Process Chemistry in the Pharmaceutical Industry

Volume 2

Challenges in an Ever Changing Climate

Process Chemistry in the Pharmaceutical Industry

Volume 2
Challenges in an Ever Changing Climate

Edited by
Kumar Gadamasetti
Tamim Braish

CRC Press
Taylor & Francis Group
Boca Raton London New York

CRC Press is an imprint of the
Taylor & Francis Group, an **informa** business

CRC Press
Taylor & Francis Group
6000 Broken Sound Parkway NW, Suite 300
Boca Raton, FL 33487-2742

First issued in paperback 2020

ISBN-13: 978-0-367-57759-9 (pbk)
ISBN-13: 978-0-8493-9051-7 (hbk)

Library of Congress Cataloging-in-Publication Data

Process chemistry in the pharmaceutical industry. Volume 2, Challenges in an ever changing climate / editors, Kumar Gadamasetti and Tamim Braish.
    p. ; cm.
Includes bibliographical references and index.
ISBN-13: 978-0-8493-9051-7 (hardcover : alk. paper)
    1. Pharmaceutical chemistry. 2. Chemical processes. I. Gadamasetti, Kumar G., 1949- II. Braish, Tamim. III. Title: Challenges in an ever changing climate.
    [DNLM: 1. Chemistry, Pharmaceutical--methods. 2. Drug Evaluation. 3. Pharmaceutical Preparations--analysis. 4. Research Design. 5. Technology, Pharmaceutical--methods. QV 744 P963 2008]

RS403.P7653 2008
615'.19--dc22
        2007020281

Dedication

KG would like to dedicate this book to his wife Vidya, daughter Stuthi, and son Pratik.

TFB would like to dedicate this book to his wife Teresa and son Fehme.

Forewords

by Mauricio Futran, Trevor Laird, and Steven Ley

The impact of novel medicines for unmet medical needs brought forward by the pharmaceutical industry in the last thirty years is tremendous: drugs for metabolic diseases (cholesterol, hypertension, diabetes), HIV treatments, new antibiotics, treatments for rheumatoid arthritis, schizophrenia, etc. These medicines have profoundly transformed the treatment paradigm in their respective fields, with a major impact of extending and enhancing human life.

The biological advances that bring such marvels, whether from academia, biotech or big pharma, are often highlighted in the media. The development of these discoveries into practical drugs is less well known and is the area where the pharmaceutical industry has excelled. Within this arena, Process Chemistry plays a pivotal role. No other development aspect can take place without a drug available to fuel it, and no drug can reach the market without a practical, robust and economical process to make it.

Process chemists and chemical engineers need to devise synthetic routes and processes based on commercially available materials and reagents, must be economical and green, and must be practical in terms of reagents, capital requirements and cost of goods.

The pressures on Process Chemistry are mounting rapidly: Industry seeks rapid entry into the clinic, postponing development spend until proof of concept, then rapid commercialization once benefit is established. External and internal trends look for more complete data, deep process understanding, absolute control of genotoxic impurities, establishing of design space and quality by design, and real time control of the process via on line sensors.

Furthermore, the need for even higher productivity and intensifying competition now pushes us towards the industrialization of the Process R&D activity. Various PR&D groups now use continuous processes and microreactors, predictive modeling of processes and properties, automation, parallel experimentation and the use of robots; DOE and multivariate analysis, etc.

Taken all together, these drivers and trends are transforming Process Chemistry profoundly and at an accelerating rate. The contributions in this volume are a good illustration of this transformation, and should shed light on the excitement, achievements, and opportunities embodied by this important element of pharmaceutical drug development. I recommend this book as essential reading not only for those of us in the industry, but to the academic community as well, given that the challenges and opportunities in Process Chemistry are much less understood than those in Medicinal Chemistry.

Mauricio Futran
Bristol-Myers Squibb Co.

Since the publication of the last volume of *Process Chemistry in the Pharmaceutical Industry* in 2000, there have been several books on this interesting topic, but there is a need for even more books that examine chemistry from an industrial viewpoint and business issues from a process R&D perspective. It is timely therefore, that a second, completely new volume should appear with lots of innovative case studies. Although there are many examples of case studies which have appeared in the pages of the *Organic Process Research and Development* journal or even in the spectacular special edition of *Chemical Reviews* devoted to process chemistry, the book format allows the authors more leeway to tell the story, to explain the problems that occurred and how they were solved, in the same way as if they were giving a lecture. A book such as this deserves a special place on the bookshelves of all process chemists and engineers.

The subtitle "Challenges in an Ever-Changing Climate" is extremely relevant for the process chemist operating today. The increasingly complex nature of API structure coupled with tighter time scales poses demanding challenges on the process chemistry community and these issues are brought to the fore by the editors in the early part of the book. Chapters also discuss key regulatory issues, globalization and business trends (including outsourcing and the rise of the influence of India and China) and new technologies that will continue to change the way in which process research and development is carried out. Process R&D issues in the manufacture of macromolecules and biologics are also highlighted.

The editors have also commissioned additional chapters on Biotransformations, reflecting the increased used of this technology in manufacturing APIs and the move towards Green Chemistry, and on Crystallization, which is so important, not only for API quality, but for controlling the physical attributes of solid forms. The discussion of these issues alongside the case studies makes this book unique.

It is therefore sad that many industrial companies now have no library in which to house such a book. This should not deter process chemists who should endeavor to build up their own library (as I have done) of books of relevance to this unique sector of industry. This latest volume, edited by Kumar Gadamasetti and Tamim Braish, which contains so many interesting chapters and case studies, not all of which have previously appeared in print, should have pride of place on the process chemistry bookshelf. The editors are to be congratulated on encouraging the contributing authors to take part and deliver the goods. Well done to you all.

Trevor Laird
Editor, Organic Process Research and Development

Process chemistry plays a key role in bringing drugs to the marketplace. However, the ever changing demands of the industry and society place tremendous pressure on all aspects of process chemistry, none more so than the future environmental demands. This, of course, creates the need for greater innovation, efficient gains, and improved technologies. This new text nicely captures many of these new trends, strategies, and developing methods to provide a dynamic view of all the opportunities that ensue during the process development of drug candidates. As a synthesis chemist, I very much welcome the detailed descriptions of the various case studies presented, and sympathize with the scientists involved in their complex task to balance all of the parameters that make the difference between success and failure.

The authors have addressed the important issues of the day as they relate to organometallic and asymmetric processes, the green chemistry and sustainability demands together with the future potential of enzymatic intervention and biotransformations.

I welcome too the coverage of the importance of new technologies and analytical techniques, and the impact of developing demographic changes in India and China particularly on how synthesis may be conducted in the future. This is an important debate and one that is well beyond simple anticipated financial savings.

The practical issues associated with crystals and morphology are nicely addressed in this new text by contrasting the art with the science involved. Similarly, two chapters, one on the developing importance of monoclonal antibodies and one on PEGylation procedures, gives a clear impression of the new challenges associated with synthesis and the need to expand horizons beyond the current comfort zone.

Likewise I was impressed by the discussion and emphasis on the importance of the knowledge-based approach particularly on reaction kinetic understanding and the use of software packages to mine information to aid the decision making process. Clearly these on-line monitoring techniques are impacting the streamlining of process operations. Inevitably these methods will find their way into continuous processing protocols and the anticipated productivity gains through telescoping techniques.

The industry faces enormous challenges to its survival owing to escalating costs and the need to provide safe and effective treatment for human diseases. Moreover, the downstream issues, especially as they relate to societal expectations and regulatory pressures, cannot be ignored.

The modern world benefits from, but often takes for granted, these crucial elements of the drug discovery process. Nevertheless, to strive to improve and solve the healthcare needs for the betterment of mankind is a truly noble goal.

Steven V. Ley
1702 Professor of Chemistry
The University of Cambridge
Cambridge, UK

Table of Contents

Preface

The pharmaceutical industry is committed to finding new ways and new medicines to alleviate the burden of disease and come up with cures to improve human life. Advancements in science and growth in technology are leading to personalized medicine that may someday be the norm rather than the exception. Symptoms may be forecasted immediately and treatment may be instantaneous to ultimately stop the progression of disease and lead to better life. For now, a paradigm shift in therapeutics from small molecules to biologics and macromolecules has become prominent toward the end of the last millennium, and this change is poised to remain and grow.

The first volume of this series has gained enormous respect and popularity in both the pharmaceutical and academic arenas. The theme, *Process Chemistry in the Pharmaceutical Industry*, led one of us (Gadamasetti) to work on a global forum to set the stage for key personnel from the pharmaceutical industry, academia, and regulatory agencies to gather to discuss the topics related to active pharmaceutical ingredients (APIs) and find solutions in chemistry, morphology, engineering, and regulatory compliance. With the help of the American Chemical Society (ACS), one of us (Gadamasetti) led the ACS ProSpectives, "Process Chemistry in the Pharmaceutical Industry," as the chairman (2002–2004). During the meetings and on several occasions over the years, we have had numerous opportunities to discuss the burning issues and the need for a second volume. Even though the first volume addressed many of these issues with real-life examples, the second volume became absolutely necessary, as the pharmaceutical industry has been faced with a magnitude of change never seen before. It is an evolution in an industry known for its stability. Small molecule therapeutics are no longer the only drugs that companies can develop, and in many cases, existing therapies are not adequate. New medicines are needed, and quickly. How can we develop medicines faster and cheaper without jeopardizing patient safety?

The purpose of this book is to highlight the importance of an area of research in the pharmaceutical industry known as process research and development and the challenges ahead. In the pharmaceutical industry, the medicinal chemist is faced with the daunting task of finding the next drug buster. This is by far the toughest job, because many medicinal chemists could work all their lives synthesizing molecules that may never make it to the market.

Drug candidates progress through development slowly, requiring 9 to 12 years before reaching the patient. What stems as an idea in the lab could become a reality, after agonizing preclinical and clinical investments, benefiting patients worldwide, which makes the journey extremely rewarding but also enormously expensive.

The process chemist must develop a commercial route for a drug candidate that addresses cost issues, environmental concerns, atom economy, and ease of synthesis, and with the specified quality attributes that will ensure patient safety during development and post-launch, underscoring the importance of process research and development disciplines in the industry.

We sincerely believe and hope that the subject material of this current volume will reach a wider array of readers and will help them understand and apply the knowledge in teaching and in solving problems at work.

Kumar Gadamasetti
Tamim Braish

Acknowledgments

We would like to thank all of the contributors to this volume for their hard work, contributions, and most importantly their patience. We want to thank the reviewers as guardians of the science for their critical feedback, their constructive criticism, and their invaluable effort hat helped shape the book.

We also want to thank colleagues within the pharmaceutical industry for their encouragement and ideas on what they thought would be a good contribution to the book that made the volume an enjoyable project and a great addition to the literature.

Additionally, we are indebted to many people who provided support and mentorship: Steve Ley (Cambridge University), Mauaricio Futran (Bristol-Myers Squibb), and Trevor Laird (Scientific Update) for their suggestions and foreword; Philip Fuchs (Purdue University) and Victor Snieckus (Queens University) for their support and guidance; Nancy Jacoby for her assistance in reviewing several manuscripts. Colleagues from Pfizer: Frank Urban, Tom Crawford, Sarah Kelly, Ed Kobelski, Roger Nosal, Jodi Gaynor, Stephane Caron, Lynne Handanyan, Jeff Blumenstein, and former colleagues (of KG) from Bristol-Myers Squibb for their critical feedback and support.

We thank the staff at Taylor and Francis for their outstanding efforts and support especially for last minute changes and additions.

Kumar & Tamim

The Editors

Kumar Gadamasetti is the founder and president of Delphian Pharmaceuticals, a cancer therapy start-up [2003] company, located in the San Francisco Bay area. Previously he held positions at Bristol Myers-Squibb Company, New Brunswick, New Jersey, as a senior research investigator in the Process Research Group and at Amgen, Thousand Oaks, California, as the head of the Process Research and Development Department. At Bristol Myers-Squibb, he worked on syntheses of anticancer drug taxol (Paclitaxel™) and monoclonal antibody drug conjugates, and at Amgen, he directed the Process Research and Development Group to advance small molecules in the areas of calcimimetics and the cardiovascular and central nervous systems, and supported other drug discovery programs. Prior to starting Delphian Pharmaceuticals, he was the chief operating officer of X-Mine, a bioinformatics company (South San Francisco, California), and senior director of chemistry research and development at Discovery Partners International Inc., located in South San Francisco. A winner of a Presidential Award at Bristol Myers-Squibb, Gadamasetti edited the first volume, entitled *Process Chemistry in the Pharmaceutical Industry* (1999). He has authored several papers and U.S. patents. He was the founder and chairman of the American Chemical Society's ProSpectives, "Process Chemistry in the Pharmaceutical Industry" (2002–2004). Currently he serves as an editorial board member of the Amercian Chemical Society journal, *Organic Process Research and Development*.

Gadamasetti has been a visiting professor at Catholic University of Louvain, Louvain-la-Neuve, Belgium (2000), and a visiting scholar at Humboldt University, Berlin, Germany (2005). He obtained his Ph.D. degree in 1987 from the University of Vermont, Burlington (Research Advisor: Martin Kuehne). He conducted his postdoctoral studies at University of Virginia, and Virginia Polytechnic Institute and State University, Blacksburg, before joining Bristol Myers-Squibb in 1991. He obtained his B.S. and M.S. from Osmania University, Hyderabad, India.

Tamim Braish started his career as a bench chemist in the process group at Pfizer in Groton, Connecticut, working in the area of quinolone antibacterials where he developed several commercial routes to drug candidates currently in production. He became a manager in 1997 in charge of the preparations laboratory, and was involved with governance at the project and the portfolio levels, and in 2003 he became a senior director responsible for the Full Development Group within the Chemical Research and Development Group at Pfizer. He serves on the editorial board of the *Organic Process Research and Development* journal and is the author of many publications and patents. His interests span both the synthetic chemistry arena as well as the regulatory arena. Braish was born in Lebanon and grew up in Indiana. He earned a B.S. degree in chemistry from Indiana University–Purdue University, Fort Wayne, and a Ph.D. from Purdue University in 1986 with Professor Philip Fuchs on the synthesis of natural products.

Contributors

David J. Ager
DSM, PMB 150
9650 Strickland Road Suite 103
Raleigh, North Carolina 27615, USA
e-mail: david.ager@dsm.com

Joseph Amato
Department of Process Research
Merck Research Laboratories
P.O. Box 2000
Rahway, NJ 07065 USA

Christopher P. Ashcroft
Chemical Research and Development
Pfizer Global Research and Development
Sandwich, CT13 9NJ, Kent, UK
e-mail: christopher.ashcroft@pfizer.com

William F. Bailey
University of Connecticut
Department of Chemistry
Storrs, CT 06269-3060 USA
e-mail: bailey@uconnvm.uconn.edu

Laura A. Bass
Pfizer Global Biologics
Pharmaceutical Sciences
700 Chesterfield Village Parkway
Chesterfield, MO 63017 USA
e-mail: laura.a.bass@pfizer.com

Donna G. Blackmond
Department of Chemistry
Department of Chemical Engineering and
  Chemical Technology
Imperial College, London SW7 2AZ UK
e-mail: d.blackmond@imperial.ac.uk

Geneviève Boice
Department of Process Research
Merck Research Laboratories
P.O. Box 2000
Rahway, NJ 07065 USA

Alfio Borghese
Solvay Pharmaceuticals
P.O. Box 9001380
DA Weesp
The Netherlands
e-mail: a.borghese@hotmail.com

Tamim F. Braish
Pfizer Global R&D
Groton, CT 06340 USA
e-mail: tamim.f.braish@pfizer.com

John J. Buckley
Pfizer Global Biologics
Pharmaceutical Sciences
700 Chesterfield Village Parkway
Chesterfield, MO 63017 USA
e-mail: john.joseph.buckley@pfizer.com

Erick M. Carreira
Laboratorium für Organische Chemie HCI
  H335
ETH Zürich
8093 Zürich, Switzerland
e-mail: carreira@org.chem.ethz.ch

Susan Casnocha
Pfizer Global Biologics
Pharmaceutical Sciences
700 Chesterfield Village Parkway
Chesterfield, MO 63017 USA
e-mail: susan.a.casnocha@pfizer.com

Stephen Challenger
Chemical Research and Development
Pfizer Global Research and Development
Sandwich, CT13 9NJ, Kent, UK
e-mail: stephen.challenger@pfizer.com

John Chung
Department of Process Research
Merck Research Laboratories
P.O. Box 2000
Rahway, NJ 07065 USA

David A. Conlon
Process Research & Development
Bristol-Myers Squibb Company
One Squibb Drive
48-1-1031
New Brunswick, NJ 08903-0191 USA
e-mail: david.conlon@bms.com

Karen Conrad
Department of Process Research
Merck Research Laboratories
P.O. Box 2000
Rahway, NJ 07065 USA

Edward Corley
Department of Process Research, Merck
 Research Laboratories
P.O. Box 2000
Rahway, NJ 07065 USA

Jotham W. Coe
Neuroscience Medicinal Chemistry
Pfizer Global Development and Research
Eastern Point Road
Groton, CT 06340 USA
e-mail: jotham.w.coe@pfizer.com

Paul Collins
Department of Process Research
Merck Research Laboratories
P.O. Box 2000
Rahway, NJ 07065 USA

Raymond J. Cvetovich
Department of Process Research
Merck Research Laboratories
P.O. Box 2000
Rahway, NJ 07065 USA

Fons De Knaep
Johnson & Johnson, Beerse B-2340
Belgium
e-mail: fdnkaep@prdbe.jnj.com

Andrew M. Derrick
Chemical Research and Development
Pfizer Global Research and Development
Sandwich, CT13 9NJ, Kent, UK
e-mail: andrew.derrick@pfizer.com

Lisa DiMichele
Department of Process Research
Merck Research Laboratories
P.O. Box 2000
Rahway, NJ 07065 USA

Peter J. Dunn
Chemical Research and Development
Pfizer Global Research and Development
Sandwich, CT13 9NJ, Kent, UK
e-mail: peter.dunn@pfizer.com

Ronald Fedechko
Pfizer Global Biologics
700 Chesterfield Village Parkway
Chesterfield, MO 63017 USA
e-mail: ronald.w.fedechko@pfizer.com

Hans-Jürgen Federsel
AstraZeneca, Process R&D
151 85 Södertälje, Sweden
e-mail: hans-jurgen.federsel@astrazeneca.com

Rory F. Finn
Pfizer Global Biologics
Pharmaceutical Sciences
700 Chesterfield Village Parkway
Chesterfield, MO 63017 USA
e-mail: rory.f.finn@pfizer.com

Mauricio Futran
Process R&D
Bristol-Myers Squibb
New Brunswick, NJ 08903 USA
Mauricio.futran@bms.com

Kumar Gadamasetti
Delphian Pharmaceuticals
Pharmaceutical R&D
Belmont, CA 94002 USA
e-mail: delphianpharma@aol.com

Yousef Hajikarimian
Chemical Research and Development
Pfizer Global Research and Development
Sandwich, CT13 9NJ, Kent, UK
e-mail: yousef.hajikarimian@pfizer.com

Steve Hannon
Davos
600 East Crescent Avenue
Upper Saddle River, NJ 07458 USA
e-mail: hannons@DAVOS.com

Ronald L. Hanson
Bristol-Myers Squibb
New Brunswick, NJ 08903-0191 USA
e-mail: ronald.hanson@bms.com

Sa V. Ho
Pfizer Global Biologics
Pharmaceutical Sciences
700 Chesterfield Village Parkway
Chesterfield, MO 63017 USA
e-mail: sa.v.ho@pfizer.com

Dave Hughes
Department of Process Research
Merck Research Laboratories
P.O. Box 2000
Rahway, NJ 07065 USA

Hiroshi Iwamura
Department of Chemistry
Imperial College
London SW7 2AZ UK

Bill Izzo
Chemical Engineering Research and
  Department of Process Research
Merck Research Laboratories
P.O. Box 2000
Rahway, NJ 07065 USA
e-mail: bill_izzo@merck.com

Shu Kobayashi
Graduate School of Pharmaceutical Sciences
The University of Tokyo,
The HFRE Division
ERATO
Japan Science and Technology Agency (JST)
Hongo, Bunkyo-ku, Tokyo 113-0033, Japan
e-mail: skobayas@mol.f.u-tokyo.ac.jp

Steven V. Ley
The University of Cambridge
Trinity College, Cambridge, UK
e-mail: Svl1000@cam.ac.uk

John Lu
Mitsui & Co., (U.S.A.)
New York, USA

Suju P. Mathew
Department of Chemistry
Imperial College
London SW7 2AZ UK

Farah Mavandadi
Discovery Drive
Charlottesville, VA 22911 USA
e-mail: farah.mavandadi@biotage.com

J. Chris McWilliams
Department of Process Research
Merck Research Laboratories
P.O. Box 2000, Rahway, NJ 07065 USA
e-mail: jmcwilli@boehringer-ingelheim.com

Paul Mensah
Pfizer Global Biologics
Pharmaceutical Sciences
700 Chesterfield Village Parkway
Chesterfield, MO 63017 USA
e-mail: paul.mensah@pfizer.com

Alain Merschaert
Eli Lilly & Company
Chemical Product R & D
Lilly Development Centre S.A.
Parc Scientifique de Louvain-la-neuve
Rue Granbonpré 11
B-1348 Mont-Saint-Guibert, Belgium

Jianming Mo
Pfizer Global Biologics
Pharmaceutical Sciences
700 Chesterfield Village Parkway
Chesterfield, MO 63017 USA
e-mail: jianming.mo@pfizer.com

Carlos. A. Mojica
Pfizer Global R&D
Groton, CT 06340 USA
e-mail: carlos.mojica@pfizer.com

John Mott
Pfizer Global Biologics
Pharmaceutical Sciences
700 Chesterfield Village Parkway
Chesterfield, MO 63017 USA
e-mail: john.e.mott@pfizer.com

Klaus Müller
F. Hoffmann-La Roche AG
Pharmaceuticals Division
4070 Basel, Switzerland
e-mail: klaus.mueller@roche.com

Jerry A. Murry
Department of Process Research
1 Amgen Center Dr.
Newbury Park, CA 91320 USA
e-mail: jmurray@amgen.com

Oscar Navarro
ICIQ
Av Paisos Catalans 16
43007 Tarragona
Spain
e-mail: onavarro@uci.edu

Sandeep Nema
Pfizer Global R&D
Groton, CT 06340 USA

Steven P. Nolan
ICIQ
Av Paisos Catalans 16
43007 Tarragona
Spain
e-mail: snolan@iciq.es

Timothy Norris
Pfizer Global R&D
Groton, CT 06340 USA
e-mail: timothy.norris@pfizer.com

Thomas C. Nugent
Department of Chemistry
School of Engineering and Science
International University Bremen
Campus Ring 1
28759 Bremen, Germany
e-mail t.nugent@iu-bremen.de

Chikako Ogawa
Eisai Research Institute
Lead Identification
4 Corporate Drive
Andover, MA 01810 USA
e-mail: chikako_ogawa@eri.eisai.com

Terry L. Rathman
t-Links Consulting
100-29 Willow Run,
Suite Li, Gastonia, NC 28056
e-mail: trathman@carolina.rr.com

Robert A. Reamer
Department of Process Research
Merck Research Laboratories
P.O. Box 2000
Rahway, NJ 07065 USA

David Robins
Davos
600 East Crescent Avenue
Upper Saddle River, NJ 07458 USA
e-mail: darobins@DAVOS.com

Mark Rogers-Evans
F. Hoffmann-La Roche AG
Pharmaceuticals Division
4070 Basel, Switzerland

Cécile Savarin
Department of Process Research
1 Amgen Center Dr.
Newbury Park, CA 91320 USA
e-mail: csavarin@amgen.com

Ichiro Shinkai
Beta Chem Inc.
1-2-1 Ohtemachi
Chiyoda-Ku
Tokyo 100-0004, Japan
e-mail: shinkai.ichiro@beta-chem.com

Robert A. Singer
Chemical R&D
Pfizer Global Development and Research
Eastern Point Road
Groton, CT 06340 USA
e-mail: robert.a.singer@pfizer.com

L. St. Pierre-Berry
Pfizer Global R&D
Groton, CT 06340 USA
e-mail: laurie.a.st.pierre@pfizer.com

Anders Sveno
AstraZeneca
Process R&D
151 85 Södertälje, Sweden
e-mail: andres.sveno@astrazeneca.com

Nicholas M. Thomson
Chemical Research and Development
Pfizer Global Research and Development
Sandwich, CT13 9NJ, Kent, UK
e-mail: nick.thomson@pfizer.com

Rajappa Vaidyanathan
Chemical R & D
Pfizer Global Research and Development
Groton, CT 06340 USA
e-mail: Rajappa.vaidyanathan@pfizer.com

Harry A. Watson Jr.
Chemical R&D
Pfizer Global Development and Research
Eastern Point Road
Groton, CT 06340 USA

Dierk Wieckhusen
Novartis Pharma AG
Process Technology
WSJ145.10.01A, Lichtstr. 35
CH-4002 Basel, Switzerland
e-mail: dierk.wieckhusen@novartis.com

Georg Wuitschik
Laboratorium für Organische Chemie HCI
  H335
ETH Zürich
8093 Zürich, Switzerland
e-mail: wuitschik@org.chm.ethz.ch

Xu Feng
Process Research Department
Merck Research Laboratories
P.O. Box 2000
Rahway, NJ 07065 USA
e-mail: feng_xu@merck.com

Natalia Zotova
Department of Chemical Engineering and
  Chemical Technology
Imperial College
London SW7 2AZ UK

Affiliations

AstraZeneca, Process R&D, 151 85 Södertälje, Sweden

Beta-Chem, Inc. 1-2-1 Ohtemachi, Chiyoda-Ku, Tokyo 100-0004, Japan

Biotage, Discovery Drive, Charlottesville, VA 22911 USA

Bristol-Myers Squibb, New Brunswick, NJ 08903-0191 USA

Davos, 600 East Crescent Avenue, Upper Saddle River, NJ 07458 USA

Delphian Pharmaceuticals, Pharmaceutical R&D, Belmont, CA 94002 USA

DSM, PMB 150, 9650 Strickland Road Suite 103, Raleigh, North Carolina 27615, USA

Eli Lilly & Company, Chemical Product R&D, Lilly Development Centre S.A., Parc Scientifique de Louvain-la-neuve, Rue Granbonpré, 11, B-1348 Mont-Saint-Guibert, Belgium

Eisai Research Institute, Lead Identification, 4 Corporate Drive Andover, MA 01810 USA

ETH Zürich, Laboratorium für Organische Chemie HCI H335, 8093 Zürich, Switzerland

F. Hoffmann-La Roche AG, Pharmaceuticals Division, 4070 Basel, Switzerland

Imperial College, Department of Chemistry, Department of Chemical Engineering and Chemical Technology, London SW7 2AZ UK

International University Bremen, Department of Chemistry, School of Engineering and Science, Campus Ring 1, 28759 Bremen, Germany

Johnson & Johnson, Beerse B-2340, Belgium

Merck Research Laboratories, Rahway, NJ 07065 USA

Mitsui & Co., New York, USA.

Novartis Pharma AG, Process Technology, WSJ145.10.01A, Lichtstr. 35, CH-4002 Basel, Switzerland

Pfizer Global Research and Development, Sandwich, CT13 9NJ, Kent, UK

Pfizer Global Biologics, Pharmaceutical Sciences, Chesterfield, MO 63017 USA

Pfizer Global R&D, Groton, CT 06340 USA

t-Links Consulting, 100-29 Willow Run, Suite Li, Gastonia, NC 28056 USA

The University of Cambridge, Trinity College, Cambridge, UK

University of Connecticut, Department of Chemistry, Storrs, CT 06269-3060 USA

University of New Orleans, Department of Chemistry, 2000 Lakeshore Dr., New Orleans, LA 70148 USA

The University of Tokyo, Graduate School of Pharmaceutical Sciences, The HFRE Division, ERATO, Japan Science and Technology Agency (JST), Hongo, Bunkyo-ku, Tokyo 113-0033, Japan

1 Process Chemistry in the Pharmaceutical Industry: Challenges in an Ever Changing Climate—An Overview

Kumar Gadamasetti

CONTENTS

1.1 INTRODUCTION

The pharmaceutical industry strives to produce medical breakthroughs to improve the quality of human life. The average cost to develop and advance a new chemical entity (NCE) from inception to market, as a successful drug, is about $1 billion.[1] There are no guarantees of sustained success or longevity for any given drug in the market as evidenced by the recent withdrawal from market of Vioxx.[2] In spite of a higher degree of attrition at various stages of development and the statistics indicating a downward trend in the number of NCEs in the past 3 to 4 years, the worldwide pharmaceutical market growth was positive to the tune of 6 to 7% in 2006, and it is projected that global sales will reach $665 billion to $685 billion in 2007.[6a] The pharmaceutical industry is going through a continuous ferment of change and reinvention to cope with attrition, high development cost, and the potential lack of substrate. Adaptation to growing change has become a norm in an industry that was known for its stability over the last several decades of the 20th century. The mission of process chemistry in the pharmaceutical industry is to provide documented, controlled, economic, green, and safer synthetic processes for development and large-scale manufacturing under regulatory guidance to support clinical trials and future commercial requirements of an active pharmaceutical ingredient (API). The main focus of this volume is twofold:

1. Outline approaches to synthetic processes and realistic solutions to some problems associated with process development and manufacturing in industry
2. Discuss changes that the pharmaceutical industry has been going through and how these changes have affected the process scientist, engineer, or technologist in tackling the challenges in this constantly evolving environment

The subject matter and the authors from academia and industry were carefully selected to make an attempt to lay out the mosaic of comprehensive information with real case studies to present to the process personnel and the individuals interested in such sciences and technology. It is hoped that the subject matter in this volume will help you to learn, develop, and eventually contribute positively by tackling challenges toward improving the quality of human life.

1.2 CHANGE AND CHALLENGES

The pharmaceutical industry has experienced an enormous array of change in terms of management, accountability, and ultimately productivity, toward the end of the 20th century. Process chemists constitute a small fraction of the whole development discipline, yet they are known to play a vital role in making significant contributions in the industry. Although changes focused toward process optimization, controls, and equipment were categorized as "anticipated change," the changes due to advances in technology and the new forms of APIs (active pharmaceutical ingredients: small molecules and biologics) led to new paradigms in the process technology disciplines. Outlined below are a few selected elements in terms of technology (small versus large molecules), global regulatory compliance, and globalization, relevant to process chemists, technologists, and engineers.

1.2.1 Technology: APIs in the 21st Century

Advances in the human genome project and the identification of genes and proteins that underlie diseases led to a plethora of targets in biologics as APIs for the treatment of human diseases. The amalgamation of fermentation experts, cell biologists, enzymologists, and process engineers/scientists has helped accelerate the recent biologic drugs to market. Process personnel have taken the task to learn, cross-train, and bridge the gap to address the needs to meet the development timelines. A paradigm shift in therapeutics from small molecules to biologics and macromolecules became prominent toward the end of the last millennium, and this change is poised to remain and grow. Roughly 60% of the revenue growth for "Big Pharma" is projected to come from biologic products through 2010.[4]

1.2.1.1 Small and Large Molecules[3,4]

Small molecule drugs, defined as drugs with typical molecular weights <500 daltons, have been the main drivers of sales growth for Big Pharma. The projections lean toward a change in this trend by the turn of this decade. The pharmaceutical industry is expected to move rapidly toward biologic products, defined usually, but not exclusively, as protein-based therapeutic agents. Therapeutic proteins, monoclonal antibodies, and vaccines, are expected to complement or replace many small molecules over the next decade or two. Herceptin®, Lucentis® from Genentech, Epogen® and Aranesp® from Amgen, Procrit® from Johnson & Johnson, Humulin® from Eli Lilly, and Novolin® from Novo Nordisk, are a few selected examples of large molecule therapeutics.

1.2.1.2   Growth Forecast for Small and Large Molecules and Generics

Datamonitor,[4b] which forecasts revenue growth, predicts the compound annual growth rate (CAGR) of 13% for Big Pharma during 2004–2010. In contrast, the CAGR for small molecule products is expected to be about 1% over the same period. On the flip side, the generics companies see small molecules as blockbuster drugs that continue to drive their market in the future with the belief that "the end of blockbusters is not upon us, despite what some analysts are saying."[4c] The global market for generic drugs is poised for strong (11 to 13%) growth by 2007. According to Datamonitor (London, UK), generic drugs are expected to reach sales of $160 billion by 2015 as drug patents continue to expire.[4a]

1.2.2   Globalization

Contract manufacturing organizations (CMOs) in Italy, Spain, Eastern bloc countries, China, and India have become major partners for the pharmaceutical and biotech companies. Cost reductions and competition to manufacture API formulations and the manufacturing of generic drugs have created this market and have led to the growth in globalization. The industry has benefited from enormous improvements in the quality of work and the timely delivery of the APIs that are well within the set parameters (cost, timeline, and quality). Outsourcing has become a part of the development strategy for most fully integrated pharmaceutical companies (FIPCOs) as well as mid- to small-size pharmaceutical and biotech companies. Most Big Pharma companies develop the processes in-house and transfer the robust processes to CMOs overseas. Globalization has necessitated that process personnel develop practical technical transfer methods, tight controls on releasing specifications, and thorough knowledge of regulatory compliance. In essence, globalization created a healthy worldwide competition in the process community. Although initially the benefits were enjoyed mutually between pharmaceutical/biotech industry and the contract research organizations (CROs)/CMOs, lately the industry seems to be facing tough competition due to the growth of generic drugs overseas.

1.2.3   Regulatory Compliance[7]

Pharmaceutical industrial globalization brought a general awareness that it was essential to make pharmaceutical manufacturing more efficient and less wasteful. The pressure is enormous on regulators globally to focus on the most critical issues affecting product quality and assuring patient safety. The International Conference on Harmonization[7d] (ICH) issued a series of quality standards (Q series: Q8, Q9, Q10, etc.) to ensure efficiency and quality assurance in drug substance (API) and drug product manufacturing. A continuous exchange of dialogue between U.S. as well as worldwide regulatory agencies with the pharmaceutical counterparts on global regulatory awareness and advancements is the mantra for today's industrial survival and longevity. The need for process development to encompass regulatory compliance, current good manufacturing practice (cGMP) guidelines, 21 CFR Part 11,[7e] utilizing process analytical technology[7c] (PAT) tools in advancing the APIs through various phases of drug development, is an added responsibility for the process scientist in the modern era of pharmaceutical industry. As the regulatory environment changes, it is likely that many small biotech companies may not have sufficient in-house expertise in good manufacturing practice (GMP) compliance and will as a result underestimate what it takes to manufacture drugs safely.[7f] However, these small biotech companies will now benefit greatly from the already established market of CROs/CMOs that have honed their process development and regulatory skills with Big Pharma over the years.

An in-depth discussion of PAT and its applications in industrial processes is presented by Merck and Pfizer process scientists in Chapters 19, 20, 21, 22, and 23. Readers are encouraged to read the plethora of information, guidance and updates provided by the U.S. Food and Drug Administration[7a] (FDA) and international agencies.[7]

1.3 GREEN CHEMISTRY

The focus for pharmaceutical industries is to find ways to develop chemical products and environmentally friendly, efficient processes that require fewer reagents and minimize the production of toxic gases and toxic waste, while being operationally safe and economical. Chapters 15 and 16 in this volume, one from academia (Ogawa and Kobayashi) and the other from industry (Dunn), are dedicated to green chemistry.

The accomplishments by Codexis and Merck, recipients of the Presidential Green Chemistry Challenge Award in 2006, are noteworthy, and they highlight a future trend in the industry (Scheme 1.1 and Scheme 1.2).[5] Some companies have created organized plans for developing environmentally friendly processes.[5c]

1.4 PRESCRIPTION PHARMACEUTICAL DRUGS

"Growth in the global pharmaceutical market is expected to moderate in 2007 although opportunities for biologics and generics loom large and Asia shifts the balance of API power."[4a] Global pharmaceutical sales are expected to increase 5 to 6% by 2007 to reach $665 billion to $685 billion down from a growth of 6 to 7% in 2006, according to recent analysis by IMS Health (Fairfield, Connecticut). Growth of biologics, like Epogen® and Neupogen® (Amgen), Herceptin® and Lucentis® (Genentech), Gardasil (Merck), Humulin® (Eli Lilly), and Novolin (Novo Nordisk), is projected

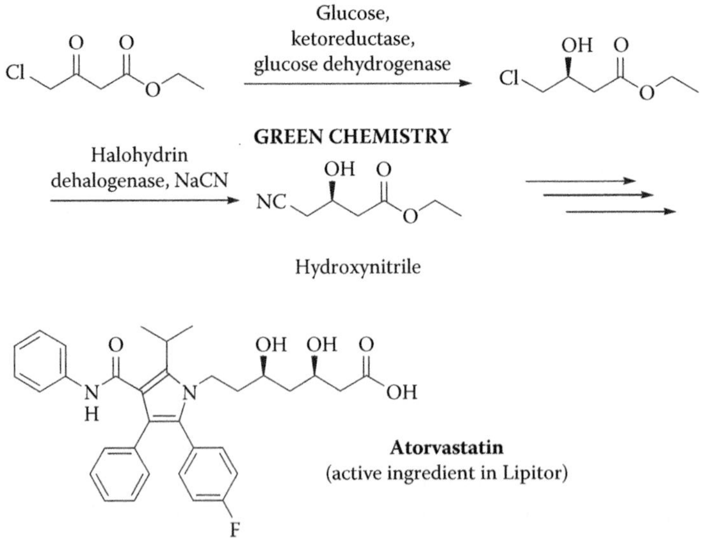

SCHEME 1.1 Green chemistry efficiency in the synthesis of atorvastatin (Lipitor®) by Codexis scientists. Codexis researchers engineered three enzymes to create a more efficient biocatalytic process to make hydroxynitrile, the intermediate that forms the chiral dihydroxy acid side chain essential to atorvastatin's synthesis. (From *Angewandte Chemie International Edition*, 44, 362, 2005. With permission.)

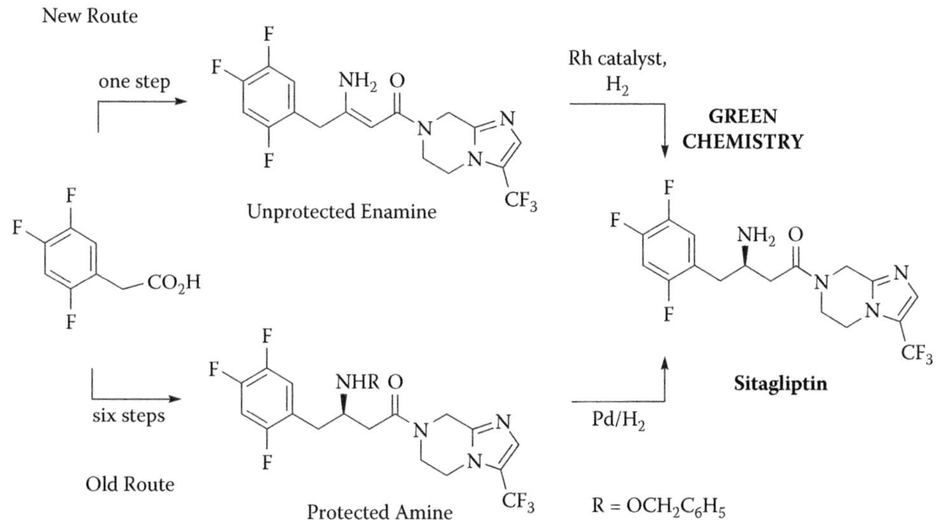

SCHEME 1.2 Green chemistry in the synthesis of sitagliptin (Januvia™) by Merck process scientists. Merck process chemists redesigned and significantly shortened the original synthesis of type 2 diabetes drug candidate sitagliptin (Januvia) to include an unprecedented efficient hydrogenation of an unprotected enamine. (From *C&EN*, July, 24–27, 2006. With permission.)

to grow more as compared to small molecules.[4a] Biologics are often very expensive, injectable medicines used for hard-to-treat diseases like cancer, severe anemia, and rheumatoid arthritis. Table 1.1 lists the 20 top-selling prescription drugs currently on the market. Macromolecules are prominent on this list (30%) and will likely continue to grow.

1.5 OUTLINE OF CONTENTS

In classifying the division of chapters and in selecting the specific contents of individual chapters, the editors made a conscientious effort to identify the changes the pharmaceutical industry has been going through during the past decade, while preserving the basic mosaic of elements (methodology, optimization, controls, scale-up, and engineering and commercial processes) on which process chemistry is built. The topics in a given chapter may cover classical case studies of process development, the manufacturing processes of current specific drugs in advanced stages of development, the drugs in market, and specific classes of compounds of significant value for the industry. With the intent of addressing the changes the industry has been going through, the chapters focus on globalization, outsourcing, regulatory compliance, and biologics.

1.5.1 CASE STUDIES

Case studies include taking the drug discovery processes to commercial processes via process research and development. An in-depth discussion on varenicline (Chapter 3) and sunitinib (Chapter 4), both recently approved drugs, originated from Pfizer's discovery and process development group by Coe and Vaidyanathan and coworkers, respectively. Savarin and coworkers from Merck, in Chapter 5, outline a detailed work on efficiency and scalable process of a potent MC4 receptor antagonist followed by the work on LY414038, a 5HT$_2$ antagonist from Eli Lilly process scientists, in Chapter 6. A contribution from AstraZeneca in Chapter 7 on the synthesis of robalzotan originated from Federsel's group. The subsequent two specific drug development projects were submitted by Pfizer—one from Challenger and colleagues from United Kingdom, on the endothelin antagonists, UK-350926 and UK-349,862. Norris outlined the details of the development and scale-up of

TABLE 1.1
Top 20 Prescription Drugs[6]

1

(optically active)

A. Lipitor
B. Atorvastatin
C. Pfizer
D. High cholesterol
E. $8.4B
F. $13.3B [1]

2

(optically active)

A. Zocor
B. Simvastatin
C. Merck & Co.
D. High cholesterol
E. $4.4B
F. $5.5B [5]

3

(optically active)

A. Nexium
B. Esomeprazole
C. AstraZeneca
D. Heartburn
E. $4.4B
F. $6.2B [3]

4

(optically active)

A. Prevacid
B. Lansoprazole
C. Abbott & Takeda
D. Heartburn
E. $3.8B
F. $4.0B

5

(optically active)

A. Advair (Seretide)
B. Fluticasone propionate
C. GlaxoSmithKline
D. Asthma
E. $3.6B
F. $5.9B [4]

TABLE 1.1 (CONTINUED)
Top 20 Prescription Drugs[6]

6	 (optically active)	A. Plavix B. Clopidogrel C. Bristol-Myers Squibb & Sanofi-Aventis D. Heart disease E. $3.5B F. $6.4B [2]
7	 (optically active)	A. Zoloft B. Sertraline C. Pfizer D. Depression E. $3.1B
8	Hormone	A. Epogen B. Erythropoietin C. Amgen D. Anemia E. $3.0B
9	Hormone	A. Procrit B. Erythropoietin C. Johnson & Johnson D. Anemia E. $3.0B
10	Recombinant protein	A. Aranesp B. Darbepoietin alfa C. Amgen D. Anemia E. 2.8B F. $4.3B [9]
11	Protein (recombinant tumor necrosis factor alfa receptor)	A. Enbrel B. Etanercept C. Amgen and Wyeth D. Rhematoid arthritis E. 2.7B F. $4.1B [10]
12		A. Norvasc B. Amlodipine C. Pfizer D. High blood pressure E. $2.6B F. $5.0B [6]

TABLE 1.1 (CONTINUED)
Top 20 Prescription Drugs[6]

13

A. Seroquel
B. Quetiapine
C. AstraZeneca
D. Schizophrenia
E. $2.6B

14

A. Effexor XR
B. Venlafaxine
C. Wyeth
D. Depression
E. $2.6B
F. $3.8B

15

A. Zyprexa
B. Olanzapine
C. Eli Lilly
D. Schizophrenia
E. $2.5B
F. 4.7B [7]

16

A. Singulair
B. Montelukast
C. Merck & Co.
D. Asthma and allergies
E. $2.5B

17

A. Protonix
B. Pentoprazole
C. Wyeth
D. Heartburn
E. $2.4B

18

A. Risperdal
B. Risperidone
C. Johnson & Johnson
D. Schizophrenia
E. $2.3B
F. $4.3B [8]

19   Glycoprotein (growth factor or cytokine)

A. Neulasta
B. Granulocyte colony-simulating factor
C. Amgen
D. Chemotherapy side effects
E. $2.2B

TABLE 1.1 (CONTINUED)
Top 20 Prescription Drugs[6]

20	Chimeric monoclonal antibody (engineered protein)	A. Remicade
		B. Infliximab
		C. Johnson & Johnson
		D. Rhematoid arthritis
		E. $2.2B

Note: A = brand name; B = chemical name; C = marketer; D = indication; E = annual sales in United States (December 2006; USD in billions); and F = worldwide 2006 annual sales in USD billions (number indicates top ten worldwide).

tesocefovecin sodium. Three specific case studies entail chiral amines (Nugent, University of Bremen, Germany), unnatural amino acids (Ager, DSM), and lithium–halogen exchange reactions (Bailey and Rathman, University of Connecticut). These constitute important building blocks and useful methodologies for the process chemist.

1.5.2 Asymmetric Synthesis, Macromolecules, and Special Topics

Single isomers have gained enormous popularity in the pharmaceutical world. Potential generic competition of select single enantiomer drugs interested the chemists as well as the legal firms. Plavix, Nexium, Zocor, Pravacol (or Mevalotin), Zoloft, Ciralex (or Lexapro), and Zithromax are among the selection of drugs that gained popularity.[3] Structures and sales are represented in Table 1.1.

A large number of drugs marketed today are chiral in nature. In 2005, single-enantiomer therapies had sales of $225 billion, representing 37% of the total final formulation pharmaceutical sales of $602 billion. Among the small molecules, single-enantiomer drugs are not only critical in new drug development, but they also can be used as a defense strategy by innovator drug companies against generic competition.[3] The success of optically active Nexium by AstraZeneca and on the downside the litigations surrounding clopidogrel by Sanofi-Aventis and Bristol-Myers Squibb are noteworthy.[8] Oxetan-3-one chemistry and synthesis by Carreira and colleagues in Chapter 13 and a discussion of C–C and C–N bond-forming reactions (Chapter 14) by Navarro and Nolan provide outstanding material.

Among the special topics, two chapters from the Pfizer process development group have focused on therapeutic monoclonal antibodies (Chapter 26) and pegylation of biological molecules (Chapter 24). Hopefully, the readers will appreciate the first-hand knowledge of the science and technology of the challenges of handling, modifying, and purifying large molecules. Microwave-mediated organic synthesis in the laboratories gained popularity in recent years. Chapter 25 enlists the merits and applications of such reactions.

1.5.3 Enzymatic Intervention, Crystallization, and Morphology and Chemical Engineering

Microbial technology enables enzymatic interventions for a given building block (e.g., β-keto esters to β-(S)-hydroxy esters) or generates the whole drug (e.g., pravachol from Bristol-Myers Squibb). Both approaches are very valuable in the pharmaceutical industry. Hanson from Bristol-Myers Squibb provided an outstanding discussion with schematics of building blocks and drugs in Chapter 17. Crystallization, in the early years, was considered more an art than a science. Chapter 18 by Wieckhusen from Novartis outlines an excellent step-by-step tutorial to achieve control in large-scale industrial crystallizations. Blackmond and colleagues from Imperial College, London, UK, portray the "reaction progress kinetic analysis" and the applications to the pharmaceutical world in Chapter 27.

1.5.4 REGULATORY COMPLIANCE, PROCESS PATENTS, AND INSTRUMENTATION

Contributions in Chapters 19, 20, 21, and 22 are focused on *in situ* reaction monitoring using mid-infrared (IR) spectroscopy, stemming from McWilliams and coworkers of the Merck process group. Detailed discussion on process analytical technology (PAT) and associated tools is outlined by Mojeca and the Pfizer group, and an outline of regulatory compliance and the effect on process scientists' outlook is summarized in Chapter 1 and in Chapter 2.

1.5.5 OUTSOURCING

Globalization in the modern era has led to worldwide outsourcing. Two chapters were dedicated to the changes in the pharmaceutical industry in terms of outsourcing. In Chapter 28, Lu and Shinkai (Beta Chem, Japan), formerly from Merck, discuss the trends. Robins and Hannon (Davos), in Chapter 29, outline the importance of sourcing pharmaceutical products to China and India.

1.6 FUTURE TRENDS

The title of this book is centered on the challenges in an ever-changing climate in the pharmaceutical industry's process development and manufacturing disciplines. Highlighted in this chapter are key factors influencing such challenges, addressing the need for change. The second chapter in this volume entitled "Emerging Trends in Process Chemistry," by Braish, De Knaep, and Gadamasetti delineates the essence of the function of the process development chemist in the pharmaceutical industry subject to a continuous state of change.

REFERENCES AND NOTES

1. (a) Tufts Center for the Study of Drug Development. The average cost to develop a novel prescription drug, including cost of postapproval release, is $897 million (in 2003); http://csdd.tufts.edu/NewsEvents/RecentNews.asp?newsid=29. (b) The average cost to develop a new biotechnology product is $1.2 billion. http://csdd.tufts.edu/NewsEvents/NewsArticle.asp?newsid=69.
2. Vioxx, the Merck blockbuster drug, was withdrawn from the market for the risk of serious cardiovascular side effects.
3. Supplement to *Pharmaceutical Technology*, October 2006, s14–18.
4. (a) Simon King, "The Evolving Pharmaceutical Value Chain: Forecasting Growth for Small and Large Molecules," Supplement to *Pharmaceutical Technology*, October 2006, s6. (b) "Big Pharma Turns to Biologics for Growing to 2010" (DMHC2190) *Datamonitor PLC*, London, May 2006. (c) "Pharma Challenged," Joanne Grimley, *C&E News*, December 4, 2006, 17–28.
5. "Going Green Keeps Getting Easier," Stephen K. Ritter, *C&EN*, July 10, 2006, 24. (a) *Angewandte Chemie International Edition*, 2005, 44, 362. (b) Hydrogenation of an unprotected enamine in sitagliptin by Merck Process Group; *C&EN*, July 2006, 24–27. (c) "Industry Takes Steps Toward Greener API Manufacturing," Kaylynn Chiarello, *Pharmaceutical Technology*, November 2005, 46; Bristol-Myers Squibb instituted a Green Chemistry Program that encourages scientists to think about minimizing environmental hazards when developing new compounds. The company uses a "process greenness scorecard" in its education and training and in every step of a compound's development to rate the impacts of certain reagents and solvents on its processes.

6. (a) IMS Health; *Pharmaceutical Technology*, January 2007, 52–58. http://open.imshealth.com/webshop2/IMSinclude/i_article_20061204a.asp; www.imshealth.com. "Competition in the World APIs Market," Chemical Pharmaceutical Generic Association, Milan, Italy, March 2006. (b) *Forbes* February 27, 2006; Top-selling drugs, www.forbes.com/2006/02/27/pfizer-merck-genentech-cx_mh_0224topsellingdrugs.html?partner=alerts.

7. (a) U.S. Department of Health and Human Services, http://www.fda.gov/. (b) "FDA Issues Advice to Make Earliest Stages of Clinical Drug Development More Efficient" (January 12, 2006); www.fda.gov/bbs/topics/news/2006/NEW01296.html. (c) Process Analytical Technology (PAT) Initiative; www.fda.gov/cder/OPS/PAT.htm. (d) International Conference on Harmonization (ICH); www.ich.org/cache/compo/276-254-1.html. (e) 21 CFR Part 11; www.fda.gov/ora/compliance_ref/part11/. (f) "Keeping Pace," Laura Bush, *Pharmaceutical Regulatory Guidance Book*, July 2006, 68. (g) CDER: Center for Drug Evaluation and Research; www.fda.gov/cder/. (h) CBER: Center for Biologics Evaluation and Research; www.fda.gov/cber/.

2 Emerging Trends in Process Chemistry

Tamim F. Braish, Fons De Knaep, and Kumar Gadamasetti

CONTENTS

2.1 INTRODUCTION

The function of the process development chemist in the pharmaceutical industry is in a state of transition. Historically, the primary functions of the process development chemist were devising efficient and commercially viable synthetic routes to advance a medicinal candidate through the preclinical and clinical development timeline, establishing and transferring a commercial synthetic route to the manufacturing division, and contributing to the chemistry, manufacturing, and controls (CMC) section of the new drug application (NDA) submission. Although these responsibilities remain prominent, the role of the process chemist has expanded to accommodate changes prompted by increased regulatory expectations and in response to pervasive pressures to improve business efficiency.

2.2 CHANGE

Internal factors that have influenced the role of the process chemist are typically driven by economics (i.e., cost-cutting measures, streamlined processes, and shortened development timelines). External factors have primarily emerged due to improved technology and changes in regulatory expectations i.e., increased emphasis on development of technological innovations like PAT [process analytical technology], implications associated with expectations for control of potential genotoxic impurities to extremely low levels, adaptation to and navigation around complicated intellectual property obligations for patented intermediates or processes claimed by third parties. The process chemist typically begins a career in the pharmaceutical industry as a synthetic organic chemist; however, addressing these internal and external factors requires additional familiarity with toxicology, patent law, analytical chemistry, supply chain, clinical pharmacology, and regulatory disciplines. Most importantly, a process chemist must understand and integrate the properties and characteristics of the active pharmaceutical ingredient (API) with the target product profile and appropriate drug product formulation criteria. In essence, the process chemist is expected to deliver a robust, efficient, economical, and adaptable manufacturing process that consistently produces quality APIs that can withstand scrutiny from global regulatory authorities.

Increased market pressure on innovator pharmaceutical companies to reduce costs of prescription medications and simultaneously reduce the time to bring new medications to market has had a tremendous impact on business development. In addition, the rate of drug candidate attrition has altered the research and development (R&D) investment paradigm to minimize development until proof of clinical viability has been demonstrated. These internal pressures are driving fundamental changes in process chemistry to streamline timelines without the added cost.

To respond to change, the resourceful process chemist has devised new modes of operation, and change has become the norm rather than the exception. This chapter will discuss factors that influence timelines, reduce cost, and influence the process of developing drugs.

2.3 TIMELINE COMPRESSIONS

Timeline compressions in the pharmaceutical industry are typically driven by how fast a clinical program can be run. This, in turn, is driven by how well a biomarker for a disease that is being treated is understood and, most importantly, validated. Hence, the clinical program determines the development timeline of the drug, once a candidate has been nominated by the discovery team. The process chemist responds to the toxicology and the clinical needs by providing the technology that will enable the medicinal candidate to be manufactured at a scale to meet the clinical supply demand as the program progresses through the development pipeline and ultimately to be launched.

One cost-cutting trend has been to minimize investments in terms of development work on a candidate about which little is known in terms of the mechanistic understanding of the disease and no precedence exists in the marketplace. These candidates typically have little chance of success, yet the reward could be huge if these candidates make it to market, as these would typically be first in class and meet a patient need not fulfilled prior to launch. The amounts of resources and investments dedicated to a candidate are directly proportional to how well a mechanism is understood. Candidates targeting a well-understood physiological mechanism would be well resourced, because the chances of making it to the market are better than average. However, when little-resourced candidates reach a proof of concept, the process chemist has some catching up to do, as the NDA filing cannot be delayed. The typical development timeline will then be compressed, and the process chemist has to make up that lost time.

2.3.1 AUTOMATION

One way to address timeline compressions is through automation. Automation can help the chemist develop high-throughput screens to quickly screen reactions to choose the right bond-forming steps. Once a bond-forming step has been selected, automation can help define the critical or noncritical process parameters to optimize the process and render it robust. Great advances have been made in the equipment and the tools to aid the chemist in making decisions faster and better. However, this does not come at a small cost. First, the equipment is quite expensive, and many sets are needed to make it a truly integrated tool for most development chemists. Although it is a one-time cost, automation adds significantly to the development cost. Second, there is the energy barrier associated with the human nature of resisting change. Once these issues are overcome, instrumentation can become a chemist's greatest ally.

Software to handle and analyze the large amounts of data generated through automation has evolved tremendously, but these tools have not fully evolved to move from the hands of the specialists and the technophiles into the hands of a typical chemist, which introduces yet another barrier to fully adopting automation. Downscaling often combined with software simulation of lab information that is translated into plant conditions has become indispensable, sometimes even allowing direct scale-up from lab to large reactor in a safe and reliable fashion. The data manipulation and simulation as well as the continued progress made on the electronic notebook will continue to evolve and will eventually become easy-to-use tools that are indispensable to every chemist.

2.3.2 Timeline Compressions versus Regulatory Hurdles

Another way to deal with timeline compressions is to manage regulatory expectations. For example, filing whatever manufacturing process data is available at the time proof of concept is achieved can reduce time. In some cases, even in the absence of a significant investment in the technology to make a molecule more efficiently, the process may be perfectly suitable for an NDA filing. This is a valuable option for highly potent drugs (low volumes), where cost of goods is not critical, and the saving opportunities, as a result, are limited. Postfiling changes allow you to spend your resources on real drugs that could contribute great savings to a manufacturing organization. However, for changes with major regulatory impact, the lead time to implement significant process modifications or optimizations or an alternate manufacturing process is long and may require revalidation and prior approval via submission of a regulatory supplement. A significant opportunity may be lost unless the work is initiated early (i.e., in parallel to the synthetic route in the initial NDA submission). Additional resources with multiple independent teams working on the same project are typically required. Postfiling changes become a part of the life-cycle management for a drug, and the International Conference on Harmonization (ICH) has started to address this topic through Q10.[1] Q10 will continue to attempt to strike a balance between the delays in regulatory reviews and oversight that can hold up the applicant's implementation of an improved manufacturing process.

2.3.3 Postfiling Changes

These postfiling changes become an expectation, and the drivers are several. These could be economic (cheaper intermediates, fewer steps) or environmental (in order to avoid the use of less-than-desirable reagents that could potentially harm the environment), or simple throughput in order to meet the market requirements. The changes typically drive the chemistry to become greener, a very common practice in process chemistry, yet the term has been coined only recently.[2] Process chemistry is driven by economics and the desire to make processes economical while using readily accessible cheap starting materials and minimizing waste.

Postapproval changes that describe alternative manufacturing processes, second-generation synthetic routes, or significant modifications pose a major challenge to the industry. Generally, a sponsor is obligated to demonstrate that the quality of the API manufactured from a new process is equivalent to the quality of the API manufactured from the registered synthetic process. For example, regulators typically expect comparative equivalence of API impurity profiles or justification and qualification for the presence of new impurities that might arise from the alternate process. This could be easy to achieve if changes to the filed process are only streamlined modifications in process conditions or parameters, and there are no changes in reagents or in bond-forming steps.

But once the bond-forming steps in the process are changed, the impurity profile of the drug substance will most definitely change. Consequently, the introduction of purification steps may be necessary to meet the filed specifications. Otherwise, an impurity qualification strategy in collaboration with safety evaluation may be required to insure the quality of the material has not changed. Generic companies that tend to change the synthetic route[3] have to go through a similar process in the United States and Europe but not necessarily in the rest of the world, where cost is more critical than quality.

2.4 TRENDS IN THE REGULATORY ENVIRONMENT

All activities of a process chemist are influenced by regulatory requirements and expectations described in regulations, guidelines, and precedents. The selection of regulatory starting materials, the control of impurities, and results from process validation and qualification are all subject to regulatory scrutiny. The regulatory environment is also continuously changing. Unfortunately, the rate and impact of change are different in Europe, the United States, Japan, and other global markets and regulatory authorities are not always aligned. Although harmonization of regulatory expectations has been initiated through the adoption of ICH guidelines, most of the differences in regulatory obligations are subjective and derived by an individual country's ability to maintain a regulatory organization and evaluate prospective medicinal candidates. As a result, the process chemist tends to assume the most conservative position in order to ensure adherence to global regulatory requirements.

Regulatory agencies typically issue a "guideline" on a certain topic and allow the public and interested parties to comment on the guidance and its impact on the pharmaceutical industry. The greater the impact of the guidance, the more the comments, and the more likelihood that the guidance will stay in draft form longer, in order to allow authors of the guidance and interested parties the time to debate and potentially incorporate useful comments within the guidance. Adoption of or changes to a guideline can take several years to finalize and usually lag behind the actual needs of the stakeholders. Health authorities are well aware of this issue and have intensified their interactions with industry in an effort to be more proactive.

Within the last few years, several guidelines governing the manufacture and control of APIs have been issued that fundamentally influence the work of the process chemist. These include the following:

- Genotoxic impurities
- Good manufacturing practices (GMPs) for the 21st century
  - ICH Q8 (Design Space)
  - ICH Q9 (Risk Management)
  - ICH Q10
  - Process analytical technology (PAT)
- FDA guidance on starting materials

2.4.1 GENOTOXIC IMPURITIES

Genotoxic impurities are chemical compounds that may be mutagenic and could potentially damage DNA.[4] ICH Q3A addresses the control and qualification strategy for many process-related impurities, but it does not really address genotoxic impurities. Most pharmaceutical companies have incurred delays and even clinical holds related to the presence of potential genotoxic impurities in APIs. Although it is the responsibility of every applicant to avoid exposing patients to these types of impurities, the development of process strategies to limit the presence of these impurities and demonstrate adequate control is a tremendous challenge for the process development chemist. Frequently, reactive reagents and other efficient chemical substrates used in the manufacture of active pharmaceutical ingredients are inherently toxic. In fact, an API by definition must trigger a physiologic response to be effective. Regulatory authorities generally expect sponsors of clinical trials and commercial marketing authorizations to demonstrate the removal of potentially genotoxic impurities or control them to minute levels in the ppm range. There are also nonalkylating agents that are classified genotoxic due to the nature of the functional groups they possess. One such example is aniline and derivatives thereof. Even intermediates in a synthesis that contain these functionalities are considered potentially genotoxic, and a strategy to control impurities containing these functional moieties either by demonstrating that these impurities are non-genotoxic or by them to the ppm level should be part of a regulatory submission strategy during development and prior to NDA submission.

In addition, final salt-forming steps can introduce genotoxic impurities. Some examples include formation of methyl chloride as a side reaction of hydrochloric acid in methanol, or esters of methanesulfonic acid as the by-product from the methanesulfonic acid salt formation step in alcoholic solvents.[5]

While process-related impurities of the API are potential sources of genotoxic impurities, excipient degradation products or impurities resulting from interactions of APIs with excipients may also be genotoxic. Aldehydes are typical examples from degradation of polyethylene glycol-based formulations. In addition, genotoxic impurities from one process that may carry over to another process must be managed and controlled by adopting appropriate plant cleaning methodology. Although most multipurpose chemical manufacturing plants that prepare APIs for pharmaceutical use are obligated to demonstrate control of cross-contamination in accordance with current good manufacturing practice (cGMP), chemical plants that manufacture precursors and other chemical reagents typically are not. Consequently, the process chemist frequently assumes responsibility for ensuring the quality of these materials is adequately controlled.

Advancements in analytical detection tools and techniques that can assess the presence of impurities to levels at or near 1 ppm or below have undoubtedly influenced increased expectations and scrutiny by regulatory health authorities.

The European Medicines Agency's (EMEA) Committee for Medicinal Products for Human Use (CHMP)[6] recently issued its final guideline with recommended limits for exposure to potential genotoxic impurities, where the threshold of toxicological concern (TTC) is 1.5 µg per day for commercially approved drugs.[7] There are, however, no direct guidelines describing recommended limits of genotoxic impurities for drugs in development. Consequently, some regulatory authorities default to application of this guideline even during the investigational stage of development. The process chemist is therefore faced with the challenge of evaluating process conditions and alternative synthetic strategies to reduce or demonstrate control for these impurities to unprecedented levels. Setting specifications for an API becomes a challenging problem that spans multiple disciplines, including chemistry, formulations, toxicologist, quality control, quality assurance, and regulatory personnel.

PhRMA published a draft position paper proposing a classification system for genotoxic impurities and associated limits during each phase of development based on the level of risk these impurities may pose on the safety of subjects participating in clinical trials.[8] This staged TTC concept was also recommended for establishing limits for commercial products.

The impact of this guidance on generic drugs, especially those with new synthetic routes, has not yet been fully realized.

2.4.2 Pharmaceutical cGMPs for the 21st Century

This guidance,[9] issued in the fall of 2004,[10] is essentially encouraging the industry to adopt new technologies (such as PAT) to control quality and introducing a risk-based approach to the science and to ensure the FDA's practices in reviews and inspections are based on state-of-the-art pharmaceutical science. It was defined elegantly by Janet Woodcock (FDA Commissioner, October 2005) as a "Mutual Goal of Industry, Society and Regulators." And it encourages the creation of a maximally efficient, agile, flexible pharmaceutical manufacturing sector that reliably produces high-quality drug products without extensive regulatory oversight.

Harmonization with the European Union started with the creation of two expert working groups (EWGs) to address Q8 and Q9.[11]

ICH Q8[12] allows an applicant to show how well a process performs over a range of variables through the use of design of experimentation, PAT, and risk management tools and "seeks to incorporate elements of risk and *quality by design* throughout the life-cycle of the product." Q8 also invokes the principle of design space.

2.4.3 DESIGN SPACE

Design space[13] is the collective knowledge gathered during the development phase of a process to understand whether certain variability in the process could affect the quality of the product. This process knowledge then defines the boundaries within which process variability is controlled. If a certain chemical step was known to make an undesired genotoxic impurity or an impurity affecting the API quality, understanding what factors produce and/or remove this impurity constitutes process understanding, which helps you set parameters for your design space. For example, if during the processing step, the temperature of the mixture increased by 10°C from the usual norm for that reaction and the level of that impurity did not increase from the normal level, then this temperature increase is not a critical process parameter. However, if the impurity level does increase under that scenario, then temperature is a critical parameter to quality and must therefore be controlled. This is clearly a simplistic example describing a single variable; in reality, the quality attributes of a process are a function of many variables that influence each other. What is important for the chemist is to understand the impact of these variables on safety and efficacy.

DOE (design of experiments) and PAT are critical tools with which to gather the relevant data that constitute the fundamentals of the design space. An interesting article sponsored by PhRMA describes a science- and risk-based approach and provides insight in the matter.[14]

The API, as an outcome of the designed processes, needs to be fit for use in the formulating of the drug product where the requirements need to be well defined and understood. Increased API complexity, lower solubility and bioavailability, less stable APIs, and more complicated "intelligent" formulations as a result impose specific form requirements (salts, particles, etc.) calling for the process chemist's creativity and expertise to find adequate solutions. Polymorphism continues to be an important aspect to be managed, and there is an excellent chapter on crystallization and polymorph controls in this book.

2.4.4 RISK MANAGEMENT

ICH Q9 defines "the principles by which risk management will be integrated into decisions regarding quality and CGMP compliance." Q9 focuses on important aspects of this risk management, risk control, communication, and review. It also speaks to the emerging trend of much closer collaboration between the regulators and the industry (see Reference 6 for an excellent overview).

The guidance sets the framework for the future and will continue to refine the concept of quality systems, encourage future cGMP international harmonization, and achieve science-based quality standards.

Most recently ICH Q10 has emerged as an approach to address quality issues of both the API and the drug product over the whole lifecycle of the drug. Currently, it is very difficult to incorporate improvements in a process that were developed over many years of learning without regulatory oversight. Q10 would improve the quality monitoring and review and would provide greater assurance that the continuous improvements in a process do not result in unintended consequences.[1]

2.4.5 Process Analytical Technology

This was a successfully launched guidance representing a very close collaboration between the industry and the regulatory agencies. The concept started in 2001, the draft guidance was issued September 2003, and it was released a year later. The guidance[15] can be viewed on the FDA's Web site, and the goal is "to understand and control the manufacturing process, which is consistent with the current drug quality system: *quality cannot be tested into products; it should be built-in or should be by design.*"

PAT equates to process understanding where sources of variability are identified, explained, and managed by the process, and the product quality attributes could be predicted by the process.[16] The other major feature of this guidance is around feedback control and intervention to immediately correct a problem during manufacturing as it arises. PAT may be applied to old products within a manufacturing facility or to new products through the NDA process, and it is not an obligation, as following the guidance is completely voluntary. The FDA actively encourages applicants in this direction.

2.4.6 FDA's Starting Materials Guidance

During the after-phase II meeting with the FDA, an applicant would usually discuss the starting material strategy. The starting materials for an API represent the point at which GMP begins in the API manufacturing process. The Starting Material Guidance issued January 2004 describes the concept of "propinquity," which means proximity or nearness.[17] The concept is that there must be several steps between a starting material and an API. Lots of comments from PhRMA and others were provided back to the FDA citing inconsistencies with ICH guidance. There have been several presentations by FDA personnel describing this guidance[18] and explaining its purpose. The industry has argued that as long as a starting material meets certain quality specifications, then it should not matter how it is made as long as vendors can meet these set specifications, regardless of whether the material has a significant or nonsignificant pharmaceutical use. Furthermore, pharmaceutical companies are expected to have an appropriate system for supplier management, i.e., to carefully evaluate process changes at their suppliers as well as evaluate change of supplier for impact on API impurity profile. The FDA sees it differently and argues that process understanding means to understand the whole process only from starting materials. Although this guidance was recently withdrawn, regulatory agencies continue to request more information about where raw materials come from and whether they were made under GMP regardless of controls and set specification you have in place.

The process chemist must again strike a balance between how much process knowledge to generate in phases of development where attrition is high and even during the later stages of development where there may still be a chance to change to a more optimal process that may require a completely new starting material.

Other current trends discussed in this book include the emergence of the Asian markets and the development of macromolecules as pharmaceuticals as opposed to small molecules. Separate chapters in this book discuss these topics at length.

2.5 EMERGING ASIAN MARKETS

Another major business trend has been the emergence of both India and China as sources of cheaper labor and as emerging pharmaceutical markets. Two excellent chapters in this book discuss their differences and the advantages and disadvantages of each country, and therefore, these topics will not be discussed here. However, the two emerging economies can no longer be ignored due to the availability of capable and well-trained process chemists and a plethora of CGM-compliant manufacturing plants. The trend to "outsource" or "off-shore" some of the development activities will likely continue as long as a risk-mitigating strategy is in place to deal with intellectual properties issues.

2.6 MACROMOLECULES

Anytime the term *process chemist* is used, it usually relates to the development of a small synthetic molecule. However, there has been a growth explosion in the development of macromolecules as of late. These may include proteins, small molecules attached to proteins, pegylated proteins, DNA and RNA fragments, oligonucleotides, monoclonal antibodies, and even personalized medicines in vaccines being made from a specific patient to treat the same patient. As this is a growing trend within the pharmaceutical industry, the term *process chemist* should include development chemists who are involved in all of the above-mentioned areas. A major difference between these areas and the development of small molecules is that the process to manufacture these complex products usually defines the product versus the other way around for small molecules, where the product defines the process.

The training of these development chemists is different, but the regulatory hurdles are similar. In addition, the outsourcing of development activities in these areas is currently limited due to limited worldwide talent in this area. However, that could change rapidly as emerging Asian markets adapt to change.

2.7 FUTURE TRENDS

In addition to the current trends highlighted in this chapter, the authors believe that the following future trends will likely affect the pharmaceutical industry for years to come. These trends will bring significant change to the industry and will bring many challenges to the process chemist. These trends include the following:

- New "intelligent formulations" developed in an effort to rescue certain candidates that may lack certain inherent "drug-like" characteristics. These intelligent formulations can improve the half-life of the drug, can enhance its absorbance, and can blunt C_{max} when needed. These formulations offer new opportunities to get candidates to patients faster and can also enhance the intellectual property position of a certain drug. On the other hand, these formulations can be challenging to the process chemist as they sometimes may require specific final forms with specific particle size distributions that may not be attainable by common means.
- Process chemists will be even more integrated with medicinal chemists as they continue to be more proactive in assisting in the synthesis and the nomination of candidates into development. This becomes even more critical as timelines continue to compress with less time to apply the right science to a synthetic problem.
- As the assault of generic companies on intellectual properties continues, more process patents will likely be filed in order to secure the intellectual space for processes and products. This not only would allow companies the freedom to operate but could potentially enhance the product life of the drug.
- Regulatory approvals of a larger number of macromolecules mentioned earlier as opposed to small molecules will likely continue in the future. This has huge impacts on the "traditional process chemists" in terms of the different skills needed. Chemists could and should branch out into the peptide chemistry and nucleotide chemistry arenas of synthetic chemistry. However, this branching out may be much more difficult in the area of monoclonal antibodies where the skill sets are very different.
- Closer collaborations between pharmaceutical companies and the regulatory agencies will continue. As new guidelines emerge, the pharmaceutical companies will be more integrated in the regulatory decision processes. The process chemist needs to be heavily involved in this dialogue in order to implement changes readily and to communicate the process chemist's point of view and challenges as well.

- The new emerging market's regulatory requirements, processes, and ways of interactions need to be mastered. They may not in all aspects be aligned with those of the United States and the European Union, hence adding another complicating factor.
- Further investments by large pharmaceutical companies in small and emerging biotech companies that are focused on specific therapeutic areas will continue for many years. This will help large companies to add more candidates in their R&D portfolios, and small companies will benefit from the infusion of money into their operations. Depending on the licensing deal signed, these candidates usually pose unusual challenges to the development organization to ensure quality and no interruptions of supplies to the market.
- New emerging low-cost markets are coming online everyday. These pose opportunities in terms of finding them early and potentially saving development costs. The opportunities are usually short-lived as the standard of living in many of these emerging economies will catch up with the rest of the world, and the cost savings will eventually go away. Companies should therefore stay on the lookout for these emerging economies and be there early to capture the savings and the market. Where will the next low-cost market be? Will South America emerge as a new economic power in the future?
- In some cases where the side effects of a certain drug candidate are well understood and can be attributed to certain physiological changes or a genetic aberration such as a lack of a certain receptor or enzyme, one can, in theory, test for this side effect before taking the drug. Developing a diagnostic tool could be one way to deal with this issue. This area of developing a diagnostic to monitor for the side effects of a drug is in its infancy, but it could potentially save many development candidates, which may otherwise not make it to the market due to the lack of tolerance to risk. Although this sounds simple, the downsides of this strategy are the cost and the timelines associated with developing yet another tool that can help in the development of a drug. Ultimately, the drug will be safer but at a much higher cost.

The job of the process chemist continues to evolve and be redefined due to internal and external changes and influences, and the job continues to be extremely rewarding, in terms of the impact a process chemist can make on an organization, especially for candidates that have achieved proof of concept and are ready to file an NDA. Their efforts are key to the advancement of new drugs to the market, to getting cost of goods and NPVs correct, and to supporting launch with a robust commercial supply chain. Second, many new challenges, as the ones discussed above, will continue to be added to the chemists' scope, increasing the diversity of the discipline and continuously making the job stimulating and exciting. Third, no two drugs are alike and no two development pathways are identical; therefore, each drug candidate poses a different challenge, keeping the process chemists challenged and focused on the job.

Change is inevitable and trends come and go, but the discipline remains grounded in the science of organic chemistry. Nonetheless, good development chemists have learned to anticipate change early and to adapt quickly to new processes and new technologies, establishing efficient ways of doing work that keep the science fresh, exhilarating, and fun.

REFERENCES

1. www.fda.gov/cder/meeting/ICH/famulare20061002.pdf; www.fda.gov/ohrms/dockets/AC/04/slides/ 2004-4052S1_04_Massa.ppt and Gerry Migliaccio, AAPS Workshop-Pharmaceutical Quality Assessment, October 6, 2005 (www.aapspharmaceutica.com/workshops/PharmaceuticalQuality100505/ migliaccio.pdf).
2. Anastas, P. and Warner, J., *Green Chemistry: Theory and Practice*, Oxford University Press, Oxford, 1998.

3. Basak, Arup K., Raw, Andre S., Al Hakim, Ali H., Furness, Scott, Samaan, Nashad I., Gill, Devinder S., Patel, Hasmukh B., Powers, Roslyn F., and Yu, Lawrence, Pharmaceutical impurities: regulatory perspective for abbreviated new drug applications, *Advanced Drug Delivery Reviews*, 59, 64–72, 2007.

4. Guidance issued by the EMEA (European Medicines Agency, Evaluation of Medicines for Human Use, London, June 23, 2004. See also Reference 6.

5. David, J. Snodin, Residues of genotoxic alkyl mesylates in mesylate salt drug substances: real or imaginary problems? *Regulatory Toxicology and Pharmacology*, 45 (1), 79–90, 2006.

6. www.emea.europa.eu/htms/human/swp/swpfin.htm; CPMP/SWP/5199/02.

7. For a 60 mg daily dose, 1.5×10^{-6} g of impurity/day/60×10^{-3} g of drug/day $\times 1000 = 25$ ppm. Hence, for a drug with a 60 mg daily dose, the guidance allows 25 ppm of one genotoxic impurity.

8. Müller, L., et al., A rationale for determining, testing, and controlling specific impurities in pharmaceuticals that possess potential for genotoxicity, *Regulatory Toxicology and Pharmacology* 44, 198–211, 2006.

9. This is not a guidance, but an approach that later appeared as a guidance in ICH Q8, Q9, and Q10.

10. www.fda.gov/cder/gmp/gmp2004/GMP_finalreport2004.htm.

11. www.rpsgb.org.uk/pdfs/qpsymp050215pg.pdf.

12. www.fda.gov/cber/gdlns/ichq8pharm.htm.

13. www.aapspharmaceutica.com/workshops/pharmaceuticalquality/10505/hussain.pdf.

14. Ganzer, W.P., Materna, J.A., Mitchell, M.B., and Wall, L.K., Current thoughts on critical process parameters and API synthesis, *Pharmaceutical Technology*, July 2005.

15. www.fda.gov/cder/OPS/PAT.htm.

16. Personal communication with Chris Watts of the FDA PAT policy team, Center for Drug Evaluation and Research (CDER), office of Pharmaceutical Science. From a talk by Chris Watts presented at Pittcon Conference, March 2, 2005, Orlando, FL.

17. www.fda.gov/cber/gdlns/drugsubcmc.htm.

18. www.fda.gov/cder/present/DIA2004/Miller.ppt.

3 Varenicline: Discovery Synthesis and Process Chemistry Developments

Jotham W. Coe, Harry A. Watson Jr., and Robert A. Singer

CONTENTS

3.1 OVERVIEW

Varenicline (**1**; Figure 3.1), a partial agonist of the $\alpha 4\beta 2$ nicotinic acetylcholine receptor (nAChR) developed specifically for smoking cessation, was submitted for approval to the U.S. Food and Drug Administration in November 2005 and approved in May 2006. This discussion describes to the synthesis of the material after its 1997 nomination as a potential new medicine. In the course of these efforts, an economic and safe synthetic process for large-scale manufacturing of varenicline was identified and is the subject of this chapter. What typically transpires as clinical candidates progress is a battle for resources to support ongoing optimization of existing discovery syntheses

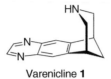

Varenicline **1**

FIGURE 3.1 Varenicline.

versus the development of novel manufacturing routes. As is described herein, the discovery route was viable for commercial synthesis and supported all development efforts through Phase 2; however, another elegant approach was developed that addressed critical issues and became the manufacturing process of choice.

3.2 INTRODUCTION: PHARMACOLOGY AND BACKGROUND

Few lifestyle choices have had so profound and lasting an effect on today's world economy and health as tobacco smoking.[1] The current toll on human health is staggering, as half of the world's 1.25 billion smokers are expected to die from smoking-related illnesses.[2] Appreciation of smoking's dire health effects led to the development of medicines such as controlled-release nicotine in gum and patch forms in the 1980s.[3] Zyban® (bupropion; GlaxoSmithKline, Philadelphia, PA)[4] was the next breakthrough and was the first nonnicotine therapy, identified for its unexpected effects on smoking rates in depressed patients for whom bupropion was intended (sold as Welbutrin® [Glaxo-SmithKline, Philadelphia, PA]). Although neither treatment achieves high long-term quit rates (<10%), both herald the possibility that nicotine dependence can be treated with medicines.[5]

Nicotine binds at high-affinity nicotinic receptors in the brain, causing the release of dopamine downstream in the mesolimbic dopamine system. This primary physiological effect from tobacco smoking creates reinforcing sensations that support dependence and maintain the habit. More important, abstinence from smoking leads to low dopaminergic tone, creating a significant urge to smoke again to compensate for low dopamine levels.[6]

The project team recognized the need for a medicine with a "dual" mechanism of action to treat nicotine dependence from tobacco smoke.[7,8] We reasoned that an agent that blocked not only the dopamine elevation from smoking but also the dopamine low, or void, from abstinence would address both sensations that lead to continued smoking. Seeking this dual effect, we specifically targeted nicotinic partial agonists of the neuronal α4β2 nAChR as medicines. The hypothesis was that "partial" activation of this receptor would elicit a moderate and sustained increase in mesolimbic dopamine that would be below the peak levels induced by nicotine but sufficient to relieve the low dopamine levels encountered in the absence of nicotine during smoking cessation attempts. Low levels of dopamine have been associated with craving for and withdrawal from nicotine and are the key syndromes that precipitate relapse to smoking behavior. Additionally, by binding to the α4β2 nAChR, a nAChR partial agonist would competitively shield the smoker from nicotine-induced dopaminergic activation in the event of renewed smoking. In theory, without the nicotine-induced elevation in mesolimbic dopamine levels, tobacco use would not produce a pharmacological reward. Thus, a partial agonist was considered uniquely suited as a treatment for this condition.[9]

3.3 DISCOVERY SYNTHESIS OF VARENICLINE

Varenicline, a small (molecular weight 211.27) achiral alkaloid, was identified in 1997 as a potential treatment for smoking cessation.[7] Its selection as our clinical candidate came after years of effort to identify an effective and safe partial agonist selective for the α4β2 nAChR subtype. A majority of the analogs explored in the course of our work were structurally related to the natural product (–)-cytisine and variations thereof.[7,9,10] The discovery synthesis shown in Scheme 3.1 is rooted in

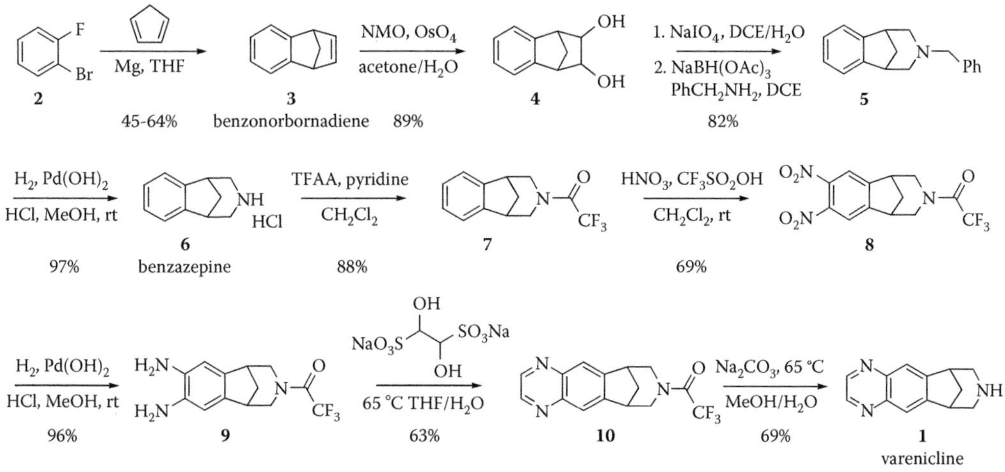

SCHEME 3.1 Discovery synthesis of varenicline (**1**).

the methodology used to make hundreds of analogs on a milligram to greater than 100-gram scale. As such, it served well throughout the discovery phase and into the early process scale-up work.

When varenicline was accepted as a candidate for development, analytically pure material was provided from our discovery laboratory in support of early development work (see Scheme 3.1). Based on the potency of varenicline, the project team anticipated that 50 grams would be sufficient for these studies. In the end, our assumption proved correct, as a mere 24 grams of varenicline supported all early toxicology and formulation work. This triggered advancement of the compound into the investigational new drug application development phase and the need for good manufacturing practice (GMP) clinical supplies.

As evident above, the discovery route to varenicline is a linear, nine-step synthesis that takes advantage of symmetry.[7] Notably, the first step of the route is the only one that involves carbon–carbon bond formation. Benzazepine **6**, a key synthetic intermediate, was known in the literature, and its preparation is the primary subject of this chapter. The latter steps have proven robust and safe on scale, providing a six-step regulatory synthesis from **6** to varenicline delivered in the desired salt form.

The original literature preparation of benzazepine **6** (Scheme 3.2) was reported in 1978 by Mazzocchi and Stahly[11] and began with benzonorbornadiene (**3**), a compound prepared by the benzyne Diels–Alder reaction of 1-bromo-2-fluorobenzene (**2**) and cyclopentadiene.[12] Mazzocchi and Stahly's preparation involved hydration of the olefin to generate **11** and sequential oxidations of **11** (Al(OtBu)$_3$, SeO$_2$, and KO$_2$) that ultimately led to intermediate diacid **13** (see Scheme 3.2). Conversion of **13** to the corresponding anhydride followed by treatment with ammonium hydroxide and thermal dehydration gave cyclic imide **14**. Lithium aluminum hydride reduction provided **6** in 2% overall yield.

The discovery synthesis of varenicline emerged from a program initially focused on the preparation of derivatives of the bicyclic lupin alkaloid (–)-cytisine, a natural product that was rigorously established to be a partial agonist of the α4β2 nAChR subtype in 1994.[13] The majority of our targets were [3.3.1]-bicyclic derivatives and [3.2.1]-bicyclic 3,5-bridged aryl piperidines,[10] each derived from key olefinic intermediates (Figure 3.2). A routine cyclopentene-piperidine conversion emerged from these synthetic efforts that, as in Mazzocchi's approach, exploited bicyclic precursors to generate the required *cis*-dicarbonyl intermediates. In our syntheses, olefins were converted to intermediate dialdehydes via an OsO$_4$-N-methylmorpholine-N-oxide (OsO$_4$-NMO) dihydroxylation[14] and oxidative cleavage,[15] followed by conversion to the corresponding N-benzyl piperidines using Maryanoff's reductive amination protocol.[16]

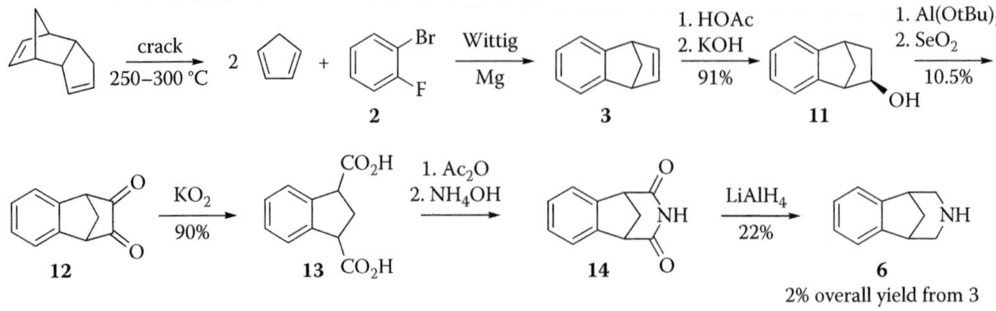

SCHEME 3.2 Mazzocchi and Stahly's synthesis of benzazepine **6**.

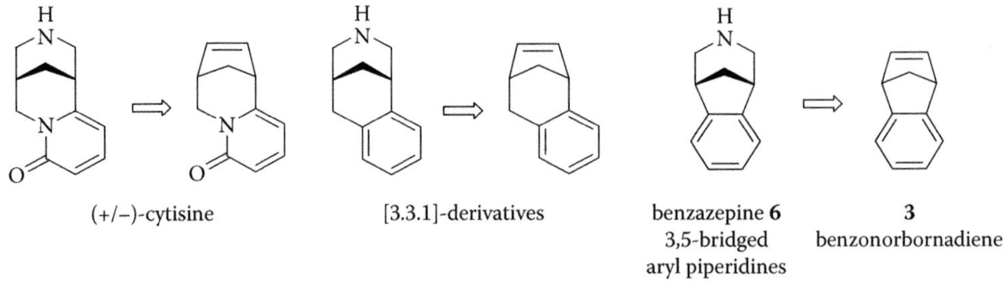

(+/−)-cytisine   [3.3.1]-derivatives   benzazepine **6**   **3**
                                                      3,5-bridged   benzonorbornadiene
                                                      aryl piperidines

FIGURE 3.2 Olefinic precursors to bicyclic cytisine derivatives.

3.4 EARLY PROCESS CHEMISTRY CAMPAIGNS: THE FEARED BENZYNE-OSMIUM ROUTE TO BENZAZEPINE FROM BENZONORBORNADIENE

From the outset, discovery and process chemists joined forces to deliver the first 150-gram campaign and to transition the existing expertise. Benzazepine **6** was prepared using the discovery approach in the discovery laboratory, and its conversion to varenicline was then completed in the process laboratory. This crucial discovery/process collaboration delivered the 150-gram campaign in 6 weeks, and the rapid delivery greatly accelerated GMP toxicology studies, early proof-of-concept studies in human clinical trials, and development work toward the final salt form and tablet formulation. In addition, as is described herein, very practical solutions to scale-up issues were incorporated, paving the way for two large scale-up campaigns over the next 10 months, all using the discovery synthetic route.

Key issues were identified that required attention before starting on kilo laboratory scale-up campaigns. These included the following:

1. Cracking of dicyclopentadiene on scale would require dedicated, specialized equipment and high thermal input (250–300°C).
2. Because cyclopentadiene is unstable, low-temperature storage and immediate use would be required.
3. The Grignard route from **2** generated highly reactive benzyne intermediates. A controlled addition rate would be necessary for safe operation.
4. Catalytic OsO$_4$ dihydroxylation of **3** risked contamination of pilot plant and kilo laboratory equipment, necessitating testing for residual osmium at part per billion levels.
5. The dinitration process posed thermal hazards requiring thorough safety assessment.

6. Nitrated intermediates are often dangerous to prepare, store, and fully purge from final active pharmaceutical ingredient.
7. Catalytic reduction of dinitrated intermediates (e.g., **8**), which is often dangerous, requires careful control.
8. Anilines must be fully purged from the final product.
9. Purity levels of discovery lots of varenicline were >99.7%, setting a high lower-limit for kilo-scale campaigns.

During the discovery synthesis phase leading to varenicline, we generated benzonorbornadiene **3** using the Grignard approach described by Wittig and Knauss in the 1950s.[12] Few challenges arose as we routinely performed the sequence on a ~100-gram scale. However, as discussed below, a number of insights allowed us to better understand the observed performance of this and other steps on the kilo scale over the course of the program.

3.4.1 CRACKING AND STORAGE OF CYCLOPENTADIENE

The source of cyclopentadiene for the benzyne reaction was from the large-scale cracking of the dimer (Figure 3.3). The casual cyclopentadiene laboratory user would typically give little thought to the challenge of scaling this process to multi-kilo scale other than knowing that the process requires constant vigilance.

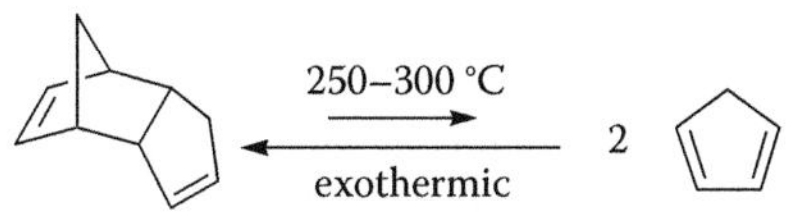

FIGURE 3.3 Cyclopentadiene cracking and dimerization.

For the 150-gram bulk reload campaign, an *Organic Synthesis* laboratory method[17] was used to prepare the required 400 grams of cyclopentadiene. On the kilo scale, the general practice of maintaining inert atmosphere and collecting cyclopentadiene at low temperatures followed by immediate use would be especially problematic without dedicated cracking and storage equipment. Cyclopentadiene storage was also a major concern, and safety laboratory chemists became engaged to evaluate the stability of this key material.

By using instruments that measure temperature versus time and pressure versus time under conditions that avoid heat loss (adiabatic conditions), rate information can be obtained to characterize the exothermic potential of chemical decompositions. Analyses of this type provide onset temperature, rates of temperature and pressure rise, and time-to-maximum-rate (TMR) data. Accelerated reaction calorimetry (ARC) is particularly sensitive at detecting onset temperature for decomposition (usually calibrated to any temperature excursion of 0.02°C/min, which will send the equipment into an exotherm tracking mode). ARC uses 1 to 5 grams of material, whereas the vent sizing package (VSP) test uses 20 to 50 grams. Because the VSP test uses a larger sample in a thin-walled container, the rates of maximum temperature and pressure rise closely simulate what would occur on a larger scale under low-heat-loss conditions. As such, the VSP data are useful in determining the size required for relief vents on reaction tanks, and these data are considered more directly applicable to large-scale reaction planning. The assessment in Figure 3.4 was obtained.

Clearly, cyclopentadiene deserved great respect. Based on these observations and additional experiments, two recommendations for cyclopentadiene storage were proposed:

• Immediately dilute the cyclopentadiene directly from the cracker with ten volumes of hexanes held cold (at around –40°C).
• Store the pure cyclopentadiene packed in dry ice at –40 to –50°C.

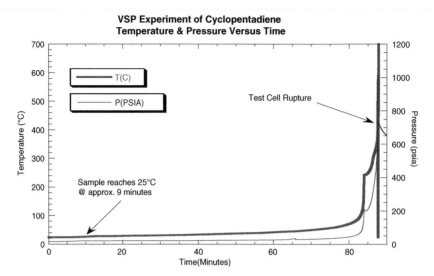

FIGURE 3.4 (See color insert following page 40). The vent sizing package (VSP) test of cyclopentadiene reveals a slow exotherm upon reaching 25°C that accelerates to a violent decomposition after 80 min. The decomposition caused the test cell to rupture. These data show that the consequences of a runaway reaction would be severe.

An accelerated reaction calorimetry (ARC) test determined the time-to-maximum rate (TMR) of the decomposition from subambient conditions. Further testing on the material diluted in hexane/heptane was performed to investigate the consequences of a runaway reaction in these mixtures. ARC showed an exotherm from the first heat–wait–search step (at 31°C), which progressed to more than 110°C. A second exotherm was seen at 160°C, progressing to 200°C. The initial decomposition showed a TMR on the order of several days from the storage temperature (–40°C) anticipated in kilo laboratory campaigns. The TMR is on the order of an hour from room temperature, however, which supports the data from the VSP test of cyclopentadiene.

The first method had a theoretical advantage in that high dilution would reduce the adiabatic temperature rise of a runaway reaction, making the consequences more manageable. The downside, however, was the need for much larger –40°C storage capacity for extended periods prior to use. The second method was considered preferable, as 3 kilograms of neat cyclopentadiene can readily be stored in a 5-liter flask packed in an insulated chest containing dry ice.

Safety testing certainly added to our vigilance for kilo-scale preparations, especially as we engaged an outside vendor. For upcoming scale-up campaigns, coolers were bought and packed with dry ice. The freshly cracked and distilled cyclopentadiene from 2-, 10-, and 11-kilogram sequential batches was stored over a 2-week period. The batches were safely carried into Grignard-induced benzyne Diels–Alder reactions to produce more than 20 kilograms of benzonorbornadiene in less than a month, only 8 weeks after the scale-up campaign began.

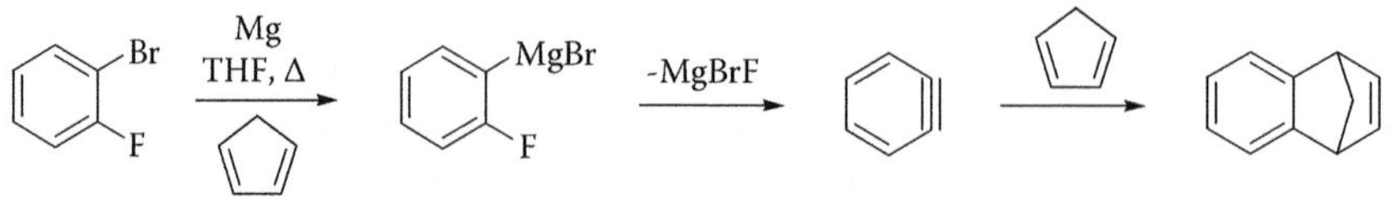

SCHEME 3.3 Grignard-based benzyne Diels–Alder reaction with cyclopentadiene.

3.4.2 Scaled Benzyne Reactions

The first "large-scale" Grignard-based benzyne reaction of 2-bromofluorobenzene (1 kg) was operated in a 12.5-liter round-bottom flask in the discovery laboratory (Scheme 3.3). A mixture of cyclopentadiene and 2-bromofluorobenzene in tetrahydrofuran (THF) was added to magnesium turnings in warm THF. Once the Grignard was initiated, a gentle reflux was maintained in a routine reaction using a controlled addition rate of the combined precursors.[12] After an aqueous workup, rather than using chromatography, we purified the crude product by distillation on a rotary evaporator; this was the first of a series of key improvements that enabled the move to process scale at an outside vendor. With this simple innovation, distilled product was collected in less than 30 min (80°C water bath with a vacuum pump attached, 445 grams), greatly simplifying future large-scale preparations and purifications of benzonorbornadiene (3).

Grignard-based benzyne reactions typically produce numerous benzyne-derived by-products.[18] For this reason, yields of benzonorbornadiene via this route are often moderate (~50%), as we also observed on large-scale. Fortunately, the by-products are higher in molecular weight, allowing purification based on boiling point. Additional removal of undesired products from the distilled crude reaction extracts (typically >90% pure) was achieved in the dihydroxylation step with OsO_4, which sequestered all benzonorbornadiene as crystalline analytically pure diol 4 (see Scheme 3.1 and Scheme 3.4).

Improvements to the efficiency, safety, and cost of operating benzyne-based benzonorbornadiene synthesis became available from concurrent studies in the discovery effort (Figure 3.5).[19] A halogen–metal exchange protocol avoided the Grignard initiation step, further simplifying large-scale reactions. By studying the reactions of readily available 1,2-dihalobenzenes (2a, b, and c), yields of benzonorbornadiene (3) were improved to 89% by treating a 0°C mixture of 1,2-dibromobenzene and cyclopentadiene (1.0 to 1.2 equiv) with n-butyllithium (1.05 equiv) in nonpolar solvents such as toluene (or hydrocarbons such as hexanes). Remarkably, in ethereal solvents,

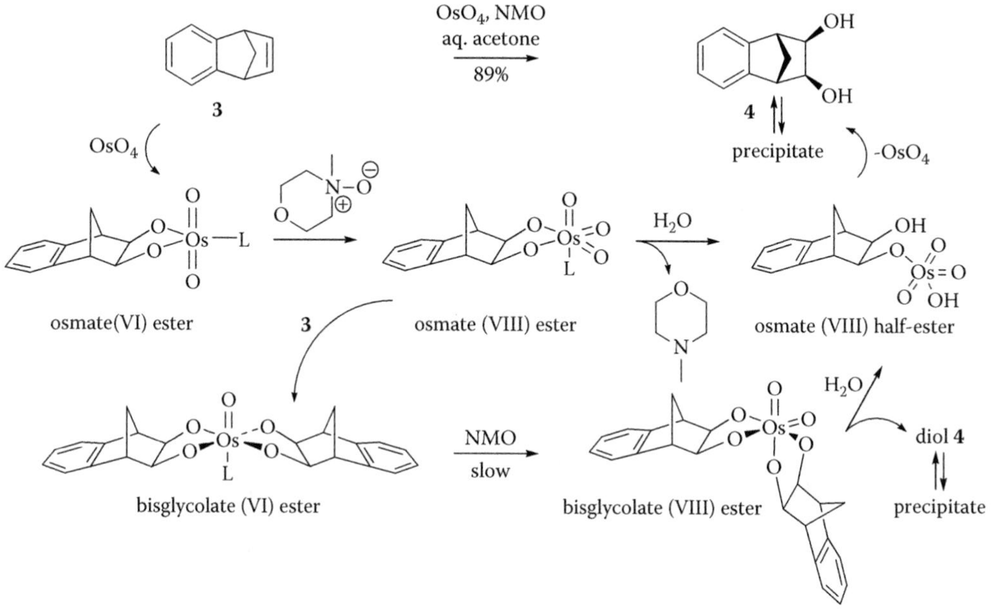

SCHEME 3.4 Catalytic cycle in the dihydroxylation of benzonorbornadiene (3).

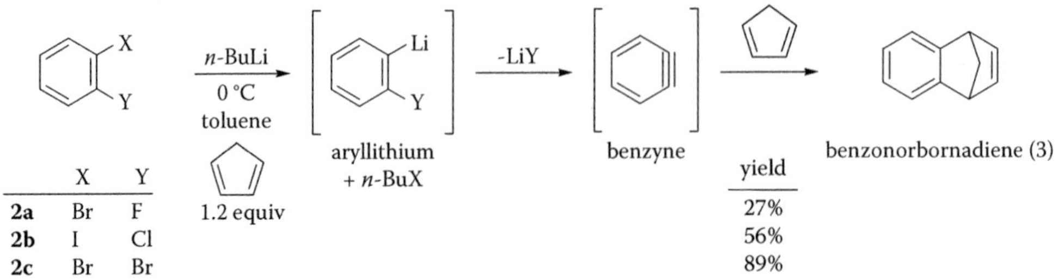

	X	Y					yield
2a	Br	F					27%
2b	I	Cl					56%
2c	Br	Br					89%

aryllithium + n-BuX   benzyne   benzonorbornadiene (3)

n-BuLi, 0 °C, toluene; −LiY; 1.2 equiv

FIGURE 3.5 Halogen–metal exchange-mediated benzyne Diels–Alder reaction with cyclopentadiene.

competing cyclopentadiene metalation with *n*-butyllithium is observed. Although the use of non-etherial solvents is an application of the conditions pioneered by Hart et al. with highly substituted 1,2-dihalobenzenes in diethylether,[20] only a few examples were known in the case of cyclopentadiene.[21] This modification improved the overall yield of bicyclic aryl piperidine **6** to >64% from 1,2-dibromobenzene **2c** and was demonstrated on a multi-kilo scale. In addition, ARC measurements by the safety lab determined that the thermal output with these modifications was easily managed under dose-controlled conditions.

3.4.3 OSMIUM-CATALYZED DIHYDROXYLATION OF BENZONORBORNADIENE

An osmium-catalyzed dihydroxylation of benzonorbornadiene **3** using a procedure developed by VanRheenan at the Upjohn Corporation was routinely applied for the preparation of diol **4** and other targets.[14] Under the standard conditions, conversion to diol **4** was often extremely sluggish below 4 to 5 mol% OsO$_4$ catalyst load. These conditions had been acceptable for most exploratory medicinal chemistry syntheses where the somewhat water-soluble diol products were isolated as oily solids, purified by traditional aqueous workup and chromatography. Fortunately, on a large scale, the isolation of **4** was greatly facilitated by crystallization. While monitoring a particularly sluggish conversion of benzonorbornadiene **3** to **4** on a large scale, it was noticed that crystals had been induced to form after several days. Apparently the glass pipette used to pull samples for thin-layer chromatography analysis provided a suitable surface to initiate the process. With vigorous stirring, precipitated product **4** accumulated, and the reaction rate accelerated.

This discovery—that the reaction was driven by precipitation of crystalline product—was an invaluable example of Le Chatelier's principle. Presumably the bicyclic scaffold of **4** stabilizes intermediate osmate esters via the rigid syn-diol, slowing hydrolysis of the osmate (VIII) ester, inhibiting catalytic turnover (Scheme 3.4). It is also likely that this catalytically active species adds to additional olefin **3**, leading to a bisglycolate (VI) ester, which is believed to be subject to slow catalytic turnover.[22] Higher concentrations of substrates (0.5 to 1.5 M) and rapid stirring induced the granular crystalline product to precipitate directly from the reaction mixture, typically driving complete conversion within 60 hours. Decanting or filtering and rinsing the product with fresh acetone afforded **4** in 89% yield using 0.13 to 0.26 mol% catalyst loading.[23] As an additional benefit, the filtrate from these reactions could be partially concentrated to give a dark mixture containing reduced osmium by-products presumed to be osmate esters of **4**. This filtrate could be reused in what we termed a "living" reaction. Simply introducing fresh benzonorbornadiene **3**, solvents, and NMO continued the osmium recycle and diol preparation; four or five recycles became routine.

3.4.4 OSMIUM-CATALYZED DIHYDROXYLATION: A PROCESS PERSPECTIVE

In appraising the overall route for commercialization, we determined that the residual osmium was the key concern, potentially risking contamination of multipurpose equipment in the kilo laboratory and pilot-plant facilities with the toxic heavy metal. We required a method to quantify and eliminate residual osmium, as only osmium-free materials should enter subsequent steps. We also evaluated alternative osmium-mediated methods. These studies revealed that sodium chlorite ($NaClO_2$) was a particularly effective alternative reoxidant (versus NMO), allowing the use of only 0.045 mol% of the potassium osmate. Although complete reaction was observed in 30 minutes, the yield (72%) and quality of the final product **4** did not provide a competitive advantage to the OsO_4-NMO/crystallization method, so this method was not used for bulk preparation campaigns.[23] We also studied stoichiometric potassium permanganate dihydroxylation of **3** as a potential replacement of the osmium-catalyzed dihydroxylation and found it to be highly pH dependent. Under basic conditions, the diol could be generated quantitatively, but because of its high crystallinity, the diol was difficult to separate from the residual stoichiometric MnO_2: pure material could be isolated in only 35% yield. Under acidic conditions, the desired dialdehyde could be generated directly. But because the required stoichiometric oxidant risked further oxidation to the dicarboxylic acid as well as isolation challenges, these conditions were not further pursued on scale.

3.4.5 RESIDUAL OSMIUM MEASUREMENTS

After a great deal of experimentation, inductively coupled plasma mass spectrometry (ICP-MS) was determined to be the most accurate and dependable analytical technique to measure residual osmium levels. This measurement was required before material could enter multipurpose production facilities. The results from ICP-MS were exceptionally accurate; even measurements from a control standard, spiked with 9.13 ppm of osmium, recorded a value of 10.0 ppm. Control samples of diol **4** (+/– heavy metals) and materials from subsequent steps were analyzed to track residue purging in later steps. These results showed that silica gel filtration of the reductive amination product **5** effectively removed all residual osmium (<0.04 ppm).

3.4.6 PIPERIDINE SYNTHESIS: OXIDATIVE CLEAVAGE/REDUCTIVE AMINATION

In the discovery laboratory, diol **4** produced by this dihydroxylation method was conveniently converted to **5** via a two-step procedure (Scheme 3.5).[14,15] Oxidative cleavage of **4** by $NaIO_4$ (1 equiv) gave dialdehyde **15** in aqueous dichloroethane. Aqueous washes effectively removed $NaIO_x$ salts, and the extracts were treated with benzylamine (1.05 equiv), dried by filtration through a cotton plug, then added to $NaBH(OAc)_3$ (3.6 equiv).[16] Conversion to **5** was complete within hours. Aqueous workup and silica gel pad filtration of the extracts gave **5** in 82 to 95% yield.

Considerable experimentation with this sequence during the discovery phase had provided little improvement in yield. Premixing the dialdehyde and benzylamine followed by addition to a slurry of $NaBH(OAc)_3$ in dichloroethane often gave yields as high as 95% on laboratory scale. Mixing

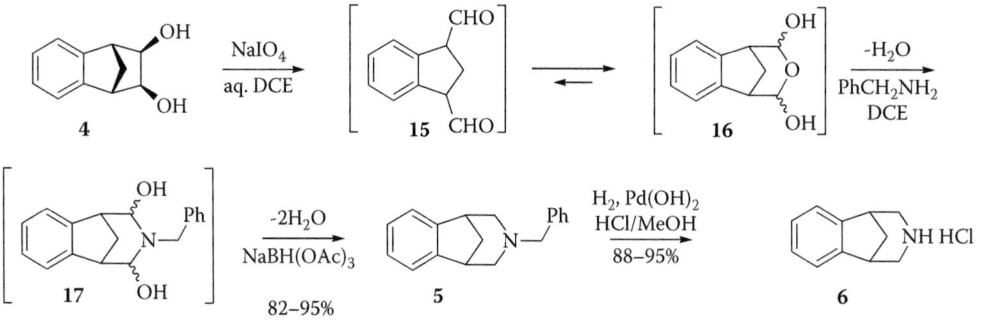

SCHEME 3.5 Oxidative cleavage/reductive amination.

of dialdehyde **15** and benzylamine generated water, which was readily separated from the chlorinated solvent by filtration through cotton. This observation suggested that dialdehyde **15** existed as hydrate **16**. Unhydrated dialdehyde would theoretically form a bridged piperidine diol (**17**) with benzylamine without liberating water upon mixing before NaBH(OAc)$_3$ reduction. Alternative explanations for water production included the formation of enamines; however, in related systems these conjugated intermediates were highly colored, which was not observed in this reaction sequence. The premixing liberated water and was accompanied by a mild exotherm, leading to biphasic mixtures that required separation, adding an operation that would potentially lead to varied hold times and yields on larger scale.

This cumbersome method was supplanted by a simpler operation for scale-up campaigns in which a mixture of crude dichloroethane extracts of dialdehyde and NaBH(OAc)$_3$ was treated with benzylamine. Yields using this method were consistent on laboratory scale (50 to 70%). The reaction reveals the impressive selectivity of NaBH(OAc)$_3$ as a reducing and dehydrating agent, as 3 equiv of water are produced from the presumed aldehyde hydrate (**16**). Even though not the highest yielding on a laboratory scale, this approach was readily scaled and returned a 91.6% yield for the process in the conversion of more than 15 kilograms of diol. Other methods such as adding dialdehyde (**15/16**) to premixed benzylamine and NaBH(OAc)$_3$ risked diamination and gave variable lower yields. Under these conditions, the predominant product was still the desired bridged adduct **5**, demonstrating the facility of intramolecular iminium ion formation and reduction.

3.4.7 FINAL OSMIUM QUESTION ON LARGE SCALE

The oxidative cleavage/reductive amination sequence was readily scaled. After standard aqueous workup, the final manipulation involved a simple filtration through silica gel to remove salts, polar by-products, and residual osmium from **5**. The success of this procedure stems from the remarkable nonpolar nature of the tertiary amine. In the favored chair conformation, the 3,5-bicyclic aryl group shields the axial nitrogen lone pair and ensures an equatorial disposition of the benzyl group (Figure 3.6). The lone pair is thus buried inside a lipophilic hydrocarbon framework, making this material particularly nonpolar. A silica gel pad filtration using the extraction solution of **5** from the reaction workup provided material with the expected low osmium levels as determined by ICP-MS (<0.04 ppm). Benzyl group removal by hydrogenolysis of the HCl salt of **5** gave benzazepine **6**.

3.4.8 QUINOXALINE RING: DISCOVERY AND SYNTHESIS: DINITRATION OF
TRIFLUOROACETYL–BENZAZEPINE

Appending the quinoxaline ring system to benzazepine **6** required amine protection before the dinitration reaction. The fused bicyclic system imposes close proximity of the nitrogen atom to the aromatic ring. As a result, the amine and sp^2-protected forms (amides and carbamates) unexpectedly inhibited electrophilic aromatic substitution (equations a and b, Scheme 3.6). We suspected that interactions of electrophilic reagents with the amine or protected amines generated competing

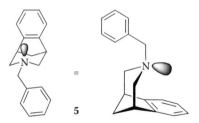

FIGURE 3.6 Nonpolar tertiary *N*-benzylbenzazepine **5**.

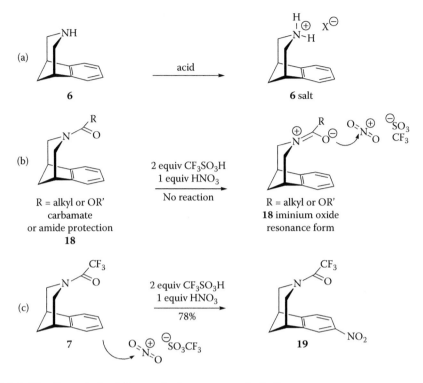

SCHEME 3.6 Bicyclic aryl piperidine: protecting groups and nitration.

cationic intermediates that inhibited electrophilic aromatic substitution chemistry. Only with the trifluoroacetamide group did electrophilic substitution chemistry proceed. We presume that a substantial decrease in basicity of the trifluoroacetamide compared to that of other protecting groups insulates the functionality from competitive interaction with electrophilic reagents.

The nitration of trifluoroacetamide **7** is at the core of the varenicline discovery story. In early studies with the unprotected free amino benzazepine **6** and carbamate-protected versions **18**, nitration of the aryl ring did not proceed (see Scheme 3.6). The unreactive nature of the ring system necessitated the study of more forcing conditions. One critical breakthrough was the use of nitronium triflate.[24] This compound is an exceedingly powerful nitrating reagent with a number of advantages over more commonly used alternatives. Chief among these are its organic solubility, allowing *in situ* generation in a number of chlorinated solvents, and its high reactivity at low temperature. In addition, as a result of its reactivity, the reagent is generally fully consumed in the process, leaving little or no unconsumed reagent and, therefore, fewer unstable by-products. When TFA amide **7** was first studied, nitronium triflate was under investigation in slight molar excess, which gave rise to rapid and clean conversion to mononitrated product **19**, which crystallized in pure form in 78% yield.

Following this development, we scaled up the reaction using the standard 30% excess of nitronium triflate (−78 to 0°C) described by Blutcher and Coon.[24] After aqueous workup, the mononitrated product nicely crystallized from the crude mixture. Upon reexamination of the filtrate using GC-MS analysis, it was noticed that small amounts of dinitrated products were present in the crude reaction mixture. Upon chromatographic isolation from the mother liquors, the major dinitrated by-product was identified as **8**, the result of vicinal dinitration (9 to 1, ortho to meta product **20**).[7] The combination of TFA protection and an exceptionally powerful nitrating agent revealed a surprisingly selective directional effect inherent to the 3,5-bicyclic aryl piperidine ring system. Although only a few percent of **8** was produced the first time as a by-product, its formation

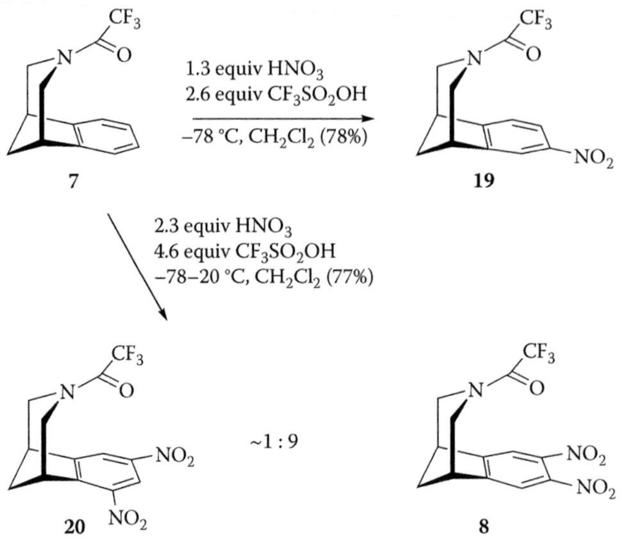

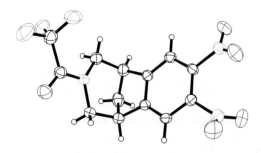

SCHEME 3.7 Dinitration of bicyclic aryl piperidine 7.

made possible the discovery of a direct approach to the quinoxaline heterocycle and, for the first time, varenicline. Optimized conditions to generate **8** simply involve exposure of benzazepine **7** to 2.3 equiv of nitronium triflate in CH_2Cl_2 (Scheme 3.7) giving **8** in 77% yield after crystallization from the crude mixture, which contained <10% of meta-isomer **20**. Other nitrating agents do not efficiently dinitrate **7** under standard conditions. Thus, the combination of the trifluoroacetamide protecting group and nitronium triflate was not only key but also essential to the varenicline discovery. This expedient chemistry remains the most efficient route to the quinoxaline ring of **1**.

Rigorous safety evaluation of the dinitration and reduction sequence was performed in the safety laboratory. The expectation was that the material would prove unsafe, but, fortunately, this proved not to be true. We presume that steric interactions inherent in **8** serve to prevent high levels of aromatic destabilization by the buttressing ortho-dinitro functionality (Figure 3.7). Apparently without readily achieving coplanarity, the nitro groups do not simultaneously delocalize the ring system, thus avoiding a serious destabilizing effect. Because **8** crystallized directly from the crude reaction workup mixture, only a simple recrystallization was required for final purification.

Early lots of varenicline were routinely generated through reduction of **8** to the corresponding diaminobenzazepine **9** and condensation of **9** with glyoxal (sodium bisulfite addition adduct) to afford crystalline quinoxaline trifluoroacetamide **10** (Scheme 3.8). Deprotection, salt formation, and crystallization completed the synthesis of **1** tartrate in 44% overall yield from **6** in early

FIGURE 3.7 (See color insert following page 40). Single-crystal X-ray structure of trifluoroacetamide **8**.

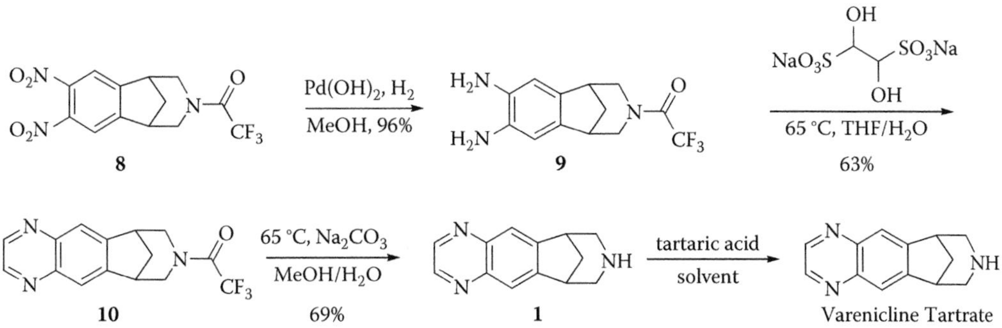

SCHEME 3.8 Conversion of dinitrated bicyclic aryl piperidine **8** to varenicline (**1**).

campaigns. Minor modifications to these steps, such as use of glyoxal and controlling pH, have improved the yield of **1** from **6** to >80%, but the original approach to varenicline from **6** remains unchanged.

Over the course of 10 months, this synthesis generated 4.7 kilograms of varenicline (17% yield overall from benzonorbornadiene) from 50 kilograms of fluorobromobenzene **2** and 20 kilograms of dicyclopentadiene. This material fueled a fast-paced development program through the 6-month toxicological evaluation, salt selection, formulation and stability studies, and Phase 2 clinical trials.

3.4.9 Process Research Reengaged for Early Development

With sufficient material in hand from the "feared benzyne-osmium sequence" to support immediate drug development activities, our process research efforts turned to establishing a preferred commercial synthetic route to varenicline (**1**). Given the extensive and highly successful development of the discovery synthesis from **6** to varenicline in early scale-up campaigns, and the supportive studies by the safety laboratory, the process research task therefore became clear—to establish a reliable supply of benzazepine **6**.

The benzonorbornadiene route established a direct and reliable preparation of **6** from the oxidative cleavage product dialdehyde **15**, but two steps remained problematic for commercial scales: (1) the preparation of benzonorbornadiene and (2) the use of osmium in the cleavage step. Although the benzonorbornadiene synthesis was amply demonstrated to be safe and efficient, the large-scale cracking and storage of cyclopentadiene was considerably more problematic, as shown by its explosive dimerization potential. All efforts to establish large-scale commercial sources of benzonorbornadiene from outside vendors failed. In the event one was established, a viable alternative to osmium oxidative cleavage was required. Although dedicated equipment or suppliers could surmount the osmium issue, avoiding it altogether was preferable, because osmium would always pose a toxicity risk and the potential for unnecessary cross-contamination of GMP equipment and facilities.

3.4.10 Ozonolysis

The discovery approach to benzazepine **6** was considered commercially viable if the osmium dihydroxylation/periodate sequence was replaced by ozonolytic cleavage of benzonorbornadiene **3** (Scheme 3.9), with the caveat that a dedicated set of vendors would need to be identified to support this conversion. We established an ozonolysis route through collaboration with vendors that used cryogenic conditions in methylene chloride and a dimethylsulfide quench of intermediate ozonides. The yield of the overall process, including subsequent reductive amination using the protocol already established with benzylamine and sodium triacetoxyborohydride, was 56%. N-Benzyl group cleavage from the derived product proved unreliable, however, presumably because of residual

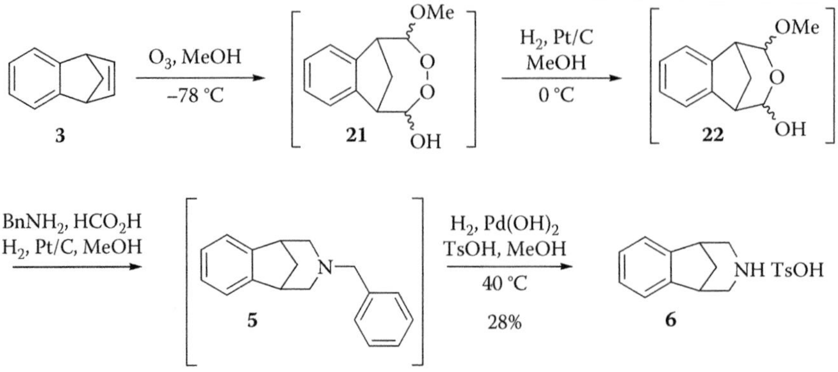

SCHEME 3.9 Ozonolysis of benzonorbornadiene **3** and subsequent reductions to benzazepine **6**.

lingering dimethylsulfide and related inhibitors. Optimally, benzazepine **6** was isolated as an HCl salt in 28% yield overall (from **3**) by transfer hydrogenolysis with 30% by weight of 5% palladium on carbon as catalyst. The low yield, and the concerns surrounding large-scale dimethylsulfide usage, led us to consider further refinements to the process.

Metal-catalyzed hydrogenolysis of the peroxide intermediates was considered an environmentally friendly and more economical approach that offered the possibility of sequential reductions and direct access to **6** from **3** in a one-pot sequence. In a promising conceptual approach, we found that the ozonide or peroxide intermediate derived from **3** could be readily reduced by an amine such as triethylamine (by *N*-oxide formation) before the reductive amination. Unfortunately, the resulting *N*-oxide complicated the reductive amination step by giving rise to side products. Using benzylamine as the peroxide reductant also led to mixtures of reductive amination side products in the hydrogenolysis step, a result of successful peroxide transfer to benzylamine but incomplete reduction of the resulting hydroxylamine. In the end, initial hydrogenolysis of peroxides coupled with subsequent reductive amination with benzylamine provided a viable process.

The sequential process developed called for treatment of benzonorbornadiene **3** with a stream of ozone in methanol at −78°C to provide methoxyhydroperoxide intermediate **21**[25] that was reduced with a platinum catalyst under a hydrogen atmosphere (see Scheme 3.9). The highly exothermic reduction required careful control. If isolated, this product was observed to exist as a mixture of methoxy-glycals **22** by NMR.[26] However, mixing crude reduced glycal **22**, with benzylamine, and formic acid (or acetic acid) allowed direct reduction under the hydrogenolysis conditions using the platinum catalyst (15 to 45 psi of hydrogen)[27] allowing complete conversion to *N*-benzylbenzazepine **5** within a few hours. After catalyst filtration, hydrogenolysis of the *N*-benzyl group of **5** was readily accomplished in the presence of *p*-toluenesulfonic acid and Pearlman's catalyst under a hydrogen atmosphere. Benzazepine **6** was isolated as the tosylate salt in 28% yield from benzonorbornadiene **3**. A single palladium catalyst-derived protocol was identified that affected reduction of the peroxide, reductive amination, and *N*-benzyl group cleavage in stages merely by adjusting reaction temperatures and pressures. Although the overall yield was somewhat lower (15 to 20%), the principal objectives of this methodology had been established and provided a manufacturing option for consideration.[23]

3.5 RETHINKING THE PROBLEM: DEVELOPMENT OF A NEW APPROACH TO BENZAZEPINE

Although our tandem ozonolysis route had the potential to provide rapid throughput to benzazepine **6** from **3** and the advantages of coupling multiple transformations in a single vessel and solvent (ozonolysis, peroxide reduction, reductive amination, and benzyl hydrogenolysis, all in methanol),

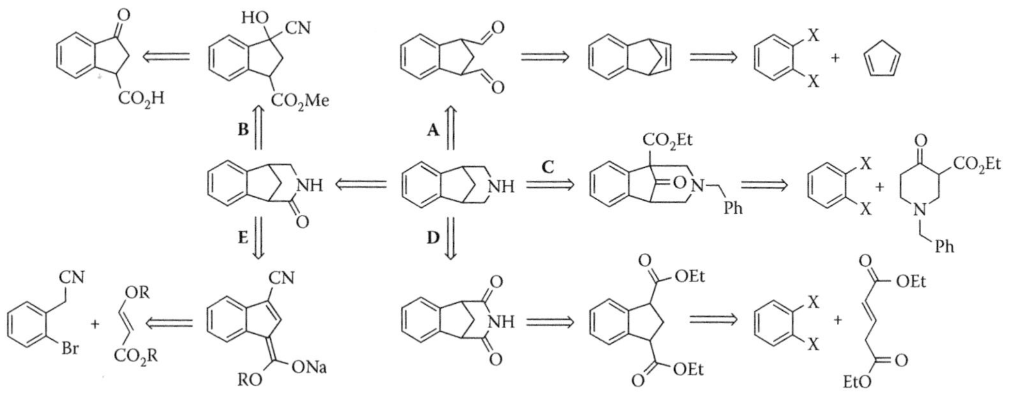

FIGURE 3.8 Retrosynthetic analyses of benzazepine **6**. A number of indane ring syntheses were envisioned to give access to benzazepine **6**. **Route A** (discussed in the discovery and ozonolysis approach above) provides rapid and direct access via dialdehyde precursors. **Route B** features lactam generation as the key step in a cyanohydrin approach. **Route C** provides a convergent nucleophilic S_NAr approach involving preformed piperidine precursors. **Route D** is the related indane-1,3-diester interception of the Mazzocchi imide. **Route E** involves a related acrylate-derived fulvene approach that incorporates all lactam elements in a convergent strategy.

the pitfalls of this route from a manufacturing perspective were still significant. Chief among these problems was catalyst control of peroxide reduction on a manufacturing scale, which posed a significant safety concern. The buildup of stoichiometric peroxide could be avoided by using bus-loop reactors that oxidize only small portions of the material at a time followed by the use of a similar process in the reductive step,[28] but this would involve significant hurdles for a tandem process. In addition, the issue of finding a commercial supplier of cyclopentadiene monomer on demand (or commercial benzonorbornadiene) remained to be resolved.

To overcome these obstacles, we turned our attention to novel approaches to benzazepine **6** that did not involve benzonorbornadiene and obviated the problems presented by cyclopentadiene generation and oxidative cleavage chemistry (Figure 3.8). Both the Mazzocchi and discovery syntheses generated 1,3-indane dicarbonyl intermediates that were converted to the bicyclic aryl piperidine nucleus. We therefore focused on alternative indane syntheses to access this functionality, targeting readily available synthetic precursors.

3.5.1 1-Indanone-3-Carboxylate/Cyanohydrin Route to Benzazepine 6 (Route B)

The first of these alternative approaches started with commercially available indanone-3-carboxylic acid **23** (derived from phenylsuccinic anhydride via Friedel–Crafts chemistry).[29] This indanone was a viable precursor to **6** through an aminomethyl homologation with cyanide or nitromethane (Scheme 3.10). Our intent was to obtain ester **25**, which after reduction would afford methylamine **28**, which in turn would cyclize to generate lactam **29**, giving benzazepine **6** upon reduction. Stereocontrol in the homologation was unnecessary, as lactam **29** would theoretically provide a thermodynamic sink under epimerizing reaction conditions, capturing the desired *cis*-indane isomer.

Indanone **23** was converted to methyl ester **24** (see Scheme 3.10).[30] We found that the alkaline conditions required for nitromethane additions or Wittig homologation of ketone **24** led to benzylic deprotonation. We therefore turned to an acid-catalyzed cyanide homologation strategy.[31] Literature procedures using trimethylsilyl cyanide were studied using either 18-crown-6 or ZnI_2 as catalysts in methylene chloride at ambient temperature.[32] The Lewis acid catalyst (ZnI_2) was favored, as crown ethers are costly and known teratogens and typically not used in development unless

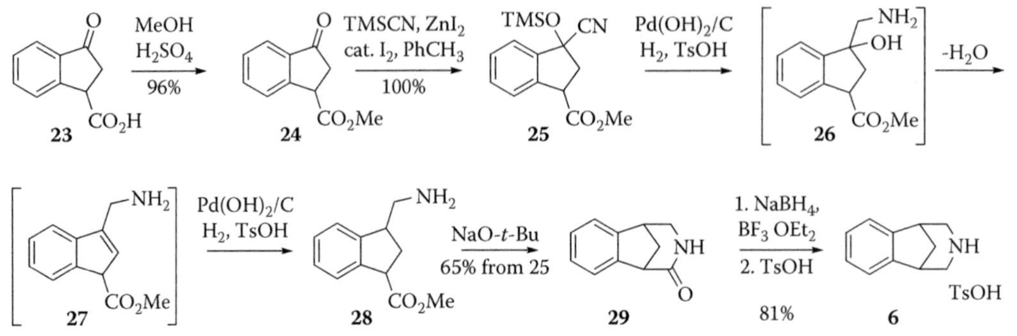

SCHEME 3.10 Route to **6** via 1-indanone-3-carboxylic acid (**23**).

absolutely necessary. Conversion of indanone **24** to the cyanohydrin **25** (a 2:1 mixture favoring the undesired trans-diastereoisomer) required heating the reaction at 50°C for several hours with trimethylsilyl cyanide (1.2 equiv) and ZnI$_2$ (0.01 equiv) in toluene.

The catalytic activity of ZnI$_2$ varied from batch to batch. By adding catalytic iodine, the performance of the catalyst greatly improved, presumably through the *in situ* formation of the more reactive and more soluble Lewis acid Zn(I$_3$)$_2$.[33] A small amount of acetonitrile was added to slightly increase the overall solvent polarity and increase catalytic activity to a level similar to that observed using methylene chloride. These modifications greatly improved this step, which provided silylated-cyanohydrin **25** in quantitative yield. For the subsequent hydrogenation of **25**, care was needed to avoid palladium catalyst poisoning by excess cyanide, which was removed by a critical aqueous workup protocol.

Hydrogenation of **25** was accomplished using Pearlman's catalyst to provide amino-ester **28** as a 10:1 mixture of diastereoisomers favoring the desired cis isomer.[34,35] Excess acid (>1 equiv of *p*-toluensulfonic or sulfuric acid) was necessary to facilitate the hydrogenation by promoting relatively rapid nitrile reduction to protonated amine **26**. This protocol prevents catalyst deactivation. Acid also catalyzes the subsequent benzylic alcohol dehydration and the final reduction to **28**. The observed increase in diastereoselectivity during the hydrogenation is consistent with acid-catalyzed elimination of water to form intermediate indene **27** followed by reduction from the least hindered face of **27** to afford the *cis* isomer of **28** as the major product. We did not observe elimination of water prior to nitrile reduction.[36]

We initially explored mild acids and bases for the cyclization of the lactam **28**, such as acetic acid, cat. tosic acid, or methylimidazole, but these reagents required high reaction temperatures (100 to 120°C) and long reaction times (over 1 day) and resulted in low yields (20 to 40%). We then explored stronger bases, such as sodium *tert*-butoxide, and discovered that cyclization could be readily accomplished under ambient conditions in good yields (50 to 65%). Eventually we found that we could suppress the main hydrolysis side product by removing water through distillation of solvent such as n-propanol, and the yields improved to 70 to 80%. Both diastereoisomers were observed during the course of the cyclization, but the *cis*-diastereomer was eventually funneled away to product and left ~5 to 10% of the *trans*-diastereoisomer remaining. Filtration to remove catalyst and treatment of the crude filtrate containing **28** with sodium *tert*-butoxide at room temperature led to rapid conversion to lactam **29**. Aqueous workup and recrystallization provided **29** in 65% yield from **24**. A minor side product, the amino acid hydrolysis product of **28**, failed to cyclize to **29** under all conditions examined. This by-product is presumably produced by water liberated in the dehydration step, from the water wet Pd/C catalyst, or potentially from water contained in the hygroscopic sodium *tert*-butoxide. Lactam reduction by *in situ*-generated borane followed by tosylate salt formation gave **6** in an isolated yield of 81%.[37] By this route, **6** was produced in five synthetic steps and 51% overall yield from indan-1-one-3 carboxylic acid **23**.

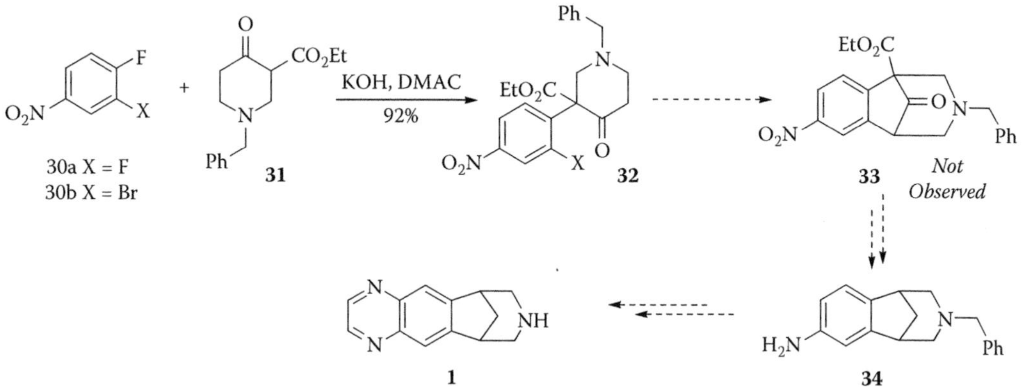

SCHEME 3.11 Nucleophilic aromatic substitution strategy: failed ring closure efforts (route C).

The favored features of the route—sequential single-pot hydrogenation, dehydration, reduction, and basic lactam cyclization steps—made it highly attractive for manufacturing. Theoretically, catalyst poisoning by residual cyanide could be avoided without isolation of crude cyanohydrin, thereby allowing direct conversion of **24** to lactam **29** in a single vessel. We were still compelled, however, to find a route that avoided the safety issues associated with benzyne, osmium, hydroperoxides, and cyanide, and hoped to incorporate the highly crystalline lactam **29** into future approaches.

3.5.2 PIPERIDINE ANNELATION/AROMATIC SUBSTITUTION STRATEGY (ROUTE C)

The nucleophilic aromatic substitution reaction of the readily available carboethoxy piperidone **31** with 1,2-dihalo-4-nitrobenzenes **30a** or **30b** provided an avenue for directly joining the piperidine and aryl components of **6** via piperidine **32** (Scheme 3.11, route C). Precedent for the subsequent conversion of **32** to **33** was limited to the preparation of fused indanyl rings. Although the [3.2.1]-bridged bicyclic systems were unknown, the [3.3.1]-bicyclic systems had been prepared via palladium-mediated intramolecular ketone arylation.[38] Under S_NAr conditions, the initial bond between **31** and **30a** or **30b** formed readily in polar solvents N,N-dimethylformamide and N,N-dimethylacetamide (DMF or DMAC) at room temperature to give **32** (92%, HPLC). Unfortunately, attempts to affect the subsequent ring closure to **33** by either S_NAr or palladium catalysis failed. We suspect that geometric constraint thwarted this intramolecular bicyclic ring formation. Attempts to reverse the bond-forming steps by first coupling at the bromide of **30b** with **31** under palladium catalysis also failed, which suggested that electron transfer between palladium and nitroarene **32** inhibited ring closure under cross-coupling conditions.[39]

3.5.3 NUCLEOPHILIC AROMATIC SUBSTITUTION STRATEGY (ROUTE D)

To reduce the strain associated with cyclization, we next studied acyclic piperidine precursors as nucleophiles to ultimately intercept the Mazzocchi, discovery, and cyanohydrin approaches. Diethyl glutaconate **35** coupled readily with fluoro-4-nitroarenes **30a** or **30b** to give **36** (95%, HPLC; Scheme 3.12).[40] Again, however, ring closure to **37** did not proceed by either S_NAr (**36a**, X = F) or palladium catalysis (**36b**, X = Br). The newly formed allylic anion of **36** presumably delocalized into the nitroaromatic ring to generate electron-rich character inhibiting subsequent nucleophilic S_NAr substitution or the oxidative addition required to enter palladium catalysis pathways.

We next attempted to accomplish the annulation with substrates lacking the deactivating nitro group. A Michael addition elimination strategy was explored with readily available fluoro-derivative

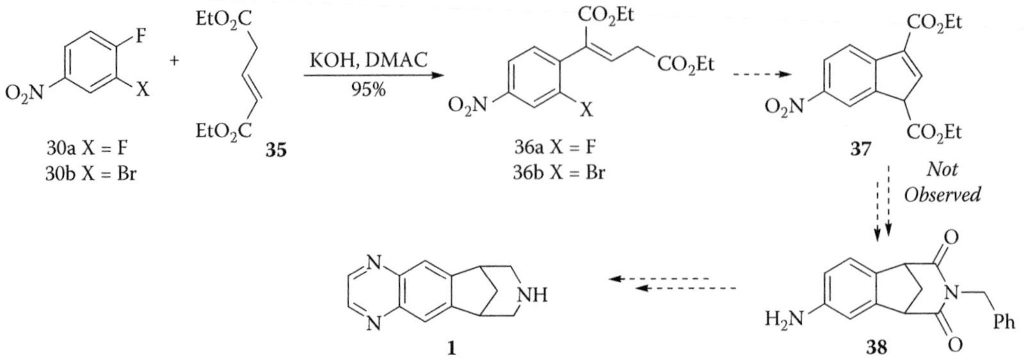

SCHEME 3.12 Diethylglutaconate strategy (route D).

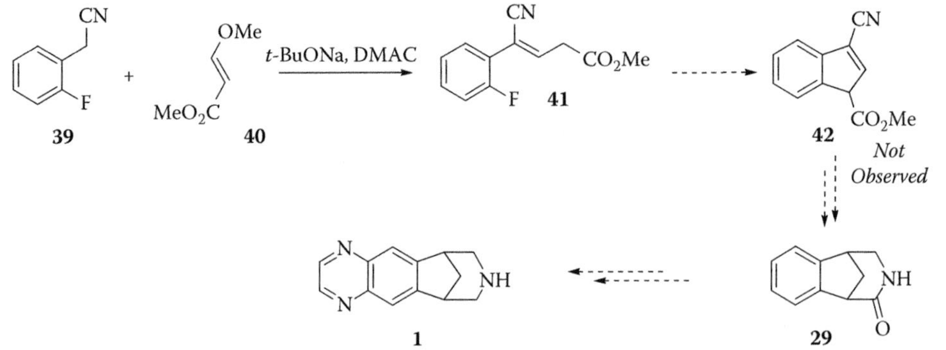

SCHEME 3.13 Unsuccessful S_NAr strategy to prepare **1**.

39 and acrylate **40** to generate cyano-substituted intermediate **41** in good yield. Again, however, cyclization of **41** to **42** did not proceed under S_NAr conditions (Scheme 3.13).

Indene ring formation seemed hopelessly inhibited using this approach. The trajectory required for the conjugated sp^2-hybridized aryl-propene anion to add to yet another sp^2-hybridized carbon atom likely imposes a tremendous energetic cost in the transition state (Figure 3.9). We therefore concluded that a significant alteration of the ring closure geometry was necessary. We reasoned that substitution of fluorine by bromine would permit palladium-catalyzed oxidative insertion at the electrophilic center, thereby promoting successful ring closure through a less-strained palladium-mediated end-on trajectory.

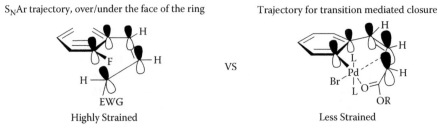

FIGURE 3.9 Trajectories to fused five-membered ring formation.

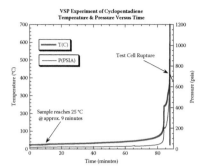

VSP Experiment of Cyclopentadiene
Temperature & Pressure Versus Time

Test Cell Rupture

Sample reaches 25 °C
@ approx. 9 minutes

Temperature (°C) / Pressure (psia) / Time (minutes)

T(C)
P(PSIA)

FIGURE 3.4 The vent sizing package (VSP) test of cyclopentadiene reveals a slow exotherm upon reaching 25°C that accelerates to a violent decomposition after 80 min. The decomposition caused the test cell to rupture. These data show that the consequences of a runaway reaction would be severe. An accelerated reaction calorimetry (ARC) test determined the time-to-maximum rate (TMR) of the decomposition from subambient conditions. Further testing on the material diluted in hexane/heptane was performed to investigate the consequences of a runaway reaction in these mixtures. ARC showed an exotherm from the first heat–wait–search step (at 31°C), which progressed to more than 110°C. A second exotherm was seen at 160°C, progressing to 200°C. The initial decomposition showed a TMR on the order of several days from the storage temperature (–40°C) anticipated in kilo laboratory campaigns. The TMR is on the order of an hour from room temperature, however, which supports the data from the VSP test of cyclopentadiene.

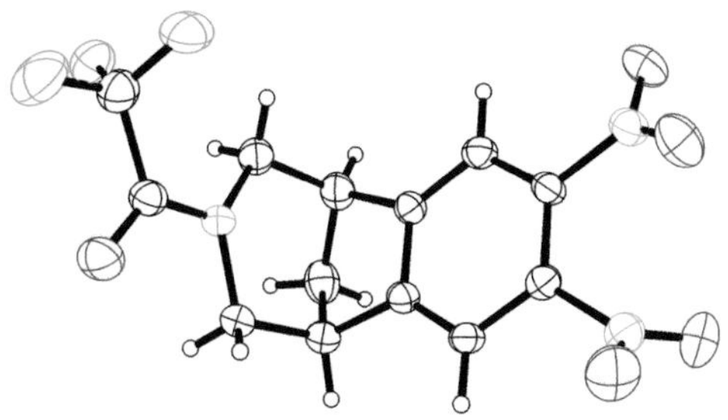

FIGURE 3.7 Single-crystal X-ray structure of trifluoroacetamide **8**.

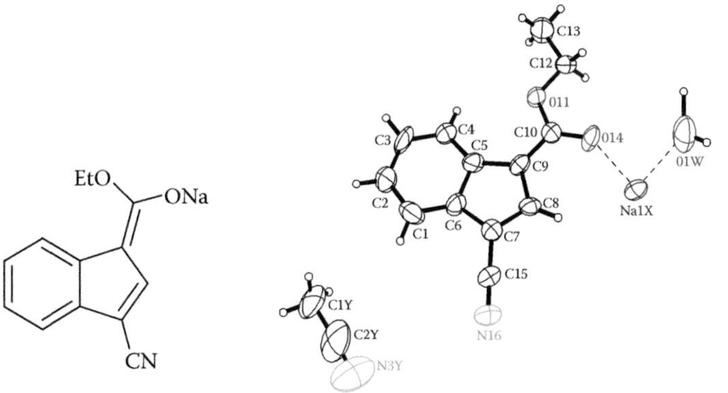

50 Na

FIGURE 3.10 Single-crystal X-ray structure of sodium enolate **50** (and acetonitrile solvate).

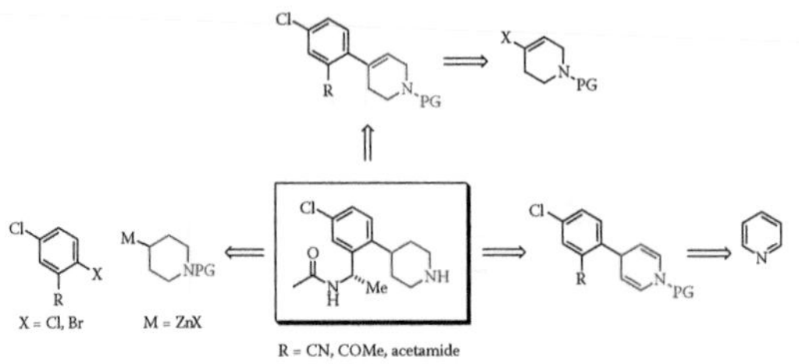

SCHEME 5.4 4-Arylpiperidine.

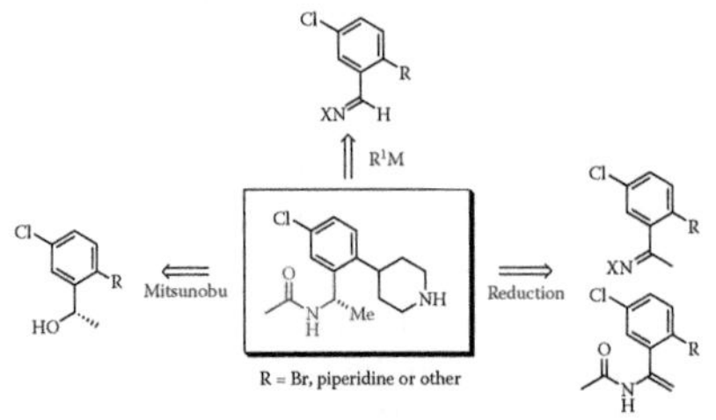

SCHEME 5.5 Introduction of the phenethylamine chiral center.

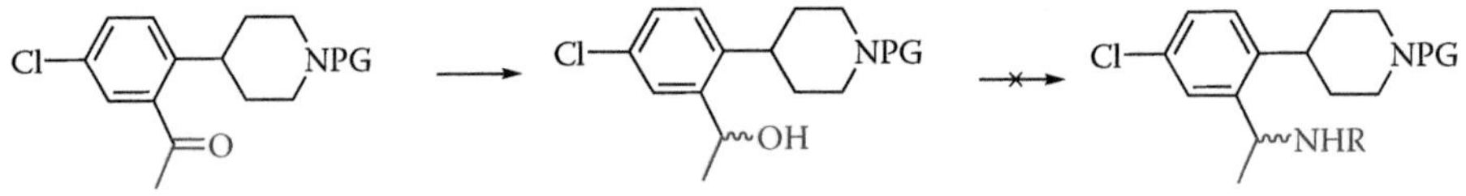

32

SCHEME 5.12 Mitsunobu approach.

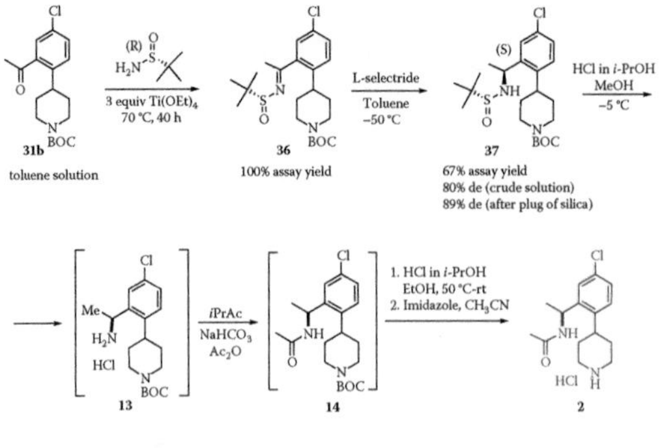

SCHEME 5.16 From ketone to phenethylamine.

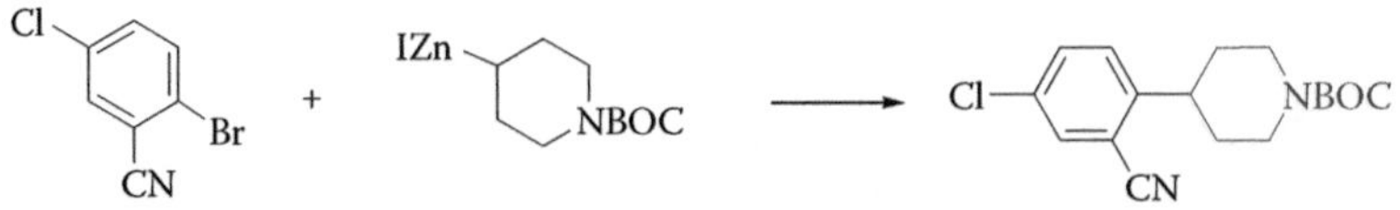

TABLE 5.2 STRUCTURE Negishi Approach

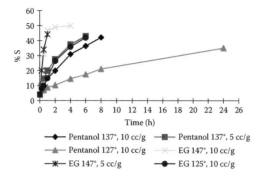

FIGURE 6.4 Solvent effect on the racemization rate.

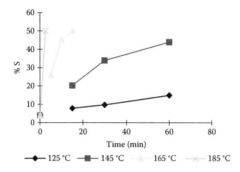

FIGURE 6.5 Racemization in ethylene glycol: T° effect.

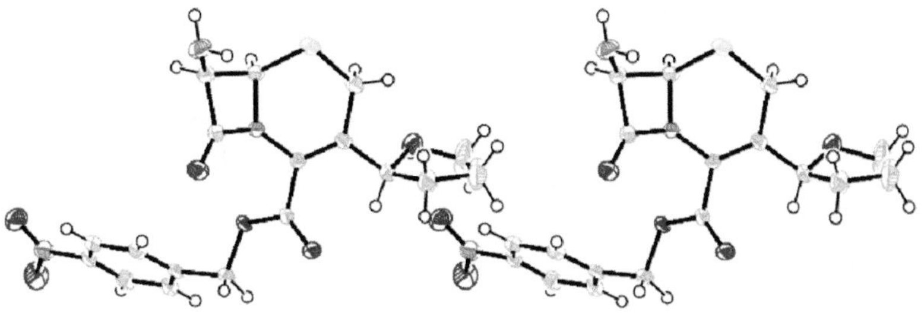

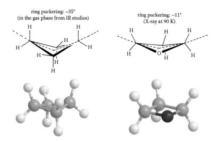

FIGURE 11.3 Stereochemistry of cephem **30** derived from single-crystal x-ray data.

FIGURE 13.1 Comparison of puckering between cyclobutane and oxetane.

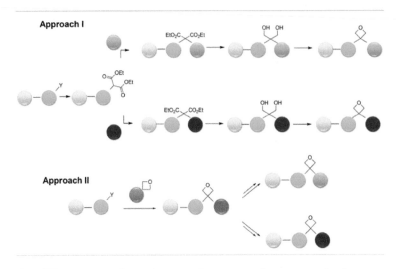

FIGURE 13.6 Two different approaches to incorporating oxetane rings onto molecules.

Legend cell:
$$M^{+n}$$
$$pK_h \quad ^a$$
$$WERC \quad ^b$$

Element	pK_h	WERC
Li $^{+1}$	13.64	4.7×10^7
Be	—	—
B $^{+3}$	—	—
C	—	—
N	—	—
Na $^{+1}$	14.18	1.9×10^8
Mg $^{+2}$	11.44	5.3×10^5
Al $^{+3}$	4.97	1.6×10^0
Si $^{+4}$	—	—
P $^{+5}$	—	—
K $^{+1}$	14.46	1.5×10^8
Ca $^{+2}$	12.85	5×10^7
Sc $^{+3}$	4.3	4.8×10^7
Ti $^{+4}$	≤2.3	—
V $^{+3}$	2.26	1×10^3
Cr $^{+3}$	4.0	5.8×10^{-7}
Mn $^{+2}$	10.59	3.1×10^7
Fe $^{+2}$	9.5	3.2×10^6
Co $^{+2}$	9.65	2×10^5
Ni $^{+2}$	9.86	2.7×10^4
Cu $^{+2}$	7.53	2×10^8
Zn $^{+2}$	8.96	5×10^8
Ga $^{+3}$	2.6	7.6×10^2
Ge $^{+4}$	—	—
As	—	—
Rb	—	—
Sr	—	—
Y $^{+3}$	7.7	1.3×10^7
Zr $^{+4}$	0.22	—
Nb $^{+5}$	(0.6)	—
Mo $^{+5}$	—	—
Tc	—	—
Ru $^{+3}$	—	—
Rh $^{+3}$	3.4	3×10^{-8}
Pd $^{+2}$	2.3	—
Ag $^{+1}$	12	$>5 \times 10^6$
Cd $^{+2}$	10.08	$>1 \times 10^8$
In $^{+3}$	4.00	4.0×10^4
Sn $^{+4}$	—	—
Sb $^{+5}$	—	—
Cs	—	—
Ba $^{+2}$	13.47	$>6 \times 10^7$
Ln $^{+3}$	7.6 – 8.5	$10^6 - 10^8$
Hf $^{+4}$	0.25	—
Ta $^{+5}$	(−1)	—
W $^{+6}$	—	—
Re $^{+5}$	—	—
Os $^{+3}$	—	—
Ir $^{+3}$	—	—
Pt $^{+2}$	4.8	—
Au $^{+1}$	—	—
Hg $^{+2}$	3.40	2×10^9
Tl $^{+3}$	0.62	7×10^5
Pb $^{+2}$	7.71	7.5×10^9
Bi $^{+3}$	1.09	—

Element	pK_h	WERC
La $^{+3}$	8.5	2.1×10^8
Ce $^{+3}$	8.3	2.7×10^8
Pr $^{+3}$	8.1	3.1×10^8
Nd $^{+3}$	8.0	3.9×10^8
Pm	—	—
Sm $^{+3}$	7.9	5.9×10^8
Eu $^{+3}$	7.8	6.5×10^8
Gd $^{+3}$	8.0	6.3×10^7
Tb $^{+3}$	7.9	7.8×10^7
Dy $^{+3}$	8.0	6.3×10^7
Ho $^{+3}$	8.0	6.1×10^7
Er $^{+3}$	7.9	1.4×10^8
Tm $^{+3}$	7.7	6.4×10^6
Yb $^{+3}$	7.7	8×10^7
Lu $^{+3}$	7.6	6×10^7

[a] $pK_h = -\log K_h$. Reference 6a,b. [b] Exchange rate constants for substitution of inner-sphere water ligands. Reference 6c.

FIGURE 15.1 Hydrolysis constants (Kh) and water exchange rate constants for substitution of inner-sphere water ligands for metal cations.

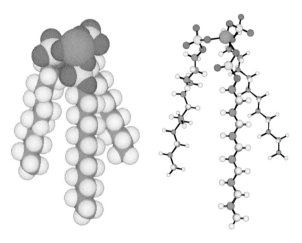

FIGURE 15.2 Sc(C$_{12}$H$_{25}$OSO$_3$)$_3$.

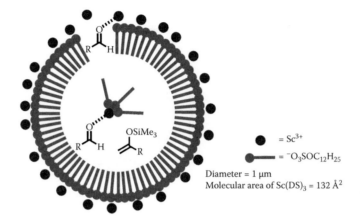

FIGURE15.3 A hydrophobic particle.

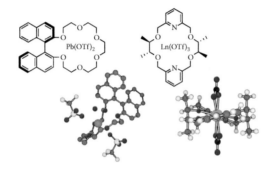

FIGURE 15.4 Chiral Pb and Ln catalysts for asymmetric Mukaiyama aldol reactions.

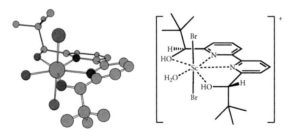

FIGURE 15.5 [3·ScBr$_2$·H$_2$O]$^+$ moiety in the X-ray structure of [3·ScBr2·H$_2$O]·Br·H$_2$O. Hydrogen atoms are omitted for clarity.

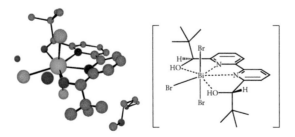

FIGURE 15.6 The X-ray crystal structure of [BiBr$_3$·1]·(H$_2$O)$_2$·dimethyl ether. Dimethyl ether is omitted for clarity.

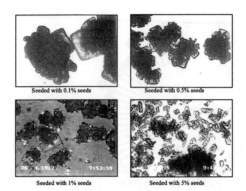

Seeded with 0.1% seeds   Seeded with 0.5% seeds

Seeded with 1% seeds   Seeded with 5% seeds

FIGURE 18.3 API particle size as a function of the amount of seeds added.

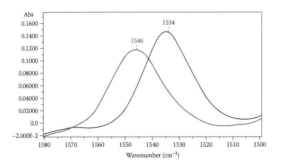

FIGURE 20.4 Infrared absorbance shift upon addition of 12-crown-4 to **11b**.

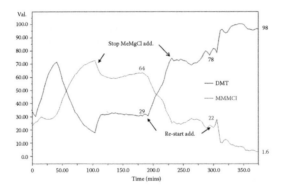

FIGURE 22.6 Dimethyltitanocene (DMT) and monomethylmonochloro titanocene concentration profiles from discontinuous addition of Grignard solution.

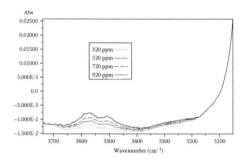

FIGURE 22.7 Water concentration in the DMT solution.

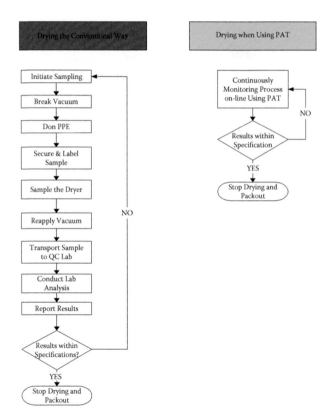

FIGURE 23.1 Process steps in a conventional drier monitoring operation versus a process analytical technology (PAT) drier application.

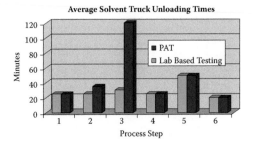

FIGURE 23.2 Average times involved in testing, release, and unloading a solvent tank truck (study conducted by Pfizer Inc., Groton, CT).

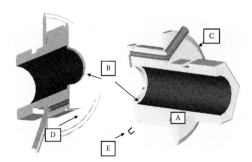

FIGURE 23.3 Typical process analytical technology (PAT) probe holder for automated filter-driers.

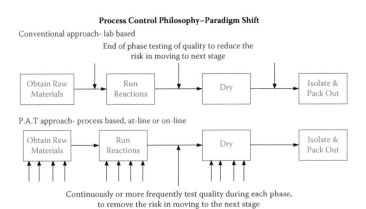

FIGURE 23.5 Scheme of conventional versus process analytical technology (PAT) process control paradigms.

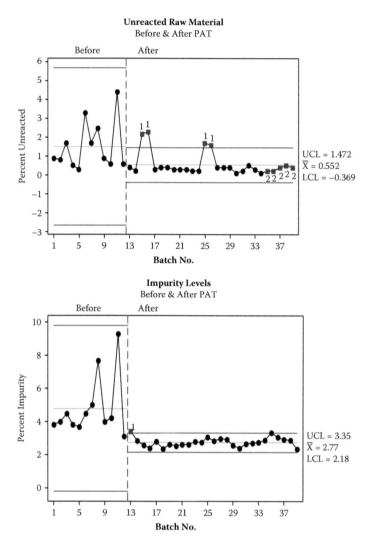

FIGURE 23.9 Control chart of percent unreacted before and PAT (top) and impurity levels before and after implementing PAT (bottom).

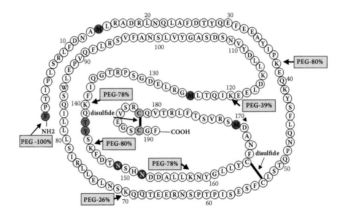

FIGURE 24.5 Effects of PEG/GHA molar ratio on yield of PEG-GHA (4 through 6). Reaction conditions: GHA = 10 mg/ml, PEG/GHA molar ratio = 7 to 11, T = 20°C. Rate constants: k_1 = 23,000, k_2 = 9000, k_3 = 3300, k_4 = 2500, k_5 = 1750, k_6 = 1150, k_7 = 700, k_8 = 350 in units of L/mol-hr; k_d = 1.3 hr^1.

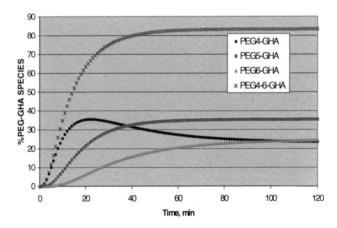

FIGURE 24.6 PEG-GHA 4-6 Formation at 25°C, GHA = 14 mg/mL, (PEG/GHA) molar ratio = 8.5.

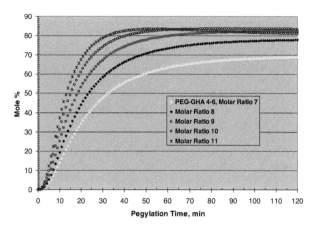

FIGURE 24.7 Effects of PEG/GHA molar ratio on yield of PEG-GHA (4 through 6). Reaction conditions: GHA = 10 mg/ml, PEG/GHA molar ratio = 7-11, T = 20°C; Rate constants: k_1 = 23000, k_2 = 9000, k_3 = 3300, k_4 = 2500, k_5 = 1750, k_6 = 1150, k_7 = 700, k_8 = 350 in units of L/mol-hr; k_d = 1.3 hr^{-1}.

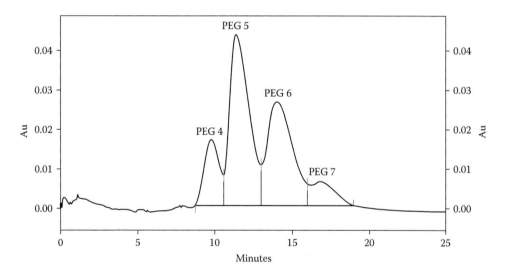

FIGURE 24.9 Detection of de-PEGylation and PEG chain truncation.

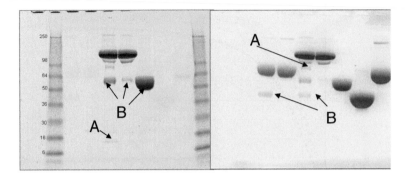

A: De-PEGylation

B: PEG Truncation

FIGURE 24.10 Detection of de-PEGylation and PEG chain truncation.

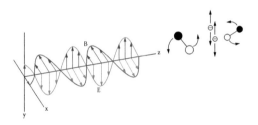

FIGURE 25.1 Dipolar molecules and ions try to move with an oscillating electric field.

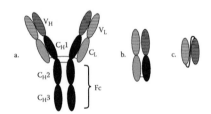

FIGURE 26.1 (a) The general structure of an IgG antibody. (b) The structure of a Fab fragment containing the variable domains and the C_H1 and C_L constant regions. (c) A single-chain Fv (scFv) fragment. The V_H and V_L domains in a scFv fragment are joined together by a flexible peptide linker encoded by the scFv gene.

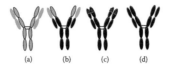

FIGURE 26.2 The differences from a murine antibody (a) to a fully human antibody (d). In (b), the constant regions of the murine antibody have been replaced with human constant domains. In (c), the amino acid sequences adjacent to the three hypervariable regions in the variable domains had been humanized by modeling amino acid substitutions that are more commonly found in human antibodies.

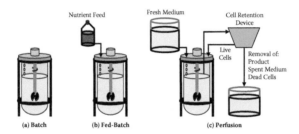

FIGURE 26.3 Modes of bioreactor operation. (A) Batch bioreactor, (B) fed-batch reactor, and (C) perfusion bioreactor.

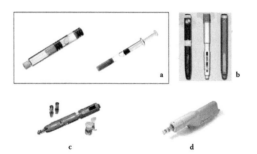

FIGURE 26.5 Current and future trends in delivery systems for mAbs include (a) dual-chamber cartridge and prefilled syringe with attached needle (courtesy Vetter), (b) pen systems for self-administration, and (c, d) needle-free injection systems (BioJect).

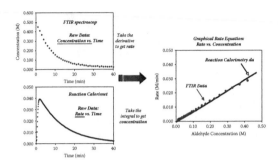

FIGURE 27.1 Monitoring reaction progress in the proline-mediated aldol reaction of Scheme 27.1 by Fourier transform infrared (FTIR) spectroscopy and reaction calorimetry. The data are manipulated in each case as illustrated in a "graphical rate equation" plotting rate versus substrate concentration for the limiting substrate.

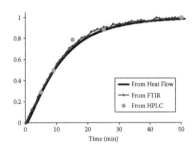

FIGURE 27.2 Comparison of conversion versus time for the reaction of Scheme 27.1 using high-performance liquid chromatography (HPLC) sampling of product concentration to *in situ* monitoring by Fourier transform infrared (FTIR) spectroscopy and reaction calorimetry.

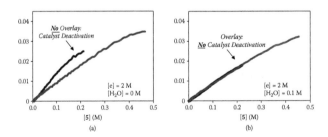

FIGURE 27.3 Comparison of rate versus [aldehyde] **2** for two reactions of Scheme 27.1 carried out at the same excess [*e*] with initial conditions as given in Table 27.1 (magenta dots: Exp. 1; blue dots: Exp. 2). (a) No water added; (b) 0.1 M water added to the reaction.

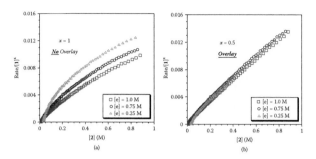

FIGURE 27.4 Comparison of rate/[**1**]^x versus [aldehyde] **2** for three reactions of Scheme 27.1 carried out at different excess [e] as shown in the figure legends. [**2**]$_0$ = 1 M, [**4**] = 0.1 M in each case. (a) $x = 1$; (b) $x = 0.5$.

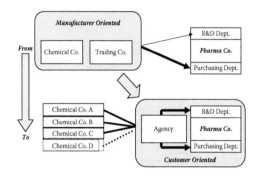

FIGURE 28.1 From trading company to agent.

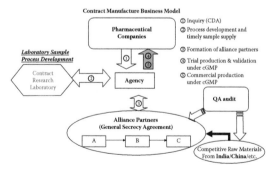

FIGURE 28.2 Contract manufacture business model (CDA, confidential disclosure agreement; cGMP, current good manufacturing practice; QA, quality assurance).

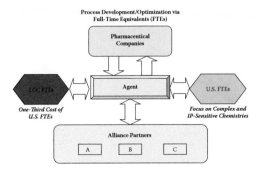

FIGURE 28.3 Process development/optimization via full-time equivalents (IP, intellectual property).

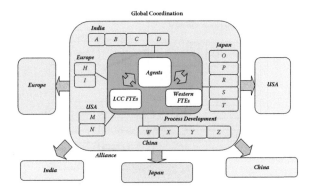

FIGURE 28.4 Global coordination (FTEs, full-time equivalents).

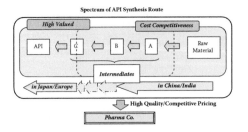

FIGURE 28.5 Spectrum of active pharmaceutical ingredient (API) synthesis route.

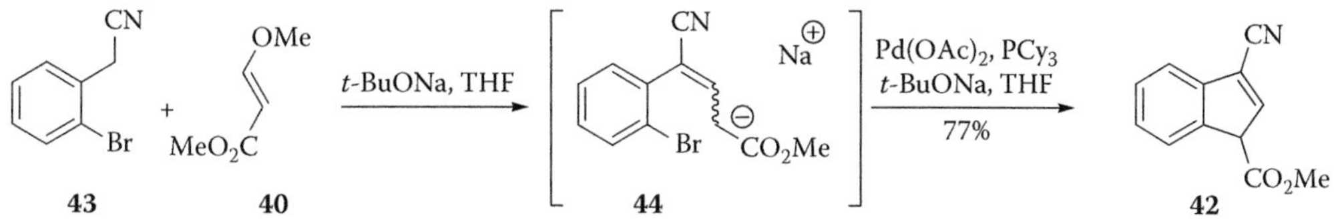

SCHEME 3.14 Palladium-catalyzed cyclization route to cyano-carboxyindene **42**.

3.5.4 PALLADIUM-CATALYZED CYCLIZATION ROUTE TO BENZAZEPINE (ROUTE E)

The required substrate (**44**) for the new palladium-catalyzed cyclization route (**44**) was prepared as shown in Scheme 3.14 (route E). This compound, isolated as a mixture of olefin isomers, was obtained by the high-yield coupling reaction of 2-bromophenylacetonitrile (**43**) and acrylate **40**, which were added concurrently to a solution of t-BuONa (2 equiv) in 1,2 dimethoxyethane or THF. Palladium catalysts were then screened in the hopes of finding a phosphine ligand that would permit oxidative addition to the electron-rich arene (despite the conjugated allylic anion) and catalyze the subsequent ring closure to indene **42**. Although the allylic anion could impede oxidative addition in the catalytic cycle, as was likely the case with the nitro-substituted systems (see Scheme 3.11 and Scheme 3.12), the anion of **44** could theoretically isomerize to the *cis* olefin geometry required for attack on palladium. However, upon heating **44** with t-BuONa, catalytic triarylphosphines (e.g., 0.1 equiv PPh$_3$, or DPPF [diphenylphosphinoferrocene]) and various palladium sources (5 mol%), no reaction occurred. We speculated that weakly coordinating phosphines in combination with Pd sources do not promote oxidative addition because they do not support electron transfer from palladium to the aryl halide. Also, we presumed that oxidative addition is impeded by the electron-rich conjugated allylic anion, but we hoped that using more electron-rich dialkylarylphosphines, such as the 2-dicyclohexyl-biphenylphosphines developed by the Buchwald group, would facilitate this electron transfer step.[41] Gratifyingly, cyclization to indene **42** proceeded with high efficiency with 2-dicyclohexyl-biphenylphosphine in 1,2-dimethoxyethane or THF at 60°C over 2 to 3 hours.[42]

With these conditions identified, we then successfully carried out the one-pot condensation of **43** with acrylate **40** and the subsequent cyclization without isolating **44** to obtain indene **42**. This was accomplished in 77% yield with stoichiometric t-BuONa in the presence of the palladium catalyst. A number of factors were found to govern the process (Scheme 3.15). Reaction efficiency was improved when methoxy acrylate (**46**, R = Me) was replaced by the ethoxy derivative (**47**, R = Et), thereby avoiding methoxide generation. Methoxide can hinder palladium catalysis through the formation of bridged dimeric palladium complexes, analogous to what has been observed with hydroxide.[43] Replacing **46** with **47** reduced the palladium catalyst loading from 5 to 10 mol% to

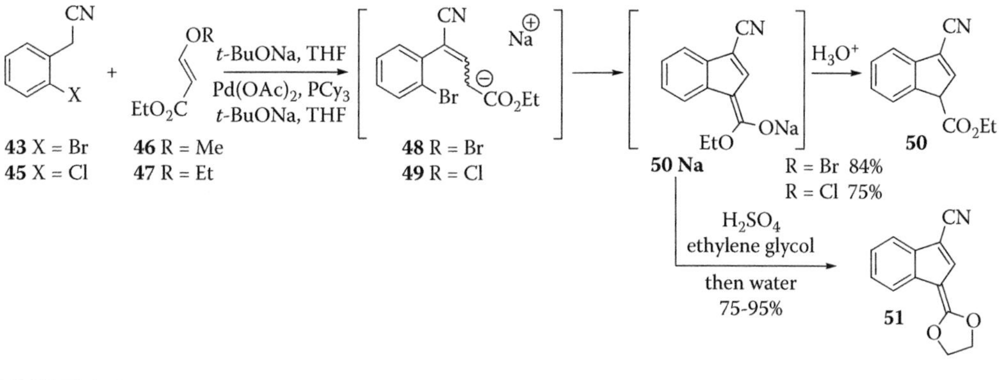

SCHEME 3.15 Tandem acrylate addition and palladium-catalyzed cyclization.

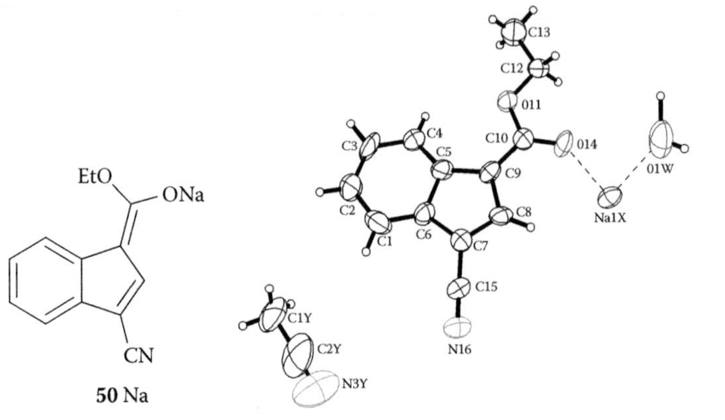

FIGURE 3.10 (See color insert following page 40). Single-crystal X-ray structure of sodium enolate **50** (and acetonitrile solvate).

0.5 to 2.0 mol%. We were also pleased to find that the relatively expensive dialkylbiarylphosphines could be replaced with readily available tri-*t*-butylphosphine or tricyclohexylphosphine. Tricyclohexylphosphine was typically used because it is an easy-to-handle crystalline solid.[44] Although we initially carried out the tandem condensation/cyclization process in 1,2-dimethoxyethane, efficient palladium catalysis was possible in THF, a preferred solvent. For economic reasons, 2-chlorophenylacetonitrile **45** was evaluated as a replacement for **43**. Unfortunately, while the corresponding aryl bromide **43** reacts at 40°C with 1 mol% palladium, the chloride **45** required heating to 60°C for at least 12 hours and 5 to 10 mol% catalyst loading. We presume the use of chloride results in less facile oxidative addition.

Neutral **50** was challenging to isolate as it was a low-melting solid, sensitive to decarboxylation. We were, however, able to isolate samples of the sodium salt by extraction at alkaline pH and subsequent solvent displacement. Structure elucidation using single-crystal X-ray diffraction revealed it to be a single olefin isomer (Figure 3.10). In practice, the sodium salt was isolated as a foamy solid or dark oil, because it was not readily crystallized from solution.[45] By studying derivatives, we recognized that cyanobenzofulvene intermediate **51** was more crystalline than **50** or its salts. This material was best generated *in situ*, by simply adding ethylene glycol and sulfuric acid (as catalyst and dehydrating agent) to the reaction mixture after the sodium salt of **50** had formed. With prolonged stirring at room temperature (12 to 48 hour), precipitated **51** was isolated in 75 to 95% yield from 2-bromophenylacetonitrile (**43**). Isolation of **51** was not only practical, it also efficiently purged phosphines and other by-products that had plagued the reliability of subsequent reactions (Scheme 3.16).

With **51** in hand, hydrogenation under acidic conditions was readily achieved in alcoholic solvents to form **28**, which was cyclized under basic conditions to give lactam **29**. Finally, conversion of lactam **29** to the desired benzazepine **6** was accomplished by borane reduction (generated *in situ*) to form benzazepine **6**, isolated as the tosylate salt in 81% yield. The reduction could also be accomplished with lithium aluminum hydride. This route was efficient and robust, as it uses

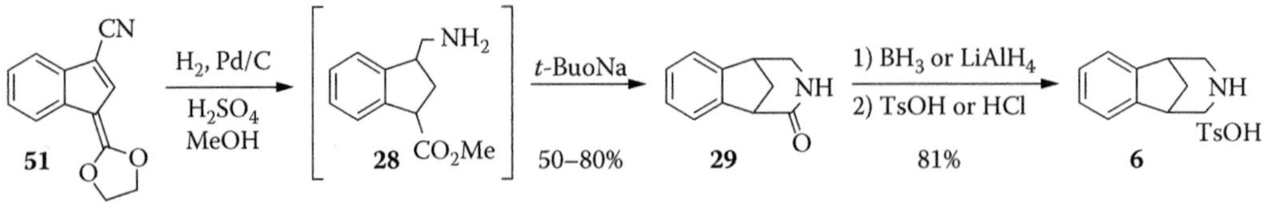

SCHEME 3.16 Streamlined approach to benzazepine **6** via benzofulvene intermediate **51**.

inexpensive commodity chemicals, avoids hazardous reagents, and provides the two highly crystalline intermediates—cyanobenzofulvene **51** and lactam **29**—both of which are readily isolated in high states of purity.[46]

3.6 CONCLUSION

We have described the many issues we faced in providing a relatively simple material (**6**), one whose conversion to the final active pharmaceutical ingredient, in this case varenicline, followed the path of its original discovery. As is often the case in early development, the process chemist must rely on the discovery assembly strategy to meet rapid development timelines. Teamwork and collaboration made these studies productive as well as enjoyable and provided the impetus to delve more deeply into the safety aspects of the original route, which was rigorously established to have an unexpectedly safe profile for a nitration-based sequence. Although multiple approaches to **6** proved highly efficient and viable for scale-up, only the final route avoided reagents that required special handling precautions and hazardous intermediates. The final route liberated simple, innocuous by-products, alkali metal salts, alcohols, or water, thereby avoiding unnecessary toxic reagents and making it possible to meet ever-increasing purity and safety specifications. As we have shown, costs, by-products, and thermodynamics became important considerations for synthesis on a large scale. The final synthetic approach addresses all of these factors in an elegantly simple and safe synthesis.

ACKNOWLEDGMENTS

We acknowledge the diligent and insightful contributions of our colleagues who have made the work described herein possible: Michael G. Vetelino, Crystal G. Bashore, Krista E. Bianco, Michael C. Wirtz, Eric Arnold, Jianhua Huang, and Mark Bundesman for being tireless and upbeat pursuers of new medicines. We are grateful for the privilege of working with many Pfizer colleagues, including Brian T. O'Neill, Robert A. Volkmann, Joseph Lyssikatos, Bertrand Chenard, Mark A. Dombroski, Martin Jefson, and Anabella Villalobos, whose advice and collaborative spirit bring enjoyment to a demanding job. Mark Hampden-Smith recommended and Jeff Kiplinger implemented ICP-MS, now the industry standard.

We thank Stephané Caron and Enrique Vasquez for process research efforts and support, and David J. am Ende, Steven J. Brenek, and Pamela J. Clifford for indispensable safety data. Mike Johnson manufactured the chromatography equipment overnight in the Building 156 machine shop. Jason McKinley contributed pivotal work to demonstrate the tandem Michael addition/palladium-catalyzed cyclization and investigations of the nucleophilic aromatic substitution strategy. Steve Massett, Tim Watson, and colleagues Bob Handfield, Mike Vetelino, and Jim McGarry worked to scale up the new route to the benzazepine, and development support for commercial scale-up was provided by Frank Busch, Greg Withbroe, Terry Sinay, Joe Rainville, Eanne O'Maitiu, and John O'Sullivan. Lisa Newell, Lewin Wint, George Quallich, Mark Delude, and Darlene Hall provided exploratory development synthesis support. Stephané Caron, Bob Dugger, and Keith DeVries made helpful suggestions regarding the ozonolysis approach. Steve Massett suggested the use of indanone carboxylic acid as a raw material. Jon Bordner provided single-crystal X-ray data. We are indebted to Bob Volkmann and Ralph Robinson for a thorough review of the manuscript.

Finally, we thank Paige R. Brooks for her inspiring approach to laboratory chemistry. Paige contributed many aspects of the discovery synthesis that made varenicline a reality. She collaborated on the initial scale-up with Karl Ng in the process kilo laboratory and with chemists Jim Arnett and Joseph Petraitis. Especially noteworthy was her discovery of the dinitration reaction that is still used in the manufacture of varenicline.

REFERENCES AND NOTES

1. "Counterblaste to Tobacco," King James, 1604.
2. (a) Ezzati, M.; Lopez, A.D. Estimates of global mortality attributable to smoking in 2000. *Lancet* 2003, *362*, 847–852. (b) Doll, R.; Peto, R.; Boreham, J., et al. Mortality in relation to smoking: 50 years' observations on male British doctors. *BMJ* 2004, *328*, 1519–1527.
3. Rose, J.E. Nicotine addiction and treatment. *Ann. Rev. Med.* 1996, *47*, 493–507.
4. (a) Hays, J.T.; Ebbert, J.O. Bupropion sustained release for treatment of tobacco dependence. *Mayo Clin. Proc.* 2003, *78*, 1020–1024. (b) Holm, K.J.; Spencer, C.M. Bupropion: a review of its use in the management of smoking cessation. *Drugs* 2000, *59*, 1007–1024.
5. (a) Fiore, M.C.; Bailey, W.C.; Cohen, S.J., et al. Treating tobacco use and dependence: Clinical practice guideline. Washington, DC: U.S. Department of Health and Human Services. June 2000. (b) Murray, R.P.; Voelker, H.T.; Rakos, R.F., et al. *Addict. Behav.* 1997, *22*, 281–286.
6. (a) Picciotto, M.R.; Zoli, M.; Rimondini, R., et al. Acetylcholine receptors containing the β2 subunit are involved in the reinforcing properties of nicotine. *Nature* 1998, 391, 173–177. (b) Watkins, S.S.; Epping-Jordan, M.P.; Koob, G.F., et al. Blockade of nicotine self-administration with nicotinic antagonists in rats. *Pharmacol. Biochem. Behav.* 1999, 62, 743–751. (c) Tapper, A.R.; McKinney, S.L.; Nashmi, R., et al. Nicotine activation of α4* receptors: sufficient for reward, tolerance, and sensitization. *Science* 2004, *5*, 1029–1032. (d) Maskos, U.; Molles, B.E.; Pons, S., et al. *Nicotine reinforcement and cognition restored by targeted expression of nicotinic receptors. Nature* 2005, *436*, 103–107. (e) Laviolette, S.R.; Van der Kooy, D. The neurobiology of nicotine addiction: bridging the gap from molecules to behavior. *Nature Rev. Neurosci.* 2004, *5*, 55–65.
7. Coe, J.W.; Brooks, P.R.; Vetelino, M.G., et al. Varenicline: a novel, potent, and α4β2 selective nicotinic receptor partial agonist for smoking cessation. *J. Med. Chem.* 2005, *48*, 3474–3477.
8. Mansbach, R.S.; Chambers, L.K.; Rovetti, C.C. Effects of the competitive nicotinic antagonist erysodine on behavior occasioned or maintained by nicotine: comparison with mecamylamine. *Psychopharmacology* 2000, *148*, 234–242.
9. (a) Coe, J.W.; Vetelino, M.G.; Bashore, C.G., et al. In pursuit of α4β2 nicotinic receptor partial agonists for smoking cessation: carbon analogs of (–)-cytisine. *Bioorg. Med. Chem. Lett.* 2005, *15*, 2974–2979. (b) Cohen, C.; Bergis, O.E.; Galli, F., et al. SSR591813, a novel selective and partial α4β2 nicotinic receptor agonist with potential to smoking cessation. *J. Pharmacol. Exp. Ther.* 2003, *306(1)*, 407–420.
10. Coe, J.W.; Brooks, P.R.; Wirtz, M.C., et al. 3,5-Bicyclic aryl piperidines: a novel class of α4β2 neuronal nicotinic acetylcholine receptor partial agonists for smoking cessation. *Bioorg. Med. Chem. Lett.* 2005, *15*, 4889–4897.
11. (a) Mazzocchi, P.H.; Stahly, B.C. Synthesis and pharmacological activity of 2,3,4,5-tetrahydro-1,5-methano-1H-3-benzazepines. *J. Med. Chem.* 1979, *22*, 455–457. (b) Mokotoff, M.; Jacobson, A.E. *J. Het. Chem.* 1970, *7*, 773–778. (c) Mazzocchi. P.H.; Harrison, A.M. Synthesis and analgetic activity of 1,2,3,4,5,6-hexahydro-1,6-methano-3-benzazocines. *J. Med Chem.* 1978, *21*, 238–240.
12. (a) Wittig, G.; Knauss, E. Dehydrobenzene and cyclopentadiene. *Chem. Ber.* 1958, *91*, 895–907. (b) Muir, D.J.; Stothers, J.B. Carbon-13 magnetic resonance studies. 145. An examination of β-enolization in benzobicyclo[2.2.2], [3.2.1], and [3.2.2] ketones. *Can. J. Chem.* 1993, *71*, 1290–1296.
13. Coe, J.W. Total synthesis of (+/)-cytisine via the intramolecular heck cyclization of activated n-alkyl glutarimides. *Org. Lett.* 2000, *2*, 4205–4208.
14. (a) Pappo, R.; Allen, D.S. Jr.; Lemieux, R.U., et al. Osmium tetroxide-catalyzed periodate oxidation of olefinic bonds. *J. Org. Chem.* 1956, *21*, 478–479. (b) VanRheenen, V.; Cha, D.Y.; Hartley, W.M. Catalytic osmium tetroxide oxidation of olefins: *cis*-1,2-cyclohexanediol. *Org. Synth.* 1988, (Coll. Vol.) 6, 342–348.

15. (a) Emerson, W.S. The preparation of amines by reductive alkylation, in *Organic Reactions*, Vol. 4, Wiley, New York, 1982, chap. 3. (b) Bols, M; Persson, M.P.; Butt, W.M., et al. T. Synthesis of a ribofuranosyl cation mimic. *Tetrahedron Lett.* 1996, *37*, 2097–2100. (c) Evans, D.A.; Illig, C.R.; Saddler, J.C. Stereoselective synthesis of (±)-cyanocycline. *J. Am. Chem. Soc.* 1986, *108*, 2478–2479. (d) Kawaguchi, M.; Ohashi, J.; Kawakami, Y., et al. Facile synthesis of morpholines and azacrown ethers by ozonolysis of cyclic olefins and reductive n-alkylation. *Synthesis*, 1985, 701–703. (e) Fray, A.H.; Augeri, D.J.; Kleinman, E.F. A convenient synthesis of 3,6-disubstituted 3,6-diazabicyclo[3.2.2]nonanes and 3,6-diazabicyclo[3.2.1]octanes. *J. Org. Chem.* 1988, *53*, 896–899. (f) Wolin, R.; Wang, D.; Kelly, J., et al. Synthesis and evaluation of pyrazolo[3,4-b]quinoline ribofuranosides and their derivatives as inhibitors of oncogenic Ras. *Bioorg. Med. Chem. Lett.* 1996, *6*, 195–200. (g) Meyers, A.I.; Nguyen, T.H. C2 symmetric amines. II. Asymmetric synthesis of C2 (3S,3′S)- and (3R,3′R)-dimethyl-4H-dinaphth[2,1-c:1′,2′-e]azepines. *Tetrahedron Lett.* 1995, *36*, 5873–5876.

16. Abdel-Magid, A.F.; Carson, K.G.; Harris, B.D., et al. Reductive amination of aldehydes and ketones with sodium triacetoxyborohydride. Studies on direct and indirect reductive amination procedures. *J. Org. Chem.* 1996, *61*, 3849–3862.

17. Moffett, R.B. Cyclopentadiene and 3-chlorocyclopentene. *Org. Syn.* 1963, Coll. Vol. IV, 238–243.

18. Chen, L.S.; Chen, G.J.; Tamborski, C. The synthesis and reactions of *o*-bromophenyllithium. *J. Organomet. Chem.* 1980, *193*, 283–292.

19. Coe, J.W.; Wirtz, M.C.; Bashore, C.G., et al. Formation of 3-halobenzyne: solvent effects and cycloaddition adducts. *Org. Lett.* 2004, *6*, 1589–1592.

20. (a) Hart, H.; Lai, C.-Y.; Nwokogu, G.C., et al. Bisannelation of arenes with bis-aryne equivalents. *J. Am. Chem. Soc.* 1980, *102*, 6651–6652. (b) Hart, H.; Shamouilian, S. New phenanthrene synthesis via *ortho* bis(aryne) equivalents. Application to permethylphenanthrene. *J. Org. Chem.* 1981, *46*, 4874–4876. (c) Nwokogu, G.C.; Hart, H. Attempts to polylithiate tetrabromoarenes. The compatibility of methyl sulfate and butyllithium. *Tetrahedron Lett.* 1983, *24*, 5725–5726. (d) Hart, H.; Nwokogu, G.C. The dilithiation mechanism in the reactions of polyhaloarenes as diaryne equivalents. *Tetrahedron Lett.* 1983, *24*, 5721–5724. (e) Blatter, K; Schlueter, A.-D. Model studies for the synthesis of ribbon-shaped structures by repetitive Diels-Alder reaction. *Chem. Ber.* 1989, *122*, 1351–1356.

21. (a) Hart, H.; Lai, C.; Nwokogu, G.C., et al. Tetrahalobenzenes as diaryne equivalents in polycyclic arene synthesis. *Tetrahedron* 1987, *43*, 5203–5224. (b) Koenig, B.; Knieriem, B.; Rauch, K., et al. 4,5,12,13-Tetrabromo[2.2]paracyclophane—a new bis(aryne) equivalent. *Chem. Ber.* 1993, *126*, 2531–2534.

22. (a) Wai, J.S.M.; Marko, I.; Svendsen, J., et al. A mechanistic insight leads to a greatly improved osmium-catalyzed asymmetric dihydroxylation process. *J. Am. Chem. Soc.* 1989, *111*, 1123–1125. (b) Kwong, H.L.; Sorato, C.; Ogino1, Y., et al. Preclusion of the second cycle in the osmium-catalyzed asymmetric dihydroxylation of olefins leads to a superior process. *Tetrahedron Lett.* 1990, *31*, 2999–3002.

23. Brooks, P.R.; Caron, S.; Coe, J.W., et al. Synthesis of 2,3,4,5-tetrahydro-1,5-methano-1h-3-benzazepine via oxidative cleavage and reductive amination strategies. *Synthesis* 2004, *11*, 1755–1758.

24. (a) Coon, C.L.; Blucher, W.G.; Hill, M.E. Aromatic nitration with nitric acid and trifluoromethanesulfonic acid. *J. Org. Chem.* 1973, *38*, 4243–4248. (b) For an additional related example of dinitration, see Tanida, H.; Ishitobi, H.; Irie, T., et al. Substituent effects and homobenzylic conjugation in benzonorbornen-2(exo)-yl *p*-bromobenzenesulfonate solvolyses. *J. Am. Chem. Soc.* 1969, *91*, 4512–4520. (c) Fonseca, T.; Gigante, B.; Marques, M.M., et al. Synthesis and antiviral evaluation of benzimidazoles, quinoxalines and indoles from dehydroabietic acid. *Bioorg. Med. Chem.* 2004, *12*, 103–112.

25. (a) Criegee, R.; Schroeder, G. A crystalline primary ozonide. *Chem. Ber.* 1960, *93*, 689–700. (b) Schreiber, S.L.; Claus, R.E.; Reagan, J. Ozonolytic cleavage of cycloalkenes to terminally differentiated products. *Tetrahedron Lett.* 1982, *23*, 3867–3870. (c) Thompson, Q.E. Ozonolysis of dihydropyran. Reactions of 4-hydroperoxy-4- methoxybutyl formate. *J. Org. Chem.* 1962, *27*, 4498–4502. For general references, see Criegee, R. Mechanism of ozonolysis. *Angew. Chem., Int. Ed. Engl.* 1975, *87*, 765–771 and Lee, D.G.; Chen, T. Cleavage Reactions. In *Comprehensive Organic Synthesis*, Vol. 7, Trost, B.M., Ed., Pergamon Press, Oxford, 1991, pp. 541–591.

26. Pollart, K.A.; Miller, R.E. Ozonolysis-reductive amination of olefins. *J. Org. Chem.* 1962, *27*, 2392–2394.

27. White, R.W.; King, S.W.; O'Brien, J.L. Catalytic reduction of ozonides. II. Synthesis of amines from olefins. *Tetrahedron Lett.* 1971, *12*, 3591–3593.

28. Kiwi-Minsker, L.; Joanette, E.; Renken, A. Loop reactor staged with structured fibrous catalytic layers for liquid-phase hydrogenations. *Chem. Eng. Sci.* 2004, *59*, 4919–4825.

29. (a) Fernandez, F.; Garcia-Mera, X.; Morales, M., et al. Synthesis of new 6-substituted purinyl-5-nor-1-homocarbanucleosides based on indanol. *Tetrahedron* 2004, *60*, 9245–9253; (b) Grunewald, G.L.; Sall, D.J.; Monn, J.A. Conformationally defined adrenergic agents. 13. Conformational and steric aspects of the inhibition of phenylethanolamine N-methyltransferase by benzylamines. *J. Med. Chem.* 1988, *31*, 433–444. (c) Baker, W.; Leeds, W.G. Attempts to prepare new aromatic systems. *J. Chem. Soc.* 1948, 974–979. (d) Hansen, H.-J.; Sliwka, H.-R.; Hug, W. On the absolute configuration of (+)-indan-1-carboxylic acid. *Helv. Chim. Acta* 1982 *65*, 325–343.

30. House, H.O.; Sauter, F.J.; Kenyon, W.G., et al. Perhydroindan derivatives. IX. Diels-Alder reaction with 7-methylindene derivatives. *J. Org. Chem.* 1968, *33*, 957–961.

31. O'Donnell, C.J.; Singer, R.A.; Brubaker, J.D., et al. A general route to the synthesis of 1,5-methano- and 1,5-ethano- 2,3,4,5-tetrahydro-1h-3-benzazepines. *J. Org. Chem.* 2004, *69*, 5756–5759.

32. (a) Gassman, P.G.; Talley, J.J. Cyanohydrins—a general synthesis. *Tetrahedron Lett.* 1978, *19*, 3773–3776. (b) Evans, D.A.; Truesdale, L.K.; Carroll, G.L. Cyanosilylation of aldehydes and ketones. A convenient route to cyanohydrin derivatives. *Chem. Commun.* 1973, 55–56. (c) Lidy, W.; Sundermeyer, W. Cleavage reactions of trimethylsilyl cyanide, a new synthesis of O-(trimethylsilyl)cyanohydrins. *Chem. Ber.* 1973, *106*, 587.

33. Singer, R.A. Pfizer Global Development and Research, Groton, CT. Unpublished work, Feb. 2000.

34. Ratio determined using HPLC and GC-MS. The major component was determined to be the cis-isomer, as it is rapidly consumed in the subsequent cyclization, whereas the minor isomer is consumed more slowly.

35. The diastereoselectivity approaches 1:1 upon increased reaction time, indicative of acid-catalyzed isomerization of the ester.

36. Support for this presumed intermediate has been reported in similar systems, see Trivedi, B.K.; Blankley, C.J.; Bristol, J.A., et al. N6-Substituted adenosine receptor agonists: potential antihypertensive agents. *J. Med. Chem.* 1991, *34*, 1043–1049.

37. (a) Brown, H.C.; Heim, P.; Yoon, N.M. Selective reductions. XV. Reaction of diborane in tetrahydrofuran with selected organic compounds containing representative functional groups. *J. Am. Chem. Soc.* 1970, *92*, 1637–1646. (b) Brown, H.C.; Korytnyk, N. Hydroboration. IV. A study of the relative reactivities of representative functional groups toward diborane. *J. Am. Chem. Soc.* 1960, *82*, 3866–3869.

38. For related palladium-catalyzed cyclizations of enolates, see Ciufolini, M.A.; Qi, H.B.; Browne, M.E. Intramolecular arylations of soft enolates catalyzed by zerovalent palladium. *J. Org. Chem.* 1988, *53*, 4149–4151. Muratake, H.; Natsume, M.; Nakai, H. Palladium-catalyzed intramolecular -arylation of aliphatic ketone, formyl, and nitro groups. *Tetrahedron* 2004, *60*, 11783–11803. Muratake, H.; Natsume, M. Palladium-catalyzed intramolecular -arylation of aliphatic ketones. *Tetrahedron Lett.* 1997, *38*, 7581–7582.

39. Gallo, E.; Ragaini, F.; Cenini, S., et al. Investigation of the reactivity of palladium(0) complexes with nitroso compounds: relevance to the palladium-phenanthroline-catalysed carbonylation reactions of nitroarenes. *Organomet. Chem.* 1999, *586*, 190–195.

40. For malonate anion additions to fluoro-nitroarenes, see (a) Selvakumar, N.; Azhagan, A.M.; Srinivas, D., et al. A direct synthesis of 2-arylpropenoic acid esters having nitro groups in the aromatic ring: a short synthesis of (±)-coerulescine and (±)-horsfiline. *Tetrahedron Lett.* 2002, *43*, 9175–9178. (b) Selvakumar, N.; Yadi Reddy, B.; Sunil Kumar, G., et al. Dimethyl malonate as a one-carbon source: a novel method of introducing carbon substituents onto aromatic nitro compounds. *Tetrahedron Lett.* 2001, *42*, 8395–8398. (c) Gurjar, M.K.; Murugaiah, A.M.S.; Reddy, D.S., et al. A new route to prepare 6-chloro-5-(2-chloroethyl)oxindole. *Org. Process Res. Dev.* 2003, *7*, 309–312.

41. Recent work has led to the use of bulk electron-rich phosphine ligands to facilitate palladium-catalyzed processes with problematic oxidative addition. For examples, see (a) Littke, A.F.; Fu, G.C. A convenient and general method for Pd-catalyzed Suzuki cross-couplings of aryl chlorides and arylboronic acids. *Angew. Chem., Int. Ed. Engl.* 1998, *37*, 3387–3388. (b) Shen, W. Palladium catalyzed coupling of aryl chlorides with arylboronic acids. *Tetrahedron Lett.* 1997, *38*, 5575–5578. (c) Old, D.W.; Wolfe, J.P.; Buchwald, S.L. A highly active catalyst for palladium-catalyzed cross-coupling reactions: room-temperature Suzuki couplings and amination of unactivated aryl chlorides. *J. Am. Chem. Soc.* 1998, *120*, 9722–9723.

42. For related applications of palladium-catalyzed arylations alpha to carbonyls, see (a) Palucki, M.; Buchwald, S.L. Palladium-catalyzed -arylation of ketones. *J. Am. Chem. Soc.* 1997, *119*, 11108–11109. (b) Fox, J.M.; Huang, X.; Chieffi, A., et al. Highly active and selective catalysts for the formation of aryl ketones. *J. Am. Chem. Soc.* 2000, *122*, 1360. (c) Nguyen, H.N.; Huang, X.; Buchwald, S.L. The first general palladium catalyst for the Suzuki-Miyaura and carbonyl enolate coupling of aryl arenesulfonates. *J. Am Chem. Soc.* 2003, *125*, 11818–11819. (d) Hamann, B.C.; Hartwig, J.F. Palladium-catalyzed direct -arylation of ketones. Rate acceleration by sterically hindered chelating ligands and reductive elimination from a transition metal enolate complex. *J. Am. Chem. Soc.* 1997, *119*, 12382–12383. (e) Kawatsura, M.; Hartwig, J.F. Simple, highly active palladium catalysts for ketone and malonate arylation: dissecting the importance of chelation and steric hindrance. *J. Am. Chem. Soc.* 1999, *121*, 1473–1478. (f) Culkin, D.A.; Hartwig, J.F. Palladium-catalyzed -arylation of carbonyl compounds and nitriles. *Acc. Chem. Res.* 2003, *36*, 234–245. (g) Satoh, T.; Kawamura, Y.; Miura, M., et al. Palladium-catalyzed regioselective mono- and diarylation reactions of 2-phenylphenols and naphthols with aryl halides. *Angew. Chem., Int. Ed. Engl.* 1997, *36*, 1740–1742.

43. It has been shown that hydroxide bridged dimeric palladium complexes can form and impede catalysis, see Hartwig, J.F. Palladium-catalyzed amination of aryl halides: mechanism and rational catalyst design. *Synlett* 1997, *4*, 329–340.

44. We observed less-efficient reactions with tricyclohexylphosphine that had been exposed to the air, presumably due to oxidation of the phosphine. For best results on scale, we used new containers of the phosphine to minimize opportunity for air oxidation before use. For improved reliability, we found that dicyclohexylphenylphosphine was less susceptible to air oxidation and provided a more stable catalyst (as did dialkylbiarylphosphine ligands; see Reference 42).

45. The sodium salt of **50** was crystallized from acetonitrile and toluene and isolated as a single olefin isomer. The structure was elucidated by single-crystal X-ray diffraction.

46. Singer, R.A.; McKinley, J.D.; Barbe, G., et al. Preparation of 1,5-methanol-2,3,4,5-tetrahydro-1*h*-3-benzazepine via pd-catalyzed cyclization. *Org. Lett.* 2004, *6*, 2357–2360.

4 The SUTENT® Story

Rajappa Vaidyanathan

CONTENTS

4.1   INTRODUCTION

Sunitinib malate (**1**; Figure 4.1) is the active pharmaceutical ingredient in SUTENT® capsules, a promising drug recently approved for the treatment of gastrointestinal stromal tumors (GIST) and advanced renal cell carcinoma. This was the first instance in which the U.S. Food and Drug Administration (FDA) granted approval to an oncology product for two indications simultaneously. Sunitinib malate is a small-molecule drug that selectively targets multiple protein receptors, called receptor tyrosine kinases (RTKs). Inhibition of these RTKs is believed to starve tumors of blood and nutrients needed for growth and simultaneously kill the cancer cells that constitute tumors.

In this chapter, the evolution of the active pharmaceutical ingredient (API) manufacturing process for sunitinib malate will be described. The different routes evaluated to prepare the API, the factors that influenced the choice of the commercial process, salt selection, scale-up experience, and regulatory issues will be discussed in detail. The benefits and challenges of developing a "one-pot" process will also be highlighted. It is hoped that this chapter will provide a sense of the spectrum of issues to be considered while designing a commercial process.

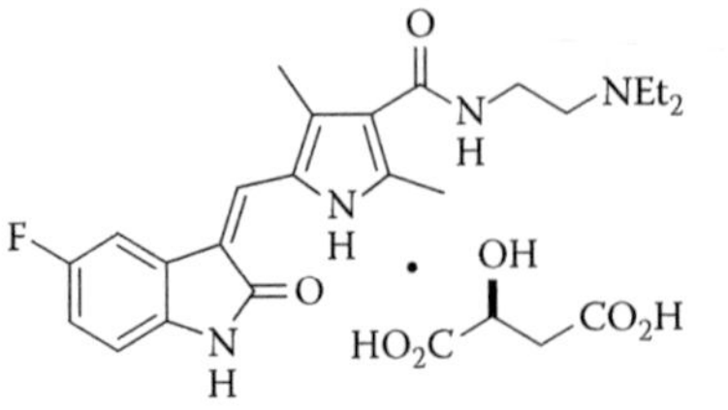

FIGURE 4.1 Suntinib Malate, the active pharmaceutical ingredient in SUTENT® Capsules.

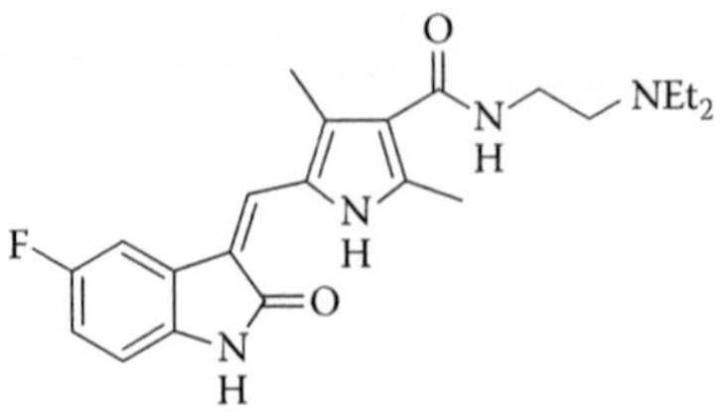

4.2 RETROSYNTHETIC ANALYSIS

The free-base form of sunitinib malate (**2**) contains a highly substituted pyrrole core bearing an amide side chain at the 3-position and an indolinone substituent at the 5-position. Two routes were originally evaluated for the synthesis of free base **2**. These were the early and late amidation routes. In the early amidation route, the pyrrole core is assembled after the amide side chain is installed, and the late amidation route involves amidation after the pyrrole core is assembled. These two approaches are depicted in Figure 4.2.

In the early amidation approach, **2** may be assembled by condensation of aldehyde-amide **4** with 5-fluorooxindole **3**. The aldehyde-amide would arise from a Knorr pyrrole reaction involving oxime **5** (derived from *t*-butyl acetoacetate) and β-ketoamide **6**, which can, in turn, be synthesized from diketene **7** and *N,N*-diethylethylenediamine **8**.

In the late amidation approach, free base **2** may be synthesized from acid-aldehyde **9**, 5-fluorooxindole **3**, and *N,N*-diethylethylenediamine **8**. Acid aldehyde **9** can be synthesized from diester **10** which would arise out of a Knorr pyrrole reaction involving oxime **5** and ethyl acetoacetate **11**.

Both of these routes were investigated in the laboratory as part of the early process research efforts. Once these routes were successfully demonstrated in the laboratory, they were critically evaluated for scalability. These efforts are discussed in detail in Section 4.3.

4.3 EARLY PROCESS RESEARCH EFFORTS

4.3.1 EARLY AMIDATION ROUTE[1]

The early amidation route commenced with the synthesis of β-ketoamide **6**. Reaction of diketene with *N,N*-diethylethylenediamine in *t*-butyl methyl ether afforded **6** in excellent yield (Scheme 4.1).[2] The β-ketoamide was prone to decomposition and had to be either used immediately or stored at −20°C for later use. It was found that **6** was typically contaminated with polymeric material that carried over from diketene. However, this did not cause any problems in the downstream chemistry.

Oxime **5** was synthesized by treatment of *t*-butyl acetoacetate with sodium nitrite in acetic acid. Reaction of **5** with β-ketoamide **6** in the presence of zinc and acetic acid according to the classic Knorr pyrrole formation conditions led to pyrrole **12**. Although this reaction worked fairly well (60 to 70% yield), workup and product isolation proved problematic as the reaction scale was increased (>10 g). Typically, the products of Knorr reactions (when β-ketoesters are used instead of β-ketoamides) are isolated by a water knock-out at the end of the reaction.[3] However, in the case of **12**, the presence of the pendant amine functionality rendered precipitation of the product from an acidic reaction mixture impossible. Precipitation of the product required basic conditions; unfortunately, at pH 9, gelatinous zinc salts crashed out of solution, making an extraction or isolation

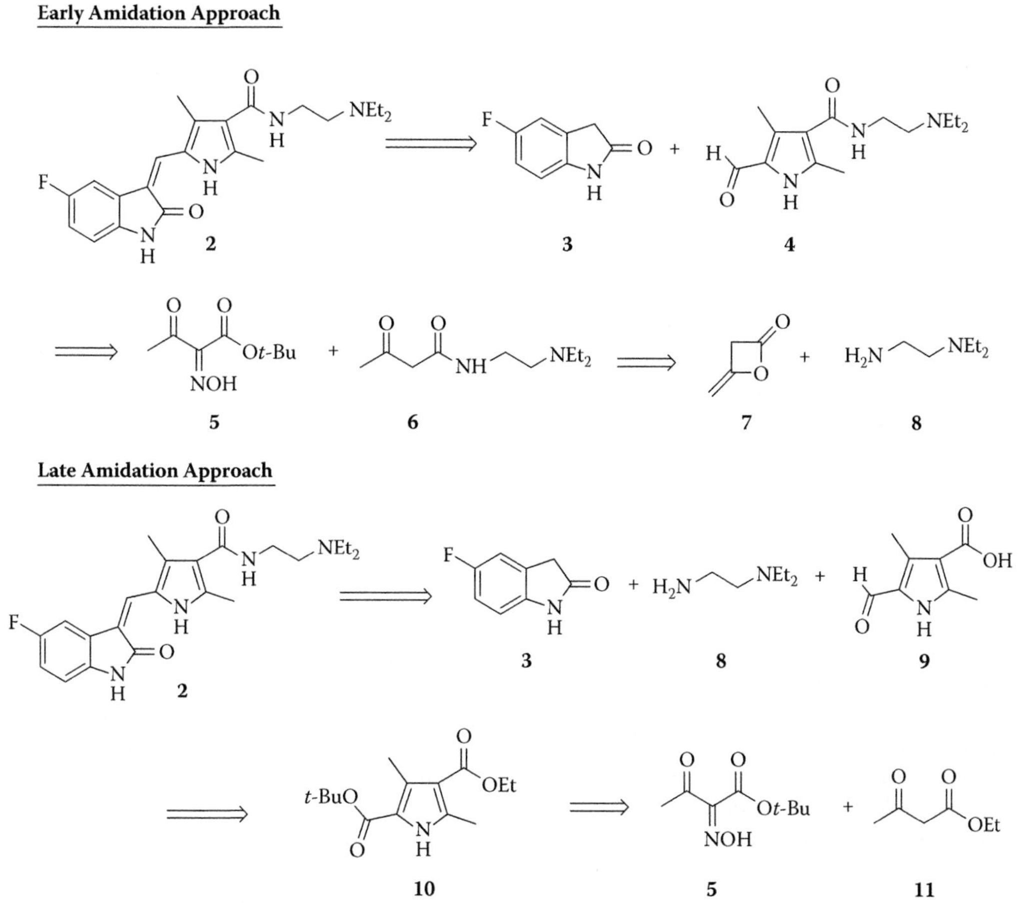

Early Amidation Approach

Late Amidation Approach

FIGURE 4.2 Retrosynthetic analysis.

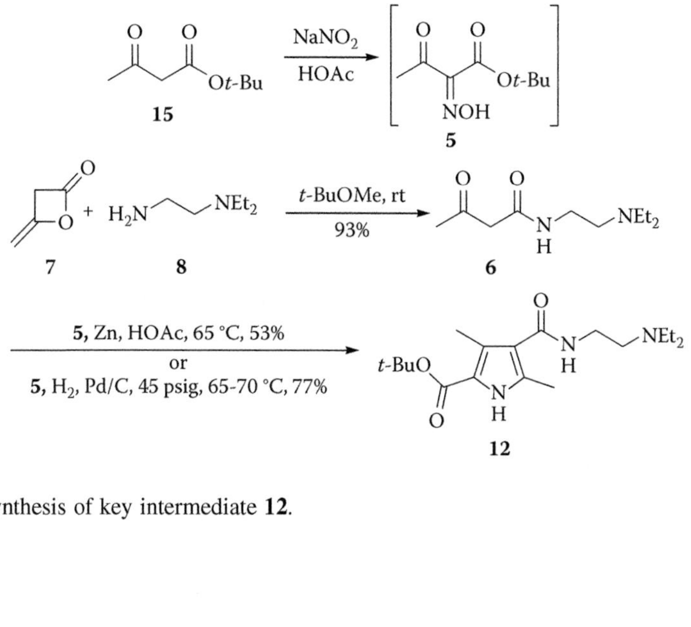

SCHEME 4.1 Synthesis of key intermediate **12**.

extremely difficult. Attempts to remove the zinc salts by filtration before proceeding with an extractive workup proved fruitless.

The workup problem was partially resolved by adding methylene chloride to the reaction mixture at the end of the pyrrole formation reaction, followed by basification with 50% NaOH until the pH was 13 to 14. The zinc salts that formed at pH 9 dissolved at pH 13 to 14. A series of water–CH_2Cl_2 extractions led to isolation of the product in 53% yield. Although this strategy worked well in the laboratory, it was not a viable proposition for scale-up, because it was impossible to avoid precipitation of zinc salts during workup even though they could be dissolved later. These gelatinous salts could potentially cause further complications during scale-up (such as binding the agitator), and therefore other alternatives to the zinc chemistry had to be explored. In addition, the environmental concerns associated with the use and disposal of stoichiometric zinc were important factors in our decision to pursue other options for the Knorr pyrrole synthesis reaction.

One attractive alternative to the zinc chemistry was to use hydrogenation conditions to form the pyrrole core. Oxime reduction[4] and pyrrole synthesis[5] under hydrogen are well documented in the literature; however, very few reports of Knorr synthesis of pyrroles bearing amide side chains under hydrogenation conditions have been published.[6] To our delight, hydrogenation of a mixture of amide **6** and oxime **5** over 10 wt% Pd/C catalyst in acetic acid (45 psig, 65 to 70°C, 7 h) led to the clean formation of pyrrole **12**. The reaction workup was simple—the catalyst was filtered off, the filtrate was basified to pH 11 to 13 and the mixture was extracted with CH_2Cl_2 to afford the product in 77% yield. These results demonstrated that the hydrogenation procedure was a viable alternative to the zinc protocol, especially for compounds that cannot be extracted or precipitated under acidic conditions.

Having assembled the pyrrole core, we endeavored to attach the oxindole side chain and complete the synthesis of **2**. Decarboxylation of **12** was achieved in quantitative yield by reaction with H_2SO_4 in methanol. The α-free pyrrole thus obtained (**13**) was treated with chloromethylene-dimethylammonium chloride in acetonitrile to form the Vilsmeier adduct **14** *in situ*.[7] Addition of 5-fluorooxindole and KOH to the reaction mixture at this stage afforded the desired product **2**, which was isolated in 74% yield upon direct filtration of the reaction mixture (Scheme 4.2). This synthetic sequence worked well in the laboratory, but further work on this route was suspended until both the early and late amidation routes were critically evaluated and the commercial route was chosen.

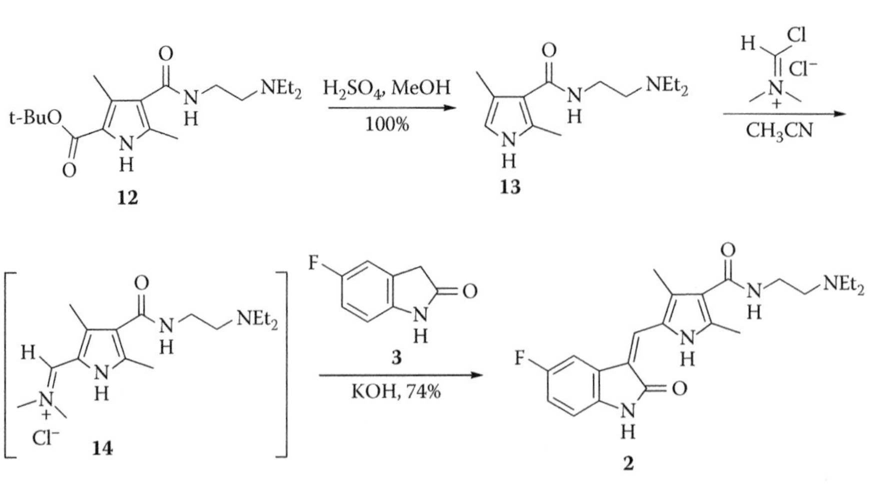

SCHEME 4.2 Early amidation route-end game.

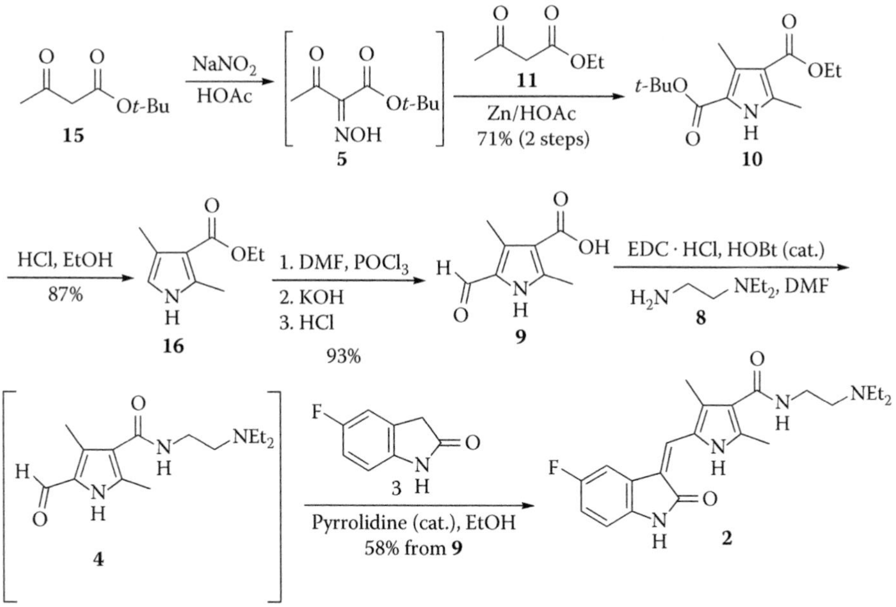

SCHEME 4.3 Late amidation route.

4.3.2 LATE AMIDATION ROUTE

In the late amidation route, the strategy was to assemble the pyrrole core before installing the amide side chain (Scheme 4.3).

Oxime **5**, derived from *t*-butyl acetoacetate, was treated with ethyl acetoacetate **11** in the presence of zinc and acetic acid according to the classic Knorr pyrrole formation conditions to afford pyrrole **10**. Unlike the early amidation approach, isolation of **10** from the zinc protocol was straightforward—addition of water to the reaction mixture precipitated the product. Further work on extending the hydrogenation conditions described above to the synthesis of pyrrole **10** was not performed in-house, because this compound was commercially available from several sources. The substitution pattern in **10** is significant because it allows for efficient removal of the *t*-butyl ester moiety under acidic conditions without affecting the ethyl ester, paving the way for introduction of the formyl group at that position.

Decarboxylation of **10** with HCl led to clean formation of the α-free pyrrole **16**. Vilsmeier formylation followed by saponification of the ethyl ester moiety and acidification led to the acid aldehyde **9**. Time constraints during the early stages of our work forced us to make **9** in-house because vendors quoted rather long lead times for delivery of this material. However, this compound is now commercially available and can be purchased from several vendors.

Pyrrole **9** was coupled with diamine **8** in the presence of EDC·HCl (1-(3-dimethylaminopropyl)-3-ethylcarbodiimide hydrochloride) and HOBt (1-hydroxybenzotriazole) to give the aldehyde-amide **4**. This intermediate was not isolated; instead, it was treated with 5-fluorooxindole **3** in the presence of catalytic pyrrolidine to give free base **2** in 58% yield.

4.4 ROUTE SELECTION

At this stage, with the limited laboratory scale experience at our disposal, it was time to make the all-important decision—choice of route for commercialization. From preliminary cost estimates, the early amidation approach appeared to be slightly more cost-effective. Moreover, the early amidation route was also more atom economical, because no coupling agents were required for amide bond formation. On the other hand, the late amidation approach had a strategic advantage in that acid-aldehyde **9** could be used as the starting material for the synthesis of a host of entities structurally related to **2**. At the time, there were several molecules in the pipeline containing the pyrrole core attached to differently substituted oxindoles and amines. The potential cost savings from the early amidation approach were not deemed to be substantial enough to offset the long-term advantages offered by the late amidation route. Therefore, from a strategic sense, it seemed prudent to adopt the late amidation approach for commercialization.

There were several issues with the late amidation approach (as it stood at the time) that needed to be addressed prior to commercial route development. A regulatory starting material (RSM) strategy needed to be in place. It was imperative to get the regulatory agency's (or agencies') approval on our choice of RSMs. Next, there were processing considerations. During the early stages of development, it was found that free base **2** had very poor crystalline properties. Hence, a salt-screening exercise had to be undertaken in order to discover an entity with better crystalline properties. EDC·HCl is relatively expensive and is a sensitizer. Moreover, HOBt poses an explosion hazard and would have to be avoided. In short, better amidation conditions needed to be developed. Furthermore, it was felt that a telescoped "one-pot" process would be the best option in the absence of crystalline or easily isolable intermediates. Even though telescoped processes are desirable, they come with a price tag—the absence of isolated intermediates puts the onus on the process chemist to design a robust process with enough quality controls and purge points to remove any process-related impurities, residual starting materials, impurities in starting materials, and derivatives of those impurities, and still routinely produce an acceptable quality API.

4.5 REGULATORY STARTING MATERIAL STRATEGY

The FDA was engaged in discussions in the early stages of commercial process development to craft a mutually acceptable RSM strategy. The CDER (Center for Drug Evaluation and Research) guidance on "Submitting Supporting Documentation in Drug Applications for the Manufacture of Drug Substances" (February 1987) states the following with regard to the definition of the starting material:

> While a definition of starting material applicable to all situations cannot be given, the following criteria for defining a starting material may be helpful:
>
> 1. It is incorporated into the new drug substance as an important structural element.
> 2. It is commercially available.
> 3. It is a compound whose name, chemical structure, chemical and physical characteristics and properties, and impurity profile are well defined in the chemical literature.
> 4. It is obtained by commonly known procedures (this applies principally to starting materials extracted from plants and animals, and to semi-synthetic antibiotics).

Compounds **3**, **8**, and **9** met the aforementioned criteria to qualify as RSMs. The FDA was provided a packet of information containing several literature reports citing the syntheses of **3**, **9**, and analogous compounds. Furthermore, several suppliers were identified for the manufacture of these compounds. Diamine **8** is a true commodity chemical available from multiple vendors. Based on the information provided, the FDA concurred with the use of these three compounds as RSMs, with the proviso that tests be developed to identify impurities in these RSMs, the fate of these impurities and the ability of the process to purge them be demonstrated, and that any impurities carried through the process above 0.1 area% be identified and quantified. This work was duly accomplished after the commercial process was developed and prior to filing the NDA.

4.6 SALT SELECTION

Early development work utilized free base **2** in some preclinical toxicology and early Phase I clinical studies. As the development program progressed, **2** was found to have unacceptably slow filtration properties, making its large-scale isolation unattractive. For instance, filtration of 13 kg of **2** required 8 h for a 2″ cake on a 4′ Nutsche filter. A salt-screening exercise was undertaken in an effort to quickly identify a crystalline and pharmaceutically acceptable salt of **2** for commercialization.

Free base **2** was treated with several readily available acids to afford the corresponding salts. The resulting salts were crystallized and evaluated for physicochemical properties including crystallinity, toxicity, hygroscopicity, animal bioavailability, solubility, stability, and morphology. As a result of this investigation, the L-malate salt (sunitinib malate, **1**) emerged as the front-runner and was therefore chosen for further development.* The following sections describe the development of a commercial process for the manufacture of sunitinib malate.

4.7 COMMERCIAL PROCESS DEVELOPMENT

Conceptually, the route chosen for commercialization was the late amidation route, and the target molecule was **1**. However, several issues (as described above) had to be addressed, the first among them being amidation conditions.

Amidation reactions are some of the most widely utilized in organic chemistry; however, several of the commonly used amidation conditions are not suitable for large-scale manufacture. For instance, the use of carbodiimides poses a problem in that the reagents are potent sensitizers and the by-product ureas are not easily removable by nonchromatographic methods. Several other activation conditions such as formation of acid chlorides followed by amidation were explored in the laboratory and met with moderate success at best. Activation of **9** with iso-butyl chloroformate gave the desired mixed carbonic anhydride (Scheme 4.4), but subsequent reaction of the amine **8** occurred at the carbonate center to afford **9** and the isobutyl carbamate of **8** (via pathway "b"), instead of amide **4** (pathway "a").

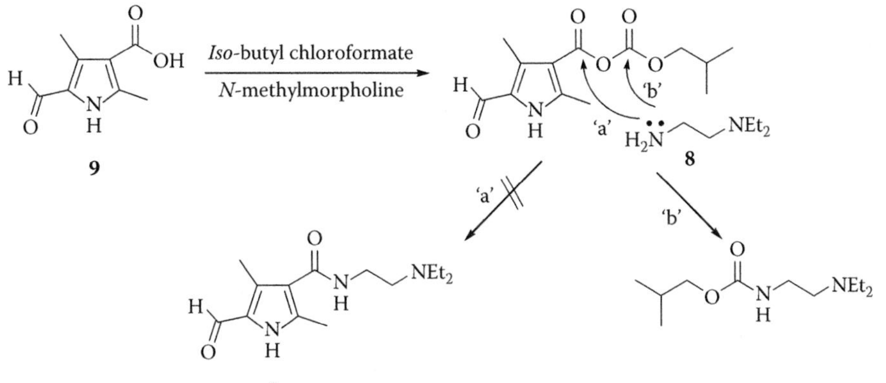

SCHEME 4.4 Attempted amidation using a mixed carbonic anhydride.

---

* The following question would be entirely justified. "Why the L-malate salt and not the D-malate or the DL-malate?" The answer is simple: there was a bottle of L-malic acid in the lab at the time of initiating the salt-screening exercise, and we did not have the D- or DL- forms. The development of the L-malate salt proceeded so rapidly that project timelines would have been delayed had a switch to the D- or DL- form been attempted.

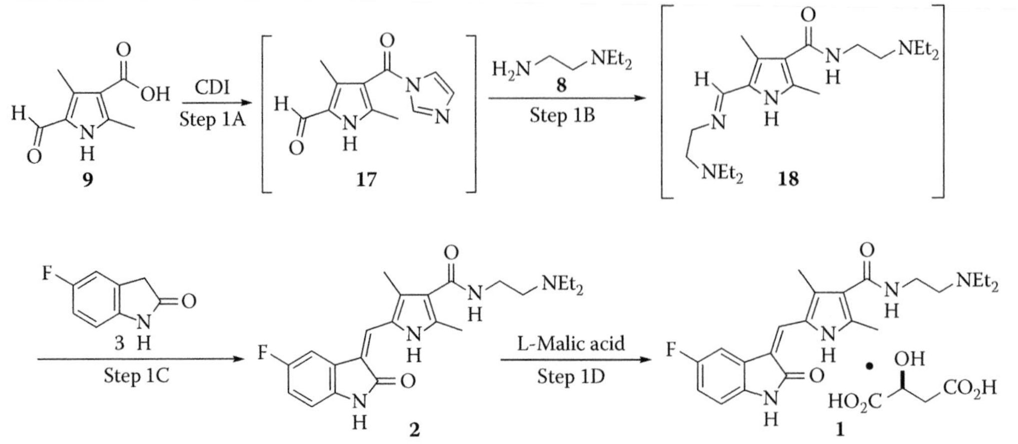

SCHEME 4.5 Late-stage amidation via activation using CDI.

However, activation of **9** with N,N′-carbonyldiimidazole (CDI) in tetrahydrofuran followed by treatment with **8** cleanly led to complete amidation. The activation reaction afforded the imidazolide intermediate **17** which, upon reaction with **8**, underwent facile amidation. The amidation reaction was accompanied by imine formation at the aldehyde end of the molecule to give imine-amide **18**. This necessitated the use of excess **8** to ensure complete amidation. Several attempts to hydrolyze the imine to the aldehyde prior to coupling with **3** were met with only partial success; but it was found that the imine could react efficiently with 5-fluorooxindole (**3**) to give desired free base **2** (Scheme 4.5).

Now, this series of chemical transformations had to be translated into a good process for the manufacture of sunitinib malate, **1**. The first factor to be considered was the solvents to be used in each of the transformations. The CDI activation and amidation reactions (Steps 1A and 1B) were found to run satisfactorily in a variety of solvents, and THF was ultimately chosen as the solvent for these steps due to its relatively low boiling point.

Likewise, several solvent systems were screened for the subsequent processing steps, that is, reaction of **18** with **3** to give **2** (Step 1C), and salt formation with L-malic acid to give **1** (Step 1D). Several factors were considered as part of this solvent-screening exercise. First, two equivalents of imidazole and excess **8** (released from the imine) would be present at the end of the free-base formation step. These would have to be removed from the reaction mixture prior to salt formation with L-malic acid (Step 1D); otherwise, they would compete with **2** to form salts with L-malic acid. The removal of these basic impurities would have to be done extractively (and not by filtration) because the free base had poor filtration properties, and the impetus for making the malate salt was to avoid isolation of the free base. The free base **2** (as well as other compounds possessing an oxindole–pyrrole–carboxy moiety) had extremely low solubility in most solvent systems. Thus, it was a challenge to design an extractive purification. Next, a solvent system from which the API could be easily and selectively recovered (crystallized) was needed. To address these issues, a systematic approach (now popularly known as the "bottom-up approach")[8] was adopted.

The solubility of **1** in a variety of solvents was examined. Solvent systems containing ICH class 1 and class 2 solvents were deliberately eschewed, especially for the final isolation step. Based on this study, 20% water/1-butanol emerged as the solvent of choice. Interestingly, **1** had a higher solubility (approximately 7 mg/mL) in 20% water/1-butanol than in pure 1-butanol (0.3 mg/mL) at room temperature. Naturally, the solubilities were much higher at higher temperatures. Thus, 20% water/1-butanol was added to **1** and heated to approximately 90°C to form a solution. This was then distilled; **1** crystallized out of solution as the azeotrope (42.5% water) was removed. The recoveries in this crystallization were typically around 90%.

Now, working backward, the solubility of free base **2** was examined. This compound had a solubility of approximately 37 mg/mL in the 20% water/1-butanol solvent system at room temperature. Any more water added to this solvent system formed a separate phase, rendering this solvent system an attractive option for the extractive removal of water-soluble impurities. Moreover, in an effort to minimize the number of solvents used in the process, Step 1C reaction was also attempted in 1-butanol and was shown to be extremely efficient.

Once the solvents were chosen, other parameters were investigated systematically, and the process was designed so as to routinely produce API of the desired quality in good yields. Because this was a one-pot process, it was imperative to build in "quality by design." Conditions such as stoichiometry, reaction time, and temperature were carefully chosen to minimize formation of impurities and afford an API of acceptable quality and yield. During development, all the reactions in Steps 1A, 1B, and 1C were monitored by high-performance liquid chromatography (HPLC) to determine progress of the reaction and impact of incomplete reactions on API yield and quality. An important factor to bear in mind is that only impurities containing the oxindole–pyrrole–carboxy moiety crystallized along with the API due to the similarity in their solubilities; other impurities were easily removed in the processing steps. Each of the steps is described in detail below.

4.7.1 STEP 1A

A significantly incomplete Step 1A reaction would lead to residual acid-aldehyde **9**, which would lead to the carboxylic acid impurity **19** (Figure 4.3) in the API. Based on careful experimentation, the Step 1A reaction was designed to routinely proceed to completion with 1.1 equiv of CDI in THF at 35 to 45°C within 2 h in the laboratory. More CDI or longer reaction times did not adversely affect API quality. The reaction mixture at the end of step 1A was carried on to the next step without any further purification.

4.7.2 STEP 1B

An incomplete Step 1B reaction would lead to residual imidazolide **17**, which could then be transformed to impurities **19** and **20** (Scheme 4.6) in the subsequent processing steps (by reaction of **17** with **3** in Step 1C followed by reaction of the imidazolide moiety with water or 1-butanol). Based on systematic investigation, the Step 1B reaction was designed to routinely proceed to completion with 3 equiv of **8** within 12 h at 35 to 45°C in the laboratory.* Addition of more **8** or longer reaction times did not have any adverse effect on API quality. At the end of the reaction, 1-butanol was added, and the THF was removed by vacuum distillation, and the mixture was carried on to the next step without further purification.

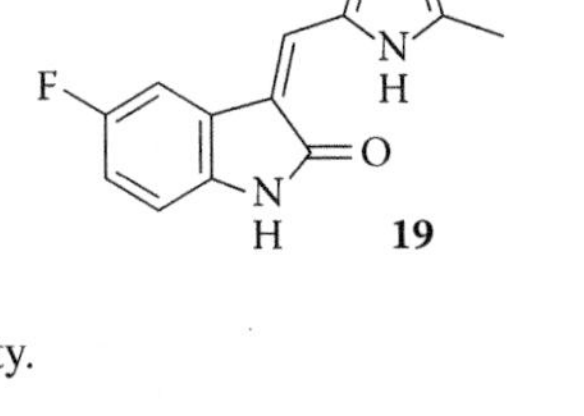

FIGURE 4.3 Carboxylic acid impurity.

---

* Theoretically, the amount of **8** required would be 2 + (2 × excess CDI), that is, 1 equiv each for the formation of the amide and imine, and 2 equiv per excess equivalent of CDI (to form the corresponding urea).

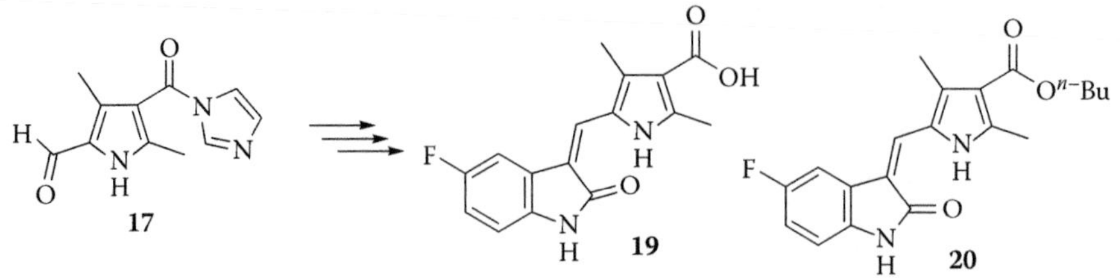

SCHEME 4.6 Formation of impurities from **17**.

4.7.3 STEP 1C

An incomplete Step 1C reaction would result in excess **18** remaining, which would get hydrolyzed to **4** during the aqueous washes that follow. Because **4** does not possess the oxindole–pyrrole–carboxy structure, it is much more soluble than the API and is easily lost to the aqueous layer during the acidic washes and to the mother liquor during the final crystallization. Nevertheless, for economic reasons, the Step 1C reaction was designed to typically be complete within 3 h at room temperature with 1.1 equiv of **3**. Any excess **3** would not cause any problems downstream, because it can be easily removed during the crystallization.

At the end of the Step 1C reaction, 1-butanol and water were added to the reaction mixture and were heated to ~80°C. The mixture was washed with water (and dilute HCl), maintaining the pH at or below pH 7. These washes ensured complete removal of residual imidazole and **8**, and worked well on scale. The organic layer was then washed with aqueous inorganic base at pH ~11 to remove any residual HCl and any other acidic impurities (such as unreacted **9**). This was then followed by a water wash to remove as much of the inorganic base as possible from the organic layer.

4.7.4 STEP 1D

Addition of L-malic acid to the reaction mixture containing free base **2** followed by concentration by vacuum distillation led to precipitation of sunitinib malate (**1**), which was isolated by filtration. The filtration of the L-malate salt took only about 30 min on a 50-kg scale (compared to 8 h for 13 kg of the free base), thereby vindicating the choice of the L-malate salt as the API.

This process, as described thus far, worked extremely well on multi-kilo scale in the pilot plant, producing API in over 80% chemical yield, with very low levels of impurities by HPLC. Nevertheless, the scale-up experience generated a few surprises and taught us a few lessons, as discussed in the next section.

4.8 SCALE-UP ISSUES AND LESSONS LEARNED

4.8.1 LESSON 1: EFFECT OF CARBON DIOXIDE ON AMIDATION RATE

During our exploratory studies on this route, the amidation reaction (Step 1B) reached 90% conversion in approximately 4 h, and was typically complete in less than 12 h. However, when attempts were made to run this chemistry in the pilot plant on a multi-kilogram scale, the amidation reaction was substantially slower, reaching only 37% completion in 4 h, and about 50% in 12 h. There was no option but to extend the reaction stir time and wait. While the reaction was slowly progressing in the pilot plant, we started thinking about what could have gone wrong. The reaction eventually reached completion in 50 h; however, the problem had to be resolved almost immediately because the next batch was slated to start in 3 days, and any delay had the potential to adversely affect project timelines.

One potential explanation for the slow reaction in the pilot plant was that the CO_2 released in the imidazolide formation step was not removed completely from the reactor in the pilot plant prior to addition of diamine **8**. This lingering CO_2 could have reacted with the amine to form the carbamate salt, thus slowing the amidation reaction.

In order to verify this hypothesis, two experiments were set up in the laboratory, starting with isolated imidazolide **17**.* In one case, CO_2 was bubbled into a slurry of imidazolide **17** and imidazole** in THF for 15 min prior to the addition of the amine, and the other was run under CO_2-free conditions. The reaction with CO_2 reached 88% conversion in 4 h, while the CO_2-free reaction was substantially slower and was only 37% complete in the same time. These experiments clearly demonstrated that CO_2 has a beneficial, rather than deleterious, effect on the reaction rate.

Based on these results, a conscious attempt was made to retain the CO_2 liberated in Step 1A in solution during Step 1B in the next pilot-plant batch. This was done by slowing the agitation in Steps 1A and 1B and by maintaining a very low nitrogen sweep of the headspace. As expected, the amidation reaction reached approximately 90% conversion in 4 h and was complete in less than 12 h without any complications. An alternative approach was to keep the vent closed during the CDI activation and amidation reactions, and this was also shown to be successful in a separate run in the plant. Thus, the problem of slow reaction was solved without any adverse impact on project timelines. In retrospect, better agitation and venting in the pilot plant in our first batch led to removal of CO_2 from the reactor, and thus slowed the amidation reaction.

In separate experiments, CO_2 was bubbled through solutions of the amine, imidazolide **17**, and the solvent prior to the reaction. Another reaction was carried out at 50 psig CO_2. Interestingly, all these reactions proceeded at the same rate as the normal reaction (one-pot conversion of the acid to the amide via the imidazolide without venting the CO_2). This suggests that only a critical amount of CO_2 in solution is needed to catalyze the reaction; additional CO_2 does not necessarily help.

It was found that this CO_2 catalysis was generally applicable to a wide range of substrates, and the results have been published.[9] The next question was, "Can one 'jump-start' a slow, CO_2-free reaction by bubbling in CO_2 mid-way through the reaction?" The reaction of imidazolide **21** with benzyl amine was chosen as the test reaction. In the experiment, a mixture of **21**, imidazole, and benzyl amine was allowed to stir at 45°C. After 1 h (approximately 1% conversion to the amide), CO_2 was sparged into the reaction mixture for 15 min, and the mixture was stirred at 45°C. As expected, the reaction rate increased dramatically and was nearly identical to that of the CO_2-catalyzed reaction (Figure 4.4). This conclusively established the fact that CO_2 catalyzes the reaction and may be used to "jump-start" slow amidation reactions.

Based on our observations and literature precedent, the following pathways were postulated to explain the catalytic effect of CO_2 on amidation reactions. Reaction of the amine with CO_2 would lead to the alkylammonium N-alkyl carbamate, **23**. Nucleophilic attack of the oxygen center of carbamate **23** on the imidazolide would give intermediate **24**, which upon extrusion of CO_2 would lead to the amide (Scheme 4.7, pathway "a"). Alternately, the nitrogen center of tautomer **25** may attack the carbonyl of the imidazolide to directly give the amide (Scheme 4.7, pathway "b").

In summary, the first lesson learned was the ability of CO_2 released during the activation of a carboxylic acid to catalyze the subsequent amidation reaction. Although the mechanistic aspects need some clarification, the utility of this phenomenon in organic synthesis, especially on large scale, is clear.

---

* This intermediate is typically not isolated in the process; however, it is an isolable, stable solid.

** Imidazole was added to these reactions in order to mimic the reaction mixture in a conventional reaction where the imidazolide would not be isolated (i.e., the mixture after CDI activation would contain the imidazolide and imidazole). However, the presence or absence of imidazole does not affect the results of the CO_2 experiments.

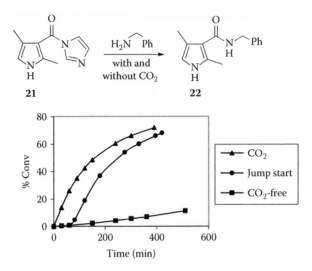

FIGURE 4.4 Jump start experiment.

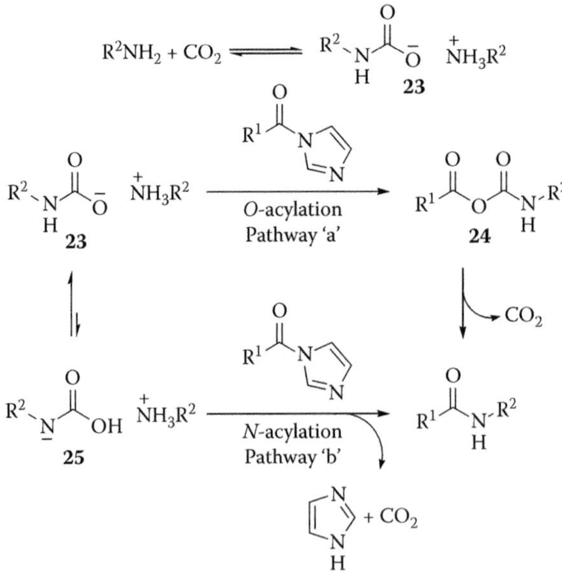

SCHEME 4.7 Possible pathways of CO_2 catalysis.

4.8.2 LESSON 2: ACETATE PROMOTED AMINAL FORMATION

The reactions in Steps 1A, 1B, and 1C were monitored by HPLC during the early runs. A peak eluting after **17** was consistently observed at 0.1 to 1 area % in the in-process HPLC analysis of step 1A during several pilot runs; however, it was observed at higher levels (5 to 20 area %) in later runs. Nevertheless, all of these batches afforded acceptable quality API, with no discernible difference in yields over the previous batches.

Extensive analytical work on a sample of the reaction mixture at the end of Step 1A identified this peak as corresponding to the aminal **26** (Scheme 4.8). Subsequently, an authentic sample of

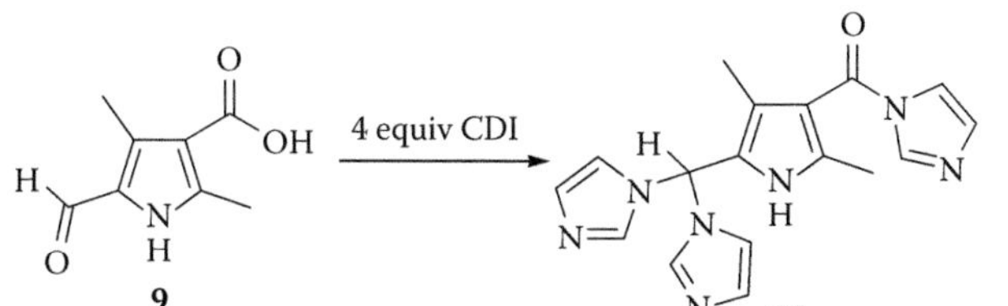

SCHEME 4.8 Synthesis of **26**.

26 was synthesized by treatment of **9** with a large excess of CDI, and its structure was confirmed by nuclear magnetic resonance (NMR) and mass spectrometry. This sample of **26** was subjected to the normal sunitinib malate processing conditions from Step 1B onwards and was completely converted to sunitinib malate with no new impurities over 0.10%. This added credence to the expectation that the presence of **26** in the HPLC chromatogram of Step 1A would not cause any API quality problems. However, the sudden increase in the levels of this peak was intriguing, and root cause for its presence needed to be investigated.

Preliminary investigations revealed that this issue (high levels of **26**) was specific to the batch of acid-aldehyde **9** used in the later runs, and that the "contaminant" that caused high levels of **26** needed to be present in only catalytic quantities to promote the side reaction. Moreover, when the implicated batch of **9** was washed with water, dried, and then used in the Step 1A reaction, the peak due to **26** dropped to typical levels (0.1 to 1 area %). This demonstrated that the "contaminant" was water soluble and removable by washing the crystals of **9** with water.

The final step in the synthesis of **9** is the saponification of the ester in **16**, followed by precipitation by acidification, and filtration (Scheme 4.3). Although HCl was used for this purpose when **9** was made in-house, the vendor utilized acetic acid. The by-product of this transformation is potassium acetate and could be a potential contaminant in **9**. Further support for this hypothesis was garnered by analysis of the filtrate obtained after washing the implicated batch of **9** with water and methanol, which revealed the presence of potassium (by qualitative elemental x-ray microanalysis) and acetate (by Fourier transform infrared [FT-IR] spectroscopy). ICP (inductively coupled plasma) tests did not show the presence of any other ionic impurities at significant levels.

Our first instinct was to spike the Step 1A reaction mixture with a catalytic quantity of potassium acetate. A sample of acetate-free **9**, which was made using HCl for the acidification step and did not result in high levels of **26** under Step 1A conditions, was subjected to Step 1A reaction conditions in the presence of a small amount of added potassium acetate. However, no increase in the level of **26** was observed in this case. This result was intriguing and suggested that our hypothesis (that potassium acetate was the culprit) was wrong. However, we rationalized that any residual potassium acetate in **9** would have to be coated on the surface; therefore, the spiking experiment described above did not really mimic the reaction conditions in the plant where elevated levels of **26** were observed.

To test whether surface-bound acetate was the cause of elevated levels of **26**, a sample of acetate-free **9** was washed with a solution of potassium acetate and dried. This sample coated with potassium acetate (estimated at 1.25 wt%) was then subjected to the Step 1A conditions. The HPLC chromatogram of this reaction showed 10 area % of **26**, demonstrating that the side reaction was indeed promoted by potassium acetate. The difference in outcomes of the potassium acetate spiked and coated experiments could be due to proximity or solubility effects, because both **9** and potassium acetate are insoluble in the reaction solvent (THF).

It was also determined that **26** was formed by reaction of **9** with CDI and not imidazole. Thus, the maximum amount of **26** that could be formed in the sequence to synthesize sunitinib malate cannot be greater than the excess amount of CDI used (assuming imidazolide formation is faster than animal formation). In any event, the formation of high levels of **26** did not cause any API quality or yield issues.

In summary, the main lesson learned here is the ability of acetate to promote the reaction of CDI with aldehydes. Additionally, this teaches us the importance of carefully planning experiments to test hypotheses—a simple spiking experiment need not reproduce reality. More experimental work aimed at understanding the mechanism and expanding the synthetic utility of this novel reaction is in progress and will be communicated in due course.

4.8.3 LESSON 3: OXIDATIVE DEALKYLATION OF AMINES

Here is one more minor lesson learned during development of the RSM impurity strategy. We had agreed with the FDA that we would identify all impurities observed at 0.1 area % or higher in our RSMs. As the potential impurities in diamine **8** were being evaluated by gas chromatography (GC), a peak (m/z 142 by GC/MS) was consistently observed at 0.2% in all lots of **8** purchased from multiple suppliers. This was intriguing and suggested that it could be an artifact or thermal degradant under the GC assay conditions. This peak seemed to grow in size when either the injection port temperature was increased or a sample of **8** was exposed to air prior to GC analysis. Literature precedent suggested that tertiary amines underwent dealkylation via the corresponding N-oxides via an oxidative degradation pathway.[10] It was rationalized that a similar pathway could be operative in this case as well. If so, the by-product would be acetaldehyde. This could then combine with another molecule of **8** to give the imine **27** (m/z 142), as depicted in Scheme 4.9. Indeed, when a sample of **8** was treated with acetaldehyde prior to analysis by GC, only the m/z 142 peak was observed in the GC trace, confirming our hypothesis.

In summary, the lesson learned here is that the oxidative dealkylation of tertiary amines can occur under GC conditions. Furthermore, it is important to do a complete analytical evaluation before an observed chromatographic peak is concluded to be an impurity in the sample.

4.9 CONCLUSIONS

In summary, two approaches to sunitinib malate were evaluated, and the late amidation route was chosen for commercialization. The commercial process was carefully developed and successfully implemented on multi-kilo scale to produce the API in over 80% chemical yield in four telescoped chemical transformations. The process was designed with adequate controls and purge points to routinely afford API containing very low levels of impurities. Several lessons were learned during the scale-up of this process, and they were discussed in this chapter. These learning points serve to emphasize the fact that no transformation can be considered trivial; even simple, mundane reactions like amidations are capable of generating surprises, and opening up new and interesting possibilities for future research. Even the most casual reader would have noticed a few gaps or unanswered questions in the story; confidentiality considerations prevented me from discussing those aspects in detail.

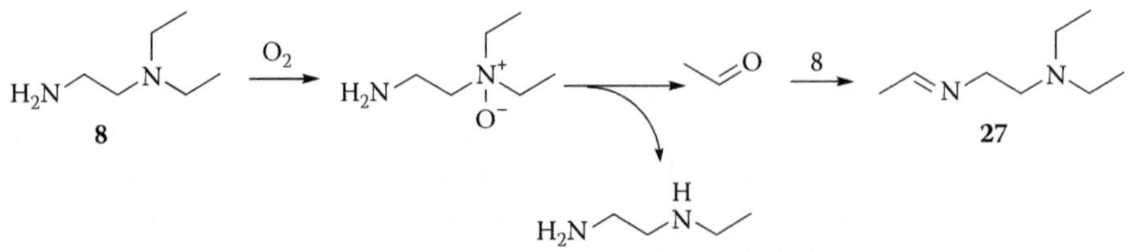

SCHEME 4.9 Oxidative dealkylation of **8**.

Given the life-threatening nature of the cancers targeted by sunitinib malate, there was a big emphasis on speeding the medicine to market, even before the FDA granted priority review status for this drug. To achieve discovery, process research and development, analytical research and development, API manufacturing, project governance, drug product development, and regulatory affairs compliance, a total of 12 sites spread across 10 time zones were involved in efforts to bring this molecule to market.

ACKNOWLEDGMENTS

This chapter represents the efforts of numerous individuals, whose substantial contributions made this work possible. Even though it is impossible to acknowledge each of them individually, I would like to name a handful of people who were most intimately involved with the work described here. Jeffrey Havens, Jerad Manley, and Monica Kalman were the key researchers on the early amidation approach. Thomas Fleck, Michael Mauragis, Alan Jin, Mark Maloney, and Vikram Kalthod made significant contributions to the late amidation route. William Snyder, Chadd Gromaski, Michael Gangwer, and James Tam provided the bulk of the analytical support. Patricia O'Driscoll, Shona Fox, and other colleagues from the manufacturing facility helped with the technology transfer. Adrian Davis, Gary Martin, and their respective groups helped with structural elucidation work. Peter Giannousis and David Waite provided strategic input at different stages of this project. Nathan Ide provided helpful suggestions during the preparation of this manuscript. A special word of thanks to Thomas Runge for providing valuable guidance throughout the course of this work, and in the preparation of this manuscript.

This paper is dedicated to Prof. V. Srinivasan, Loyola College, Chennai, India, on the occasion of his 60th birthday.

REFERENCES

1. Manley, J.M.; Kalman, M.J.; Conway, B.G.; Ball, C.C.; Havens, J.L.; Vaidyanathan, R. *J. Org. Chem.* 2003, 68, 6447.
2. Beholz, L.G.; Benovsky, P.; Ward, D.L.; Barta, N.S.; Stille, J.R. *J. Org. Chem.* 1997, 62, 1033.
3. (a) Fischer, H. *Org. Synth. Coll.* Vol. 2 1943, 202. (b) Treibs, A.; Hintermeier, K. *Chem. Ber.* 1954, 86, 1167.
4. (a) Itoh, K.; Sugihara, H.; Miyake, A.; Tada, N.; Oka, Y. *Chem. Pharm. Bull.* 1978, 26, 504. (b) Doyle, T.W.; Belleau, B.; Luh, B.-Y.; Conway, T.T.; Menard, M.; Douglas, J.L.; Chu, D.T.-W.; Lim, G.; Morris, L.R.; Rivest, P.; Casey, M. *Can. J. Chem.* 1977, 55, 484.
5. (a) Winans, C.F.; Adkins, H. *J. Am. Chem. Soc.* 1933, 55, 4167. (b) Ochiai, E.; Tsuda, K.; Ikuma, S. *Chem. Ber.* 1935, 68, 1710.
6. Moon, M.W.; Church, A.R.; Steinhards, A. Ger. Offen. (1973) DE 2235811.
7. Hafner, K.; Bernhard, C. *Angew. Chem.* 1957, 69, 533.
8. Chen, C.-K.; Singh, A.K. *Org. Process Res. Dev.* 2001, 5, 508.
9. Vaidyanathan, R.; Kalthod, V.G.; Ngo, D.P.; Manley, J.M.; Lapekas, S.P. *J. Org. Chem.* 2004, 69, 2565.
10. Grossi, L. *Free Rad. Res. Comms.* 1990, 10, 69.

5 An Efficient and Scalable Process for the Preparation of a Potent MC4 Receptor Agonist

Cécile Savarin, John Chung, Jerry A. Murry,
Raymond J. Cvetovich, Chris McWilliams, Dave Hughes,
Joseph Amato, Geneviève Boice, Karen Conrad,
Edward Corley, Robert Reamer, and Lisa DiMichele

CONTENTS

5.1 INTRODUCTION

Drug candidate **1** is a potent melanocortin receptor agonist with the potential for the treatment of obesity. This chapter will describe our efforts toward developing a robust synthesis of the target molecule. We will begin with an overview of the medicinal chemistry approach, then describe an interim route that was used to prepare initial kilogram quantities and finally present our long-term solution to the synthesis of MC4R agonist **1**.

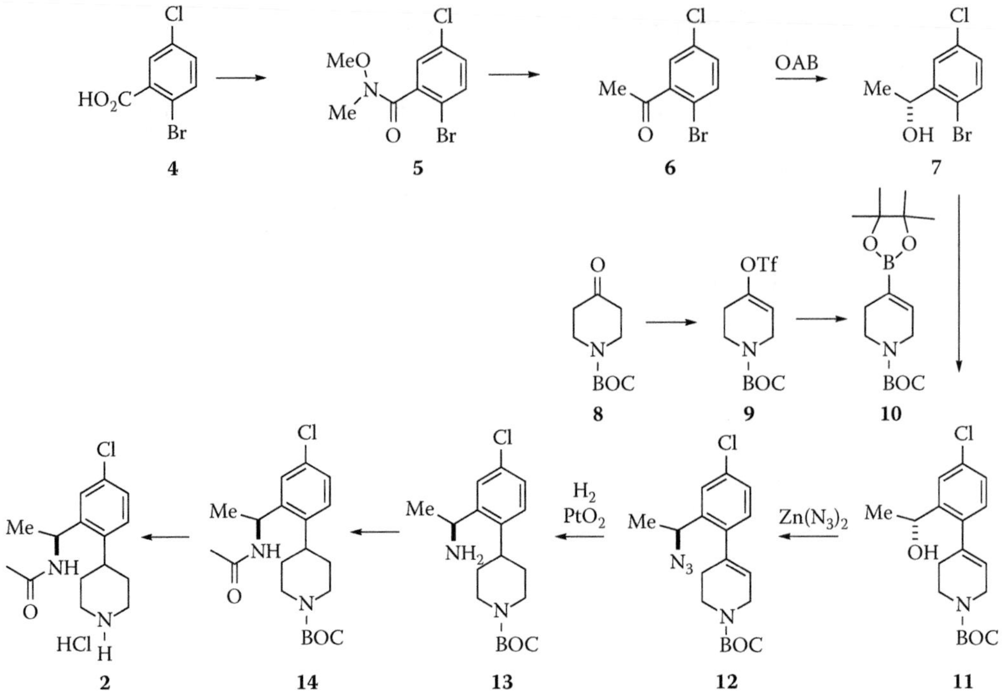

SCHEME 5.1 Medicinal approach to the piperidine fragment.

5.1.1 MEDICINAL CHEMISTRY

The original synthesis[1] involved a final stage coupling of piperidine amine **2** and acid **3** fragments to provide the amorphous HCl salt of **1**. The synthesis to amine **2** began with 2-bromo-5-chlorobenzoic acid **4** that was elaborated to the phenethyl alcohol **7** via asymmetric reduction of the corresponding acetophenone utilizing catalytic oxazaborolidine reduction. The resulting aryl bromide **7** was then coupled to vinyl boronate **10**, which was prepared from the requisite piperidinone **8**. The phenethyl alcohol **11** was then converted to azide **12** utilizing zinc azide under Mitsunobu conditions. The trisubstituted olefin and alkyl azide were simultaneously reduced under hydrogenation conditions to produce the desired piperidine ring. Amine **13** was then acetylated and the BOC (*t*-butoxycarbonyl) group was removed to provide piperidine amine **2** in >99% enantiomeric excess (ee) (Scheme 5.1).

This ten-step approach has several impediments to a long-term efficient synthesis:

1. The preparation of the boronate ester utilizes pinacol diborane, which is expensive and available in limited quantities.
2. The hydrogenation of the trisubstituted olefin of the piperidine results in various amounts of dechlorination.
3. The preparation and use of zinc azide.
4. The need for chiral high-performance liquid chromatography (HPLC) separation to upgrade the ee.

Acid fragment **3** was rapidly prepared by 1,3 dipolar cycloaddition of the requisite cinnamate **17** with *tert*-butyl azomethine ylide **16**. However, the synthesis of the azomethine ylide **15** required rather harsh conditions in the alkylation step and suffered low yield in the methoxymethylation due to its inherent instability. This provided a racemic mixture of 3,4-*trans*-disubstituted pyrrolidine

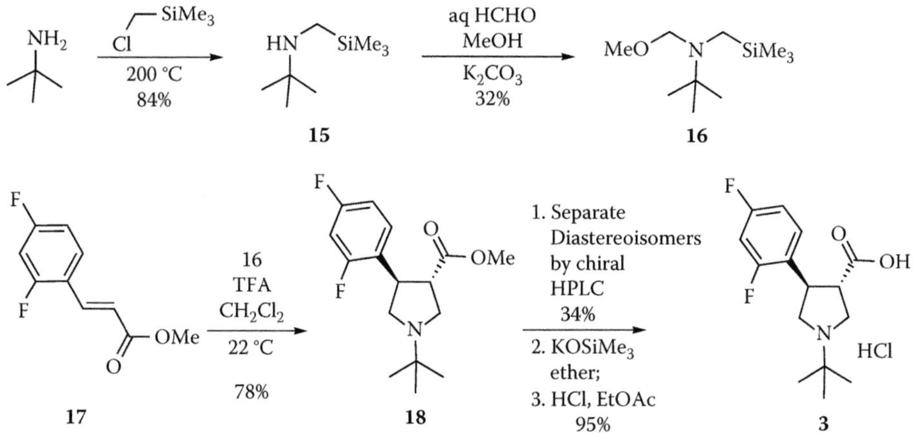

SCHEME 5.2 Early synthesis of pyrrolidine acid **3**.

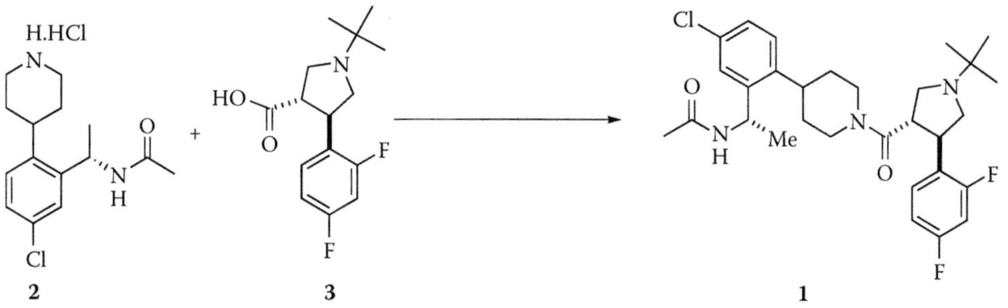

SCHEME 5.3 Amide formation.

18 (Scheme 5.2). This material was purified by chiral chromatography to provide the desired *trans* ester in high ee. Hydrolysis provided crystalline acid **3** as the hydrochloride salt.

The last step of the medicinal chemistry protocol consisted of coupling amine **2** and acid **3** using HATU (*O*-(7-Azabenzotriazol-1-yl)-N,N,N′,N′-tetramethyluronium hexafluorophosphate) in DMF (N,N-Dimethylformamide). After aqueous workup, the free base of final product **1** was first purified by column chromatography and then treated with HCl to afford drug candidate **1** as an amorphous HCl salt (Scheme 5.3).

5.1.2 INTERIM ROUTE DEVELOPMENT

The first goal of the project team was to develop an interim route that addressed some of the major hurdles to large-scale (kilogram-scale) preparation that could be utilized to make kilogram quantities of the target molecule in a timely manner.

5.1.2.1 Piperidine

As mentioned in the previous section, there were several concerns about the initial synthesis to the amine fragment. As a result, the team did not feel that simple variations of this approach could be implemented for the preparation of initial quantities of material. Fortunately, there exists a plethora of methods in the literature for preparing these moieties as outlined in retrosynthetic fashion in Scheme 5.4 and Scheme 5.5.

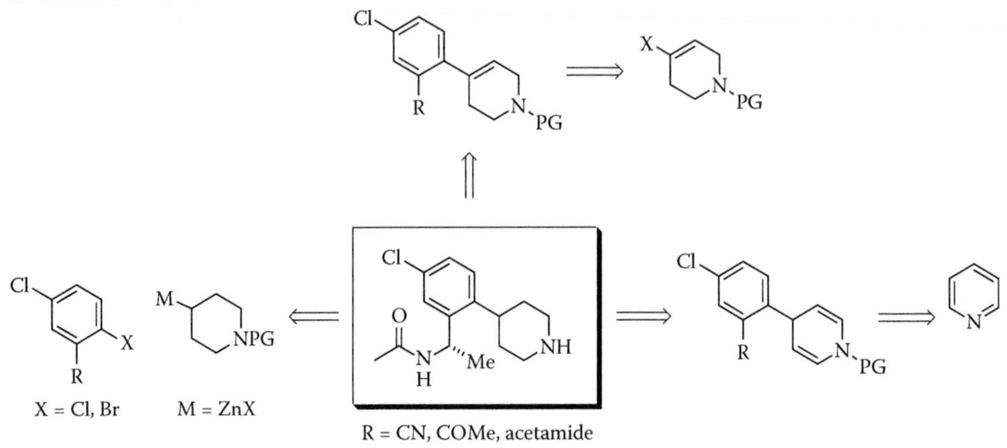

SCHEME 5.4 (See color version following page 40). 4-Arylpiperidine.

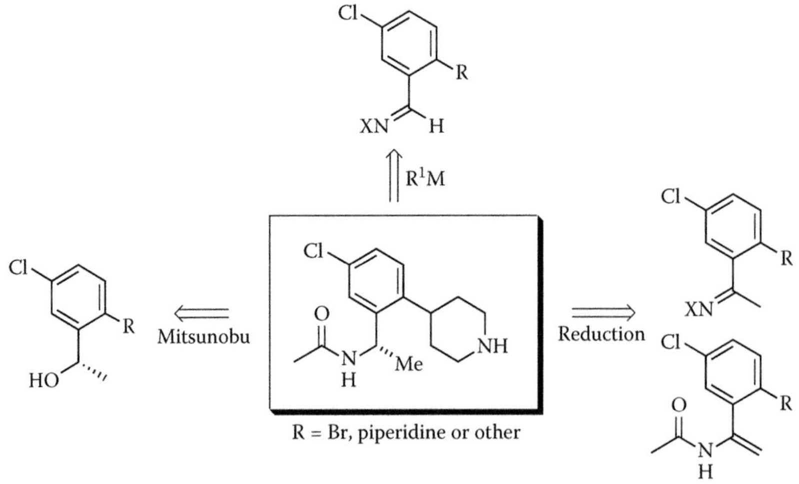

SCHEME 5.5 (See color version following page 40). Introduction of the phenethylamine chiral center.

4-Arylpiperidines are ubiquitous pharmacophores that can be prepared via transition metal-catalyzed coupling reactions or nucleophilic additions to activated pyridinium ions. Although sp^2–sp^2 coupling, as exemplified by the vinyl boronate coupling, is extensively precedented, more recent methodology has outlined the potential for sp^3–aryl couplings that directly affords the desired product at the right oxidation state and avoids the risk of dechlorination observed in catalytic hydrogenation conditions. Alternatively, the additions of aryl Grignard reagents to activated acyl pyridinium ions in the presence of Cu(I) provide a variety of 4-aryl piperidine derivatives (see Scheme 5.4).[2]

Likewise, the preparation of phenethyl amines is well precedented. The benzylic amine functionality can be introduced by either displacement of the phenethyl alcohol (Mitsunobu), Grignard addition to an aldimine derivative, or reduction of an enamine and ketimine precursors (see Scheme 5.5). Due to the steric environment presented by the *ortho* piperidine group, displacement of the benzylic alcohol proved problematic. Although displacement of the alcohol could be effected when R was less sterically imposing, this route was less desirable due to the need to tolerate a benzylic amine during introduction of the piperidine fragment. The most efficient protocol for introduction

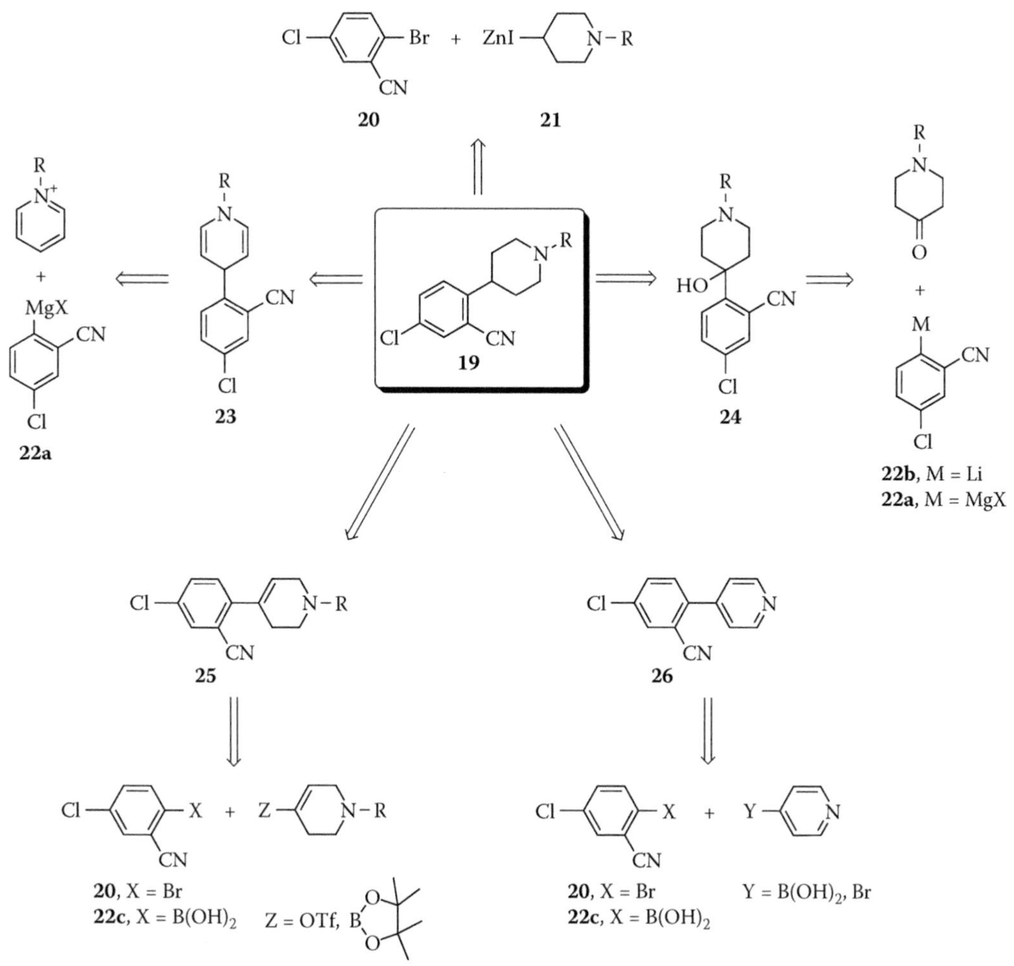

SCHEME 5.6 Retrosynthetic analysis to BOC-(*t*-butoxycarbonyl) aryl nitrile.

of this functional group is the asymmetric hydrogenation of enamides; however, these reductions are very sensitive to steric environment, and substantial optimization would be needed prior to the identification of effective reaction conditions. Direct reductive amination has recently been dramatically improved by the advent of the *tert*-butyl sulfinamide by Ellman.[3]

These methodologies provided numerous potential pathways to examine. We desired a quick and reliable method that could make kilograms of material, was not necessarily cost effective, but would provide the rapid production of material. As a result, we decided to introduce the piperidine fragment first, followed by the phenethylamine. This would require a functional group that could later be transformed to the phenethylamine. A nitrile is a robust carboxylic acid equivalent that can survive numerous transformations, does not suffer from solubility and reactivity problems that esters and acids do, and can be transformed at the appropriate time via hydrolysis or alkyl Grignard addition. With that in mind, several possible approaches to the piperidine ring introduction are presented in Scheme 5.6. We proceeded to prepare these various benzonitrile intermediates and investigated which of these would be most appropriate for the downstream functional group conversion of nitrile to phenethylamine.

We first needed to select an appropriately substituted benzonitrile. 3-Chlorobenzonitrile is readily available and cheap; however, introduction of a functional group in the 6 position could be difficult. Metalation of this material occurs predominantly in the 2 position. Electrophilic

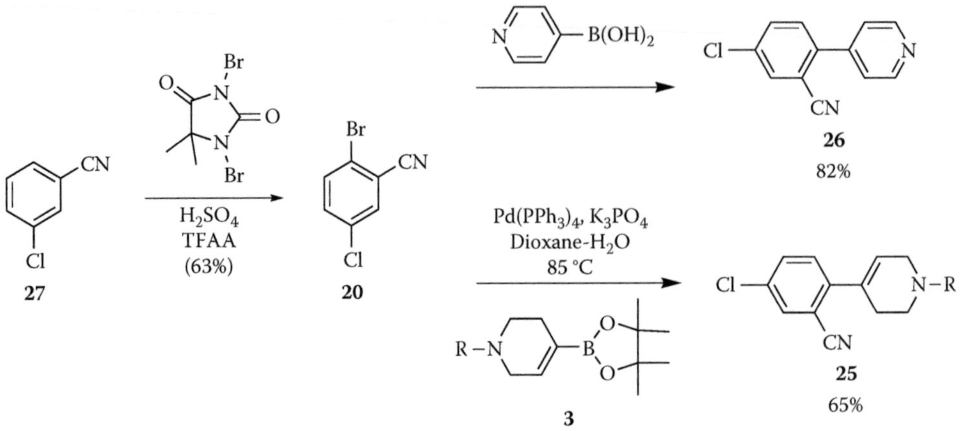

SCHEME 5.7 Suzuki approach.

bromination of benzonitrile occurs at the meta position, so one may predict that the C5 position would be reactive unless the chloride directed the electrophile *para* to the C6 position. Gratifyingly, exposure of 3-chlorobenzonitrile to brominating reagents in the presence of strong acid provided the desired product, although in moderate yield, along with regioisomers. Careful optimization of the reaction parameters, including the brominating reagent, solvent, and nature of the acid, improved the reaction yield from 50 to 63%. Optimally, 2-bromo-5-chlorobenzonitrile **20** was prepared from 3-chlorobenzonitrile using dibromodimethylhydantoin/TFA (trifluoroacetic acid)/sulfuric acid, followed by a heptane swish of the crude product to remove trace amounts of regioisomers and dibrominated products. Employing this material in Suzuki couplings both with vinyl boronate and pyridine boronate esters provided the desired products **25** and **26** in good yields (Scheme 5.7).

The electronic nature of the coupling partners could be reversed by introducing a boronic acid fragment on the benzonitrile, such as boronic acid **22c**, and affecting the coupling with an aryl triflate of pyridinium bromide. Because the aryl boronate could be prepared using the aryl lithium in the presence of triisopropyl borate, this would obviate the need for the expensive pinacol diborane. Reduction of the intermediate pyridine or tetrahydropyridines would provide the desired products (see Scheme 5.8).

However, as previously noted, the sp^2–sp^2 coupling routes suffered from the necessity of hydrogenating the double bond with the liability of dechlorination. Thus, alternate protocols for introducing this fragment were investigated. One method that has been extensively used is the addition of aryl Grignard reagents to acyl pyridinium salts (see Scheme 5.9).[4] This can provide addition products either at the 2 or 4 positions. The 4-substituted isomer can preferentially be formed in the presence of copper I salts. Formation of the aryl Grignard could be effected at low temperature utilizing Knochel conditions with *iso*-propylmagnesium chloride in THF. Attempts to form Grignard **22a** using magnesium metal resulted in extensive decomposition, presumably through reaction with the nitrile. Addition of this Grignard to a preformed suspension of the arylpyridinium salt (formed from the corresponding chloroformate addition to pyridine), provided the desired dihydropyridine **23** in good yield. We also observed that maintaining an inert atmosphere was critical in preventing the formation of Ullmann dimerization product **29**. The dihydropyridine was subsequently reduced using mild "Wilkinson's catalyst" conditions (Scheme 5.9). Interestingly, the intermediate isolated from incomplete reactions was monohydrogenated product **30**, and we did not observe any isomerization of the double bond to the more substituted position in conjugation with the aryl ring. Importantly, dechlorination was not observed under these conditions.

Of interest, it was possible to effect protecting group exchange from the benzyl or phenyl carbamate to the *tert*-butyl derivative by exposure of this material to potassium *tert*-butoxide in

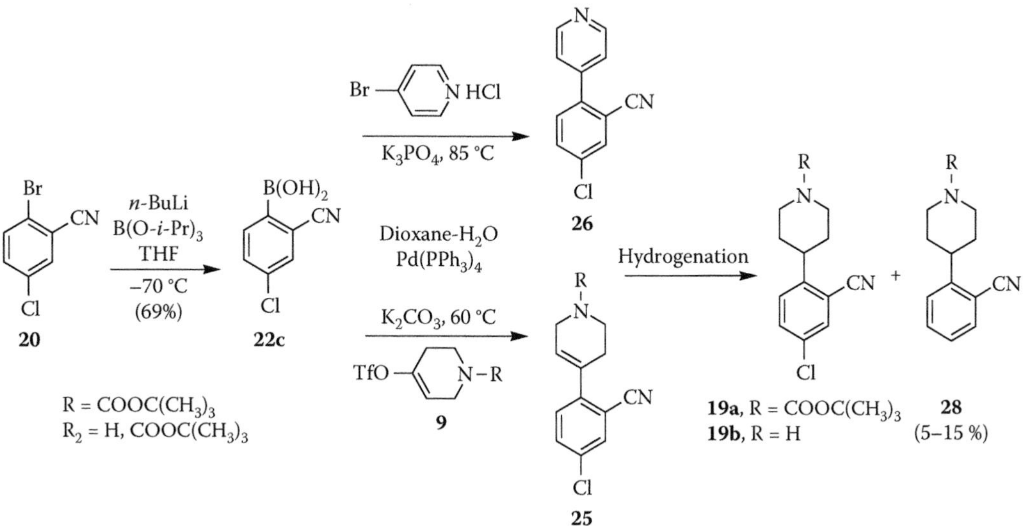

SCHEME 5.8 Suzuki alternatives.

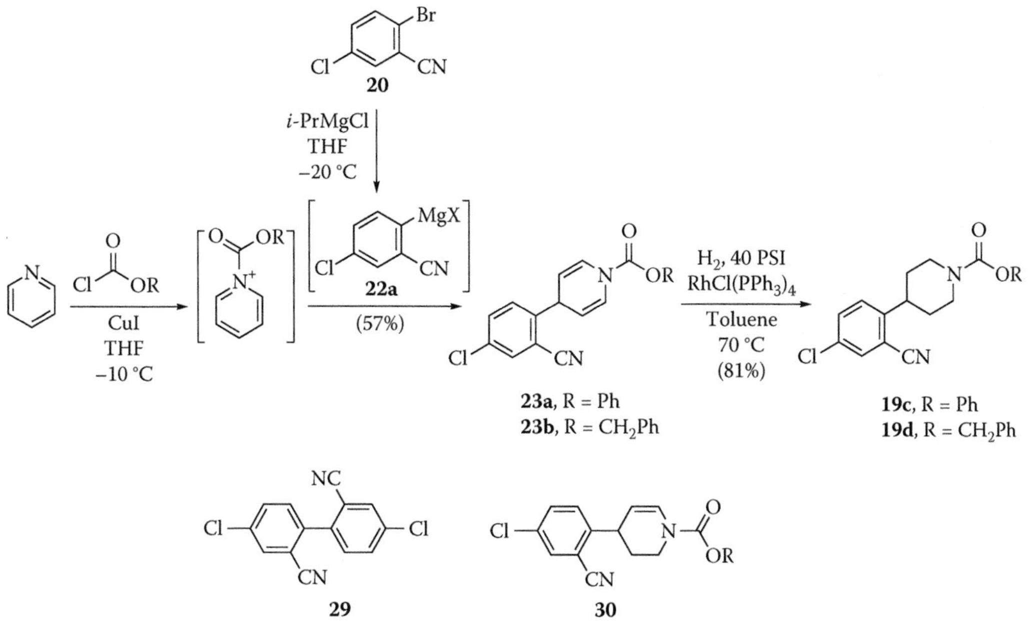

SCHEME 5.9 The acyl pyridinium route.

THF giving access to CBZ (benzyloxycarbonyl), BOC, and phenyl carbamate piperidine intermediates (Scheme 5.10).

With piperidine nitriles in hand, we next investigated the functional group interconversion of the benzonitrile to the phenethylamine. Methyl Grignard addition to the aryl nitrile occurred with concomitant loss of the protective group. The addition of copper salts did not improve the reaction. However, using an excess of methyl Grignard, followed by treatment with BOC$_2$O, we were able to obtain the desired BOC-protected ketone in 65% isolated yield. Because a late-stage deprotection

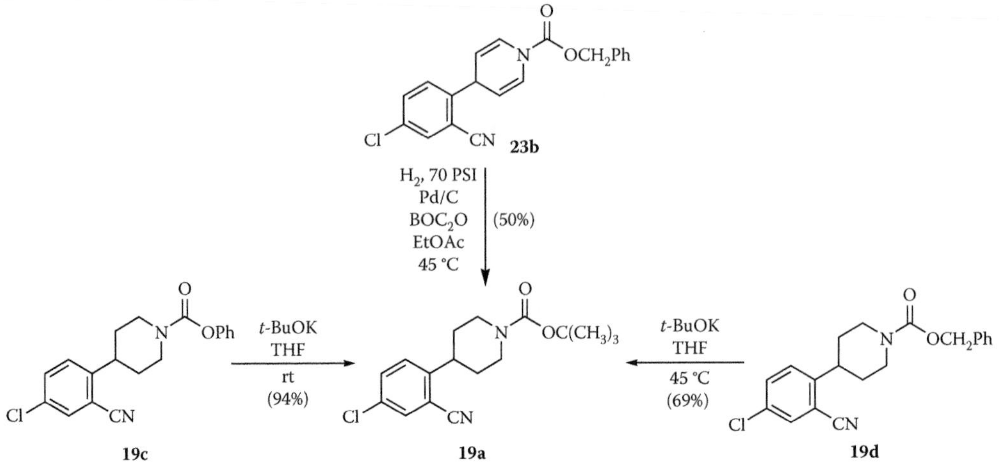
O
N
OCH2Ph
Cl
CN
23b
H2, 70 PSI
Pd/C
BOC2O
EtOAc
45 °C
(50%)
OPh
t-BuOK
THF
rt
(94%)
OC(CH3)3
45 °C
(69%)
19c
19a
19d

SCHEME 5.10 Preparation of BOC nitrile.

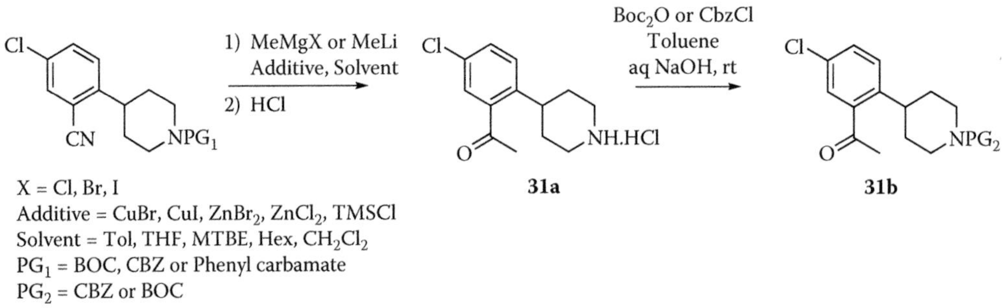
1) MeMgX or MeLi
Additive, Solvent
2) HCl
Boc2O or CbzCl
Toluene
aq NaOH, rt
NPG1
NH.HCl
NPG2
31a
31b
X = Cl, Br, I
Additive = CuBr, CuI, ZnBr2, ZnCl2, TMSCl
Solvent = Tol, THF, MTBE, Hex, CH2Cl2
PG1 = BOC, CBZ or Phenyl carbamate
PG2 = CBZ or BOC

SCHEME 5.11 Ketone synthesis.

in the presence of a phenethyl amine and aromatic arylchloride would be required, the BOC protective group appeared to be the best choice (see Scheme 5.11).

At this point, we investigated the reduction of the ketone to the phenethyl alcohol followed by Mitsunobu displacement as well as direct reductive amination. We were able to efficiently reduce the ketone racemically using $NaBH_4$ in MeOH to alcohol **32**. However, the subsequent Mitsunobu reaction proved problematic and no desired product was observed (Scheme 5.12).

The asymmetric reduction of less-hindered acetophenone **6** utilizing oxazaborolidine was sluggish but did provide a good yield of alcohol **7** in 88% ee. Investigating other chiral-reducing agents revealed that Noyori's DiPen transfer hydrogenation conditions provided 90% ee and good yield. Next we investigated the Mitsunobu displacement with a variety of nitrogen nucleophiles. The

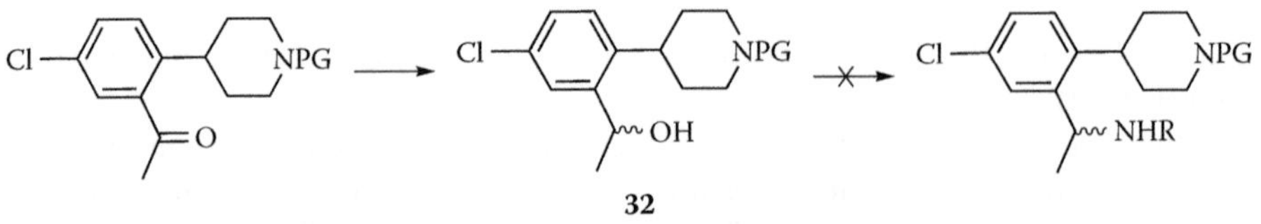
Cl
NPG
O
OH
NHR
32

SCHEME 5.12 (See color version following page 40). Mitsunobu approach.

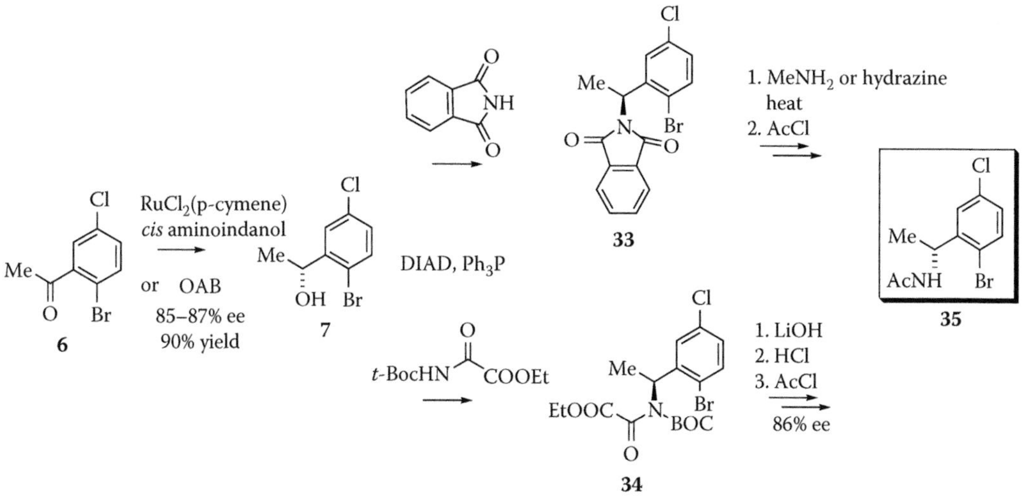

Other reagents used: $(PhO)_2P(O)N_3$/DBU; AcNH-COCOOEt; *t*-BOC-NHAc DIAD/Ph_3P/HN(*t*BOC)$_2$, *t*-BOC-NH-COCOOEt, CF_3CONH_2, Phthalimide, CF_3CONH-$COCF_3$

SCHEME 5.13 Alternative Mitsunobu approach.

reduced DEAD (diethylazodicarborylate) by-product was competitive with most nucleophiles that were employed, but phthalimide proved to be the best nucleophile and provided acceptable yields of the desired product. This route could potentially be used as the pathway to product, but the low productivity of these steps encouraged us to investigate the direct reductive amination scenario (see Scheme 5.13).

The reduction of sulfinyl imines is well precedented. Recently, Ellman demonstrated that the *tert*-butyl derivative is a highly selective reagent for this transformation that can also be removed under very mild conditions. Formation of the imine using titanium ethoxide followed by reduction under standard conditions (sodium borohydride in methanol) provided the desired product as a 5:1 mixture of diastereomers (Scheme 5.14).

We decided to investigate this reduction further and probe the effect of solvent, temperature, and reducing agent on the diastereoselectivity. Remarkably, changing the solvent to THF reversed the sense of selectivity. More bulky L-selectride showed improved selectivity over that observed with sodium borohydride. Lowering the temperature and moving to more nonpolar solvents increased the selectivity to 10:1 (Table 5.1).

We presume the increase in selectivity is due to the cyclic transition state in the nonpolar solvent which is broken up in more polar media. This type of behavior was previously observed in the addition of Grignard reagents to aldimines (Scheme 5.15).

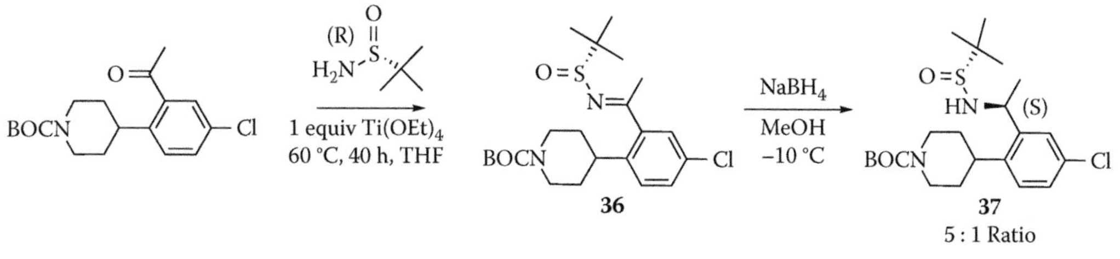

SCHEME 5.14 Ellman's chemistry.

TABLE 5.1
Change in Selectivity of the Imine Reduction

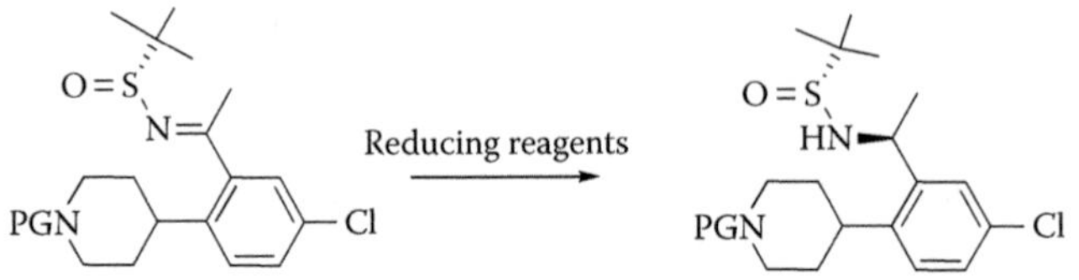

Protective Group	Solvent (T°C)	Reagents	Ratio (S/R)
	THF (−40)	NaBH₄	1:3
	MeOH (−40)		6:1
CBZ	THF (−20)	L-selectride	7:1
	THF (−15)	LiAl[OC(CH₃)]₃H	1:5
	THF (−50)		7:1
	MTBE (−50)		8:1
t-BOC	CH₂Cl₂ (−50)	L-selectride	9:1
	Toluene (−50)		10:1

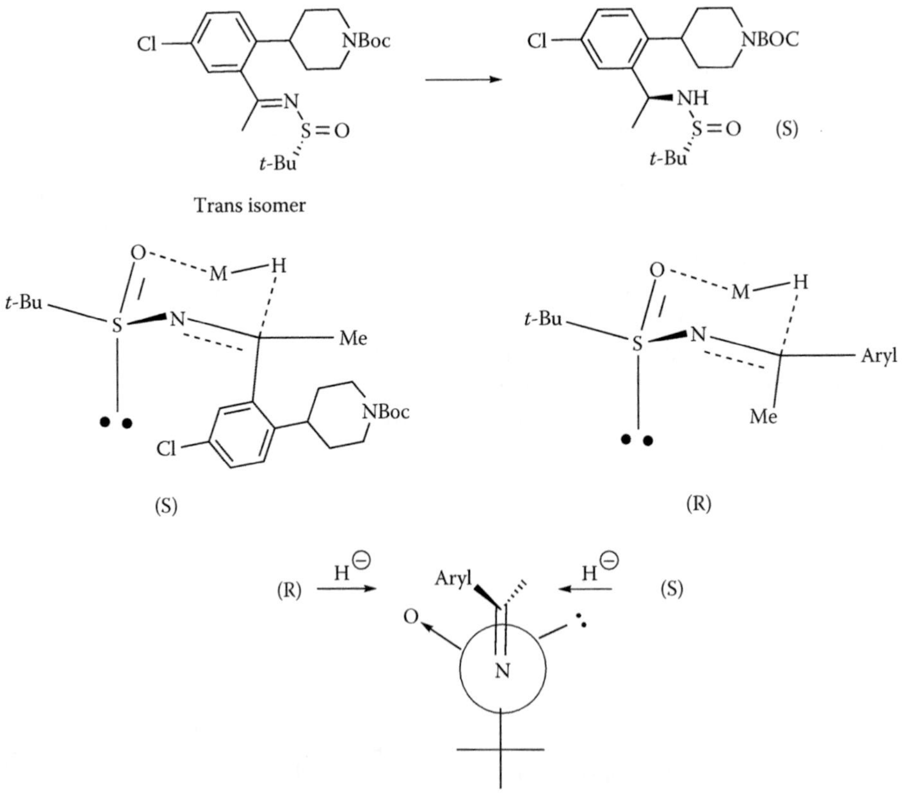

Trans isomer

SCHEME 5.15 Proposed transition state.

The final optimized protocol involved the formation of the imine in toluene using titanium ethoxide as the Lewis acid, followed by quenching with water and filtering off the titanium oxide products. The resulting imine solution was then reduced to provide 10:1 diastereoselectivity. This

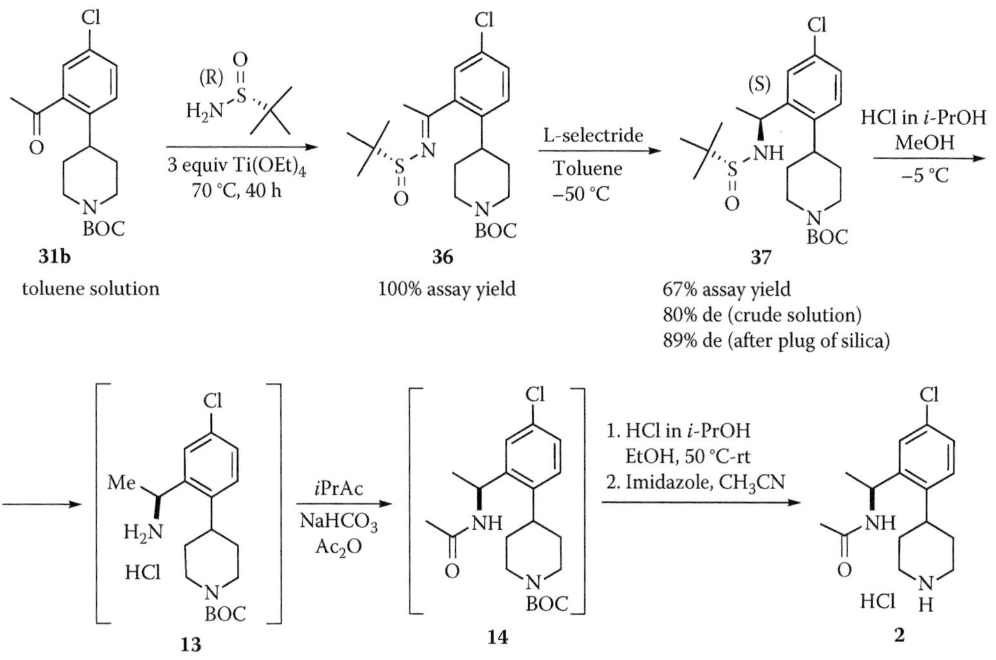

SCHEME 5.16 (See color version following page 40). From ketone to phenethylamine.

material was carried directly through to the BOC deprotection, and the resulting amine could be isolated as a crystalline solid (Scheme 5.16).

5.1.2.2 Pyrrolidine Acid

The initial asymmetric synthesis (see Scheme 5.2) of pyrrolidine acid **3** suffered from a chiral HPLC bottleneck. As a result, chiral salt resolution was investigated. The rapid discovery of a crystalline di-*p*-toluoyl-D-tartaric acid salt provided the necessary means to resolve and purify the desired diastereomer. Using 0.65 equivalent of the acid in methyl *tert*-butyl ether (MTBE), the crystallized salt was shown to be a 92:8 ratio of (3S,4R):(3R,4S) diastereomers. The resolved tartaric acid salts were then recrystallized from *n*-butanol to ratios of >99:1 in a 42% overall yield on a 2-kg scale. Further improvements were made in the preparation of the azomethine ylide precursor **38**. In step 1, using dimethyl sulfoxide (DMSO) as the solvent, the reaction temperature of trimethylsilylmethylation of *tert*-butylamine was lowered from 200°C used in the original synthesis to 80°C. In step 2, substituting *n*-butanol in place of methanol reduced the amount of oligomerization observed and increased the yield to 69%. Overall, these improvements allowed for the preparation of pyrrolidine acid **3** in 22% overall yield in 99% ee from cinnamate **39** (Scheme 5.17).

5.1.3 FINAL COUPLING

With the two fragments in hand, we investigated the final amide coupling and product isolation. Because final product **1** was not crystalline at this time, it was imperative that the coupling reaction provide product as pure as possible so as not to introduce any by-products that would necessitate the use of chromatography for purification. Exposing the acid and amine to EDC (*N*-(3-Dimethylaminopropyl)-*N*′-ethylcarbodiimide) and HOBT (1-Hydroxybenzotriazole) resulted in clean conversion. Quenching with water and extraction with *i*PrAc provided the desired amide. A concentrated solution of the API in CH₃CN was lyophilized under high vacuum to provide an amorphous powder.

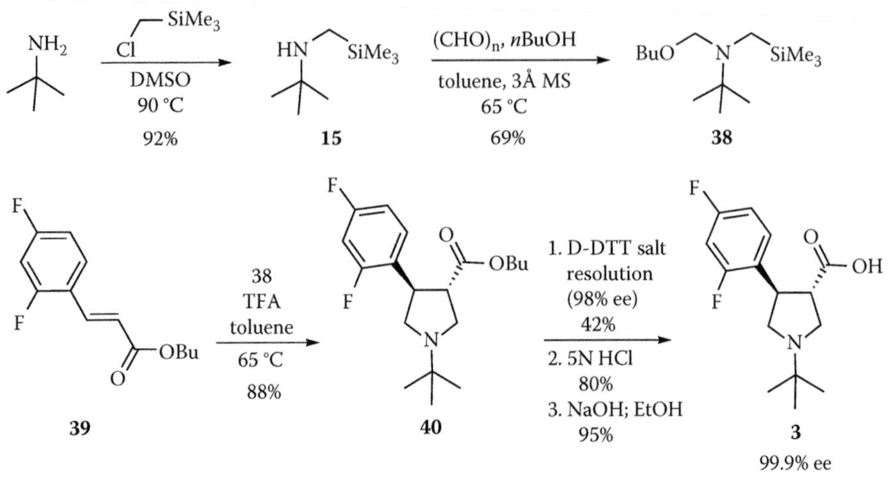

SCHEME 5.17 First-generation process pyrrolidine acid synthesis.

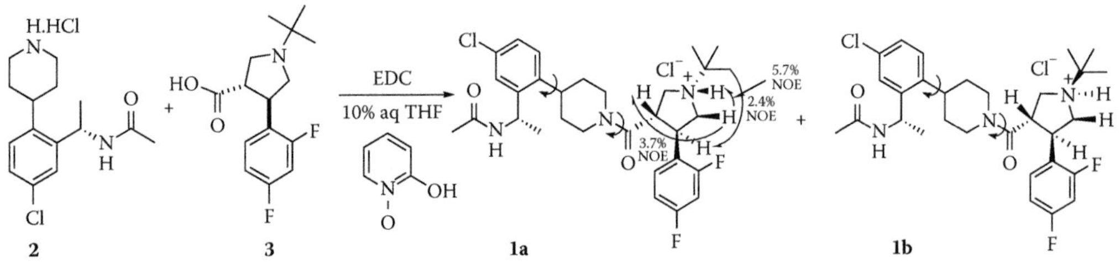

SCHEME 5.18 EDC (*N*-(3-Dimethylaminopropyl)-*N*′-ethylcarbodiimide) coupling.

NMR analysis revealed that this material exists as a complex mixture of amide rotamers as well as diastereomers due to the new chiral center created by the protonated amine. Shown in Scheme 5.18 are the major and minor diastereomers, **1a** and **1b**, with rotation indicated about the amide linkage. The stereochemistry of the major diastereomer, **1a**, was determined based on the observed NOEs shown with the structure. In characterizing this compound, it was observed that the dynamic equilibrium of diastereomers was significantly affected by solvent polarity. A crystalline perchlorate salt was obtained, and the single-crystal X-ray of this salt revealed that the piperidine ring is positioned directly over the aryl ring which results in a significant upfield shift of the axial protons.

Thus, an interim route had been identified to prepare the desired product. The acid fragment was prepared in racemic fashion utilizing a dipolar cycloaddition and resolved via chiral salt formation. The piperidine piece was prepared using Comins chemistry to put in the 4-arylpiperidine piece and Ellman chemistry to introduce the phenethylamine. Although this route was useful for preparing initial quantities, it would not be a long-term cost-effective route. Other methods would need to be investigated.

5.2 LONG-TERM ROUTE SELECTION

The goals of the long-term route development were to replace any stoichiometric chiral auxiliaries or resolutions with catalytic asymmetric reagents as well as replace any unstable reactive organometallic reagents requiring cryogenic conditions with room temperature-stable intermediates. These changes would affect overall throughput, cycle time, and cost effectiveness.

5.2.1 THE AMINE FRAGMENT

The previous synthesis of the amine fragment relied on Comins methodology to install the piperidine ring and Ellman methodology to install the phenethylamine. The Comins chemistry, although efficient, relied on cryogenic conditions to control the unstable benzonitrile Grignard reagent. Employing a more stable coupling that afforded the desired oxidation state could improve the efficiency and throughput for this process. Sp^3–aryl couplings were reported by Knochel.[5] The zinc reagents are prepared in high yield and are stable even at elevated temperatures. Formation of the piperidine zinc reagent **21** was effected utilizing the secondary alkyl iodide and zinc metal. After filtration of the reaction solution, the zinc reagent could be stored for extensive periods without loss of reactivity. Exposing this reagent to the aryl bromide under typical conditions did not result in any appreciable coupling. Control reactions with commercially available cyclohexyl zinc demonstrated that the aryl bromide was an effective coupling partner. Thus, additives were investigated in an attempt to accelerate this transformation. Addition of copper I salts to this reaction resulted in complete consumption of the starting material. Changing to Pd-dppf further increased the yield of this reaction. Thus, this procedure allowed for the direct introduction of the piperidine ring onto the aryl fragment without using cryogenic conditions.

4-Iodo-N-BOC-piperidine **42** was obtained by iodination of 4-hydroxy-*N*-BOC-piperidine **41** with iodine/triphenylphosphine/imidazole in THF. An extractive workup removed the triphenylphosphine oxide, allowing a direct crystallization of the pure iodide from ethanol/water that typically contained <1 mol% of triphenylphosphine oxide and other by-products. DMAC was later identified as a better solvent to avoid the formation of THF-polymer or dimer impurities resulting from the interaction of I_2/THF.

The Negishi coupling started with the synthesis of the zinc iodide reagent (Table 5.2, Scheme 5.19).[6a] The zinc was activated by a modification of the Knochel procedure, then reacted with a solution of iodide **42** in DMAC to provide the desired alkyl zinc species. The zinc insertion was run in the presence of 5 weight% of CelPure P65 to aid in the filtration of excess zinc from the insertion reaction mixture. The resulting solution was stable at 22°C under N_2 for >6 months.

The zinc reagent was then coupled with aryl bromide **20** affording piperidine **19a** in 72% yield after aqueous workup, Ecosorb treatment to remove metals, and crystallization from ethanol/water.

An efficient introduction of the phenethylamine moiety was envisioned from reduction of an enamide intermediate, but the formation of the enamide proved somewhat problematic. Initial attempts to form the enamide from the oxime **43** resulted in closure to the cyclic imine **45**, presumably via the intervention of a nitrene intermediates (Scheme 5.20).[6b]

TABLE 5.2
Negishi Approach

Catalyst	Additive	Conditions	Yield
(2-furyl)$_3$P, Pd$_2$(dba)$_3$	None	THF, 65°C	0
	CuI	THF, 65°C	35%
	CuI	DMAC, 80°C	35%
Cl$_2$Pd(dppf)	CuI	THF, 65°C	85%
	CuI	DMAC, 80°C	95%
	None	DMAC, 80°C	Trace

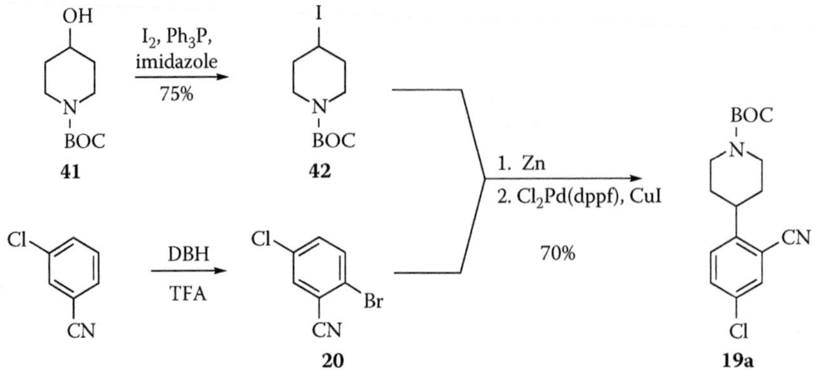

SCHEME 5.19 Negishi coupling.

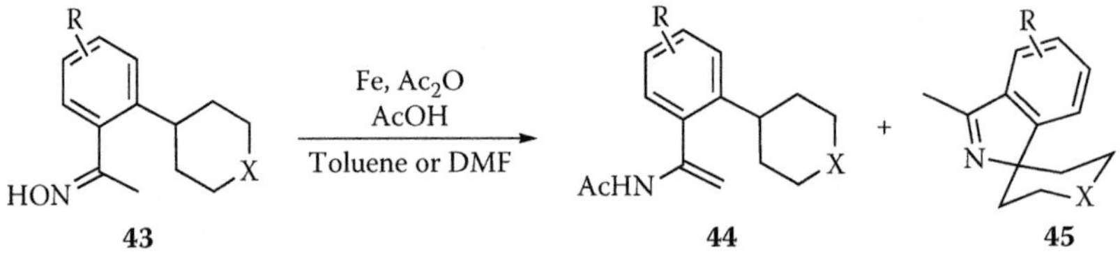

SCHEME 5.20 Enamide from oxime.

Attempts to directly form enamide **44** from methyl Grignard addition and trapping with acetic anhydride resulted in a variety of products including C-acylation as well as the imine (or the ketone **31**, the result of hydrolysis of this intermediate). An investigation into this reaction revealed that both the addition and the acylation were sensitive to both solvent and temperature. The effect of additives was investigated, and it was discovered that the use of MeLi with lithium bromide had a dramatic effect on the reaction profile (Scheme 5.21).

The enamide formation was first run using commercially available MeLi·LiBr in ether (86% assay yield).[7] The reaction was then improved and optimized for pilot plant using a new source of MeLi (MeLi in cumene-THF), adding LiBr separately as a solid. However, upon scale-up, a 1:1 BOC-nitrile·LiBr complex crystallized out of the MTBE solution at ambient temperature (17°C). The insolubility of the complex in MTBE led to a slow reaction with MeLi, requiring higher temperatures which resulted in the formation of by-product which could not be rejected by crystallization, and a lower overall assay yield (70%). Keeping nitrile **19a** and LiBr in solution prior to the MeLi addition was critical to successful enamide formation, and it was found that adding THF was key at preventing the crystallization of the complex (at 30°C, the complex's solubility in a 16.6% THF in MTBE solution was 68 g/L). However, adding THF also promoted the formation

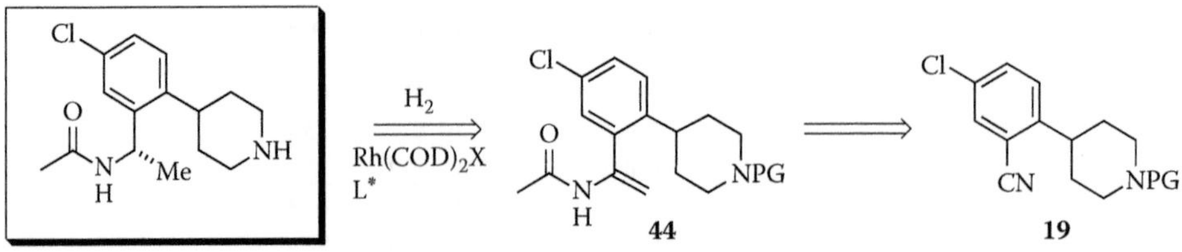

SCHEME 5.21 Enamide synthesis and hydrogenation.

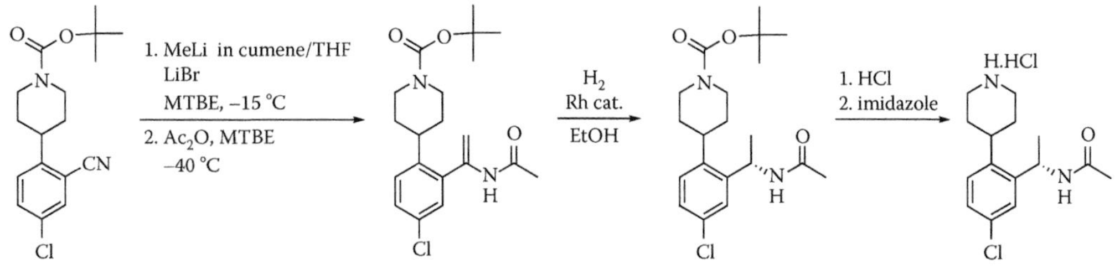

SCHEME 5.22 Enamide formation and hydrogenation.

of a significant amount of diacetylated enamide during the Ac_2O quench (25 to 40% compared to enamide). After aqueous workup and solvent-switching to EtOH, it was found that the addition of base (NaOH, Na_2CO_3, or NH_4OH) effectively converted the diacetylated compound to the desired enamide **44**. Further study found that adding aqueous NaOH until the apparent pH was >12 was required for complete conversion of diacetylated compound to the enamide within an hour. The crude enamide could then be crystallized out of ethanol/water. The enamide was submitted to hydrogenation conditions using 0.2% Rh(S,S-Me-BPE)Rh(COD)BF_4, under 90 psi H_2 at room temperature in denatured ethanol. The reaction showed complete conversion within 4 to 5 h with 87% ee (hydrogen uptake stopping after 2 to 3 h) in 100% assay yield. The deprotection step begins with an Ecosorb treatment of the ethanol solution after hydrogenation (98.4% recovery).

The ethanol solution was then treated with HCl (5 to 6 N in i-PrOH) at 50°C for 6 h (completion of the reaction was observed) and stirred overnight at room temperature. The excess HCl was then quenched with imidazole and after solvent-switching to CH_3CN (solvent composition CH_3CN:EtOH, 10 to 20:1), the desired HCl salt of piperidine amine **2** crystallized out in >99% ee (81% isolated yield from before Ecosorb treatment; >99 wt%, >99.7 % ee) (Scheme 5.22).

5.2.2 THE ACID FRAGMENT

Our initial improvement in the synthesis of pyrrolidine acid **3** relied on a racemic 1,3 dipolar cycloaddition followed by resolution. Attempts to devise asymmetric protocols of this reaction using chiral auxiliaries were not productive. The results from our laboratories were consistent with literature findings,[8] with a moderate diastereoselectivity of 3 to 4:1 at best obtained even when double chiral auxiliaries were used.[9] Several other approaches, such as Aza-Cope/Mannich reaction, intramolecular C–H insertion, and asymmetric aryl 1,4 addition, did not bear fruit.

A fundamentally different approach was undertaken.[10] In designing a new synthesis, we made the following considerations: (1) because direct installation of *tert*-butyl group onto pyrrolidine nitrogen is not a trivial reaction, we would install the *tert*-butylamine as a single entity; (2) the *trans* configuration of the two adjacent stereogenic centers in which the C3 center bearing the carboxylate group being epimerizable could simplify the synthesis to a one chiral center problem. The later consideration was derived from the observation that pure *trans* ester when treated with NaOH/MeOH, gave initially a 97:3 *trans/cis* mixture of the esters, and then proceeded to give only *trans* acid.

Intramolecular displacement of benzylic chlorides by nitrile anions was previously demonstrated as an efficient method making substituted pyrrolidines;[11] however, the question of enantioselectivity was not addressed. Recent works on the synthesis of substituted cyclopentane showed that intramolecular enolate displacement of benzylic phosphate proceeded in clean inversion.[12] Thus, it was envisioned that an appropriately substituted cyano phosphonate could be used to construct the pyrrolidine ring (Scheme 5.23). This intermediate could potentially be produced by the addition of a *tert*-butylamine to acrylonitrile. The desired amine could be produced via reaction

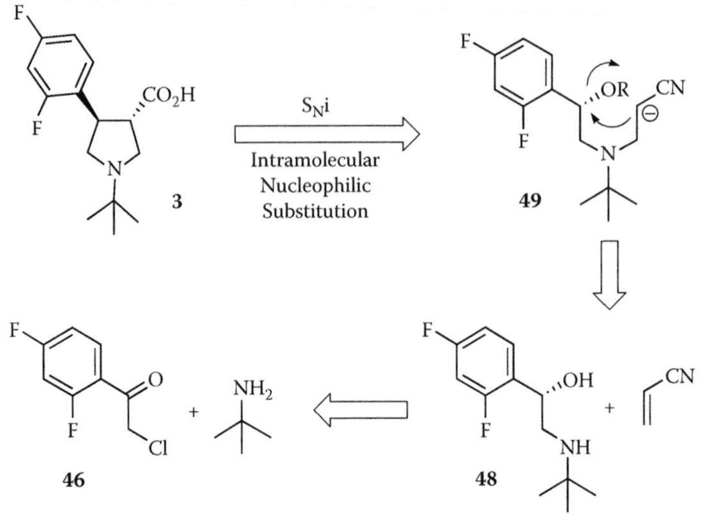

SCHEME 5.23 Retrosynthetic analysis of pyrrolidine acid **3**.

of the *tert*-butylamine with the chiral epoxide. The epoxide could be derived from the chlorohydrin *in situ*. The chlorohydrin could come from reduction of the appropriate chloroketone.

5.2.2.1 Asymmetric Reduction of Chloroketone

2-Chloro-1-(2,4-difluorophenyl)ethanone **46** was identified as a cost-effective[13] ($30/kg on metric ton scale) starting material toward the synthesis of amino alcohol **48**. Many options were evaluated for this asymmetric reduction, and ultimately, an oxazaborolidine-catalyzed reduction was chosen based on the availability, ease of operation, and high reproducibility. Aryl chloromethyl-ketones have also been asymmetrically reduced using catalytic (*S*)-MeCBS, and Burkhardt[14a] describes the advantages of using borane-diethylaniline as the borane source. In addition, adding ketone to the reducing agents at 31 to 32°C improves the enantioselectivity.[14] Building on this literature precedent, we determined that the controlled addition of ketone **46** to the reducing complex at 40°C afforded alcohol **47** in >98% ee's, and allowed the (*S*)-MeCBS catalyst loading to be reduced to as low as 0.1 mol% from the typical 5 to 10 mol% used in these reductions (Table 5.3).

Because it was observed that alcohol **47** tends to co-distill with toluene during *in vacuo* solvent removal, the reaction solvent was switched to MTBE. Using 0.5 mol% (*S*)-MeCBS, ketone **46** was reduced with 1 equiv of borane–diethylaniline complex in MTBE at 40°C, with the controlled addition of the ketone over a 10-h period. Alcohol **47** was isolated in 98% yield and 98.9% ee as the *S*-enantiomer on multi-kilogram scale.

5.2.2.2 *Tert*-Butylamine Displacement

Conversion of crude chloro alcohol **47** to amino alcohol **48** was accomplished by dissolving **47** in a methanol/*tert*-butylamine mixture and heating to reflux (56 to 60°C) in the presence of 1 equiv of solid NaOH. The use of NaOH allows the rapid formation of intermediary epoxide **51**, whose formation was confirmed by NMR and HPLC. Initially, with 1:1 MeOH/*tert*-butylamine, the displacement produced an 80:20 ratio of amino alcohols **48**:**52**. Decreasing the amount of methanol led to reduced levels of the undesired regioisomer, with only 4% **52** being formed with a 1:5 ratio of MeOH/*tert*-butylamine. Further reduction in the amount of MeOH led to much slower reaction rates. After workup of the latter conditions, crystallization from heptane afforded an 89.8% isolated yield of amine **48** (from ketone **46**) that was 99.9% pure with >99.9% ee as the (*S*)-enantiomer.

TABLE 5.3
(S)-MeCBS-Catalyzed Reduction of Chloroketone

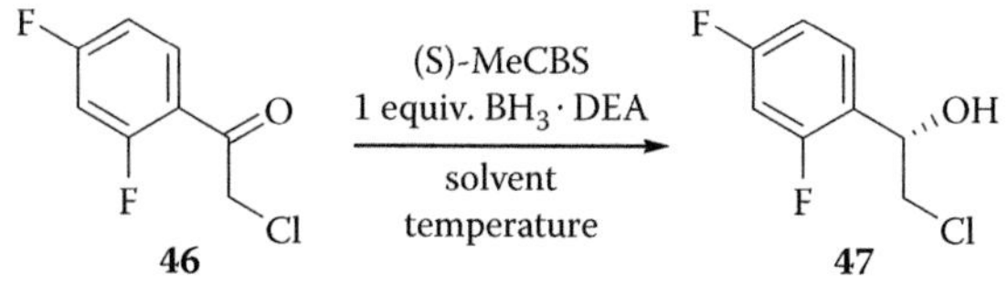

Solvent	Catalyst (mol%)	Temperature	Additional Time of 46	Optical Purity of 47[a]
Toluene	10	20°C	2.5 h	83.8% ee[b,c]
	10	27°C	4 h	98.2% ee
	1	36°C	6.5 h	97.6% ee
MTBE	1	36°C	5 h	98.6% ee
	0.5	36°C	1.3 h	95.4% ee
	0.5	36°C	1.3 h	93.7% ee[c]
	0.5	40°C	10 h	98.9% ee
	0.1	40°C	8 h	94.2% ee

[a] Determined by chiral high-performance liquid chromatography (HPLC) analysis.
[b] $BH_3.SMe_2$.
[c] 0.67 eq borane.

Source: Chung, John Y.L. et al. *The Journal of Organic Chemistry*, Vol. 70(9), 3592–3601, ACS, 2005.

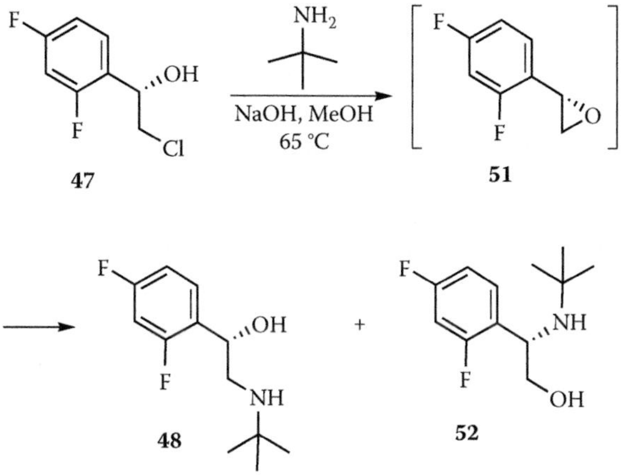

MeOH : tBuNH$_2$	48 : 52
1 : 1	80 : 20
1 : 5	96 : 4

SCHEME 5.24 Conversion to amino alcohol **48**. *Source*: Chung, John Y.L. et al. *The Journal of Organic Chemistry*, Vol. 70(9), 3592–3601, ACS, 2005.

This crystallization rejected all of isomer **52** and increased the enantiomeric purity of isolated amine **48** (98.9% ee in the reaction mixture, see Scheme 5.24).

TABLE 5.4

Effect of Additives in the Conjugate Addition of Amino Alcohol to Acrylonitrile

	Additive			24 h		48 h	
Entry	EtOH (5 eq)	HCONH$_2$ (5 eq)	Mont. K10 (10 wt%)	LCAP[a] Conversion 49/(48 + 49)	49% yield	LCAP Conversion 49/(48 + 49)	49% yield
1				90.6	83.7	95.8	92.9
2	+			92.8	88.7	94.8	88.7
3		+		97.3	92.1	97.2	90.4
4	+	+		96.6	93.7	96.4	97.1
5	+		+	91.6	86.2	93.9	88.7
6		+	+	97.2	87.0	90.0[b]	70.3[b]
7	+	+	+	96.4	90.4	96.4	95.4

[a] HPLC area % at 210 m.
[b] Uncertain data due to solvent evaporation.

Source: Chung, John Y.L. et al. *The Journal of Organic Chemistry*, Vol. 70(9), 3592–3601, ACS, 2005.

5.2.2.3 Acrylonitrile Conjugate Addition

Initial experiments on the conjugate addition of amino alcohol **48** to acrylonitrile found the reactions were sluggish and gave incomplete addition due to the steric hindrance of the amine and retro-Michael reaction. Employing excess acrylonitrile in acetic acid–methanol mixtures at reflux for 20 h typically gave a 75% assay yield and 65% isolated yield. Further investigation revealed that acetic acid actually prevented complete conversion (75:25 equilibrium mixture). Refluxing in neat acrylonitrile without acetic acid gave an improved 85 to 90% conversion after 24 h, but the remaining 10 to 15% was slow to convert. Various additives and physical activation methods were explored in hopes of driving the reaction to completion. After screening a number of additives, we focused on two. As shown in Table 5.4, both ethanol and formamide exhibited positive effects. Ultimately, the inclusion of one equivalent each of ethanol and formamide, introduced at the latter stages of the reaction, cleanly drove the reaction to completion. The strong hydrogen-bond donor formamide provided activation toward conjugate addition on the acrylonitrile, while ethanol minimized impurity formation.

Thus, heating a mixture of amino alcohol **48** and acrylonitrile at 77°C for 20 h, followed by addition of one equivalent each of formamide and ethanol and continued heating for another 12 h, afforded nitrile **49** in 98% assay yield. After crystallization from heptane, nitrile **49** was isolated in 92% yield and >99.9% ee (Scheme 5.25). The structure of **49** was also confirmed by single-crystal X-ray analysis.

5.2.2.4 Nitrile Anion Cyclization

Initial attempts for the cyclization of **49** using Ts$_2$O or Ms$_2$O and Li-hexamethyldisilazide (LiH-MDS) (2 equivalents) gave only 15% and 25% of desired pyrrolidine nitrile **50**, respectively. The reactions produced many by-products with large amounts of starting material remaining. We then explored less-reactive activators. Diethyl chlorophosphate emerged as an excellent choice, which, with the use of two equivalents of NaHMDS produced pyrrolidine nitrile **50** in 90% yield as an

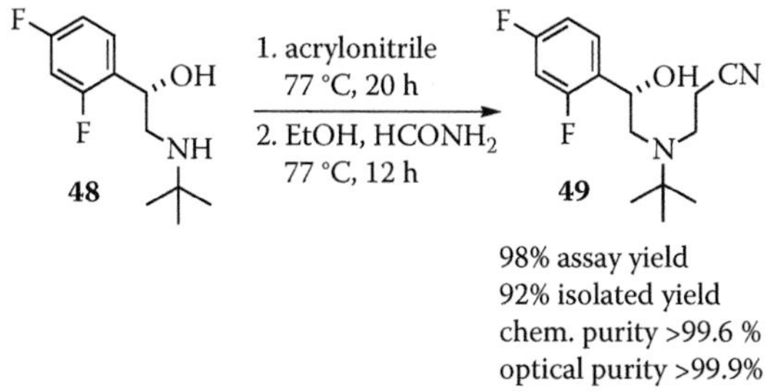

SCHEME 5.25 Optimized conjugate addition of amino-alcohol **48** to acrylonitrile. *Source*: Chung, John Y.L. et al. *The Journal of Organic Chemistry*, Vol. 70(9), 3592–3601, ACS, 2005.

	yield
TsOTs/LiHMDS	15%
MsOMs/LiHMDS	25%
ClPO(OEt)$_2$/NaHMDS	90% (t:c 80:20)
ClPO(OEt)$_2$/LiHMDS	95% (t:c 80:20)

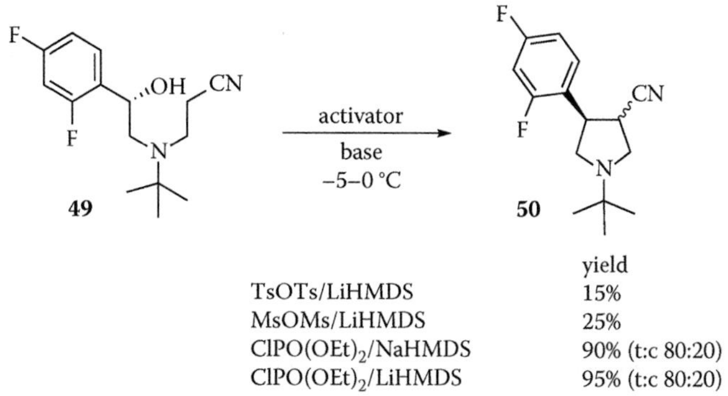

SCHEME 5.26 Effect of activator in the cyclization. *Source*: Chung, John Y.L. et al. *The Journal of Organic Chemistry*, Vol. 70(9), 3592–3601, ACS, 2005.

80:20 *trans:cis* mixture (Scheme 5.26). Further optimization resulted in 95% yield of pyrrolidine nitrile **50** using LiHMDS as the base, which significantly reduced the formation of impurities **53** to **56** (Figure 5.1). Importantly, the stereocenter at the aryl ring underwent complete inversion and only a single isomer was produced.

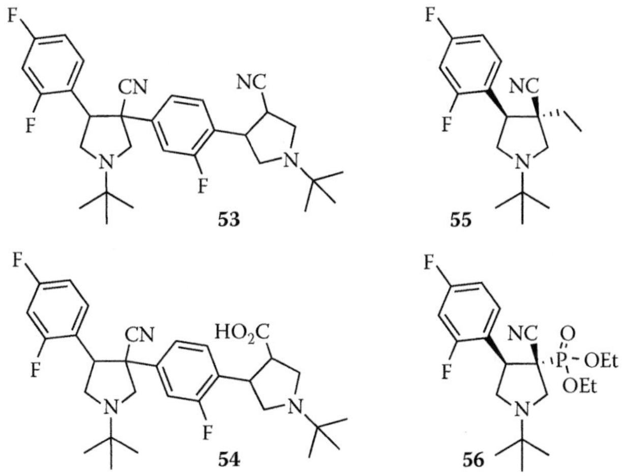

FIGURE 5.1 Impurities from cyclization.

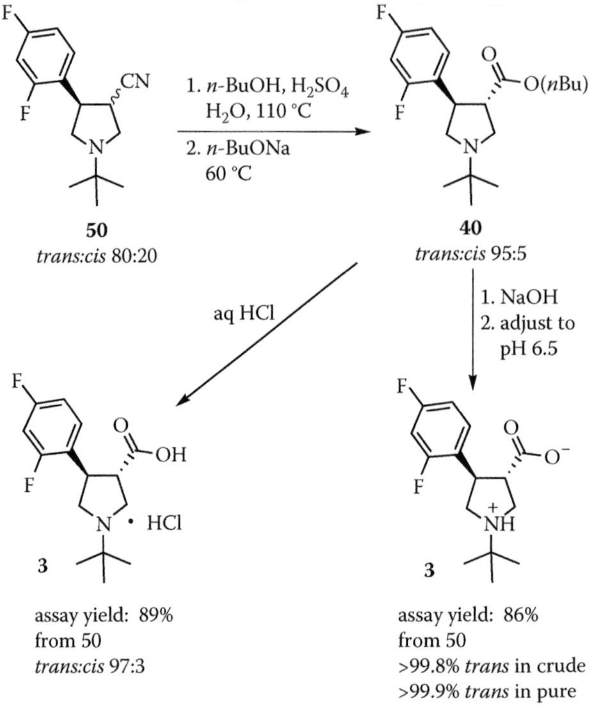

SCHEME 5.27 Hydrolysis/epimerization of pyrrolidine nitrile via ester **40**. *Source*: Chung, John Y.L. et al. *The Journal of Organic Chemistry*, Vol. 70(9), 3592–3601, ACS, 2005.

5.2.2.5 Epimerization/Saponification of Pyrrolidine Nitrile

Initially, the 80:20 *trans:cis* mixture of pyrrolidine nitrile **50** was converted to a mixture of *n*-butyl esters under sulfuric acid catalysis, followed by epimerization with *n*-BuONa, and then hydrolysis to the acid (Scheme 5.27). The best *trans:cis* ratio of *n*-butyl esters achieved was 95:5. Hydrolysis of esters with HCl afforded the HCl salt of **3** in 89% overall yield, which led to a minor upgrade in the *trans:cis* ratio (97:3). On the other hand, hydrolysis of the *n*-butyl ester by NaOH and subsequent pH adjustment to 6.5 afforded pyrrolidine acid **3** in 86% yield and >99.9 chemical and optical purities.

The basic hydrolysis of nitrile **50** with aqueous NaOH in ethanol was examined, which proceeded through intermediate amides **57**, and reached 98% conversion to acid **3** within 4 hr with <1% of amides remaining. Both *cis*- and *trans*-amides were observed by liquid chromatography/mass spectrometry (LC/MS) during reaction, and the structure of *trans*-amide **57t** was confirmed by LC/MS and independent synthesis (by treatment of *trans*-pyrrolidine acid with CDI and ammonium hydroxide). It is reasonable to postulate that the hydrolysis of both *cis*-nitrile **50** and *cis*-amide **57c** to the corresponding *cis*-acids was slow relative to epimerization, which provided a mechanism for complete conversion of *cis*-nitrile **50** to *trans*-acid **3**.

This procedure produced crude pyrrolidine acid **3** that was >99.6% *trans* with an optical purity of 99.5% ee, indicating that there was no chirality leakage in the cyclization–hydrolysis–epimerization process. This **3** was isolated by crystallization from IPA/MTBE in 95% yield with 99.97 LCAP purity and >99.9% ee on multi-kg scale. Thus, an efficient direct one-step basic epimerization–hydrolysis of the pyrrolidine nitrile mixture to the *trans*-pyrrolidine acid **3** was achieved (Scheme 5.28).

In summary, the final synthesis of the piperidine fragment **2** and acid fragment **3** is summarized in Scheme 5.29.

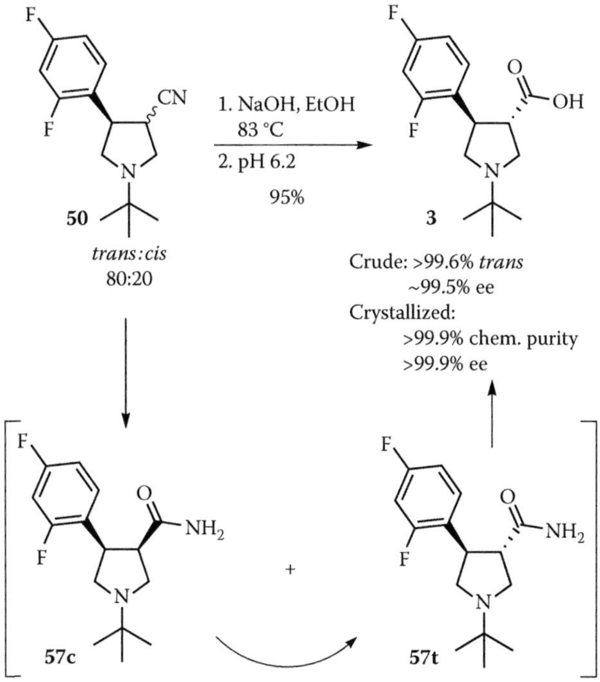

SCHEME 5.28 One-step conversion of *trans:cis* mixture pyrrolidine nitrile **50** to *trans*-pyrrolidine acid **3**. *Source*: Chung, John Y.L. et al. *The Journal of Organic Chemistry*, Vol. 70(9), 3592–3601, ACS, 2005.

5.2.3 FINAL COUPLING

Three procedures for amide formation were demonstrated: EDC coupling of the acid, Schotten–Bauman with the acid chloride, and CDI-promoted formation in acetonitrile. The optimum end-game coupling consisted of using a CDI coupling of the amine hydrochloride salt **2** and pyrrolidine acid **3** affording final API in 91% yield. The free base of final drug **1** was then treated with HCl in IPA to afford the corresponding bis-HCl salt of product **1** in 73% overall yield. Alternatively, the bis-HCl salt was further treated with NaOH to afford the mono-HCl salt after crystallization in 94% yield (Scheme 5.30).

5.3 CONCLUSION

In summary, BOC-nitrile **19** was synthesized through the Negishi coupling of BOC-iodopiperidine **42** and aryl bromide **20**. The nitrile was then transformed to enamide **44** via addition of methyl lithium, and *N*-acetylation. Asymmetric hydrogenation of the enamide, followed by deprotection, resulted in the desired amine piece **2** in 87% ee. Crystallization of the HCl salt of the amine provided **2** in >99% ee. Pyrrolidine acid **3** was prepared by a highly efficient nitrile anion cyclization strategy in 72% overall yield and excellent enantioselectivity. Diethyl phosphate and LiHMDS were shown to be the optimum benzylic leaving group and base, respectively, in the key displacement–cyclization reaction that proceeded with clean inversion. Another key feature is the kinetic control hydrolysis–epimerization of the *cis*-pyrrolidine nitrile via the *cis*-pyrrolidine amides which epimerized faster relative to hydrolysis and funneled all *cis*-nitrile to the *trans*-pyrrolidine acid. We have also shown that simple additives such as formamide accelerated the conjugate addition of hindered amine to acrylonitrile. Amine **2** and acid **3** were coupled using CDI in acetonitrile, and then final product **1** was isolated first as the bis-HCl salt, then finally as the mono-HCl salt. This chemistry was successfully implemented in a pilot plant to deliver the desired product.

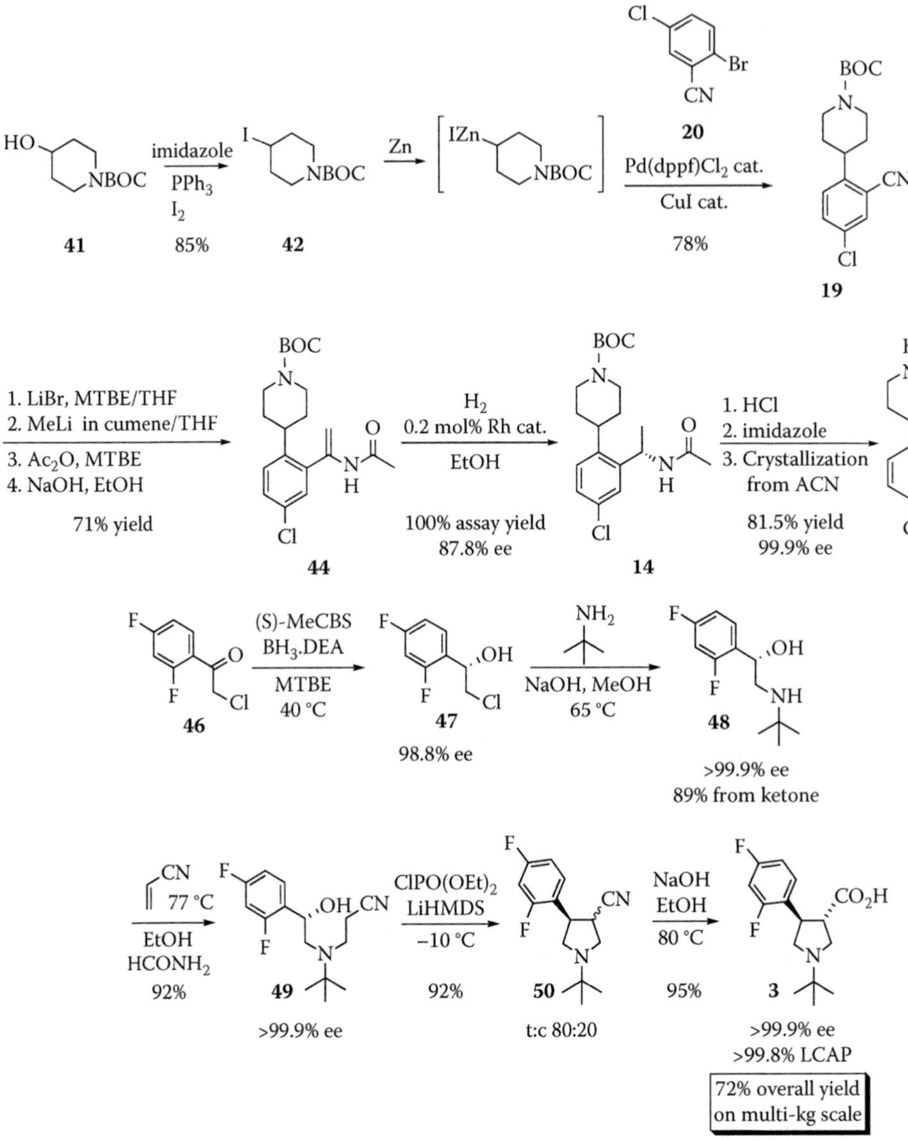

SCHEME 5.29 Long-term route.

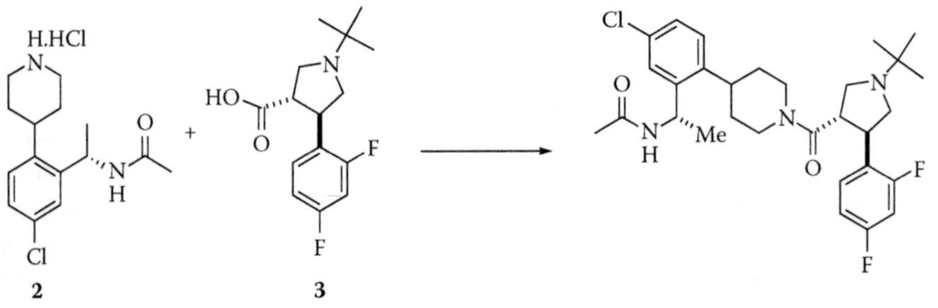

SCHEME 5.30 Amide coupling.

REFERENCES

1. (a) Ujjainwalla, F.; Chu, L.; Goulet, M.T.; Lee, B.; Warner, D.; Wyvratt, M.J. PCT Int. Appl. WO 2002068388 A2 20020906, 2002. (b) Sings, H.; Ujjainwalla, F. PCT Int. Appl. WO 2005009950 A2 20050203, 2005. (c) Ujjainwalla, F.; Chu, L.; Goulet, M.T.; Lee, B.; Warner, D.; Wyvratt, M.J. Manuscript in preparation.

2. (a) Savarin, C.; Boice, G.; Matty, L.; Murry, J.; Corley, E.; Conrad, K.; Hughes, D. Process and Intermediates for the Preparation of 4-Aryl Piperidines. PCT Int. Appl. 6998488 B2, 2006. (b) Boice, G.N.; Savarin, C.G.; Murry, J.A.; Conrad, K.; Matty, L.; Corley, E.; Smitrovich, J.; Hughes, D. *Tetrahedron* 2004, *60*, 11367–11374.

3. (a) Borg, G.; Cogan, D.A.; Ellman, J.A. *Tetrahedron Lett.* 1999, *40*, 6709–6712. (b) Liu, G.; Cogan, D.A.; Owens, T.D.; Tang, T.P.; Ellman, J.A. *J. Org. Chem.* 1999, *64*, 1278–1284. (c) Ellman, J.A.; Owens, T.D.; Tang, T.P. *Accounts Chem. Research* 2002, *35*, 11, 984–995.

4. Comins, D.L.; Abdullah, A.H. *J. Org. Chem.* 1982, *47*, 4315–4319.

5. (a) Knochel, P.; Calaza, M.I.; Hupe, E. Carbon-carbon bond-forming reactions mediated by organozinc reagents, in *Metal-Catalyzed Cross-Coupling Reactions* (2nd ed.), de Meijere, A. and Diederich, F., Eds., Wiley, New York, 2004, pp. 619–670. (b) Knochel, P.; Jones, P.; Langer, F. Organozinc chemistry: an overview and general experimental guidelines, in *Organozinc Reagents: A Practical Approach*, Knochel, P. and Jones, P., Eds., Oxford University Press, Oxford, 1999, pp. 1–21. (c) Knochel, P.; Perea, J.; Jones, P. *Tetrahedron* 1998, *54*, 29, 8275–8319.

6a. Corley, E.; Conrad, K.; Murry, J.; Savarin, C.; Holko, J.; Boice, G. *J. Org. Chem.* 2004, *69*, 5120–5123.

6b. Savarin, C.G.; Grise, C.; Murry, J.A.; Reamer, R.A.; Hughes, D.L., *Org. Lett.* 2007 9, 6, 981–983.

7. (a) Savarin, C.G.; Boice, G.N.; Murry, J.A.; Corely, E.; DiMichele, L.; Hughes, D. *Org. Lett.* 2006, *8*, 3903–3906. (b) Boice, G.; McWilliams, C.; Murry, J.; Savarin, C.; Stereoselective Preparation of 4-Arylpiperidine Amides by Asymmetric Hydrogenation of a Prochiral Enamide and Intermediates of This Process. PCT Int. Appl. 2006, WO 2006057904 A1 20060601.

8. Karlsson, S.; Hogberg, H.-E. *Tetrahedron Asymmetry* 2001, *12*, 1977–1982.

9. Belyk, K.M.; Beguin, C.D.; Palucki, M.; Grinberg, N.; DaSilva, J.; Askin, D.; Yasuda, N. *Tetrahedron Lett.* 2004, *45*, 3265–3268.

10. Chung, J.Y.L.; Cvetovich, R.; Amato, J.; McWilliams, J.C.; Reamer, R.; DiMichele, L.; *J. Org. Chem.* 2005, *70*, 9, 3592–3601.

11. Achini, R. *Helv. Chim. Acta* 1981, *64*, 2203–2218.

12. Song, Z.J.; Zhao, M.; Desmond, R.; Devine, P.; Tschaen, D.M.; Tillyer, R.; Frey, L.; Heid, R.; Xu, F.; Foster, B.; Li, J.; Reamer, R.; Volante, R.; Grabowski, E.J.J.; Dolling, U.H.; Reider, P.J.; Okada, S.; Kato, Y.; Mano, E. *J. Org. Chem.* 1999, *64*, 26, 9658–9667.

13. This intermediate is common to other pharmaceuticals (e.g., fluconazole and voriconazole).

14. (a) Salunkhe, A.M.; Burkhardt, E.R. *Tetrahedron Lett.* 1997, *38*, 1523–1526. (b) Xu, J.; Wei, T.; Zhang, Q. *J. Org. Chem.* 2003, *68*, 10146–10151.

6 Process Research and Development of LY414197, a 5HT$_{2B}$ Antagonist

Alfio Borghese and Alain Merschaert

CONTENTS

6.1 INTRODUCTION

Systematic structure activity relationships (SARs) around the scaffold of the natural alkaloid yohimbine led to the discovery of 1,2,3,4-tetrahydro-β-carboline (THβC) derivatives that have been shown to be very potent and highly selective 5HT$_{2B}$ receptor antagonists.[1] Of particular interest are 1-substituted-THβC compounds (**1** to **4**) that were selected for further development (Figure 6.1).

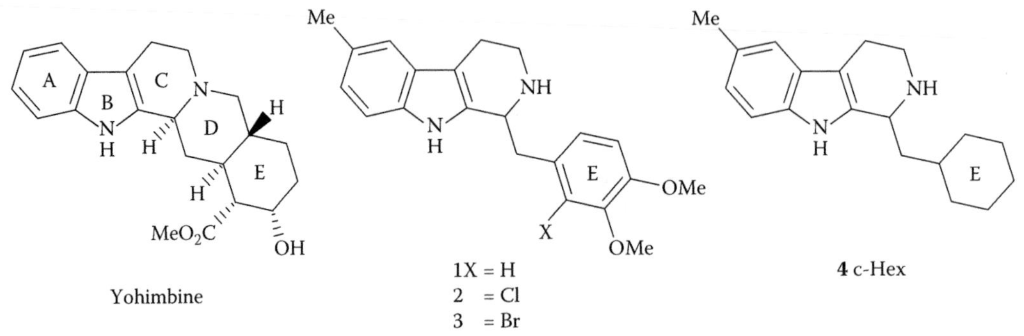

FIGURE 6.1 From yohimbine to 1,2,3,4-tetrahydro-β-carboline (THβC) derivatives.

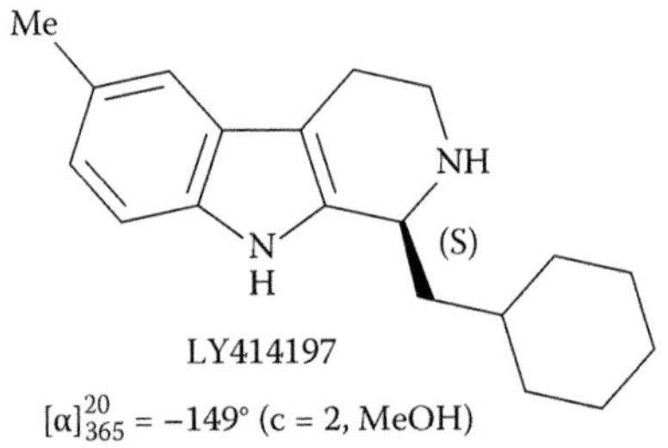

LY414197

$[\alpha]_{365}^{20} = -149°$ (c = 2, MeOH)

FIGURE 6.2 THβC selected for development.

The enantiomers of these four THβC derivatives **1**, **2**, **3**, and **4** were resolved and systematically tested in the 5HT$_{2B}$ receptor binding assay. Invariably, one enantiomer of each THβC was found to be active. The (*S*) enantiomer of 1-cyclohexylmethyl-THβC **4** (LY414197) showed the best profile (Ki = 0.7 nM) and was therefore selected for clinical evaluation as a potential drug candidate for the prophylactic treatment of migraine (Figure 6.2).

6.2 RETRO-SYNTHETIC ANALYSIS

The retro-synthetic analysis outlined in Scheme 6.1 shows that in addition to the Pictet–Spengler route previously used by discovery, several others are also possible. The Bischler–Napieralski condensation of 5-methyltryptamine and 2-cyclohexylacetic acid chloride as well as the Friedel–Crafts reaction of the N-protected 5-methyltryptamine allows preparation of the prochiral imine derivative **7**. The 2-cyclohexylmethyl side chain may also be introduced via alkylating an unsubstituted THβC by 2-cyclohexylmethyl chloride. As shown in Scheme 6.1, all of these routes involve 5-methyltryptamine as a common building block.

To reach our target molecule, the routes involving the cyclization step looked particularly attractive. The Bischler–Napieralski or related Friedel–Crafts route offers the potential for an asymmetric synthesis through enantio- or diastereoselective reduction of the imine penultimate intermediate, whereas the Pictet–Spengler approach is the shortest possible pathway.

6.3 PROCESS EVALUATION

The starting point of the 5HT$_{2B}$ project at the Chemical Product Research and Development Department was the evaluation of the synthetic method used by medicinal chemists to prepare the initial small quantities of material. In general, the synthetic strategy designed by the medicinal

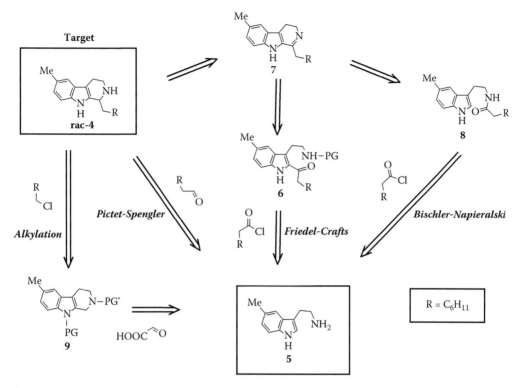

SCHEME 6.1 Retro-synthetic analysis.

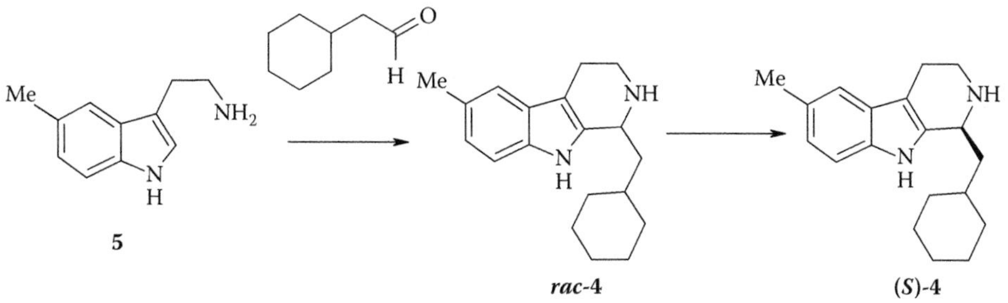

SCHEME 6.2 Discovery route to LY414197.

chemist is used primarily to support SAR work by providing an easy access to common intermediates readily convertible to the various related structures. On the other hand, the synthetic approach for the large-scale production of LY414197 will have to accommodate economic, safety, and environmental constraints.

The discovery approach to LY414197 involves a traditional Pictet–Spengler condensation of 2-cyclohexylacetaldehyde with 5-methyltryptamine (Scheme 6.2). At this stage of research, the desired (*S*)-enantiomer obtained by chiral preparative chromatography was not considered as an option for large-scale preparation.

Optimization of this synthetic route to LY414197 positioned it as a reasonable candidate as a production process, provided a scalable resolution methodology is developed. Much effort was thus expended in finding an acceptable resolution process of the THβC.

In parallel, we also decided to focus our route selection work on the research and development of an asymmetric synthesis articulated around the asymmetric reduction of an imine penultimate intermediate.

In this chapter, we will discuss the various approaches investigated for the synthesis of LY414197 and the development of the route selected for the preparation of the first kilogram for toxicological evaluation. A particular emphasis will be put on comparing a synthesis of racemic 1-cyclohexylmethyl-THβC **4** followed by a classical resolution with an asymmetric approach in terms of economy, safety, and environmental constraints. The discussion will thus essentially be articulated around a scale-up perspective, but mechanistic aspects will also be covered.

6.3.1 SYNTHESIS OF 5-METHYLTRYPTAMINE

6.3.1.1 Route Selection

As 5-methyltryptamine **5** was the key starting material for all possible routes discussed above and because it was expensive and not easily available in bulk at the time of development of LY414197, a large part of our process development efforts focused on the preparation of this key intermediate. The literature concerning the synthesis of tryptamines or indoles is abundant.[2] We focused our development effort on several synthetic routes using Fischer-type indole methods of 4-methylphenylhydrazine with various suitable aldehydes or equivalents (Scheme 6.3).[3] This methodology to construct the indole synthon is one of the more efficient and shortest ones. Also, from an economical and operational point of view, it is the most advantageous (one step). Other potential routes from 5-methylindole (e.g., via gramine,[4] glyoxamide,[5] or nitrovinyl[6] derivatives) or via metal-catalyzed reactions might also be envisaged but were not tested. All of these approaches suffered from the necessity to perform two or three steps and the use of hazardous, expensive reagents and 5-methylindole, which at that time had limited availability.

After evaluating each of the proposed routes, we came to the conclusion that the one-step Fischer indole-type reaction **b** was the most promising as a scalable process despite a low yield. Ready availability of the starting raw materials (*p*-tolylhydrazine and 4,4-diethoxy-butylamine) at low cost compensates for the low productivity of that transformation.

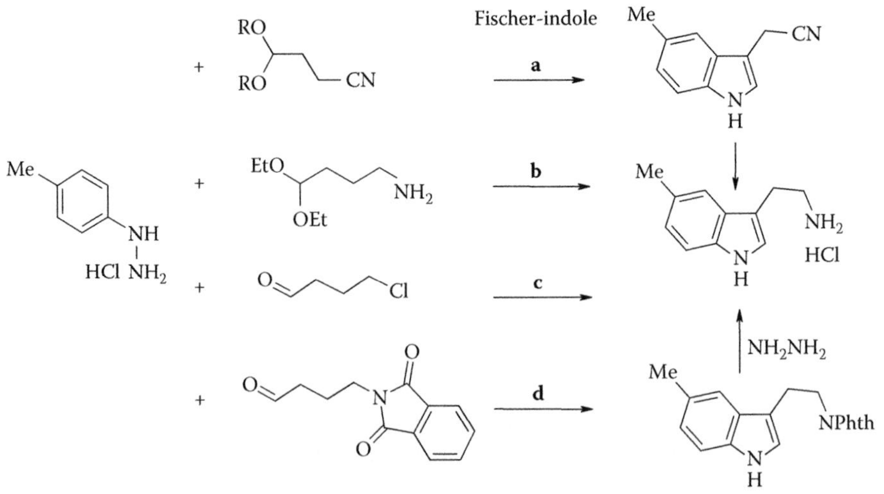

SCHEME 6.3 Fischer indole tryptamine synthesis.

TABLE 6.1
Effect of H$_2$O/AcOH Ratio on 5-Methyltryptamine Yield

H$_2$O/AcOH %w/%w	Reaction Time (h)	5-Methyltrypt./a %HPLC area	Yield 5-Methyltrypt. (%)
96/4	3	96/5	17
85/15	3	84/16	24
75/25	3	84/16	41
0/100	0.5	57/43	5

6.3.1.2 Large-Scale Preparation of 5-Methyltryptamine

The Fischer indole reaction between 4-methylphenylhydrazine hydrochloride and 4,4-diethoxy-butylamine affords 5-methyltryptamine with a low yield (average isolated yields = 35%).

Attempts to optimize the yield were unsuccessful. The main parameters affecting the yield of the Fischer indole reaction are the added acid (AcOH > TFA > HCl > H$_2$SO$_4$) as well as the solvent composition. In pure water, no reaction was observed, whereas in pure acetic acid, we mainly observed the degradation of the 4-methylhydrazine as evidenced by hydrogen nuclear magnetic resonance (^{1}H NMR) and high-performance liquid chromatography (HPLC) analysis. The best compromise was to perform the Fischer indole reaction in a 75/25 AcOH/H$_2$O mixture at 80°C (Table 6.1).

Compound **a** is the major impurity formed during the reaction as a result of the Pictet–Spengler reaction between 5-methyltryptamine **5** and 4,4-diethoxy-butylamine (Scheme 6.4). We found that using 0.87 equiv of 5-diethylacetal aminobutane minimizes the formation of **a**.

Scale-up results (Table 6.2) show that the process was very reproducible from laboratory scale to pilot-plant scale in terms of yield and quality of the isolated 5-methyltryptamine hydrochloride.

6.3.1.3 Workup Procedure

The 5-methyltryptamine is a suspected hazardous material. Therefore, to avoid worker exposure to the material during the pilot-plant operations, a process was developed by which that material was directly used in the next step (Pictet–Spengler) without isolation and drying of the solid. The resulting wet technical-grade 5-methyltryptamine hydrochloride was redissolved in methanol and used directly in the next step. Ethanol, in which the next Pictet–Spengler reaction is performed, would obviously have been a more appropriate solvent. However, for solubility and limited through-put per batch reasons (the 5-methyltryptamine being four-fold less soluble in ethanol than in methanol), we opted to dissolve the tryptamine in methanol and perform a methanol/ethanol solvent exchange.

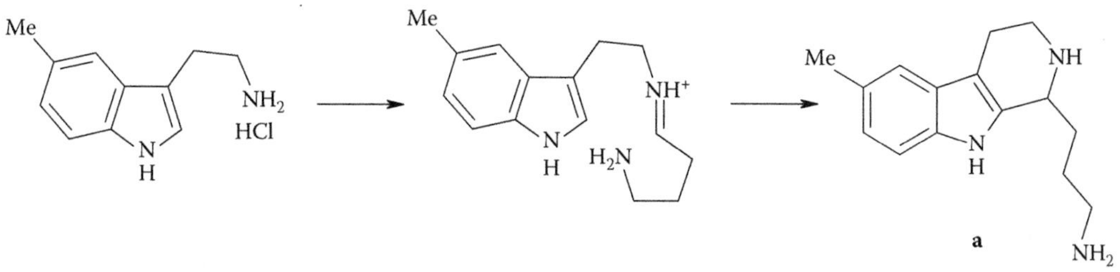

SCHEME 6.4 Major by-product of the 5-methyltryptamine synthesis.

TABLE 6.2
Scale-Up of 5-Methyltryptamine Synthesis

Scale	4-Methylphenylhydrazine (g)	Yield versus Ketal (%)	Purity HPLC % Area
Lab.	15	39	96.2
Kilo lab.	350	33	96.8
Kilo lab.	670	34	96.2
Pilot plant	7,800	34[a]	97.7
Pilot plant	39,200	36[a]	97.5

[a] Based on the methanolic solution assay.

6.3.1.3.1 Typical Large-Scale Procedure

A typical procedure to prepare the 5-methyltryptamine hydrochloride at the kilo lab scale and at the pilot-plant scale is as follows.

A mixture of 4-methylphenylhydrazine hydrochloride (350 g, 2.21 mol), 4,4-diethoxybutylamine (310 g, 1.92 mol), H_2O (7 kg), and acetic acid (2.52 kg) was heated at 80°C for 150 min. The reaction mixture was then concentrated under reduced pressure to a weight of 1.8 kg. Then, toluene (2 × 5 L) was added to the residue and the mixture concentrated again under reduced pressure until a weight of 2.2 kg was achieved. Toluene (3.5 L) was added to the residue and the pH adjusted to 9.5 by addition of 25% NH_4OH (400 mL) and H_2O (300 mL). The resulting suspension was stirred for 2 h and the crystals filtered and rinsed twice with toluene (250 mL). The cake was dissolved in MeOH (4.9 L) and used in the next Pictet–Spengler step. A methanolic solution yield assay of 33% was obtained based on the limiting starting material (4,4-diethoxybutylamine).

6.3.1.4 Alternate Synthesis

A mechanistic insight by ^{1}H NMR kinetic studies shows that only 40 to 50% of the precursor hydrazone is formed at 30°C (Scheme 6.5). Some literature data suggest that the 4,4-diethoxybutylamine in acidic media might be in equilibrium with a cyclic 1-pyrroline derivative as well as its trimeric form, thus partly explaining the low percentage of conversion into the hydrazone intermediate along with the degradation of the 4-methylphenylhydrazine at higher temperature.[7] We hypothesized that a protected 4,4-diethoxybutylamine might give better results.

The N,N-dibenzyl protected 4,4-diethoxybutylamine (Scheme 6.6) gave a better yield (77%) of the desired 5-methyltryptamine. From a process point of view, the obvious disadvantage of such a synthetic route is the need for additional protection and deprotection steps making such an approach less desirable for manufacturing. Given that, we choose not to further develop this route.

6.3.2 Synthesis of 2-Cyclohexylacetaldehyde

6.3.2.1 Route Selection

The discovery route utilized the pyridinium chlorochromate (PCC) oxidation of 2-cyclohexylethanol in CH_2Cl_2 in presence of molecular sieves.[8] It is a simple process, as the aldehyde is simply isolated by filtration of the reaction mixture through silica gel. However, this process was proven to be difficult to scale up due to difficulties of filtration of the chromium salts. Furthermore, the environmental issues created by the large amount of toxic chromium salts make this process unsuitable for large-scale synthesis. Two other processes (Scheme 6.7) were therefore developed and tested to prepare the required 2-cyclohexyl acetaldehyde at the pilot-plant scale:

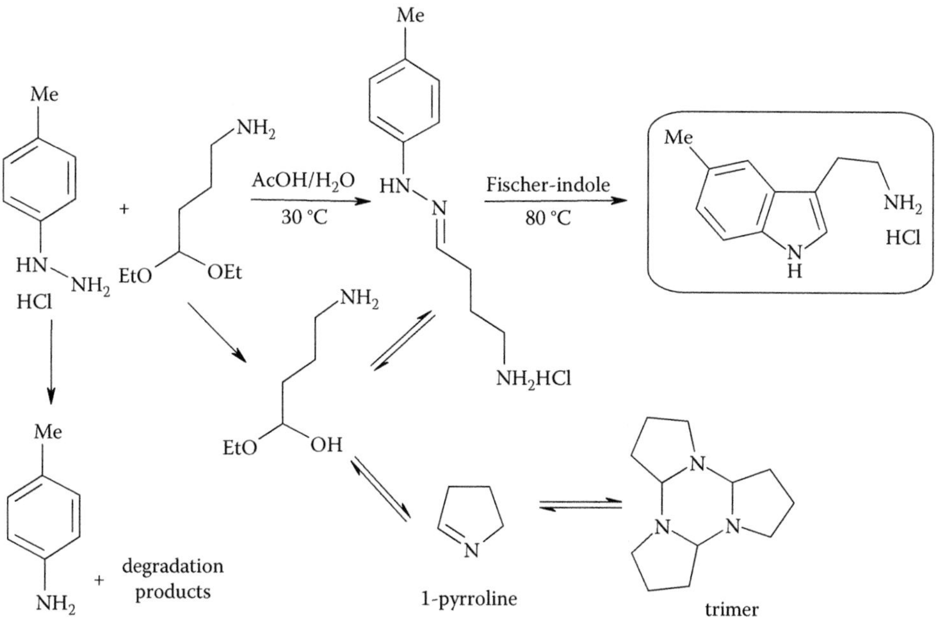

SCHEME 6.5 Fischer indole synthesis of 5-methyltryptamine.

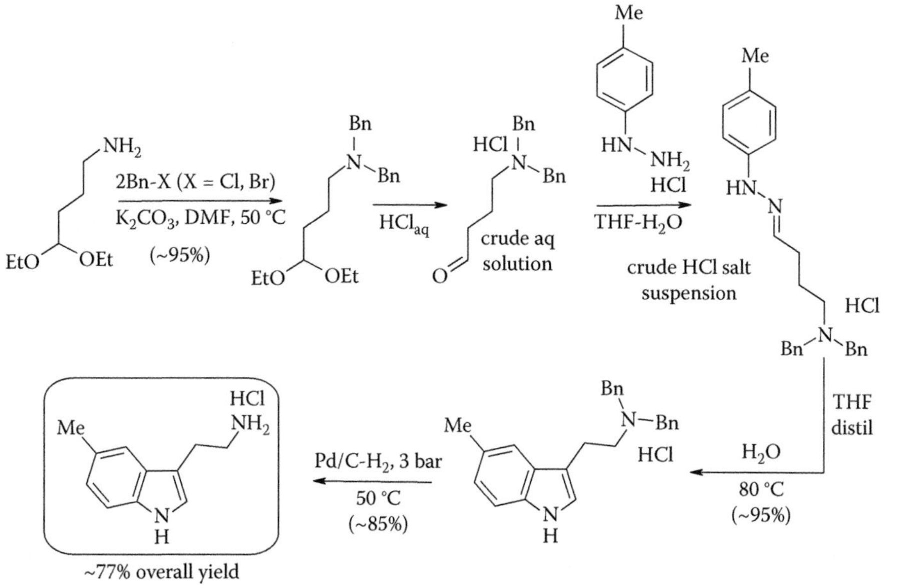

SCHEME 6.6 Synthesis of 5-methyltryptamine via the *N,N*-dibenzyl-protected 4,4-diethoxybutylamine.

1. Oxidation of 2-cyclohexylethanol using the 2,2,6,6-tetramethyl-1-piperidinyloxy (TEMPO)/KBr/NaOCl system in CH$_2$Cl$_2$ performed at room temperature[9]
2. Reduction of 2-cyclohexylethylacetate with DIBAL-H at 70°C in methyl *tert*-butyl ether (MTBE)[10]

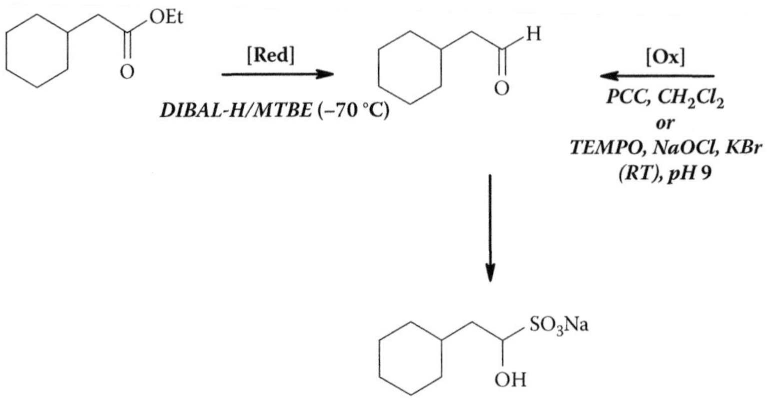

SCHEME 6.7 Synthetic routes to 2-cyclohexylacetaldehyde.

Both starting raw materials (2-cyclohexylethanol and ethyl cyclohexylacetate) were available on large scales and did not represent a sourcing issue for future manufacturing. Both processes developed gave the cyclohexylaldehyde in high yield (80%).

However, it became obvious that the oxidation route would be the preferred one as it uses safer reactants than the reduction route (DIBAL-H). In addition, the oxidation reaction performed at room temperature represented a major advantage compared to the reduction route that has to be performed at low temperature to avoid overreduction by-products.

The oxidation of 2-cyclohexylethanol is performed by adding the bleach (13% NaOCl) at a controlled rate to a buffered (TEMPO/KBr/NaHCO$_3$) biphasic H$_2$O/CH$_2$Cl$_2$ reaction mixture. The oxidation is completed at the end of the addition of the NaOCl. Separation of the organic layer affords after concentration to dryness of the 2-cyclohexylacetaldehyde as a colorless oil.

The 2-cyclohexylacetaldehyde is too air sensitive and needs to be protected to be further handled and stored. The aldehyde was protected either as a dimethylacetal or as a sodium bisulfite adduct. However, we discovered that the dimethylacetal derivative was unreactive in the reaction conditions (acidic) in the Pictet–Spengler reaction generating the tetrahydro-β-carboline derivative. In addition, the dimethylacetal derivative is an oil, making its isolation difficult on the plant scale. Therefore, we choose to isolate the 2-cyclohexylacetaldehyde as a bisulfite adduct, yielding a white crystalline solid.

6.3.2.2 Large-Scale Preparation

A typical procedure to prepare the sodium bisulfite adduct of 2-cyclohexylacetaldehyde is as follows. A 10-L reactor is charged with 2-cyclohexylethanol (350 g), CH$_2$Cl$_2$ (1.37 L), and 2,2,6,6-tetramethyl-1-piperidinyloxy (TEMPO) (4.27 g) under inerted atmosphere. A solution of KBr (32.8 g) in water (137 mL) is added to this mixture. The biphasic reaction mixture is cooled to about 0°C, and a solution of NaOCl (188 g, 14.7% active Cl) and sodium bicarbonate (93.1 g) in water (1.1 L) is slowly added (35 min) to the reaction mixture. After a post-stirring period of 2 min, the organic layer is separated, washed with a solution of 37% HCl (75 mL) and KI (17.5 g) dissolved in water (980 mL), and charged into the reactor together with MeOH (3.43 L). Then, a solution of NaHSO$_3$ (313 g) dissolved in water (630 mL) is added to the solution. The resulting suspension is stirred for 16 h at 20°C, filtered, washed with MeOH (875 mL), and the cake dried under reduced pressure to give 432 g of the sodium bisulfite adduct of 2-cycloacetaldehyde (69%).

Surprisingly, this TEMPO-mediated oxidation process, followed by the formation of the crystalline sodium bisulfite adduct scaled up in the pilot plant (100 L and 600 L) afforded the sodium bisulfite adduct with variable yields (48% and 71%). Careful investigation of what happened during the pilot-plant synthesis reveals that the lower yield was linked to the contact time between the

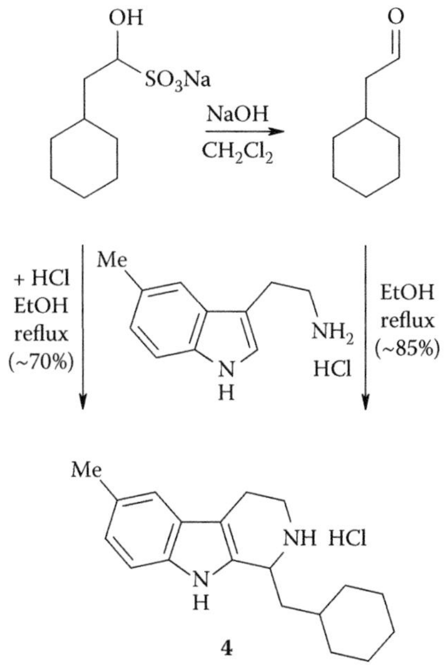

SCHEME 6.8 Pictet–Spengler synthesis of *rac*-THβC (**4**).

MeOH and the 2-cyclohexylacetaldehyde, forming the dimethylacetal derivative, before addition of the NaHSO$_3$. We solved this issue by simply replacing MeOH by EtOH, thus yielding an 83% sodium bisulfite adduct.

6.3.3 PICTET–SPENGLER REACTION

6.3.3.1 Route Selection

The acid-catalyzed Pictet–Spengler reaction between tryptamine derivatives and aldehydes is a well-established method for preparing tetrahydro-β-carboline (THβC) derivatives.[11] Our first trials were aimed at using the bisulfite adduct as the carbonyl source in order to minimize the number of process steps. The reaction was performed by reacting the 5-methyltryptamine hydrochloride with an excess (1.3 equiv) of the sodium bisulfite adduct in EtOH at reflux, in the presence of one extra equivalent of HCl. The *rac*-THβC was simply isolated as a hydrochloride salt by filtration of the reaction mixture (Scheme 6.8).

The addition of an extra equivalent of nonaqueous HCl is necessary to obtain an acceptable yield of 5-methyltryptamine hydrochloride. Without this extra HCl, the isolated yields dropped to the 15 to 36% range. However, we soon reached the conclusion that this process had some robustness issues.

The best result obtained using the bisulfite process affords the *rac*-THβC.HCl in 70% yield. Attempts to complete the reaction by adding more HCl led to the formation of a major impurity **b** (Figure 6.3), at the expense of 5-methyltryptamine. This impurity was very easily removed as it was very soluble in EtOH. High HCl concentration led to the quantitative formation of impurity **b**. Use of aqueous HCl causes the appearance of an additional impurity **c**, which is very insoluble, and contaminates the isolated *rac*-THβC (Table 6.3). The process using the sodium bisulfite adduct was first scaled up to 250 L to produce the first large batch (10 kg) of *rac*-THβC.

Given the issues associated with the bisulfite process, a new process was developed in which the free aldehyde was first liberated from its bisulfite salt under basic (NaOH) conditions in

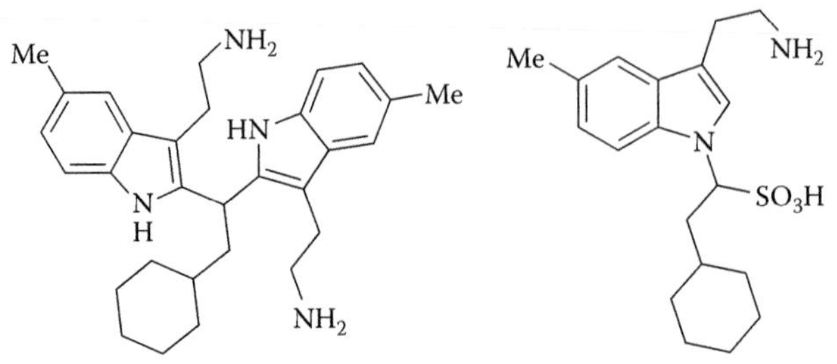

FIGURE 6.3 Main Pictet–Spengler by-products using the sodium bisulfite adduct.

TABLE 6.3
Pictet–Spengler Reaction Using Sodium Bisulfite Adduct or Free Aldehyde

Carbonyl Source	Experimental Conditions	b (%)	c (%)	THβC (%)
Bisulfite add.	+1 equiv HCl/EtOH	4	ND	70
	+1.7 equiv HCl/EtOH	95	ND	5
	Solvent = H$_2$O, 1 equiv HCl		50	50
Free aldehyde	No extra HCl added	1	ND	85
	+1 equiv HCl/EtOH	100	ND	ND
	Solvent = 3:1 EtOH:H$_2$O	1	ND	85
	No extra HCl added			

H$_2$O/CH$_2$Cl$_2$. The CH$_2$CH$_2$ layer containing the aldehyde was used in the Pictet–Spengler reaction after a solvent exchange to EtOH to give after isolation the *rac*-THβC with 85% yield.

In this case, addition of 1 equiv of HCl/EtOH yields impurity **b** quantitatively. The presence of water has no effect on the yield but significantly slowed the reaction (Table 6.3).

6.3.3.2 By-Products Mechanism Hypothesis (Scheme 6.9)

The formation of impurity **b** is clearly linked to the presence of an excess of HCl, which presumably blocks the formation of the imine intermediate for the Pictet–Spengler reaction by shifting the equilibrium toward the protonated form of the tryptamine. Under these circumstances, reaction at the other nucleophilic site (i.e., the position 2 of the indole nucleus) occurs to ultimately yield **b**.

To explain the occurrence of the other major by-product **c**, we hypothesized an initial step involving protonation of 5-methyltryptamine **5** at position 3, ultimately leading to a known hexahydropyrrolo[2,3-b]indole tricyclic nucleus **e**.[12] This aminal derivative can react on either nitrogen with the bisulfite adduct to yield either the target THβC derivative **4** or the by-product **c**. Again, we assume that the ratio **4/c** is mostly affected by the pH and therefore by the amount of HCl used in the reaction, although **4** was not formed under anhydrous conditions.

6.3.4 Optical Resolution of *rac*-THβC

The desired (*S*)-isomer was obtained by resolution of *rac*-THβC performed by crystallization of its diastereomeric salts formed with (*S*)-(+)-camphorsulfonic acid in EtOH. This resolving agent was selected by screening a collection of chiral acids ((*S*)-(-)-malic acid, (+)-o,o-*p*-toluyl-D-tartaric

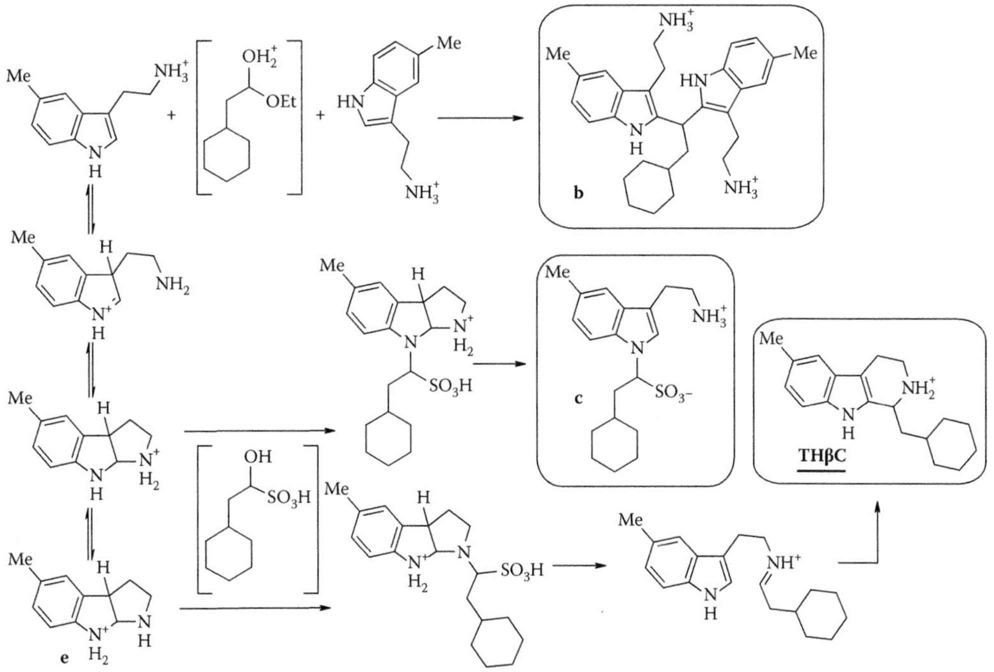

SCHEME 6.9 Formation of by-products **b** and **c** in Pictet–Spengler step.

acid, (+)-10-camphorsulfonic acid, L-(-)-tartaric acid monohydrate, L-(+)-tartaric acid, and (S)-(+)-mandelic acid) in EtOH. These experiments were carried out under concentrated conditions in order to assess the eutectic composition by chiral analysis of the mother liquors.[13] Only the (S)-(+)-camphorsulfonic acid yielded enriched mother liquors in the opposite diastereomeric salt. The measured eutectic composition (27:73 S:R) allowed us to calculate a theoretical yield of resolution of 31.5%.

Experimentally, we did the resolution by using 1 equiv of (S)-(+)-camphorsulfonic acid added to the *rac*-THβC free base in EtOH and refluxed for 2 h. Filtration of the resulting slurry yielded the (S)-THβC camphorsulfonate salt with 39.1% yield and a 82.8% d.e. Because the camphorsulfonate salt was not acceptable as a final pharmaceutical salt, we had two options:

1. Upgrade the camphorsulfonate salt to an accepted optical purity followed by a salt exchange to prepare the final form
2. Perform the salt exchange after the first resolution and crystallize the final form to reach an acceptable optical purity

To optically upgrade the obtained chiral camphorsulfonate salt, two crystallizations in EtOH were necessary (y = 87.7%, 96.8% d.e. and y = 90%, >99.9% d.e.). It is to be noted that the overall experimental yield (30.8%) is very close to the expected theoretical yield. The HCl and the citrate salt displayed acceptable biopharmaceutics characteristics and were selected as potential final forms.

6.3.4.1 Issues with Final Salt Preparation

We observed that the salt exchange process from an optically pure diastereomeric salt to an enantiomeric salt decreased the optical purity of the THβC. The exchange process from a 100% d.e. (S)-THβC.(S)-CSA to the (S)-THβC.HCl gave a final HCl salt with 99.1% e.e. One possible explanation for such phenomenon was the racemization of the optically pure THβC derivative under strong acidic conditions (HCl) via a ring opening (retro-Pictet–Spengler)/re-closure reaction.

Another situation arose during the preparation of the citrate salt as a result of the particular shape of the solubility phase diagram (two eutectics) and mainly the eutectic composition.

The salt exchange from the 84% d.e. (S)-THβC.(S)-CSA to (S)-THβC.citrate was even worse, yielding crystals of the citrate salt with a 40% d.e. The mother liquors were 96% d.e., indicating that (S)-enriched material was obtained in solution. In that situation, to obtain optically enriched crystals by crystallization requires that the minimal optical purity of (S)-THβC.(S)-CSA before the salt exchange to the citrate salt be higher than 96% d.e. The percentage of racemization due to the acidity of the medium was not known in this case.

Given these issues, the HCl and citrate salts were abandoned as final pharmaceutical salts.

Finally, we discovered that L-(+)-tartaric acid was suitable for preparing the final salt and optically upgrading the enriched (S)-THβC. This will have the advantage to eliminate the issue associated with the optical purification of an enriched enantiomeric mixture ((S)-THβC.HCl or (S)-THβC.citrate) (solubility phase diagram with two eutectics) if the minimum required optical purity is not achieved.

6.3.4.2 Pilot-Plant-Scale Resolution (Scheme 6.10)

In a typical resolution procedure, the rac-THβC.HCl (3.199 kg, 91.6% w/w) and K_2CO_3 are slurried under reflux in EtOH (58.6 L) for 8 h. After cooling to room temperature, the suspension is filtered and the resulting ethanolic solution of the free base (rac-THβC) is percolated through a silica gel bed (5.15 kg). The silica gel is washed with EtOH (4.16 L). Both ethanolic solutions of rac-THβC are pooled together and concentrated until a concentration of 1 g rac-THβC per 26 mL EtOH is obtained. To that ethanolic solution is added (S)-(+)-camphorsulfonic acid (2.135 kg, 1 equiv). The reaction mixture (suspension) is refluxed for 2 h and stirred for another 2 h at room temperature. The crystals are filtered, washed with EtOH (4.6 L), and dried between 40 to 50°C under reduced pressure to yield the diastereomeric salt (S)-THβC.(S)-CSA (1.849 kg, 39.1%, 82.8% d.e.).

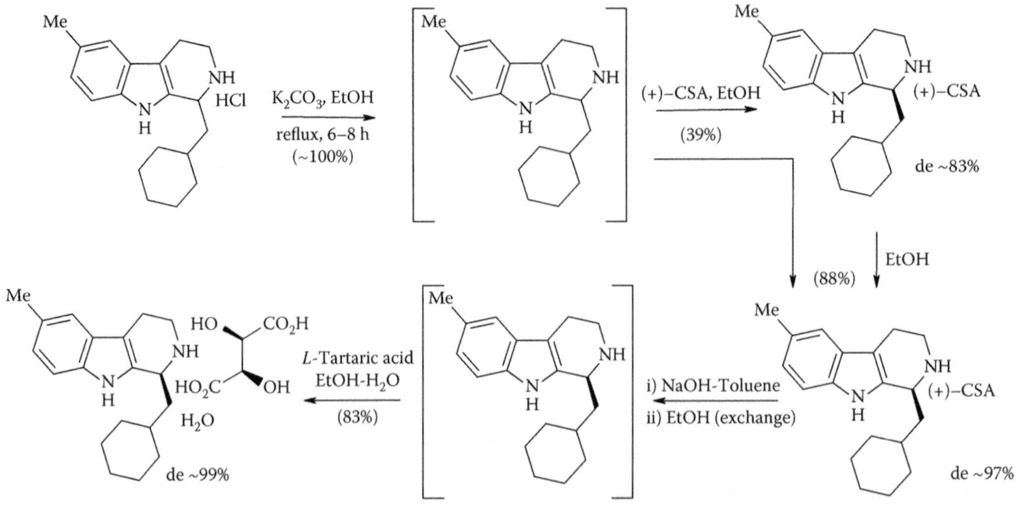

SCHEME 6.10 Optical resolution of rac-THβC and final salt formation.

6.3.4.3 Purification of (S)-THβC.(S)-CSA

A suspension of 1.839 kg of (S)THβC.(S)-CSA (82.8% d.e.) in 9.1 kg of anhydrous EtOH is heated to reflux for 2 h. After cooling and stirring to room temperature for 2 h, the crystals are filtrated, washed with 1.6 kg of EtOH and dried between 40 and 50°C under reduced pressure, to afford 1.613 kg of purified material (yield = 87.7%, 100.7% w/w, 96.8% d.e.).

6.3.4.4 Final Salt Formation and Purification

A suspension of 6.207 kg of (S)-THβC.(S)-CSA (95% d.e.) in toluene (124 L), water (9.9 L), and 30% NaOH (2.5 L) was stirred at 40°C until dissolution (40 min). After cooling to room temperature, the organic layer was separated and polish filtrated. The toluene was stripped off under reduced pressure until a minimum volume. Then, EtOH (74 kg) was added to the reaction mixture and the toluene/EtOH azeotropic mixture (bp: 76.7°C; azeotropic composition: toluene:EtOH 32:68) distilled at ambient pressure until a residual toluene content of maximum 1% was achieved (gas chromatography [GC] analysis). A mass adjustment (1g (S)-THβC/5 g reaction mixture) was then performed by addition of EtOH (7.2 kg), followed by the addition of L-tartaric acid (1.986 kg) dissolved in water (305 L). The mixture was then heated for 1 h at reflux and stirred for 16 h at room temperature. The crystals were filtered, rinsed twice with water (17 L), and dried at 50°C under reduced pressure for 20 h to afford 4.595 kg of (S)-THβC-(L)-tartrate monohydrate (yield: 86%, 99% d.e.).

6.3.5 RACEMIZATION AND RESOLUTION

As described in the previous sections, we discovered and developed an efficient resolution procedure for the synthesis of LY414197 and decided to move forward with that approach into the pilot plant. However, in order to improve the overall yield of the synthesis, it became critical to recycle the mother liquors coming from the resolution step. We therefore focused on the research and development of an efficient racemization method.

The racemization of chiral 1-substituted-1,2,3,4-tetrahydro-β-carbolines is well known.[14] This reaction is generally performed either under strong acidic (Scheme 6.11, path A) or basic (Scheme 6.11, path B) conditions as well as via oxidation–reduction processes (Scheme 6.11, path C). From a literature review, no optimal method has been developed for a large-scale synthesis. Most methods involve harsh conditions, long reaction times, and expensive reagents, and generally afford poor yields.

6.3.5.1 Racemization through Oxidation-Reduction

As we were not only interested in the development of a racemization method but also wanted to evaluate an asymmetric synthesis articulated around the imine intermediate **7** (*vide infra*), we initially investigated its controlled preparation by oxidation of the unwanted (R)-THβC **4** obtained from the mother liquors. Among the various methods initially tested, good results (approximately 75 to 80% *in situ* yield of imine **7**) were obtained with NaOCl in methanol/THF at 0 to 5°C for 3 h.[15] The major by-product is the overoxidized β-carboline derivative (5 to 10%), although in some experiments, low levels of the unstable N-chloroamine intermediate were also detected. Later on, approximately 68% *in situ* yield was obtained with tetra-*n*-propylammonium perruthenate (0.05 equiv) as catalyst with *N*-methylmorpholine oxide (1.5 equiv) as cooxidant in acetonitrile at room temperature.[16] However, in this latter method, up to 16% of totally oxidized β-carboline was also formed. The imine **7** was then directly reduced with sodium borohydride to produce the racemic material in approximately 50% isolated overall yield. Although the aromatic β-carboline by-product was easily removed upon salt formation, the above approach suffered from several major drawbacks: difficulty to control the overoxidation of the desired dihydro-β-carboline to the β-carboline on

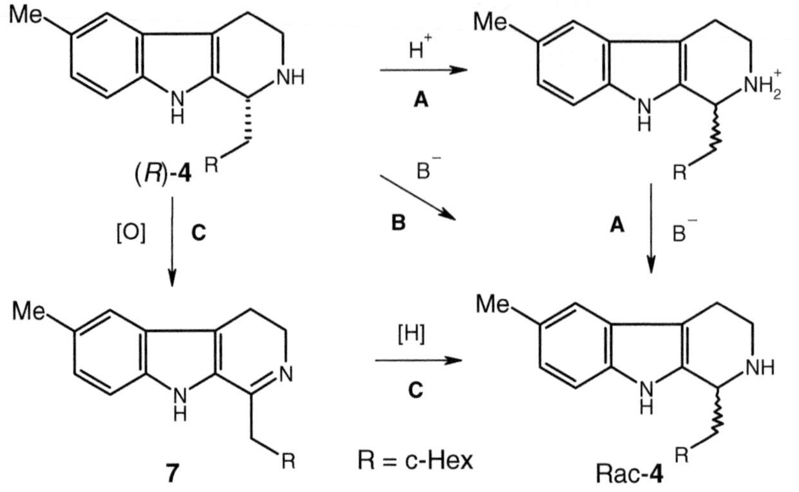

SCHEME 6.11 Racemization pathways of 1-substituted-THβC.

scale, moderate overall yield, and difficulty getting rid of the boron salts after reduction. We therefore turned our attention to another racemization method involving transition metals. Platinium oxide (Adam's catalyst) is claimed to be the catalyst of choice for the racemization of isoquinoline alkaloids.[20] Thus, according to the standard method reported in the literature, we treated a 25/75 S/R mixture with PtO$_2$ (100% w/w) in 95% ethanol under a hydrogen atmosphere (40 psi) for 16 h which led to racemic THβC **4** in 74% yield accompanied with 7% imine **7** and 4% of overoxidized β-carboline. Various attempts to lower the catalyst loading (down to 10% w/w) with Pt/C led to significantly longer reaction time as well as increased amounts of imine **7** and overoxidized β-carboline by-product. Interestingly, however, in contrast to the literature on isoquinoline alkaloids, we obtained faster reaction rates with palladium in the absence of hydrogen atmosphere.[20] An 8/92 S/R mixture was racemized within 2 h in refluxing methanol with 10% Pd/C (1% w/w Pd versus substrate). Although we only obtained a 50% *in situ* yield of racemic THβC **4**, 32% of imine **7** was also formed under these conditions along with approximately 10% of overoxidized β-carboline. The crude reaction mixture was then submitted to a hydrogen atmosphere to reduce the imine intermediate, therefore affording about 80% overall of racemic THβC **4**. From the large amount of imine formed in these last experiments, it is reasonable to think that racemization is not only caused by homolytic cleavage of the C1–H bond which affords a radical that equilibrates. With a secondary amine, the zerovalent metal (M = Pt or Pd) may induce dehydrogenation via oxidative addition into the C–H or N–H bond followed by β-elimination of MH$_2$ leading to the imine **7**, which is then, in the presence of M(0) and H$_2$ (regenerated after reductive elimination of MH2), reduced to the racemic THβC derivative.[17]

Unfortunately, several attempts to selectively access imine **7** by adding hydrogen acceptors such as cyclohexene to the mixture to trap H$_2$ were not successful.

6.3.5.2 Racemization under Acid and Basic Conditions

In addition to the above methods involving the oxidation of **4** to the imine **7**, we tested basic and acidic conditions for the racemization of the mother liquors. We quickly abandoned the basic conditions as significant degradation, and very little if any racemization was observed even after prolonged treatment with *n*-BuLi or LDA.

Fortunately, a much faster racemization was obtained using acidic catalysts. Complete racemization of the mother liquors was achieved within 16 h and with up to 95% yield using aqueous HCl in dioxane at reflux. However, with that procedure as well as with all the procedures discussed

TABLE 6.4
Racemization of the (*R*)-Enriched (+)-CSA Salt

Entry	Initial Composition[a] S/R (%/%)	Solvent (Vol)	T (°C) (time (h))	Final Composition[a] S/R (%/%)	Yield[b] (%)
1	22/78	*n*-BuOH (2.5)	117 (72)	50/50	62
2	22/78	*n*-BuOH (15)	117 (72)	50/50	71
3	25/75	*n*-BuOH (8)	117 (72)	47/53	82
4	22/78	*n*-BuOH (7.5)	137 (28)	50/50	33

[a] Determined *in situ* by chiral high-performance liquid chromatography (HPLC).
[b] *In situ* yield determined by HPLC.

above, the free basing of the mother liquors must be performed before the racemization reaction, and a subsequent resolution step using an additional amount of expensive resolving agent must be performed. Based on some early observations like the slight decrease of the optical purity of highly enriched (*S*)-enantiomer during the development of the resolution process (*vide supra*), we hypothesized that camphorsulfonic acid should be strong enough to induce racemization of the unwanted (*R*)-enriched mother liquors (*R:S* ~ 80:20).

After some unsuccessful preliminary experiments directly conducted on the ethanolic mother liquors, we turned our attention to higher boiling alcohols such as *n*-butanol and *n*-pentanol (Table 6.4).

As indicated in Table 6.4, the complete racemization of mother liquors was obtained in refluxing *n*-butanol at various concentrations (entries 1 through 3). On the other hand, significant degradation was observed at a higher temperature (entry 4) as indicated by the *in situ* yield determined by HPLC. From these initial attempts, we noticed, however, that after cooling the solution to room temperature, crystallization occurred, resulting in the obtention of (*S*)-enriched material (Table 6.5). A major by-product was isolated and characterized as structure **d** (Scheme 6.11). Based on the results described in Table 6.5, the overall yield of the optical resolution step could thus be improved from 35 to ~70% using *n*-butanol as solvent for the racemization.

TABLE 6.5
Optical Resolution after Racemization of the (+)-CSA Salt

Entry	Chiral Composition[a] (%/%)	Solvent (vol)	Crystals[b] S/R (%/%)	Yield[c] (%)	ML S/R (%/%)	Yield[d] (%)
1	50/50	*n*-BuOH (15)	84/16	31	24/76	40
2	47/53	*n*-BuOH (8)	82/18	40	18/82	42[e]

[a] *In situ*, determined by chiral high-performance liquid chromatography (HPLC) after racemization.
[b] Determined by chiral HPLC on isolated crystals.
[c] Isolated yields corrected for potency by HPLC w/w assay using an external reference.
[d] *In situ* yields determined by HPLC w/w using an external reference.
[e] A second recycling of these mother liquors afforded a further 20% yield of 93/7 *S/R* crystals.

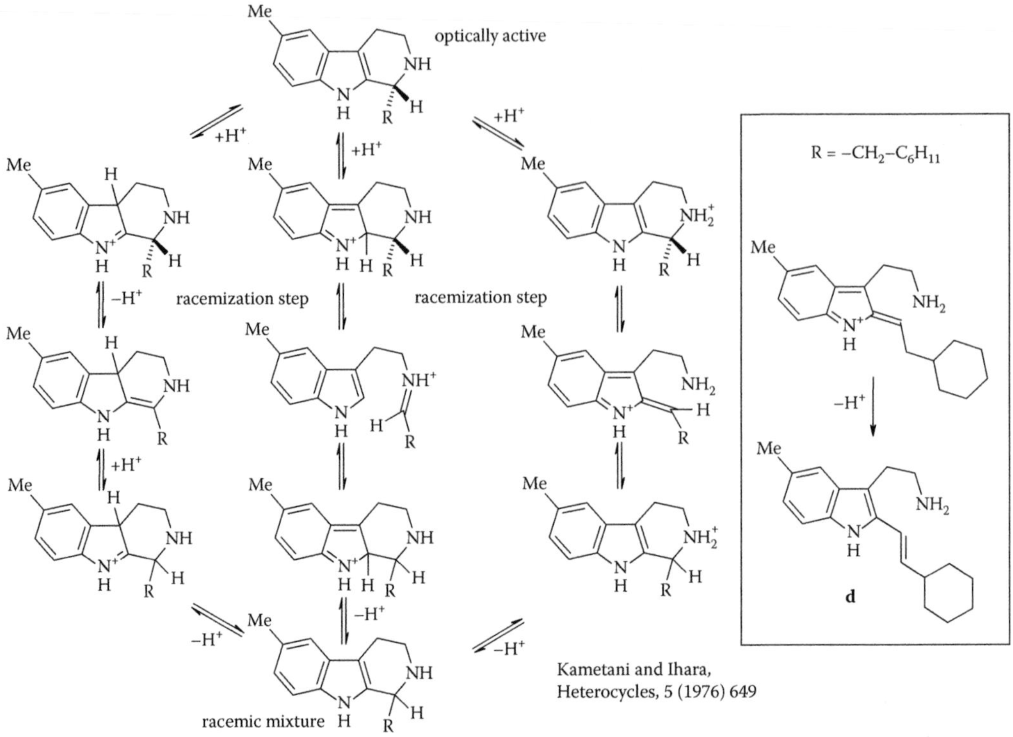

SCHEME 6.12 Mechanisms of acid-catalyzed racemization of 1-substituted THβC and major by-product **d**.

Acid-catalyzed racemization involves protonation at various positions of the THβC structure (Scheme 6.12) as reported in the literature. It is believed that several mechanisms might be operating, depending on reaction parameters such as acid pK_a, concentration and solvent properties (e.g., H-bonding, chelating strength). The formation of impurity **d** in our process indicates that at least the right-hand pathway involving cleavage of the C1–N bond is operating with camphorsulfonic acid in *n*-butanol.

Therefore, we tested other solvents in order to speed up the reaction without causing the degradation of the THβC derivative. As shown in Figure 6.4, a significant rate acceleration was obtained in ethylene glycol (EG). We assume that this acceleration could be due to the better stabilization of the positively charged intermediates by the 1,2-diol (see mechanisms in Scheme 6.12).

The reaction in ethylene glycol was then performed at several temperatures ranging from 125 to 185°C (Figure 6.5). Starting from a 5:95 *S*:*R* mixture, the racemization was achieved in less than 8 h at 125°C and within 1 min at 185°C. A mass temperature of about 150 to 155°C was found to be a good compromise involving reaction time, degradation control, and process scale-up. Under these conditions, racemization of the mother liquors was typically achieved in less than 30 min.

Scheme 6.13 summarizes the different acid-catalyzed processes that we have successfully developed.

6.3.5.2.1 Racemization–Resolution: A Semicontinuous Process

The very fast and clean racemization obtained in ethylene glycol at higher temperature also opened the door for the development of a semicontinuous process. The concept, schematized in Figure 6.6, was successfully demonstrated at the laboratory scale. After 13 consecutive resolution–racemization

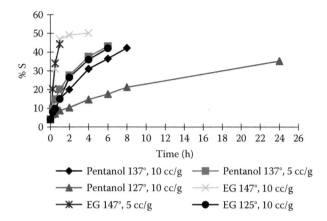

FIGURE 6.4 (See color insert following page 40). Solvent effect on the racemization rate.

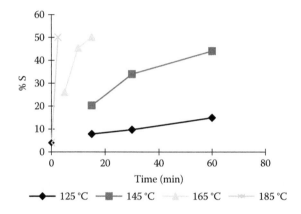

FIGURE 6.5 (See color insert following page 40). Racemization in ethylene glycol: T° effect.

cycles performed in approximately 24 h, the 27:73 *S:R* mother liquors (eutectic mixture) yielded crystals with a 79:21 *S:R* composition in 66% yield.

6.3.5.2.2 *Representative Laboratory-Scale Procedure*

A 27:73 mixture of (*S*)-4:(*R*)-4 is suspended in eight volumes ethylene glycol and stabilized at 30°C for 2 h. The reaction mixture is filtered, and the filtrate is transferred into a second reactor, preheated to 170°C. After 15 min, the solution is allowed to flow back to the first reactor and is stabilized at 20 to 25°C for about 1.5 h. The suspension is filtered, and the filtrate is submitted to the next racemization step. The racemization–resolution process is repeated 13 times to afford a 79:21 mixture of crystals (*S*)-4:(*R*)-4 in 66% yield.

Alternatively, we have also achieved the direct one-pot resolution–racemization of the racemic mixture coming from the Pictet–Spengler by the progressive addition of triglyme to a suspension of the CSA salt in *n*-butanol in order to progressively lower its solubility (Table 6.6). At the end of the process, filtration of the suspension afforded a 70% yield of (*S*)-**4**.CSA (*S:R* 94:6) (Scheme 6.14). Although it was not formally demonstrated in the laboratory, we assume that similar results could be obtained in a shorter reaction time using an ethylene glycol/anti-solvent mixture.

Me
NH
(+)–CSA
NaOH/EtOH
HO(CH2)2OH, 145 °C, <30 min
(100%)
THβC.(+)–CSA (R > S)
M-L (~20/80)
(95%)
M-L in ethylene glycol
THβC (R > S)
(~20/80)
Dioxane/HClaq
reflux
(>95%)
precipitation
with water
20 °C
THβC.(+)–CSA (R = S)
(~50/50) in glycol
resolution
in glycol
20 °C
(>90%)
(25–30%)
NH HCl
salt exchange
THβC.HCl (R = S)
(~50/50)
THβC.(+)–CSA (R = S)
(~50/50)
THβC.(+)–CSA (S > R)
(~90/10)
+ M–L
resolution in EtOH

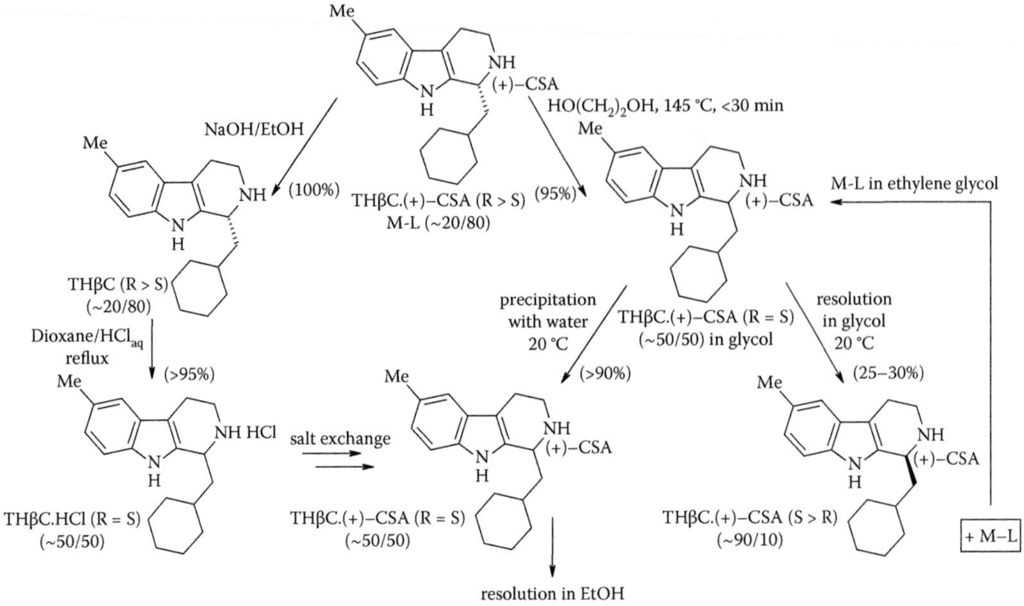

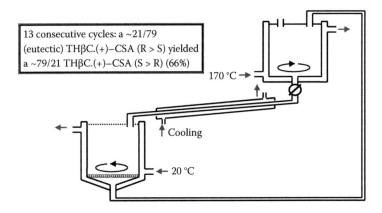

SCHEME 6.13 Acid-catalyzed racemization processes.

13 consecutive cycles: a ~21/79
(eutectic) THβC.(+)–CSA (R > S) yielded
a ~79/21 THβC.(+)–CSA (S > R) (66%)
170 °C
Cooling
20 °C

FIGURE 6.6 Laboratory-scale semicontinuous process.

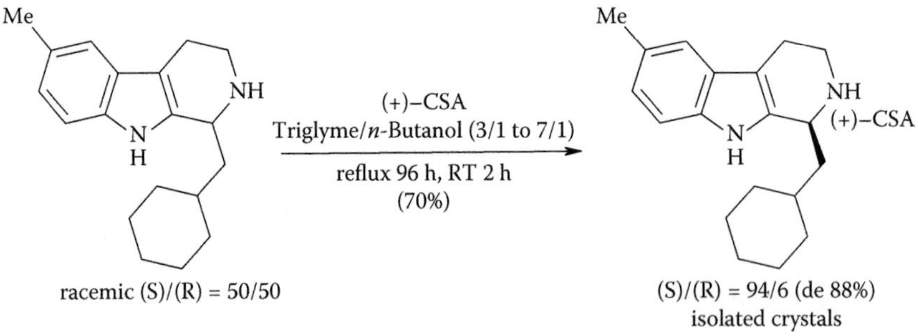

Me
NH
(+)–CSA
Triglyme/n-Butanol (3/1 to 7/1)
reflux 96 h, RT 2 h
(70%)
racemic (S)/(R) = 50/50
(S)/(R) = 94/6 (de 88%)
isolated crystals

SCHEME 6.14 One-pot racemization–resolution.

TABLE 6.6
One-Pot Racemization-Resolution

Entry	Time (h)	Trglyme/n-BuOH (v/v)	(S)/(R) (%/%)[a]
1	0	3/1	50/50
2	24	3/1 → 4/1	58/42
3	48	4/1 → 5/1	72/28
4	72	5/1 → 7/1	80/20
5	96	7/1	82/18

[a] Percent area of each enantiomer determined by chiral high-performance liquid chromatography (HPLC) on the crude reaction mixture.

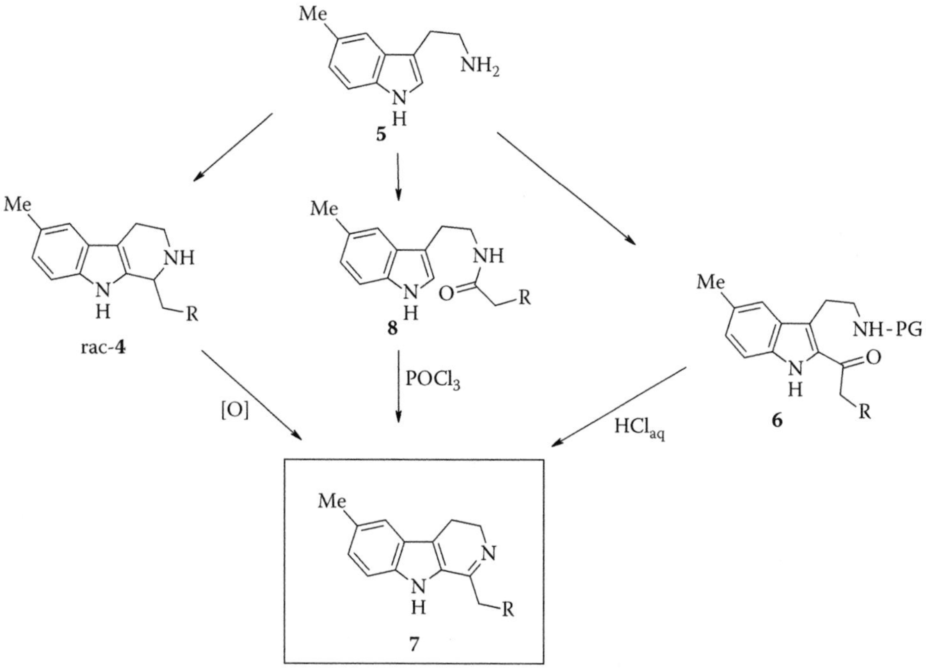

SCHEME 6.15 Access to prochiral imine **7**.

6.3.6 ASYMMETRIC SYNTHESIS

As mentioned in the retro-synthetic analysis, the imine derivative **7** is the key intermediate of our enantioselective synthesis strategy.[18] It was obtained by various pathways (Scheme 6.15):

1. Oxidation of the tetrahydro-β-carboline derivative obtained by Pictet–Spengler reaction (*vide supra*, 75 to 80%)
2. Bischler–Napieralski cyclization of amide **8**, readily obtained by reaction of 5-methyl-tryptamine with 2-cyclohexyl acetyl chloride (75%)
3. Acid-catalyzed cyclization of ketone **6**, obtained by Friedel–Crafts acylation of *N*-acetyl-tryptamine with 2-cyclohexyl acetyl chloride (45 to 50%)

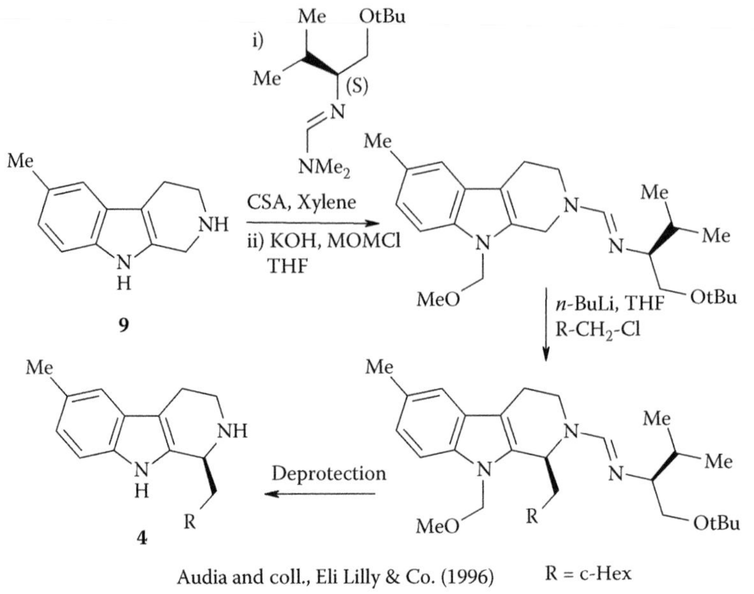

SCHEME 6.16 Asymmetric synthesis of racemic THβC through alkylation.

An alternative stereoselective approach involving a C-1 alkylation of the THβC **9** (Scheme 6.16) was not considered, although we previously demonstrated its usefulness for the synthesis of the benzyl-THβC derivatives **1** to **3**.

The Bischler–Napieralski synthesis of imine **7** was compared to the Pictet–Spengler/oxidation pathway previously used to racemize the mother liquors. The overall yield, although unoptimized, was higher than 85% using POCl₃ in toluene for the cyclization step.[19] Here also, the major issue was to minimize the level of overoxidation to the β-carboline. Alternative procedures involving PCl₅ or POCl₃/P₂O₅ mixtures were not successful, resulting in incomplete reactions, important stirring issues, or higher amounts of by-products.

A related approach toward **7** involves the Friedel–Crafts acylation of *N*-acetyltryptamine catalyzed by AlCl₃ (45 to 50% yield).[20] This latter procedure offers the advantage of better control of the acid-catalyzed cyclization step, therefore resulting in lower levels of overoxidation but suffering from the need to perform an additional protection step.

With several viable methods in hand to access **7**, we investigated its asymmetric reduction. The well-known NaBH₄/CBZ-(*S*)-proline method previously used with success for 1-benzyl-THβC analogs **1** to **3** was investigated first.[21] Although an 82% e.e. was obtained without any optimization, the process was significantly limited by the need to use 3 equiv of an expensive chiral reagent and the usual difficulty in removing the boron salts upon workup.[22]

We therefore quickly turned our attention to the ruthenium-catalyzed asymmetric transfer hydrogenation recently reported by Noyori.[23] Without any optimization, 95% yield and 96% e.e. were obtained with 0.25 mol% catalyst and formic acid-triethylamine 5:2 azeotropic mixture (2.5 mL/g) in CH₂Cl₂ at room temperature for 8 h (Scheme 6.17).[24,25] Apart from the high yield, enantiomeric excess, and turnover, this procedure is particularly simple to carry out. It also allows an easy recovery of the optically active amine by filtration, as its formiate salt at the end of the reaction, if needed, would offer an additional improvement in optical purity.

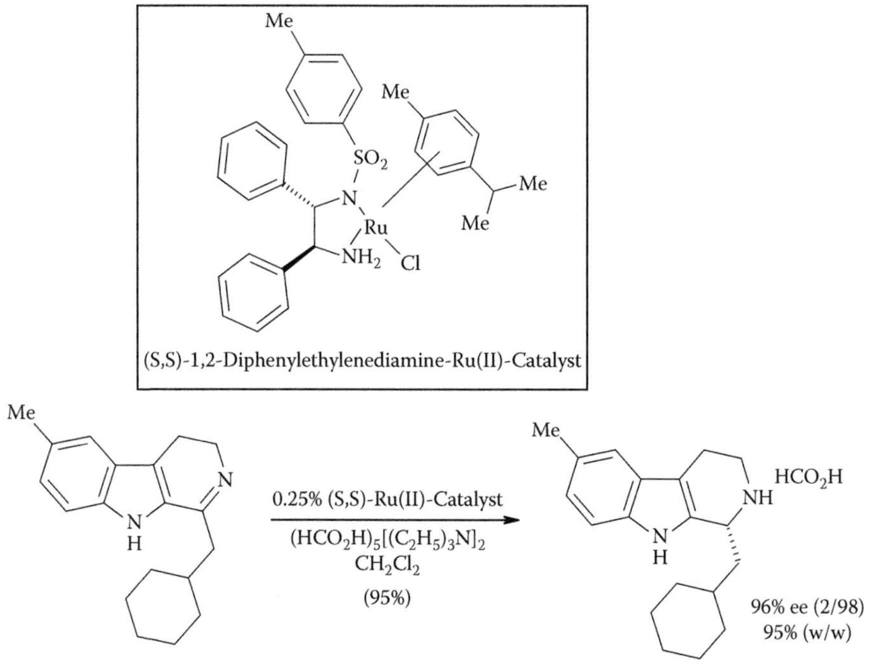

SCHEME 6.17 Catalytic asymmetric synthesis of racemic THβC.

6.4 CONCLUSIONS

We developed two highly efficient approaches toward the 5HT$_{2B}$ receptor antagonist LY414197. The first route utilized a Pictet–Spengler reaction to construct the racemic THβC derivative. Access to the most active (*S*)-enantiomer was then achieved in up to 70% yield through a one-pot resolution–racemization process. The second approach is based on the chiral *N*-arenesulfonylated-1,2-diamine/ruthenium(II)-catalyzed asymmetric reduction of a 3,4-dihydro-β-carboline precursor, obtained from a Bischler–Napieralski or related Friedel–Crafts methodology.

Both routes involved 5-methyltryptamine as key starting material. As this compound was not easily accessible in bulk at that time, we also put substantial effort into the development of a robust and inexpensive process based on a classical Fischer indole synthesis.

ACKNOWLEDGMENTS

This chapter represents the outstanding contributions of many scientists within Lilly Research Laboratories, discovery chemists, process chemists, analytical chemists, and chemical engineers. Among these, the authors would like to particularly acknowledge H. Gorissen for his leadership and guidance throughout the development of this project. We are also indebted to the important contributions made by J.P. van Hoeck, L. Antoine, M. Vanmarsenille, F. Driessens, P. Boquel, C. Rypens, X. Lemaire, C. Masson, Ch. Stevens, K. Stockman, A. Mockel, Cl. Delatour, P. Delbeke, B. Bonnier, O. David, M. Rinaldi, F. Baudoux, J. Audia, D. Nelson, and V. Libert.

REFERENCES AND NOTES

1. Audia, J.E., et al. *J. Med. Chem.* 1996, *39*, 2773–2780.
2. (a) Abramovitch, R.A.; Shapiro, D. *J. Chem. Soc.* 1956, 4589; (b) Güngor, T.; Malabre, P. *J. Med. Chem.* 1994, *37*, 4307–4316; (c) Szantay, C.; Szabo, L. *Synthesis* 1974, 5, 354; (d) Speeter, M.E.; Anthony, W.C. *J. Am. Chem. Soc.* 1954, *76*, 6208; (e) Freund, M.E.; Mahboobi, S. *Helv. Chem. Acta* 1990, *73*, 439; (f) Bûchi, G.; Mak, C.P. *J. Org. Chem.* 1977, *42*, 1784; (g) Flaugh, M.E.; Crowel, T.A. *J. Med. Chem.* 1979, *22*, 63; (h) Kozikowski, A.P.; Chen, Y.Y. *J. Org. Chem.* 1981, *46*, 5248; (i) Audia, J.E., et al., European Patent application 0 620 222 A2.
3. Grandberg, I.I., et al. *Proc. Acad. Sci., USSR* 1967, *176*, 828; Bretherick, L., et al., *J. Chem. Soc* 1961, 2919; Keglevic, D., et al. *Croatia Chemica Acta* 1961, *33*, 83; Zapeda, L.C., et al., *Tetrahedron* 1989, *45*, 6439.
4. Albright, J.D.; Snyder, H.R. *J. Am. Chem. Soc.* 1958, *81*, 2239; Martin, S.F.; Liras, S. *J. Am. Chem. Soc.* 1993, *115*, 10450.
5. Kametami, T.; Takano, S.; Hibino, S.; Takeshita, M. *Synthesis* 1972, 9, 475.
6. Barrett, A.G.M.; Graboski, G.G. *Chem. Rev.* 1986, *86*, 751–762.
7. Kraus, G.A.; Neuenschwander, K. *J. Org. Chem.* 1981, *46*, 4791.
8. Righi, G.; Rumbolt, G.; Bonini, C. *J. Org. Chem.* 1996, *61*, 3557.
9. (a) Anelli, P.; Montanari, F.; Quici, S. *Org. Synth.* 1993, Collect. Vol. VIII, 367. (b) De Nooy, A.; Besemer, A.; Van Bekkum, H. *Tetrahedron* 1995, *51*, 8023.
10. Kobayashi, Y.; Takemoto, Y.; Kamijo, T.; Harada, H.; Ito, Y.; Terashima, S. *Tetrahedron* 1992, *48*, 1853.
11. Cox, E.D.; Cook, J.M. *Chem. Rev.* 1995, *95*, 1797–1842 and references cited therein. Czerwinski, K.M.; Cook, J.M. *Adv. Het. Nat. Prod. Synth.* 1996, *3*, 217–277. Love, B.E. *Org. Prep. Proc. Int.* 1996, *28*, 1–64.
12. Swaminathan, N. *Diss. Abstr. Int., B* 1996, *57*(4), 2572.
13. (a) Jacques, J.; Collet, A.; Willen, S. *Enantiomers, Racemates and Resolution*, Wiley, New York, 1981; (b) Borghese, A.; Libert, V.; Zhang, T.; Alt, Ch.A. *Org. Process Res. and Dev.* 2004, *8*, 532–534.
14. For a general reference, see Kametami, T.; Ihara, M. *Heterocycles* 1976, *5*, 649–668 and references cited therein.
15. Akimoto, H.; Okamura, K.; Shioiri, T.; Kuramoto, M.; Kikugawa, Y.; Yamada, S.I. *Chem. Pharm. Bull.* 1974, *22*, 2641–2623.
16. Goti, A.; Romani, M. *Tetrahedron Lett.* 1994, *35*, 6567. Kamal, A.; Howard, P.W.; Reddy, B.S.N.; Reddy, B.S.P.; Thurston, D.E. *Tetrahedron* 1997, *53*, 3223.
17. Murahashi, S.; Yoshimura, N.; Tsumiyama, T.; Kojima, T. *J. Am. Chem. Soc.* 1983, *105*, 5002.
18. For a related oxidation–reduction approach, see Tietze, L.F.; Zhou, Y.; Töpken, E. *Eur. J. Org. Chem.* 2000, 2247–2252.
19. For a review, see *Org. Prep. Proc. Int.* 1996, 1–64.
20. Orito, K.; Matsuzaki, T.; Suginome, H. *Heterocycles* 1988, *27*, 2403–2412.
21. Yamada, K.; Takeda, M.; Iwakuma, T. *J. Chem. Soc. Perkin Trans. I* 1983, 265 and references cited therein.
22. Even though we have shown that recovery of the CBZ-(*S*)-proline is feasible.
23. Uematsu, N.; Fujii, A.; Hashiguchi, S.; Noyori, R. *J. Am. Chem Soc.* 1996, *118*, 4916–4917.
24. The experiment was conducted with the (*S,S*) chiral catalyst that afforded the unwanted (*R*)-4.
25. Narita, K.; Sekiya, M. *Chem. Pharm. Bull.* 1977, *25*, 135–140.

7 To Overcome the Hurdles: Coping with the Synthesis of Robalzotan, a Complex Chroman Antidepressant*

Hans-Jürgen Federsel and Anders Sveno

CONTENTS

7.1 INTRODUCTION: ENTRY INTO THE FIELD OF CNS ACTIVE CHROMANS

When a molecule has been identified as possessing interesting pharmacological activities, the search for structural analogues of the parent compound that might demonstrate a different and preferably better set of characteristics usually begins. These features could, for example, be lower acute toxicity, higher stability, more potency, easier synthesis, changes to more optimal physical properties, or improved bioavailability; it all comes down to what is judged to be most important when aiming for a specific application or use. This stimulating and highly competitive search for new

---

* A part of this work has been presented in lecture format at two conferences: Sveno, A., The development of an effective process for a challenging chiral chroman drug molecule, 13th International Conference on Organic Process Research and Development, Nice, France, 4th–6th April 2006; Federsel, H.-J., Chromans: Innocent targets or tough opponents? Experiences from process development of two neuroscience drug candidates, 13th FECHEM Conference on Heterocycles in Bioorganic Chemistry, Sopron, Hungary, 28th–31st May 2006.

members of a structural class is by no means novel. Instead this hunt for identifying successively better and better compounds has been practiced for decades with the ultimate goal to find what in today's language is called the "best-in-class" molecule. Noticeable compound families in which such an iterative process has been successfully demonstrated, are, in roughly chronological order, β-lactam antibiotics (penicillins and cephalosporins), tricyclic antidepressants, [1,4]-benzodiazepines for use as anxiolytics and sedatives, benzimidazole-sulfoxide proton pump inhibitors to treat gastrointestinal disorders, and, most recently, statins for treatment of high blood cholesterol.

In the late 1980s and early 1990s, there were dedicated efforts to explore the suitability of using chromans as agents to alleviate symptoms of depression and anxiety. Especially, derivatives belonging to the 3-aminochroman series had demonstrated a pronounced affinity to the 5-HT$_{1A}$ receptor, and the goal was to identify compounds that would be highly selective either as agonists or antagonists.[1,2] As a matter of fact, the core part constitutes a rather simple oxygen-containing heterocycle—benzo[b]pyran—and this compound class incorporates several members with pronounced biological activity. Representative examples would be the lipophilic antioxidant vitamin E and its many derivatives, as well as structurally diverse natural products that display properties such as prevention of HIV replication in human T-lymphoblastic cells, estrogen antagonism implying usefulness as antifertility agents, anticancer activities for treatment of malignant tumors, antihypertensives, and antifungals.[3]

The first candidate drug (CD) to emerge from an extensive in-house pharmacological screening of a wide variety of chroman derivates was NAE-086 (internal Astra code) that subsequently was given the generic name ebalzotan.[4,5] This compound, which is a potent 5-HT$_{1A}$ agonist, was eventually given to healthy human volunteers in Phase I studies—but the observation of unwanted side effects brought the project to a halt, and further development was abandoned. In parallel to this work, the search for a follow-up based on the chroman structural platform aimed at improvement of efficacy and tolerance. The outcome of these efforts led to NAD-299, later to be named robalzotan, a candidate with selective antagonist properties capable of increasing neuronal transmission.[6] As had been anticipated, this new molecule demonstrated clear advantages in early studies in humans and the candidate in progressed to the stage of proof-of-concept studies. However, in this case, there were difficulties as trials on a small number of patients suffering from depression showed a clear lack of efficacy; therefore, this project was also discontinued without delivering any of the medical advantages that had been outlined as the ultimate goal. In Figure 7.1, the structures of these molecules are displayed to allow side-by-side comparison.

The setbacks experienced in these two cases from the medical point of view did not, however, eliminate the need for robust syntheses that could be scaled up and used to produce the required amounts of active pharmaceutical ingredient (API). When embarking on the ebalzotan project, the only route available for making this nontrivial molecule consisted of 13 linear steps that were utilized by the medicinal chemistry department when preparing pre-CD quantities; see Scheme 7.1 for details on what reactions were used to synthesize this product.[4] This rather long sequence was required as no suitable advanced building block was found to be commercially available. Thus, the

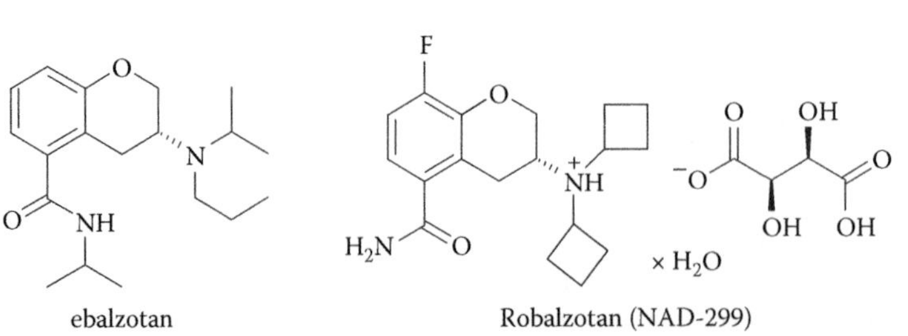

FIGURE 7.1 Structural comparison of ebalzotan (free base) and robalzotan (L-tartrate monohydrate).

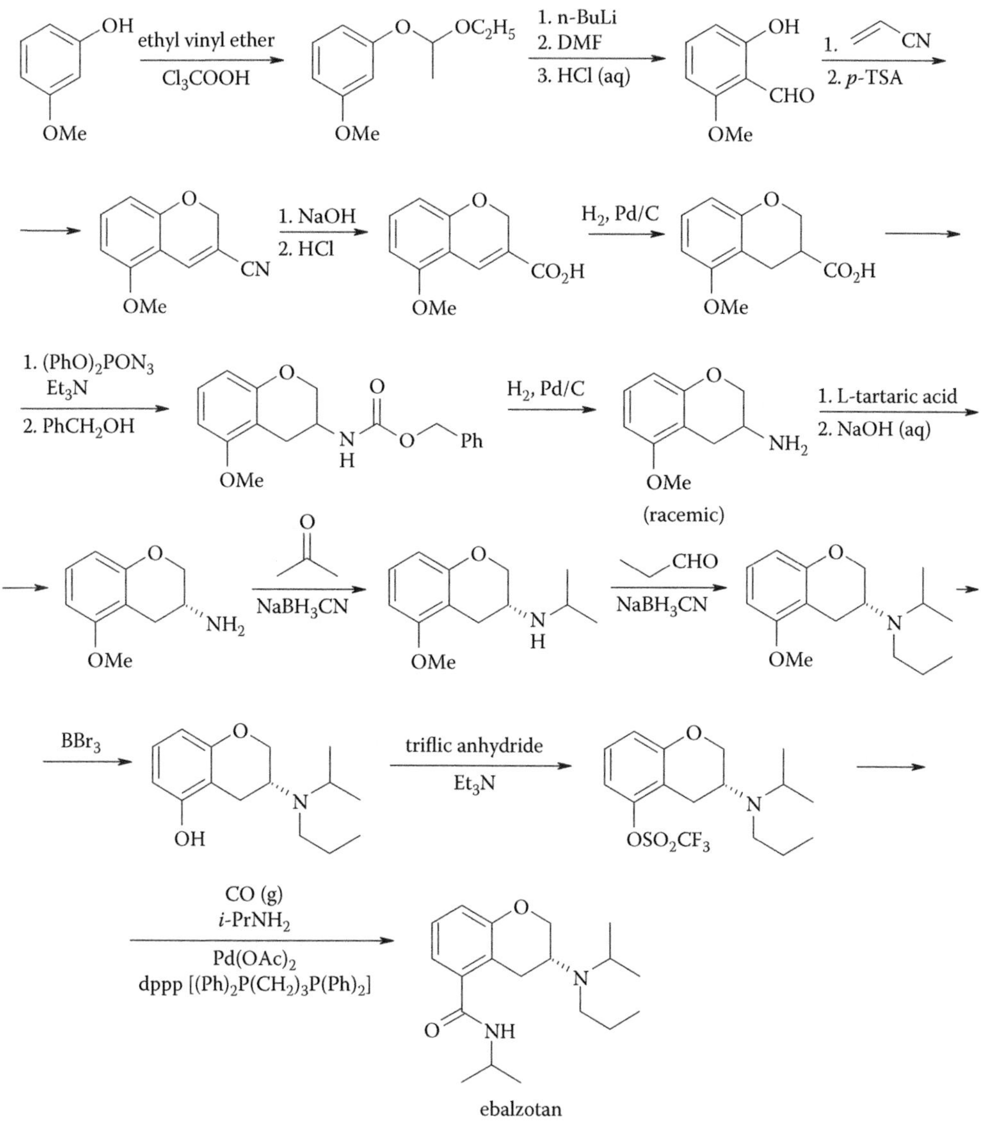

SCHEME 7.1 The synthetic sequence to reach ebalzotan in 13 linear steps incorporating a *de novo* assembly of the chroman nucleus and a diastereomer resolution (abbreviations: DMF = *N,N*-dimethylformamide; *p*-TSA = para-toluenesulfonic acid).

synthesis had to be back-integrated to the stage of a fairly trivial starting material (resorcinol monomethyl ether). Furthermore, a major drawback was the fact that the introduction of the correct stereochemistry (*R*-enantiomer) did not occur until halfway through the process, and this meant losing 50% of valuable material in the form of the wrong isomer during resolution. Notwithstanding, this was the chemistry that was developed, scaled up, and eventually optimized, allowing the production of multi-kilogram batches of the desired product.[7,8]

At the time of starting the robalzotan project, a well-developed process to prepare the forerunner ebalzotan was already in hand to a level that allowed the production of multiple kilograms of all intermediates, inasmuch as the starting material was freely accessible. Not surprising then that the search for a feasible first synthesis of the new target compound attempted to build on as much as possible of the experience and knowledge that was previously achieved. Racemic aminochroman (see Scheme 7.1) was quickly identified as a key building block, which brought with it the advantage that a *de novo* method for synthesizing the chroman skeleton was not needed. Furthermore, after the resolution (again a known method from ebalzotan), the desired product could be conceived in a series of transformations constituting the changing of substituents around the core heterocyclic moiety. When trying to validate this approach, it was, however, easily recognized that this process of making the desired product was far from being ideal for the long-term supply of larger quantities. The major weakness was the requirement to introduce the fluorine atom on the aromatic ring, which turned out to be technically extremely complicated. Moreover, this feature, together with other constraints, led to a cost of goods that was orders of magnitude higher than expected. In this chapter, the efforts that eventually led to the development of a robust and reliable manufacturing process that met predefined criteria of scalability, quality, and cost will be described.

7.2   AN UPHILL STRUGGLE: THE FIRST ATTEMPT TO PRODUCE MATERIAL AT SCALE

At the time the candidate was nominated into development—in the mid-1990s—the interactions between the process research and development (PR&D) and medicinal chemistry parts of the organization were relatively weak, and this was to a large extent dependent on the organizational framework that was in place at that time and put these two interlinked disciplines too far away from each other. Furthermore, there was a lack of understanding of the challenges and complexities in other functional areas as well as an inability to communicate in an efficient manner.[9–11] Ignoring the reasons, the outcome of this was a late involvement of the PR&D department resulting in a situation in which the formal requirement of the first delivery at scale was not realized until the point where the acceptance of the new candidate drug had been formally made by the business. This meant that the activities focusing on robalzotan and how to produce this molecule did not really start until considerable amounts were urgently needed in order to kick off other streams of work, notably in safety assessment, enabling continued studies on toxicological properties and pharmaceutical development so that a suitable formulation could be designed and tested. Fortunately, the situation has improved considerably over the last decade as PR&D has become much more integrated into the discovery activities; an achievement that to a large extent is due to enhanced communication across departmental interfaces.[12]

Upon closer examination of the 11-step synthetic route (see Scheme 7.2) provided by medicinal chemistry, several features were noteworthy:

- The resolution of the aminochroman described earlier carries with it considerable cost implications, because at least 50% of the material is wasted.
- After the resolution the (*R*)-configuration is retained throughout the remainder of the synthesis.
- Introducing the fluorine atom via an electrophilic fluorination reagent requires a two-stage bromination and low-temperature metalation procedure that is followed by a chromatographic purification.
- The transformation of the phenolic OH group to the carboxamide ($ArCONH_2$) requires five discrete steps starting with activation (to the triflate), a Pd-catalyzed carbon monoxide insertion followed by ester hydrolysis, acid chloride formation, and amidation.
- Cyclobutyl substitution of the amino group under reductive conditions is effected early on in the chain by using cyclobutanone which is by far the highest cost-contributing chemical in the whole process.

Introduction of fluorine substituents is generally seen as a technically complicated and demanding procedure due to, for example, the often extreme reaction conditions, the handling of

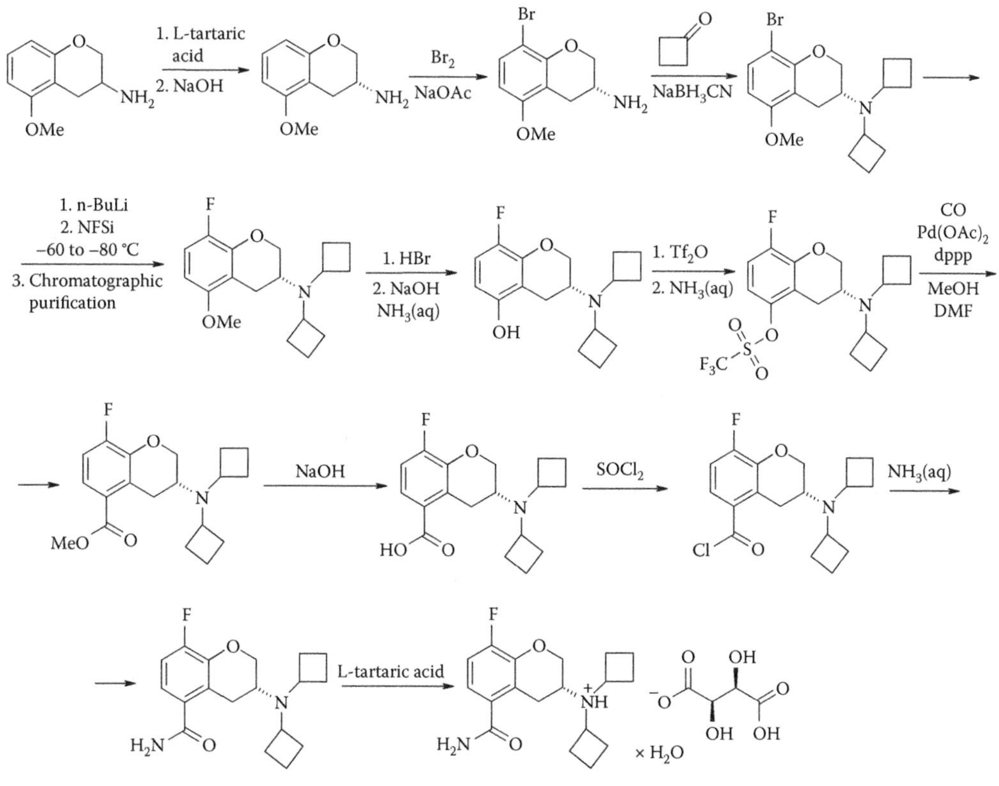

SCHEME 7.2 In 11 steps to robalzotan from racemic 3-aminochroman including an electrophilic fluorine introduction as a key transformation (abbreviations: Tf$_2$O = (CF$_3$SO$_2$)$_2$O; dppp = [(Ph)$_2$P(CH$_2$)$_3$P(Ph)$_2$]).

fluorine-containing reagents, and the high risk of excessive corrosion in many standard-type construction materials (glass, stainless steel). The procedure applied to incorporate the fluorine substituent in the early step of this reaction sequence builds on the use of an electophilic reagent, NFSi (*N*-fluorobenzenesulfonimide)[13] that became commercially available on larger scale in the early 1990s. This approach allows a carbanion that is generated via a metalation procedure (using BuLi in this case) to be neatly transformed to a C-F moiety, albeit a reaction that often requires very low temperatures (<−50°C).[14,15] Besides being a very atom-inefficient transformation in which only about 6% of the atomic weight of NFSi is utilized in the form of one fluorine equivalent, this reaction is severely hampered by its sensitivity toward proton-donating agents. Thus, protic solvents, for example, alcohols or water, residues of starting materials, upstream synthetic intermediates with active hydrogen atoms, or even the reagent NFSi by virtue of the aromatic ortho-protons can act as an apparent proton source under the strongly basic conditions generated in the system.

In our case, this side reaction was particularly undesired as the corresponding *H*-analogue of robalzotan (see Scheme 7.3) displayed a pronounced central nervous system (CNS) activity. Assessment of its potency showed that the affinity toward the target receptor was ten times higher than the parent compound, and moreover, it exhibited the properties of an agonist—the mechanism of action was opposite to that of robalzotan (antagonistic profile). This conflicting pharmacological situation had to be addressed from a quality point of view as the projected efficacy studies might lose their relevance and reliability if too high a level of the hydrogen impurity was present. In order to avoid this potentially negative impact, the challenging level of 0.1% was set for the *H* compound as its maximum content in the final product. Considering the close similarity, both in chemical and physical terms, of *H*- and *F*-containing compounds in general, it is obvious that separating mixtures

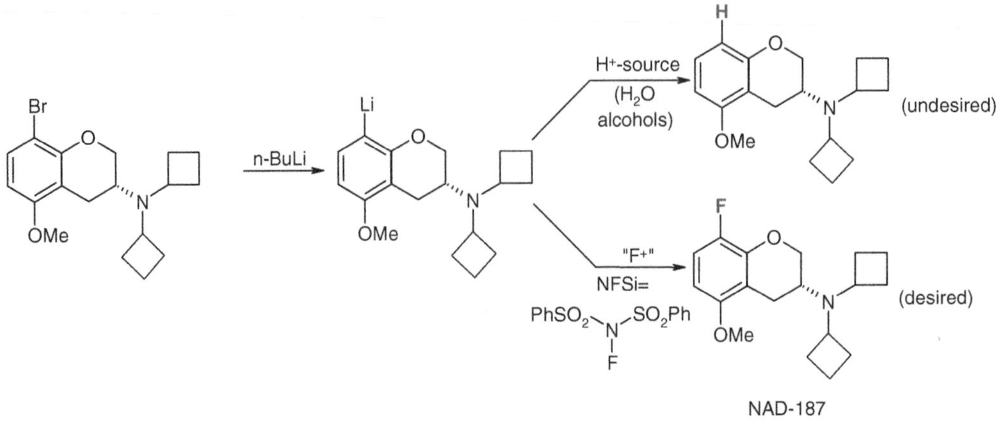

SCHEME 7.3 Generation of the highly undesired H analogue. The presence of electrophilic species, such as water or alcohols, will be captured by the reactive Li intermediate and form by-products.

of them will be a complicated task that requires sophisticated techniques. Even with access to these, there is no guarantee that a complete separation, meaning no trace of the unwanted analogue, will ever be attained. As a knock-on effect, this leads to a considerable risk that residual amounts of the des-*F* impurity, irrespective of how small that might be, are cross-contaminating stages in the downstream process or could even be enriched during crystallization of intermediates or in the isolated end product to a nonacceptable level.

For the PR&D team, this meant that the outcome of the fluorination was crucial to the entire process and the ability to deliver a material with the desired quality features. Given the late start of the work to develop the medicinal chemistry method, there was not much time available for familiarization with this novel chemistry in order to scrutinize boundaries and options. A relatively small number of laboratory experiments had to suffice, and luckily, they showed consistent results with what had been reported confirming that the formation of *H* analogue could be controlled at a low enough level (*F* versus *H* analogue ~7:1) so that an adequate purification would not be a major issue. However, the outcome from the run in the pilot plant was disastrous; the reaction behaved far from what had been expected, and the crude product showed the shocking result of an almost 40:60 mixture of the two components (*H* and *F* analogues, respectively). This result was attributed to a number of factors that had come into play during the scale-up of this step in a 600-L vessel size. Some of these factors may include inferior quality of the aromatic bromide starting material, inability of the cooling equipment to secure the low temperatures needed, inefficient mixing on adding the solid NFSi reagent, and presence of protic impurities (notably H_2O) in the industrial-grade solvents used. A far from satisfactory control of the sensitive reaction conditions and process critical parameters had, thus, resulted in something that was close to being described as catastrophic, and the imminent question raised was "can a method be quickly developed that will allow our desired product to be salvaged?"

In line with what was stated previously on the resemblance of properties between the *H* and *F* analogues, the prime option for purification became preparative column chromatography. Together with an expert and instrumentation company specializing in this field, the conditions—type of stationary phase, composition of eluant system, flow rate—that would deliver the best performance that allows as close to full separation of the two major components as possible, preferably in quantitative yield, were investigated and screened. It was found that a straight-phase system on a standard silica column using dichloromethane (CH_2Cl_2) for elution was optimal. In order to produce 1 kg of the desired fluorinated intermediate (NAD-187), a high-performance liquid chromatography (HPLC) column of 30-cm width was utilized, and the eluant consumption mounted to 6 m^3 CH_2Cl_2. Suffice it to say that this purification exercise required the input of about 200 man-hours of work, albeit with a successful outcome resulting in controlling the by-product at below the 0.1% level; and it is obvious that the value of the product obtained had increased substantially.

The rest of the synthesis proceeded in a largely nondramatic fashion allowing for the isolation of 0.5 kg robalzotan in the form of its tartrate salt. The stereochemical purity of this material was >99% enantiomeric excess (ee) but, unfortunately, containing a somewhat higher level of the *H* derivative (0.3%) than expected. Starting from roughly 17 kg of the racemic aminochroman, the total yield over the entire synthetic sequence (11 steps) was only 1%, which equates to an average yield/step of around 65%. From a chemistry point of view, however, two steps stand out as the weakest links in the chain—the resolution of the racemic starting material with a yield that by default is <50% and the just described fluorination in which a disappointingly low yield of 35% was achieved after chromatography. Comparing cost figures is always sensitive and might lead to wrong conclusions, but in this case, it seems highly relevant as an illustration of the bad shape in which the process was: the cost of robalzotan after the first pilot-plant campaign extrapolated to 1 kg was $200,000, and this figure needs to be compared with the predefined target price of "merely" $10,000.

Irrespective of the far from ideal process that was at our disposal, a second batch was quickly orchestrated to respond to the fact that not enough material had been produced in the first attempt. Using the NAD-187 that had already been chromatographically purified, but the largest portion of which had been shelved for later conversion, it was possible, based on learning from the first run, to successfully make a second 7-kg batch of the API. This enabled other parts of the project to start their activities, but for PR&D, it brought the realization to the forefront that this process of synthesizing robalzotan was not sustainable, and as a consequence, the route had to be redesigned.

7.3 ENVISAGING A NEW ROUTE AND ITS STRATEGIC IMPLICATIONS

When extrapolating the stage-by-stage performance of the process and the achieved delivery from the first campaign—that is, without taking any improvements into account—to meet the forthcoming need of 15 kg of the API, the scenario was rather discouraging. In particular, this was true for the fluorination step which required a chromatographic operation. It was estimated that the contemplated batch size would consume over 100 m^3 of solvent (CH$_2$Cl$_2$) for elution purposes and numerous hours of labor. Hence, it is not far-fetched that the project was taken back to the drawing board in a trial that aimed at unraveling alternative synthetic options. Applying the principles of retrosynthesis to the target molecule, coupled with some prerequisites that had to be fulfilled, notably the absence of a fluorination step and as late an introduction of the cyclobutyl substituents as possible, a new route was conceived; see Scheme 7.4.

The major characteristic features in this novel sequence are the following:

- There is a total of 13 linear steps.
- The crucial *F* substituent is present in the starting material.
- The pyran ring required to form the chroman skeleton is generated *de novo*.
- An array of three transformations comprised of a consecutive olefin nitration/reduction protocol generates a racemic primary amine.
- The desired configuration is established via a classical resolution.
- Dual introduction of the cyclobutyl units is performed in the last chemical step.

Switching from the old synthetic route to this second-generation approach had successfully addressed a number of the previous shortcomings. However, new problems that stood out as being crucial to the survival of the new sequence were created. First, the generation of the chromene system by using a Claisen rearrangement reaction in the third step required heating to temperatures

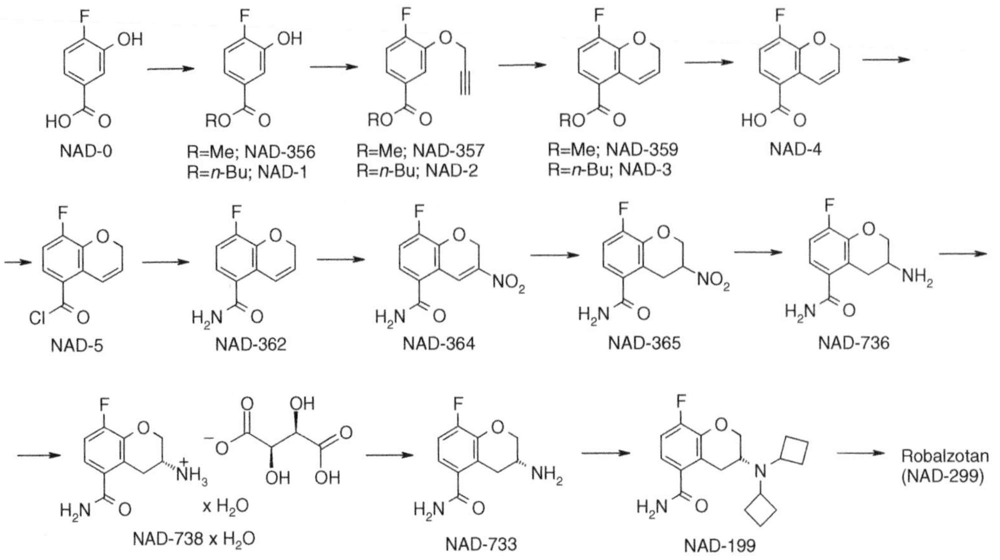

SCHEME 7.4 Second generation synthesis of robalzotan comprised of 13 steps. Key features are the presence of the fluorine substituent in the starting material, a *de novo* construction of the chroman skeleton, and a late stage diastereomer resolution.

well above 200°C, which is clearly higher than many standard-configured process plants are designed for (normally operated at maximum of 150 to 160°C). Second, given the intrinsic safety concerns implicit in nitrations, due to harsh reaction conditions and the often high-energy content of the product, step 7 forming a 3-nitrochromene intermediate was seen as a potential bottleneck. This uncertainty was enhanced by the low yields obtained in this transformation, from preliminary laboratory experiments, despite using a huge excess of reagents. Continued scrutiny and exploration of the other steps in this route revealed some further weaknesses—the reduction of the saturated nitrochroman (step 9) and the reductive alkylation (step 12) responsible for the introduction of the cyclobutyl substituents.

History has a tendency to repeat itself, which in this case meant that due to a compressed delivery schedule, there was only very limited time available to gain a good understanding of the new route and, furthermore, to improve on some of the underperforming steps. A quick scale-up to satisfy the needs for API, to initiate clinical phase I studies, resulted in a yield of only 0.3%, which did not compare favorably to the laboratory yield of 5%. This batch produced 2 kg of robalzotan of outstanding quality with an enantiomeric purity >99% ee. The major flaws in the chemistry (expressed as poor yields) occurred exactly where they had been projected as mentioned previously, and it was obvious from working on pilot scale that virtually all steps were hampered by low volume efficiency (amount of material produced per unit reactor volume) and complicated workup procedures.

After having conducted the first campaign in-house adopting the new sequence in its entirety, the accumulated hands-on experience was used to analyze the prospects for successful process development. The pros and cons of different scenarios were identified and amalgamated into strategic options that had to be prioritized. Due to the poor shape of the process in this initial stage, it became quite clear early on that some sort of outsourcing solution would have to be devised, if only to address the strained internal capacity situation. But in this case, there was also a strong technical driver to go to external sources, namely the aforementioned high-temperature transformation (Claisen step), and this brought the project in contact with the Italy-based R&D contractor Zambon. Based on its capability to work at elevated reactor temperatures (>200°C) coupled with the experience from conducting more conventional chemical reactions on scale, it was most beneficial to both parties to extend the collaboration beyond just the ring-closing step. Instead of restricting the work to a particular key transformation, the agreement was made to outsource the first six steps of the synthesis leading up to the fluorinated carboxamido-chromene (NAD-362), a satisfactory crystalline and stable compound. After a short while, however, it became obvious that further external support would be required to meet the project targets, and this was addressed by including the nitration reaction in the overall deal. This was an excellent tactical move, as Zambon belonged to the relatively small group of companies that had the technology to operate nitration chemistry on large scale; therefore, it was decided that it should take responsibility for developing a sustainable and safe process that could be optimized and implemented in its manufacturing facilities to allow production of large amounts of material. There was one drawback with this plan, namely the unsuitability of shipment of the nitro-chromene thus produced due to safety concerns and stability issues. These limitations were also valid in the case of the next intermediate in the sequence, the nitrochroman, and so the options were to either stop at the aforementioned chromene moiety or to add the subsequent three steps and pursue the synthesis through to the racemic aminochroman. Not surprisingly, it was this latter scenario that was chosen but with the caveat that the nitro-to-amine reduction conditions would be developed internally and then transferred to Zambon to conduct the scale-up.

7.4 TURNING THE NEW ROUTE INTO A WORKABLE PROCESS

The newly designed sequence had successfully tackled the challenge to identify an imaginative and considerably improved synthesis for robalzotan. A number of problematic issues had been dealt with, and it seemed that the target compound could be assembled in a clever way. As witnessed by the results obtained in the first scaled-up batch, we were still far from a fully viable process. Realizing that many of the separate steps were in poor shape and required a substantial makeover to bring them to an acceptable level, one of the transformations—the chromene nitration—stood out as being more critical to the survival of the synthetic approach than all the others. It was blatantly obvious that considerable innovative power would be needed to make this crucial step feasible for operation on plant scale, but the same pressure would have to be exerted on other parts along the reaction path. The following step-by-step description will show, in a consecutive order, how this was managed.

7.4.1 STEP 3: CONTROLLING THE CLAISEN REARRANGEMENT

At elevated temperatures, unsaturated alkyl aryl ethers undergo an intriguing reaction where the olefinic appendage is transposed onto either the *ortho* or *para* positions of the aromatic ring, depending on the substitution pattern. This transformation, normally referred to as the Claisen rearrangement, constitutes, in mechanistic terms, a [3,3]-sigmatropic shift, and when applied to our aryl propargyl ether substrate, this provided an elegant direct entry to the chroman skeleton. In the course of the first pilot campaign, the decision was made to distribute the ether intermediate that had been prepared as starting material for the rearrangement into five sub-batches as a measure to minimize risk of failure, as the reaction had not been extensively mapped at that point. Thermal heating to 210°C in *N,N*-diethylaniline (DEA) as a solvent offered a much lower yield and more by-products than observed in the laboratory. The major by-product was the five-membered furan congener, which had been observed in laboratory experiments at a 3 to 5% level. The level of this impurity in the pilot plant varied considerably from 8 to 10% in the first batches to >80% in the final batch; see Scheme 7.5.

The failure to successfully repeat the laboratory results was attributed to the prolonged heating times when operating in large-scale equipment, which meant a longer residence time at temperatures below 200°C leading to the side reactions observed. Later, however, it was identified that an even more crucial parameter was the presence of base (notably in the form of K_2CO_3) that had

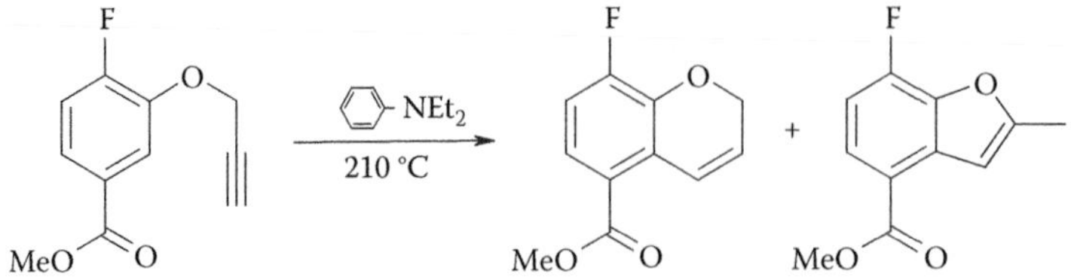

Batch #	Furan (%)
1	10
2	8
3	11
4	17
5	>80

SCHEME 7.5 Conducting the Claisen rearrangement of the aryl propargyl ether in diethylaniline at elevated temperatures. Formation of the desired chroman product is accompanied by generation of furan by-product in successively increasing amounts.

inadvertently been left in the process stream from the previous step. Furthermore, better analytical control revealed that the reaction also produced dimers and oligomers as well as provided a more detailed insight into the stability properties of the starting ether. Thus, the data for the methyl ester (NAD-356)—the form in which the substrate was used initially—pointed at a significantly lower stability when handled in solution. As a consequence, the ester functionality was shifted from methyl to the *n*-butyl analogue (NAD-1), and with this measure the stability issues were effectively controlled under the actual range of reaction parameters. At this stage, this process step did not meet even modest expectations of robustness, and this was further augmented by the inefficiency seen in the workup where the solvent (DEA) had to be washed off in a repeated high-volume extraction procedure.

The increased understanding of the intrinsic characteristics and peculiarities of this reaction was seen as a logical next step. It was obvious that there were at least two pathways in place, one leading to the six-membered ring and the other to the five-membered ring. Our early experiments had shown that rather small changes in reaction conditions had an immediate impact on the selectivity, albeit without our understanding the exact reasons for this anomaly. Preliminary data suggested a solvent effect in this reaction because the formation of chroman was favorable in DEA while the formation of the furan was favorable in ethylene glycol. Another variable that caught our attention was the presence of various basic components in the system and their possible influence. In order to come to grips with this complex problem, it was felt that a mechanistic understanding would be essential, and fortunately, there had been some rather detailed investigations into this area during the 1960s and 1970s.[16,17] An allenic intermediate is implicated in this reaction, which, depending on the exact structural arrangement, can undergo a [2 + 4] cycloaddition or an enolization-driven nucleophilic ring closure. Applying these principles to the current problem and expanding the investigation to analogous substrates, an interesting picture emerges; see Scheme 7.6.

In addition to the pronounced solvent effect mentioned previously (entries 1-4), there is a clear electronic influence exerted by the substituent on the aromatic ring in controlling the ratio between five- and six-membered ring products being formed. The differences observed are most likely due to the relative electron-donating or -withdrawing properties of the substituent groups (entries 1 and 2 vs. entries 3 and 4). An even more dramatic impact is seen when comparing the product composition obtained in the presence or absence of a basic additive. Performing the rearrangement in the same solvent (DEA), the addition of only 1 to 2% of K_2CO_3 will completely reverse the ratio between chroman and furan and deliver the latter as the predominant product as opposed to the former when no base is included (entries 3 and 5). A similar outcome was seen when using

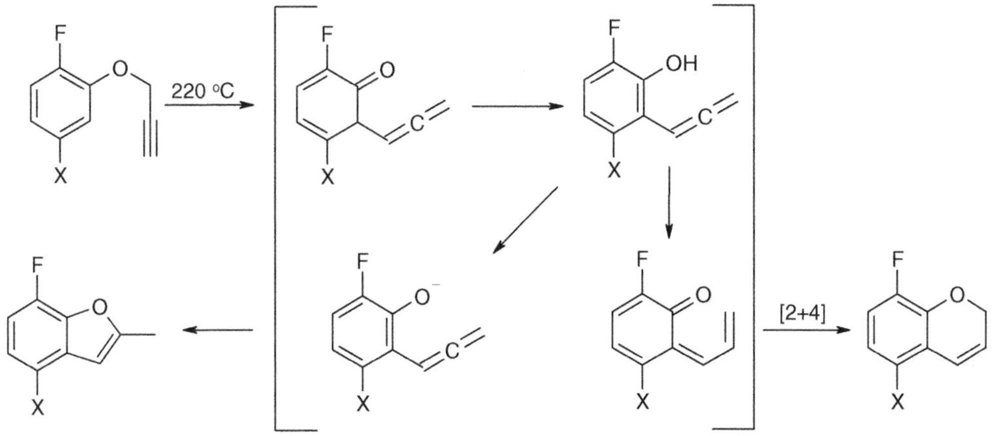

		Diethylaniline		PEG	
Entry #	X	6-ring (%)	5-ring (%)	6-ring (%)	5-ring (%)
1	OMe	100	-	66	33
2	Me	99	1	40	60
3	CO$_2$Me	97	3	-	100
4	CF$_3$	94	6	-	100
5	CO$_2$Me + K$_2$CO$_3$ (1-2%)	0-20	80-100	-	-

SCHEME 7.6 Results from investigation of factors—type of substituent (X) on aromatic ring, presence of base additive (K$_2$CO$_3$), and solvent (DEA or PEG)—influencing the outcome of the Claisen rearrangement recorded as amounts of chroman and furan products formed, respectively.

other bases such as potassium acetate. We quickly realized that the base is an important controlling factor in this reaction, as is the extent to which the enolate intermediate is stabilized.

With much increased understanding of the reaction and its controlling factors, the next challenge to address was the search for a solvent that incorporated the good performance shown by DEA in creating the chroman product, but, at the same time, one that offered a considerably easier workup combined with a less-pronounced toxicity profile (aromatic amines are notoriously toxic). When screening a variety of options, diphenyl ether was quickly identified as a suitable replacement.

Above 210°C, the reaction would result in the formation of an ever-increasing amount of the indanone by-product; see Scheme 7.7. Interestingly, this side reaction turned out to be far from novel, as the mechanism behind forming 2-indanones on flash vacuum pyrolysis of phenyl propargyl ethers had already been thoroughly investigated and unraveled.[18-20]

The conclusion was drawn that there was a maximum temperature that clearly should not be superseded as long as the chroman was the desired product. On the other hand, temperatures that were too low started to increase the reaction time, which would constitute a drawback from a processing point of view as well as a risk of product decomposition. Therefore, it was judged that a set temperature of 195°C would represent an ideal compromise between these counteracting effects. Before large-scale testing of this newly developed procedure for running the reaction, another feature was addressed: the intrinsic instability presented by 2*H*-chromene esters. The elegant solution to this problem was an immediate hydrolysis of the product obtained on ring closure without prior isolation, and this afforded the chroman as the carboxylic acid (NAD-4), which displayed considerably improved stability properties.

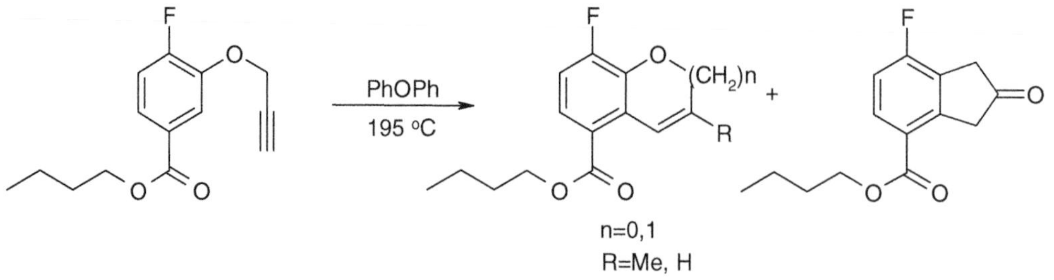

n=0,1
R=Me, H

Temp (°C)	Indanone (%)
195	10
240	35

SCHEME 7.7 Competing reactions in the Claisen rearrangement. The formation of unwanted 2-indanone by-product is strongly temperature dependent, which makes control of the amount possible.

Running the second campaign proved much more successful, and the yield achieved for the desired chroman was in the range of 80 to 82% (>99% product conversion in reaction solution) on an approximately 55-kg scale and operating in 500-L vessel size, with only small amounts of furan being formed. However, another phenomenon was observed—severe corrosion on the inside of the stainless steel reactor used for the thermal rearrangement. It was immediately suspected that hydrogen fluoride (HF) might be released in small amounts during the reaction, and reports in the literature[21] describing that as little as 5 ppm would suffice for eliciting corrosive behavior strengthened this view. After extensive investigations, the formation of HF in the system was confirmed at a level just below 5 ppm, and analysis of the headspace in the reactor definitely showed that the atmosphere was acidic. The best way to prevent this unwanted situation from causing potentially harmful damage to our reactors was to envisage the use of a suitable base, unless it was deemed economically viable to invest in a reactor made of entirely corrosion-resistant material such as titanium or other sophisticated alloy-based alternatives (Hastelloy®). The search was quickly narrowed down to N-methylmorpholine (NMM) that seemed ideally suited for this purpose. Unfortunately, however, testing NMM in different reaction systems revealed that the ability to perform as an effective base was better in other solvents than in the one currently used in the synthesis (diphenyl ether). Therefore, as it was seen as unavoidable to have a presence of a base in the reaction mixture, the solvent had to be changed for the third time, now shifting to xylene. Given the considerably lower boiling point of this solvent (around 140°C), in order to reach the required reaction temperature, the rearrangement had to be performed under an elevated pressure (3 bar). Another downside of relying on NMM as a crucial ingredient in this step was the concomitant formation of small amounts of the furan by-product. On balance, it was, nonetheless, considered appropriate to accept this degree of undesired side reaction just to avoid detrimental corrosion. When scaling this third-generation synthesis to pilot manufacture, the outcome was satisfactory with an isolated yield of 80 to 82% (at a 55-kg scale in 500-L reactor equipment), and only 2 to 3% of furan was generated. Unfortunately, however, the annoying issue with corrosion on the reactor walls (pitting) remained despite running the reaction in the presence of base. The best way to deal with this severe drawback in a production system was, evidently, to operate in an entirely corrosion-proof environment. Thus, conducting field trials with various construction materials in authentic pilot equipment showed that Hastelloy 276 suffered the least corrosive stress. This measure also eliminated both the need for a base to be present and the accompanying yield penalty due to formation of furan.

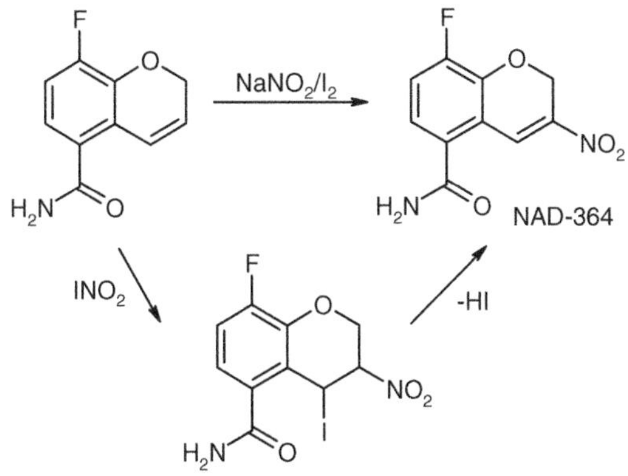

SCHEME 7.8 Nitration of 2-chromenes under stoichiometric conditions using $NaNO_2$ and large excess of elemental iodine.

7.4.2 STEP 7: AN INNOVATIVE CATALYTIC OLEFIN NITRATION

At a position about halfway through the entire synthesis, this step was seen as probably one of the most critical when judging the potential for survival of the current route. If anything, this reflects the respectfulness with which nitrations are viewed from an industrial perspective, both from their potential hazardous reaction conditions and from the risk of generating high-energy-containing products. The method delivered from the medicinal chemistry group was actually built on a literature precedent[22] but suffered from very low yields being achieved (25 to 30% on laboratory scale). Furthermore, the overall conditions—$NaNO_2$ in the presence of excess iodine (a known free radical reaction[22])—were judged not to be suitable for scale-up, at least not beyond initial pilot manufacture. Thus, after the initial formation of nitryliodide (INO_2), a homolysis takes place leading to the attack of an NO_2 species onto the olefin. The resulting intermediate, a 3-nitro-4-iodo adduct, is formed in a completely regiospecific manner that then undergoes a facile elimination of HI to generate the nitro-substituted 3-chromene product (NAD-364); see Scheme 7.8. An apparent feature of this synthetic step is that iodine acts as a reaction promoter, and, on completion of its mission, departs in reduced form as iodide.

Time constraints on the project forced this more laboratory-like procedure to be probed in the pilot plant on a 1000-L scale. Surprisingly, given the far from optimal method that was more or less a translation of the small-scale procedure, the outcome at 45% yield was only 5% lower than the best laboratory results. Due to the massive excess of both I_2 (3 mol eq) and $NaNO_2$ (3 mol eq), copious amounts of inorganic salts were generated with consecutive workup challenges to be dealt with. The low yield was blamed on the instability of the product in the reaction medium. Moreover, the low solubility of the nitro-chromene necessitated the use of large volumes of solvent in order to ensure good stirring of the reaction mixture, thus making the process less efficient and the workup extremely complex. Dedicated and continued efforts to improve on these shortcomings, unfortunately, did not provide better yields or reduce the amount of reagents needed.

At this point, it was painfully realized that this methodology was not sustainable for further use, and the attention was directed toward finding a totally novel approach, with the proviso that the chromene should constitute the starting material and an aminochroman should be the ultimate product. Several alternatives were considered—S_N2 displacement on position 3 of the chroman (gave mostly elimination), aziridination (resulted in allylic amination), epoxide ring opening (wrong regioisomer)—but all had to be discarded as not feasible for further development, with one exception, however; the conjugate addition of nitrogen nucleophiles to the chromene double bond in the presence of $CsCO_3$ and CsF. As promising as this might have appeared, a breakthrough was suddenly achieved in the old I_2-mediated nitration reaction that changed the scenario completely.

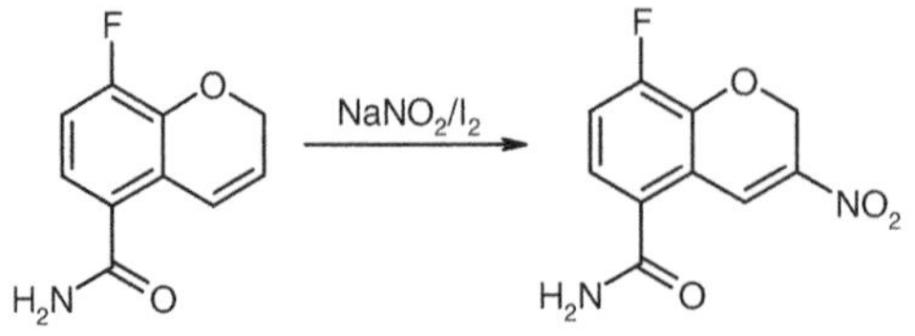

NAD-362 NAD-364

$$I_2 + NO_2^- + \text{NAD-362} \longrightarrow 2\,I^- + H^+ + \text{NAD-364}$$
$$2\,NO_2^- + 2\,I^- + 4\,H^+ \longrightarrow I_2 + 2\,NO + 2\,H_2O$$
$$3\,NO_2^- + 3\,H^+ + \text{NAD-362} \longrightarrow 2\,NO + 2\,H_2O + \text{NAD-364}$$

SCHEME 7.9 A catalytic breakthrough. Balanced reaction formulae show that, at least in theory and partly based on redox potentials of the species involved, the application of I_2 in catalytic amounts is possible as the iodide generated will be oxidized back to iodine.

Thus, as indicated previously, with the iodine→iodide reduction taking place in conjunction to the nitration creating pseudo-catalysis conditions, albeit operating at stoichiometric quantities, it was questioned whether the NO_2 substitution could be transformed into a truly catalytic process with respect to I_2. Listing all components used as ingredients to build up the matrix required for conducting the nitration and, based on this, setting up balanced reaction formulae showed that I_2 has the potential of being present as a catalyst; see Scheme 7.9.

Inasmuch as this finding was in accordance with our expectations, it was more surprising to realize the way in which the catalytic cycle was closed—offering a reoxidation path for iodide back to iodine. As it turned out, the nitrating agent in the form of free nitrite anion furnishes the oxidant, and this capability was confirmed in the current reaction system by comparing the redox potentials of the species involved. Therefore, using excess $NaNO_2$ under acidic conditions should suffice to alter the process into catalytic mode with an incredibly enhanced performance and potential to be competitive on production scale. Testing of this hypothesis, however, resulted in a disappointing failure as no catalytic cycle could be established. Being so close to something really groundbreaking, it was decided to continue the search for other oxidants, as it was believed that insufficient oxidation power of nitrite was the root cause of the reaction failure. This endeavor was rewarded when it was shown that a mixture of peracetic acid and hydrogen peroxide was successful where nitrate had failed; iodide underwent smooth reoxidation to iodine. In fact, the best conditions were realized with a commercial reagent (Oxystrong™) that in addition to the peracid (15%) and H_2O_2 (24%) also contains acetic acid (18%) and water (43%). Optimizing the process under these circumstances reduced the I_2 loading to merely 20 mol%, and furthermore, an important feature of obtaining the best outcome was the slow charging of the reoxidizing agent.[23] Operating in this way gave a consistent and reproducible yield on large scale of 70 to 75%; see Scheme 7.10.

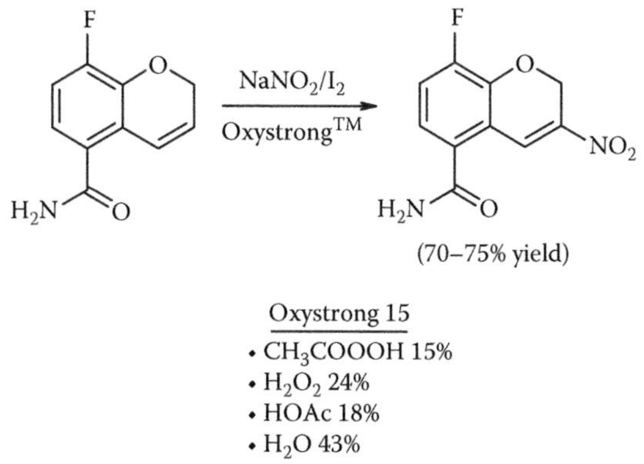

(70–75% yield)

Oxystrong 15
- CH$_3$COOOH 15%
- H$_2$O$_2$ 24%
- HOAc 18%
- H$_2$O 43%

SCHEME 7.10 An elegant nitration. The final process turned out to be an efficient way of conducting the reaction in high yield.

7.4.3 COMBINING A NITRO GROUP REDUCTION WITH A RACEMATE RESOLUTION IN ONE STEP

With an elegant and effective nitration process in hand, the next task was to develop a reliable procedure for conducting consecutive reductions of the olefinic bond and the nitro functionality. Performing these in a one-pot process was, for obvious reasons, very tempting, but all attempts in this regard led to failure. It was believed that the reason for this failure is the fact that the nitro group was reduced prior to the olefin, and this created an unstable enamine intermediate, which most likely hydrolyzed to the ketone. Therefore, the initial method identified to reduce the C–C double bond using NaBH$_4$ was retained without modification, and that offered a smooth transformation to the nitro-chroman. We then turned our attention to the nitro reduction as we needed to address two issues. First, we wanted to avoid the use of RaNi (>1 mole eq. used) as a reducing agent, which was the way medicinal chemistry conducted the reaction, and second we needed to address the high dilution issue associated with the low solubility of the nitro-chroman intermediate. It was quickly found that changing from a stoichiometric to a catalytic process was possible. Initially, the intent was to use a heterogeneous platinum catalyst (Pt/C) in a solvent mixture composed of EtOAc/EtOH in combination with small amounts of a suitable acid—HCl at first but later switching to *p*-TSA (*p*-toluenesulfonic acid) due to better performance—to prevent an unfavorable formation of by-products. In spite of this precautionary measure, a lot of problems were experienced with impurities, starting with the *N*-ethyl-substituted aminochroman analogue as the most prevalent; see Scheme 7.11.

A simple, yet efficient way to counteract this unwanted contamination was to replace EtOH, the progenitor of the ethyl substituent, with *iso*-propanol with its considerably lower reactivity to act as an alkylating agent. Another problem that cropped up during the hydrogenation was the appearance of the 8-*H* analogue of the aminochroman (Scheme 7.11). Recall that this is an analogous impurity to that seen in the fluorination step in the first-generation chemistry that required the tedious and costly chromatographic purification (Scheme 7.3). Its appearance here, as a consequence of the sensitivity of the C–F bond toward reducing conditions, in relatively close vicinity of the final product definitely could not be ignored. Instead, a targeted search for alternative procedures that would, preferably, eliminate the risk of it being generated was immediately launched. This proved, however, to be a tough task, and screening different solvents, acids, and catalysts did not provide a solution to this problem even if some trends were visible. As it had been observed that Pd/C offered the best selectivity of the catalysts investigated, the drastic approach was taken to perform the reaction in a solvent, toluene, in which the nitro-chroman is virtually insoluble. By adding a small quantity of *N,N*-dimethylformamide (DMF), the solubility properties were improved

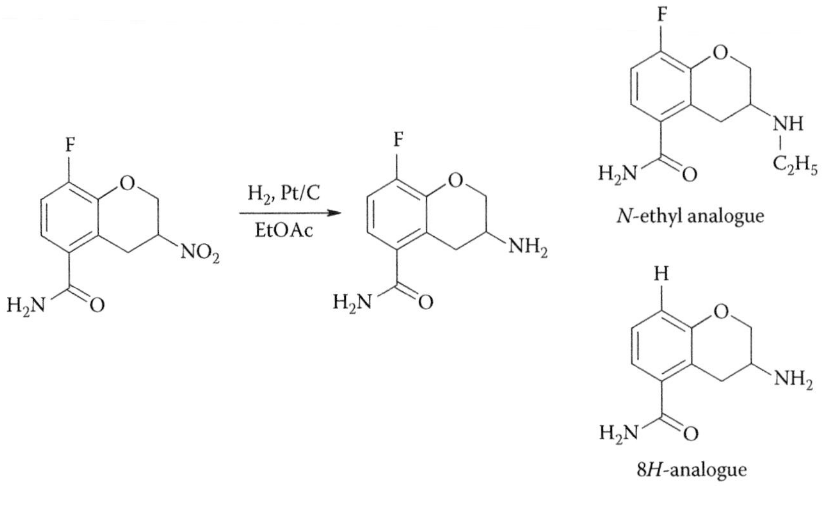

SCHEME 7.11 Appearance of two unexpected by-products at a late stage of the process required targeted countermeasures.

to a level where the reaction was still fast and the resulting product displayed an excellent impurity profile. Contrary to our optimistic expectations, the scale-up of this promising laboratory method turned into a veritable catastrophe in the pilot plant. Unsatisfactory stirring combined with undissolved starting material and precipitated product depositing on the catalyst surface rendered the conversion at a very slow rate, offering a yield of less than 40% compared to twice as much obtained in laboratory experiments. This scenario for a viable process was bound to go wrong as the stability lacking properties of the nitro-chroman at the reaction temperature (70°C) required the reduction to proceed as fast as possible.

The idea was conceived that, based on all these negative results, a more effective solvent system had to be devised where both starting material and product had to be soluble. Only under these circumstances would it be possible to avoid catalyst coating. As is surprisingly often the case, the inclusion of water in systems where organic molecules are undergoing reactions can have a miraculous impact, especially when the compounds involved contain both polar and nonpolar domains. And this experience proved to be true again in this case, as a mixture of THF and H_2O was found to provide the best conditions for this transformation with good solubilities being achieved for both components (nitro- and aminochroman, respectively). A direct benefit emanating from this was the fact that the reaction could be more concentrated (lowering the solvent volume) and, furthermore, the operating temperature could be reduced significantly (from 70 to 50°C). The advantageous influence exerted by p-TSA, that repeatedly had been manifested, was found to be valid under the new conditions also and, hence, was kept intact. Putting all these parameters in place gave a very clean conversion to the racemic aminochroman in high yield (quantitative) in which it was almost impossible to find even traces of the feared 8-*H* contaminant. An extra bonus was delivered when investigating the possibility to perform the resolution *in situ* during the actual workup. Thus, driven by the desire to avoid unnecessary isolations, in part at least due to the complex solubility properties alluded to previously, it was found that adding natural tartaric acid led to precipitation of the wanted (*R*)-enantiomer as the diastereomeric L-tartrate salt. This telescoping of two steps—nitro reduction and resolution—including a base treatment (NEt_3) to liberate the free amine, proved to be a highly efficient and robust method for obtaining the product in very good yield (90% in theory; 45% from racemate) and of excellent quality; see Scheme 7.12.

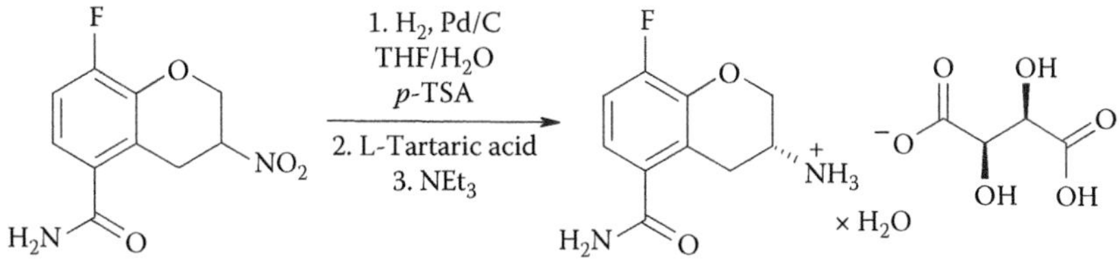

SCHEME 7.12 Successful telescoping of two steps—a nitro group reduction and a diastereomer resolution—into one.

7.4.4 THE END GAME: ADDING THE TWO CYCLOBUTYL GROUPS

Attempts to integrate the formation of the free aminochroman from its tartrate salt with the consecutive cyclobutyl additions in one pot failed, as the process was disturbed due to crystallization of tartrate salts. Instead, a procedure was developed using calcium hydroxide, as base, which, when operating in H_2O, generated a precipitate of calcium tartrate that could be removed, leaving the aminochroman in the aqueous phase. Normally, textbooks claim that water is far from being the preferred solvent in effecting an efficient conversion of ketones to imines as its presence drives the equilibrium in the opposite direction. In this case, however, the formation of the alkylimino adduct was so favored that obtaining the mono-cyclobutylamino intermediate after concomitant hydride reduction proceeded smoothly. Unfortunately, this straightforward process design was not applicable to the succeeding second reductive amination step. Luckily, the mono-*N*-alkyl moiety showed considerable improved solubility properties in organic solvents, which allowed the reaction to be conducted in another system after swapping H_2O with methanol; see Scheme 7.13 for the first method.

In spite of lending itself to successful scale-up in the pilot plant, this procedure presented several drawbacks. First, the massive excess of cyclobutanone (5.5 eq compared to the theoretical amount of 2.0 eq needed), an overly expensive building block in bulk quantities, generated a very

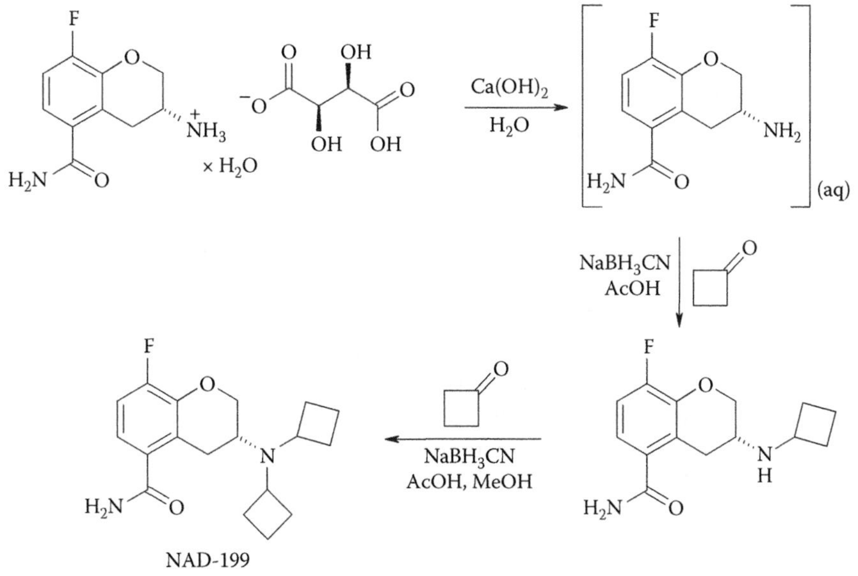

NAD-199

SCHEME 7.13 The first-generation dialkylation procedure: an inefficient method requiring excess cyclobutanone and extremely long reaction time (7 days).

negative impact on the overall price of the final product. Furthermore, the reducing agent, $NaBH_3CN$, was also applied in large excess (5.5 eq), which, due to its relatively high cost, adds to the unfavorable economics. And with the intrinsic risk that this reagent releases toxic HCN, its use on a large scale was not very desirable. Finally, the need to keep a stable pH in this highly pH-sensitive reaction called for the consumption of large volumes of acetic acid. Following this methodology, the conversion at the end was still below 90%, despite using vast amounts of costly reactants and reagents and operating in keeping with a tedious and laborious procedure that required their introduction portion-wise over an extended period of time; otherwise, running a one-pot process would drop the yield to just around 50%.

Investigating the reasons for the sluggish second alkylation revealed that the bulkiness of the system was a major factor. Studies performed during the lead optimization phase established that the four-membered alicyclic ring was the largest that could be introduced to successfully create the *bis*-cycloalkyl motif. Another factor to take into account was the risk for being exposed to a competing reaction where cyclobutanone would undergo reduction to the corresponding alcohol. This behavior actually dominated when running under catalytic conditions (H_2, Pd/C) in the sense that after a smooth introduction of the first cyclobutyl group, the second equivalent could not be attached, and instead, the preferred reaction pathway led to the consumption of cyclobutanone merely forming various side products. A further scrutiny of the factors governing this sensitive reaction balancing between two main transformations—the wanted reductive dialkylation versus the unwanted cyclobutanone reduction—was obviously needed. Literature data[24] suggested that reduction of ketones (and aldehydes) in the presence of $NaBH_3CN$ should be negligible at pH >5. However, in our case, this statement did not seem to apply, as we were facing a major problem with reduction of our precious reactant even when operating at pH 5 to 6. This issue was addressed in a comparative study where the outcome from the separate exposure of cyclobutanone and the mono-cyclobutylamino intermediate, respectively, as a function of pH under the current reaction conditions was monitored; see Table 7.1.

The beneficial effect of higher pH on reducing the formation of cyclobutanol is clearly evident; however, this desired effect is counteracted by a gradual but significant decrease in the rate of adding the second alkyl substituent. Fortunately, the relative change in the propensity for keto-group reduction is greatest between pH 5 and 6, where, at the same time, the loss in reaction rate for the desired cyclobutyl addition is modest. As a consequence, performing the transformation at pH 6 would seem to be the best compromise possible under the given circumstances.

Moving on in the search for an improved second version of the reductive dialkylation, it was decided to retain the conditions as initially established for the first two steps (formation of the free amine base and the synthesis of the mono-cyclobutyl intermediate). Focusing instead on the final part—the addition of the second cyclobutyl equivalent—the pH was rigorously controlled to stay within a narrow window between 5.7 and 6.0 by continuously adding HOAc via a pump connected

TABLE 7.1
Effect of pH on Achieving Undesired Reaction (Reduction of Cyclobutanone) versus Desired Transformation (Reductive Alkylation to Afford NAD-199) (See Scheme 7.13)

pH	Cyclobutanol (%)	NAD-199 (%)
5.0	62	12
6.0	28	10
7.0	21	5

to the pH meter. Furthermore, in order to minimize cyclobutanone reduction, the NaBH$_3$CN reagent was introduced slowly over an extended period of time (10 h in total) using a pumping device. Putting these improvements in place resulted in a tremendous shortening in reaction time, from 7 days to 12 h. When tested on 600-L scale, these modifications also impacted positively on the overall yield of this step (increase from 60 to 70%) and lowered the amounts of cyclobutanone and NaBH$_3$CN consumption (from 5.5 to 4.5 and 4.0 eq, respectively).

The outcome achieved after these changes in the method was seen as a big step forward, albeit not enough to satisfy the needs of a commercially acceptable process. Major drawbacks that required continued attention were the procedure applied to liberate the free amine from its tartrate salt, which involved extractions and a solvent swap, the handling of an expensive and toxic reducing agent (NaBH$_3$CN), and the demand for explosion-proof special equipment for pH control. As a first measure, an extensive investigation was launched aiming at identifying a suitable replacement for NaBH$_3$CN. It was quickly found that most alternatives considered—for example, NaBH(OAc)$_3$, NaBH(OCOCF$_3$)$_3$, borane morpholine, and borane tetrabutylamine—were not suitable for the intended purpose. They either did not reduce the initially formed iminium ion (formed after condensing aminochroman with cyclobutanone) or caused a reduction of the cyclobutanone reactant. Identified in the search was only one agent, pyridine borane (C$_5$H$_5$N·BH$_3$), that displayed a similar reactivity to NaBH$_3$CN. Fortuitously, this chemical is available in bulk (as a concentrated solution in pyridine) at a cost of only about 20% (on a weight basis) of that of NaBH$_3$CN, and its handling is considerably easier as the dissolution step can be omitted. A downside that posed some concern was related to safety, health, and environment (SHE) aspects and the fact that it had been classified as toxic, albeit non-HCN evolving as with the previous reagent. The major disadvantage, however, were connected with the explicit exothermic properties, which, when heating pure pyridine borane >45°C, was exposed through a very rapid decomposition (polymerization) concomitant with H$_2$ evolution. This vigorous and hazardous event is further worsened by the fact that it takes place in an autocatalytic mode. Due to the way in which the experimental protocol was designed—a slow and controlled addition of the reducing agent was prescribed—this obvious risk could be mitigated in the sense that no exothermic reaction occurred at as high a temperature as 65°C. With this margin, it was considered that the process using pyridine borane would meet the safety criteria but at the same time, it was acknowledged that large quantities of this material should not be kept in stock.

The endeavors thus described were crowned by incorporating all improvements exploited into a third process, which demonstrated a reliable performance on scale-up; see Scheme 7.14.

Starting with the neutralization of the tartrate salt, the use of Ca(OH)$_2$ in H$_2$O was replaced by a much easier methodology utilizing K$_2$CO$_3$ in MeOH. This change allowed the free aminochroman base to be used directly as a methanol solution (after filtering off solid potassium tartrate), omitting both an extraction and a solvent swap. The major advantage, however, was offered by the opportunity to conduct the consecutive alkylations in one pot. A further upside with this procedure, in addition to shortening the reaction time from 12 to 8 h, was that the amount of cyclobutanone required could be slightly reduced to 4 eq. (from 4.5 eq.), leaving the process with a ratio of 2:1 for each alkylation reaction. Implementation of these changes resulted in an appreciable yield improvement up to 80%, which was largely attributed to performing the sequence in a one-pot scenario. It also turned out that switching to pyridine borane as reducing agent enabled the removal of the strict pH control that was required in earlier versions of this reaction. After adding the required amount of acid (HOAc) in one aliquot at the beginning of the reaction, the pyridinium acetate formed *in situ* proved to buffer the system at exactly the right pH value throughout the entire transformation—a truly beautiful spin-off from the development and optimization of a step that at first sight was regarded as a major challenge.

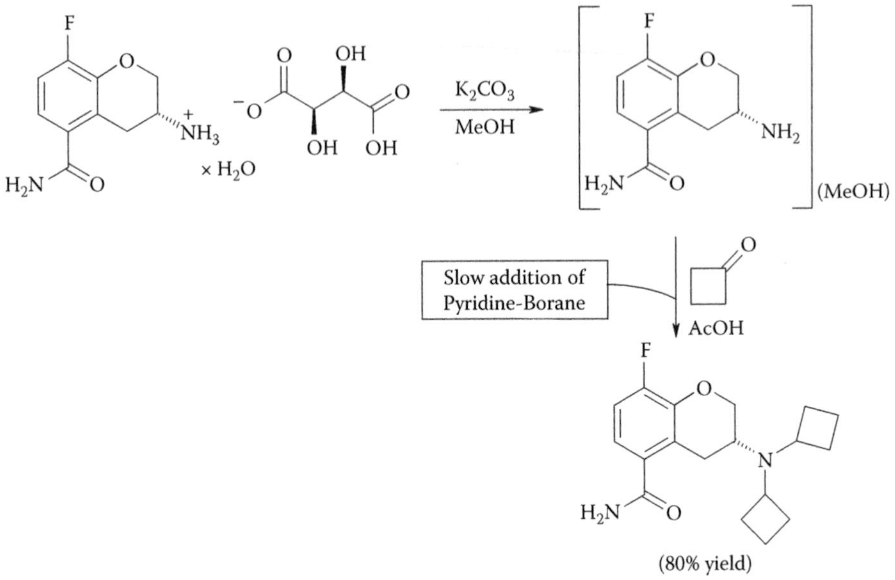

SCHEME 7.14 The final procedure. After development and refinement, the dialkylation could be performed in one pot using a considerably smaller excess of cyclobutanone and a reaction time of only 8 h.

7.5 REACHING THE TARGET: A COMMERCIALLY VIABLE PRODUCTION PROCESS

After several years of hard development work that required input of considerable resources in man-hours and economic investment as well as a high degree of innovative thinking, the initially rather poor and barely scalable synthesis had been turned into a process entirely viable for large-scale manufacture. Although the number of steps is still fairly high, considering the synthetic challenge of constructing the core heterocyclic system, the result is acceptable. Additionally, all steps operated at high to excellent yields (excluding, for obvious reasons, the resolution) and proceeded in a robust and predictable manner producing the various intermediates and the final product in excellent quality. A flowchart of the entire process including the normal batch sizes in each step is depicted in Scheme 7.15.

Thus, the output from one manufacturing campaign offered about 140 kg of the active substance robalzotan at a steady-state scenario, which equates to an overall yield of 13 to 14%. However, considering the situation from a strictly commercial point of view with requirements in the multi-tonne range per annum, then some major drawbacks appear. The current method requires three discrete steps for the conversion of the chromene ester (NAD-3) to the corresponding amide (NAD-362), and this is definitely an area for improvement. Initial laboratory trials showed that it was entirely feasible to conduct this transformation directly, but changes in project status did not allow these ideas to be pursued further.

A more significant weakness was the late stage resolution that, by default, caused half of the valuable content in the process stream to be discarded. At the time, technological limitations (mostly in the form of unsuitable catalysts and ligands) did not allow a stereoselective reduction of the 3-nitrochromene to be successfully conducted. Hence, the reactions had to be performed in a racemic mode up to the point where the 3-aminochroman was isolated as an L-tartrate salt. This approach left half of the material processed through to this late stage intermediate in the mother liquor, albeit represented by the unwanted (S)-enantiomer. Maybe this is not a big deal when considering small-scale laboratory experiments, but it is of major concern when trying to address

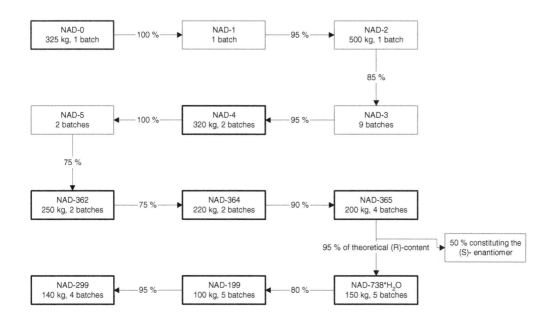

SCHEME 7.15 The process flowchart for manufacture of robalzotan (NAD-299); the structures representing the different compound codes are specified in Scheme 7.4. Boxes in bold indicate isolated stages. Batch sizes for each step are given inside the boxes together with the number of discrete batches optimized to achieve a good materials throughput in the process. Yields are as depicted and represent the status at closure of the project.

the loss of precious compounds in various effluent streams. And in this case, there were strong drivers at hand—mainly a call for improving the overall process yield and reducing the amount of waste generated—demanding that options would be investigated. Thus, a variety of strategies were explored and tested that eventually would lead either directly to the (R)-configuration via an inversion process or, alternatively, to a racemate that subsequently could be resolved using the already existing protocol or something similar. Unfortunately, several of the methods probed showed no hints of success, for example, displacing the amino functionality by an azide group in an S_N2-mode,[25] converting the amine to an imine (reaction with an aldehyde) and then hoping for an isomerization that after reduction would re-create the racemic aminochroman, or trying to oxidize the amino group to the nitro analogue which only resulted in complex mixtures of products. At the end, however, one chemical approach proved to be successful—the exhaustive alkylation of the aminochroman substrate leading to a quarternary ammonium species that could be subjected to a Hofmann elimination leading to a 3-chromene moiety that, fortuitously, was identical to a process-specific intermediate. This elegant loop was found to work very well and allowed the re-entry of salvaged material stemming from the previously undesired stereoisomer into the normal manufacturing stream. Key features of this recycle were the isolation of the (S)-enantiomer in the form of its solid D-tartrate from the mother liquor as the L-tartaric acid salt does not crystallize in this system followed by neutralization to liberate the free amine. Following is the attachment of three methyl groups effected by dimethyl sulfate to render an ammonium intermediate that is then subjected to treatment with NaOH at pH > 12. A smooth elimination (of Me_3N) occurs leading to the aforementioned chromene at a yield which on scale was >90% and with a quality that even exceeded what is normally observed in our process; see Scheme 7.16. Integrating this measure in the production process (see Scheme 7.15) was shown to increase the total yield to 19% after applying this recycling strategy five to six times.

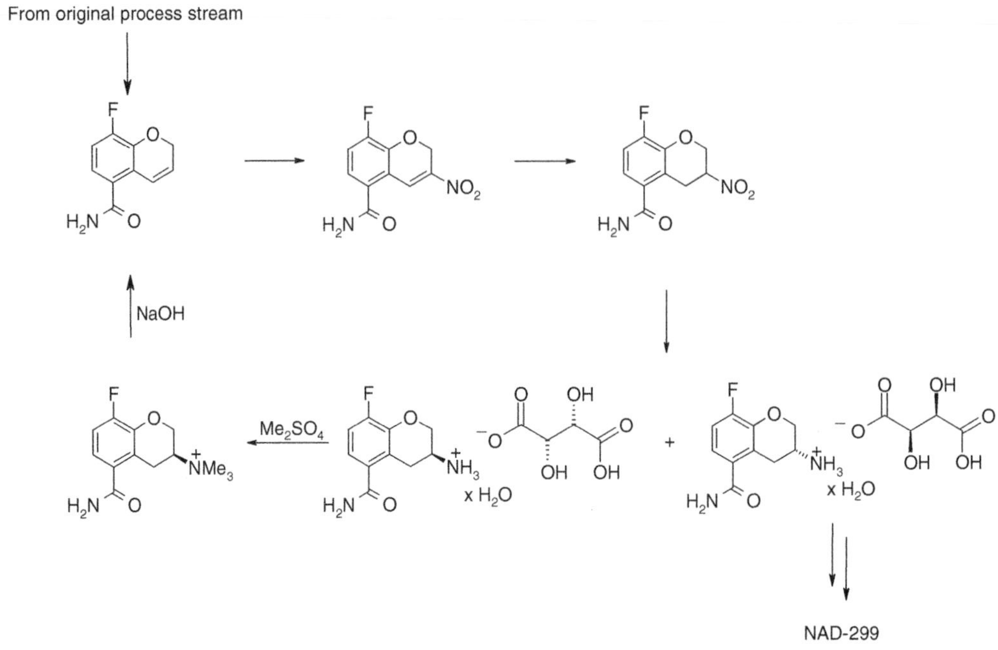

From original process stream

NAD-299

SCHEME 7.16 Closing the loop. Improving the overall economy by recirculating the wrong enantiomer via conversion to a 3-chromene intermediate specific to the synthetic route allowing its introduction into the normal process stream.

7.6 SMALL MOLECULE—BIG TROUBLE

It was already mentioned that cyclobutanone was a significant contributor to the cost of the finished API (see Section 7.2). Superficially seen, this might seem a bit strange, but when looking closer at the synthetic procedure, it becomes very clear why this is the case. The literature method[26] most commonly utilized to make this material uses an acid-catalyzed ring expansion of cyclopropyl carbinol at elevated temperature (100°C) forming cyclobutanol followed by a chromium(VI) oxidation (using excess CrO_3). Fractionated distillation affords the desired product in overall yields in the range of 30 to 50% depending on the exact procedure. This is not an environmentally friendly synthesis, and furthermore, it is tainted by obvious weaknesses from a scale-up perspective. Nonetheless, as other approaches[27] to prepare this molecule did not offer any appreciable advantages (nasty chemistry, complex workup, similar yields), this was the route on which our contractor decided to focus.

In addition to developing and trimming this procedure to enable a sustainable production on commercial scale, a few items presented some challenge to our project and, consequently, had to be dealt with. First, it was observed that handling, shipping, and storage of cyclobutanone of a high-purity grade (>99%) generated a white solid precipitate. This phenomenon was ascribed to a thermally induced radical process (homolytic cleavage of the $CO-CH_2$ bond), an entirely feasible mechanism given the high energy due to ring strain in the four-membered cyclic system. Structural elucidation of this impurity was not conclusive, but on analyzing samples it was apparent that one or even two carbonyl moieties were present. Based on this and other findings, the assumption was made that we were dealing with a polyketone or possibly a polyester. An even more important question based on the intended use of cyclobutanone as a late-stage building block was how the unwanted by-product formation could be prevented. As it turned out, the addition of a radical scavenger in ppm amounts—for example, BHT (2,6-di-*tert*-butyl-4-methyphenol)—effectively inhibited the polymeric impurity from being formed.

Second, when disseminating the spectrum of low molecular weight impurities in normal batches of cyclobutanone a few striking features became apparent—namely, that there are a variety of carbonyl-containing products present capable of competing in the reductive amination of the aminochroman substrate at an equal rate or faster than cyclobutanone. Perhaps the most obvious of these is cyclopropylcarboxaldehyde—formed in the second step of the process oxidizing remaining amounts of the starting material cyclopropyl carbinol—and typical levels are a few tenth of a percent. In order to avoid these by-products in the API, which could be potentially difficult (if not impossible) to separate, an extremely tight specification was set, requiring the content of any impurity to not exceed 0.1 area% (except for cyclobutanol, for which considerably higher levels could be tolerated). By formulating the quality criterion in this fashion, there was virtually no risk that an unknown carbonyl compound would be able to "sneak" into the final chemical reaction and produce unacceptable contamination of robalzotan analogues. Because it was known that the distillation step required to purify the cyclobutanone typically produced a product containing 0.3 to 0.6% of the aforementioned "detrimental" aldehyde, a purification method was urgently needed. It was found that reacting the "crude" product with a suitable amine (notably diethanolamine) at an elevated temperature for several hours reduced the content to about one tenth the original amount and this was far below the set specification.

Additionally, the manufacturing process of cyclobutanone produced a series of aliphatic ketones. Most of these were easily eliminated in the final purification (column distillation) except for 2-pentanone, whose boiling point was too close to the desired cyclobutanone. Therefore, reaching the very stretched specification requirement ($\leq 0.1\%$) was a technical challenge that forced several of the fractions to be redistilled. Not surprisingly, this extra labor had a direct impact on the cost of this raw material; hence, there was an economic driver to try and loosen the demand on how much residual 2-pentanone could be accepted without compromising the quality of the API. Exploring the reactivity of the 2-pentanone in the reductive aminations, we were pleasantly surprised to observe that it was less reactive than cyclobutanone. We attributed this rate difference to steric factors, or in other words, the bulkiness of 2-pentanone is much higher than that of cyclobutanone. With this information in hand, the specification of the 2-pentanone was eased to 0.4% which was easily met by our contractor. But most importantly, the bulk price of cyclobutanone came down by as much as 10%. This case, if anything, shows the importance of working closely with your suppliers.

7.7 THE LAST STUMBLING BLOCK: DEALING WITH A LATE-STAGE SAFETY ISSUE

In the early part of the robalzotan process, when transforming the phenol NAD-1 to the corresponding propargyl ether (NAD-2), the reaction was conducted using the propargyl chloride reagent as a toluene solution (70% w/w). Unfortunately, due to the highly exothermic properties of this compound together with a pronounced volatility, a number of incidents with this compound were reported by our vendor. This resulted in a decision taken on their behalf that this material would no longer be made available on the market. Instead, customers would only be offered a much less concentrated product at a content of only 30% w/w. The repercussions this had on our process were considerable; this low level of concentration and the fact that the reaction had to be performed in a polar solvent (acetone) reduced the rate of this transformation significantly. In actual fact, it was estimated that conducting this step under the modified conditions would not be possible for practical purposes, and hence, an alternative had to be rapidly sought.

The basic idea was to change the method as little as possible at this late stage of the project in order to avoid introducing too many unknowns that could have an impact further downstream. Shifting to other activated propargyl alcohols as new reactants was the obvious approach; thus, the corresponding tosylate, mesylate, and bezenesulfonate were tested. None was intrinsically safe—they all displayed exothermic properties—but because the volatilities of the activated esters were considerably lower, their handling properties became much safer. It was found that using the tosylated moiety was advantageous as the reaction went to completion in just $4^1/_2$ h (compared to 24 h when operating in the old fashion with propargyl chloride). However, initial formation of thick slurry and a problematic workup pointed out some weaknesses, and therefore, a new process had to be designed. Moving to a "classical" two-phase scenario using H_2O and xylene as the solvents in the presence of a phase-transfer catalyst (tetrabutylammonium bromide) and a base (K_2CO_3) proved to be the right measure. Under these circumstances and after some further optimization, the resulting reaction performed very well (at a set temperature of 65°C), rendering an almost quantitative yield in combination with pleasingly small amounts of impurities. An elegant spin-off opened up by the change in synthetic method was the realization that in switching to xylene in this step, coupled with fact that the succeeding reaction (the Claisen rearrangement; see Section 7.4.1) also utilized this solvent, there was an opportunity to streamline the process and telescope the first three steps. Changing from toluene in the esterification reaction (NAD-0→NAD-1) to xylene was quite facile, and this ultimately eliminated the need for conducting tedious solvent swaps, allowing three steps to be nicely telescoped with each other—a key feature of good processes.

7.8 LEARNING AND CONCLUSIONS

The main intent of this account is to provide insight into the way that a viable chemical manufacturing method for a complex drug substance has been developed. It is hoped that it will create a feeling for the kinds of problems one might be facing when starting with a medicinal chemistry route and the high degree of creativity and innovative thinking that will be required to solve them. Furthermore, this case study should convey a sense that due to the dynamic nature of this business, there is always a chance for unforeseen late-stage technical issues or other challenges and what kind of countermeasures these could provoke. A full-fledged process is the sum of myriad chemical and technological components, and it describes the way these are linked together with the aim to allow production of a given product featuring a set of predetermined attributes.

In this particular instance, the general view was that building on previous experience from an analogous compound would be relatively easy. This assumption proved to be completely false, and the original synthetic route had to be abandoned in favor of a new and totally redesigned process. To make each of the 13 linear steps work in an optimal fashion meant solving a large number of diverse problems, and the input of resources was substantial. All of these efforts were rewarded with success, as an economically viable, robust, chemically and technically feasible process capable of delivering robalzotan in bulk quantities—in 13 to 14% or 19% yield, respectively, depending on whether the unwanted enantiomer was incorporated or not—resulted.

ACKNOWLEDGMENTS

Big and complicated projects will always require a wholehearted engagement from people willing to contribute with their knowledge and expertise. The robalzotan case is no exception in this respect. As the work described above has enjoyed a high degree of external collaboration with colleagues in Zambon Group S.P.A. in Italy—dispersed over the two sites in Milan and Vicenza—we would like to extend our deep gratitude and appreciation to all their staff members who have been involved, notably, to Drs. Livius Cotarca, Marco Villa, Vincenzo Cannata, Francesco Ponzini, and Aldo Belli. Likewise, a large number of coworkers at Astra and later AstraZeneca have been engaged along the timeline described by this project starting in early 1995 and not ending until a decade later. The list of internal colleagues from various disciplines within the PR&D organization in Södertälje is far too long to be included in its entirety. However, there are a few names that warrant being specifically highlighted—Marika Lindhagen, Tibor Musil, Anders Nilsson, Lars Lilljequist, Göran

Fredriksson, Marie Tysk-Rönnqvist, Nicklas Westerholm, Ann-Sofi Kindstedt Danielsson, Dr. Tesfai Sebhatu, and Dr. Jörgen Alvhäll—due to their outstanding support and skills to make robalzotan a success story from a PR&D perspective.

REFERENCES

1. Al Neirabeyeh, M. et al., Methoxy and hydroxy derivatives of 3,4-dihydro-3-(di-*n*-propylamino)-2*H*-1-benzopyrans: new synthesis and dopaminergic activity, *Eur. J. Med. Chem.*, 26, 497, 1991.
2. Liu, Y. et al., Derivatives of 2-(dipropylamino)tetralin: effect of the C-8 substituent and the interaction with 5-HT$_{1A}$ receptors, *J. Med. Chem.*, 36, 4221, 1993.
3. Trost, B.M. et al., Synthesis of chiral chromans by the Pd-catalyzed asymmetric allylic alkylation (AAA): scope, mechanism, and applications, *J. Am. Chem. Soc.*, 126, 11966, 2004.
4. Thorberg, S.-O. et al., Aminochromans: potent agonists at central dopamine and serotonin receptors, *Acta Pharm. Swed.*, 24, 169, 1987.
5. Hammarström, E. et al., Synthesis of novel 5-substituted 3-amino-3,4-dihydro-2*H*-1-benzopyran derivatives and their interactions with the 5-HT$_{1A}$ receptor, *J. Med. Chem.*, 43, 2837, 2000.
6. Sorbera, L.A., Leeson, P., and Castañer, J., Robalzotan tartrate hydrate, *Drugs Fut.*, 24, 740, 1999.
7. Federsel, H.-J., Development of a process for a chiral aminochroman antidepressant: a case story, *Org. Process Res. Dev.*, 4, 362, 2000.
8. Federsel, H.-J., Logistics of process R&D: transforming laboratory methods to manufacturing scale, *Nature Rev. Drug Discov.*, 2, 654, 2003.
9. Federsel, H.-J., Building bridges from process R&D: from a customer-supplier relationship to full partnership, *Pharm. Sci. Technol. Today*, 3, 265, 2000.
10. Federsel, H.-J., The role of process R&D in drug discovery: evolution of an interface, *Chim. Oggi/Chem. Today*, 22(3/4), 9, 2004.
11. Potoski, J., Timely synthetic support for medicinal chemists, *Drug Discov. Today*, 10, 115, 2005.
12. Federsel, H.-J., The integration of process R&D in drug discovery: challenges and opportunities, *Comb. Chem. High Throughput Screen.*, 9, 79, 2006.
13. Differding, E. and Ofner, H., *N*-Fluorobenzenesulfonimide: a practical reagent for electrophilic fluorinations, *Synlett*, 187, 1991.
14. Snieckus, V. et al., Directed ortho metalation-mediated F$^+$ introduction. Regiospecific synthesis of fluorinated aromatics, *Tetrahedron Lett.*, 35, 3465, 1994.
15. Briner, P.H. et al., Practical synthesis of 2-amino-5-fluorothiazole hydrochloride, *Org. Process Res. Dev.*, 10, 346, 2006.
16. Zsindely, J. and Schmid, H., Sigmatropische Umlagerungen von Aryl-propargyläthern: Synthese von 1,5-dimethyl-6-methylen-tricyclo[3,2,1,0^{2,7}]-oct-3-en-8-on-Derivaten, *Helv. Chim. Acta*, 51, 1510, 1968.
17. Koch-Pommeranz, U., Hansen, H.-J., and Schmid, H., Die durch Silberionen katalysierte Umlagerung von Propargyl-phenyläther, *Helv. Chim. Acta*, 56, 2981, 1973.
18. Hansen, H.-J. and Schmid, H., Aromatic sigmatropic rearrangements, *Chem. Brit.*, 5, 111, 1969.
19. Trahanovsky, W.S. and Mullen, P.W., Formation of 2-indanone and benzocyclobutene from the pyrolysis of phenyl propargyl ether, *J. Am. Chem. Soc.*, 94, 5911, 1972.
20. Al-Sader, B.H. and Al-Fekri, D.M., On the mechanism of flash vacuum pyrolysis of phenyl propargyl ether. Intramolecular deuterium kinetic isotope effect on Claisen rearrangement, *J. Org. Chem.*, 43, 3626, 1978.
21. Rowe, D., Tantalising materials, *Chem. Engineer*, June 24, 19, 1999.
22. Hassner, A., Kropp, J.E., and Kent, G.J., Addition of nitryl iodide to olefins, *J. Org. Chem.*, 34, 2628, 1969.
23. Paiocchi, M. et al. PCT WO 02-00575, 2002; [to Zambon group, Italy].
24. Lane, C.F., Sodium cyanoborohydride: a highly selective reducing agent for organic functional groups, *Synthesis*, 135, 1975.
25. Johansen, C. and Fiksdahl, A., Inversion of chiral α-methylbenzylamine, *Chirality*, 6, 161, 1994.
26. Krumpolc, M. and Rocek, J., Cyclobutanone, *Org. Synth. Coll. Vol.VII*, 114, 1990.
27. *Org. Synth. Coll. Vol. VI*, 316, 1988 provides descriptions of three different approaches.

8 Chiral Amine Synthesis—Strategies, Examples, and Limitations

Thomas C. Nugent[1]

CONTENTS

8.1 INTRODUCTION

Extensive reviews on the general subject of chiral amine synthesis are available,[2] but none overwhelmingly focus on the industrial feasibility of synthesizing unfunctionalized chiral amines.[3] In this light, a focused overview pertaining to the synthesis of structurally diverse aliphatic and aromatic α-chiral primary amines is presented.[4]

Strategies allowing facile access to a wide variety of α-chiral primary amines, from commodity chemicals, are central to the material presented here. The unspoken goal of this aspiration is a one-step technology capable of α-chiral amine synthesis from an inexpensive starting material which itself must be represented by a structurally diverse number of members. While we are not yet there, some of the methods developed in the last 5 to 10 years are on the cusp of this ambition, making it technically feasible to synthesize these compounds with high enantioselectivity, but often without industrial applicability. The intent here is to show methods that will help define the next trends in process research. Examples are provided and help define the challenges that remain, and hopefully inspire the development of the next generation of concepts and solutions. Of particular note, aliphatic α-chiral primary amines are discussed because of a historical lack of success regarding their efficient synthesis.

8.2 CHIRAL AMINES AS ELEMENTS OF DIVERSITY FOR DRUG DEVELOPMENT

Living systems contain enormous molecular diversity and in many instances demand similar diversity in structure from external medicinal agents to modify their processes. Regarding the identification of new chemical entities for this purpose, the types of structural diversification available would appear endless, but most of the simple templates and diversification elements have already been investigated. In this regard access to α-chiral amines remains limited, yet they are arguably powerful leverage points for defining future small molecule therapies because they contain a high density of structural information and the inherent capability for hydrogen bonding. For these reasons methods allowing facile access to these advanced building blocks, in high enantiomeric excess, are desirable for maintaining successful drug discovery programs.

8.3 DEFINING THE FOCUS—THE CHALLENGING ASPECTS OF α-CHIRAL AMINE SYNTHESIS

α-Chiral amine synthesis represents many challenges because it falls into the category of functional group interconversion at the highest level of synthetic expression. The challenge becomes formidable when expressed as the wish to introduce a heteroatom (nitrogen) with simultaneous and complete chemo-, regio-, diastereo-, and enantiocontrol, into a commodity chemical during a single organic transformation. This represents an important and unmet goal, its realization would allow medicinal chemists to more rapidly meet the compound diversity and screening goals of drug discovery, and allow process chemists to efficiently synthesize α-chiral amines on scale.

α-Chiral tertiary amines rarely contain more than two α-stereocenters and would be considered the result of intended diversification (Figure 8.1). Secondary amines containing one or even two α-stereocenters, while more interesting from the viewpoint of further chemical modification, would also generally be considered a resting point of earlier planned diversification. These realities make secondary and tertiary amines less attractive from the standpoint of the diversification needs of medicinal chemists, and the more focused and specialized starting points of process research chemists.

These considerations bring us to our focal point: *primary amine synthesis and the point of most consternation, defining the first stereocenter alpha to a nitrogen atom.* This problem is the focus of the remainder of this writing, as methods for converting primary amines into secondary or tertiary amines or amides are already well documented in the racemic manifold of organic synthesis.[5]

Within the hierarchy of defining the challenge at hand, the task of distinguishing the different types of α-chiral primary amines and the corresponding strategies for their synthesis requires addressing. Simple disconnection analysis reveals that tertiary α-chiral primary amine synthesis requires carbanion chemistry,[6] while secondary α-chiral primary amine synthesis allows greater flexibility (Table 8.1). A direct consequence has been the development of fewer methods for tertiary

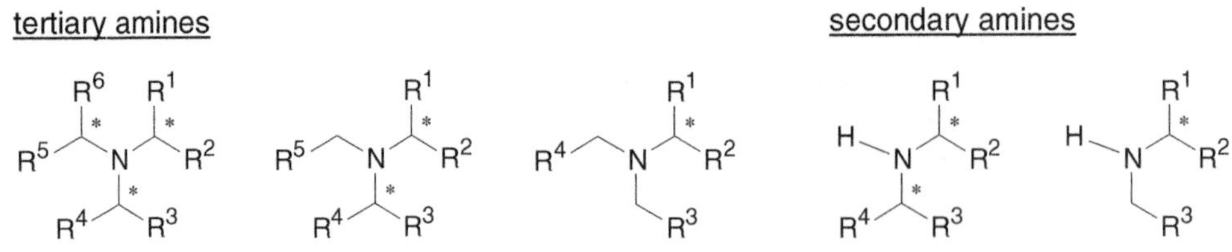

tertiary amines   secondary amines

FIGURE 8.1 Generic permutations of α-stereocenters in tertiary and secondary amines.

TABLE 8.1
General Overview of α-Chiral Primary Amine Synthesis

Key Step[a]	Secondary α-Chiral Amines			Tertiary α-Chiral Amines	
	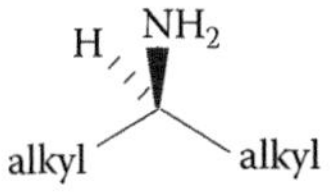 H, NH₂ alkyl–alkyl	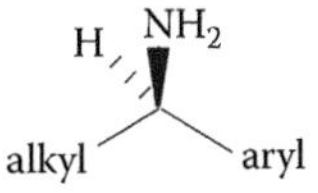 H, NH₂ alkyl–aryl	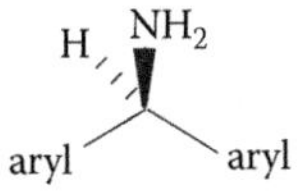 H, NH₂ aryl–aryl	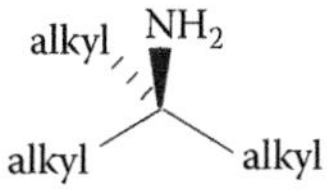 alkyl, NH₂ alkyl–alkyl	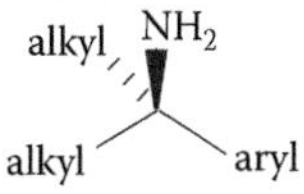 alkyl, NH₂ alkyl–aryl
Carbanion addition to imine derivative	2–3 steps	2–3 steps	3 steps	3 steps	3 steps
N-Acetylenamide hydrogenation	No examples[b]	4 steps	Not possible	Not possible	Not possible
Transfer hydrogenation of N-phosphinoylimines	4 steps	4 steps	Not investigated	Not possible	Not possible
Titanocene imine reduction	3 steps	3 steps	Not investigated	Not possible	Not possible
Reductive amination-transfer hydrogenation	No examples[c]	1 step	Not investigated	Not possible	Not possible
Reductive amination-hydrogenation	2 steps	2 steps	Not investigated	Not possible	Not possible

[a] Prochiral ketone or aldehyde starting materials are used by all methods.
[b] Pinacolone is the only demonstrated example.
[c] Examples exist, but with very low yields and ee (<30%) and discussed later.

α-chiral primary amine synthesis, which has further curbed their use as advanced building blocks in favor of the more accessible secondary α-chiral primary amines.

Although the strategies represented in Table 8.1 are comprehensive, but restricted based on their industrial promise or brevity, they do not represent cohesive general solutions to the problem of α-chiral primary amine synthesis. Instead they offer substrate specific solutions, and their individual limitations will be shortly discussed. Regardless, these methods represent significant advancement in know how regarding the number of reaction steps, yield, and ee, from just 5 to 10 years ago.

Another intimately connected concept is that of introducing the second α-chiral center next to a nitrogen atom. Although only passively addressed here, a partial answer to the state of the art can be taken from the fact that as recently as 2004, industrial chemists still considered a general synthetic strategy to the large class of ACE inhibitor drugs as viable through enantioselective ketoester reduction (and related strategies) instead of the direct route using reductive amination (Scheme 8.1).[7] The former three step strategy is lengthy and highlights the lack of knowledge concerning diastereoselective reductive amination verses the well-described success of reducing

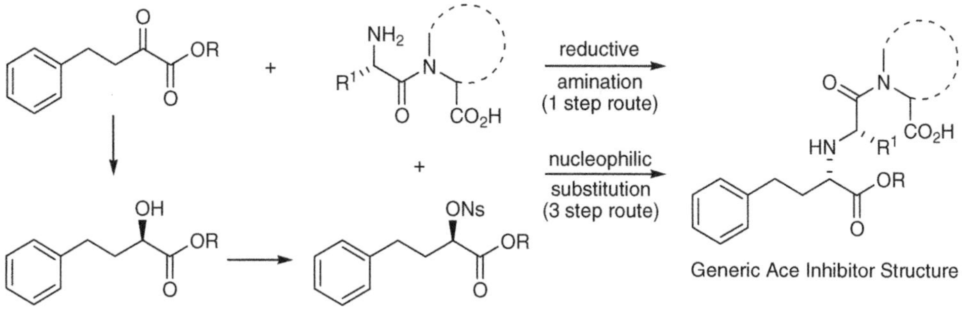

SCHEME 8.1 Reductive amination (1 step) vs hydroxyester (3 step) strategy.

many classes of ketones to alcohols with high enantioselectivity. Regardless, the example is a specialized case, and the shown reductive amination pathway has been investigated and recently improved upon.[9]

8.4   GENERAL INFORMATION BEFORE THE SPECIFIC EXAMPLES

The strategies for α-chiral amine synthesis are large in number, but few are practiced on an industrial scale. The question of what technology to use, is better defined by asking which one can deliver the required α-chiral amine within the requisite timeline, at the specified quantity, quality, and price? An overview of the most useful and promising new methods follows from the earlier defined strategies (Table 8.1). The number of reaction steps quoted is always that from the starting aldehyde or ketone to the α-chiral primary amine, and the majority of the methods examined here have the potential for scale-up. The remaining methods might be more amenable to combinatorial discovery needs, but are included to provide perspective and knowledge of present trends in α-chiral amine methodology.

The following sections focus on methods of industrial promise and/or those offering brevity in synthesis and were chosen based on reviews of the pertinent literature prior to June of 2006 and the author's opinion.[1,10] As a consequence, the provided justifications and conclusions are supported by extensive references to allow the reader to make a more critical and individual analysis as required. It is additionally important for the reader to be aware of the subject matters and strategies not included in this review. Thus, recent and significant advancements have been made for the diastereoselective and enantioselective: hydrogenation of primary enamine esters,[11] hydrogenation of α- or β-N-acetylenamide esters,[12] 1,4-addition of amines to enones,[13] chemical[14] and enzymatic[15] reductive amination of α-ketoacids, and remote amination via C-H insertion.[6] These methods often rely on pendant substrate functionality to induce high stereoselectivity and are not further discussed here, additionally biotransformations are better covered by an expert in that respective field. Furthermore, the hydroamination of olefins is early in its developmental cycle and not included here because of the present limitations concerning industrial applications.[16] Instead, the focus here is on methods enabling the use of unfunctionalized carbonyl starting materials (ketones and aldehydes), thus α- and β-amino acid products are not discussed.

The strategies presented in Table 8.1 can be generalized in the following manner: (1) carbanion addition to aldimine or ketimine derivatives; (2) sequential amination-alkylation of aldehydes (carbanion addition to in situ formed aldimine derivatives); (3) transfer hydrogenation or hydrogenation of imines; (4) reductive amination of ketones; and (5) N-acetylenamide reduction. Because of the difficulty of their synthesis, α-alkyl,-alkyl substituted amines are highlighted whenever possible.

8.5   THE CARBANION CHEMISTRY OF TERTIARY α-CHIRAL PRIMARY AMINE SYNTHESIS

8.5.1   The Method of Ellman

Organic chemists have previously produced tertiary α-chiral amines via metal carbanion additions to chiral ketimine derivatives (to our knowledge no one has ever accomplished the feat of enantioselective carbanion addition to a prochiral ketimine or ketimine derivative thereof), but these attempts pail in comparison (yield and *ee*) to the methods developed by Ellman.[17] Using a three-step reaction sequence starting from prochiral ketones,[18] high overall yields and *ee* are obtained for the primary amine (Scheme 8.2). The second, and key step, is capable of using a large variety of commercially available alkyl or aromatic organolithium reagents for addition to a diverse number of N-*tert*-butanesulfinyl ketimines.[19]

The method relies on the use of the chiral ammonia equivalents (R)- or (S)-N-*tert*-butanesulfinamide, and Ellman has optimized their synthesis via the enantioselective catalytic oxidation of

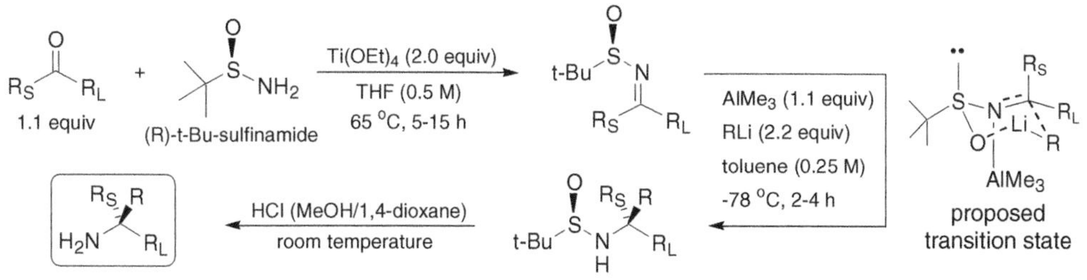

SCHEME 8.2 Ellman's general approach to tertiary α-chiral primary amines.

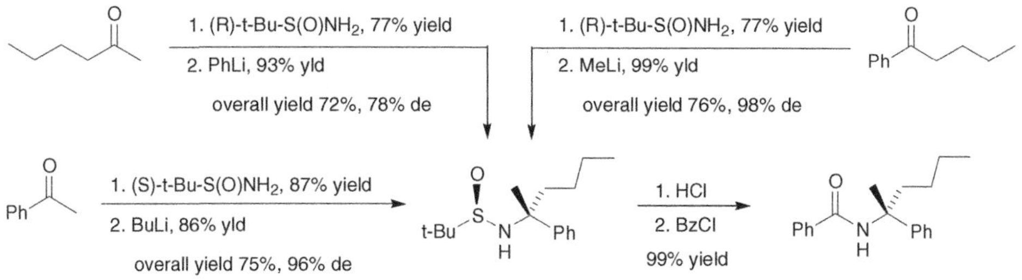

SCHEME 8.3 Three approaches to the same α-chiral primary amine.

tert-butyl disulfide.[18b] The high diastereoselectivity observed during step two (Scheme 8.2) has been rationalized by invoking a six-membered transition state complex and nicely explains the differences in stereoselectivity observed for the three different ketone starting materials shown in Scheme 8.3. Note that to arrive at the same enantiomeric series of the amine product, the acetophenone substrate requires the use of the (S)-auxiliary, the stereoconfiguration on the sulfur atom of the product is not represented in the diagram for convenience. A limited number of α-alkyl,-aryl,-aryl' and α-aryl,-aryl',-aryl'' chiral primary amines have been investigated.[19a,20]

The method calls first for the formation of a sulfinyl ketimine. The solvent, THF, is acceptable and so are the required reflux temperatures, additionally the titanium alkoxides are readily available in bulk, are inexpensive, and have been previously used for manufacturing purposes. The requirement for distillation (Kugelrohr) or chromatography of the sulfinyl ketimine product is a potential limitation, with distillation being preferred. The use of organolithium reagents (2.2 equiv), while highly reactive, may not cause as much concern as the need for −78°C reaction temperatures. Finally, the main concern would be the cost of the (R)- or (S)-N-*tert*-butanesulfinamide auxiliary. If inroads can be made into reducing the cost of these chiral auxiliaries, as compared to others (Table 8.2), then the method would likely be acceptable for industrial use because of the high overall reported yields and ee.

TABLE 8.2

Sigma-Aldrich Quote May 2006 for Two Common Chiral Ammonia Equivalents

Chemical Name	Quantity (kg)	Price (US dollars)
(S)-N-*tert*-butanesulfinamide	1.0	13,125.00
(R)-N-*tert*-butanesulfinamide	1.0	13,125.00
(S)-α-methylbenzylamine [(S)-α-MBA]	1.0	798.00
(R)-α-methylbenzylamine [(R)-α-MBA]	1.0	2,220.00

8.6 STRATEGIES FOR SECONDARY α-CHIRAL PRIMARY AMINE SYNTHESIS

The assessment of strategies for α-disubstituted chiral primary amine synthesis requires more time and some initial explanation. First it needs to be expressed that when the same amine product can be arrived at in similar quality and quantity using 'hydrogen' addition vs carbanion addition, the 'hydrogen' methods are economically preferred from a scale-up perspective. This is based on the principle of avoiding high energy starting materials which are also invariably used in excess. Regardless it will soon become apparent from our discussion that the two methods only intermittently overlap, concerning the amine product formed, and are thus complementary. This brings us to another realization; several secondary α-chiral amine substitution patterns (substituents at the chiral carbon adjacent to nitrogen) are only presently possible in reasonable yield and high ee using the carbanion strategies, but at this time lack industrial feasibility. The carbanion methods of Ellman (in this category we are referring to his aldehyde starting material methods) are an exception, but again the cost of his auxiliary will require evaluation. Furthermore, the carbanion protocols have not been extended to cyclic systems, unlike the "hydrogen" methods.

8.6.1 CARBANION CHEMISTRY

Very recently two competing methods have appeared for the enantioselective sequential amination-alkylation of aldehydes (enantioselective carbanion addition to *in situ* formed aldimine derivatives), by Hoveyda/Snapper (chiral Zr and Hf catalysts) and Charette (chiral Cu catalysts) using Et_2Zn. These methods offer high ee, diversity, brevity, and good reaction temperatures (0-22°C), but at this stage of development would not be considered for scale-up. So, why include them here? These methods, in the future, may be further improved upon by the aforementioned research groups or perhaps inspire another chemist to do so. Importantly these methods accept the presence of a large number of functional groups (e.g., alkenes, alkynes, aryl-NO_2, esters, and alkyl-bromides) on the aliphatic- or aromatic-aldehyde starting material. Both methods extensively examine the use of Et_2Zn, while occasionally showing the use of alternative carbanions R_2Zn. Only one alkyl group is transferred, thus the development of inexpensive nontransferable 'dummy' group would greatly benefit these methods when expensive R groups are required. Furthermore, it would have to be evaluated if the chromatographic purifications could be replaced by acid-base work-up conditions for the isolation of qualitatively pure amines.

The Hoveyda/Snapper method uses a 1:1 equiv ratio of the aldehyde and o-anisidine and many examples have been demonstrated (Scheme 8.4).[21] Scale-up limitations are the need for a large excess of Et_2Zn (≥6 equiv), high loading of the chiral ligand, low reactor volume efficiency, and long reaction times (minimum 24 h). Good yields and high ee are noted for step one, but the overall yields (aldehyde to primary amine) tend to be mediocre (40-60%). The deprotection step, oxidative N-Ar bond cleavage (not shown), presently requires 4.0 equiv of $PhI(OAc)_2$.[22] The authors note that preforming the imine (which adds an additional step) alternatively allows 3 equiv of Et_2Zn to be used.[23]

The method of Charette,[15b] as outlined in Scheme 8.5, is what he refers to as his one-pot process. The source of nitrogen, diphenylphosphinoylamide, is exceedingly expensive and therefore synthesized from the oxime of acetone.[24] The method reliably provides overall yields of >60% for the primary amine product, when employing Et_2Zn, and high ee. The molarity for these reactions is generally low (0.20 M, occasionally 0.1 M), and the requirement for excess aldehyde and Et_2Zn (Scheme 8.5), long reaction times (≥24 h), and high chiral ligand loadings make the method industrially restrictive. The method has been previously evaluated by industrial chemists.[25]

Charette offers a way around some of these problems by demonstrating two alternative three-step methods. For aromatic aldehydes, an N-diphenylphosphinoyl imine can be preformed and isolated. Subsequent treatment with 2.0 equiv of Et_2Zn in the presence of catalytic $Cu(OTf)_2$

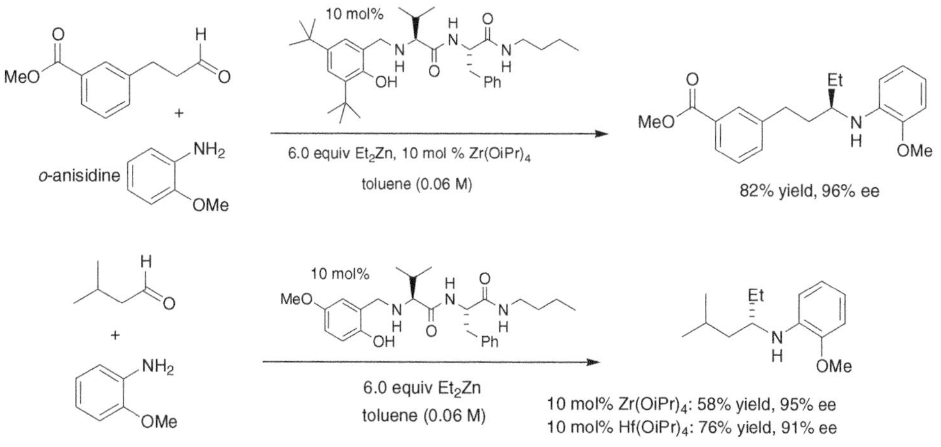

SCHEME 8.4 Hoveyda/Snapper approach.

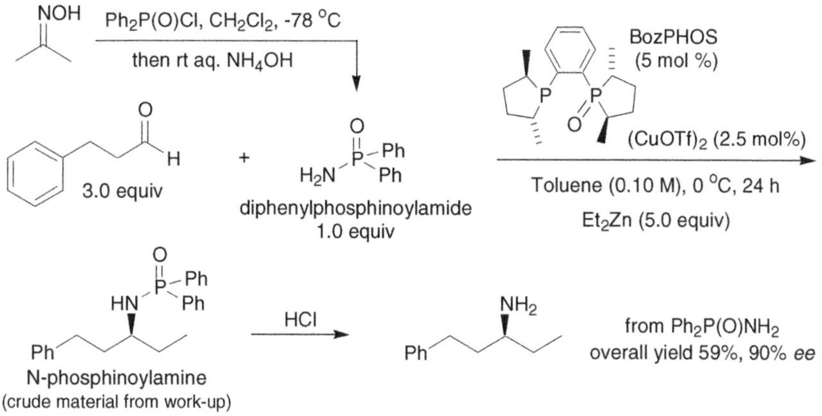

SCHEME 8.5 Charette protocol example.

(6 mol %) and the BozPHOS ligand (3 mol %), provides the N-phosphinoylamines in good and more often in excellent yield (>90%), with high ee (>90%).[26] For aliphatic aldehydes, a sulfinic acid/N-diphenylphosphinoyl amide complex of the aldehyde is required. This preformed and iso-latable compound is then treated with 2.5 equiv of Et$_2$Zn in the presence of Cu(OTf)$_2$ (4.5 mol%) and BozPHOS (5 mol %) to give the N-phosphinoylamines, again, in high yield (>86%) and ee (>90%).[27] It should be noted that removal of the diphenylphosphinoyl moiety is considered a high yielding deprotection step.

In our laboratories, we have recently developed a one-pot asymmetric sequential amination-alkylation reaction. The new method consists of prestirring an aldehyde (1.0 equiv), (R)- or (S)-α-methylbenzylamine (1.05 equiv), and 5 mol % Ti(OiPr)$_4$ for 30 min. This mixture, at room temperature, is then added dropwise to a solution of a cuprate in THF or Et$_2$O at −78°C (Scheme 8.6).[28] The method may be of interest because it uses the least expensive chiral ammonia equivalent available (Table 8.2) in essentially equimolar quantities (1.05 equiv) as compared to the limiting reagent, the aldehyde. Additionally, this one-pot process has significant yield advantages over the two-step process (formation of the (R)- or (S)-N-α-methylbenzyl aldimine, isolation of this chiral aldimine, followed by carbanion addition).[29] The reactions are fast (3 h at −78°C), but the low temperature requirements, low reactor volume to product ratio, and the need for 3 equiv of a cuprate (CuBr based) will be considered restrictive. Many times the cuprate can be reduced to 2.0 equiv

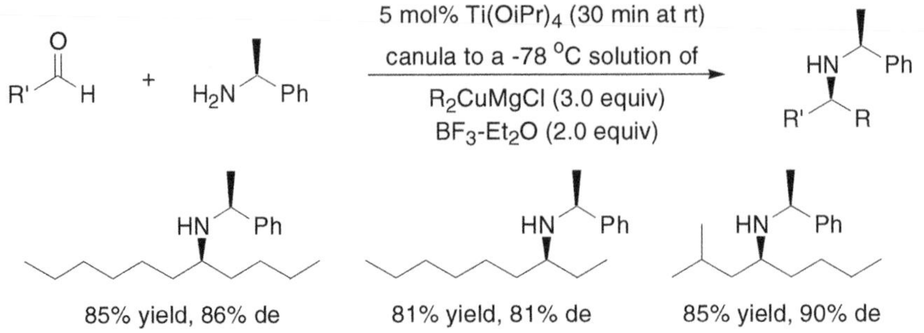

SCHEME 8.6 Nugent protocol and representative examples.

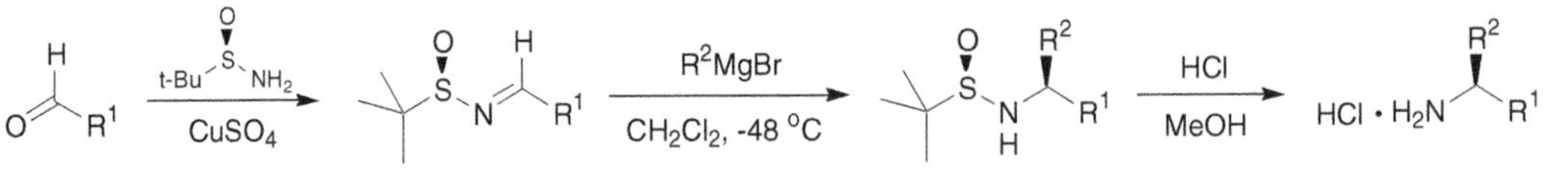

SCHEME 8.7 Ellman general strategy and examples.[19a]

TABLE 8.3
Examples of Ellman's Method as Delineated in Scheme 8.7

R¹	R²	Overall Yield (%)	ee
Et	Me	89	86
Et	iPr	86	96
Et	Ph	86	92
iPr	Me	85	96
iPr	Et	84	94
iPr	Ph	80	78
iPr	Vinyl	63	76
Ph	Me	77	94
Ph	Et	84	84
p-MeOPh	Et	71	98
Ph	iPr	29	Not recorded
Ph	Vinyl	67	88
Bn	Me	67	90
Bn	Et	66	84
Bn	Vinyl	62	82
Bn	Ph	63	90

by using CuI as a source of Cu(I), but the yield and ee are both often lower by 4-5%. Hydrogenolysis of the auxiliary provides the corresponding primary amine with the same ee as the de installed during the carbanion addition step. The relative configuration at the new chiral center is assumed based on analogous data comparison with similar compounds.

Finally we arrive at the method of Ellman, this time for secondary α-chiral primary amines (Scheme 8.7).[17b,c] It is analogous to his earlier delineated method for tertiary α-chiral primary amines but with industrial advantages. (R)- or (S)-N-*tert*-butanesulfinamide (1.0 equiv) is added to an aldehyde (1.1 equiv), and CuSO₄ (2.0 equiv), instead of Ti(OiPr)₄, allowing formation of the

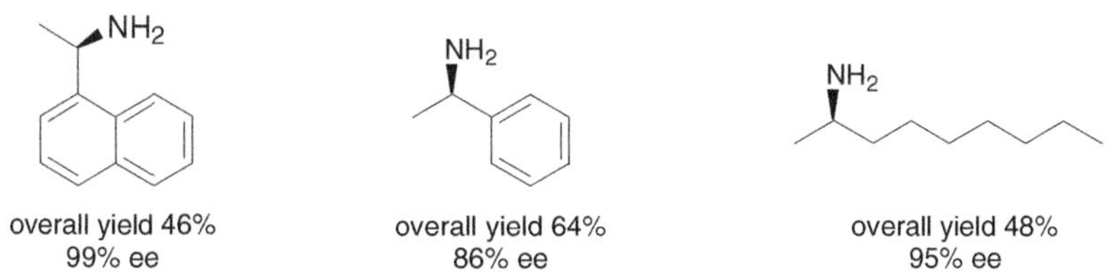

overall yield 46%
99% ee

overall yield 64%
86% ee

overall yield 48%
95% ee

FIGURE 8.2 Avecia limited examples.

TABLE 8.4
Avecia Limited Method[32]

			Yield Data (%)		
Substrate	Preferred Metal[a]	Oxime	Phosphinoyl Imine	Phosphinoyl Amine	Primary Amine
1-Acetyl Naphthalene	Rh[b]	65	94	95	80, 99% ee
Acetophenone	Rh[c]	95	80	95	88, 86% ee
2-Octanone	Ir[d]	87	84	95	69, 95% ee

[a] Refers to the enantioselective step (reduction of the phosphinoyl imine), requires 24 equiv of Et$_3$N/HCO$_2$H (2:5 ratio), reactions performed at room temperature.
[b] 0.25 mol% [RhCp*Cl$_2$]$_2$ with 0.5 mol% (R,R)-N-tosyl-1,2-diamino-1,2-diphenylethane.
[c] 1.0 mol% [RhCp*Cl$_2$]$_2$ with 2.0 mol% (R,R)-N-tosyl-1,2-diamino-1,2-diphenylethane.
[d] 1.0 mol% [IrCp*Cl$_2$]$_2$ with 2.0 mol% (R,R)-N-tosyl-1,2-diamino-1,2-diphenylethane.

N-t-butanesulfinyl aldimines at room temperature.[18a] Unlike his tertiary α-chiral primary amine method, no AlMe$_3$ is required, and 1.2 to 2.0 equiv of a Grignard reagent are used instead of an organolithium reagent. Additionally, the reaction temperature is significantly improved (–48°C, instead of –78°C). This asymmetric method surpasses all other enantioselective methods in this genre and consistently provides high overall yield and high enantiomeric excess. Points for further consideration would be the requirement of three reaction steps and the expense of the chiral ammonia equivalent (R)- or (S)-N-*tert*-butanesulfinamide (2-methyl-2-propanesulfinamide).

Finally, Ellman has recently developed a method for the synthesis of α-aryl,-aryl' substituted primary amines using the corresponding N-t-butanesulfinyl aldimine (chiral auxiliary approach) or via an N-diphenylphosphinoyl aldimine (catalytic enantioselective method) using an arylboronic acid in combination with chiral Rh catalysts.[30]

8.6.2 ENANTIOSELECTIVE IMINE REDUCTION

Regarding the reduction of ketimines and N-derivatives thereof, the enantioselective transfer hydrogenation,[31,32] or hydrogenation[33] of N-phosphinylimines, preferably over other N-functionalized imines, e.g. -OR, -SO$_2$R, or -NR$_2$, can impart high enantioselectivity. The hydrogenation method uses very high H$_2$ pressure (70 bar = 1015 psi) and thus the transfer hydrogenation method is focused on here instead. In the Avecia Limited patents regarding the use of the CATHyTM (Catalytic Asymmetric Transfer Hydrogenation) catalysts, three substrates are demonstrated (Figure 8.2 and Table 8.4). A four step process is used [prochiral ketone → oxime → phosphinoyl imine → phosphinoylamine → primary amine] and industrially acceptable improvements in catalyst loading (vs those quoted in Table 8.5 which were obtained from the patents) are discussed in a review.[31] The method has been demonstrated on an industrial scale.

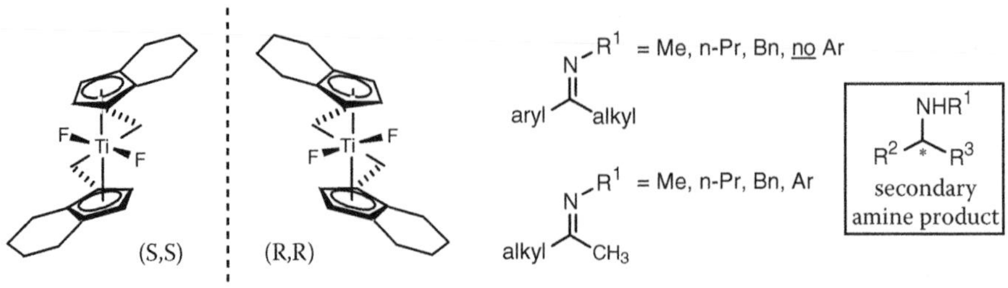

FIGURE 8.3 Buchwald's titanocene method – precatalysts, substrates, general product.

The enantioselective imine hydrogenation[34] and hydrosilylation[35] methods of Buchwald are distinct from all other methods for several reasons. The first unique feature is the unparalleled breadth of N-alkyl or N-aryl imines that can act as substrates for this reaction (Figure 8.3). If α-chiral primary amines are the final goal, N-benzyl or N-aryl imines are necessarily required, and Buchwald extensively examined these substrates, but he did not examine the subsequent deprotection step which would allow a primary amine to be realized. Based on literature precedent, hydrogenolytic cleavage of N-benzyl amine products would be expected to be facile and high yielding, while reductive oxidation of N-aryl products would be expected to be low yielding.[22] Beyond these considerations, the diversity of the α-chiral primary and secondary amine products afforded by this method are truly distinctive, making it the only method that could be considered as general in nature.

The hydrogen-based titanocene method was developed first, but suffers from high catalyst loading (5 mol %) and the need for very high H_2 pressure (generally 138 bar = 2000 psi). For the hydrosilylation methods, the specific silane can vary, but usually PMHS (polymethylhydrosiloxane) or $PhSiH_3$ is used, with the molar quantity ranging from reasonable (1.5 equiv) to excessive verses the imine. An enormous number of examples have been demonstrated and depending on the substrate examined the catalyst loading usually varies between 0.1 to 2.0 mol %. Yields and ee for the amine products are generally good to high, but the imine (required for the stereodetermining step) formation yields are generally lacking.

Only a cursory mention is made here of the much heralded success achieved while using chiral Ir catalysts (chiral ferrocenylphosphine derivatived ligands) for the enantioselective hydrogenation process leading to (S)-metolachor (Figure 8.4); simply because this strategy is not considered a general solution for primary amine synthesis. Regardless, this general class of chiral transition metal catalysts has found broad application for the enantioselective reduction of N-aryl imines and the N-acetylenamides of α- and β-ketoesters. The development of the process for metolachor itself, which requires an N-aryl imine precursor, is a great story of perseverance and know-how at Solvia and is mandatory reading to fully appreciate the dynamics of developing a catalytic enantioselective process for industrial scale production.[36] It additionally demonstrates that high hydrogen pressures (80 bar = 1160 psi) can be used for large scale manufacturing.

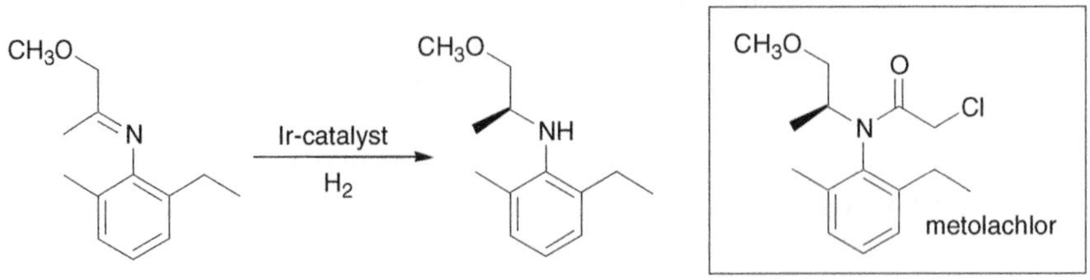

FIGURE 8.4 Enantioselective reduction of the imine precursor for metolachlor.

TABLE 8.5
Degussa AG Chemistry[37]

Substrate	Yield of Primary Amine (%)	Enantiomeric Excess (%)
Acetophenone	92	95
Phenyl ethyl ketone	89	95
3'-Methylacetophenone	74	89
4'-Methylacetophenone	93	93
4'-Methoxyacetophenone	83	95
4'-Chloroacetophenone	93	92
4'-Bromoacetophenone	56	91
4'-Nitroacetophenone	92	95
1-acetylnaphthalene	69	86
2-acetylnaphthalene	91	95
2-octanone	44	24
2-methylcyclohexanone	63 (64 cis, 36 trans)	17 cis, 64 trans

8.6.3 REDUCTIVE AMINATION VIA TRANSFER HYDROGENATION

Concerning reductive amination, the independent and combined efforts of the industrial group of Riermeier (Degussa AG),[37,38] and the academic group of Börner[38] have perhaps produced the most significant advancements in α-chiral amine synthesis to date! They have reported the enantioselective reductive amination of acetophenone (and substituted derivatives), acetylnaphthalenes, and phenyl ethyl ketone in high yield and ee (Table 8.5). The particular transfer hydrogenation conditions are of the Leuckart-Wallach reaction type, using excess NH_3/HCO_2NH_4 (80°C, ~20 h) with 0.5 to 1.0 mol % of a (R)- or (S)-BINAP (or a derivative thereof) $RuCl_2$ complex, for the one-step synthesis of α-aryl,-alkyl substituted chiral primary amines. The reaction actually produces a mixture of the primary amine and the corresponding formyl derivative (RHNC(O)H), thus the crude product is treated with HCl (EtOH/H$_2$O) at reflux to give exclusively the primary amine in good to excellent yield.

As of this writing, the method provides low yields and ee for other aromatic substrates, e.g. 1-indanone (6% yield, no ee reported, chiral Ru catalyst),[38] and aliphatic ketones, e.g., 2-octanone (44% yield, 24% ee, chiral Ru catalyst[37] or 37% yield with an achiral Rh catalyst[38]). The same authors have also recently made inroads, concerning the use of aromatic ketones and hydrogen,[39] although the presented transfer hydrogenation method appears to be superior as of now. The development of these methods for alkanone substrates would be another precedent setting break through.

8.6.4 REDUCTIVE AMINATION WITH H$_2$

Reductive amination is historically defined as the combination of a Brønsted acid, a ketone, an amine, and a coexisting reductant.[40] The recent evolution of titanium (IV) alkoxides, as mild Lewis acid replacements for Brønsted acids, owes its genesis to a clever modification of titanium amide chemistry[41] demonstrated by Mattson et al. in 1990.[42] By prestirring a ketone, amine, and Ti(OiPr)$_4$ (neat), followed by the addition of EtOH and NaBH$_3$CN, the desired reductive amination product was afforded (Scheme 8.8). The same authors suggested that these reductive aminations proceeded through a hemiaminal titanate intermediate, based on IR spectroscopy before addition of the reductant.

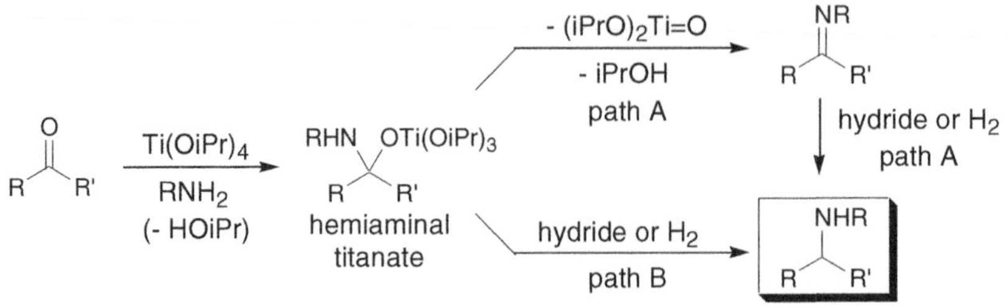

SCHEME 8.8 Ti(IV) mediated reductive amination – potential reaction pathways.

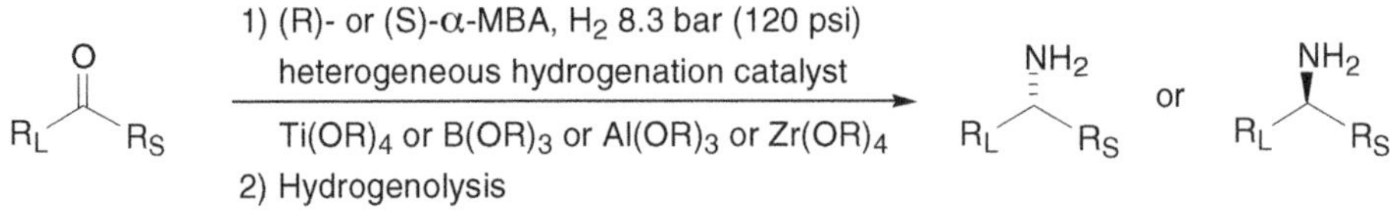

SCHEME 8.9 Two-step strategy for α-aryl,-alkyl and α-alkyl,-alkyl′ primary amines.

The general reaction has been further developed using different hydride reagents,[43] and we were the first to expand the reducing systems to those capable of using hydrogen and simultaneously demonstrated its application to asymmetric reductive amination. Our insights came from exploring the use of Ti(OiPr)$_4$/Pt-C/H$_2$ for reductive amination of an enantiopure quinuclidinone, and culminated in the most efficient synthesis of quinuclidines to date.[44]

More recently we have contributed to the goal of α-chiral primary amine synthesis by developing a two-step strategy relying on the asymmetric reductive amination of prochiral ketones with the chiral ammonia equivalent (R)- or (S)-α-methylbenzylamine (Scheme 8.9).[45,46] Step one, the key step, is the hydrogenation of a prochiral ketone substrate in the presence of (R)- or (S)-α-methylbenzylamine (α-MBA) (1.1 equiv), Ti(OiPr)$_4$ (1.2 equiv), and a heterogeneous hydrogenation catalyst, providing the amine diastereomers in good yield and diastereoselectivity (Figure 8.5). Use of reductive amination allows the normally stepwise excessive procedures of chiral auxiliary approaches to be avoided by incorporating a nitrogen atom (chiral auxiliary) and a new stereogenic center at a ketone carbonyl carbon simultaneously.

Depending on the ketone examined (acyclic vs cyclic, alkyl-alkyl vs. aryl-alkyl, sterically encumbered vs unencumbered), the correct combination of heterogeneous hydrogenation catalyst ((Raney Ni, Pt-C, or Pd-C)), solvent (0.5 M), and temperature is crucial for allowing high yield and de with practical reaction times (generally 10 h) at 8.3 bar (120 psi) of hydrogen. The reductive amination products themselves can be isolated in qualitative purity after standard acid-base work-up procedures, the residual α-MBA (3-5%) is removed from the crude product after washing with sat. NH$_4$Cl. Step two, hydrogenolysis, allows removal of the chiral auxiliary providing the α-chiral primary amines in good overall yield (5 examples 71 to 78%, 1 example 64%, Table 8.6). Particularly noteworthy is the ability of aliphatic prochiral ketones to serve as good to excellent substrates. Historically, and to this day, the efficient synthesis of α-alkyl,-alkyl′ chiral primary amines has been difficult and laborious.

Three recent findings have been critical for improving the above described method. First, we have introduced the use of B(OiPr)$_3$ or Al(OiPr)$_3$, to replace Ti(OiPr)$_4$.[46] This is important because on work-up Ti(OiPr)$_4$ can sometimes produce finely dispersed TiO$_2$, forcing a celite filtration on-scale. To avoid this the less expensive B(OiPr)$_3$ of B(OMe)$_3$ can be used, albeit in greater equivalents to have a similar effect as Ti(OiPr)$_4$.[48] Second, the de can be significantly improved upon by using 50-100 mol % Yb(OAc)$_3$ instead of the aforementioned Lewis acids. The most dramatic effect is

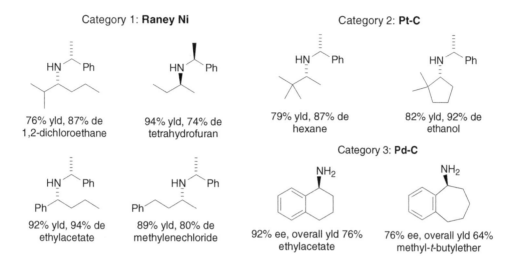

FIGURE 8.5 Correlation of heterogeneous hydrogenation catalyst with ketone structure and product examples.

observed when examining straight chain 2-alkanones, e.g. the reductive amination product of 2-butanone increases from 74% to 80% de, 2-hexanone from 71% to 85% de, and 2-octanone from 72% to 87% de.[49] Finally, a limited number of Lewis acids, Yb(OAc)$_3$ [10 mol %], Y(OAc)$_3$ [15 mol %], and Ce(OAc)$_3$ [15 mol %], have been identified as catalytically useful. These Lewis acids, in contrast to many others, are capable of suppressing alcohol by-product formation (ketone reduction), and allow good yields and reaction rates to be achieved during the reductive amination step. If no Lewis acid or the wrong one is present, large quantities of the alcohol by-product can be expected.

Finally, the strategy should be of interest because it only requires two reactions steps. This is possible because ketimine isolation is not required, and while rarely discussed can be time consuming, may provide mediocre yield, and unnecessarily lengthens the synthesis of amines.[50] Furthermore, all the reagents are already in use by the pharmaceutical industry, a broad range of ketone substrates are suitable (*even aliphatic ketones*), either enantiomeric form of the α-chiral amine product can be produced, and the process has been demonstrated on a 20 g scale. The method is compatible with acetonides, ethers, silyl ethers, bulky esters, secondary amides, tertiary amides, carbamates, urethanes, etc. With these beneficial qualities noted, the method suffers when non-branched 2-alkanones are used (product des <75%). In these cases, HCl salt formation allows further enrichment via crystallization, alternatively stoichiometric Yb(OAc)$_3$ can be used during the reductive amination to allow enhanced de.[49] Both of these solutions require additional processing time and/or cost and require consideration before scale-up.

8.6.5 N-Acetylenamides

The synthesis of N-acetylenamides proceeds in mediocre overall yield after two reaction steps from ketones,[51,52] so it is fortunate that their lack of geometric purity generally does not hinder their

TABLE 8.6
Examples of α-Chiral Primary Amines Synthesized from Ketones[a]

Primary Amine	Regioselectivity[b]	Pd Source[c]	Additive	Overall Yield (%)[d]	ee (%)[e]
(structure, NH₂)	>99:1	Pd(OH)₂-C	AcOH	71	74
(structure, NH₂)	>99:1	Pd-C	None	71	87
(structure, Ph…NH₂)	8:1	Pd-C	None	74	90[f]
(structure, Ph…NH₂)	>99:1	Pd-C	None	78	80
(structure, NH₂)	>4:1	Pd-C	AcOH	76	92
(structure, NH₂)	>40:1	Pd-C	AcOH	64	76

[a] All debenzylations were performed in CH₃OH at 22°C with H₂ 4 bar (60 psi).
[b] Regioselectivity for step 2 (hydrogenolysis), desired/undesired benzylic cleavage ratio.
[c] Hydrogenolysis step.
[d] Overall isolated yield of the primary based on the starting ketone.
[e] Determined by chiral gas chromatography (trifluoroacetyl derivative) or high-performance liquid chromatography (benzoyl derivative) analysis of the of the primary amine.
[f] The de of the starting material before hydrogenolysis was 94%.[47]

conversion to N-acetylated α-chiral amines with high ee (Scheme 8.10). Viable ketone starting materials are pinacolone and aryl-alkyl ketones.[52,53] Recently Merck scientists[11b,c] were able to alleviate the need for N-acetylenamides, their enantioselective method only requiring simple primary enamines. The ketone precursors for those studies were β-ketoesters, but the focus here is on unfunctionalized ketone substrates and the corresponding non-acylated enamine substrates are not yet acceptable substrates for enantioselective hydrogenation. N-acetyl-α-arylenamides represent the overwhelming majority of substrates examined and have been extensively investigated by Zhang.[54]

The examples shown do not include the many examples demonstrating different substituent substitution patterns on the phenyl ring (Scheme 8.10), where R is held constant and equal to H. Instead the substrates shown try to demonstrate the diversity of substitution possible for R (Scheme 8.10). For the Table 8.7 entries 1, 2, 4, and 5, reactions times vary between 48-60 h at room temperature under 10 bar (145 psi) of H₂.[55] Earlier, 1999, Zhang described very similar substrates

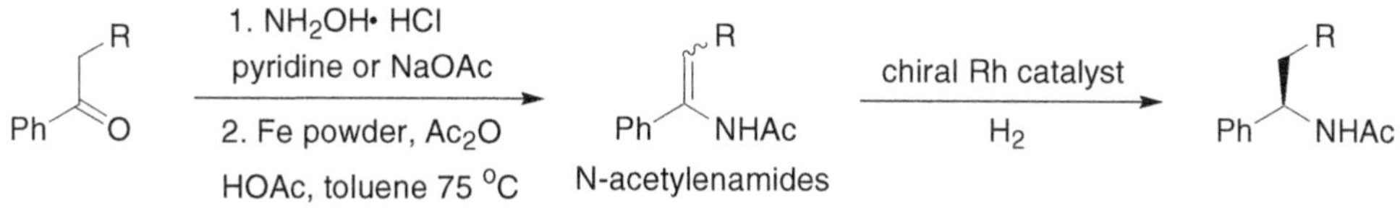

SCHEME 8.10 Typical *N*-acyl-α-arylenamides substrates.

TABLE 8.7
Zhang Catalytic Enantioselective N-acyl-α-Aryl-Enamide Reductions

Entry	R	Catalyst-Chiral Ligand	ee
1	H	2 mol% [Rh(COD)$_2$]SbF$_6$-(*R,S,S,R*)-DIOP*	98
2	CH$_3$	2 mol% [Rh(COD)$_2$]SbF$_6$-(*R,S,S,R*)-DIOP*	97
3	Et	1 mol% Rh(COD)$_2$PF$_6$-(*R,R*)-Binaphane	97
4	iPr	2 mol% [Rh(COD)$_2$]SbF$_6$-(*R,S,S,R*)-DIOP*	99
5	Bn	2 mol% [Rh(COD)$_2$]SbF$_6$-(*R,S,S,R*)-DIOP*	99

under the influence of 1 mol % Rh(COD)$_2$PF$_6$-(R,R)-Binaphane (Figure 8.6), which after 24 h at 1.4 bar (20 psi) of H$_2$ provided ee's ranging from 1-4% less than the results shown in Table 8.7.[56] In this reference he also provides ee data for the N-acetyl-α-phenylenamide substrate when R = Et (Table 8.7, entry 3). In 1998, he showed results approximately 4-8% lower in ee from those in Table 8.7 using 1 mol % [Rh(COD)$_2$]OTf-(R,R)-BICP (Figure 8.6) after 24 h at 2.8 bar (40 psi) of H$_2$.[57] Additionally he has reported on benzocyclic (aryl-alkyl cyclic) N-acetylenamide examples.[51,58] Several examples using monodendate phosphorous based ligands (phosphoramidites[53] and phosphites,[59] see Figure 8.6) have recently begun to appear and are starting to approach the ees achieved by the Zhang research group. Isolated yield data for the enantioselective N-acetylenamide hydrogenations are overwhelming not reported in the literature. When yields are mentioned, they are quoted as quantitative. If we assume every enantioselective reduction is quantitative, then overall yields for

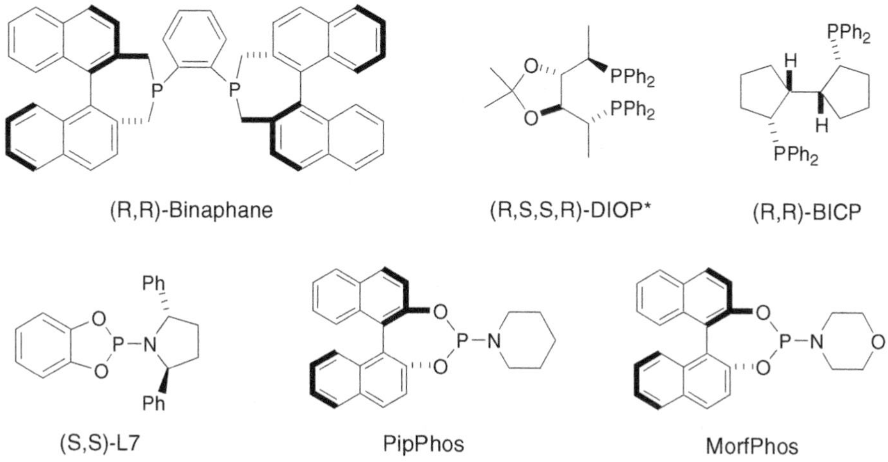

FIGURE 8.6 Examples of some highly efficient and new ligand types.

this approach are still lower than 55% in general, based on the four steps required to obtain an α-chiral primary amine. This is due in large part to the low yield of N-acetylenamide formation from ketones (Scheme 8.10). Additionally, the last step, amide hydrolysis is by no means trivial and would have to be compatible with any incidental functionality. Another relevant analysis would have to involve the Rh to chiral ligand ratio and overall catalyst loading to determine industrial feasibility. Regardless, it is clearly evident that the enantiodetermining step is high yielding and proceeds with very high stereoselectivity.

8.7 CORRELATION OF α-CHIRAL AMINE STRUCTURE TO BEST METHODS AND SCALE-UP POTENTIAL

Table 8.8 and Figure 8.7 can be used as an initial guide to gain quick perspective regarding a specific α-chiral primary amine substitution pattern and the likely method of choice to arrive at it. The first entry represents a generic tertiary α-chiral primary amine, access to these structures is presently only possible using the method of Ellman. The individual references in Table 8.8 represent good starting points for further detailed evaluation (also see the earlier relevant sections of this chapter for more references) and are based on literature published before June of 2006.

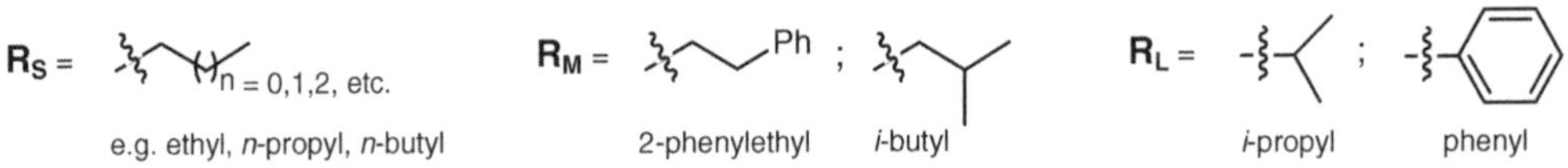

FIGURE 8.7 Examples of R_S, R_M, and R_L for Table 8.8.

CONCLUSION

It should be clear by now that no functional group holds the overwhelming dominance that the carbonyl moiety does when discussing amine synthesis. Depending on the subclass of α-chiral amine desired, a particular carbonyl substrate and strategy will be evident based on the strategies presented here. Methods using molecular hydrogen or transfer hydrogenation reagents generally excel at producing α-chiral primary amine products with sterically dissimilar α-,α-substituents in high ee. Thus ketone starting materials of the general structure $R_S C(O)CH_3$, $R_M C(O)CH_3$, $R_L C(O)CH_3$, $R_L C(O)R_S$, are excellent substrates (Figure 8.7). Carbanion addition methods are overwhelmingly superior (vs the 'hydrogen' methods) at enabling high ee when two similarly sized substituents are on the stereocenter adjacent to the nitrogen atom. Thus the combination of a carbanion source (R_S or R_M) with an aldehyde of general structure: $R_S C(O)H$ or $R_M C(O)H$ or $R_L C(O)H$, has a good probability of being controlled with high stereoselectivity. At the present stage of the developmental cycle, it can be stated that the two main strategies ('hydrogen' addition vs carbanion addition) complement one another. When the two methods compete directly, which is not often, the 'hydrogen' based methods are more cost effective.

From the process chemistry standpoint, assessment of synthetic design can often be an intricate matter shaped by the early stage needs of salt screening, formulation, toxicology, and/or clinical trial studies. Confidentiality, and the quality and quantity of material an outside vendor can reliably deliver are additional factors that are intertwined with the above noted themes. In a situation where the α-chiral amine product is of high enough value, time constraints may allow a higher cost methodology to be seriously considered over the short term. For long term manufacturing solutions, an in depth scientific evaluation, free of non-scientific interruptions, is required and necessitates the evaluation of several synthetic strategies and likely as many corresponding technologies.

TABLE 8.8
Correlation of α-Chiral Primary Amine Structure and Viable Strategies

Amine	Recommended Strategy	Industrial Feasibility	Reaction Steps
R^1,,, NH$_2$ / R^2 — R^3	Carbanion addition (Ellman)[17b,c]	Possible	Three
NH$_2$ / R_S — CH$_3$	1. Reductive amination (H$_2$)[45,46] 2. Transfer H$_2$ (Avecia Ltd)[31,32]	1. Possible 2. Yes	1. Two 2. Four
NH$_2$ / R_M — CH$_3$	1. Reductive amination (H$_2$)[45,46] 2. Carbanion (Ellman)[17c]	1. Possible 2. Possible	1. Two 2. Three
NH$_2$ / R_L — CH$_3$	1. Red. amin. transfer H$_2$[37,38] 2. Reductive amination (H$_2$)[45,46] 3. Carbanion (Ellman)[17c]	1. Yes 2. Possible 3. Possible	1. One 2. Two 3. Three
NH$_2$ / R_S — R_S	Carbanion addition 1. Hoveyda/Snapper[21b,22a] 2. Charette[27]	1. Not likely 2. Not likely	1. Two 2. Three
NH$_2$ / R_M — R_S	Carbanion addition 1. Hoveyda/Snapper[21b,22a] 2. Charette[15b,27]	1. Not likely 2. Not likely	1. Two 2. One or two
NH$_2$ / R_L — R_S	1. Red. amin. transfer H$_2$[37,38] 2. Reductive amination (H$_2$)[45,46] 3. Carbanion (Ellman)[17c]	1. Yes 2. Possible 3. Possible	1. One 2. Two 3. Three
NH$_2$ / R_L — R_S	Not enough examples	–	–
NH$_2$ / R_M — R_M	Carbanion addition (Ellman)[17c]	Possible	Three
NH$_2$ / R_M — R_L	Carbanion addition (Ellman)[17c,30]	Not likely	Three

Finally, few building blocks hold as much potential for aiding future drug development as α-chiral amines and are as inaccessible for general use as α-chiral amines are. Consequently, their incorporation into the drug discovery process will not become a commonplace reality until commercial access to diverse sets of α-chiral amines is made possible or when expedient, inexpensive, and robust preparations of them are forthcoming.

ACKNOWLEDGMENTS

Trained in industry as a process research chemist, I thank my prior industrial supervisors and colleagues for their insights over the years, but in particular my first supervisor, Dr. Robert Seemayer, for his patience and willingness to discuss all of the nuances of process chemistry that a younger scientist will naturally be curious about — invaluable!

REFERENCES

1. The author worked as a process research chemist for three years with Catalytica/DSM (Mountain View, CA) and then 2 years with Pharmacia/Pfizer (South San Francisco, CA) before joining Jacobs University (Bremen, Germany) as an assistant professor of organic chemistry in October 2003. E-mail address: t.nugent@jacobs-university.de.

2. (a) Chen, B.; Dingerdissen, U.; Krauter, J. G. E.; Rotgerink, H. G. J. L; Möbus, K.; Ostgard, D. J.; Panster, P.; Riermeier, T. H.; Seebald, S.; Tacke, T.; Trauthwein, H. *Appl. Catal. A: General* 2005, 280, 17–46; (b) Vilaivan, T.; Bhanthumnavin, W.; Sritana-Anant, Y. *Curr. Org. Chem.* 2005, 9, 1315–1392; (c) Breuer, M.; Ditrich, K.; Habicher, T.; Hauer, B.; Keßeler, M.; Stürmer, R.; Zelinski, T. *Angew. Chem.* 2004, 116, 806–843; *Angew. Chem., Int. Ed.* 2004, 43, 788-824; (d) Spindler, F.; Blaser, H.-U. Enantioselective reduction of C=N bonds and enamines with hydrogen. In *Transition Metals for Organic Synthesis*, 2nd Edition (Ed.: Beller, M.; Bolm, C.), Wiley-VCH, Weinheim, 2004, pp. 113–123; (e) Kukula, P.; Prins, R. *Topics in Catalysis* 2003, 25, 29–42; (f) Besson, M.; Pinel, C. *Topics in Catalysis* 2003, 25, 43–61; (g) Blaser, H.-U.; Malan, C.; Pugin, B.; Spindler, F.; Steiner, H.; Studer, M. *Adv. Synth. Catal.* 2003, 345, 103–151; (h) Juaristi, E.; León-Romo, J. L.; Reyes, A.; Escalante, J. Tetrahedron: *Asymmetry* 1999, 10, 2441–2495; (i) Kobayashi, S.; Ishitani, H. *Chem. Rev.* 1999, 99, 1069–1094.

3. Unfunctionalized amines here implies the exclusion of substrates requiring functional groups that are intimately involved in the transition state for production formation, e.g., ester chelation to a metal center.

4. The term 'chiral amine' is used in the title, but refers to an α-chiral amine, i.e., an α-carbon stereocenter adjacent to a nitrogen atom.

5. The time honored adage, especially in sophomore organic chemistry courses, of teaching students that over alkylation of amines makes their substitution chemistry unfit for consideration is outdated; though limited the strategy should not be ignored for the mono-alkylation of amines, see, for example, Hayler, J. D.; Howie, S. L. B.; Giles, R. G.; Negus, A.; Oxley, P. W.; Walsgrove, T. C.; Whiter, M. *Org. Process. Res. Dev.* 1998, 2, 3–9.

6. The remote insertion of nitrogen into the unfunctionalized backbone of an alkane is a true break through in technology, see: (a) Kim, M.; Mulcahy, J. V.; Espino, C. G.; Du Bois, J. *Org. Lett.* 2006, 8, 1073–1076; (b) Wehn, P. M.; Du Bois, J. *Org. Lett.* 2005, 7, 4685–4688; (c) Espino, C. G.; Fiori, K. W.; Kim, M.; Du Bois, J. *J. Am. Chem. Soc.* 2004, 126, 15378–15379.

7. (a) Blaser, H.-U.; Eissen, M.; Fauquex, P. F.; HungerBühler, K.; Schmidt, E.; Sedelmeier, G.; Studer, M. Comparison of Four Technical Syntheses of Ethyl (R)-2-Hydroxy-4-Phenylbutyrate. In *Asymmetric Catalysis on Industrial Scale: Challenges, Approaches, and Solutions*; Blaser, H.-U.; Schmidt, E. Eds.; Wiley-VCH Verlag GmbH & Co. KGaA: Weinheim, Germany, 2004, pp 91–103; (b) Blaser, H.-U.; Burkhardt, S.; Kirner, H. J.; Mössner, T.; Studer, M. *Synthesis* 2003, 1679–1682.

8. Similar examples of ketone reduction, followed by alcohol activation, and nucleophilic displacement thereof by an amine, have been demonstrated for other drug classes, see Reference 5 and Noyori, R.; Ohkuma, T. *Angew. Chem.* 2001, 113, 40–75; *Angew. Chem., Int. Ed.* 2001, 40, 40–73.

9. (a) Huffman, M. A.; Reider, P. J. *Tetrahedron Lett.* 1999, 40, 831–834; (b) Stereoselective process for Enalapril. Huffman, M. A.; Reider, P. J.; Leblond, C.; Sun, Y.; Merck & Co., Ltd. Patent number US6025500, 2000.

10. Note that very specialized syntheses, i.e., specialized for one substrate in particular, are not considered here.

11. (a) Bunlaksananusorn, T.; Rampf, F. *Synlett* 2005, 17, 2682–2684; (b) Ikemoto, N.; Tellers, D. M.; Dreher, S. D.; Liu, J.; Huang, A.; Rivera, N. R.; Njolito, E.; Hsiao, Y.; McWilliams, J. C.; Williams, J. M.; Armstrong, J. D.; Sun, Y.; Mathre, D. J.; Grabowski, E. J. J.; Tillyer, R. D. *J. Am. Chem. Soc.* 2004, 126, 3048–3049; (c) Hsiao, Y.; Rivera, N. R.; Rosner, T.; Krska, S. W.; Njolito, E.; Wang, F.; Sun, Y.; Armstrong, J. D.; Grabowski, E. J. J.; Tillyer, R. D.; Spindler, F.; Malan, C. *J. Am. Chem. Soc.* 2004, 126, 9918–9919.

12. (a) Hu, X.-P.; Zheng, Z. *Org. Lett.* 2005, 7, 419–422; (b) Zhang, Y. J.; Kim, K. Y.; Park, J. H.; Song, C. E.; Lee, K.; Lah, M. S.; Lee, S.-G. *Adv. Synth. Catal.* 2005, 347, 563–570; (c) You, J.; Drexler, H.-J.; Zhang, S.; Fischer, C.; Heller, D. *Angew. Chem.* 2003, 115, 942–945; *Angew. Chem., Int. Ed.* 2003, 42, 913–916.

13. Phua, P. H.; de Vries, J. G.; Hiia, K. K. *Adv. Synth. Catal.* 2005, 347, 1775–1780.

14. Kadyrov, R.; Riermeier, T. H.; Dingerdissen, U.; Tararov, V. I.; Börner, A. *J. Org. Chem.* 2003, 68, 4067–4070.

15. (a) Menzel, A.; Werner, H.; Altenbuchner, J.; Groeger, H. *Eng. in Life Sci.* 2004, 4, 573–576; (b) See references cited within: Côté, A.; Charette, A. B. *J. Org. Chem.* 2005, 70, 10864–10867.

16. (a) Qin, H.; Yamagiwa, N.; Matsunaga, S.; Shibasaki, M. *J. Am. Chem. Soc.* 2006, 128, 1611–1614; (b) Karshtedt, D.; Bell, A. T.; Tilley, T. D. *J. Am. Chem. Soc.* 2005, 127, 12640–12646; (c) Zulys, A.; Dochnahl, M.; Hollmann, D.; Loehnwitz, K.; Herrmann, J.-S.; Roesky, P. W.; Blechert, S. *Angew. Chem.* 2005, 117, 7972–7976; *Angew. Chem. Int. Ed.* 2005, 44, 7794–7798.

17. (a) McMahon, J. P.; Ellman, J. A. *Org. Lett.* 2004, 6, 1645–1647; (b) J. A. Ellman, *Pure Appl. Chem.* 2003, 75, 39–46; (c) Ellman, J. A.; Owens, T. D.; Tang, T. P. *Acc. Chem. Res.* 2002, 35, 984–995.

18. a) Liu, G.; Cogan, D. A.; Owens, T. D.; Tang, T. P.; Ellman, J. A. *J. Org. Chem.* 1999, 64, 1278–1284; b) Cogan, D. A.; Liu, G.; Kim, K.; Backes, B. J.; Ellman, J. A. *J. Am. Chem. Soc.* 1998, 120, 8011–8019.

19. (a) Cogan, D. A.; Liu, G.; Ellman, J. *Tetrahedron* 1999, 55, 8883–8904; (b) Cogan, D. A.; Ellman, J. A. *J. Am. Chem. Soc.* 1999, 121, 268–269.

20. (a) Shaw, A. W.; deSolms, S. J. *Tetrahedron Lett.* 2001, 42, 7173–7176.

21. (a) Akullian, L. C.; Snapper, M. L.; Hoveyda, A. H. *Angew. Chem.* 2003, 115, 4376–4379; *Angew. Chem. Int. Ed.* 2003, 42, 4244–4247; (b) Akullian, L. C.; Porter, J. R.; Traverse, J. F.; Snapper, M. L.; Hoveyda, A. H. *Adv. Synth. Catal.* 2005, 347, 417–425.

22. (a) Porter, J. R.; Traverse, J. F.; Hoveyda, A. H.; Snapper, M. L. *J. Am. Chem. Soc.* 2001, 123, 10409–10410, (b) For further N-aryl oxidative cleavage protocols, see: Kobayashi, S.; Ueno, M.; Suzuki, R.; Ishitani, H.; Kim, H.-S.; Wataya, Y. *J. Org. Chem.* 1999, 64, 6833–6841.

23. No experimental description could be found.

24. See the Supporting Information section of Ref. 15b.

25. McLaughlin, M.; Rubio, S. G.; Tilstam, U.; Antunes, O. A. C.; Laird, T.; Yadav, G. D.; Andrei Zlota, A. *Org. Process Res. Dev.* 2006, 10, 168–183. See specifically pages 169–170.

26. (a) Côté, A.; Boezio, A. A.; Charette, A. B. *Angew. Chem.* 2004, 116, 6687–6690; *Angew. Chem. Int. Ed.* 2004, 43, 6525–6528; (b) Boezio, A. A.; Pytkowicz, J.; Côté, A.; Charette, A. B. *J. Am. Chem. Soc.* 2003, 125, 14260–14261; (c) Boezio, A. A.; Charette, A. B. *J. Am. Chem. Soc.* 2003, 125, 1692–1693.

27. Côté, A.; Boezio, A. A.; Charette, A. B. *Proc. Natl. Acad. Sci. U.S.A.* 2004, 101, 5405–5410.

28. Wakchaure, V. N.; Mohanty, R. R.; Shaikh, J. A.; Nugent, T. C. *Eur. J. Org. Chem.* 2007, 959–964.

29. (a) Bandini, M.; Cozzi, P. G.; Umani-Ronchi, A.; Villa, M. *Tetrahedron* 1999, 55, 8103–8110; (b) Alvaro, G.; Savioa, D. Valentinetti, M. R. *Tetrahedron* 1996, 52, 12571–12586.

30. Weix, D. J.; Shi, Y.; Ellman, J. A. *J. Am. Chem. Soc.* 2005, 127, 1092–1093.

31. Blacker, J.; Martin, J. Scaleup Studies in Asymmetric Transfer Hydrogenation. In *Asymmetric Catalysis on Industrial Scale: Challenges, Approaches, and Solutions*; Blaser, H.-U.; Schmidt, E. Eds.; Wiley-VCH Verlag GmbH & Co. KGaA: Weinheim, Germany, 2004, pp 201–216.

32. (a) Transfer hydrogenation process. Martin, J.; Campbell; L. A.; Avecia Limited, Great Britain. Patent publication number WO0112574, 2001; (b) Transfer hydrogenation process. Martin, J.; Campbell, L. A.; Avecia Limited, Great Britain. Patent publication number US6696608, 2004.

33. (a) Spindler, F.; Blaser, H.-U. Enantioselective reduction of C=N bonds and enamines with hydrogen. In *Transition Metals for Organic Synthesis*, 2nd Edition (Ed.: M. Beller, C. Bolm), Wiley-VCH, Weinheim, 2004, pp. 113-123; (b) Spindler, F.; Blaser, H.-U. *Adv. Synth. Catal.* 2001, 343, 68–70.

34. (a) Willoughby, C. A.; Buchwald S. L. *J. Am. Chem. Soc.* 1994, 116, 8952–8965; (b) Willoughby, C. A.; Buchwald, S. L. *J. Am. Chem. Soc.* 1994, 116, 11703–11714.

35. (a) Verdaguer, X.; Lange, U. E. W.; Reding, M. T.; Buchwald, S. L. *J. Am. Chem. Soc.* 1996, 118, 6784–6785; (b) Reding, M. T.; Buchwald, S. L. *J. Org. Chem.* 1998, 63, 6344–6347; (c) Hansen, M. C.; Buchwald, S. L. *Org. Lett.* 2000, 2, 713–715; (d) Yun, J.; Buchwald, S. L. *J. Org. Chem.* 2000, 65, 767–774.

36. Blaser, H.-U. *Adv. Synth. Catal.* 2002, 344, 17–31.

37. Kadyrov, R.; Riermeier, T. H. *Angew. Chem.* 2003, 115, 5630–5632; *Angew. Chem., Int. Ed.* 2003, 42, 5472–5474.

38. Method for the production of amines by reductive amination of carbonyl compounds under transfer-hydrogenation conditions. Börner, A.; Dingerdissen, U.; Kadyrov, R.; Riermeier, T. H.; Tararov, V.; Degussa AG, Germany. Patent publication number US2004267051, 2004.

39. Method for producing amines by homogeneously catalyzed reductive amination of carbonyl compounds. T. Riermeier, K.-J. Haack, U. Dingerdissen, A. Boerner, V. Tararov, R. Kadyrov; Degussa AG, Germany. Patent publication number US6884887, 2005.

40. (a) Emerson, W. S. *Org. React.* 1948, 4, 174–255; (b) Moore, M. L. *Org. React.* 1949, 5, 301–330.

41. (a) Reetz, M. T. *Organotitanium Reagents in Organic Synthesis*; Reetz, M. T. Ed.; Springer-Verlag: Berlin, 1986; specifically see p 107; (b) Reetz, M. T.; Westermann, J.; Steinbach, R.; Wenderoth, B.; Peter, R.; Ostarek, R.; Maus, S. *Chem. Ber.* 1985, 118, 1421–1440.

42. Mattson, R. J.; Pham, K. M.; Leuck, D. J.; Cowen, K. A. *J. Org. Chem.* 1990, 55, 2552–2554.

43. (a) Miriyala, B.; Bhattacharyya, S.; Williamson, J. S. *Tetrahedron* 2004, 60, 1463–1471; (b) Bhattacharyya, S.; Kumpaty, H. *J. Synthesis* 2005, 2205–2209.

44. (a) Nugent, T. C.; Seemayer, R. *Org. Process. Res. Dev.* 2006, 10, 142–148 and references cited therein for earlier work regarding the use of hydrides for Lewis acid based reductive amination; (b) Process for the Preparation of (*S,S*)-*cis*-2-Benzhydryl-3-benzylaminoquinuclidine; Nugent, T. C.; Seemayer, R. (Pfizer Products, Inc. and DSM Pharmaceuticals, Inc.), publication number: WO2004035575, 29 April 2004.

45. Nugent, T. C.; Wakchaure, V. N.; Ghosh, A. K.; Mohanty, R. R. *Org. Lett.* 2005, 7, 4967–4970.

46. Nugent, T. C.; Ghosh, A. K.; Wakchaure, V. N.; Mohanty, R. R. *Adv. Synth. Catal.* 2006, 348, 1289–1299.

47. We recently became aware of Pd-C catalysts specifically for debenzylations and suppression of racemization, see: V. Farina, K. Grozinger, H. Müller-Bötticher, G. P. Roth, Ontazolast: The Evolution of a Process. In *Process Chemistry in the Pharmaceutical Industry*; K. G. Gadamasetti, Ed.; Marcel Dekker, Inc.: New York, 1999, pp 107–124.

48. The Ellman method for sulfinyl ketimine synthesis would similarly suffer, but perhaps can be alleviated by using the Lewis acids B(OiPr)$_3$ or Al(OiPr)$_3$.

49. Nugent, T. C., El-Shazly, M.; Wakchaure, V. N. submitted for publication in 2007.

50. For a brief overview regarding imine formation, see page 1291 of Ref. 46.

51. Zhang, Z.; Zhu, G.; Jiang, Q.; Xiao, D.; Zhang, X. *J. Org. Chem.* 1999, 64, 1774–1775.

52. Burk, M. J.; Casy, G.; Johnson, N. B. *J. Org. Chem.* 1998, 63, 6084–6085.

53. (a) Bernsmann, H.; Van den Berg, M.; Hoen, R.; Minnaard, J. A.; Mehler, G.; Reetz, M. T.; De Vries, J. G.; Feringa, B. L. *J. Org. Chem.* 2005, 70, 943–951; (b) Hoen, R.; Van den Berg, M.; Bernsmann, H.; Minnaard, A. J.; De Vries, J. G.; Feringa, B. L. *Org. Lett.* 2004, 6, 1433–1436.

54. Chi, Y.; Tang, W.; Zhang, X. Rhodium-Catalyzed Asymmetric Hydrogenation. In *Modern Rhodium-Catalyzed Organic Reactions*; Evans, P. A. Ed.; Wiley-VCH Verlag GmbH & Co. KGaA: Weinheim, Germany, 2005, pp 1–31.

55. Li, W.; Zhang, X. *J. Org. Chem.* 2000, 65, 5871–5874.

56. Xiao, D.; Zhang, Z.; Zhang, X. *Org. Lett.* 1999, 1, 1679–1681.

57. Zhu, G.; Zhang, X. *J. Org. Chem.* 1998, 63, 9590–9593.

58. Tang, W.; Chi, Y.; Zhang, X. *Org. Lett.* 2002, 4, 1695–1698.

59. Huang, H.; Liu, X.; Chen, H.; Zheng, Z. *Tetrahedron: Asymmetry* 2005, 16, 693–697.

9 Unnatural Amino Acids

David J. Ager

CONTENTS

9.1 INTRODUCTION

Amino acids have become firmly established as building blocks for a wide variety of chemicals ranging from absorbents to pharmaceuticals. Nature, for the most part, uses 20 amino acids to build proteins. In addition to these proteinogenic amino acids, there is still diversity for specific applications and uses ranging from changes in the side chains of α-amino acids, a reversal of stereochemistry at the center, to positioning of the amino functionality at a site removed from the carboxylate group as in β- and γ-amino acids. These nonproteinogenic amino acids are commonly referred to as "unnatural," even though some can be isolated from natural sources.

In order to limit the size of this chapter, α-amino acids will be the emphasis.[1] The major problem for the synthesis of unnatural amino acids revolves around control of the stereogenic center, especially when large-scale synthesis is required. Inspection of the structure of an α-amino acid (**1**) allows a number of potential disconnections and approaches (Figure 9.1). The desired stereogenic center can be obtained by a resolution of a racemic mixture either in a kinetic or dynamic manner (route a). The reagent to achieve the separation can be chemical, biological, or a combination of the two. To achieve a dynamic resolution, only one substituent can be present (i.e., $R^2 = H$). Resolutions are often achieved by formation or transformation of just one enantiomer of a derivative, such as an ester (route b) or amide (route c).

The stereogenic center can be introduced by the use of a chiral auxiliary or template. Once again, this control unit can be attached to the nitrogen or carboxylate moieties (routes d and e). The plethora of chiral auxiliaries available also allows for the carboxylate and amino groups to be masked during a number of reactions. In most cases, the side chain is introduced to form the stereogenic center, although the nitrogen and carboxylate groups are also possibilities. The cost of the auxiliary used relegates this approach to small-scale work, but as the chemistry is well known,

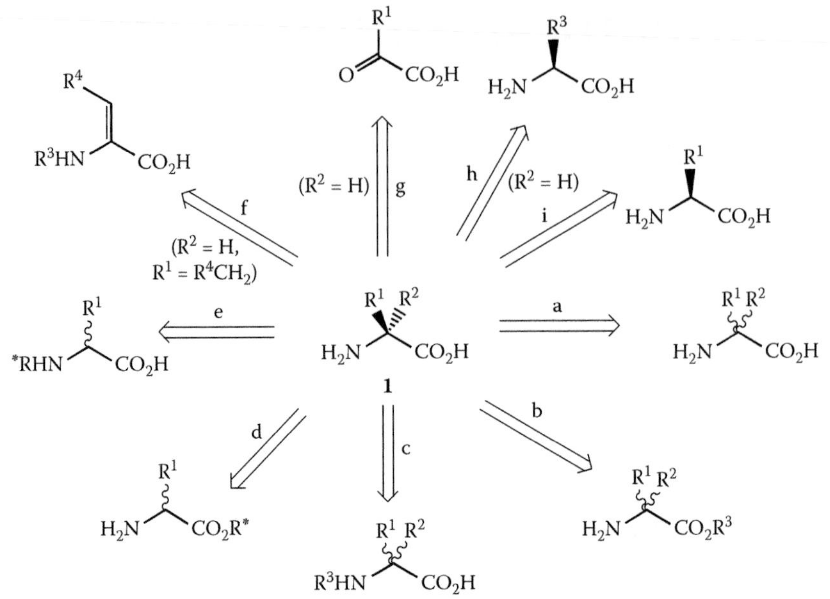

FIGURE 9.1 Approaches to α-amino acids.

it does have the advantage of working for a wide range of products with little process development being required.

Another approach that has been well documented and has been used at scale is asymmetric hydrogenation. In this case, an enamide is reduced to afford a derivative of the amino acid. Because hydrogen is added across the unsaturation, the method can only be applied to monosubstituted α-amino acids (route f).

Biological systems can use fermentative methods to produce amino acids. In some cases a metabolic pathway has been changed, but this is usually for natural amino acids, while in other cases an enzyme has been applied to a specific transformation. An example is the use of transaminases (route g) to produce unnatural amino acids such as 2-aminobutyrate and *tert*-leucine.

The final approaches rely on the transformation of a readily available, natural amino acid where the side chain is manipulated to provide an unnatural analogue (route h) or the stereogenic center of the natural amino acid is used to control the introduction of a second side chain (route i). This last method is closely related to the use of chiral templates and auxiliaries.

9.2 RESOLUTIONS

It is possible to resolve amino acids through either the amino or carboxylate functionality. If salt formation is the method employed, then usually the corresponding functionality in the substrate is protected or masked to avoid interference. The more common approach is to use an enzyme such as an esterase or lipase to hydrolyze just one enantiomer of an ester or amide substrate (Scheme 9.1).[2] An alternative is to use an amidase to convert an *N*-amide to the amine (Scheme 9.2).[3] Of

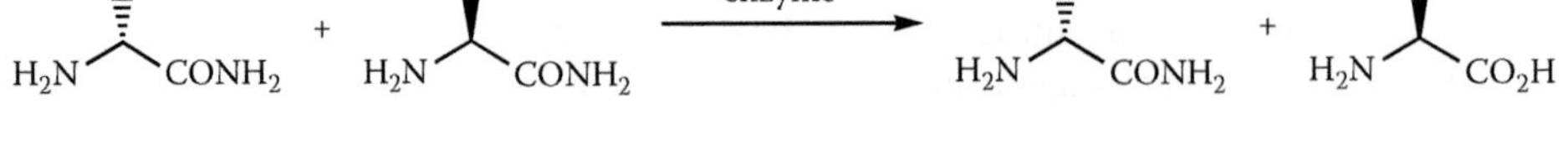

SCHEME 9.1

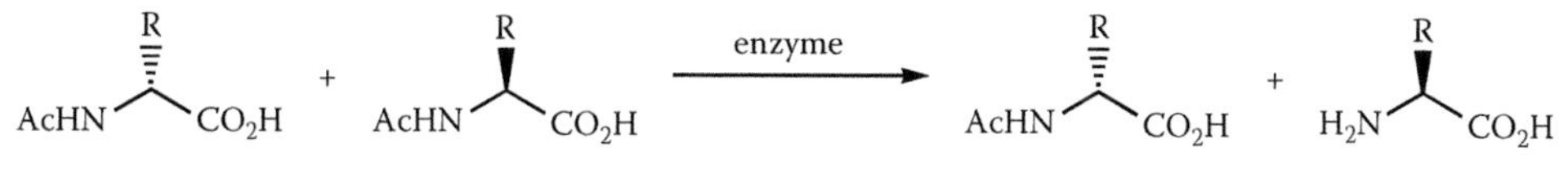

SCHEME 9.2

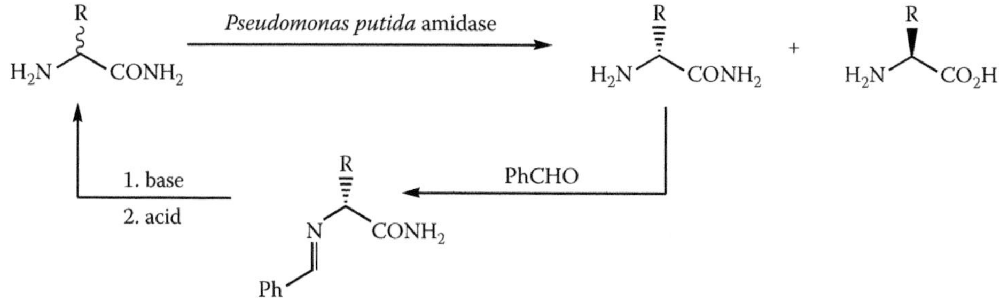

SCHEME 9.3

course, enzymes can be employed that accept the natural isomer as the unnatural one is left untouched. As it is relatively simple, by acid–base chemistry, to separate the hydrolyzed product from the untouched substrate, the resolution is achieved. A chemical transformation can then be performed to convert the substrate to the corresponding amino acid in a subsequent and separate transformation.

DSM has developed a general, industrial-scale process for the production of either L- or D-amino acids through the hydrolysis of the amide (Scheme 9.3).[4–6] The product amino acid and untouched amide are easily separated. The amide can be recycled by racemization of the benzaldehyde imine. The resolution method has been extended to α-disubstituted amino acids of which L-methyl-Dopa (**2**) is an example although the undesired isomer is not simple to recycle through the process due to the absence of an α-hydrogen.[7]

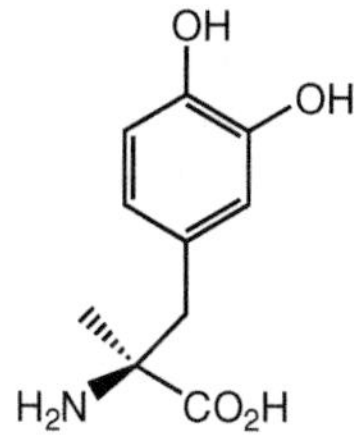

2

Another example of an α,α-disubstituted α-amino acid to be accessed by resolution is (*R*)-2-amino-2-ethylhexanoic acid (**3**), a building block for the anticholesterol drug 2164U90 (**4**) (Scheme 9.4).[8]

Attempts to resolve the racemic acid of **3** and its ester through classic resolution failed. In the early stages of development, a process based on Schöllkopf's asymmetric synthesis was developed (see Section 9.3). Large-scale development work was aimed at finding a biocatalytic process to resolve the amino acids. Racemic α,α-disubstituted α-amino esters were synthesized by standard chemistry through alkylation of the Schiff's bases formed from the amino esters (Scheme 9.5),[9] or through formation of hydantoins.[10]

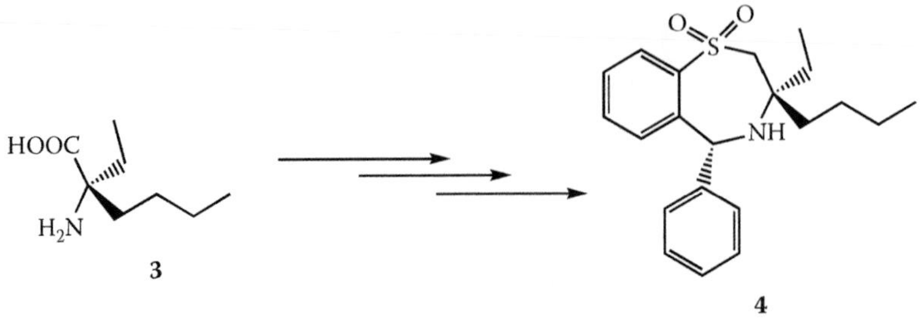

3

4

SCHEME 9.4

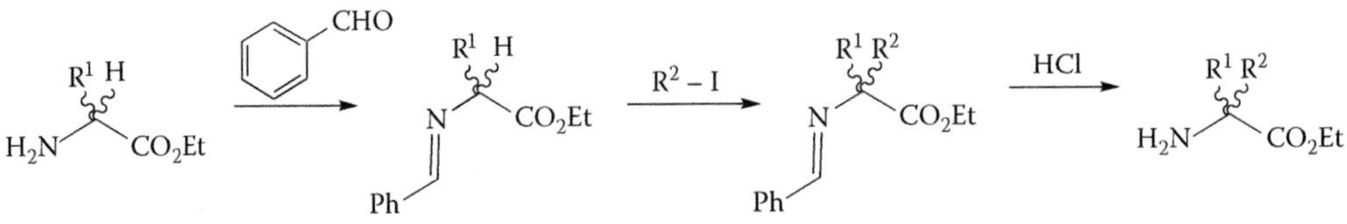

SCHEME 9.5

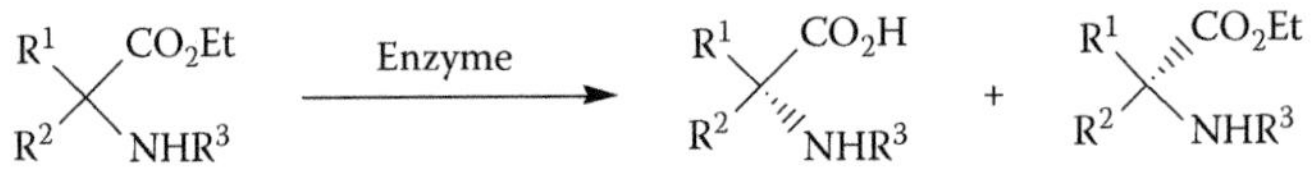

SCHEME 9.6

Typical commercial enzymes reported for resolution of amino acids were tested. Whole cell systems containing hydantoinase were found to produce only α-monosubstituted amino acids[11–17]; the acylase-catalyzed resolution of *N*-acyl amino acids had extremely low rates toward α-dialkylated amino acids[18,19]; and the nitrilase system obtained from Novo Nordisk showed no activity toward the corresponding 2-amino-2-ethylhexanoic amide.[20,21] Finally, a large-scale screening of hydrolytic enzymes for enantioselective hydrolysis of racemic amino esters was carried out. Of all the enzymes and microorganisms screened, pig liver esterase (PLE) and *Humicola langinosa* lipase (Lipase CE, Amano) were the only ones found to catalyze the hydrolysis of the substrate (Scheme 9.6).

Resolutions have been used to prepare β- and higher amino acids by a variety of resolution techniques.[22] In these cases, the lack of an epimerizable stereogenic center does not allow for dynamic resolutions to be employed.

The use of an enzymatic system to prepare δ-dicarboxylic amino acids is discussed in Section 9.5.2.

9.2.1 DYNAMIC RESOLUTIONS

The resolutions described above result in a maximum 50% yield, as the undesired isomer has to be discarded. Although the unwanted product may be racemized and resubjected to the resolving agent, this can prove costly. Only monosubstituted α-amino acids can be racemized as they have an exchangeable, acidic α-hydrogen. In a number of examples, it is possible to perform a racemization in the presence of the resolving agent. In these cases, an almost quantitative yield is attainable, as the unwanted enantiomer is converted to the desired one during the course of the reaction. The synthesis of D-proline (**5**) is an example where butyraldehyde is used to bring about

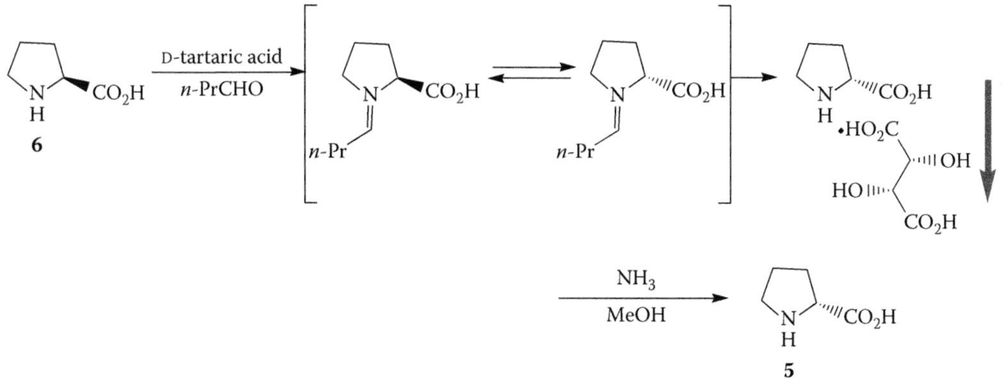

SCHEME 9.7

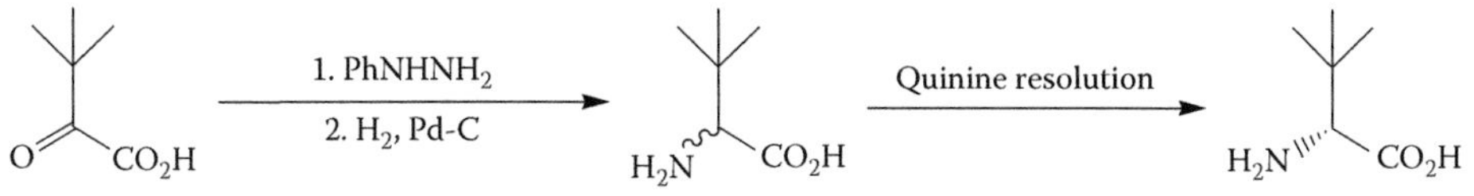

SCHEME 9.8

the racemization through imine formation (Scheme 9.7).[23,24] Treatment of L-proline (**6**) with one equivalent of D-tartaric acid and a catalytic amount of *n*-butyraldehyde in butyric acid, causes racemization due to the reversible formation of the proline-butyraldehyde Schiff base. The newly generated D-proline then forms an insoluble salt with the D-tartaric acid that precipitates out of the solution, while the soluble L-proline is continuously being racemized. The net effect is the continuous transformation of the soluble L-proline to the insoluble D-proline•D-tartaric acid complex resulting in near complete conversion. Treatment of the D-proline•D-tartaric acid complex with concentrated ammonia in methanol liberates the D-proline (**5**) (99% ee, with 80 to 90% overall yield from L-proline).

Another example is provided by D-*tert*-leucine (**7**) (Scheme 9.8) where an asymmetric hydrogenation approach cannot be used due to the lack of a hydrogen at the β-position. In this example, quinine was used as the resolving agent.[1,25]

An example of a very efficient asymmetric transformation is the preparation of (*R*)-phenylglycine amide (**8**) where the synthesis of the amino acids, by a Strecker reaction, and racemization are all performed at the same time (Scheme 9.9).[26,27] This offers a good alternative to the enzymatic resolution of phenylglycine amide with an (*S*)-specific amidase.[28]

The use of an enzyme to generate the stereogenic center while a racemization is occurring is discussed in Section 9.5.3.

9.3 CHIRAL AUXILIARIES

The use of chiral auxiliaries or templates is usually performed at smaller scale. The cost of the chiral auxiliary or template usually prohibits large-scale applications. If the chiral agent is destroyed during its removal, then, obviously, the agent cannot be recycled. Even for auxiliaries where a recycle might be feasible, good manufacturing practice (GMP) and other purification factors can make this a costly exercise. The advantage of using an auxiliary is that the stereochemical outcome

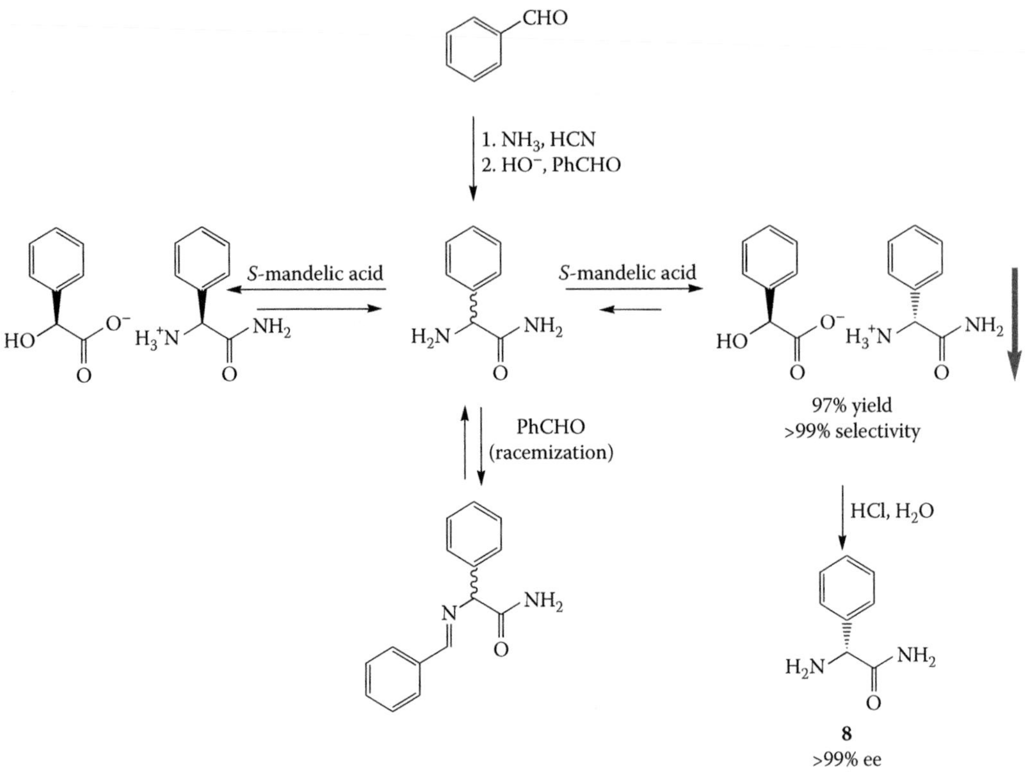

SCHEME 9.9

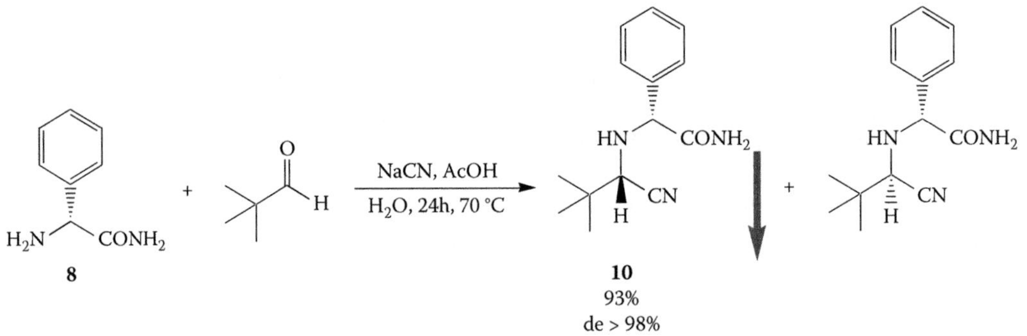

SCHEME 9.10

of the reaction is predictable and, therefore, can be applied to "new" systems. A wide variety of auxiliary systems are available, but for the synthesis of α-amino acid derivatives, many are derived from amino acids themselves.

An approach to L-*tert*-leucine (**9**) is based on the use of a cheap chiral template, D-phenylglycine amide (**8**) in an asymmetric Strecker reaction with pivaldehyde, and HCN generated *in situ* from sodium cyanide and acetic acid in water gave the aminonitrile **10** in high yield and diastereoisomeric excess (Scheme 9.10).[29]

The amide **10** can be converted to (*S*)-*tert*-leucine (**9**) as outlined in Scheme 9.11.

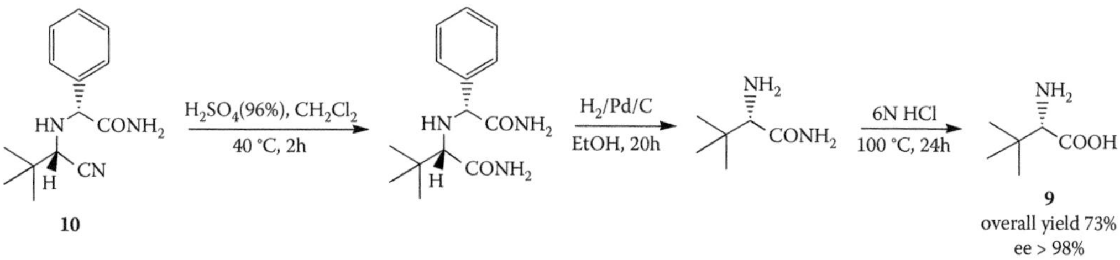

SCHEME 9.11

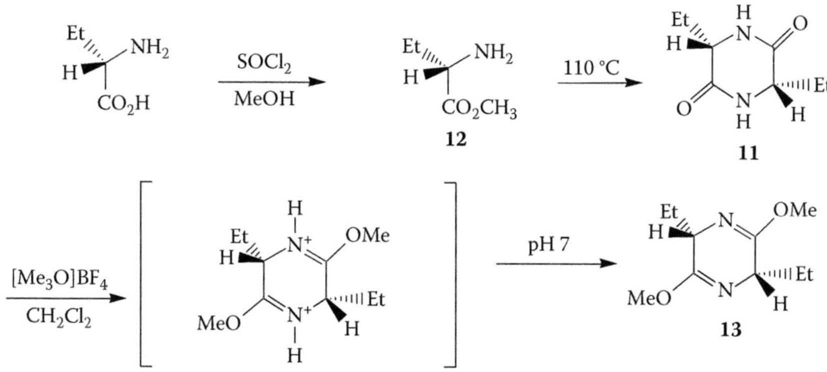

SCHEME 9.12

A process based on Schöllkopf's methodology was developed for the synthesis of (R)-2-amino-2-ethylhexanoic acid (3).[8,30] Formation of piperazinone **11** through dimerization of methyl (S)-(+)-2-aminobutyrate (**12**) was followed by enolization and methylation to give (3S,6S)-2,5-dimethoxy-3,6-diethyl-3,6-dihydropyrazine (**13**) (Scheme 9.12).

Diastereoselective alkylation of **13** with n-butyl bromide produced derivative **14**, which was then hydrolyzed by strong acid to afford the dialkylated (R)-amino acid **3** (~90% ee), together with partially racemized 2-aminobutyric acid (Scheme 9.13). Although this process provided gram quantities of the desired amino acids as analytical markers and test samples, it was neither practical nor economical for large-scale production.

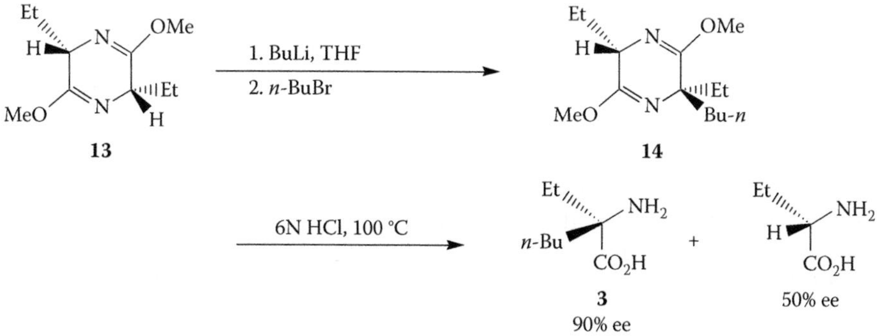

SCHEME 9.13

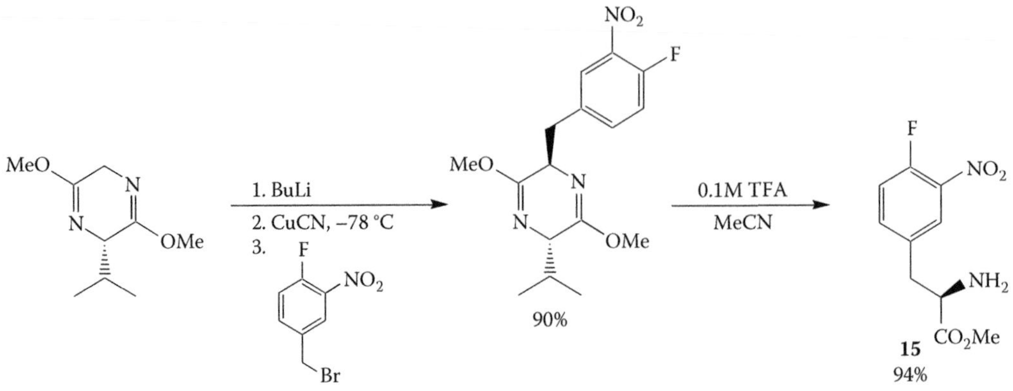

SCHEME 9.14

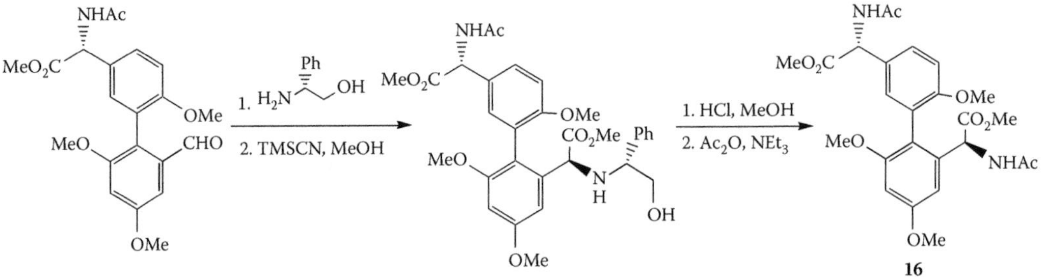

SCHEME 9.15

Another example of the use of Schöllkopf's method is for the synthesis of 4-fluoro-3-nitrophe-nylalanine methyl ester (**15**) (Scheme 9.14).[31] In this example, valine is the amino acid used as a chiral auxiliary.

In addition to amino acids, 1,2-amino alcohols can be used as chiral auxiliaries.[32] The synthesis of the amino acid derivative **16** outlined in Scheme 9.15 is a variation of a Strecker reaction.[33,34]

Evans' oxazolidinone chemistry is well documented,[32] which allows for a wide variety of reactions to be performed with a high degree of predictability. In addition to alkylation reactions to introduce the side chain of an amino acid, the nitrogen group can also be introduced in a variety of ways, one of which is illustrated in Scheme 9.16.[35]

For β-amino acids, the Michael addition of a nitrogen nucleophile to an α,β-unsaturated system has been advocated as a method to prepare β-amino acid derivatives (Scheme 9.17).[36–39] However, the method is not atom economical with regard to nitrogen delivery, and low temperatures are also required, making the approach a poor candidate for large-scale reactions. An additional problem can be encountered during the removal of the template group that induces the asymmetry, as this transformation can be low yielding due to side reactions.[40]

This template removal problem was encountered during studies to prepare the $\alpha_v\beta_3$ integrin antagonist **17**. The route that emerged used phenylglycinol as a template in a Reformatsky approach (Scheme 9.18). Unfortunately, lead tetraacetate had to be used to remove this template, so there are still problems to be overcome in this area.[41]

In many cases, chiral auxiliaries have a rich chemistry that allows them to be used at small scale in a rapid manner, as the precedents exist. However, at larger scale, the lack of atom economy and costs associated with recycle or destruction of the unit can become major issues.

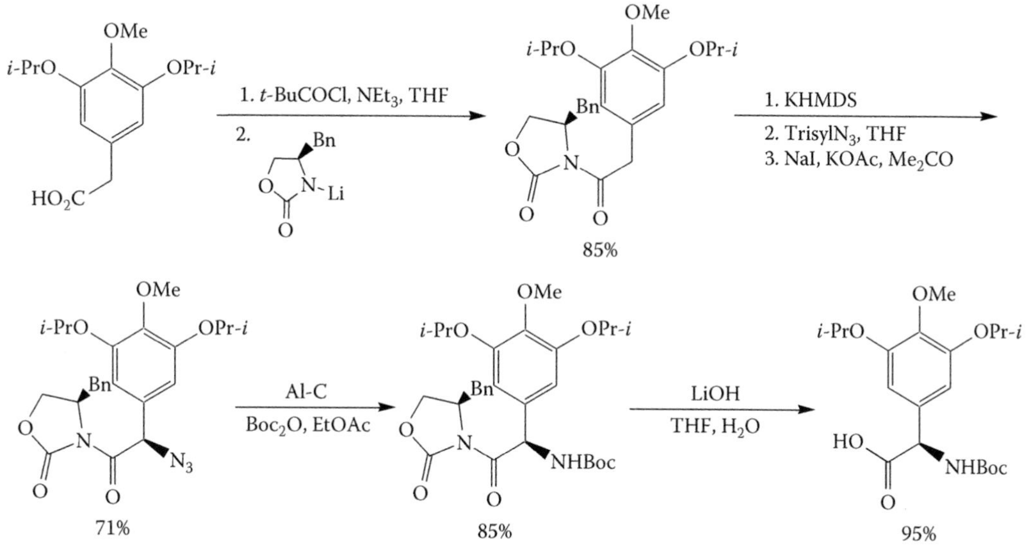

SCHEME 9.16

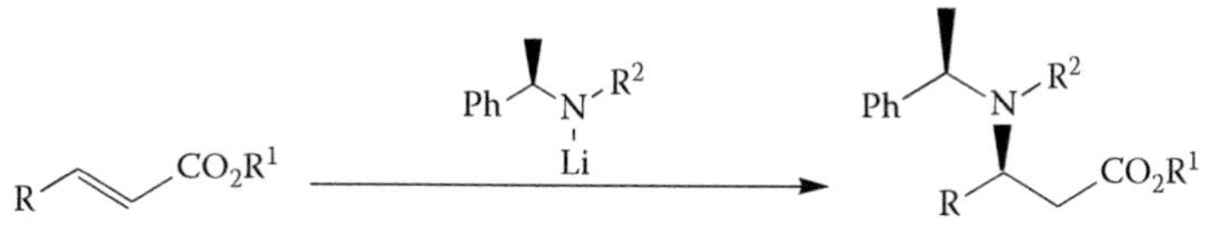

SCHEME 9.17

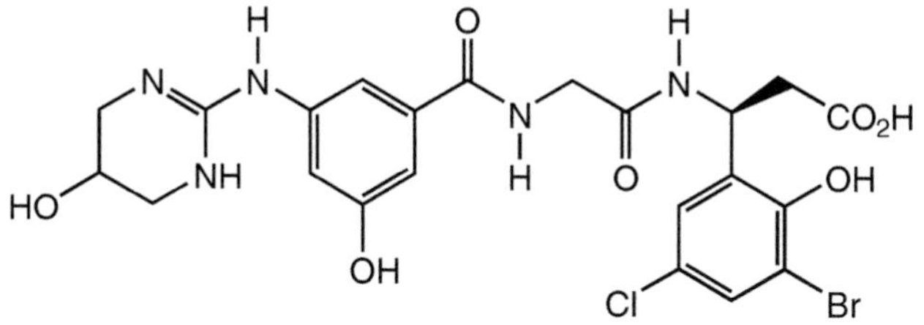

17

9.4 ASYMMETRIC HYDROGENATIONS

There are literally thousands of hydrogenation catalysts and ligand systems available for the preparation of unnatural amino acids that can be used at scale.[42] In many cases, the ligands were developed to circumvent legal and patent issues and to allow a company freedom to operate. Despite the plethora of ligands in the literature, few are available in bulk quantities, and thus, most of these have arisen from companies applying their own technology.

Perhaps the most famous example of the use of asymmetric hydrogenation at scale for the product of an "unnatural" amino acid is the Monsanto synthesis of L-dopa, a drug used for the treatment of Parkinson's disease (Scheme 9.19).[43–46] The methodology with the Knowles' catalyst system has been extended to a number of other unnatural amino acids.[46,47]

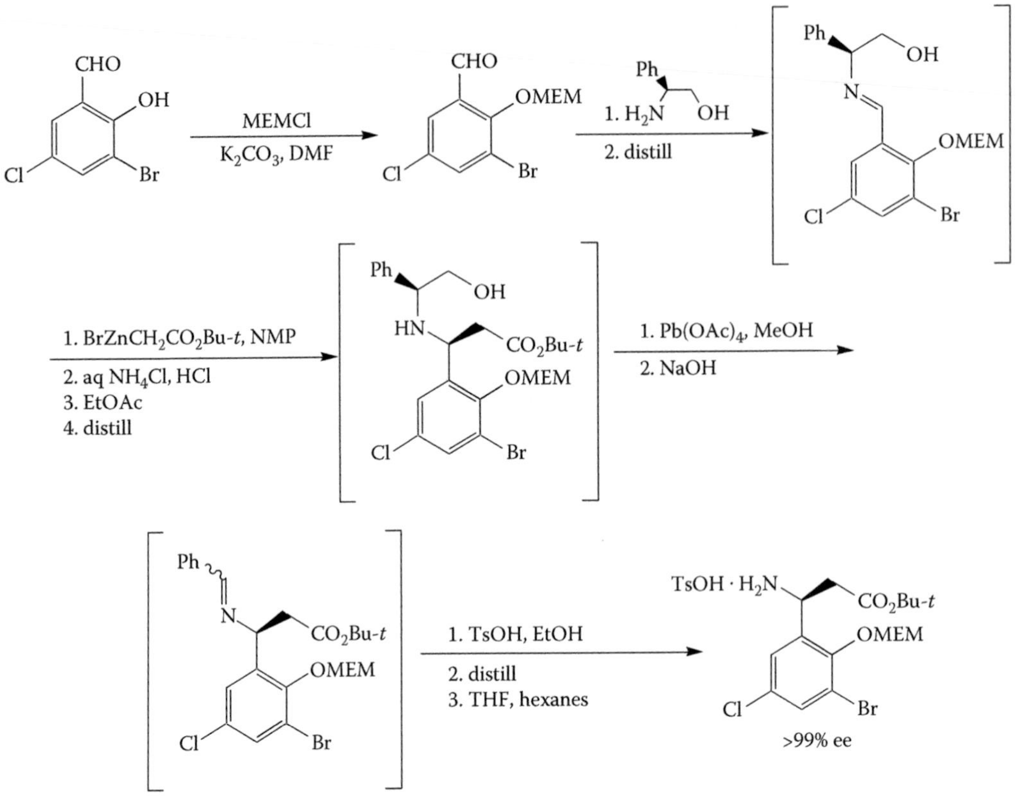

SCHEME 9.18

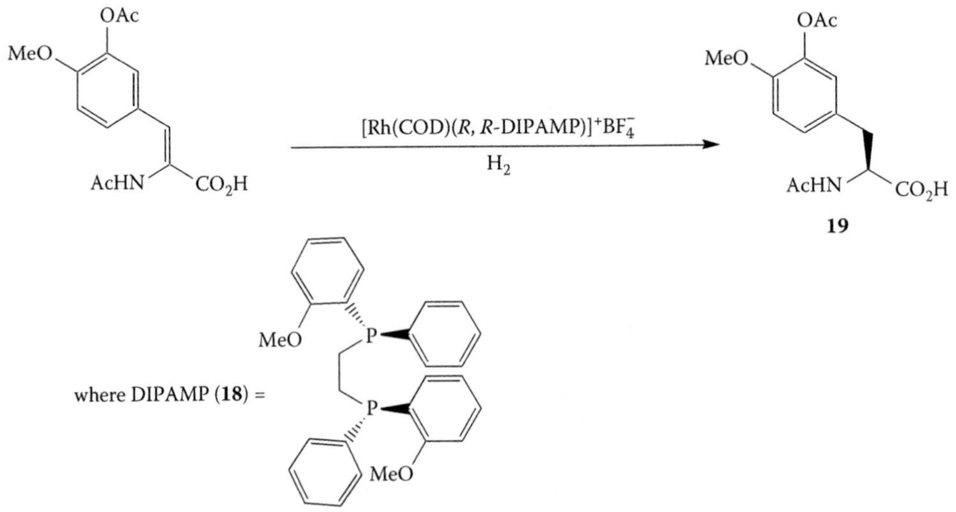

where DIPAMP (**18**) =

SCHEME 9.19

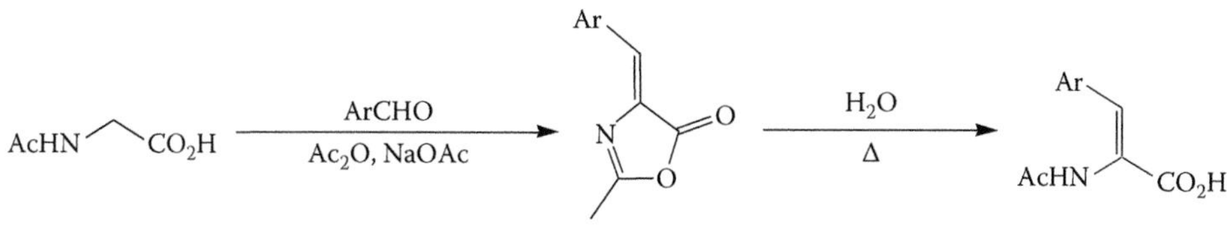

SCHEME 9.20

The asymmetric catalyst is based on the chiral bisphosphine, *R,R*-DIPAMP (**18**), that has chirality at the phosphorus atoms and can form a five-membered chelate with rhodium. The asymmetric reduction of the Z-enamide proceeds in 96% ee (Scheme 9.19).[44] The pure isomer of the protected amino acid intermediate **19** can be obtained upon crystallization from the reaction mixture as it is a conglomerate.[43] Although the catalyst system is amenable to the preparation of a wide variety of amino acids, especially substituted phenylalanine derivatives,[46,47] a major shortcoming of the approach is the need to have just the Z-enamide isomer as the substrate.

Fortunately, the general method for the preparation of the enamides is stereoselective to the Z-isomer when an aryl aldehyde is used (Scheme 9.20).[48] Other approaches to the enamide are also available, such as the Heck or Suzuki reactions (cf. Scheme 9.36).[49]

In addition to DIPAMP, which is expensive to synthesize, DuPhos and MonoPhos ligands have been employed to prepare unnatural amino acids at scale.

The MonoPhos™ family of ligands for the reduction of C–C double bonds, including enamides, is based on the 2,2-bisnaphthol (BINOL) backbone. These phosphoramidite ligands are comparatively inexpensive to prepare compared to bisphosphine ligands. For enamide reductions with MonoPhos (**20**) as ligand, it was found that the reaction is strongly solvent dependent.[50] Very good enantioselectivities were obtained in nonprotic solvents.[51,52]

The phosphoramidite ligands are surprisingly stable, even in protic solvents. Whereas, in general, P–N bonds are not stable in the presence of acids, the hydrogenation of *N*-acetyldehydroamino acids proceeded smoothly. A large range of substituted *N*-acetyl phenylalanines and their esters could thus be prepared by asymmetric hydrogenation with Rh/MonoPhos with enantioselectivities ranging from 93 to 99% (Scheme 9.21).[51,53] The configuration of the amino acid products always is the opposite of that of the BINOL in the ligand used in the hydrogenation.

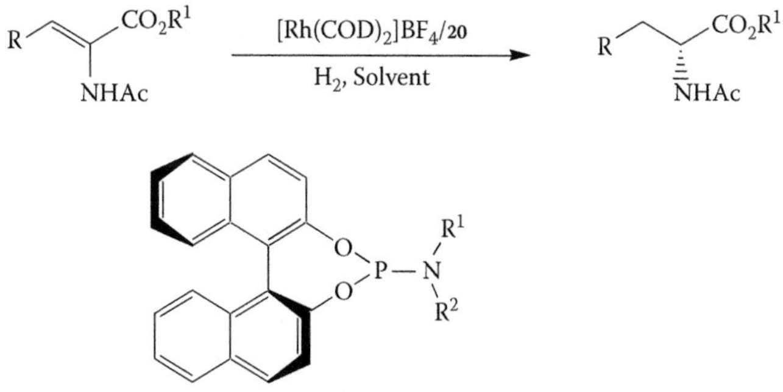

20 a $R^1 = R^2 = Me$ (MonoPhos)
b $R^1, R^2 = -(CH_2)_5-$
c $R^1 = R^2 = Et$
d $R^1 = (R)$-CHMePh, $R^2 = H$

SCHEME 9.21

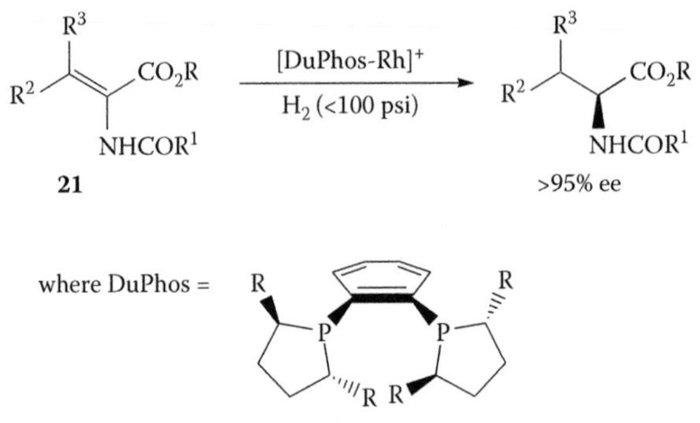

SCHEME 9.22

Due to the relative ease of synthesis of the phosphoramidites, it is simple to introduce variations within the ligand, especially the substitution on the nitrogen atom. In turn, this allows for rapid screening of ligands to find the best one for a specific substrate. The screen can be used to not only increase enantioselectivities but also to turn over numbers and frequencies. In a number of cases, the piperidine analogue (**20b**) has been found to be superior for α-amino acid and ester production over MonoPhos (**20a**).[51,54–58] The best ligands in terms of enantioselectivity are **19c**[58] and **20b**.[57] The rate of the hydrogenation is retarded by bulky groups on the nitrogen atom and by electron-withdrawing substituents on the BINOL.

Rhodium complexes of the type [(COD)Rh(DuPhos)]$^+$X$^-$ (X = weakly or noncoordinating anion) have been developed as one of the most general classes of catalyst precursors for efficient, enantioselective low-pressure hydrogenation of enamides (**21**) (Scheme 9.22).[59,60] The DuPhos approach overcomes some of the limitations of the DIPAMP system as the substrates may be present as mixtures of *E*- and *Z*-geometric isomers. For substrates that possess a single β-substituent (e.g., R^3 = H), the Me-DuPhos-Rh and Et-DuPhos-Rh catalysts were found to give enantioselectivities of 95 to 99% for a wide range of amino acid derivatives.[60,61]

As with the DIPAMP and MonoPhos ligand systems, a variety of *N*-acyl protecting groups such as acetyl, Cbz, or Boc may be employed, and enamides **21** may be used either as carboxylic esters or acids. The high catalyst activities with the DuPhos ligands and catalyst productivities are an advantage for use in large-scale reactions.[61,62]

The Me-DuPhos-Rh and more electron-rich Me-BPE-Rh catalysts can be used to access - branched amino acids (Scheme 9.22, R^2, R^3 H).[63,64] Symmetric and dissymmetric β-substituted enamides were found to hydrogenate with very high enantioselectivities. In the latter case, hydrogenation of *E*- and *Z*-enamide isomers in separate reactions allowed the generation of the second, new β-stereogenic center with high selectivity.

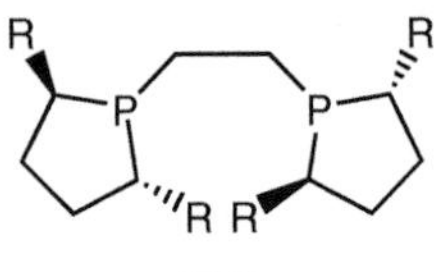

BPE

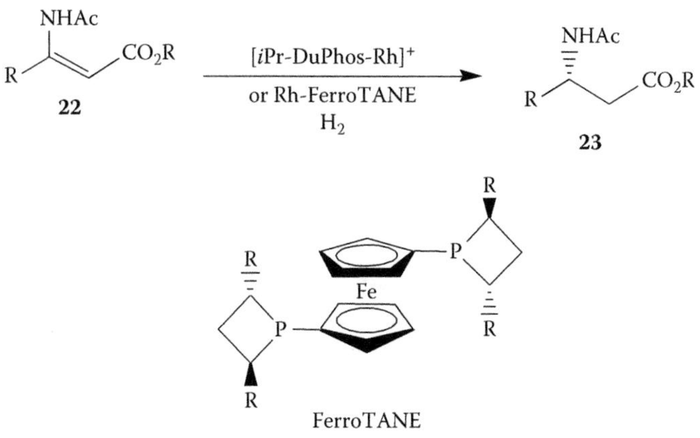

SCHEME 9.23

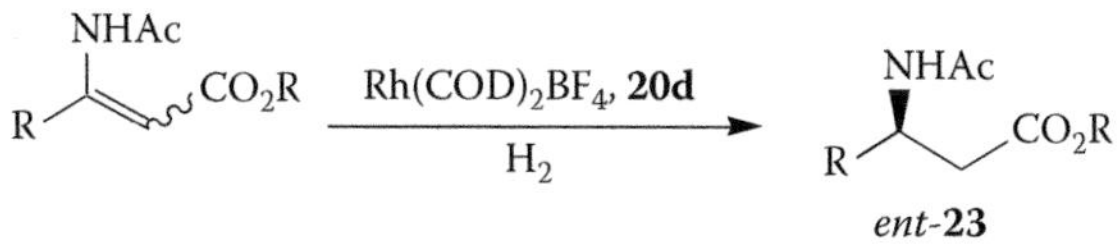

SCHEME 9.24

9.4.1 β- AND HIGHER AMINO ACIDS

Reduction of β-enamides can provide straightforward access to β-amino acids. It was demonstrated that the Me-DuPhos-Rh catalysts are very effective for hydrogenation of a series of (E)-β-enamides **22** to afford the desired β-amino acid derivatives **23** with high enantioselectivities (Scheme 9.23).[65] In addition, the Me-and Et-FerroTANE-Rh catalysts hydrogenate (E)-β-enamides with high enantioselectivities (>98%) and rates.[66] Unfortunately, neither catalyst system was able to reduce the corresponding (Z)-β-enamides with the same high level of absolute stereocontrol.[60]

The MonoPhos family of ligands does provide a means to reduce both the E- and Z-enamides to give the desired β-amino acid derivative with good enantioselectivity (Scheme 9.24).[50,67–69]

The need for acylation of the nitrogen has been circumvented by use of the JosiPhos family of ligands. The trifluoromethyl ligand (**24a**) gives high enantioselectivities with the enamine ester, while the phenyl analogue (**24b**) is better for the amide substrates (Scheme 9.25).[70]

Pfizer developed a new ligand, Trichickenfootphos (TCFP), for the production of pregabalin (**25**), an anticonvulsant related to the inhibitory neurotransmitter γ-aminobutyric acid, by asymmetric hydrogenation (Scheme 9.26).[71,72]

9.5 BIOLOGICAL APPROACHES

Enzymes can be used to introduce the stereogenic center by action on or hydrolysis of just one isomer of an amino acid derivative (Section 9.2). In addition, an enzymatic system can be used to destroy one enantiomer of an amino acid, leaving the required one untouched, as with decarboxylases.

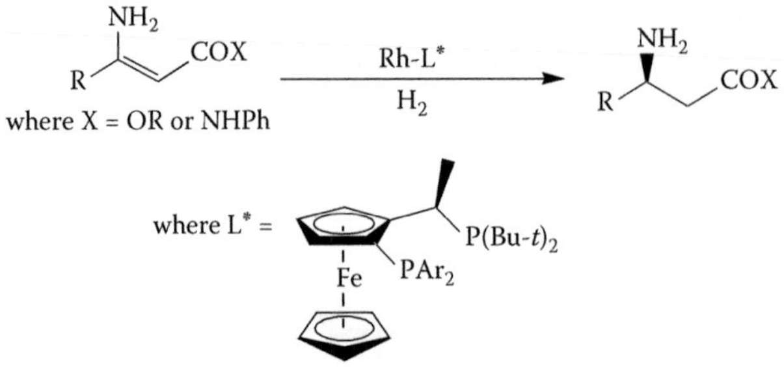

24 a Ar = 4-$CF_3C_6H_4$
b Ar = Ph

SCHEME 9.25

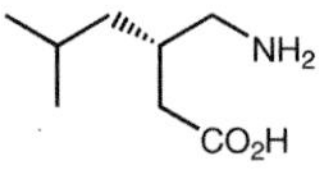

25

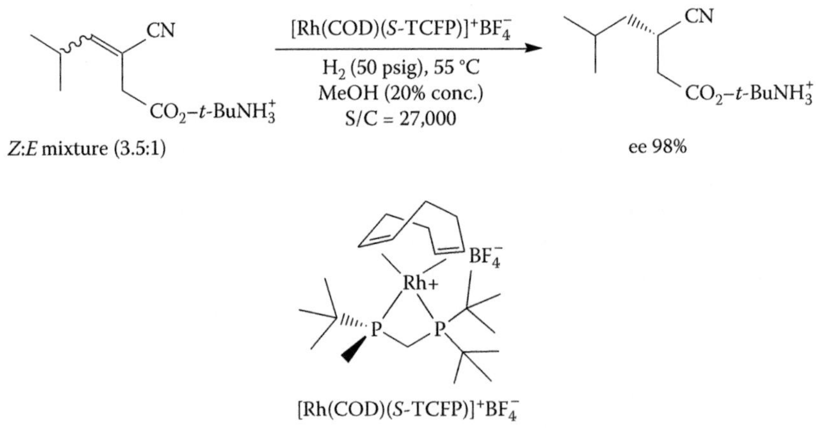

SCHEME 9.26

9.5.1 Formation of a Stereogenic Center

The aminotransferase class of enzymes, also known as transaminases, has been used to prepare a wide range of amino acids as they accept a wide range of substrates.[1,3,73] The L-aminotransferases catalyze the reaction in which an amino group from one L-amino acid is transferred to an α-keto acid (Scheme 9.27). Those enzymes most commonly used as industrial biocatalysts have been cloned, overexpressed, and generally used as whole cell or immobilized preparations.[3]

D-Amino acid transaminases (DATs) are also available and catalyze the analogous reaction to Scheme 9.27 but need a D-amino acid donor (alanine, aspartate, or glutamate) to produce another D-amino acid (Scheme 9.28).[3,74–76]

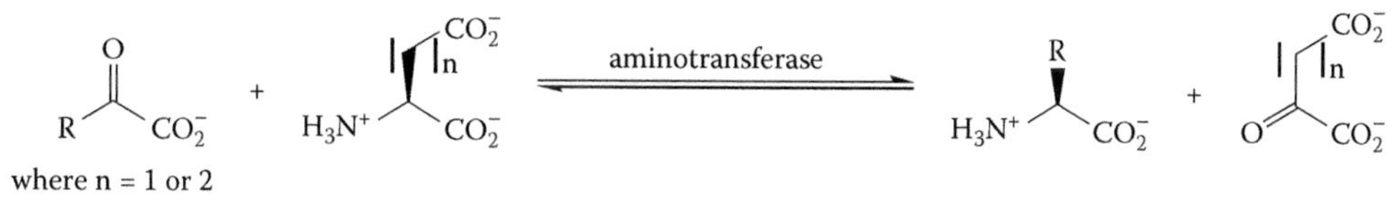

SCHEME 9.27

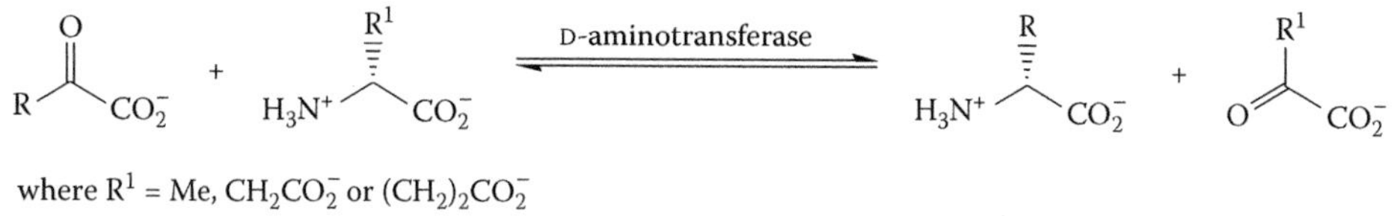

where n = 1 or 2

SCHEME 9.28

where R^1 = Me, $CH_2CO_2^-$ or $(CH_2)_2CO_2^-$

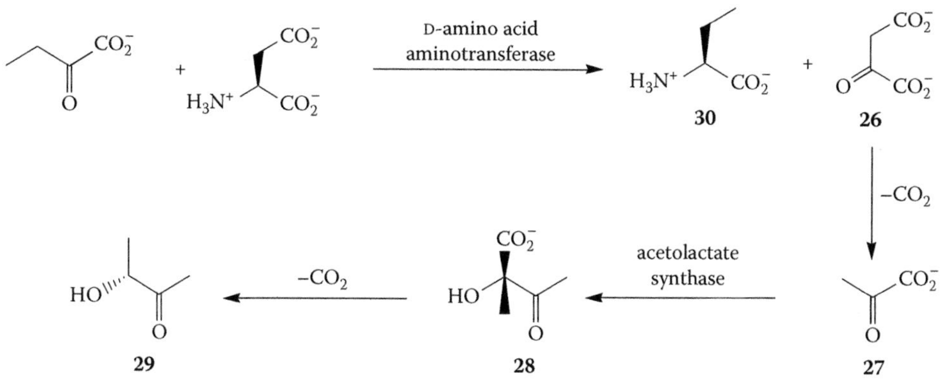

SCHEME 9.29

Although the utility of transaminases has been widely examined, one such limitation is the fact that the equilibrium constant for the reaction is around one. A shift in this equilibrium is necessary, therefore, for the reaction to be synthetically useful.[74,77] Aspartate, when used as the amino donor, is converted into oxaloacetate (**26**) upon reaction (Scheme 9.29). Because **26** is unstable, it decomposes to pyruvate (**27**) and thus favors product formation. However, because pyruvate is also an α-keto acid, it could be a substrate and be transaminated into alanine. The enzyme acetolactate synthase, which condenses two moles of pyruvate to form (S)-acetolactate (**28**), is, therefore, included in the reaction. The (S)-acetolactate can undergo decarboxylation either spontaneously or by the enzyme catalysis with acetolactate decarboxylase to give the final by-product, (R)-acetoin (**29**). This process, for example, can be used for the production of both L- and D-2-aminobutyrate (**30** and **31**, respectively) (Scheme 9.29 and Scheme 9.30).[3,73,78–80] Other than for the transaminase enzyme, the reactions cascade is the same for by-product removal for both enantiomers.

Other examples of the use of transaminases to synthesize unnatural amino acids have also been described in the literature, including L-*tert*-leucine (L-Tle) (**9**), L-2-amino-4-(hydroxymethylphosphinyl)butanoic acid (phosphinothricin) and L-thienylalanines.[3,78,81,82] Not all unnatural amino acids can be accessed by this technology. Although it works well for L-*tert*-leucine, D-*tert*-leucine remains elusive.

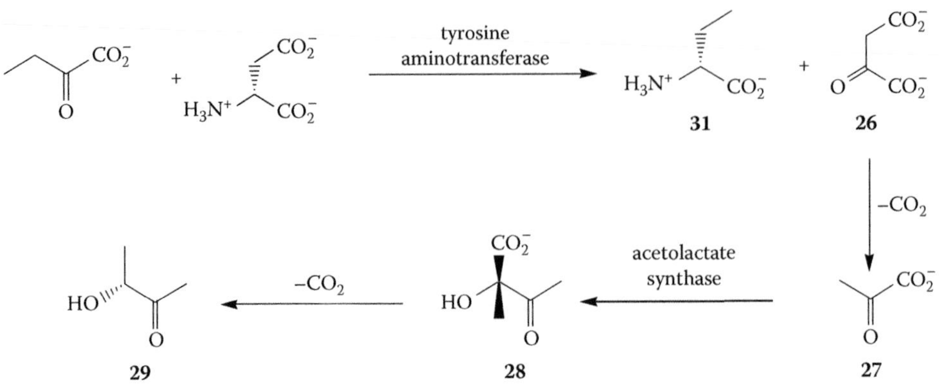

SCHEME 9.30

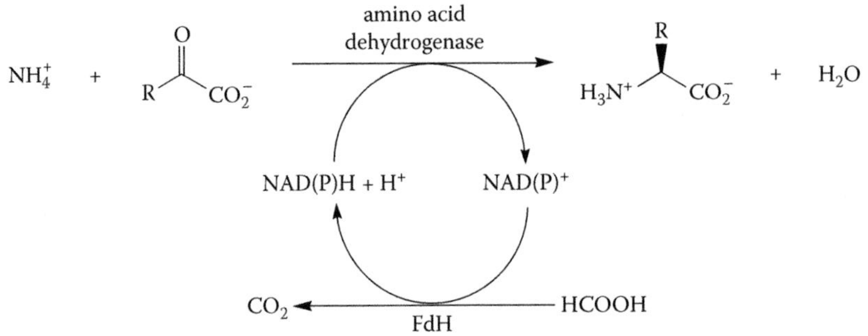

SCHEME 9.31

A variation on the transamination approach that also starts with an α-keto acid substrate is to perform a reductive amination catalyzed by amino acid dehydrogenases (dHs) (Scheme 9.31) in combination with the formate dH cofactor recycling system, although other reducing systems can be used.[3,83,84] The generation of carbon dioxide from formate drives the coupled reactions to completion.

Leucine dehydrogenase is the enzyme used as the biocatalyst in the process commercialized by Degussa to produce L-*tert*-leucine (**9**).[85–92] Similar to phenylalanine dH, leucine dH has been used to prepare numerous unnatural amino acids because of its broad substrate specificity.[93–95] For example, cloned, thermostable alanine dH has been used with a coupling enzyme system to prepare D-amino acids.[3,74]

9.5.2 DECARBOXYLASES

Amino acid decarboxylases can be used to catalyze the resolution of several amino acids. The decarboxylases are ideally suited to large-scale industrial application due to their robust nature. D-Aspartate (**32**, $n = 1$) and D-glutamate (**32**, $n = 2$) can be made economically by use of the corresponding L-amino acids as starting material through an enzymatic racemization prior to the resolution with decarboxylase (Scheme 9.32).[96,97] The combination of both activities is possible because glutamate decarboxylase is inactive at pH 8.5 during the racemization step, but regains full activity when the pH is 4.2. Conversely, glutamate racemase is active at pH 8.5, but it is completely inhibited at pH 4.2. Of course, it is also possible to use a single enzyme system if the racemic amino acid is generated by a cheap chemical process.

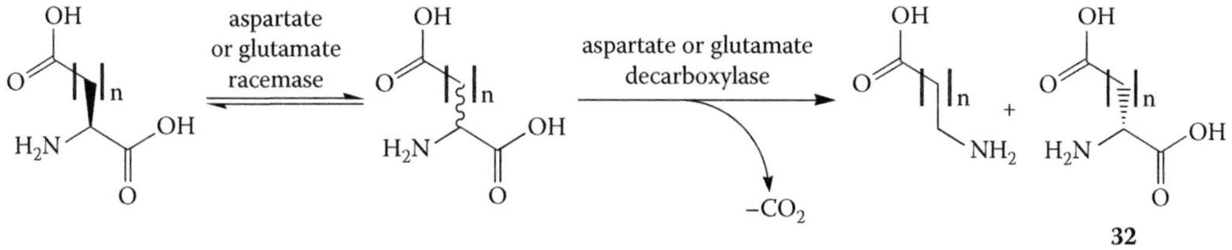

SCHEME 9.32

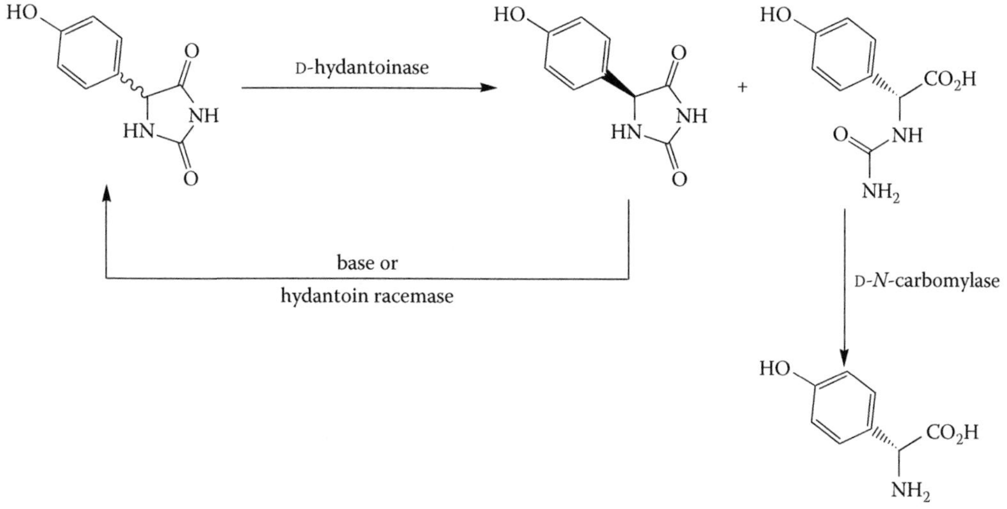

SCHEME 9.33

9.5.3 DYNAMIC PROCESSES

Hydantoins are readily epimerized, which allows access to either D- or L-amino acids.[98,99] The coupling of enzymatic steps allows equilibria to be pushed in the desired direction. Degussa and DSM both use the approach, although they have slightly different variations.[100–102] Scheme 9.33 shows the DSM route to D-*p*-hydroxyphenylglycine (**33**).[1]

The enzyme, *N*-carbamoyl D-amino acid amidohydrolase can be incorporated with hydantoinase to produce D-amino acids in one step from the corresponding racemic hydantoins.[8,103] Because many of the five-substituted hydantoins undergo spontaneous racemization under the enzymatic reaction conditions (pH > 8), complete conversion from racemic hydantoin to the D-amino acid can be achieved. The starting hydantoins are prepared by the Bucherer–Bergs reaction (Scheme 9.34),[104] the simple condensation of potassium cyanide and ammonium carbonate with the

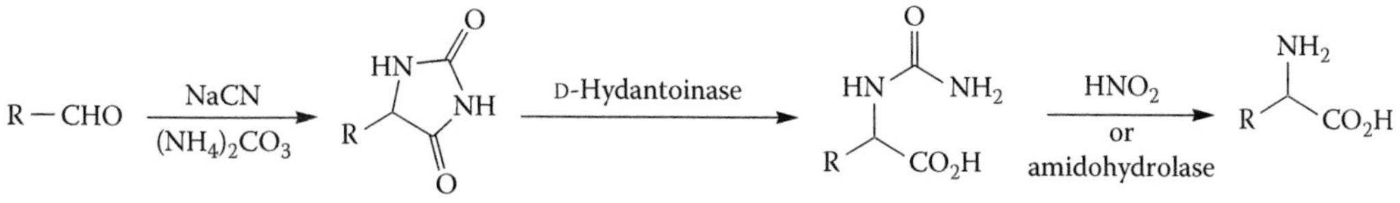

SCHEME 9.34

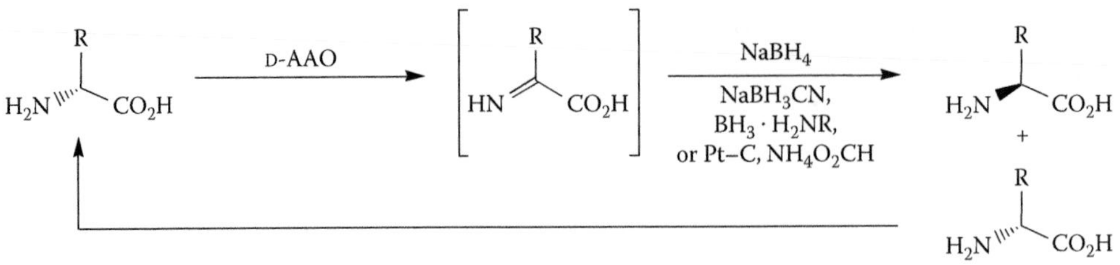

SCHEME 9.35

corresponding aliphatic or aromatic aldehydes. Many of the D-hydantoinases have broad substrate specificities and can be used for production of a variety of D-amino acids.[15,105]

D-Valine is produced from isobutyraldehyde through a Bucherer–Bergs reaction followed by a one-step enzymatic conversion of the hydantoin to the free D-amino acid.[106]

One resolution approach that is showing promise is the use of an enzyme, such as an amino acid oxidase (AAO) to convert one isomer of an amino acid to the corresponding achiral imine. A chemical reducing agent then returns the amino acid as a racemic mixture (Scheme 9.35). As the enzyme acts on only one enantiomer of the amino acid, the desired product builds up. After only seven cycles, the product ee is >99%.[107] The key success factor is to have the reducing agent compatible with the enzyme.[108,109] Rather than use the antipode of the desired product, the racemic mixture can be used as the starting material. The enzymes are available to prepare either enantiomeric series of the amino acid.

The approach can be coupled with other methods to prepare amino acids, such as to access β-substituted α-amino acids. The methodology gives a way to prepare all four possible isomers of β-aryl α-amino acids by a combination of asymmetric hydrogenation and the use of the deracemization process to invert the α-center (Scheme 9.36).[49,110]

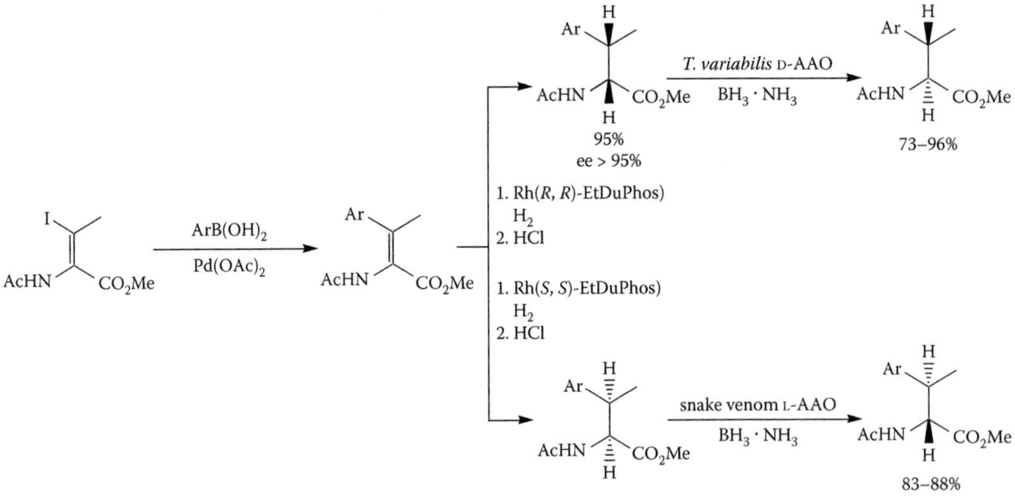

SCHEME 9.36

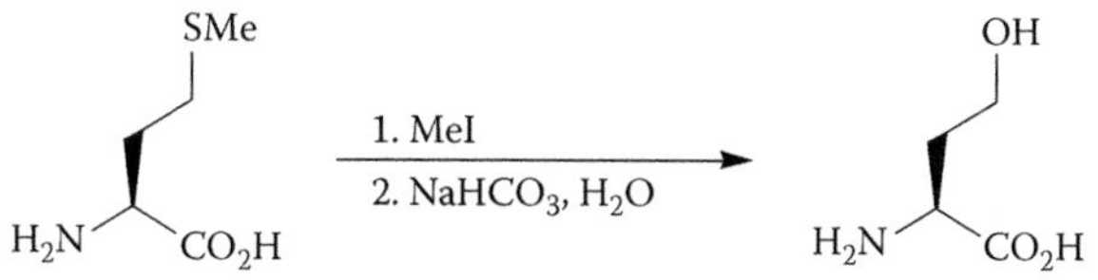

SCHEME 9.37

9.6 MODIFICATION OF OTHER AMINO ACIDS

As noted in the introduction, unnatural amino acids are available from more readily available amino acids. A wide variety of reactions are available in the literature. An example is provided by the synthesis of L-homoserine (**34**) from methionine (Scheme 9.37).[111]

A wide variety of simple transformations on chiral pool materials can lead to unnatural amino acid derivatives. This is illustrated by the Pictet–Spengler reaction of L-phenylalanine, followed by amide formation and reduction of the aromatic ring (Scheme 9.38).[1] The resultant amide **35** is an intermediate in the synthesis of a number of commercial HIV protease inhibitors.

The side chain of an amino acid can also be manipulated by biological catalysts.[73] One example takes advantage of the naturally broad substrate specificity of O-acetylserine sulfhydrylase, the final enzyme in L-cysteine biosynthesis. This enzyme accepts a variety of alternative nucleophiles, in addition to the usual sulfide required for cysteine biosynthesis,[112] to provide S-phenyl-L-cysteine (**36**), S-hydroxyethyl-L-cysteine (**37**), or phenylseleno-L-cysteine (**38**).

β-Amino acids have been prepared by the homologation of α-amino acids, but the method tends to be low yielding as well as being a relatively long sequence (Scheme 9.39),[113] or diazo compounds and a Wolff rearrangement have to be employed,[114–116] and these are not amenable to large-scale application.[22]

9.7 SUMMARY

A wide variety of methods are available for the large-scale synthesis of unnatural amino acids. In a few cases, chiral templates or auxiliaries can be used, but usually cost becomes an issue with this approach. Resolutions can be performed by either a chemical or biological agent. These methods are not cost effective because the maximum yield of the reaction is 50%. Methods in which the undesired isomer can be recycled *in situ* offer a much better solution. Biological and chemical

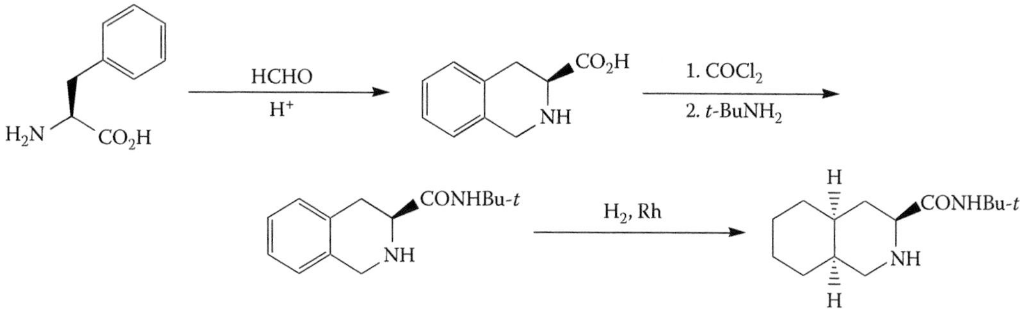

SCHEME 9.38

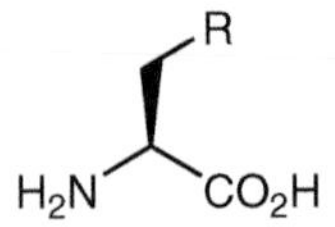

36, R = SPh
37, R = S(CH$_2$)$_2$OH
38, R = SePh

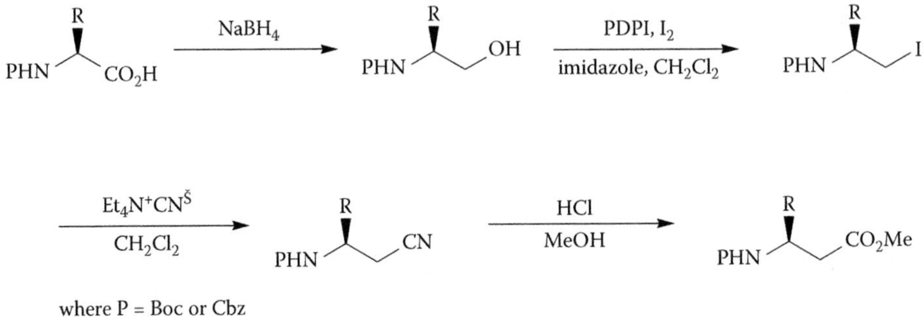

where P = Boc or Cbz

SCHEME 9.39

approaches are available to implement this dynamic resolution methodology. To generate just one enantiomer of an α-amino acid, asymmetric hydrogenation provides the tried and tested pathway. Although many catalyst systems are available, few have been applied at scale. The most common biological approach to unnatural amino acids uses a transaminase reaction, and a number of methods are available to drive the desired reaction to completion. In many cases, the large investment required to optimize a biotransformation approach can only reap rewards for large-scale applications. The exceptions are when a library of enzymes is already available to screen so that a rapid answer is found.

REFERENCES

1. Ager, D.J. In *Handbook of Chiral Chemicals*; Ager, D.J. Ed.; CRC: Taylor & Francis: Boca Raton, FL, 2005; p. 11.
2. Sonke, T.; Kaptein, B.; Boesten, W.H.J.; Broxterman, Q.B.; Kamphuis, J.; Formaggio, F.; Toniolo, C.; Rutjies, F.P.J.T.; Schoemaker, H.E. In *Stereoselective Biocatalysis*; Patel, N.K. Ed.; Marcel Dekker: New York, 2000; p. 23.
3. Pantaleone, D.P. In *Handbook of Chiral Chemicals*; Ager, D.J. Ed.; CRC: Taylor & Francis: Boca Raton, FL, 2005; p. 359.
4. Kamphius, J.; Boesten, W.H.J.; Kaptein, B.; Hermes, H.F.M.; Sonke, T.; Broxterman, Q.B.; van den Tweel, W.J.J.; Schoemaker, H.E. In *Chirality in Industry*; Collins, A.N.; Sheldrake, G.N.; Crosby, J. Eds.; Wiley: Chichester, 1992; p. 187.
5. Rutjies, F.P.J.T.; Schoemaker, H.E. *Tetrahedron Lett.* 1997, 38, 677.
6. Schoemaker, H.E.; Boesten, W.H.J.; Broxterman, Q.B.; Roos, E.C.; Kaptein, B.; van den Tweel, W.J.J.; Kamphius, J.; Meijer, E.M. *Chimica* 1997, 51, 308.
7. Kamphius, J.; Hermes, H.F.M.; van Balken, J.A.M.; Schoemaker, H.E.; Boesten, W.H.J.; Meijer, E.M. In *Amino Acids: Chemistry, Biology, Medicine*; Lubec, G.; Wagner, F. Eds.; Hanser: Munich, 1990; p. 75.

8. Liu, W. In *Handbook of Chiral Chemicals*; Ager, D.J. Ed.; CRC: Taylor & Fancis: Boca Raton, FL, 2005; p. 75.

9. Stein, G.A.; Broner, H.A.; Pfister, K. *J. Am. Chem. Soc.* 1955, 77, 700.

10. Ware, E. *Chem. Rev.* 1950, 46, 403.

11. Chevalier, P.; Roy, D.; Morin, A. *Appl. Microbiol. Biotechnol.* 1989, 30, 482.

12. Gross, C.; Syldatk, C.; Mackowiak, V.; Wagner, F.J. *Biotechnology* 1990, 14, 363.

13. Nishida, Y.; Nakamicho, K.; Nabe, K.; Tosa, T. *Enzyme Micro. Technol.* 1987, 9, 721.

14. Runser, S.; Chinski, N.; Ohleyer, E. *Appl. Microbiol. Biotechnol.* 1990, 33, 382.

15. Shimizu, S.; Shimada, H.; Takahashi, S.; Ohashi, T.; Tani, Y.; Yamada, H. *Agric. Biol. Chem.* 1980, 44, 2233.

16. Syldatk, C.; Wagner, F. *Food Biotechnol.* 1990, 4, 87.

17. West, T.P. *Arch. Microbiol.* 1991, 156, 513.

18. Chenault, H.K.; Dahmer, J.; Whitesides, G.M. *J. Am. Chem. Soc.* 1989, 111, 6354.

19. Keller, J.W.; Hamilton, B.J. *Tetrahedron Lett.* 1986, 27, 1249.

20. Kruizinga, W.H.; Bolster, J.; Kellogg, R.M.; Kamphius, J.; Boesten, W.H.J.; Meijer, E.M.; Schoemaker, H.E. *J. Org. Chem.* 1988, 53, 1826.

21. Schoemaker, H.E.; Boesten, W.H.J.; Kaptein, B.; Hermes, H.F.M.; Sonke, T.; Broxterman, Q.B.; van den Tweel, W.J.J.; Kamphius, J. *Pure Appl. Chem.* 1992, 64, 1171.

22. Liu, M.; Sibi, M.P. *Tetrahedron* 2002, 58, 7991.

23. Shiraiwa, T.; Shinjo, K.; Kurokawa, H. *Chem. Lett.* 1989, 1413.

24. Shiraiwa, T.; Shinjo, K.; Kurokawa, H. *Bull. Chem. Soc. Jpn.* 1991, 64, 3251.

25. Miyazawa, T.; Takashima, K.; Mitsuda, Y.; Yamada, T.; Kuwota, S.; Watanabe, H. *Bull. Chem. Soc. Jpn.* 1979, 52, 1539.

26. Boesten, W.H.J.: Eur. Pat, EP442,584, 1991.

27. Kaptein, B.; Vries, T.R.; Nieuwenhuijzen, J.W.; Kellogg, R.M.; Grimbergen, R.F.P.; Broxterman, Q.B. In *Handbook of Chiral Chemicals*; Ager, D.J. Ed.; CRC: Taylor & Francis: Boca Raton, FL, 2005; p. 97.

28. Sonke, T.; Kaptein, B.; Boesten, W.H.J.; Broxterman, Q.B.; Kamphius, J.; Formaggio, F.; Toniolo, C.; Rutjies, F.P.J.T.; Schoemaker, H.E. In *Comprehensive Asymmetric Catalysis*; Jacobsen, E.N.; Pfaltz, A.H.; Yamamoto, H. Eds.; Springer: Berlin, 1999; p. 23.

29. de Lange, B.; Boesten, W.H.J.; van der Sluis, M.; Uiterweerd, P.G.H.; Elsenberg, H.L.M.; Kellogg, R.M.; Broxterman, Q.B. In *Handbook of Chiral Chemicals*; Ager, D.J. Ed.; CRC: Taylor & Francis: Boca Raton, FL, 2005; p. 487.

30. Sch^llkopf, U.; Hartwig, W.; Groth, U.; Westphalen, K.-O. *Liebigs Ann. Chem.* 1981, 696.

31. Bois-Choussy, M.; Neuville, L.; Beugelsman, R.; Zhu, J. *J. Org. Chem.* 1996, 61, 9309.

32. Ager, D.J.; Prakash, I.; Schaad, D.R. *Chem. Rev.* 1996, 96, 835.

33. Chakraborty, T.K.; Reddy, G.V.; Hussain, K.A. *Tetrahedron Lett.* 1991, 32, 7597.

34. Rao, A.V.R.; Chakraborty, T.K.; Joshi, S.P. *Tetrahedron Lett.* 1992, 33.

35. Vergne, C.; Bouillon, J.-P.; Chasanet, J.; Bois-Choussy, M.; Zhu, J. *Tetrahedron: Asymmetry* 1998, 9, 3095.

36. Bunnage, M.E.; Burke, A.J.; Davies, S.G.; Goodwin, C.J. *Tetrahedron: Asymmetry* 1995, 6, 165.

37. Davies, S.G.; Dixon, D.J.J. *Chem. Soc., Perkin Trans. I* 1998, 2629.

38. Davies, S.G.; Ichihara, O. *Tetrahedron: Asymmetry* 1991, 2, 191.

39. Davies, S.G.; Ichihara, O. *Tetrahedron: Asymmetry* 1996, 7, 1919.

40. Bull, S.D.; Davies, S.G.; Delgrado-Ballster, S.; Kelley, P.M.; Kotchie, L.J.; Gianotti, M.; Laderas, M.; Smith, A.D.J. *Chem. Soc., Perkin Trans. I* 2001, 3112.

41. Clark, J.D.; Weisenburger, G.A.; Anderson, D.K.; Colson, P.-J.; Edney, A.D.; Gallagher, D.J.; Klein, H.P.; Knable, C.M.; Lantz, M.K.; Moore, C.M.; Murphy, J.B.; Rogers, T.E.; Ruminski, P.G.; Shah, A.S.; Storer, N.; Wise, B.E. *Org. Proc. Res. Dev.* 2004, 8, 51.

42. Blaser, H.-U.; Schmidt, E. In *Asymmetric Catalysis on Industrial Scale*; Blaser, H.-U.; Schmidt, E. Eds.; Wiley-VCH: Weinheim, 2004; p. 1.

43. Knowles, W.S. *Acc. Chem. Res.* 1983, 16, 106.

44. Knowles, W.S. *Angew. Chem., Int. Ed.* 2002, 41, 1998.

45. Vineyard, B.D.; Knowles, W.S.; Sabacky, M.J.; Bachman, G.L.; Weinkauff, D.J. *J. Am. Chem. Soc.* 1977, 99, 5946.

46. Ager, D.J.; Laneman, S.A. In *Asymmetric Catalysis on Industrial Scale*; Blaser, H.-U.; Schmidt, E. Eds.; Wiley-VCH: Weinheim, 2004; p. 259.

47. Laneman, S.A.; Froen, D.E.; Ager, D.J. In *Catalysis of Organic Reactions*; Herkes, F.E. Ed.; Marcel Dekker: New York, 1998; p. 525.

48. Herbst, R.M.; Shemin, D. *Org. Synth.* 1943, Coll. Vol. II, 1.

49. Roff, G.; Lloyd, R.C.; Turner, N.J. *J. Am. Chem. Soc.* 2004, 126, 4098.

50. de Vries, J.G. In *Handbook of Chiral Chemicals*; Ager, D.J. Ed.; CRC: Taylor & Francis: Boca Raton, FL, 2005; p. 269.

51. van den Berg, M.; Minnaard, A.J.; Haak, R.M.; Leeman, M.; Schudde, E.P.; Meetsma, A.; Feringa, B.L.; de Vries, A.H.M.; Maljaars, C.E.P.; Willans, C.E.; Hyett, D.; Boogers, J.A.F.; Henderick, H.J.W.; de Vries, J.G. *Adv. Synth. Catal.* 2003, 345, 308.

52. van den Berg, M.; Minnaard, A.J.; Schudde, E.P.; van Esch, J.; de Vries, A.H.M.; de Vries, J.G.; Feringa, B.L. *J. Am. Chem. Soc.* 2000, 122, 11539.

53. Willans, C.E.; Mulders, J.M.C.A.; de Vries, J.G.; de Vries, A.H.M. *J. Organomet. Chem.* 2003, 687, 494.

54. de Vries, A.H.M.; Lefort, L.; Boogers, J.A.F.; de Vries, J.G.; Ager, D. *J. Chimica Oggi* 2005, 23(2) Supplement on Chiral Technologies, 18.

55. Zeng, Q.; Liu, H.; Cui, X.; Mi, A.; Jiang, Y.; Li, X.; Choi, M.C.K.; Chan, A.S.C. *Tetrahedron: Asymmetry* 2002, 13, 115.

56. Zeng, Q.; Liu, H.; Mi, A.; Jiang, Y.; Li, X.; Choi, M.C.K.; Chan, A.S.C. *Tetrahedron* 2002, 58, 8799.

57. Bernsmann, H.; van den Berg, M.; Hoen, R.; Minnaard, A.J.; Mehler, G.; Reetz, M.T.; de Vries, J.G.; Feringa, B.L. *J. Org. Chem.* 2005, 70, 943.

58. Jia, X.; Li, X.; Xu, L.; Shi, Q.; Yao, X.; Chan, A.S.C. *J. Org. Chem.* 2003, 68, 4539.

59. Burk, M.J. *J. Am. Chem. Soc.* 1991, 113, 8518.

60. Burk, M.J.; Ramsden, J.A. In *Handbook of Chiral Chemicals*; Ager, D.J. Ed.; CRC: Taylor & Francis: Boca Raton, FL, 2005; p. 249.

61. Burk, M.J.; Bienewald, F. In *Transition Metals for Organic Synthesis and Fine Chemicals*; Bolm, C.; Beller, M. Eds.; VCH: Weinheim, 1998; p. 13.

62. Burk, M.J.; Feaster, J.E.; Nugent, W.A.; Harlow, R.L. *J. Am. Chem. Soc.* 1993, 115, 10125.

63. Burk, M.J.; Gross, M.F.; Martinez, J.P. *J. Am. Chem. Soc.* 1995, 117, 9375.

64. Hoerrner, R.S.; Askin, D.; Volante, R.P.; Reider, P.J. *Tetrahedron Lett.* 1998, 39, 3455.

65. Zhu, G.; Chen, Z.; Zhang, X. *J. Org. Chem.* 1999, 64, 6907.

66. You, J.; Drexler, H.-J.; Zhang, S.; Fischer, C.; Heller, D. *Angew. Chem., Int. Ed.* 2003, 42, 913.

67. Peña, D.; Minnaard, A.J.; Boogers, J.A.F.; de Vries, A.H.M.; de Vries, J.G.; Feringa, B.L. *Org. Biomol. Chem.* 2003, 1, 1087.

68. Peña, D.; Minnaard, A.J.; de Vries, A.H.M.; de Vries, J.G.; Feringa, B.L. *Org. Lett.* 2003, 5, 475.

69. Peña, D.; Minnaard, A.J.; de Vries, J.G.; Feringa, B.L. *J. Am. Chem. Soc.* 2002, 124, 14552.

70. Hsiao, Y.; Rivera, N.R.; Rosner, T.; Krska, S.W.; Njolito, E.; Wamg, F.; Sun, Y.; Armstrong III, J.D.; Grabowski, E.J.J.; Tillyer, R.D.; Spindler, F.; Malan, C. *J. Am. Chem. Soc.* 2004, 126, 9918.

71. Hoge, G.; Wwu, H.-P.; Kissel, W.S.; Pflum, D.A.; Greene, D.J.; Bao, J. *J. Am. Chem. Soc.* 2004, 126, 5966.

72. Laneman, S.A. In *Handbook of Chiral Chemicals*; Ager, D.J. Ed.; CRC: Taylor & Francis: Boca Raton, FL, 2005; p. 185.

73. Fotheringham, I.; Taylor, P.P. In *Handbook of Chiral Chemicals*; Ager, D.J. Ed.; CRC: Taylor & Francis: Boca Raton, FL, 2005; p. 31.

74. Galkin, A.; Kulakova, L.; Yamamoto, H.; Tanizawa, K.; Tanaka, H.; Esaki, N.; Soda, K. *J. Ferment. Bioeng.* 1997, 83, 299.

75. Tanizawa, K.; Asano, S.; Masu, Y.; Kuramitsu, S.; Kagamiyama, H.; Tanaka, H.; Soda, K. *J. Biol. Chem.* 1989, 264, 2450.

76. Taylor, P.P.; Fotheringham, I.G. *Biochim. Biophys. Acta* 1997, 1350, 38.

77. Crump, S.P.; Rozzell, J.D. In *Biocatalytic Production of Amino Acids and Derivatives*; Rozzell, J.D.; Wagner, F. Eds.; Hanser: Munich, 1992; p. 43.

78. Ager, D.J.; Fotheringham, I.G.; Laneman, S.A.; Pantaleone, D.P.; Taylor, P.P. *Chimica Oggi* 1997, 15 (3/4), 11.

79. Ager, D.J.; Li, T.; Pantaleone, D.P.; Senkpeil, R.F.; Taylor, P.P.; Fotheringham, I.G. *J. Mol. Catal. B: Enzymatic* 2001, 11, 199.
80. Taylor, P.P.; Pantaleone, D.P.; Senkpeil, R.F.; Fotheringham, I.G. *Trends Biotechnol.* 1998, 16, 412.
81. Bartsch, K.; Schneider, R.; Schultz, A. *Appl. Environ. Microbiol.* 1996, 62, 3794.
82. Meiwes, J.; Schudok, M.; Kretzschmar, G. *Tetrahedron: Asymmetry* 1997, 8, 527.
83. Hummel, W.; Kula, M.-R. *Eur. J. Biochem.* 1989, 184, 1.
84. Ohshima, T.; Soda, K. *Trends Biotechnol.* 1989, 7, 210.
85. Bommarius, A.S.; Schwarm, M.; Drauz, K. *Chimica* 2001, 55, 50.
86. Kragl, U.; Kruse, W.; Hummel, W.; Wandrey, C. *Biotechnol. Bioeng.* 1996, 52, 309.
87. Wichmann, R.; Wandrey, C.; Buckmann, A.F.; Kula, M.-R. *Biotechnol. Bioeng.* 1981, 23, 2789.
88. Bommarius, A.S.; Drauz, K.; Hummel, W.; Kula, M.-R.; Wandrey, C. *Biocatalysis* 1994, 10, 37.
89. Bommarius, A.S.; Schwarm, M.; Drauz, K. *J. Mol. Catal. B: Enzymatic* 1998, 5, 1.
90. Bommarius, A.S.; Schwarm, M.; Stingl, K.; Kottenhahn, M.; Huthmacher, K.; Drauz, K. *Tetrahedron: Asymmetry* 1995, 6.
91. Drauz, K.; Jahn, W.; Schwarm, M. *Chem. Eur. J.* 1995, 538.
92. Grˆger, H.; Drauz, K. In*Asymmetric Catalysis on Industrial Scale*; Blaser, H.-U.; Schmidt, E. Eds.; Wiley-VCH: Weinheim, 2004; p. 131.
93. Hanson, R.L.; Singh, J.; Kissick, T.P.; Patel, R.N.; Szarka, L.J.; Mueller, R.H. *Bioorg. Chem.* 1990, 18, 118.
94. Krix, G.; Bommarius, A.S.; Drauz, K.; Kottenhahn, M.; Schwarm, M.; Kula, M.-R. *J. Biotechnol.* 1997, 53, 29.
95. Sutherland, A.; Willis, C.L. *Tetrahedron Lett.* 1997, 38, 1837.
96. Archer, J.A.C.; Sinskey, A.J. *J. Gen. Microbiol.* 1993, 139, 753.
97. Jetten, M.S.M.; Sinskey, A. *J. Crit. Rev. Biotechnol.* 1995, 15, 73.
98. Batt, C.A.; Follettie, M.T.; Shin, H.K.; Yeh, P.; Sinskey, A.J. *Trends Biotechnol.* 1985, 3, 305.
99. Shiio, I.; Sugimoto, S.; Kawamura, K. *Biosci. Biotechnol. Biochem.* 1993, 57, 51.
100. Ishida, M.; Sato, K.; Hashiguchi, K.; Ito, H.; Enei, H.; Nakamori, S. *Biosci. Biotechnol. Biochem.* 1993, 57, 1755.
101. Jetten, M.S.M.; Follettie, M.T.; Sinskey, A.J. *J. Appl. Microbiol. Biotechnol.* 1995, 43, 76.
102. Shiio, I.; Yoshino, H.; Sugimoto, S. *Agric. Biol. Chem.* 1990, 54, 3275.
103. Olivieri, R.; Fascetti, E.; Angelini, L.; Degen, L. *Biotechnol. Bioeng.* 1981, 23, 2173.
104. Bucherer, H.T.; Steiner, W. *J. Prakt. Chem.* 1934, 140, 291.
105. Takahashi, S.; Ohashi, T.; Kii, Y.; Kumagai, H.; Yamada, H. *J. Ferment. Technol.* 1979, 57, 328.
106. Battilott, M.; Barberini, U. *J. Mol. Catal.* 1988, 43, 343.
107. Turner, N.J. *TIBTECH* 2003, 21, 474.
108. Alexandre, F.-R.; Pantaleone, D.P.; Taylor, P.P.; Fotheringham, I.G.; Ager, D.J.; Turner, N.J. *Tetrahedron Lett.* 2002, 43, 707.
109. Beard, T.M.; Turner, N.J. *J. Chem. Soc., Chem. Commun.* 2002, 246.
110. Enright, A.; Alexandre, F.-R.; Roff, G.; Fotheringham, I.G.; Dawson, M.J.; Turner, N.J. *J. Chem. Soc., Chem. Commun.* 2003, 236.
111. Boyle, P.H.; Davis, A.P.; Dempsey, K.J.; Hoskin, G.D. *Tetrahedron: Asymmetry* 1995, 6, 2819.
112. Maier, T.H.P. *Nature Biotechnol.* 2003, 21, 422.
113. Caputo, R.; Cassano, E.; Longobardo, L.; Palumbo, G. *Tetrahedron* 1995, 51, 12337.
114. Ellmermer-Müller, E.P.; Brössner, D.; Maslouh, N.; Takó, A. *Helv. Chim. Acta* 1998, 81, 59.
115. Leggio, A.L.; Liguori, A.; Procopio, A.; Sindona, G. *J. Chem. Soc., Perkin Trans. I* 1997, 1969.
116. Seebach, D.; Overand, M.; Kühnle, F.N.M.; Martinoni, B.; Oberer, L.; Hommel, U.; Widmer, H. *Helv. Chim. Acta* 1996, 79, 913.

10 The Chemical Development of a Potential Manufacturing Route to the Endothelin Antagonists UK-350,926 and UK-349,862

*Christopher P. Ashcroft, Stephen Challenger,
Andrew M. Derrick, Yousef Hajikarimian, and
Nicholas M. Thomson*

CONTENTS

10.1 INTRODUCTION

The endogenous endothelins 1 (ET-1, -2, and -3), a family of homologous 21-amino acid isopeptides, possess exceptionally potent vasoconstrictory activity and play an important role in the control of vascular smooth muscle tone and blood flow. Characterization of elevated endothelin levels in a variety of disease states has promoted an intense effort by a number of pharmaceutical companies to identify potent and selective nonpeptide endothelin antagonists of different subtype selectivity for the two-endothelin receptors ETA and ETB. Indications that have been targeted include congestive heart failure, pulmonary hypertension, angina, chronic renal failure, restenosis, and cancers.[1] The first compound in this class to reach the market is Bosentan (Roche Holdings AG), which was launched (Actelion Pharmaceuticals, Tracleer®) in the United States in December 2001 for the treatment of pulmonary hypertension. A number of endothelin antagonists exhibit the common structural features of typically two or more substituted aromatic or heteroaromatic rings and an acidic group, either an acid, acylsulfonamide, or sulfonamide.[2] The potent (ET$_A$ K$_i$ 4.2 nM) and selective (200- to 500-fold selectivity for ET$_A$) endothelin antagonist UK-350,926 **1** was discovered at Pfizer's Sandwich Discovery Laboratories[3] and entered development for the potential treatment and prophylaxis of acute renal failure as an intravenous product. An orally bioavailable prodrug UK-349,862 **2**, containing a hydroxymethyl group at the six position of the indole, also was nominated as a development candidate at the same time for the restenosis indication. The acylsulfonamide group of **1** and **2** are bioisosteres of carboxylic acids with a similar pK_a. This chapter

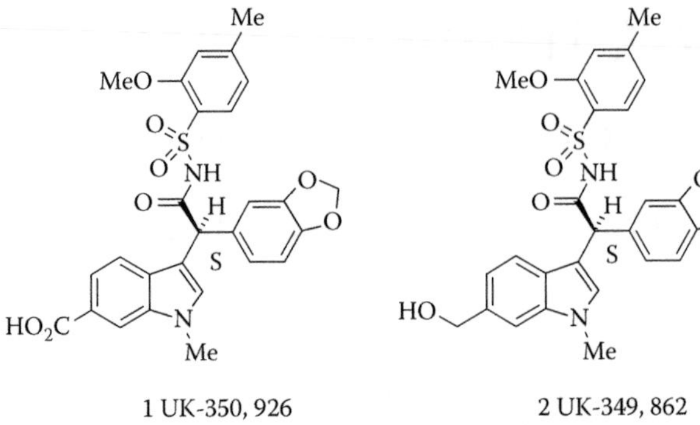

FIGURE 10.1 Endothelin antagonists (*S*)-UK-350,926 and (*S*)-UK-349,862.

describes the process research that led to the development of an efficient and robust synthetic route capable of preparing kilogram quantities of UK-350,926 **1** and UK-349,862 **2** (Figure 10.1).

10.2 MEDICINAL CHEMISTRY ROUTE

A key synthetic challenge of the early discovery routes to this series of endothelin receptor antagonists was the construction of the acylsulfonamide functionality without loss of stereochemical integrity in compounds containing a labile stereogenic center. Initially, the Discovery group prepared a series of compounds in racemic form and selected examples were resolved by chiral high-performance liquid chromatography (HPLC).[4] An alternative convergent and more process-friendly synthesis of optically pure UK-350,926 **1** was also developed by the Discovery chemists prior to the compound entering development, involving a classical resolution to access the desired (*S*)-enantiomer.[5] In the discovery chiral synthesis, the acylsulfonamide functionality was introduced by constructing the *N*-sulfonyl bond (disconnection B in Figure 10.2) by reaction of sulfonyl chloride **3** with carboxamide **4**. The alternative strategy of constructing the acylsulfonamide by making the N-carbonyl bond (disconnection A in Figure 10.2), involving the coupling of sulfonamide **5** with an activated chiral diarylacetic acid derivative **6** (Figure 10.2), was unsuccessful, resulting in racemization of the chiral carbon.[4]

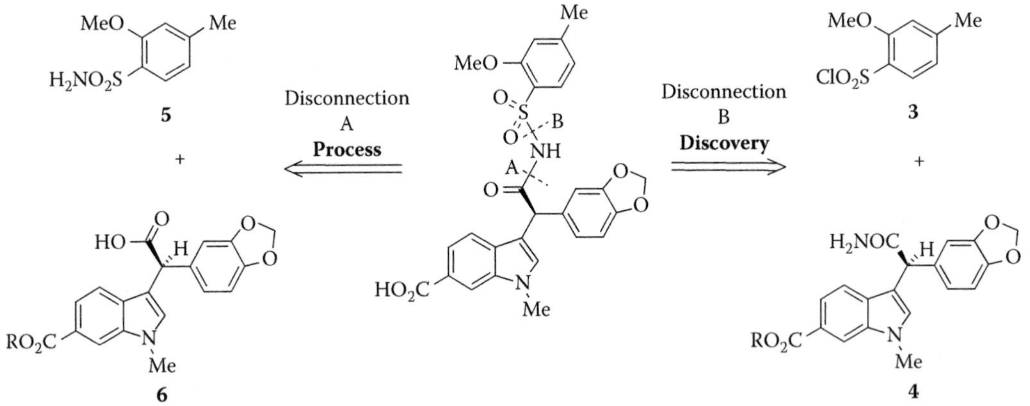

FIGURE 10.2 Synthetic strategies to construct the acylsulfonamide functionality.

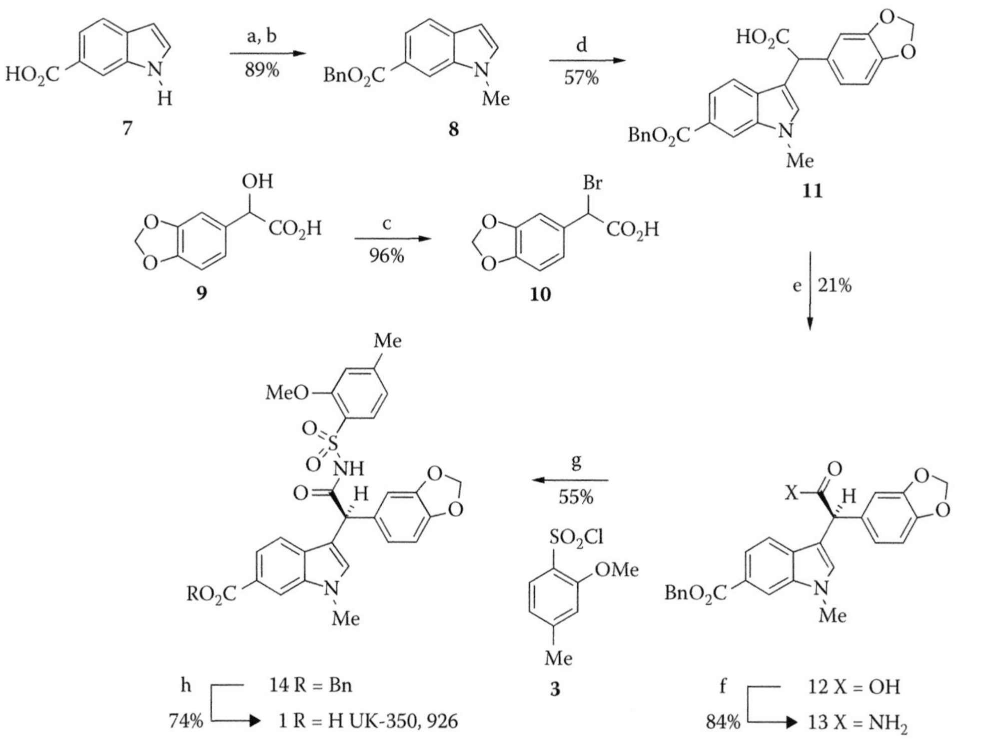

SCHEME 10.1 Discovery route to UK-350,926. Reagents and conditions: (a) 7, BnOH (1.1 equiv), 1-(3-dimethylaminopropyl)-3-ethylcarbodiimide hydrochloride (EDCI, 1.2 equiv), N,N-dimethylaminopyridine (1.3 equiv), CH$_2$Cl$_2$, rt, 12 h, chromatography, 99%; (b) NaH, THF, 0°C, 2h, MeI (1.5 equiv), 0°C, 12 h, 90%; (c) 9, 62% Aq HBr, toluene, CH$_2$Cl$_2$, rt, 4 h, 96%; (d) 8, 10 (1.08 equiv), DMF, 90°C, 4 h, chromatography, 57%; (e) 11, (R)-(+)-1-phenylethylamine (1.0 equiv), EtOAc, 21%; (f) (i) (R, S)-salt 11 with (R)-(+)-1-phenylethylamine, 2M Aq HCl, EtOAc; (ii) 12, 1-hydroxy-7-azabenzotriazole (HOAt, 1.3 equiv), 1-(3-dimethylaminopropyl)-3-ethylcarbodiimide hydrochloride (EDCI, 1.5 equiv), CH$_2$Cl$_2$, 0°C, 90 min; (iii) Aq NH$_3$ (3 equiv), 0°C, 10 min., chromatography, 84%; (g) (i) 13, NaHMDS (1.0 equiv), THF, 60°C to 40°C; (ii) 3 (1.0 equiv), 40°C, 30 min., 55%; (h) 14, 5% Pd/C. H$_2$ (60 p.s.i.), ethanol/water (9:1), recrystallization 74%. S. Challenger/Organic Process Research and Development/ACS/2005.

The successful discovery synthesis of UK-350,926 **1**, outlined in Scheme 10.1, began with the two-step esterification and methylation of commercially available indole-6-carboxylic acid **7** to give the benzyl-protected indole **8**.[3] Benzyl ester protection was chosen to avoid strongly acidic or basic conditions in the final deprotection step leading to UK-350,926 **1**. The indole derivative **8** was then alkylated at C-3 with the bromoacid **10** to give the diarylacetic acid derivative **11** in acceptable yield (57 to 61%) after column chromatography.[3,5] The bromoacid **10** was available in a high-yielding step from the corresponding commercially available mandelic acid derivative **9** using 62% aqueous hydrobromic acid in a mixture of toluene and dichloromethane. The racemic acid **11** was then resolved in a classical chemical resolution process with the chiral amine (R)-(+)-1-phenyl ethylamine. In an inefficient and unoptimized process,[5] the desired (R),(S)-salt was prepared in ethanol and selectively crystallized from ethyl acetate. The diastereoisomeric purity was upgraded to >98% d.e. after three successive recrystallizations from ethyl acetate. Salt cleavage with dilute hydrochloric acid completed the synthesis of the desired chiral acid **12**. This key intermediate was available in five steps from the two commercially available starting materials **7** and **9** in 11% overall yield (four linear steps). The chiral acid **12** was then converted into the acylsulfonamide **14** in a two-step process involving the coupling of a carboxamide anion with sulfonyl chloride **3**. The Discovery group was able to demonstrate that a 1-hydroxy-7-azabenzot-riazole (HOAt) active ester[5] of chiral acid **12** could be prepared without significant loss in optical purity, and reacted cleanly with aqueous ammonia to give the carboxamide **13** in 84% yield and >98% optical purity. It was critical that the basic urea by-product from the dehydrating reagent, 1-(3-dimethylaminopropyl)-3-ethylcarbodiimide hydrochloride (EDCI) was completely removed with a citric acid wash prior to the addition of ammonia to avoid racemization. The Discovery chemists reported an 8% loss in optical purity in one coupling reaction in which the basic urea by-product was incompletely removed prior to the addition of ammonia. Deprotonation of the amide **13** and sulfonation with sulfonyl chloride **3**[7] were examined with a limited number of bases and reaction conditions. The best results were achieved using a slight excess of sodium bis(trimethyl-silyl)amide in THF at –60°C for the deprotonation step followed by the addition of sulfonyl chloride **3** at –40°C. Stereochemical integrity was preserved, and the acylsulfonamide product **14** was obtained in a respectable 55% yield after chromatography. The synthesis of UK-350,926 **1** was completed with the deprotection of the benzyl ester **14** by hydrogenolysis over palladium on carbon. The sequence developed by the Discovery chemists was capable of producing gram quantities of material for candidate nomination studies. It provided UK-350,926 **1** in a convergent manner and in approximately 4% overall yield for the sequence of seven linear steps (11 steps in total from commercially available materials).

Although the discovery synthesis was used to prepare gram quantities of UK-350,926 **1**, there were a number of issues associated with the route for rapid scale-up and preparation of kilogram quantities. Our major concern with the synthesis was the sensitivity and lack of robustness in preparing the carboxamide **13**, with the potential to lose optical purity during the preparation of the HOAt active ester. On scale-up with increased processing times the potential optical sensitivity was more likely to become a significant problem. The discovery route also suffers from a low overall yield, largely due to an inherently inefficient classical resolution (unoptimized yield 21%). The preparation of the acylsulfonamide involved cryogenic conditions (–60°C), and the synthesis required chromatography at three steps. In addition, bromoacid **10** was lachrymatory and showed hydrolytic and thermal sensitivity.

10.3   PROCESS ROUTE TO UK-350,926 AND UK-349,862[8]

In designing a new route to this class of endothelin antagonists, a synthetic strategy that proceeds through common intermediates allowing access to both compounds was considered to be highly desirable. Early indications were that both candidates would be toxicologically bland, and it was estimated that a total of 7 kg of API would be required to progress both candidates to phase I clinical studies. Our target was to progress both candidates to phase I studies within 12 months of candidate nomination. It was envisaged that UK-349,862 **2** would be available in a final step reduction of UK-350,926 **1**. An alternative strategy was developed, which was designed to overcome some of the limitations of the discovery route, and involved the preparation of racemic UK-350,926 and resolution in the final step. On economic grounds, it is usual practice to attempt a classical resolution of optical isomers as early as possible in a synthetic pathway. However, in this case, we considered that a late-stage resolution approach would be attractive on paper due to the perceived ease of racemate synthesis and the possibility of a crystallization-induced asymmetric transforma-tion (dynamic resolution) process. In addition, UK-350,926 contains two acidic functional groups (acid pK_a = 5.01, acylsulfonamide pK_a = 5.38) and has the potential of forming mono- or di-salts with chiral amines. A small batch of the racemate of UK-350,926 was prepared and sufficient encouragement obtained in a limited resolution screen for us to make the decision to embark on the proposed new route for the first campaign in laboratory glassware.

The process synthesis of UK-350,926 **1** is outlined in Scheme 10.2. Indole-6-carboxylic acid was converted in two steps to the *N*-methylindole derivative **15** in 73% overall yield. Methyl ester protection was chosen for the acid on the grounds of atom economy, crystallinity, and ease of

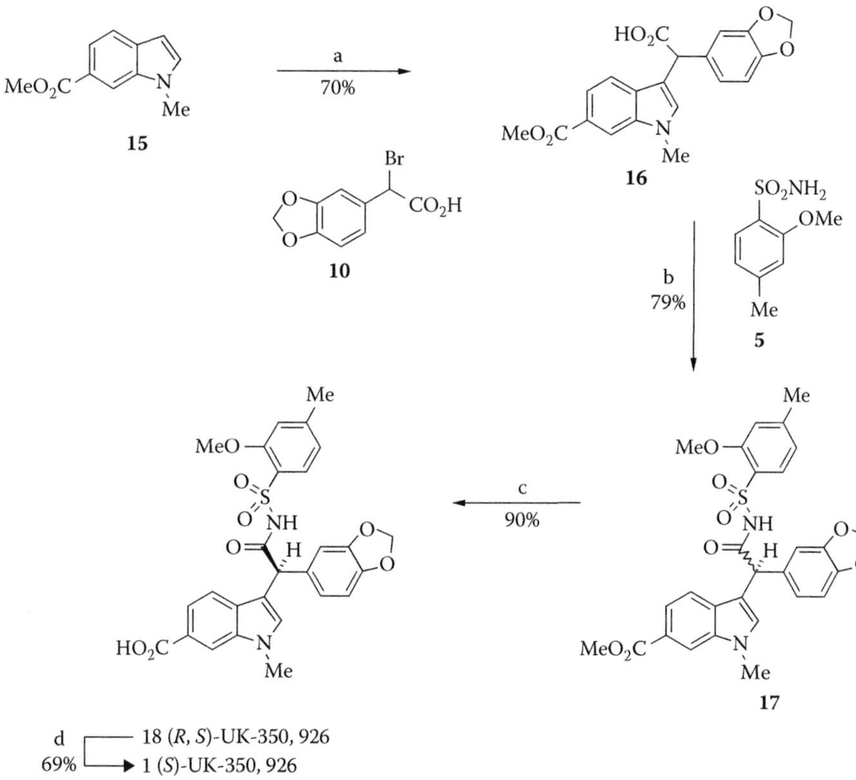

SCHEME 10.2 Process Route to UK-350,926. Reagents and conditions: (a) 15, 10 (1.1 equiv), DMF, 80°C 4 h, 70%; (b) (i) 16, 1,1-carbonyldiimidazole (CDI, 1.1 equiv), THF, reflux 1.5 h; (ii) 5 (1.1 equiv), 1,8-diazabicyclo[5.4.0]undec-7-ene (DBU, 1.1 equiv), THF reflux 4.5 h, 79%; (c) 17, NaOH (7 equiv), MeOH, water, 90%; (d) (i) 18, (S)-(-)-1-phenethylamine (2.0 equiv), THF, DME 60 to 45°C, 72 h, 76%; (ii) 1M Aq HCl , EtOAc, THF, n-hexane, 91%. S. Challenger/Organic Process Research and Development/ACS/2005.

deprotection by base hydrolysis. Carbonyl diimidazole (CDI) was used as an activating reagent instead of the more expensive carbodiimide (EDCI) in the esterification reaction, and due to safety concerns, a combination of potassium *tert*-butoxide and dimethyl sulfate was used instead of the discovery process using sodium hydride and iodomethane for the indole *N*-alkylation. The intermediate **15** is a crystalline compound and could be prepared without chromatography. Initially, **15** was prepared in-house for the first two campaigns, but for the third pilot-plant campaign, it was purchased from external vendors.[9] The discovery conditions[3] were initially used to prepare the racemic acid **16** with some minor modifications. Concerns over the stability of bromoacid **10** (Scheme 10.1) (DSC showed mild exothermic activity above 70°C) led to the substitution of toluene in the original bromination process with the lower boiling solvent dichloromethane and the imposition of a temperature limit of 40°C during the solvent strip. The crude bromoacid **10** was used in the following alkylation reaction without further purification. In the methyl ester series, the racemic acid **16** was isolated as a crystalline solid in 70% yield after an extractive workup and crystallization from dichloromethane, thus eliminating the need for purification by chromatography. This process was used to prepare approximately 22 kg of **16** from two scale-up campaigns. An alternative, more efficient acid-catalyzed direct coupling of 3,4-methylenedioxymandelic acid **9** with indole **15**, was developed in time for the third pilot-plant campaign (Scheme 10.3). Reaction of an equimolar quantity of **9** and indole **15** in acetonitrile in the presence of trifluoroacetic acid gave the desired product **16** in 86% yield. The product was insoluble in the reaction mixture and was isolated by filtration in a direct drop process and used in the next step without further

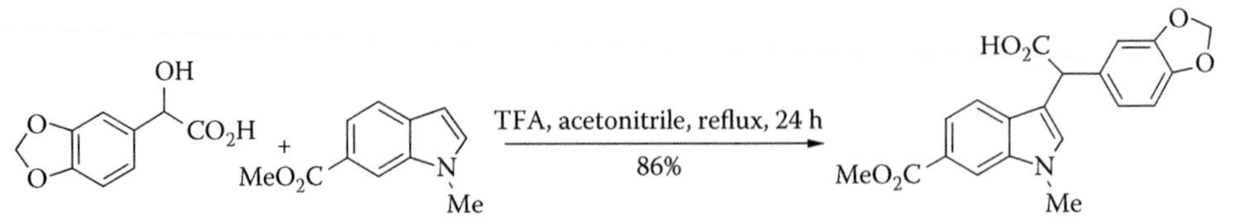

SCHEME 10.3 Direct coupling to prepare racemic acid 16. S. Challenger/Organic Process Research and Development/ACS/2005.

purification. This new process was successfully scaled up in the pilot plant on a maximum scale of 24.9 kg of **9**, avoiding the need to prepare the lachrymatory and thermally labile bromoacid **10**. A total of 78 kg of **16** was produced using a combination of the two methods without the need for chromatography.

A literature method[10] was used to prepare the acylsulfonamide **17** involving the coupling of an acyl imidazolide derived from acid **16** with sulfonamide **5** in a one-pot process. 1,8-Diaza bicycle [5.4.0]undec-7-ene (DBU) was required to catalyze the reaction. The acylsulfonamide product **17** was obtained in 79% yield after an acidic workup and crystallization from acetonitrile. This crystallization was a key purification point in the synthesis and produced material of high purity. Hydrolysis of the methyl ester **17** with excess sodium hydroxide in aqueous methanol completed the synthesis of racemic UK-350,926 **18**. The racemate was highly crystalline, and crystallization from dichloromethane gave high-quality material containing the ester **17** (0.1 to 0.5%) as the only significant impurity. A total of 18.5 kg of racemic acid **18** was prepared in the first two scale-up campaigns.

The sulfonamide **5** required for this synthesis was not commercially available at the time. We developed a four-step route (Scheme 10.4) from commercially available 4-bromo-3-methylanisole.

Initial results on the attempted resolution screen of racemic UK-350,926 with our in-house collection of chiral amines were not encouraging due to the propensity of the highly crystalline racemic free acid to preferentially precipitate from solution. The only success in the initial screen was with the chiral amine brucine hydrate **19**. Crystallization of a stoichiometric brucine salt from hot acetone gave the desired diastereoisomer in 92% recovery and 40% d.e (Scheme 10.5).

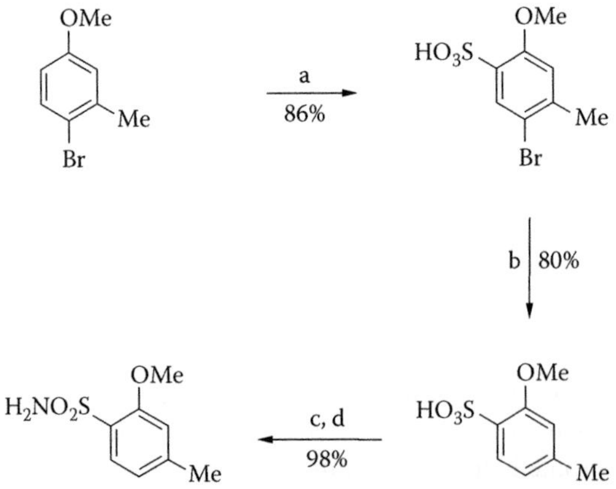

SCHEME 10.4 Preparation of sulfonamide 5. Reagents and conditions: (a) H_2SO_4 (98%), rt, 16 h; EtOAc (3 mL/g) recrystallization; (b) H_2, Pd/C (5%), MeOH, 60 psi, 60°C, 16 h; (c) $SOCl_2$, reflux 1 h, rt 16 h; (d) 0.88 M NH_3. S. Challenger/Organic Process Research and Development/ACS/2005.

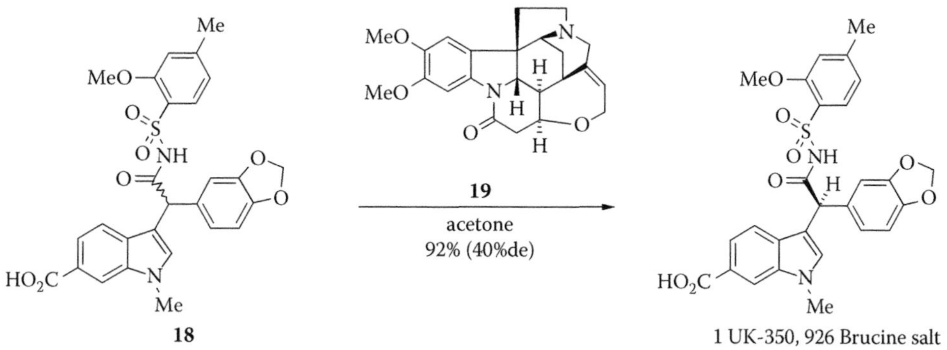

SCHEME 10.5 Resolution of racemic UK-350,926 with brucine hydrate. S. Challenger/Organic Process Research and Development/ACS/2005.

Fortunately, the natural enantiomer of brucine hydrate **19** preferentially crystallized a salt with the desired *S*-enantiomer of the acid. During the first campaign, a total of 4.77 kg of racemic acid and 3.82 kg of brucine hydrate were converted to 7.65 kg of salt. The liquors from the salt crystallization were shown to be racemic, suggesting that epimerization had occurred in solution and that a dynamic resolution process was possible in this system. The optical purity of the salt could not be upgraded by recrystallization, but advantage was taken of the highly crystalline nature of the racemic free acid. The brucine salt was broken and the racemic free acid was preferentially crystallized from ethyl acetate, leaving the desired optically enriched *S*-enantiomer in the mother liquors. Concentration of the mother liquors and crystallization from dichloromethane gave the desired *S*-enantiomer **1** in acceptable optical purity (>94% ee) and in 28% step yield. Although this process to upgrade optical purity was inefficient, it did allow for the preparation of the first batch of UK-350,926 **1** (1.34 kg) to support initial toxicology studies. In addition, 2.51 kg of good quality racemate was recovered for recycle. The overall yield for the first laboratory campaign using the new synthesis was approximately 10% (six linear steps), which compared favorably with the original discovery synthesis (4%) and vindicated our decision to switch routes early in the development program. The high toxicity of brucine hydrate and the sacrificial nature of the final step dictated that we needed an alternative resolution process in time for the second kilo laboratory campaign.

A solvent screen indicated that ethers, and in particular tetrahydrofuran (THF), were good solvents for solubilizing the racemic acid. This information led us to complete a salt screen using optically pure (*S*)-enantiomer **1** and our in-house collection of chiral amines (1 molar equivalent) in THF, which gave two crystalline salts, one with (1*R*, 2*R*)-2-amino-1-(4-nitrophenyl)-1,3-pro-panediol and the other with (*S*)-(-)-1-phenylethylamine. Interestingly, the stoichiometry of the precipitated (*S*)-(-)-1-phenylethylamine salt was 1:1.5 (acid:amine) by ^{1}H NMR analysis. In view of the low cost and bulk availability of (*S*)-(-)-1-phenylethylamine, we focused all our development efforts on this salt. Resolution of the racemic acid of **1** with (*S*)-(-)-1-phenylethylamine (1.5 molar equivalents) in THF (20 mL/g) gave the desired (*S*, *S*)-salt in 55% yield and 80% d.e. This was the starting point for further development work on this process, which led to the discovery of an efficient dynamic resolution process. The key parameters were found to be solvent composition, crystallization temperature, and amine stoichiometry. A mixture of THF and 1,2-dimethoxyethane (DME) was selected to optimize the recovery of the desired (*S*, *S*)-salt. Temperatures in the 45 to 55°C range and an excess of the chiral amine were used to ensure an acceptable rate of epimerization of the undesired (*R*)-enantiomer in solution. A combination of the above observations led to the development of the conditions chosen for scale-up. (*S*)-(-)-1-Phenylethylamine (two molar equiv-alents) was added to the racemic acid in THF (5.5 mL/g) and DME (5.5 mL/g), and the mixture was heated to 55°C for 24 h, 50°C for 24 h and 45°C for 24 h. The temperature gradient and heavy

TABLE 10.1
Scale-Up of Dynamic Resolution Process

Entry	Scale	Conditions	Yield (%)	Optical Purity (%de)
1	1 g	(S)-(-)-1-Phenylethylamine (1.5 equiv) THF	55	80
2	1.174 kg	(S)-(-)-1-Phenylethylamine (2 equiv) THF/DME, 55 to 45°C	84	94
3	6.52 kg	(S)-(-)-1-Phenylethylamine (2 equiv) THF/DME, 55 to 45°C	82	90

seeding were essential to prevent material oiling out during the initial phase of the crystallization. This process was successfully scaled-up twice in laboratory glassware on a 1-kg scale and twice in the kilo laboratory on 6.25 and 7.0 kg scales as part of the second bulk campaign (Table 10.1, Scheme 10.6) to give 9.66 kg of UK-350,926 **1** (>90% ee). The laboratory pilot on 1-kg scale gave the desired *S*, *S*-salt in 84% yield and acceptable optical purity (94% d.e.). In the kilo laboratory, the yields and optical purity were slightly lower than the laboratory batches (Table 10.1, entries 2 and 3). Fortunately, we were able to upgrade optical purity of material of 90% ee to greater than 98% by crystallization of a disodium salt of the free acid from aqueous ethanol.

The overall yield for the second campaign was increased to approximately 22% as a result of the improvements to the resolution step. During the third pilot-plant campaign, the development of both endothelin antagonists **1** and **2** was stopped due to adverse toxicology before the dynamic resolution process could be scaled further. Had the compounds continued in development, further work on the dynamic resolution would have focused on optimizing the selectivity to deliver the desired optical purity without the need for the salt upgrade, and to reduce the unacceptable processing time.

In the discovery chemistry route to racemic UK-349,862 **2**, a benzyl ester derivative of racemic acid **1** was reduced with lithium aluminium hydride (LAH) or di-isobutyl aluminium hydride (DIBAL-H). An alternative literature method,[11] avoiding LAH, was used to reduce **1** to the alcohol **2** involving the *in situ* formation of an intermediate acylimidazolide with 1,1-carbonyldiimidazole (CDI) followed by reduction with sodium borohydride in aqueous THF (Scheme 10.7). On a pilot scale in laboratory glassware, this gave UK-349,862 **2** in 86% yield and without significant loss in optical purity. However, on scale-up to 0.5 kg scale, this process resulted in 10% loss in optical purity. Quenching the reaction at the intermediate imidazolide stage and chiral purity assay of the recovered acid by HPLC indicated that optical purity was lost during the activation stage.

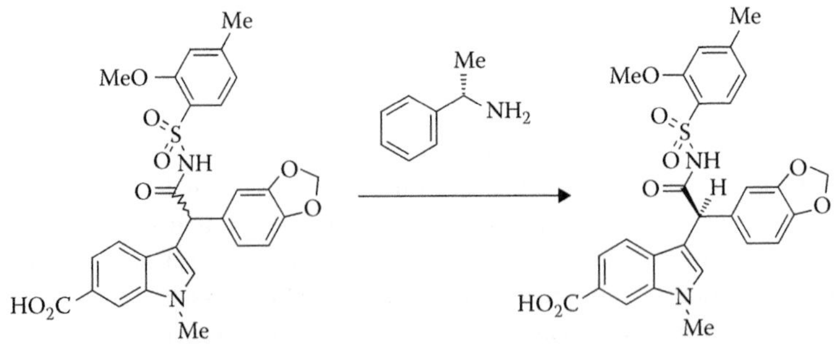

SCHEME 10.6 Dynamic resolution of racemic UK-350,926 with (S)-(-)-1-phenylethylamine. S. Challenger/Organic Process Research and Development/ACS/2005.

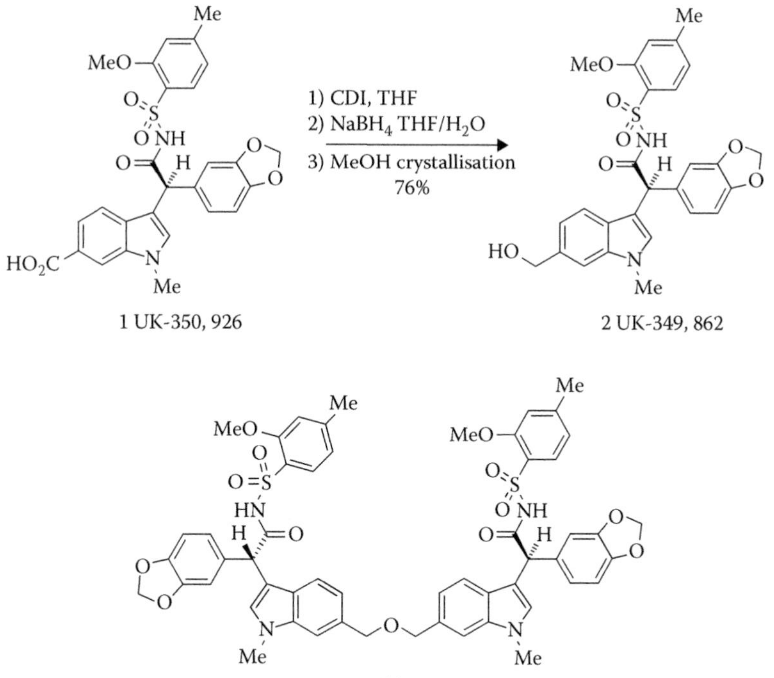

SCHEME 10.7 Synthesis of UK-349,862 from UK-350,926. S. Challenger/Organic Process Research and Development/ACS/2005.

Two reductions in laboratory glassware were successfully completed on a 0.75- and 1.5-kg scale. Provided the activation step was restricted to 1 h, the loss in optical purity could be minimized. A significant issue in this process was the stability of **2** to acid during the quench and subsequent solvent strip. A symmetrical ether dimer impurity **21** was observed when dilute hydrochloric acid was used to quench the reaction. The level of this dimer could be controlled to less than 5% by switching to a citric acid quench (pH = 3) and restricting the temperature in the solvent strip to less than 45°C. Optical purity in this series could also be upgraded by selectively removing the unwanted (*R*)-enantiomer as the racemate by crystallization from methanol. The modified reduction process was used to prepare 2 kg of UK-349,862 **2**, which was sufficient for initial toxicological evaluation.

10.4 CONCLUSIONS

We developed an efficient scalable route that allows access to the two endothelin antagonists UK-350,926 and UK-349,862 from three common commercially available starting materials. The key step involves a dynamic resolution of a late-stage intermediate with (*S*)-(-)-1-phenylethylamine. Furthermore, the new process avoids handling optically sensitive intermediates, a hazardous intermediate, expensive coupling reagents, and a cryogenic step. All of the intermediates are crystalline solids, removing the need for purification by chromatography in the original discovery synthesis. The new process was implemented on kilogram scale to prepare approximately 12.5 kg of UK-350,926 and 2 kg of UK-349,862. The overall process yield, for the four-step synthesis of UK-350,926 **1** from commercially available indole **15**, was improved from approximately 4% for the discovery synthesis to 42% through a combination of outsourcing (two steps) and improved synthetic efficiency.

ACKNOWLEDGMENTS

We thank Nick Haire and his analytical team, David Clifford, John R. Williams, and Terry V. Silk, and our Discovery colleagues David Rawson and David Ellis for excellent support during this project. We would also like to thank Mike Williams (Pfizer, Sandwich) and Steven V. Ley (University of Cambridge) for their helpful discussions.

REFERENCES AND NOTES

1. Yanagisawa, M.; Kurihara, H.; Kimura, S.; Tomobe, Y.; Kobayashi, M.; Mitsui, Y.; Yazaki, Y.; Goto, K.; Masaki, T. *Nature* 1988, 332, 411–415.
2. (a) Clark, W.M. *Current Opinion in Drug Discovery and Development* 1999, 2, 565–577. (b) Wu, C.; Holland, G.W.; Brock, T.A.; Dixon, R.A.F. *Investigational Drugs* 2003, 6, 232–239.
3. Challenger, S.; Dack, K.N.; Derrick, A.M.; Dickinson, R.P.; Ellis, D.; Hajikarimian, Y.; James, K.; Rawson, D.J. WO 99/20623, 1999.
4. (a) Rawson, D.J.; Dack, K.N.; Dickinson, R.P.; James, K. *Biorg. Med. Chem. Lett.* 2002, 12, 125–128. (b) Rawson, D.J.; Dack, K.N.; Dickinson, R.P.; James, K.; Long C.; Walker, D. *Medicinal Chemistry Research* 2004, 13, 149–157.
5. Ellis, D. *Tetrahedron Asymmetry* 2001, 12, 1589–1593.
6. Carpino, L.A. *J. Am. Chem. Soc.* 1993, 115, 4397–4398.
7. For the three-step preparation of sulfonyl chloride 14 from 4-bromo-3-methylanisole see Reference 3.
8. Ashcroft, C.P.; Challenger, S.; Clifford, D.; Derrick, A.M.; Hajikarimian, Y.; Slucock, K.; Silk, T.V.; Thomson, N.M.; Williame, J.R. *Org. Process Res. Dev.* 2005, 9, 663–669.
9. Peakdale Molecular and EMS-Dottikon AG supplied N-methylindole-6-carboxylic acid methyl ester.
10. Drummond, J.T.; Johnson, G. *Tet. Lett.* 1988, 29, 1653.
11. Sharma, R.; Voynov, G.H.; Ovaska, T.V.; Marquez, V.E. *Synlett* 1995, 839–840.

11 Cefovecin Sodium: A Single-Dose Long-Acting Antibiotic for Use in Companion Animals

Timothy Norris

CONTENTS

11.1 INTRODUCTION

Briefly, the history of the evolution of cephalosporin antibiotics starts in 1945 with the isolation of *Cephalosporium acremonium* from seawater by Professor Giuseppe Brotzu at the University of Sardinia. This organism produced material with strong antibacterial activity, one of the components of which was later isolated and identified as cephalosporin C **1**. This molecule showed relatively weak activity compared to penicillin; however, it had two potentially useful properties:

1. It showed resistance to hydrolysis by certain staphylococci penicillinases.
2. It was more stable to acid than benzyl penicillin.

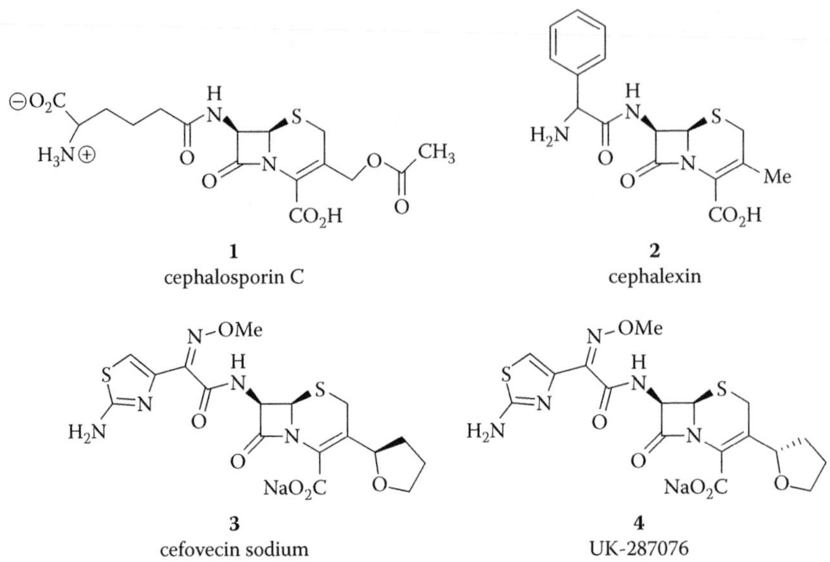

FIGURE 11.1 Comparison of Cefovecin structure with cephalosporin and cephalexin.

The disadvantage of cephalosporin C is that it is not absorbed in the gastrointestinal tract, and it took until 1969 before the first oral cephalosporin, cephalexin **2** came to be used broadly in the clinic. However, the potential was sufficient to justify at least two decades of research and development by many pharmaceutical companies. Modifications at C-3 and C-7 led to a series of improved antibiotic compounds suitable for use in mammals, including domestic animals, companion animals, and humans, either administered as a solution by injection or as a solid using an oral formulation. More detailed accounts are available for the interested reader.[1]

Cefovecin sodium, **3** the subject of this research, was selected for use in companion animals as a single-injection formulation that could be administered in a veterinarian's office because of its long-acting therapeutic effectiveness at low doses.

11.2 OVERVIEW OF SYNTHETIC STRATEGY

The product mandate for this project required that the active pharmaceutical ingredient (API) needed to be a single isomer. Methodology reported in the literature,[2] for the general class of which cefovecin is a member, provided an epimeric diastereoisomer mixture of **3** (which became cefovecin sodium) and **4** (UK-287076 sodium). Both compounds are potent, but the (S) absolute configuration at the tetrahydrofuran (THF) C-2 center was selected for commercial development (Figure 11.1).

Consideration of the annual metric ton requirement for commercial quantities of API meant that the synthesis would have to be efficient in terms of number of isolated intermediates and would have to exhibit significant atom economy. In addition, quantity of solvent used would have been minimized. Also, despite the complexity of the molecular architecture that gives rise to isomeric pairs of intermediates at multiple points in the synthesis, use of chromatography would have to be avoided to provide a commercially viable product.

Cephalosporin C **1** was first synthesized[3] by Woodward's group at Harvard in 1966 and, the chiral pool starting material for this synthesis was L-(+)-cysteine. The approach is noted below in abbreviated form to the Δ³ cephem nucleus **5** via the key 4-thia-2,6-diazabicyclo[3.2.0]heptane nucleus, **6**, Scheme 11.1. Nucleus **5** contains the same stereochemistry at C-6 and C-7 as natural cephalosporin C.

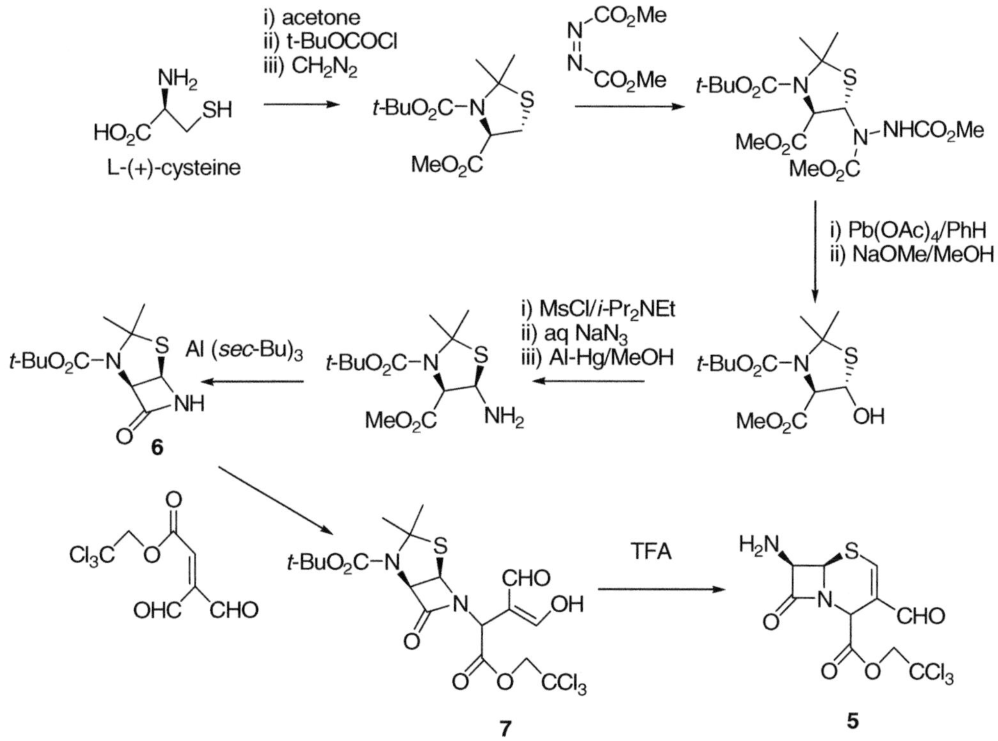

SCHEME 11.1

In more modern commercial synthesis that requires significant variation at cephem C-3 position, a convenient starting material obtained from the chiral pool is penicillin G. This also allows the process chemist to shortcut the introduction of the second chiral center, which eventually resides at C-6 and would be necessary if the L-(+)-cysteine approach is used. Use is still made of the 4-thia-2,6-diazabicyclo[3.2.0]hept-2-ene systems closely related to **7** in Woodward's synthesis. This allows access to the cephem nucleus, which can now have a wide variety of directly introduced substituents with complicated molecular architecture at C-3. This leads to more convergent and flexible approach than building a complicated molecular appendage on the C-3 aldehyde of **5** in the original approach.

Finally, the process design will require crystalline intermediates and the use of telescoped reaction sequences that are robust enough to provide clean intermediates with defined impurity profiles. This is an essential for commercial viability.

11.3 STARTING MATERIALS AND SYNTHESIS OF INTERMEDIATES USED AT CONVERGENT POINTS IN SYNTHESIS

11.3.1 PRELIMINARY SELECTION OF SUITABLE BLOCKING GROUP

The choice of the carboxylic acid blocking group (R in Scheme 11.2) is very important, as it not only has to survive multiple reactions, it also has to be readily and cleanly removed at the end of the synthesis during the formation of the penultimate intermediate. In addition, there is an important third factor. The physical properties of the isolated intermediates need to be considered. The ideal blocking group will confer sufficient crystallinity on the key intermediates that they can easily be isolated as fairly pure defined compounds from the telescoped reaction sequences needed for commercial efficiency.

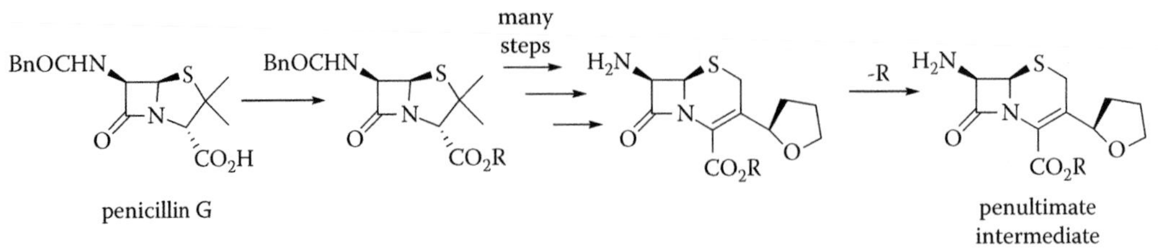

SCHEME 11.2

Many substituted benzyl and benzoyl esters were considered along with some alkyl esters. After screening, the choice came down to either allyl or 4-nitrobenzyl ester, both of which can be removed by well-defined methodology. The decision was made to use 4-nitrobenzyl ester for the commercial synthesis because of the crystalline properties it conferred on the isolated intermediates.

11.3.2 PENICILLIN G AND ITS REARRANGEMENT TO 4-THIA-2,6-DIAZABICYCLO[3.2.0]HEPT-2-ENES

Penicillin G, **8**, the 4-nitrobenzyl ester, **9**, and derived sulfoxide, **10**, are all commercially available compounds. The methodology employed to make **9** is atom efficient, requiring only a slight excess of 4-nitrobenzyl bromide to drive the reaction to completion in the presence of a phase-transfer catalyst such as TBAB. Isolation of **9** is optional and it is converted into the sulfoxide **10** by the action of peracetic acid generated from acetic acid and hydrogen peroxide. This reaction is carried out between 0 to 5°C for optimum performance and generation of minimally colored impurities. Washing with an aqueous reducing agent, typically sodium metabisulfite dissolved in water, destroys residual quantities of peroxide and peracids. The product **10** is crystallized by concentration of the dichloromethane reaction solvent and addition of isopropanol. After a period of granulation, **10** is isolated by filtration and carefully dried under vacuum. There are other variations of this well-known chemistry that give similar results, but **10** can usually be obtained from penicillin G in the range of 75 to 85% yield or greater.

The next step in the synthesis is somewhat more interesting and key to its flexibility and commercial viability. In 1970, Cooper and Jose reported[4] a novel rearrangement of penicillin V sulfoxide (Scheme 11.3). This rearrangement is quite general. When penicillin sulfoxides are treated with trialkyl phosphites, they are transformed into the 4-thia-2,6-diazabicyclo[3.2.0]hept-2-ene system in good yield. Sulfoxide **10** is heated in toluene at 105 to 110°C in the presence of 30 to 40% molar excess of triethylphosphite, the penicillin sulfur is reduced from S^{IV} to S^{II} oxidation state, and triethylphosphite (P^{III}) is oxidized to triethyl phosphate (P^V). During this redox process, C–S fission occurs in the thiazolidine ring system, but the resultant azetidinone intermediate **11** does not lose the natural chirality of the original penicillin. The sulfide anion on former penicillin C-5 attacks the carbonyl carbon of the former penicillin side chain, and a new 1,3-thiazoline ring is formed by elimination of water to yield 4-thia-2, 6-diazabicyclo[3.2.0]hept-2-ene **12**. The exocyclic olefin on **12** is rearranged by treatment with triethylamine to give the desired isolated product **13** in the range of 80 to 84% yield. Intermediates **11** and **12** are not isolated, but **12** was isolated and characterized. Intermediate **11** is inferred from the known behavior of 4-thia-2,6-diazabicyclo[3.2.0]hept-2-enes. The mechanism of the initial thermal ring opening of penicillin sulfoxides is thought to proceed by sulfenic acid-ene intermediates. Sulfenic acids are very transient species, but Barton et al. showed that the sulfenic acid-olefin intermediates could be trapped during thermolysis of penicillin sulfoxides,[5] and that they are the intermediates involved in the thermal isomerization of penicillin G (R) sulfoxides into penicillin G (S) sulfoxides.[6]

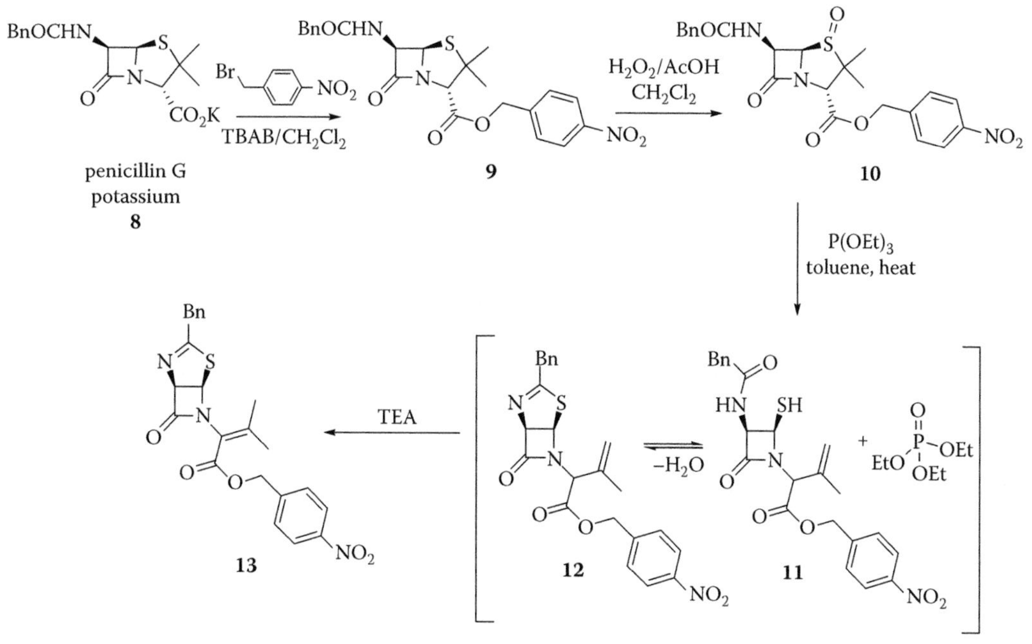

SCHEME 11.3

11.3.3 OPTICALLY ACTIVE 2-ACETYL-THF

The starting material for the chiral tetrahydrofuran side chain in the commercial cefovecin sodium synthesis is 2-acetyl-(*S*)-THF, **18**. This material is now commercially available and is manufactured from (*S*)-THF-2-carboxylic acid **14** obtained by classical resolution technology. Compound **18** provides a convenient stable starting material that can be introduced directly into the synthesis as explained in a later section.

Briefly, carboxylic acid **14** which must be an essentially pure enantiomer (ratio 98:2) is converted to acid chloride **15** which is an oil that needs to be directly reacted with Meldrum's acid **16** in the presence of 2 moles of pyridine to yield **17** as a stable solid; however, a slurry of **17** is treated with glacial acetic acid to provide **18** directly and 17 is not isolated. The process is telescoped in practice, and all the reactions are carried out in dichloromethane, which is recovered and recycled. Thus, **14** is converted into **15** using only a 10% molar excess of oxalyl chloride using 10 mole% of dimethylformamide as catalyst. The reaction is complete after 3 h reaction at room temperature (20 to 25°C); concentration and removal of the dichloromethane yields **15** as a crude oil that is suitable for direct reaction with Meldrum's acid and pyridine at 0 to 5°C in the relative molar ratio 1:2 in fresh dichloromethane. Workup with dilute hydrochloric acid and layer separation followed by water washes results in a solution of **17** in dichloromethane. The solution is concentrated to a slurry, which is treated with acetic acid at about 70 to 80°C to form **18**. The reaction mixture is worked up by adding dichloromethane and washing with aqueous sodium carbonate followed by water washes. Concentration of the washed dichloromethane layer yields enantiomer **18** that is equal in optical purity to that of the starting carboxylic acid **14**. The concentrate is vacuum distilled to provide a chemically stable and chemically pure ketone **18**. This material has been shown to have a shelf life of 5 years at room temperature (Scheme 11.4).

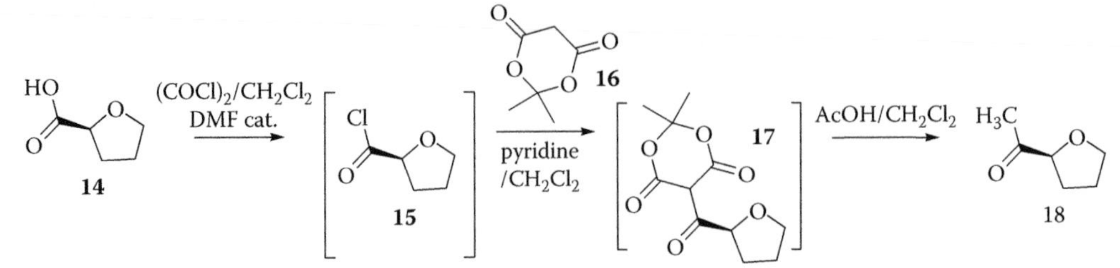

SCHEME 11.4

11.3.4 THIAZOLE SIDE CHAIN, ATMAA, AND ACTIVATED DERIVATIVE, DAMA

Aminothiazol methyloxime acetic acid[7] **19** (ATMAA) is commercially available as is the proprietary activated form of ATMAA known as DAMA **20**.[8] ATMAA side chain and its close relations are critically important structural features of third- and fourth-generation commercially viable cephalosporin-based antibiotics related to cefovecin, such as cefetamet pivoxil, cefpodoxime proxetil, cefuroxime axetil, cefpirome, cefotiam hexetil, cefdinir, ceftbuten, ceftiofur, and cefixime (Scheme 11.5).

ATMAA was first reported and prepared[9] using the methodology noted in Scheme 11.6. Its production has no doubt been significantly improved over the years, although accounts of an optimized synthesis have not been reported in the literature as far as is known.

Activated form DAMA is manufactured by reacting ATMAA with diethyl chlorothiophosphate in the presence of 1,8-dizabicyclo[5.4.0]undec-7-ene (DBU) and a simple trialkylamine such as triethylamine or tri-*n*-butylamine. The product is a crystalline solid that is a convenient reactive form of ATMAA that must, however, be stored below 5°C. Further discussion on DAMA and the activation strategy appear later in the chapter.

11.4 OZONOLYSIS AND REDUCTION

Compound **13** contains the key 4-thia-2,6-diazabicyclo[3.2.0]hept-2-ene ring system that has the two chiral centers that will form the final cephem ring nucleus. The exo-cyclic olefin is ozonolyzed to yield the ketone **20** *in situ*, which is then reduced with sodium borohydride under slightly acidic conditions. Great care is needed to ensure these operations are carried out safely on large scale. Ozone is bubbled through a mixture of dichloromethane, isopropanol, and **13** at –60 to 50°C. This temperature range is considered critical for control of operational safety and important for optimum yield. At very low temperatures, ozone can build up to explosive concentrations. After ozonolysis is complete, the reaction mixture is treated with sodium borohydride and acetic acid to yield an epimeric diastereoisomer mixture of the alcohols **21**. The workup of the reaction mixture is straightforward: initially, quenching is accomplished with aqueous sodium bisulfite, followed by acidification with hydrochloric acid and brine washes of the orgainic layer. The essentially dichloromethane layer is concentrated, and **21** is crystallized from aqueous isopropanol in 75 to 80% yield. The isolated product is an epimeric mixture of alcohols containing 8 to 11% other impurities, but this material is suitable for introduction into the next series of telescoped reactions (Scheme 11.7).

11.5 1-AZETIDINEACETIC ACID DERIVATIVES AND INTRODUCTION OF CHIRAL THF SIDE CHAIN

The next series of telescoped reactions is carefully coordinated to produce the highly substituted β-lactam **24**. The thiazoline ring in **21** is easily cleaved to yield the thiol **22**, which features a

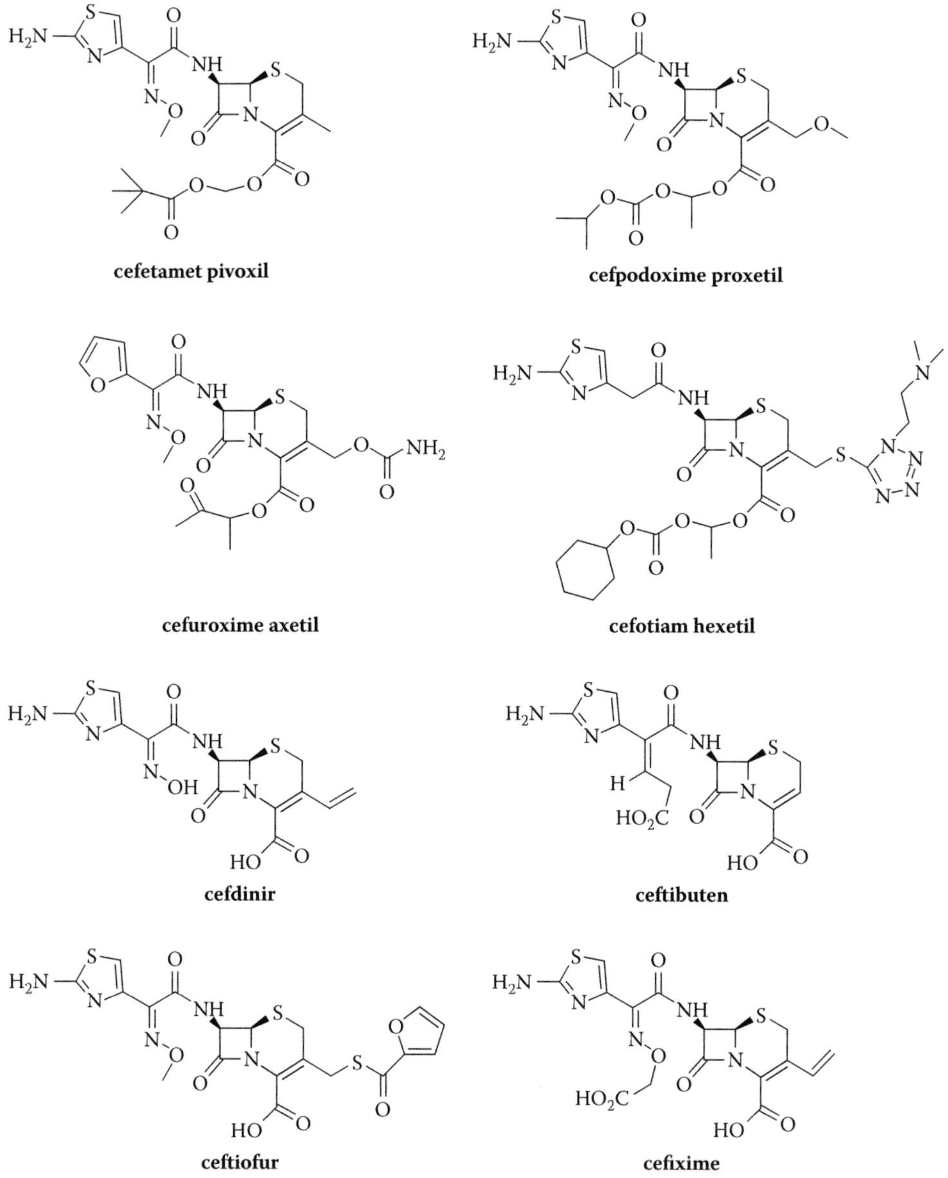

cefetamet pivoxil

cefpodoxime proxetil

cefuroxime axetil

cefotiam hexetil

cefdinir

ceftibuten

ceftiofur

cefixime

SCHEME 11.5

four-membered azetedone ring system. Compound **22** consists of a pair of epimers due to lack of stereospecificity at the α-hydroxy ester position. The thiol functional group is readily alkylated with α-bromomethylketone **23** and its derived dimethyl-ketal, which is generated *in situ* from the chiral ketone **18**. The reaction mixture on workup yields the epimeric pair of alcohols **24**. Compound **24** is crystalline due to the carefully chosen 4-nitrobenzyl protecting group and is isolated in the range of 76 to 88% yield as a diastereomeric alcohol mixture. The mixture ratio ranges from 90:10 to 70:30, with the average about 80:20. The purity of the isolated solid based on both diastereoisomers is in the range of 76 to 82%. The exact identities of the major and minor epimers were not elucidated. Compound **24** is carried forward as a mixture containing up to 8% other impurities as determined by high-performance liquid chromatography (HPLC) (Scheme 11.8).

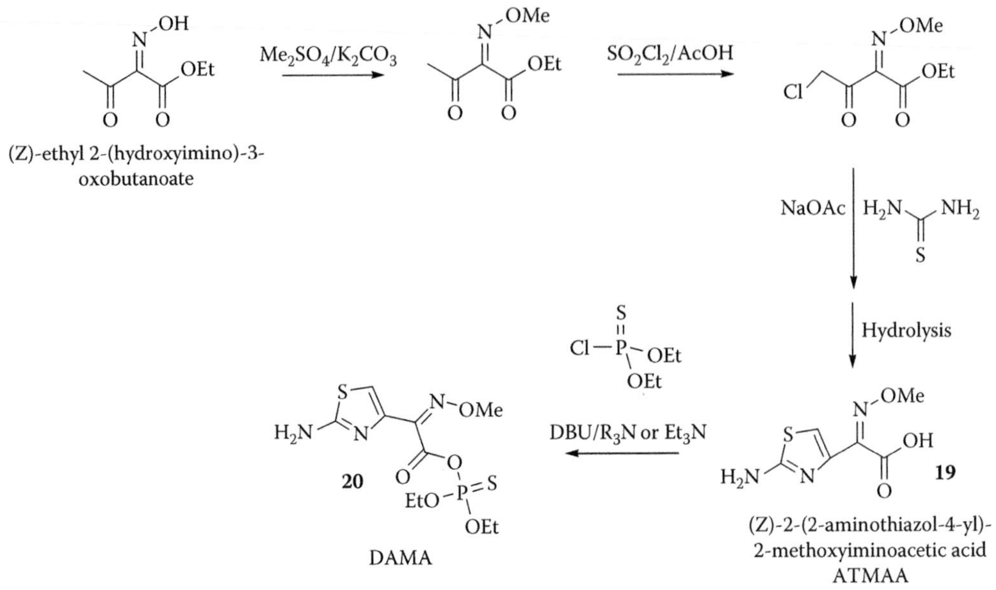

(Z)-ethyl 2-(hydroxyimino)-3-oxobutanoate

(Z)-2-(2-aminothiazol-4-yl)-2-methoxyiminoacetic acid
ATMAA

DAMA

SCHEME 11.6

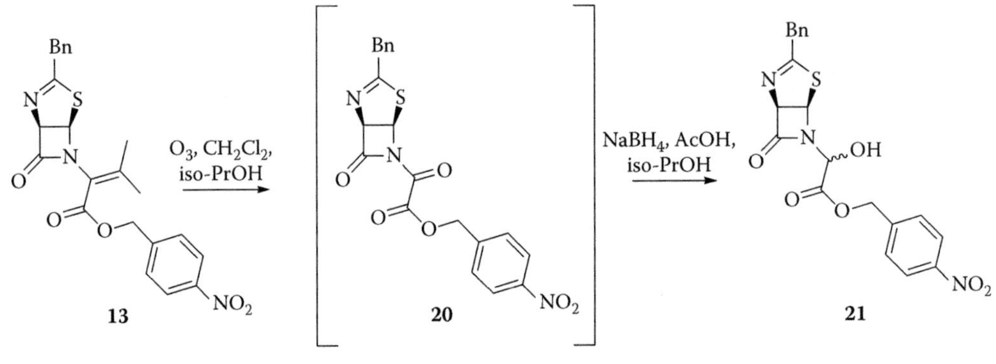

SCHEME 11.7

11.6 FORMATION OF 2*H*-DIHYDRO-1,3-THIAZINE RING

The focus of the next synthetic tactic is to convert the nonstereospecific alcohol center of **24** into either a chloride **25** or an iodide **26** in preparation of the 2*H*-dihydro-1,3-thazine ring formation using either Wittig-based chemistry or Horner–Wadsworth–Emmons-based chemistry. The alcohol function of **24** is converted into an epimeric mixture of chlorides **25**. The chlorine atom is readily introduced by treatment with sulfuryl chloride and 2-picoline, resulting in a 40:60 mix of isomers. The epimeric mixture of chloro compounds **25** is used directly in the Wittig chemistry to achieve six-membered ring closure. Transformation to the iodo analog **26** is necessary to utilize Horner–Wadsworth–Emmons chemistry for the same purpose. Thus, chloro compounds **25** are converted *in situ* to a 50:50 mixture of the iodo derivatives **26**, using the Finkelstein reaction, which uses sodium iodide in acetonitrile for best results. Yield from **24** to **26** is ~77%. In preparation for ring cyclization, **25** and **26** are generated *in situ* and used directly with the appropriate phosphorus

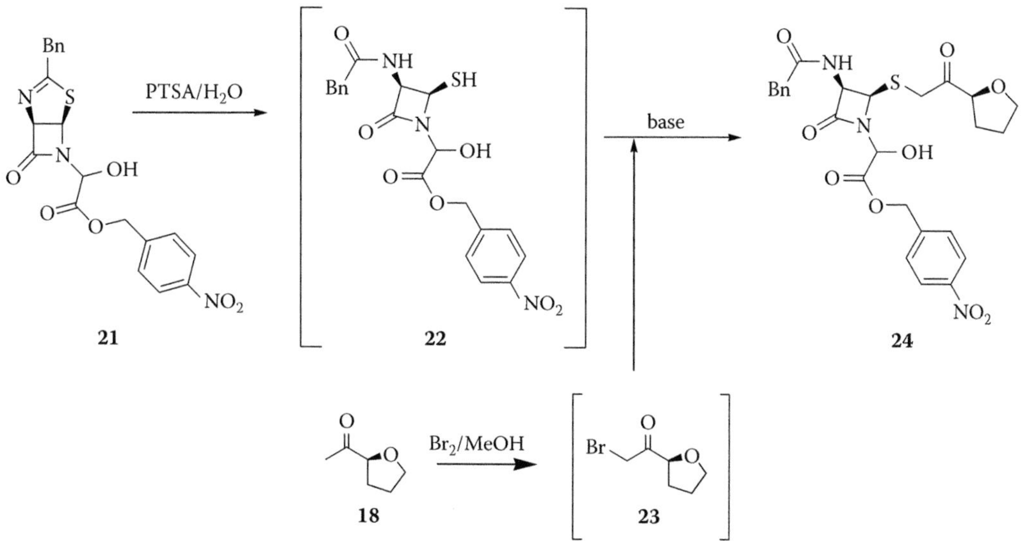

SCHEME 11.8

reagent in preparation for the cyclization and formation of the six-membered 2*H*-dihydro-1,3-thiazine ring (Scheme 11.9).

11.6.1 WITTIG CHEMISTRY

In the case of chloro compound **25**, reaction with trimethylphosphine and base (NaHCO$_3$) yields the phosphorus ylide **27**. Yield from **24** is ~80%. This is readily cyclized by gentle warming at ~30°C to form compound **29**. Optimum cyclization yield ~76% is achieved between pH 7.5 and 8.0. The bicyclic entity **29** is a doubly protected form of the cephalosporin ring system with all optically active stereochemical centers in place.

11.6.2 HORNER–WADSWORTH–EMMONS CHEMISTRY

Similarly, **29** can be formed by the Horner–Wadsworth–Emmons modification of the Wittig reaction via the phosphonate **28** rather than the phosphorus ylide **27**. In this case, the iodo compound mixture **26** is treated with trimethyl phosphite. Cyclization requires the use of lithium chloride/Hunig's base,[10] and cyclization proceeds smoothly at room temperature in similar yield to the Wittig chemistry. Details of this process with isolation and characterization of the intermediates **26**, **27**, and **29** can be found in a recently published world patent application.[11] Yields for the conversion of **24** into **27** are ~77% overall; cyclization yield from **27** to **29** is ~85%.

11.6.3 DEPROTECTION OF CEPHEM 29 AND FORMATION OF 7-AMINO-CEPHEM 30

The removal of the penicillin G group was carefully developed by chemists so that modification of the resultant exposed free amine can be derivatized with a wide variety of side chains, in a search to improve the biological activity of both cephalosporin and penicillin cores. This knowledge was used in the next stage of the synthesis. The cephem nucleus **29** is doubly protected, and the benzylcarbonyl moiety (Pen G group) is removed by treatment with phosphorus pentachloride at low temperature-40/30°C in the presence of a weak base such as 2,6-lutidine or 2-picoline. The reaction proceeds via the iminoyl chloride derivative **31** (Figure 11.2), which is decomposed by treatment of *iso*-butyl alcohol also at low temperature as the reaction is highly exothermic. The

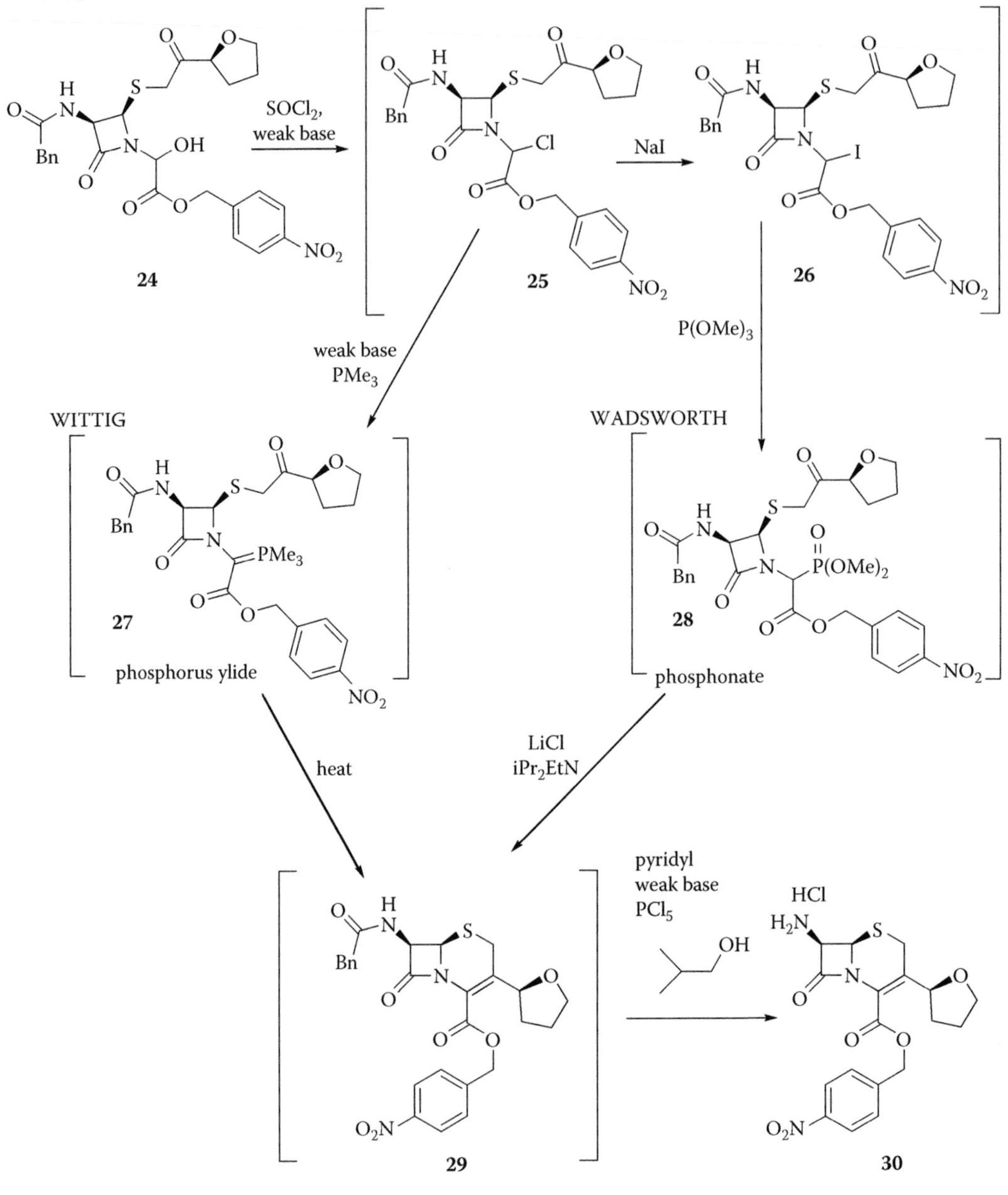

SCHEME 11.9

reaction is best carried out in dilute dichloromethane. Step yields for the conversion of **29** into **30** can be in the region of 85% if the phosphorus pentachloride reaction and subsequent decomposition of the iminoyl chloride intermediate with *iso*-butyl alcohol are carefully controlled at low temperature. Excursions into temperatures above ambient adversely affect the yield, but commercial operations have been devised in which the conversion of **24** into **30** has been achieved without isolating any intermediate compounds in 55% overall yield[12] and in high purity containing <2% of the THF (*R*) epimer **32** (Figure 11.2). A single-crystal x-ray structure was determined for cephem **30**, which confirms that the bicyclic ring stereochemical centers are identical with cephalosporin and that the THF ring has (*S*) configuration required for cefovecin (Figure 11.3). In commercial

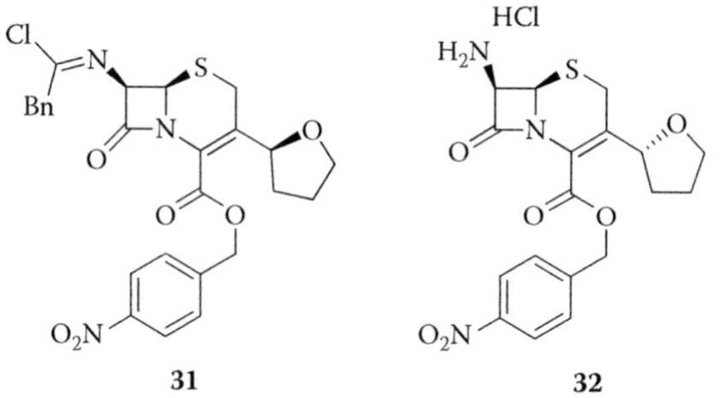

FIGURE 11.2 Structures of compounds **31** and **32**.

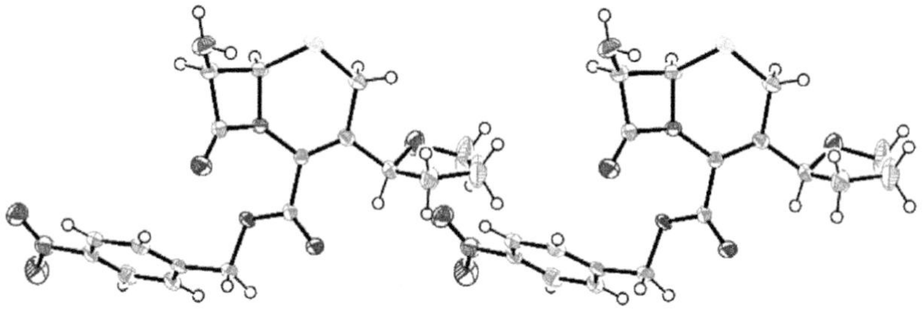

FIGURE 11.3 (See color insert following page 40). Stereochemistry of cephem **30** derived from single-crystal x-ray data.

practice, **30** is isolated as the hydrochloride salt and is a stable crystalline solid normally stored at 5°C for extended periods over 1 year.

11.7 SYNTHESIS OF CEPHEM ZWITTERION, 33

Prior to attachment of the ATTMA side chain **19**, removal of the 4-nitrobenzyl ester protecting group must be achieved using mild conditions and neutral conditions to prevent epimerization at C-7. The 4-nitrobenzyl group was introduced at the very beginning of the synthesis to confer crystallinity to the isolated intermediates. This allowed the synthesis to proceed with purging of unwanted by-products along the way and also provided storage and hold points to suit a commercial supply chain. The removal of 4-nitrobenzyl esters can be achieved in aqueous acetone at ambient temperatures (10 to 35°C) with sodium hydrosulfite[13] adjusted to pH 6.5 to 8.0 with aqueous ammonium hydroxide.[14] On completion of the reaction, the pH is adjusted to pH 3.5 to 4.5 to precipitate the product that is further purified by activated charcoal treatment in water to yield the purified cephem zwitterion **33** (Scheme 11.10). The overall yield from **30** to **33** is >75%.

11.8 USE OF THIOPHOSPHATE ESTER ACTIVATION TO FORM API

The final step of the cefovecin synthesis requires the formation of an amide bond between ATMAA **19** and cephem zwitterion **33**. There are many ways of achieving this end game; however, nothing is so elegant and high yielding as the activation of ATMAA by conversion to a thiophosphate ester to yield DAMA **20** (Scheme 11.6). In a more traditional synthesis, ATMAA might be converted to an acid chloride, but in this case, a suitable blocking group that is easily removed must be deployed. This is illustrated by the synthesis of ceftiofur (Scheme 11.11). ATMAA **19** is converted into its N-trityl derivative **34** and then to the acid chloride **35**. The acid chloride is reacted with the

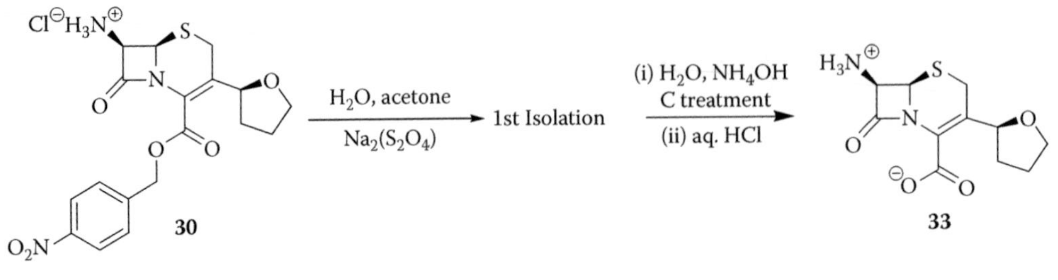

SCHEME 11.10

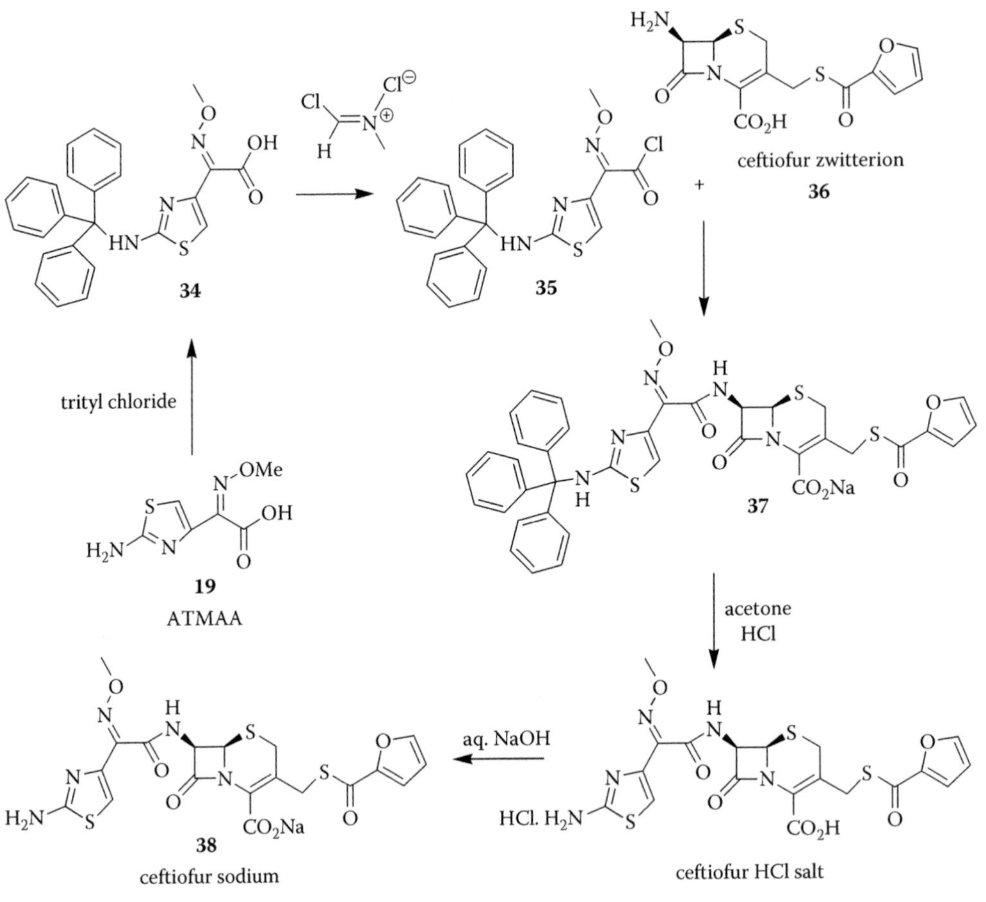

SCHEME 11.11

intermediate cephem **36** forming the amide bond to yield trityl-protected ceftiofur **37**, which then requires deprotection to yield ceftiofur, **38**.

In the cefovecin synthesis DAMA **20** is reacted directly with cephem zwitterion **33** in aqueous acetone. During the reaction, the pH of the reaction media is maintained essentially neutral or slightly alkaline by the addition of dilute aqueous sodium hydroxide in the pH range of 6.8 to 8.2. During the course of the reaction that is completed at ambient temperature after about 1 to 2 h, cefovecin sodium **3** is precipitated from the reaction media in pure form ready for formulation operations. The by-product of the reaction is diethylthiophosphate (DETP), a liquid that is washed

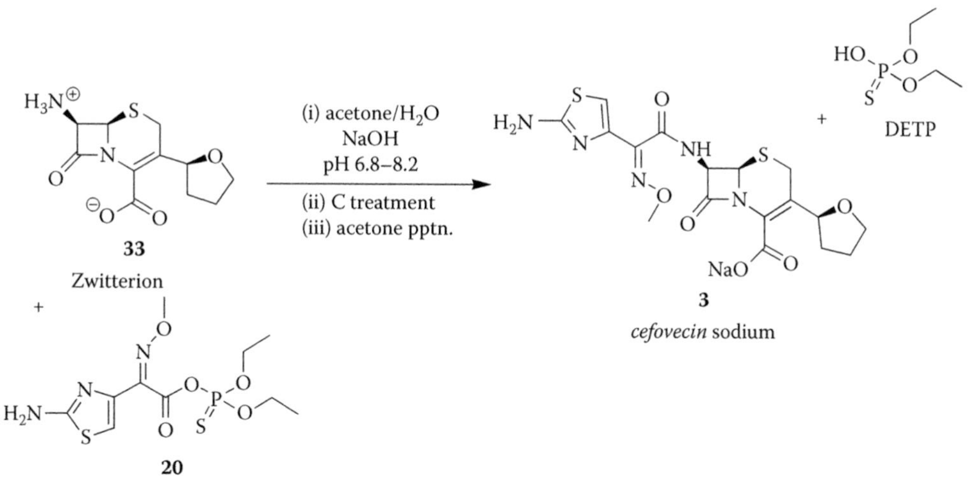

SCHEME 11.12

free from the cefovecin sodium with acetone and water (Scheme 11.12). Details of practical procedures can be found in the patent literature.[15] Yields are ~80 to 85% after purification.

11.9   FINAL API FORM AND CLINICAL USE

Cefovecin sodium is an amorphous solid, typically stored at –20°C for long-term storage of 1-year or more, and for short-term storage, 5°C is adequate. In the clinic, cefovecin sodium is supplied in vials as a sterile lyophile, which is made ready for parenteral use by injection into the companion animal by addition of sterile aqueous-based diluent. The dose is calibrated according to the size of the animal, and the parenteral solution is available for multiple surgery sessions. Normally, most skin infections are treated successfully with a single dose, which reduces the stress factor for both pet and pet owner.

ACKNOWLEDGMENTS

Development of a commercial synthesis of this complexity requires both time and dedication for a number of years. The work described here occurred in the time frame 1999 to 2005, and contributors to the science, synthetic methodology, and execution at commercial scale include the following colleagues, members of a multinational team from Australia, England, Italy, Japan, and the United States: Juan Colberg, Mark Delude, Joe Desneves, Alessandro Donadelli, Christian Dowdeswell, Fogliato Giovanni, Grant Mclachlan, Hiromasa Morita, Isao Nagakura, Russell Shine, John Tucker, Bunzo Wada, and Maurizio Zennoni. In addition, many other colleagues worked enthusiastically as managers, shift operators, and technicians to bring cefovecin sodium to life.

REFERENCES AND NOTES

1. Moellering, R.C., *Oral Cephalosporins, Antibiotics and Chemotherapy*, Vol. 47, Schönfeld, H., Series Ed., Karger, Basel, 1995.
2. (a) Bateson, J.H., Burton, G., and Fell, S.C.M., *US Patent* 6001997, December 14, 1999. (b) Bateson, J.H., Burton, G., and Fell, S.C.M., *US Patent* 6020329, February 1, 2000. (c) Bateson, J.H., Burton, G., and Fell, S.C.M.,*US Patent* 6077952, June 20, 2000.

3. Woodward, R.B. et al., *J. Amer. Chem. Soc.*, 1966, 88, 852.
4. Cooper, R.D.G. and Jose, F.L., *J. Amer. Chem. Soc.*, 1970, 92, 2575–2576.
5. Barton, D.H.R. et al., *Chem. Comm.*, 1970, 1683.
6. Barton, D.H.R. et al., *Chem. Comm.*, 1970, 1059.
7. (Z)-2-(2-aminothiazol-4-yl)-2-methoxyiminoacetic acid.
8. Lim, J.C. et al., *European Patent Specification*, EP 0 620 228 B1, March 11, 1998.
9. Takaya, T. et al., *UK Patent Application*, 2 025 933 A, January 30, 1980.
10. Di-*iso*-propylethylamine also referred to as DIPEA in some literature references.
11. Morita, H., Nagakura, I., and Norris, T., WO 2005/092900 published October 6, 2005.
12. This is true for both cyclization options; however, in the Horner–Wadsworth–Emmons chemistry, the cephem **29** is isolated before conversion into **30**.
13. Sodium dithionite.
14. Guibe-Jampel, E. and Wakselman, M., *Synth. Comm.*, 1982, 12(3), 219–223.
15. (a) Colberg, J.C. et al., WO 2002046199. (b) Colberg, J.C. et al., WO 2002046198.

12 The Lithium–Halogen Exchange Reaction in Process Chemistry

William F. Bailey and Terry L. Rathman

CONTENTS

12.1 INTRODUCTION

Organolithium chemistry had a rather late start: the first preparation of a lithium-containing organic molecule dates back to the 1917 report by Schlenk and Holtz describing the preparation of ethyllithium from diethylmercury and lithium metal.[1] It was 13 years later that Ziegler and Colonius developed a practical, general route to simple organolithiums via direct treatment of an organohalide with lithium in a hydrocarbon solvent.[2] The Ziegler–Colonius route to organolithiums, which is essentially the approach used today for the industrial production of alkyl-, vinyl-, and aryllithiums, resulted in a rapid and fruitful exploration of the chemistry of organolithium compounds during the 1930s and 1940s. Much of the progress made during this period is due to the work of the research groups of Wittig at Heidelberg and Gilman at Iowa State. One of the many pioneering contributions from these groups was their independent and virtually simultaneous discovery in 1938 of the rapid metathesis reaction between an organic halide and an organolithium (Figure 12.1) that has come to be known as the lithium–halogen interchange, or exchange, reaction.[3,4] The exchange is a reversible process that leads to an equilibrium mixture favoring the organolithium best able to accommodate the formal negative charge on the lithium-bearing carbon.[5]

The aim of this chapter is to summarize some of the more recent developments in the application of the lithium–halogen exchange reaction and to highlight the practical considerations relevant to that chemistry. The choice of topics and examples is necessarily subjective and incomplete. However, the vast primary literature, detailing the synthetic utility and mechanistic vagaries of the exchange reaction, has been extensively reviewed.[6] In this connection, it should be noted that some of what follows has been taken from two *Lithium Link* publications; one on the use of

$$R-Li \ + \ X-R' \ \rightleftharpoons \ R-X \ + \ Li-R'$$

FIGURE 12.1 Lithium-halogen exchange.

tert-butyllithium (*t*-BuLi) in organic synthesis,[7] and another on optimization of reactions involving organolithiums.[8,9]

12.2 MECHANISTIC CONSIDERATIONS

The lithium–halogen exchange, which is most readily accomplished with bromides or iodides, only rarely with chlorides, and virtually never with fluorides,[6] is mechanistically intriguing. Most chemists would share Wakefield's succinct observation that the reaction "appears electronically wrong,"[10] because the negative end of the C–Li bond dipole becomes attached to the halogen of the organic halide rather than to the partially positive carbon atom of the C–X bond. The "electronically correct" course would be expected to produce a hydrocarbon through formation of a C–C bond. This Wurtz-type coupling is, in fact, only one of a number of possible competing reactions, including β-elimination and α-metalation, that can result when an organic halide is treated with an organolithium. Despite the interest in the interchange as a synthetic method, the mechanism of the reaction was, until rather recently, something of an enigma.[11] The lack of consensus as to mechanistic detail is attributable (in retrospect) to several factors including the capricious behavior of organic halides when treated with organolithiums and the failure of early studies to take adequate account of the fact that organolithium compounds do not necessarily exist as monomers. Although organolithiums are often, as here, depicted as monomeric, they are known to exist as aggregates whose degree of association is affected by such experimental variables as solvent, concentration, and temperature.[12]

It is now generally recognized that the exchange between an organolithium and an iodide or bromide often involves nucleophilic attack of the organolithium on the halogen atom of the organohalide to give an ate complex (Figure 12.2), either as a transition state or as an intermediate.[13] Not surprisingly then, the more easily polarized iodides undergo the exchange more rapidly than do bromides; for example, low-temperature lithium–iodine exchange between *t*-BuLi and a primary alkyl iodide can be accomplished selectively in the presence of an aryl bromide.[14]

12.3 LITHIUM–HALOGEN EXCHANGE OF ARYL HALIDES AND VINYL HALIDES

Because aryllithiums, vinyllithiums, and cyclopropyllithiums are more stable than are typical alkyllithiums, the equilibrium that is established when an aryl, vinyl, or cyclopropyl halide is treated with an alkyllithium is usually quite favorable[5]: indeed, the lithium–halogen exchange was first observed in reactions of *n*-BuLi with aryl bromides.[3,4] Consequently, the lithium–halogen exchange is particularly useful for the preparation of an organolithium in which the formally anionic carbon has a high s-character and many examples of this chemistry have been reported.[6,15] Summaries of tested experimental procedures may be found in the excellent monographs by Brandsma.[16]

The exchange of an aryl bromide or iodide with an alkyllithium is typically a high-yield process that proceeds rapidly at low temperature, but side reactions, including coupling of the aryllithium with the cogenerated alkyl halide and formation of a benzyne intermediate via *ortho*-metalation of the aryl halide,[17] are sometimes observed. For all practical purposes, the rate of an exchange reaction between an aryl bromide or iodide and an alkyllithium is essentially diffusion controlled.[18]

On the industrial scale, aryllithiums are most often prepared from aryl bromides, or occasionally from aryl iodides, by low-temperature exchange with slightly more than one molar equivalent of

$$R-Li \ + \ X-R' \ \rightleftharpoons \ [R-X-R']^- \ Li^+ \ \rightleftharpoons \ R-X \ + \ Li-R'$$

ate complex

FIGURE 12.2 Ate-complex mediated exchange.

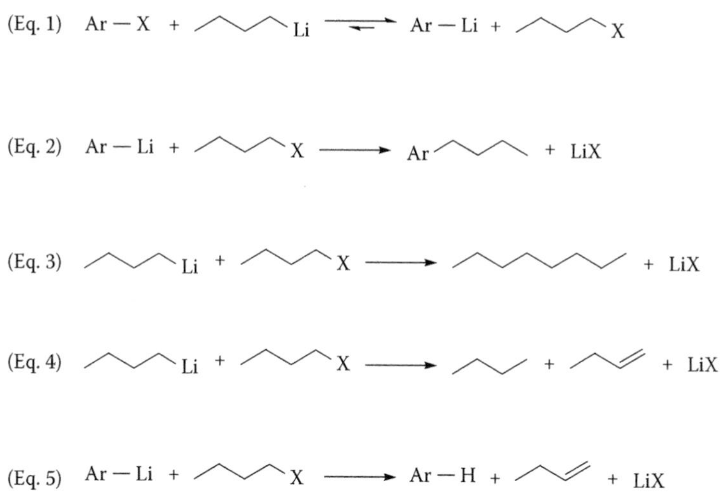

(Eq. 1) $Ar-X$ + ⌒⌒⌒Li ⇌ $Ar-Li$ + ⌒⌒⌒X

(Eq. 2) $Ar-Li$ + ⌒⌒⌒X ⟶ Ar⌒⌒⌒ + LiX

(Eq. 3)   ⌒⌒⌒Li + ⌒⌒⌒X ⟶ ⌒⌒⌒⌒⌒ + LiX

(Eq. 4)   ⌒⌒⌒Li + ⌒⌒⌒X ⟶ ⌒⌒ + ⌒⫽ + LiX

(Eq. 5) $Ar-Li$ + ⌒⌒⌒X ⟶ $Ar-H$ + ⌒⫽ + LiX

SCHEME 12.1

n-BuLi (Scheme 12.1, eq. 1).[15,19] Even though exchange equilibrium between an aryl bromide or iodide and *n*-BuLi is quite favorable, several side reactions, resulting from the presence of the *n*-butyl halide cogenerated in the exchange, may consume the aryllithium product or the *n*-BuLi reagent. These are illustrated in Scheme 12.1 (eq. 2 to eq. 5). For example, Wurtz–Fittig-type coupling between the aryllithium and the cogenerated *n*-butyl halide (Scheme 12.1, eq. 2) is a well-recognized side reaction that often ensues when the reaction mixture is warmed, and, as Negishi noted some time ago,[20] this is often a serious problem when the exchange is run in tetrahydrofuran (THF) solution. Because both coupling of *n*-butyl halide with the aryllithium and the reaction of *n*-butyl halide with *n*-BuLi to give octane (Scheme 12.1, eq. 3) are much slower than is the exchange reaction, complications due to the coupling of halides with organolithiums can be minimized by attention to process design.

Unintended quench of the aryllithium product by proton abstraction from the *n*-butyl halide coproduct via β-elimination to give 1-butene and reduced starting material (Scheme 12.1, eq. 5) is another potentially troubling side reaction that is often overlooked in process design. The formation of Ar-H following an exchange reaction between Ar-X and *n*-BuLi is most often attributed to traces of various proton sources in the reaction medium, such as adventitious water or THF. It has been our experience that *n*-butyl halide is often responsible for formation of "protonated" product. It should also be noted that the *n*-BuLi reagent may also be consumed in reactions with *n*-butyl halide to give octane in a coupling reaction (Scheme 12.1, eq. 3) or butane and 1-butene via a -elimination (Scheme 12.1, eq. 4). Such side reactions are often detected indirectly; to the extent that the *n*-BuLi reagent is consumed in reactions with *n*-butyl halide, more than one molar equivalent will be required to complete the exchange. In short, 1-butene and butane, which are quite soluble in the reaction medium at low temperatures and become part of the vapor effluent during the much higher workup temperatures, are often overlooked side products from a lithium–halogen exchange between an aryl halide and *n*-BuLi. These coupling and elimination side reactions also generate quantities of lithium halide. Depending on the type of substrate (Ar-X), side reactions can often be minimized by conducting the exchange at lower temperatures or by modification of the order of addition of the substrate, *n*-BuLi, and electrophile.

The first large-scale organic synthesis using *n*-BuLi was a lithium–bromine exchange reaction, developed in the early 1970s, for the preparation of the Eli Lilly fungicide, Fenarimol (Scheme 12.2, eq. 1).[21] In order to minimize various side reactions, the preferred process required dissolution of the 5-bromopyrimidine substrate and the dichlorobenzophenone electrophile in THF–ether

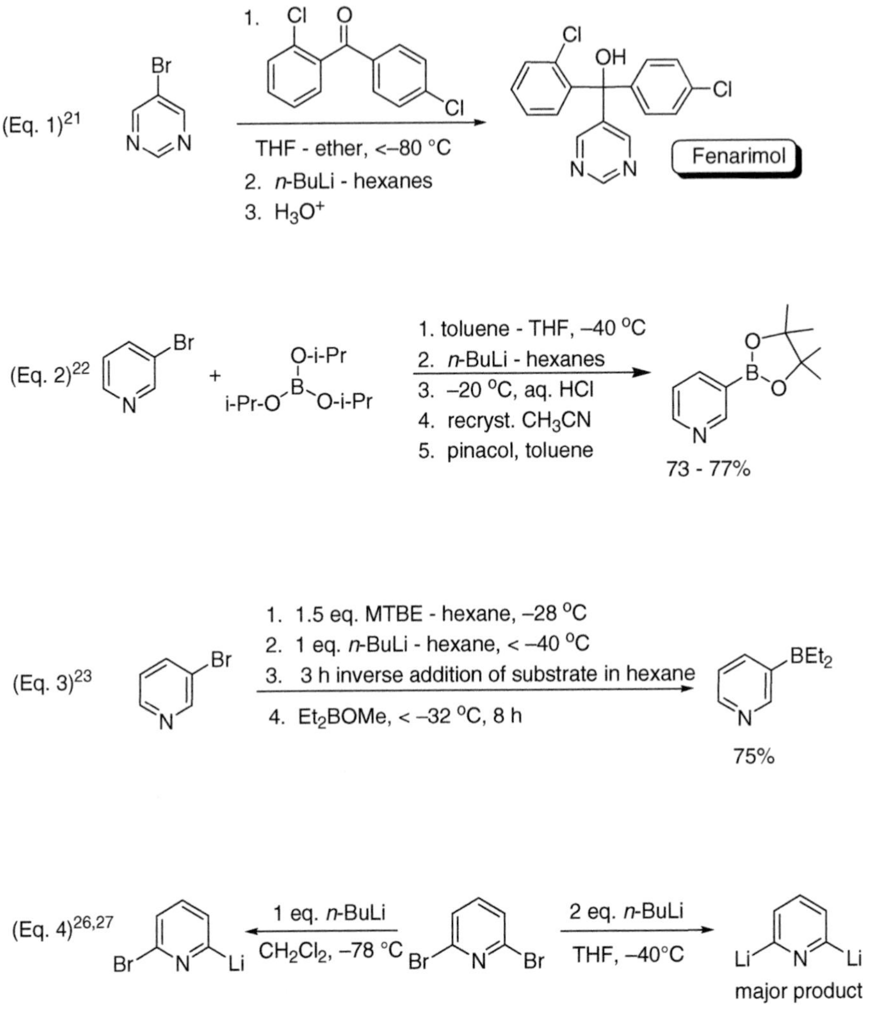

SCHEME 12.2

solution prior to the addition of *n*-BuLi in hexanes under cryogenic conditions (<–80°C). This procedure also minimized the undesirable reaction of 5-lithiopyrimidine with 5-bromopyrimidine. The alternative Eli Lilly route to this fungicide (not shown) involved "inverse addition" of a solution of 5-bromopyrimidine in THF–ether to a solution of *n*-BuLi in hexanes followed by addition of the dichlorobenzophenone. As noted below, virtually the same procedures are used today. It might also be noted that a continuous process can also be employed for the lithium–bromine exchange step leading to this antifungal.

An experiment 40 years after the Lilly procedure was disclosed, a crystalline borate ester was prepared by an analogous method. Dissolution of 3-bromopyridine and triisopropyl borate in toluene–THF followed by the addition of *n*-BuLi in hexanes at –40°C delivered the borate ester in good yield (Scheme 12.2, eq. 2).[22] The temperature used for this reaction illustrates that it is often not necessary to conduct the exchange at –78°C. It might also be noted that the use of toluene as a cosolvent is often beneficial for optimization of organolithium reactions, particularly those involving aryl or benzylic substrates.

An optimized temperature of <–40°C, employed for the generation of 3-pyridyllithium used for the preparation of diethyl-3-pyridyl borane on the pilot scale, was achieved by careful selection of solvent and addition order (Scheme 12.2, eq. 3).[23] The optimized procedure employed an "inverse addition" of 3-bromopyridine in hexane to a solution of *n*-BuLi in hexane containing a quantity of methyl *tert*-butyl ether (MTBE) to afford 3-pyridyllithium that was sparingly soluble in the reaction medium. The limited amount of MTBE used in the optimized process (1.5 molar equivalents) was responsible for the low solubility of 3-pyridyllithium. This appears to be a general phenomenon in lithium–halogen exchange reactions. Incorporation of limited quantities of a Lewis base solvent often results in cleaner chemistry at higher reaction temperatures, but the trade-off is frequently decreased solubility of the aryllithium product resulting in longer times for addition of the electrophile as well as other mixing issues. Addition of the diethylmethoxy borane electrophile to the 3-lithiopyridine (Scheme 12.2, eq. 3) took almost three times as long as the time needed to prepare the aryllithium.

It is instructive to consider in a general sense the heat evolved in a lithium–halogen exchange reaction. The heat evolved upon deaggregation and solvation of *n*-BuLi in a Lewis basic solvent[24] is generally greater than that evolved in the exchange step;[25] needless to say, any subsequent reactions that produce LiX may be quite exothermic.[25] Overall then, the heat evolved during the reaction of *n*-BuLi with an aryl halide may be summarized as lithium–halogen exchange < Lewis base solvation of *n*-BuLi « LiX formation. The inverse addition procedure (addition of the substrate to a solution of *n*-BuLi containing a quantity of Lewis basic solvent) minimizes the effects of heat generation during all steps of the exchange sequence.

Solvent selection will often predetermine the order of addition. For example, although inverse addition of 2,6-dibromopyridine to one equivalent of *n*-BuLi at –78°C has been reported to provide clean monolithiation via reequilibration of the initially formed 2,6-dilithiopyridine,[26] the use of dichloromethane, a much less basic solvent, allowed addition of *n*-BuLi to 2,6-dibromopyridine to produce the desired 2-bromo-6-lithiopyridine directly (Scheme 12.2, eq. 4).[27]

Lithium–bromine exchange was preferred over the Grignard option for the preparation of the ketone shown in Scheme 12.3, eq. 1; the solvent of choice for the exchange was MTBE rather than THF because the aryllithium was found to be unstable in THF.[28] Although 5 to 10% of quenched product was observed, it was easily removed by a re-slurry of the workup concentrate in hexanes.

In a similar fashion, a process group from SmithKline Beecham Pharmaceuticals reported the preparation of a functionalized quinoline ketone by lithium–bromine exchange as illustrated in Scheme 12.3, eq. 2.[29] The bromoquinoline substrate contains a fairly acidic amine hydrogen. Conversion to the dianion was achieved at –78°C by employing an inverse addition of the bromo-quinoline in THF to a solution of two equivalents of *n*-BuLi–hexanes in THF. This addition order ensured that *n*-BuLi would be present in excess during the addition, thus preventing competitive quenching of the initially formed 3-quinolinyllithium via proton exchange with the acidic proton of the amine moiety.

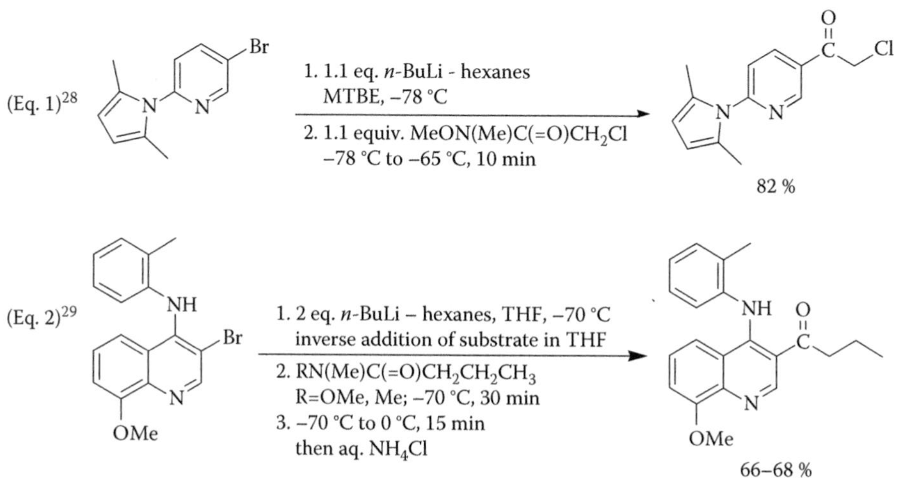

(Eq. 1)[28]

1. 1.1 eq. *n*-BuLi - hexanes
   MTBE, –78 °C

2. 1.1 equiv. MeON(Me)C(=O)CH$_2$Cl
   –78 °C to –65 °C, 10 min

82 %

(Eq. 2)[29]

1. 2 eq. *n*-BuLi – hexanes, THF, –70 °C
   inverse addition of substrate in THF

2. RN(Me)C(=O)CH$_2$CH$_2$CH$_3$
   R=OMe, Me; –70 °C, 30 min

3. –70 °C to 0 °C, 15 min
   then aq. NH$_4$Cl

66–68 %

SCHEME 12.3

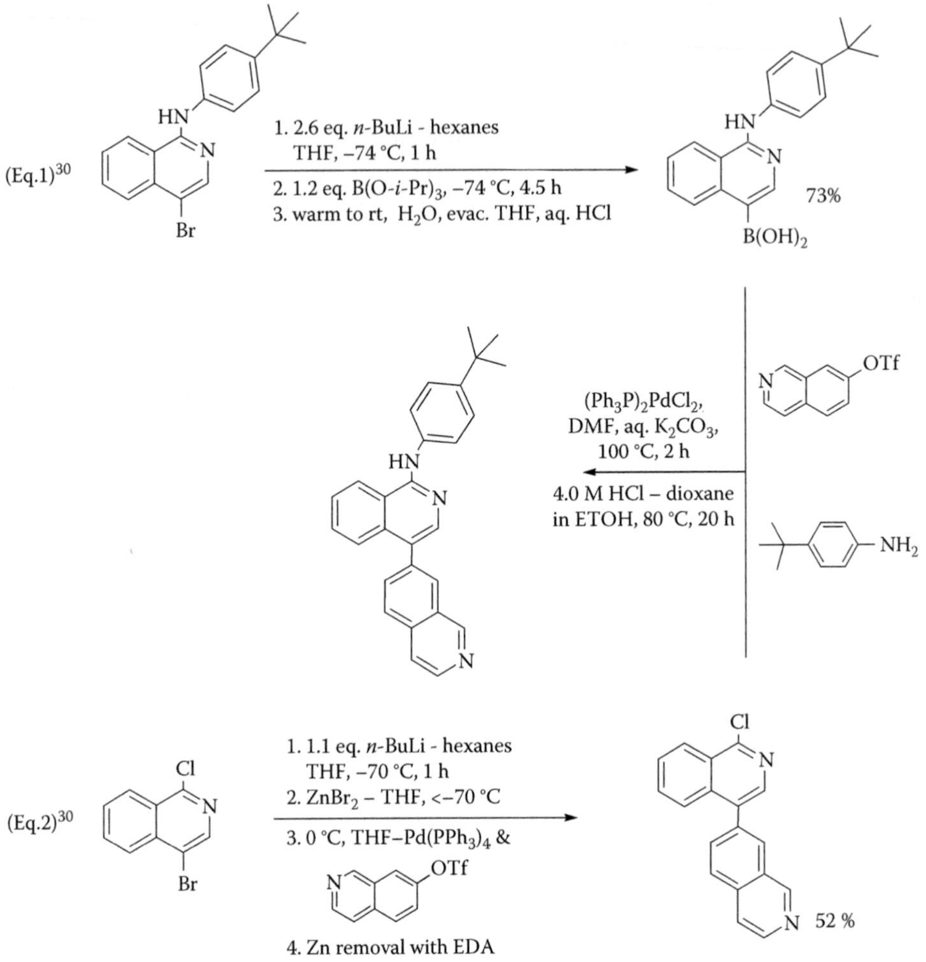

SCHEME 12.4

Various coupling methodologies are often interwoven with the lithium–halogen exchange reaction. Examples of such strategies from Novartis,[30] are shown in Scheme 12.4. The options for route selection markedly increase as the number of aryl to aryl linkages increases in the target molecule. Because the route selected for preparation of the active pharmaceutical ingredient (API) can be determined by in-house expertise, availability, or delivery dates of starting materials, it is sometimes necessary to develop different routes to an API as illustrated by the chemistry depicted in Scheme 12.4, eq. 1 and eq. 2.

The penultimate boronic acid required for the final Suzuki–Miyaura coupling was prepared in 73% yield via a standard lithium–bromine exchange followed by trapping with triisopropyl borate (Scheme 12.4, eq. 1). In contrast to seemingly similar preparations discussed above, much lower temperatures were required for this preparation to prevent unwanted side reactions between excess n-BuLi and electrophile. The scaled Negishi sequence (Scheme 12.4, eq. 2) delivered the coupling product in good yield (79% for the crude material),[30] but an extra step had to be added to remove complex bound Zn giving a final yield of 52% for pure product. Also troublesome was the 2 to 5% of unreacted 4-bromo-1-chloroisoquinoline that remained even when 1.5 equivalents of n-BuLi were used for the exchange. It is conceivable that an inverse addition of substrate to the n-BuLi might have solved the problem of recovered starting material.

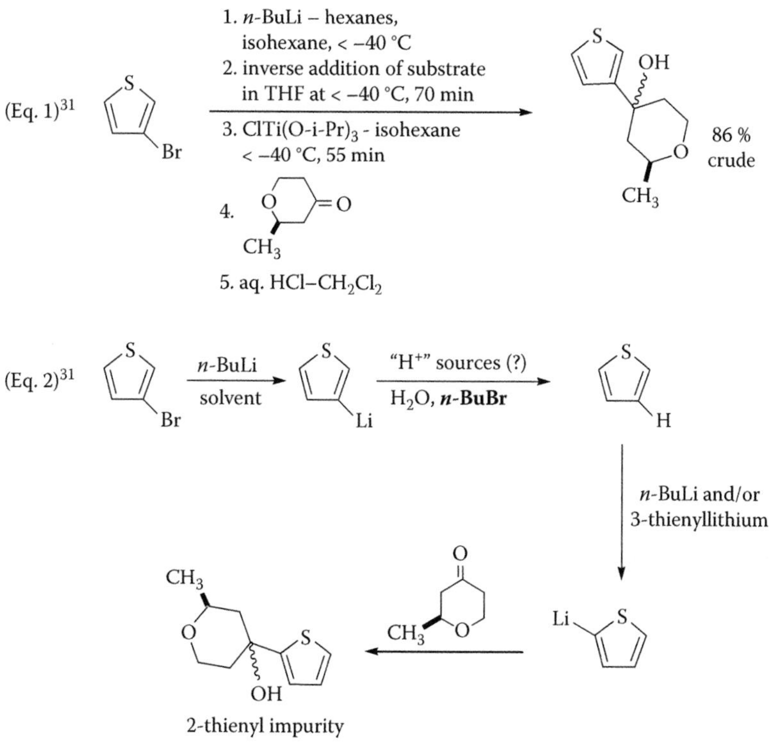

(Eq. 1)[31]

1. n-BuLi – hexanes, isohexane, < –40 °C
2. inverse addition of substrate in THF at < –40 °C, 70 min
3. ClTi(O-i-Pr)$_3$ - isohexane < –40 °C, 55 min
4.
5. aq. HCl–CH$_2$Cl$_2$

86 % crude

(Eq. 2)[31]

n-BuLi / solvent

"H$^+$" sources (?) / H$_2$O, n-BuBr

n-BuLi and/or 3-thienyllithium

2-thienyl impurity

SCHEME 12.5

Transmetalation of an organolithium to generate a less basic organometallic is quite common in processes involving the addition of ketone electrophiles that possess acidic protons. An example of such chemistry is summarized in Scheme 12.5.[31] This AstraZeneca route to a precursor of AZD4407, a 5-lipoxygenase inhibitor, involved inverse addition at –40°C of 3-bromothiophene in THF to a solution of n-BuLi in isohexane to generate 3-thienyllithium. Transmetalation of the 3-thienyllithium with chlorotitanium triisopropoxide limited enolization of the pyranone added in the next step (Scheme 12.5, eq. 1).[31] Temperature optimization studies on a laboratory scale revealed that all reactions could be conducted at approximately –20°C; however, scale-up revealed the formation of 5 to 7% of a crucial 2-thienyl impurity (Scheme 12.5, eq. 2) that had been present to the extent of less than 1% on the laboratory scale. The impurity apparently results from protonation of 3-thienyllithium, to give thiophene, followed by deprotonation at C(2) to afford the more thermodynamically stable 2-thienyllithium. It was suggested, quite reasonably, that traces of unidentified protic substances were responsible for this equilibration,[31] but it may well be that the culprit was the n-butyl bromide generated in the exchange reaction. As noted above (Scheme 12.1, eq. 5), β-elimination to give 1-butene and quenched substrate is a common, though often overlooked, side reaction when n-BuLi is used to accomplish a lithium–bromine exchange.

The difficulties associated with the presence of an alkyl halide coproduct from an exchange reaction may be avoided altogether by the use of t-BuLi rather than n-BuLi. To date, however, there are no reports in the open literature of the use of t-BuLi on large scale for the production of aryllithiums.

The principal advantage to the use of t-BuLi in a lithium–halogen exchange is the ability to render the exchange operationally irreversible by employing two molar equivalents of the reagent in a system containing a quantity of ethereal solvent such as diethyl ether, MTBE, or even THF; there is essentially no reaction of t-BuLi with an organohalide in pure hydrocarbon solution.[32] As

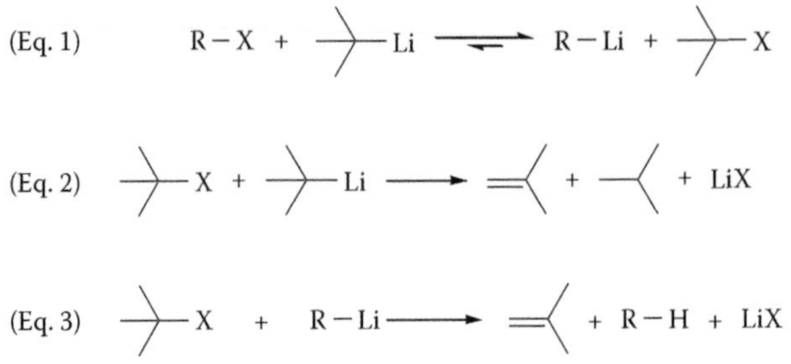

(Eq. 1)   R−X + $\rangle$−Li $\rightleftharpoons$ R−Li + $\rangle$−X

(Eq. 2) $\rangle$−X + $\rangle$−Li $\longrightarrow$ $=\!\!\langle$ + $-\!\!\langle$ + LiX

(Eq. 3) $\rangle$−X + R−Li $\longrightarrow$ $=\!\!\langle$ + R−H + LiX

SCHEME 12.6

detailed in Scheme 12.6, the first equivalent of *t*-BuLi establishes the exchange equilibrium (eq. 1), and a second equivalent rapidly consumes the *t*-butyl halide generated in the exchange to give isobutane, isobutylene, and LiX (eq. 2). A typical literature procedure involves addition of a solution of *t*-BuLi in a hydrocarbon solvent (although solutions of *t*-BuLi in pentane have commonly been used, the commercially available *t*-BuLi in heptane is by far the best reagent for the purpose) to a solution of the aryl bromide or iodide in an ethereal solvent. This procedure, which we arbitrarily term "normal addition," is not necessarily the best approach. As the *t*-BuLi is added to the halide solution, there is a competition between the product organolithium and *t*-BuLi for abstraction of a proton from the cogenerated *t*-butyl halide (eq. 3). This formal reduction of the aryl halide (eq. 3) is easily avoided by reversing the method of addition and the "inverse addition" of the halide to a solution of *t*-BuLi ensures that there is always an excess of *t*-BuLi available to consume the *t*-butyl halide (eq. 2). Consequently, the yield of an organolithium prepared by this method often approaches quantitative as demonstrated by several of the examples presented below. The chief disadvantage to the use of *t*-BuLi in the exchange reaction is the fact that an excess of *t*-BuLi is required to consume the *t*-butyl halide, and one must balance the extra cost of reagent against the yield benefit. It should also be noted that the overall conversion of an organohalide to an organolithium using two equivalents of *t*-BuLi is usually a very exothermic process, because a full equivalent of LiX is produced. For this reason, the exchange is normally conducted at very low temperatures; this too is a potential disadvantage.

Recently, it was reported that the lithium–bromine exchange between an aryl bromide and *t*-BuLi may be accomplished at 0°C, a temperature significantly higher than that normally used for this reaction, in yields exceeding 97%.[33] The presence of very small quantities (1 to 10% by volume) of any of a variety of simple ethers such as Et$_2$O, THF, tetrahydropyran (THP), or MTBE in a predominantly hydrocarbon solution of *t*-BuLi apparently results in an ether-solvated *t*-BuLi aggregate that is responsible for the exchange reaction. For example, as shown below, the lithium–bromine exchange between 1-bromo-4-*t*-butylbenzene and *t*-BuLi may be accomplished in 98% yield at 0°C by addition of a solution of the bromide in heptane to solution of *t*-BuLi in heptane containing only 1% (by volume) of THP.[33]

Lithium–halogen exchange with *t*-BuLi may also be used to prepare vinyllithiums from vinyl bromides or iodides with complete retention of configuration. Pioneering studies by the Seebach

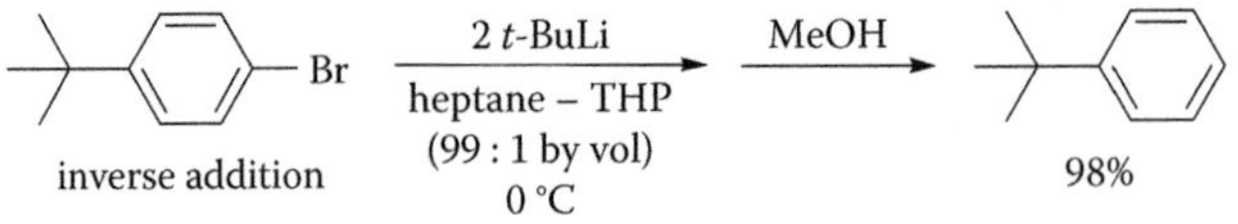

$$\text{inverse addition} \qquad \xrightarrow[\substack{\text{heptane} - \text{THP} \\ (99:1 \text{ by vol}) \\ 0\,°C}]{2\ t\text{-BuLi}} \xrightarrow{\text{MeOH}} \quad 98\%$$

FIGURE 12.3 Lithium-bromine exchange.

group used *t*-BuLi in the Trapp solvent (a 4:4:1 by volume mixture of THF, diethyl ether and pentane)[34] for the generation of vinyllithiums at very low temperatures (typically –110°C).[35] Subsequently, it was found that vinyllithiums may be generated in equally good yield at –78°C provided that the solvent system is free of THF.[36]

12.4 LITHIUM–HALOGEN EXCHANGE OF PRIMARY ALKYL IODIDES

Even though it is a fairly simple matter to prepare organolithiums having lithium bonded to an sp^2-hybridized carbon, the preparation of alkyllithiums by the lithium–halogen exchange reaction appears to be restricted to the generation of primary alkyllithiums from an alkyl iodide precursor.[32] Primary alkyllithiums may be generated very rapidly and in virtually quantitative yield by adding a solution of a primary alkyl iodide to 2.0 to 2.2 equivalents of *t*-BuLi at –78°C in a hydrocarbon–ether solvent system. Any residual *t*-BuLi remaining in solution may be easily removed prior to reaction of the primary alkyllithium by taking advantage of the differential reactivity of the two organolithiums toward ethers; allowing the reaction mixture to warm and stand at room temperature for ~1 h serves to preferentially destroy the remaining *t*-BuLi through rapid proton abstraction from the ether solvent. A detailed, general procedure for the generation of primary alkyllithiums by lithium–iodine exchange is available in a recent volume of *Organic Syntheses*.[37]

The success of this general route to primary alkyllithiums often depends crucially on the appropriate choice of halide and solvent. The more easily polarized alkyl iodides, rather than bromides or chlorides, must be used to ensure that the reaction proceeds cleanly via the ate complex: treatment of an alkyl bromide with *t*-BuLi often initiates radical-mediated reactions, and alkyl chlorides are essentially inert to the action of *t*-BuLi at low temperatures.[32,38] The exchange appears to involve dimeric *t*-BuLi. Consequently, solvent systems containing THF, TMEDA, or other highly lithiophilic Lewis bases that render the *t*-BuLi monomeric should be avoided, because elimination and coupling reactions can compete with exchange under these conditions. The best general medium for the preparation of primary alkyllithiums by lithium–iodine exchange is a predominantly hydrocarbon solvent system that contains a small quantity of a simple alkyl ether, such as Et$_2$O, MTBE, dibutyl ether, or the like, in which the *t*-BuLi is predominantly dimeric.

It might be noted that it is possible to conduct the exchange between *t*-BuLi and a primary iodide at temperatures significantly higher than those commonly used in academic research laboratories. A study of the reactions of 1-iodooctane, a representative primary alkyl iodide, with *t*-BuLi at 0°C in a variety of solvent systems composed of heptane and various ethers demonstrated that an optimal ether–heptane ratio, which varied for each of the ethers studied, was found to maximize the extent of lithium–iodine exchange and minimize side reactions such as coupling and elimination.[39] Numerous examples of the utility of this general route to primary alkyllithiums may be found elsewhere,[32,38,40] and a few additional examples are summarized in Scheme 12.7.[41–43]

12.5 SUMMARY

Laboratory-scale procedures are available for conducting lithium–halogen exchange reactions with either *n*-BuLi or *t*-BuLi; however, as noted above, there are no reports in the literature of exchange reactions conducted using *t*-BuLi on a large scale. It seems likely that *n*-BuLi will continue to be the reagent of choice for large-scale lithium–halogen exchange reactions, but this situation may change should *t*-BuLi ever become cost competitive with *n*-BuLi. Thus, the scale-up challenge is to minimize the various competitive side reactions associated with the use of *n*-BuLi in the exchange (Scheme 12.1). Once starting material availability is established, the following interdependent process parameters deserve consideration:

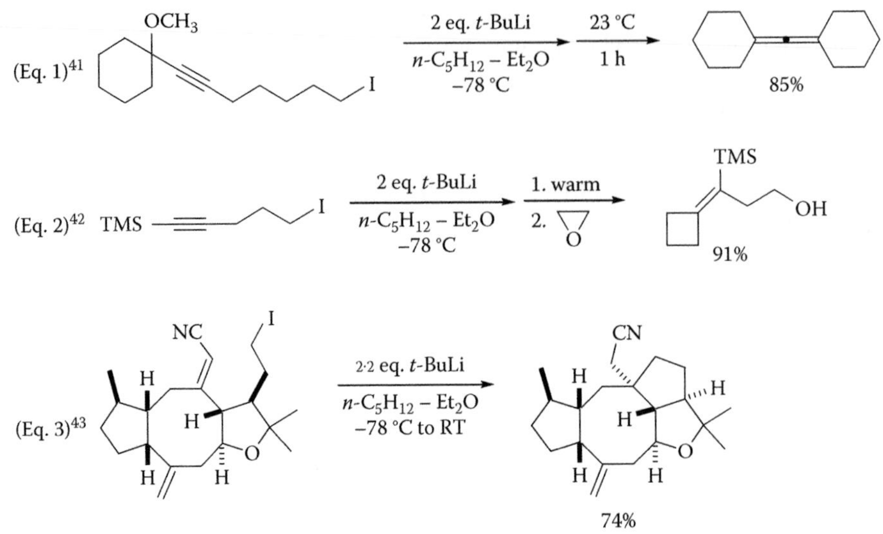

SCHEME 12.7

Order of Addition—Variation in the order of addition of substrate, alkyllithium, and electrophile are possible. The method of choice will be dictated by the chemistry. "Inverse addition" of the substrate to a solution of *n*-BuLi in the appropriate solvent may be a particularly useful approach, because it ensures that there is an excess of the alkyllithium at all times during the addition. Clearly, potential problems (HAZOP review) that may result from starting with a reactor charged with *n*-BuLi need to be considered.

Medium—The use of various Lewis bases, either as solvent or as component of a predominantly hydrocarbon medium, offers advantages. When trying to improve solubility, even hydrocarbon cosolvent selection is important, keeping in mind that aromatic solvents, especially toluene, can be beneficial. The option of using limited solubility of the desired ArLi to advantage should be considered; however, the trade-off may be mixing issues and longer addition times (Scheme 12.2, eq. 3).

Temperature—It is not always necessary to conduct the exchange reaction at temperatures typically used in a laboratory setting (<–70°C). Large-scale exchange reactions have been conducted at temperatures as high as –40°C (Scheme 12.2, eq. 2 and eq. 3). Reactions conducted at these higher temperatures may require presolvation of the *n*-BuLi (Scheme 12.2, eq. 3), good mixing, and full understanding of chemistries involved.

Substrate—Aromatic bromides or iodides (Ar-X) are rapidly converted to the corresponding aryllithium (Ar-Li) when treated with 1 molar equivalent of *n*-BuLi or 2 molar equivalents of *t*-BuLi. However, some substrates may rearrange to a more thermodynamically stable aryllithium via abstraction of a proton from a more acidic site, and this can lead to the formation of unwanted impurities (Scheme 12.5, eq. 2). Similarly, proton abstraction from aryl bromide substrates that contain an *ortho*-directing metalation group (DMG) may lead to impurities in the final product.[44]

Electrophile—It might be noted that it is sometimes possible to perform the lithium–bromine exchange in the presence of the electrophile used to trap the aryllithium product (Scheme 12.2, eq. 2).

REFERENCES AND NOTES

1. Schlenk, W. and Holtz, J., *Chem Ber.*, 50, 262, 1917.
2. Ziegler, K. and Colonius, J., *Liebigs Ann. Chem.*, 479, 135, 1930.
3. Wittig, G., Pockels, U., and Dröge, H., *Chem. Ber.*, 71, 1903, 1938.
4. (a) Gilman, H. and Jacoby, A.L., *J. Org. Chem.*, 3, 108, 1938. (b) Gilman, H., Langham, W., and Jacoby, A.L., *J. Am. Chem. Soc.*, 61, 106, 1939.
5. Applequist, D.E. and O'Brien, D.F., *J. Am. Chem. Soc.*, 85, 743, 1963.
6. (a) Jones, R.G. and Gilman, H., *Chem. Rev.*, 54, 835, 1954. (b) Gilman, H. and Jones, R.G., *Org. React. (NY)*, 6, 339, 1951. (c) Schölkopf, U., *Methoden der Organischen Chemie*, Vol. 13, Georg Thieme, Stuttgart, 1970. (d) Wakefield, B.J., *The Chemistry of Organolithium Compounds*, Pergamon Press, New York, 1974. (e) Wardell, J.L., *Comprehensive Organometallic Chemistry*, Vol. 1, Wilkinson, G., Ed., Pergamon Press, New York, 1982, p. 43. (f) Wakefield, B.J., *Organolithium Methods,* Pergamon Press, New York, 1988. (g) Schlosser, M., *Organometallics in Synthesis: A Manual*, Schlosser, M., Ed., John Wiley & Sons, New York, 2002, p. 101. (h) Sapse, A.M. and Schleyer, P.v.R., *Lithium Chemistry: A Theoretical and Experimental Overview*, John Wiley & Sons, New York, 1995. (i) Clayden, J., *Organolithiums: Selectivity for Synthesis,* Pergamon Press, New York, 2002, p. 111.
7. Bailey, W.F., *tert-Butyllithium in Organic Synthesis*, *FMC Lithium Link*, Fall 2000.
8. Rathman, T.L., *Fine Tune Your Carbanion, FMC Lithium Link*, Winter 2006.
9. Publications cited as References 7 and 8 are available free of charge from FMC Corporation, Lithium Division, on its Web site. FMC Corporation, Lithium Division, Seven Lake Pointe Plaza, 2801 Yorkmont Rd., Charlotte, NC 28208; www.fmclithium.com (accessed January 2006).
10. Wakefield, B.J., *Lithium: Current Applications in Science, Medicine, and Technology*, Bach, R.O., Ed., John Wiley & Sons, New York, 1985, p. 257.
11. A review of the literature detailing the historical development of mechanistic studies of the lithium–halogen exchange is available, see Bailey, W.F. and Patricia, J.J., *J. Organomet. Chem.*, 352, 1, 1988.
12. Ogle, C., *The Structure and Reactivity of Organolithium Reagents*, *FMC Lithium Link*, Summer 1995.
13. Wiberg, K.B., Sklenak, S., and Bailey, W.F., *J. Org. Chem.*, 65, 2014, 2000, and references therein.
14. Beak, P. and Allen, D.J., *J. Am. Chem. Soc.*, 114, 3420, 1992.
15. Slocum, D.W., *Metal–Halogen Exchange Metalations (M ↔ X) of Aryl and Heteroaryl Ring Systems Using Alkyllithiums*, FMC Lithium Link, Winter 1993.
16. (a) Brandsma, L. and Verkruijsse, H., *Preparative Polar Organometallic Chemistry*, Vol. 1, Springer-Verlag, New York, 1987. (b) Brandsma, L., *Preparative Polar Organometallic Chemistry*, Vol. 2, Springer-Verlag, New York, 1990.
17. Hoffmann, R.W., *Dehydrobenzene and Cycloalkynes*, Academic Press, New York, 1967.
18. (a) Narasimhan, N.S., Sunder, N.M., Ammanamanchi, R., and Bonde, B.D., *J. Am. Chem. Soc.*, 112, 4431, 1990. (b) Gallagher, D.J. and Beak, P., *J. Am. Chem. Soc.*, 113, 7984, 1991.
19. For recent reviews, see (a) Mealy, M.J. and Bailey, W.F., *Organomet. Chem.*, 646, 59, 2002. (b) Sotomayor, N. and Lete, E., *Current Org. Chem.*, 7, 275, 2003.
20. Merrill, R.E. and Negishi, E., *J. Org. Chem.*, 39, 3452, 1974.
21. Taylor, H.M., Davenport, J.D., and Hackler, R.E., U.S. Patent 3,887,708, 1975; U.S. Patent 3,869,456, 1975; U.S. Patent 3,868, 244, 1975; U.S. Patent 3,818,009, 1974.
22. Wenjie, L., Nelson, D.P., Jensen, M.S., Hoerrner, R.S., Cai, D., Larsen, R.D., and Reider, P.J., *J. Org. Chem.*, 67, 5394, 2002.
23. Lipton, M.F., Mauragis, M.A., Maloney, M.T., Veley, M.F., VanderBor, D.W., Newby, J.J., Appell, R.B., and Daugs, E.D., *Org. Process Res. Dev.*, 7, 385, 2003.
24. (a) Quirk, R.P. and Kester, D.E., *J. Organomet. Chem.*, 72, C23, 1974. (b) Quirk, R.P., Kester, D.E., and Delaney, R.D., *J. Organomet. Chem.*, 59, 45, 1973. (c) Quirk, R.P. and Kester, D.E., *J. Organomet. Chem.*, 127, 111, 1977.
25. Ende, D.J.am. and Braish, T., The heat generated during the *in situ* formation of LiBr in a THF–hydrocarbon solvent mixture typically used for the exchange reaction is approximately twice as large as the heat measured during the lithium–bromine exchange, personal communication, 2006.
26. Cai, D., Hughes, D.L., and Verhoeven, T.R., *Tetrahedron Lett.*, 37, 2537, 1996.

27. (a) Peterson, M.A. and Mitchell, J.R., *J. Org. Chem.*, 62, 8237 1997. (b) The selective monolithiation of 2,5-dibromopyridine has also been reported, see Wang, X., Rabbat, P., O'Shea, P., Tillyer, R., Grabowski, E.J.J., and Reider, P.J., *Tetrahedron Lett.*, 41, 4335, 2000.

28. Scott, R.W., Fox, D.E., Wong, J.W., and Burns, M.P., *Org. Process Res. Dev.*, 8, 587, 2004.

29. Atkins, R.J., Breen, G.F., Crawford, L.P., Grinter, T.J., Harris, M.A., Hayes, J.F., Moores, C.J., Saunders, R.N., Share, A.C., Walsgrove, T.C., and Wicks, C., *Org. Process Res. Dev.*, 1, 185, 1997.

30. Denni-Dischert, D., Marterer, W., Banziger, M., Yusuff, N., Batt, D., Ramsey, T., Geng, P., Michael, W., Wang, R.-M.B., Taplin Jr., F., Versace, R., Cesarz, D., and Perez, L.B., *Org. Process Res. Dev.*, 10, 70, 2006.

31. (a) Hutton, J., Jones, A.D., Lee, S.A., Martin, D.M.G., Meyrick, B.R., Patel, I., Peardon, R.F., and Powell, L., *Org. Process Res. Dev.*, 1, 61, 1997. (b) Alcaraz, M.-L., Atkinson, S., Cornwall, P., Foster, A.C., Gill, D.M., Humphries, L.A., Keegan, P.S., Kemp, R., Merifield, E., Nixon, R.A., Noble, A.J., O'Beirne, D., Patel, Z.M., Perkins, J., Rowan, P., Sadler, P., Singleton, J.T., Tornos, J., Watts, A.J., and Woodland, I.A., *Org. Process Res. Dev.*, 9, 555, 2005.

32. Bailey, W.F. and Punzalan, E.R., *J. Org. Chem.*, 55, 5404, 1990.

33. Bailey, W.F., Luderer, M.R., and Jordan, K.P., *J. Org. Chem.*, 71, 2825, 2006.

34. (a) Köbrich, G. and Trapp, H., *Chem. Ber.*, 99, 680, 1966. (b) Köbrich, G., *Angew. Chem., Int. Ed. Engl.*, 6, 41, 1967.

35. (a) Seebach, D. and Neumann, H., *Chem. Ber.*, 107, 847, 1974. (b) Neumann, H. and Seebach, D., *Chem. Ber.*, 111, 2785, 1978.

36. Bailey, W.F., Wachter-Jurcsak, N.M., Pineau, M.R., Ovaska, T.V., Warren, R.R., and Lewis, C.E., *J. Org. Chem.*, 61, 8216, 1996.

37. Bailey, W.F., Luderer, M.R., Mealy, M.J., and Punzalan, E.R., *Organic Syntheses*, Vol. 81, John Wiley & Sons, New York, 2004, p. 121.

38. Bailey, W.F., *Metal–Halogen Exchange (M ↔ X) Involving Aliphatic Substrates Using Alkyllithiums*, FMC Lithium Link, Spring 1994.

39. Bailey, W.F., Brubaker, J.D., and Jordan, K.P., *J. Organomet. Chem.*, 681, 210, 2003.

40. Bailey, W.F. and Jiang, X., *Tetrahedron*, 61, 3183, 2005.

41. Bailey, W.F. and Aspris, P.H., *J. Org. Chem.*, 60, 754, 1995.

42. Bailey, W.F. and Ovaska, T.V., *J. Am. Chem. Soc.*, 115, 3080, 1993.

43. Jamison, T.F., Shambayati, S., Crowe, W.E., and Schreiber, S.L., *J. Am. Chem. Soc.*, 119, 4353, 1997.

44. Manley, P.W., Acemoglu, M., Marterer, W., and Pachinger, W., *Org. Process Res. Dev.*, 7, 436, 2003.

13 Oxetan-3-one: Chemistry and Synthesis

Georg Wuitschik, Erick M. Carreira,
Mark Rogers-Evans, and Klaus Müller

CONTENTS

13.1 INTRODUCTION

There are numerous aspects of organic chemistry that are highly relevant in the drug discovery process.[1] This is particularly the case for the science of chemical synthesis. One notable feature of the discipline is its ability to continue to evolve in new directions in ways that redefine the synthesis of molecules and, consequently, provide access to entities not previously readily accessible. This, in turn, leads to the tangible result of providing novel scaffolds or building blocks, which can serve as starting points for the preparation of new drugs. In addition to providing novel molecular launching points for drug discovery, it is possible to consider access to small molecular units that may be appended onto scaffolds of interest, which in turn can impart or modulate the physico-chemical properties of a molecule.

In general, an important focus in the field of chemical synthesis has been on the development of methods that provide access to optically active building blocks via catalytic asymmetric synthesis.[2] However, building block methodology should be expansive and include approaches to both chiral and achiral structures, *inter alia* acyclic and cyclic as well as saturated and unsaturated. We have been interested in the development of synthetic methods that provide access to building blocks for complex molecule assembly.[3–18] Our focus has been in large part on methods that provide access to small units, which are amenable to subsequent extensive synthetic elaboration. Recently, one particular area of interest to us has been the identification of small-ring structures with unusual properties that may, in turn, be grafted onto larger scaffolds to modulate biological activity or pharmacological properties.[19] As synthetic chemists, we have been especially enamored with small-molecule entities that present challenging synthesis targets.

Examination of various small-molecule databases reveals the wide use of saturated oxygen-based heterocyclic structures such as pyrans and furans in drug discovery. These are generally employed in either of two capacities: as scaffolds or as add-ons to optimize pharmacological properties. Interestingly, the second smallest saturated heterocyclic ring systems (namely, the oxetanes) appear to have been shunned, possibly because of the perceived proclivity of these rings to undergo opening reactions. However, substituted oxetanes can be quite stable to both alkali and acid. Moreover, we have shown that 3,3-disubstituted oxetanes can be utilized as add-ons in the optimization of molecular properties, and that these are sufficiently robust in a number of chemical and biological settings. Although the saturated larger rings (e.g., oxepanes and oxecanes) in principle can serve as scaffolding for the generation of molecular diversity, their utility as add-on substitutents for the modulation of physical properties is not obvious, and any potential benefits may be counterbalanced by their increased lipophilicity or susceptibility to metabolic degradation. Thus, oxetanes that are largely unexplored represent fertile ground for exploratory investigations. In addition to the perception that the strain associated with these rings leads to high instability, there would also appear to be a general lack of practical synthetic procedures that permit their ready access.

We recently disclosed that 3-substituted and 3,3-disubstituted oxetanes can impart useful pharmacological properties when incorporated into molecular scaffolds.[19] In this chapter, we provide a discussion of this ring system and a *raison d'être* for the development of methods for their synthesis and ready incorporation into drug candidates. We focused on the preparative chemistry of 3-substituted oxetanes for a number of reasons. The positioning of substituents at C-3 coincides with the molecular plane of symmetry perpendicular to the ring plane: C-3 substitution *does not* lead to the introduction of a stereogenic center. Consequently, the two substituents can be widely varied without incurring stereochemical complications in any synthetic route. Additionally, 3,3-disubstitution leads to significant increase in the stability of the ring system.

13.2 BACKGROUND

Trimethylene oxide, the simplest oxetane, was first prepared by Reboul in 1878.[20] Subsequent analysis of the structure of this heterocycle revealed some interesting features. Thus, in contrast to what is observed with cyclobutanes, the oxetane ring is less puckered (Figure 13.1). The replacement of a methylene unit by an oxygen reduces the otherwise unfavorable eclipsing interactions that are minimized by out-of-plane distortion in the cyclobutanes.[21,22] As small saturated heterocycles, oxetanes display chemical as well as physical characteristics whose origins can be traced to their

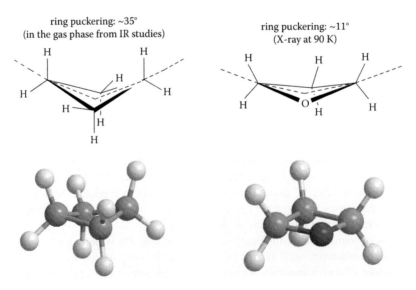

ring puckering: ~35°
(in the gas phase from IR studies)

ring puckering: ~11°
(X-ray at 90 K)

FIGURE 13.1 (See color insert following page 40). Comparison of puckering between cyclobutane and oxetane.

inherent ring strain. In trimethylene oxide, the ring strain has been determined to be 106 kJ/mol, only 1 kJ/mol less than for oxirane and 20 kJ/mol more than tetrahydrofuran.[23] Interestingly, for 3,3-disubstituted oxetanes, the ring strain is reduced by 14 kJ/mol.[23] Additionally, 3,3-disubstituted oxetanes display reduced susceptibility to ring opening, because any ring cleavage via a displacement reaction would suffer from unfavorable nonbonded interactions that are analogous to those observed at neopentyl centers.

The simplest unsubstituted oxetane has been shown to be prone to undergo acid-catalyzed ring-opening reactions with sulfuric or perchloric acid in aqueous dioxane almost as rapidly as ethylene oxide.[24] Under alkaline conditions, however, ring opening of trimethylene oxide is very slow: comparison of the hydrolyses under basic conditions of oxirane and oxetane leads to a rate difference of three orders of magnitude.[24] Organometallic reagents, such as Grignards or organolithiums, participate in ring-opening reactions with oxetanes at elevated temperatures, and these reactions often require the use of strong Lewis acids for ring activation (e.g., BF_3OEt_2).[25] The introduction of β-substitution leads to considerable steric congestion (despite reduced ring strain?), making oxetanes more resistant toward opening.[26] Oxetanes undergo reductive opening with lithium aluminum hydride at the least hindered position. It has been observed that 3,3-disubstituted oxetanes react much more slowly than their monosubstituted counterparts, with the former requiring prolonged exposure to $LiAlH_4$ in refluxing tetrahydrofuran (THF).[27] Oxetanes undergo polymerization in solvents such as chloromethane catalyzed by BF_3, forming polyethers of high molecular weight. The polymer of 3,3-bis(chloromethyl)oxetane has found wide application under the brand names Pentaplast® or Penton®.[28,29]

Substituted oxetanes are found embedded in natural products such as taxol,[30] thromboxane A$_2$,[31] oxetanocin,[32] and oxetin[33] (Figure 13.2). Taxol was first isolated from the bark of the western yew (*Taxus brevifolia*)[30] and is, together with several derivatives, used presently in cancer chemotherapy (Taxol®).[34] The structural consequences of the oxetane in taxol were subject of a computational study, from which it was concluded that the oxetane leads to the rigidification of the overall structure[35] and acts as a hydrogen bond acceptor partner for a threonine-OH in the putative binding pocket.[36] Replacement of the oxetane in taxol with azetidine, thietane, and selenetane invariably resulted in lower activity.[37–39] However, the role of the oxetane moiety remains controversial regarding the bioactivity of taxol.[36] Thromboxane A$_2$ is a compound predominantly synthesized by platelets and promotes vasoconstriction, platelet aggregation, and bronchoconstriction of the lung. It has plasma half-life of only 30 sec, before the oxetane ring hydrolyzes to give inactive thromboxane B$_2$.[31] Oxetanocin A inhibits the reverse transcriptase of HIV by mimicking adenosine.[40,41] Oxetin has been found to have herbicidal as well as antibacterial effects, but further investigation of its biological activity is ongoing.

There are anthropogenic small molecules (Figure 13.3) that also incorporate oxetane rings both as scaffold (EDO; 2,2-bis(4-ethoxyphenyl)-3,3-dimethyloxetane) and side chain (oxasulfuron). The insecticide EDO is 25 times more potent than DDT, which is also active against DDT-resistant strains of *Musca domestica*. In contrast to the notorious environmentally persistent DDT, EDO is

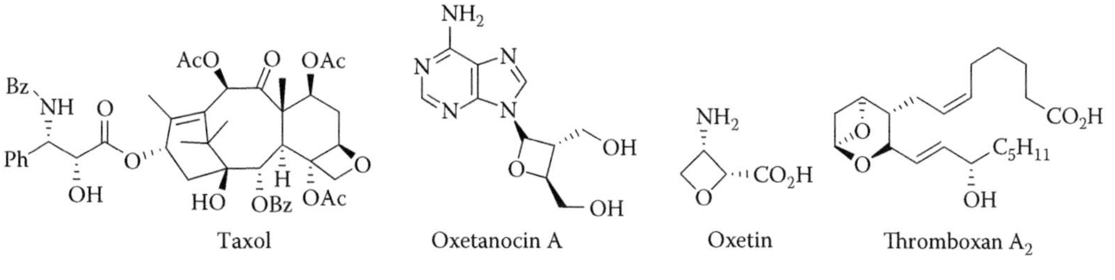

FIGURE 13.2 Natural products containing oxetanes.

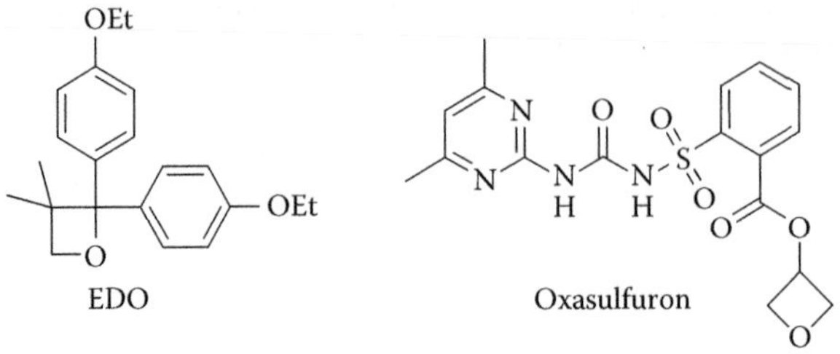

FIGURE 13.3 Oxetane-containing pesticides.

biodegradable.[42] Oxasulfuron[43] acts by inhibiting the biosynthesis of valine and isoleucine in organisms. It is used, for example, in the cultivation of soybeans to keep weeds under control. These agrochemical agents are selective against their target, because the plant is not affected due the rapid metabolic degradation of oxasulfuron in the crop.[44]

13.3 SYNTHESIS OF OXETANES

This class of saturated oxygen heterocycles has been made in a variety of different ways (Figure 13.4). Those that have found repeated use in the literature include intermolecular versions of the Williamson ether synthesis, the addition of sulfonium ylids to aldehydes, and the Paterno–Büchi cycloaddition. Among these, the Williamson ether synthesis is the most general.[26] It is interesting to note that the rate of closure of different β- and γ-chloroalcohols in aqueous base differs considerably, with epoxide formation from β-chlorohydrin favored by two orders of magnitude over the homologue.[45] It is often the case that formation of by-products and a variety of competing intermolecular reactions effectively compete with desired ring formation, lowering the yield.[45,46] The one-pot conversion of aldehydes or ketones with sulfoxonium-ylides **1** to give 2-substituted oxetanes provides an alternate route.[47] This reaction is considered to proceed via an epoxide intermediate that subsequently undergoes ring opening by a second equivalent of the ylid. The resulting γ-alkoxy sulfonium ylid then participates in an intramolecular displacement reaction to furnish a 2-substituted oxetane.[48] The same class of substituted oxetanes can be accessed via the Paterno–Büchi reaction.[49,50] This cycloaddition reaction between an aldehyde or a ketone and an electron-rich alkene

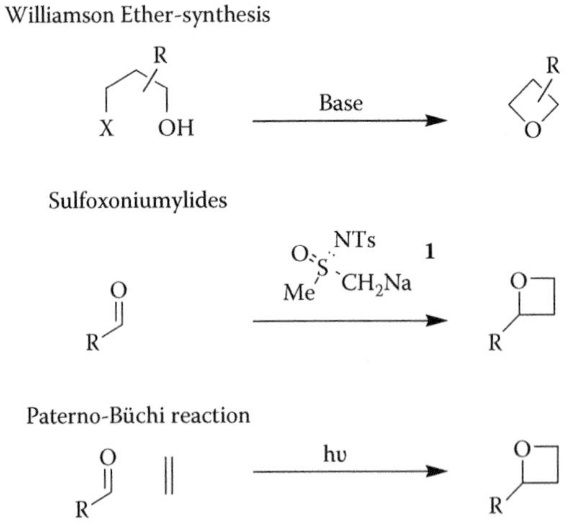

FIGURE 13.4 Common methods to synthesize oxetanes.

affords regioselectively the corresponding oxetanes in good yield.[51] An important difference between the various approaches is that in the Williamson approach, stereochemical issues are addressed separately from the ring-closing event. By contrast, in the processes that commence with carbonyl substrates, control must be exercised over the generation of stereocenters during the ring closure step. Although the approach involving intramolecular Williamson ether synthesis permits access to a range of substituted oxetanes, the latter two methods necessarily lead to oxetanes that incorporate substitution at C-2.

13.4 OXETANES IN DRUG DISCOVERY

The calculated van der Waals volumes of trimethylene oxide and propane are essentially the same. Consistent with this feature, the measured partial molar volumes in water are similar.[52,53] This leads to the hypothesis that it may be a fruitful exercise to consider the replacement of *gem*-dimethyl groups with oxetanes (Figure 13.5). This would lead to geometrically similar structures with markedly different pharmacokinetic properties. In a related fashion, the *tert*-butyl group can be considered a *gem*-dimethyl substituted ethyl group. Thus, in a model construct wherein oxetanes would function as *gem*-dimethyl group equivalents, 3-methyl 3-oxetanyl linked units would serve as *tert*-butyl surrogates. Both *gem*-dimethyl as well as *tert*-butyl groups can be found in many pharmaceutical compounds. A search of the *World Drug Index* reveals 714 entries for molecules incorporating a *tert*-butyl group, and 69 such compounds are currently in the market. *Tert*-butyl groups are often used to fill hydrophobic pockets of a certain target; *gem*-dimethyls are often introduced to protect metabolically labile positions. In both cases, however, their introduction is associated with decreased water solubility and elevated lipophilicity. Higher lipophilicity in general makes the molecule as a whole a better substrate for oxidative metabolic enzymes. However, the oxetane group is considerably reduced in lipophilicity. In a series of studies, we were able to demonstrate that introduction of the oxetane into a model compound can lead to increased water solubility in excess of 4000-fold over the parent compound. Moreover, the 3,3-disubstituted oxetanes were shown to be stable in a pH range from 1 to 10 and to enhance metabolic stability of the parent molecules significantly in standard liver microsomal assays.[19]

In order to make use of these potential beneficial properties of oxetanes as surrogates of *gem*-dimethyl and *tert*-butyl groups, synthetic methods had to be developed that allow access to a variety of 3,3-disubstituted oxetanes. In principle, there are at least two approaches that could be pursued

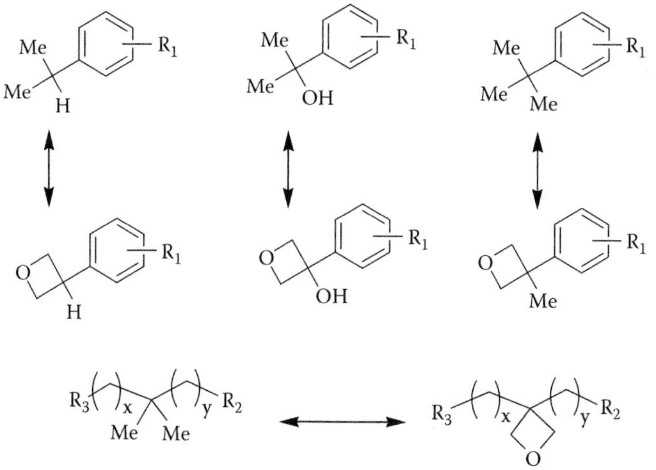

FIGURE 13.5 Relevant structural motifs for the replacement of a *gem*-dimethyl group by an oxetanyl unit.

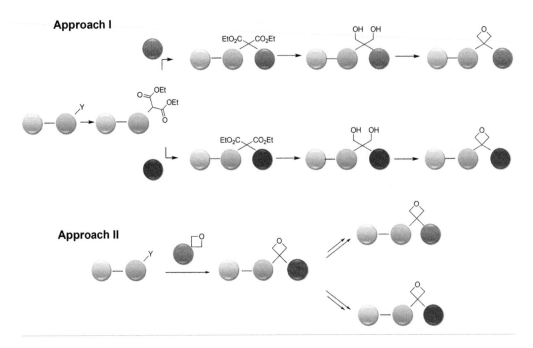

FIGURE 13.6 (See color insert following page 40). Two different approaches to incorporating oxetane rings onto molecules.

in the design of routes for the incorporation of oxetanes (Figure 13.6). In the first of these (approach I), oxetanes can be introduced from an acyclic precursor (e.g., a 1,3-diol). If the oxetane links two different fragments of a structure of interest, then in this approach the linking of the two subunits would typically precede oxetane formation. The steps associated with ring closure (i.e., activation of a diol precursor and base-induced ring formation) would necessarily need to be compatible with other elements of the molecule, and in this more linear strategy, the efficiency of ring closure may be highly substrate dependent. In a second route (approach II), units or fragments already displaying an oxetane ring would be incorporated into an existing scaffold. This second access route has the potential advantage of being inherently more flexible in providing access to a wider range of oxetane structures and streamline, or make more convergent, any synthesis of such compounds. The second sequence, however, necessitates the development of novel methods for the synthesis, manipulation, and elaboration of oxetanes. This is a proposition that made the synthetic aspects of the project intriguing.

13.5 SYNTHESIS

In order to meet the synthetic objectives outlined previously, we sought an effective means of generating oxetan-3-one.[54] The synthesis of this compound was first described by Marshall in 1952 (Scheme 13.1).[55] The route commences with treatment of chloroacetyl chloride with diazomethane to afford **2**, which upon exposure to alkali undergoes ring closure to the desired compound. Due to numerous side reactions, these investigators encountered difficulties in purifying the oxetan-3-one from the reaction mixture. Thus, they resorted to establishing the presence of oxetan-3-one by formation of its dinitrophenyl hydrazone derivative. The use of diazomethane along with the isolation problems make this sequence less than optimal for the generation of the simple oxetan-3-one.[54]

Pure oxetan-3-one was first obtained through the oxidative cleavage of 3-methylene oxetane (Scheme 13.2).[56] The synthesis of methylene oxetane begins with the Diels–Alder adduct of

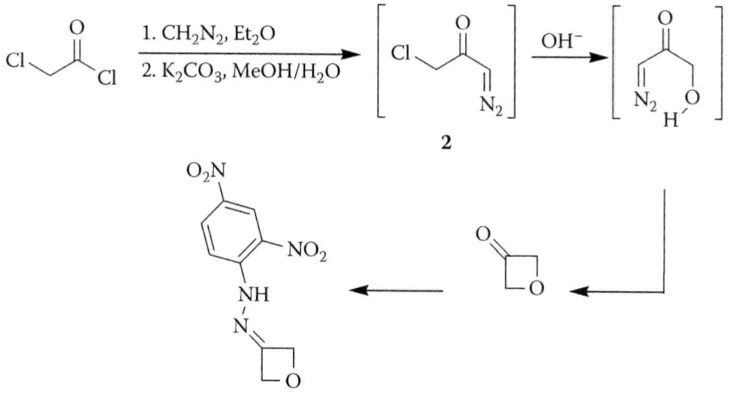

SCHEME 13.1

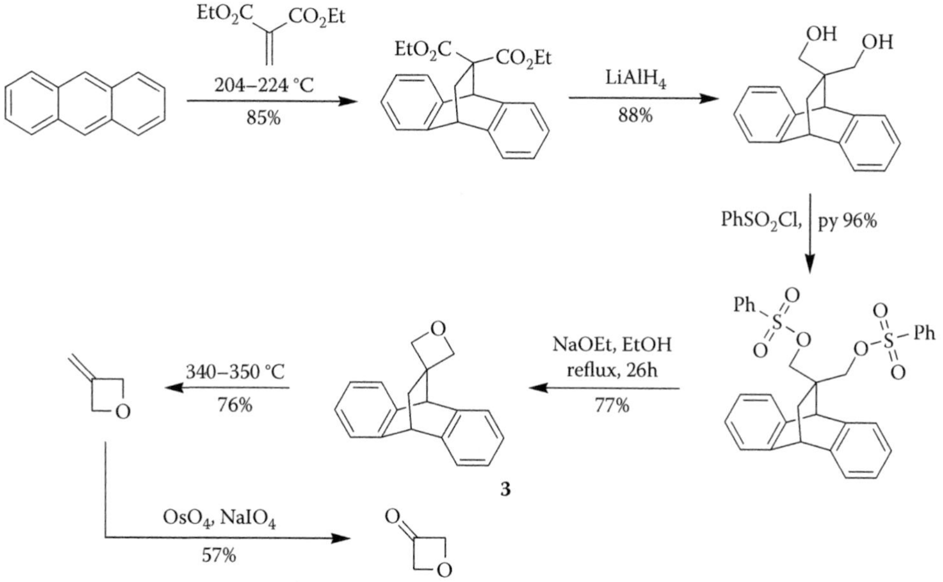

SCHEME 13.2

methylene malonate and anthracene. Diester reduction, generation of the bis-sulfonate, followed by ring closure furnished oxetane **3**. In a subsequent step, **3** undergoes retro-Diels–Alder reaction at 340 to 355°C to release 3-methylene oxetane. This low-boiling liquid can be distilled from the reaction mixture and trapped. It is important to note that the authors highlighted the fact that 3-methylene oxetane needs to be handled under inert gas atmosphere, because it very easily undergoes autoxidation to form a peroxide.[57] The safety issues along with the nature of the steps required in the sequence render, in our estimation, the approach unsuitable for the production of preparative quantities of oxetan-3-one.

A more practical method relies on the preparation of oxetane-3-ol and its subsequent oxidation to give oxetan-3-one (Scheme 13.3).[58–60] This procedure suffers from the necessity to use Pyridinium chlorochromate (PCC) for oxidation of oxetane-3-ol, because separation of oxetan-3-one from the reaction by-products could only be effected by preparative gas chromatography (GC). It was furthermore reported that other oxidizing conditions failed for this substrate.[60] Having to rely on

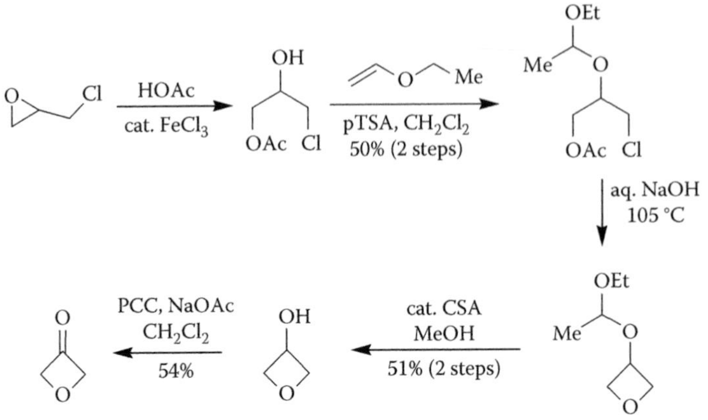

SCHEME 13.3

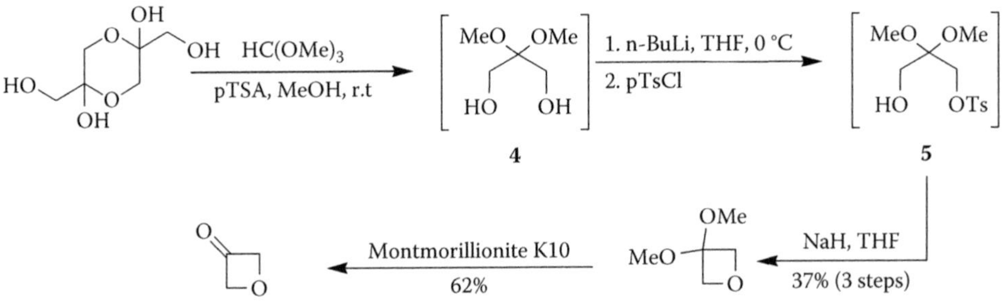

SCHEME 13.4

preparative GC poses a serious limitation on applicability and scalability of this route. To get access to large quantities of oxetan-3-one, we therefore decided to investigate alternative ways to make this compound.

In light of the limitations of the previously reported approaches, we developed a process that commences with commercially available dihydroxyacetone dimer (Scheme 13.4). This is an inexpensive starting material that is conveniently handled. When it is treated with trimethyl orthoformate and catalytic amounts of p-toluene sulfonic acid in methanol, dihydroxy acetone dimethyl ketal is readily formed over the course of approximately 8 h.[61,62] Following neutralization of the reaction mixture with basic ion exchange resin (Ambersep 9000H®), filtration, and evaporation of solvent, ketal **4** can be isolated and employed without additional purification. The diol can then be subjected to deprotonation in THF (1 equiv n-BuLi at 0°C), followed by quenching the resulting alkoxide with p-toluene sulfonyl chloride to furnish monotosylate **5**.[63] A simple aqueous workup conveniently effects removal of the lithium salts formed as by-product. The unpurified monotosylate **5** can then be subjected to NaH in THF to afford 3,3-dimethoxy oxetane. In our initial foray, the dimethyl ketal proved highly recalcitrant toward hydrolysis. Several attempts to effect cleavage failed, leading to recovery of either the starting material or decomposition products. These included exposure to mineral acid in a variety of solvents,[64] treatment with strongly acidic ion-exchange resin,[65] and H$_2$SO$_4$ on silica.[66] Ultimately, we observed that ketal hydrolysis could be effected by refluxing a solution of the ketal in dichloromethane in the presence of Montmorillonite K10 for 60 h to give the desired compound in 62% yield.

The advantages of the procedure we have developed include the relatively few steps from a convenient starting material (dihydroxy acetone dimer), the use of "low-tech" conditions throughout,

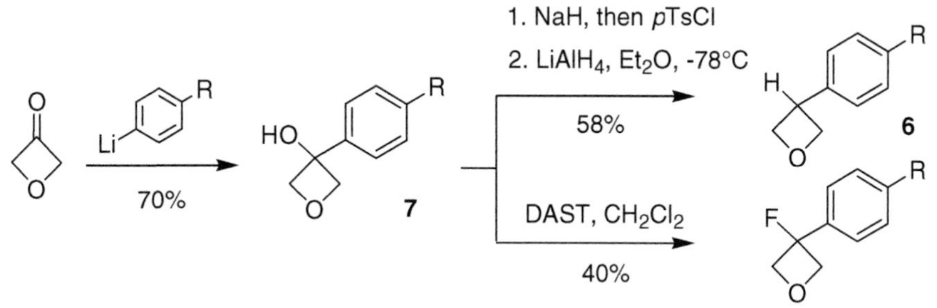

SCHEME 13.5

and, importantly, the purification of the final product by distillation. The sequence has been employed to provide more than 0.4 mole of oxetan-3-one which can be stored neat in a freezer over months without decomposition. Safety tests with differential scanning calorimetry show that oxetan-3-one can decompose violently at elevated temperature (2090 J/g liberated, onset temperature 180°C). The maximum permissible safe temperature for processes involving oxetan-3-one is recommended as 80°C. For longer processes (> 72 h), a temperature of 60°C should not be exceeded.

13.6 APPLICATIONS

We prepared a number of key compounds as benchmarks in order to investigate the effect of oxetane introduction onto some key structures (Scheme 13.5). In the first series of compounds, we explored the addition of organometallic reagents to oxetan-3-one. We demonstrated that oxetan-3-one participates in a number of nucleophilic additions by reagents including aryllithiums to give the corresponding 3-hydroxy-3-aryl substituted oxetanes. These compounds in turn can be conveniently converted to the reduced monosubstituted oxetane **6** or converted to the fluorinated analogue. Thus, treatment of **7** with tosyl chloride and then LiAlH$_4$ at low temperature afforded **6**. The tertiary alcohol **7** undergoes direct displacement to the fluoride upon its exposure to DAST in methylene chloride at −78°C.

We have also been engaged in the development of synthetic chemistry of building blocks derived from oxetan-3-one (Scheme 13.6). In this regard, oxetan-3-one reacts with commercially available stabilized ylids to give unsaturated ester **8** and aldehyde **9** at remarkably rapid rates as well as 89% and 81% yield, respectively. This was surprising given the strain of the olefin produced and the problems that had been previously noted with 3-methylene oxetane, as well as the fact that

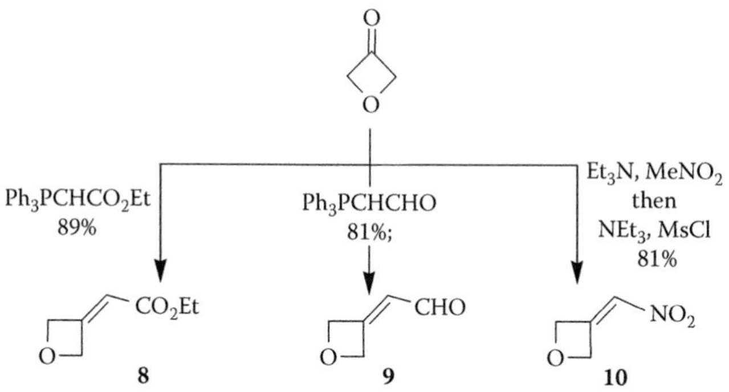

SCHEME 13.6

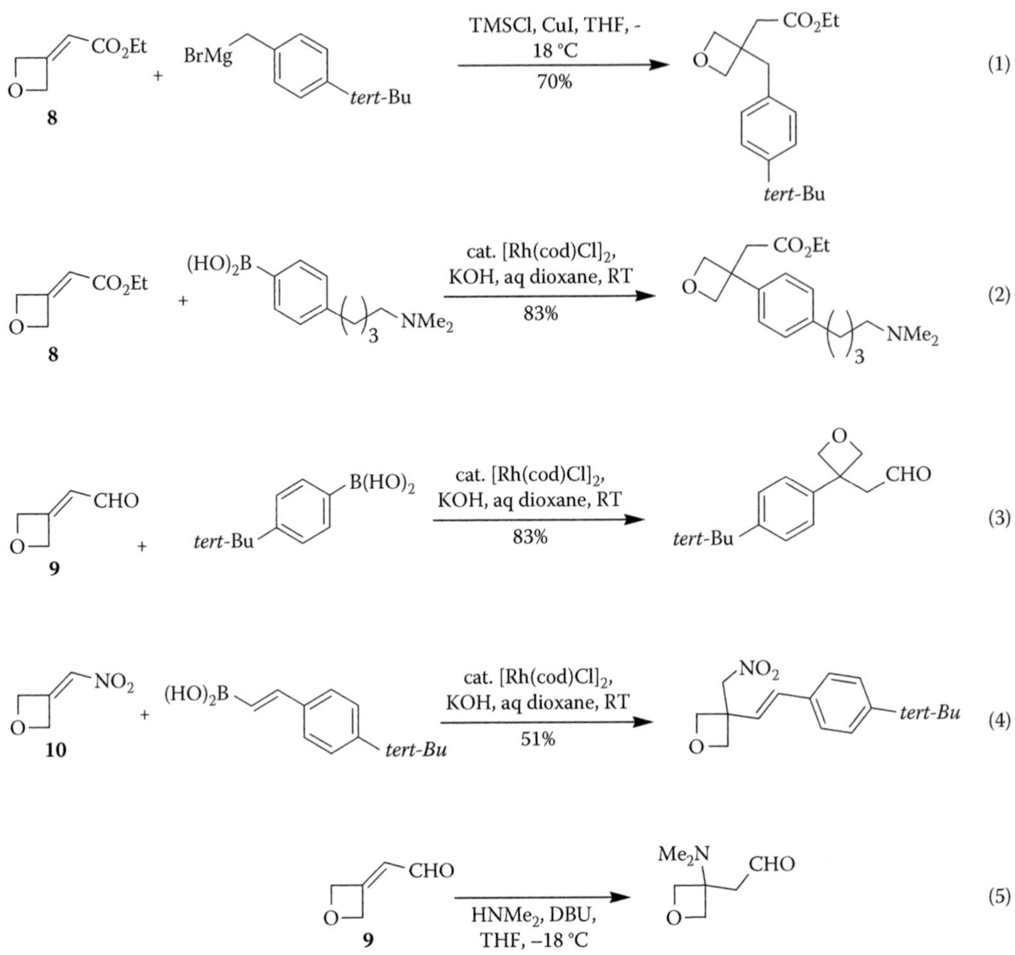

SCHEME 13.7

oxetan-3-one has been suggested to undergo self-condensation or polymerization under basic conditions. Additionally, condensation of oxetan-3-one with nitromethane furnishes nitromethylene oxetane **10** in 81% yield. Surprisingly, none of these compounds has previously been reported, yet they constitute tractable, stable entities amenable to storage.

The α,β-unsaturated ester (**8**), aldehyde (**9**), and nitro (**10**) compounds participate as electrophiles in a number of useful conjugate addition reactions. Copper-catalyzed addition of Grignard reagents provides access to aryl butanoic acid derivatives substituted with an oxetane (equation 1 in scheme 13.7). The ester, aldehyde, and nitro electrophiles also undergo mild Rh-catalyzed additions of aryl and vinyl boronic acids (equation 2 through equation 4 in scheme 13.7). Interestingly, the unsaturated aldehyde participates readily in an amine conjugate addition to afford oxetane substituted 3-amino-acetaldehyde derivatives.

13.7 CONCLUSION

Improved access routes to relatively small building blocks can open up new opportunities in drug discovery. We detailed the case of oxetan-3-one; this simple ketone is rich in the chemical reactions it and its derivatives participate in. We anticipate additional studies focused on this building block, demonstrating its unique features and the benefits that can be harnessed by its grafting onto molecular targets of interest (Scheme 13.7).

REFERENCES

1. Bleicher, K.H., Böhm, H.J., Müller, K., and Alanine, A.I., Hit and lead generation: beyond high-throughput screening, *Nat. Rev. Drug Discov.* 2 (5), 369, 2003.
2. Jacobsen, E.N., Pfaltz, A., Yamamoto, H., and Editors, *Comprehensive Asymmetric Catalysis I–III*, Springer Berlin, 1999.
3. Mish, M.R., Guerra, F.M., and Carreira, E.M., Asymmetric dipolar cycloadditions of Me3SiCHN2. Synthesis of a novel class of amino acids: azaprolines, *J. Am. Chem. Soc.* 119 (35), 8379, 1997.
4. Kim, Y., Singer, R.A., and Carreira, E.M., Total synthesis of macrolactin A with versatile catalytic, enantioselective dienolate aldol addition reactions, *Angewandte Chemie-International Edition* 37 (9), 1261, 1998.
5. Starr, J.T., Baudat, A., and Carreira, E.M., Nucleophilic additions to a spiro[2,4]hepta-4,6-diene 4-nitrile: synthesis of 1,2-disubstituted cyclopentenes, *Tetrahedron Lett.* 39 (32), 5675, 1998.
6. Alper, P.B., Meyers, C., Lerchner, A., Siegel, D.R., and Carreira, E.M., Facile, novel methodology for the synthesis of spiro[pyrrolidin-3,3'-oxindoles]: catalyzed ring expansion reactions of cyclopropanes by aldimines, *Angewandte Chemie-International Edition* 38 (21), 3186, 1999.
7. Sasaki, H. and Carreira, E.M., Efficient enantioselective synthesis of C--methyl aspartic acid and 3-amino-3-methylpyrrolidin-2-one, *Synthesis-Stuttgart* (1), 135, 2000.
8. Guerra, F.M., Mish, M.R., and Carreira, E.M., Versatile, diastereoselective additions of silyl ketene acetals, allyl tributylstannane, and Me3SiCN to N-acyl pyrazolines: asymmetric synthesis of densely functionalized pyrazolidines, *Org. Lett.* 2 (26), 4265, 2000.
9. Bode, J.W., Fraefel, N., Muri, D., and Carreira, E.M., A general solution to the modular synthesis of polyketide building blocks by Kanemasa hydroxy-directed nitrile oxide cycloadditions, *Angewandte Chemie-International Edition* 40 (11), 2082, 2001.
10. Anand, N.K. and Carreira, E.M., A simple, mild, catalytic, enantioselective addition of terminal acetylenes to aldehydes, *J. Am. Chem. Soc.* 123 (39), 9687, 2001.
11. Czekelius, C. and Carreira, E.M., Catalytic enantioselective conjugate reduction of ,-disubstituted nitroalkenes, *Angewandte Chemie-International Edition* 42 (39), 4793, 2003.
12. Ritter, T. and Carreira, E.M., 1,2,4-Oxadiazolidinones as configurationally stable chiral building blocks, *Angewandte Chemie-International Edition* 44 (6), 936, 2005.
13. Lohse-Fraefel, N. and Carreira, E.M., A modular approach to polyketide building blocks: cycloadditions of nitrile oxides and homoallylic alcohols, *Org. Lett.* 7 (10), 2011, 2005.
14. Knopfel, T.F., Zarotti, P., Ichikawa, T., and Carreira, E.M., Catalytic, enantioselective, conjugate alkyne addition, *J. Am. Chem. Soc.* 127 (27), 9682, 2005.
15. Paquin, J.F., Defieber, C., Stephenson, C.R.J., and Carreira, E.M., Asymmetric synthesis of 3,3-diarylpropanals with chiral diene-rhodium catalysts, *J. Am. Chem. Soc.* 127 (31), 10850, 2005.
16. Ritter, T., Kvaerno, L., Werder, M., Hauser, H., and Carreira, E.M., Heterocyclic ring scaffolds as small-molecule cholesterol absorption inhibitors, *Org. and Biomol. Chem.* 3 (19), 3514, 2005.
17. Waser, J., Gonzalez-Gomez, J.C., Nambu, H., Huber, P., and Carreira, E.M., Cobalt-catalyzed hydro-hydrazination of dienes and enynes: access to allylic and propargylic hydrazides, *Org. Lett.* 7 (19), 4249, 2005.
18. Aschwanden, P., Stephenson, C.R.J., and Carreira, E.M., Highly enantioselective access to primary propargylamines: 4-piperidinone as a convenient protecting group, *Org. Lett.* 8 (11), 2437, 2006.
19. Wuitschik, G., Rogers-Evans, M., Müller, K., Fischer, H., Wagner, B., Schuler, F., Polonchuk, L., and Carreira, E.M., Oxetanes as promising modules in drug discovery, *Angew. Chem., Int. Ed. Engl.*, 45 (46), 7736, 2006.
20. Reboul, M., *Ann. Chim. (Paris)* 14 (5), 496, 1878.
21. Fernandez, J., Myers, R.J., and Gwinn, W.D., Microwave spectrum and planarity of the ring of trimethylene oxide, *J. Chem. Phys.* 23 (4), 758, 1955.
22. Gwinn, W.D., Information pertaining to molecular structure, as obtained from the microwave spectra of molecules of the asymmetric rotor type, *Discuss. Faraday. Soc.* (19), 43, 1955.

23. Ringner, B., Sunner, S., and Watanabe, H., Enthalpies of combustion and formation of some 3,3-disubstituted oxetanes, *Acta Chem. Scand.* 25 (1), 141, 1971.

24. Pritchard, J.G. and Long, F.A., The kinetics of the hydrolysis of trimethylene oxide in water, deuterium oxide and 40-percent aqueous dioxane, *J. Am. Chem. Soc.* 80 (16), 4162, 1958.

25. Searles, S., The reaction of trimethylene oxide with Grignard reagents and organolithium compounds, *J. Am. Chem. Soc.* 73 (1), 124, 1951.

26. Searles, S., Jr., Oxetanes, *Chem. Heterocyclic Compds.* 19 (Pt. 2), 983, 1964.

27. Searles, S., Tamres, M., and Lippincott, E.R., Hydrogen bonding ability and structure of ethylene oxides, *J. Am. Chem. Soc.* 75 (11), 2775, 1953.

28. Farthing, A.C. and Reynolds, R.J.W., Synthesis and properties of a new polyether—poly-3,3-bis(chloromethyl)-1-oxabutene, *J. Polym. Sci.* 12 (67), 503, 1954.

29. Nair, J.K., Reddy, T.S., Satpute, R.S., Mukundan, T., and Asthana, S.N., Synthesis and characterization of energetic thermoplastic elastomers (ETPEs) based on 3,3-bis(azidomethyl)oxetane(BAMO)-3-azidomethyl-3-methyloxetane (AMMO) copolymers, *J. Polym. Mater.* 21 (2), 205, 2004.

30. Wani, M.C., Taylor, H.L., Wall, M.E., Coggon, P., and Mcphail, A.T., Plant antitumor agents. 6. Isolation and structure of Taxol, a novel antileukemic and antitumor agent from *Taxus brevifolia*, *J. Am. Chem. Soc.* 93 (9), 2325, 1971.

31. Roberts, S.M., Scheinmann, F., and Editors, *New Synthetic Routes to Prostaglandins and Thromboxanes*, Academic Press, London, New York, 1982.

32. Shimada, N., Hasegawa, S., Harada, T., Tomisawa, T., Fujii, A., and Takita, T., Oxetanocin, a novel nucleoside from bacteria, *J. Antibiot.* 39 (11), 1623, 1986.

33. Omura, S., Murata, M., Imamura, N., Iwai, Y., Tanaka, H., Furusaki, A., and Matsumoto, T., Oxetin, a new antimetabolite from an actinomycete. Fermentation, isolation, structure and biological activity, *J. Antibiot.* 37 (11), 1324, 1984.

34. Farina, V. and Editor, *The Chemistry and Pharmacology of Taxol and Its Derivatives.* [In: *Pharmacochem. Libr.*, Elsevier, Amsterdam, 1995: 22], 1995.

35. Boge, T.C., Hepperle, M., Vander Velde, D.G., Gunn, C.W., Grunewald, G.L., and Georg, G.I., The oxetane conformational lock of paclitaxel: structural analysis of D-secopaclitaxel, *Bioorg. Med. Chem. Lett.* 9 (20), 3041, 1999.

36. Wang, M., Cornett, B., Nettles, J., Liotta, D.C., and Snyder, J.P., The oxetane ring in taxol, *J. Org. Chem.* 65 (4), 1059, 2000.

37. Gunatilaka, A.A.L., Ramdayal, F.D., Sarragiotto, M.H., Kingston, D.G.I., Sackett, D.L., and Hamel, E., Synthesis and biological evaluation of novel paclitaxel (Taxol) D-ring modified analogs, *J. Org. Chem.* 64 (8), 2694, 1999.

38. Marder-Karsenti, R., Dubois, J., Bricard, L., Guenard, D., and Gueritte-Voegelein, F., Synthesis and biological evaluation of D-ring-modified taxanes: 5(20)-azadocetaxel analogs, *J. Org. Chem.* 62 (19), 6631, 1997.

39. Fenoglio, I., Nano, G.M., Vander Velde, D.G., and Appendino, G., Chemistry and occurrence of taxane derivatives. XXV. Synthesis of azetidine-type taxanes, *Tetrahedron Lett.* 37 (18), 3203, 1996.

40. Hoshino, H., Shimizu, N., Shimada, N., Takita, T., and Takeuchi, T., Inhibition of infectivity of human immunodeficiency virus by oxetanocin, *J. Antibiot.* 40 (7), 1077, 1987.

41. Huryn, D.M. and Okabe, M., AIDS-driven nucleoside chemistry, *Chem. Rev.* 92 (8), 1745, 1992.

42. Holan, G., Rational design of degradable insecticides, *Nature (London)* 232 (5313), 644, 1971.

43. Meyer, W., Application: EP 92-810027 92-810027 496701, 1992, N-[[2-[[(3-oxetanyl)oxy]carbonyl]phenyl]sulfonyl]urea derivs. and N-[[2-[[(3-thietanyl)oxy]carbonyl]phenyl]sulfonyl]urea derivs., methods for their preparation and their use as herbicides. 19920116.

44. Koeppe, M.K. and Brown, H.M., Sulfonylurea herbicide plant-metabolism and crop selectivity, *Agro Food Ind. Hi-Tech* 6 (6), 9, 1995.

45. Forsberg, G., Rate constants and reaction products of the alkaline hydrolysis of ethylene and trimethylene chlorohydrins with alkyl substituents, *Acta Chem. Scand.* 8 (1), 135, 1954.

46. Ruzicka, L., On the understanding of carbon rings I. On the constitution of zibetone, *Helv. Chim. Acta* 9, 230, 1926.

47. Welch, S.C. and Rao, A.S.C.P., Convenient one-step synthesis of 2,2-disubstituted oxetanes from ketones, *J. Am. Chem. Soc.* 101 (20), 6135, 1979.

48. Fitton, A.O., Hill, J., Jane, D.E., and Millar, R., Synthesis of simple oxetanes carrying reactive 2-substituents, *Synthesis-Stuttgart* (12), 1140, 1987.
49. Büchi, G., Inman, C.G., and Lipinsky, E.S., Light-catalyzed organic reactions. 1. The reaction of carbonyl compounds with 2-methyl-2-butene in the presence of ultraviolet light, *J. Am. Chem. Soc.* 76 (17), 4327, 1954.
50. Paterno, E. and Chieffi, G., Synthesis in organic chemistry using light. Note II. Compounds of unsaturated hydrocarbons with aldehydes and ketones, *Gazz. Chim. Ital.* 39, 341, 1909.
51. Bach, T., Stereoselective intermolecular [2+2]-photocycloaddition reactions and their application in synthesis, *Synthesis-Stuttgart* (5), 683, 1998.
52. Edward, J.T., Farrell, P.G., and Shahidi, F., Partial molar volumes of organic compounds in water. Part 1. Ethers, ketones, esters, and alcohols, *J. Chem. Soc. Lond. Faraday Trans. 1* 73 (5), 705, 1977.
53. Moore, J.C., Battino, R., Rettich, T.R., Handa, Y.P., and Wilhelm, E., Partial molar volumes of gases at infinite dilution in water at 298.15 K, *J. Chem. Eng. Data* 27 (1), 22, 1982.
54. Dejaegher, Y., Kuz'menok, N.M., Zvonok, A.M., and De Kimpe, N., The chemistry of azetidin-3-ones, oxetan-3-ones, and thietan-3-ones, *Chem. Rev.* 102 (1), 29, 2002.
55. Marshall, J.R. and Walker, J., Synthesis of simple C-substituted derivatives of dihydroxyacetone, *J. Chem. Soc.* 467, 1952.
56. Berezin, G.H., Application: U.S. Patent 3297719, 1967, 3-Hydroxymethyl-3-hydroxyoxetane. 19641127.
57. Applequist, D.E. and Roberts, J.D., Small-ring compounds. XV. Methylenecyclobutene and related substances, *J. Am. Chem. Soc.* 78, 4012, 1956.
58. Baum, K., Berkowitz, P.T., Grakauskas, V., and Archibald, T.G., Synthesis of electron-deficient oxetanes-3-azidooxetane, 3-nitrooxetane, and 3,3-dinitrooxetane, *J. Org. Chem.* 48 (18), 2953, 1983.
59. Kozikowski, A.P. and Fauq, A.H., Synthesis of novel 4-membered ring amino-acids as modulators of the N-methyl-D-aspartate(Nmda) receptor complex, *Synlett* (11), 783, 1991.
60. Wojtowicz, J.A. and Polak, R.J., 3-Substituted oxetanes, *J. Org. Chem.* 38 (11), 2061, 1973.
61. Charmantray, F., El Blidi, L., Gefflaut, T., Hecquet, L., Bolte, J., and Lemaire, M., Improved straightforward chemical synthesis of dihydroxyacetone phosphate through enzymatic desymmetrization of 2,2-dimethoxypropane-1,3-diol, *J. Org. Chem.* 69 (26), 9310, 2004.
62. Ferroni, E.L., DiTella, V., Ghanayem, N., Jeske, R., Jodlowski, C., O'Connell, M., Styrsky, J., Svoboda, R., Venkataraman, A., and Winkler, B.M., A three-step preparation of dihydroxyacetone phosphate dimethyl acetal, *J. Org. Chem.* 64 (13), 4943, 1999.
63. Picard, P., Leclercq, D., Bats, J.P., and Moulines, J., An efficient one-pot synthesis of oxetanes from 1,3-diols, *Synthesis-Stuttgart* (7), 550, 1981.
64. Allen, W.S., Bernstein, S., Heller, M., and Littell, R., Steroidal cyclic ketals. 15. 17,21-Oxido-steroids. 1. Preparation, *J. Am. Chem. Soc.* 77 (18), 4784, 1955.
65. Coppola, G.M., Amberlyst-15, a superior acid catalyst for the cleavage of acetals, *Synthesis* (12), 1021, 1984.
66. Huet, F., Lechevallier, A., Pellet, M., and Conia, J.M., Wet silica-gel—convenient reagent for de-acetalization, *Synthesis-Stuttgart* (1), 63, 1978.

14 Well-Defined (NHC)Pd (II) Complexes and Their Use in C–C and C–N Bond-Forming Reactions

Oscar Navarro and Steven P. Nolan

CONTENTS

14.1 INTRODUCTION

Research focusing on palladium compounds and their use in catalysis at both industrial and laboratory scales has exponentially increased during the last 10 years.[1,2] Although ligandless systems are also known,[2a] it is well understood that the ancillary ligation to the metal center plays a crucial role in dictating the efficiency of a catalytic system.[3] Bulky, electron-rich phosphine ligands such as P^{t}Bu$_3$ are now commonly used to stabilize the Pd(0) intermediates, thereby avoiding the precipitation of the metal in homogeneous catalysis.[4] However, the most common phosphine ligands possess several drawbacks: (1) they often are prone to air oxidation and therefore require air-free handling; (2) when these ligands are subjected to higher temperatures, significant P–C bond degradation occurs and the use of an excess of the phosphine is required; and (3) they often react with Pd precursors such as Pd(OAc)$_2$ in a reduction process forming Pd(0)P$_n$ and phosphine oxide.[5]

    N-Heterocyclic carbenes (NHCs)[6] have become increasingly popular in the last few years as they represent an attractive alternative to tertiary phosphines in homogeneous catalysis. NHC complexes exhibit reaction behavior different from phosphines, especially displaying high thermal stability and tolerance to oxidation conditions. We developed several systems based on the combination of imidazolium salts (air-stable precursors for NHCs) and Pd(0) or Pd(II) sources, to generate catalytically active species *in situ*, and these mediate numerous organic reactions, principally cross-coupling reactions.[7] These preliminary systems by us and others[8] showed the importance of the NHC/Pd ratio on the efficiency of the reactions, pointing to an optimum 1:1 ligand-to-metal ratio in most cases. From there, we focused our efforts on the development of monomeric NHC-bearing Pd(II) complexes and the study of their catalytic activity. Generally, shorter reaction times are observed in these well-defined systems, because the carbene is already coordinated to the palladium center. Also, the use of a well-defined precatalyst allows for a better knowledge of the amount of ligand-stabilized palladium species in solution, by reducing the possibility of side reactions leading to ligand or palladium precursor decomposition prior to the coordination of the ligand.

We reported on the synthesis of monomeric (NHC)Pd(allyl)Cl,[9] (NHC)Pd(acac)Cl,[10] and (NHC)Pd(carboxylate) complexes,[11] among many architectures,[12] and we studied their activation mechanisms and catalytic activities. The synthesis of most of these complexes is directly related to a successful *in situ* generation and use of NHC and the appropriate palladium source. All these complexes display very high activity as precatalysts for C–C and C–N bond formation reactions. As an added advantage, all of these Pd(II) complexes are air and moisture stable, and some are already commercially available. Here, we focus on the synthesis and applications of two of those well-defined complexes: (NHC)Pd(allyl)Cl and (NHC)Pd(acac)Cl. The multigram syntheses described herein are straightforward and afford the desired complexes in excellent yields.

14.2 (NHC)Pd(allyl)Cl COMPLEXES

The first series of well-defined (NHC)Pd(II) complexes we developed has the general formula (NHC)Pd(allyl)Cl.[9] The easy and scaleable synthesis and the versatility of these as precatalysts have resulted in the commercialization of two of those complexes:[13] (IPr)Pd(allyl)Cl (**1**)[14] and its saturated analogue (SIPr)Pd(allyl)Cl (**2**) (IPr = (*N*,*N'*-bis(2,6-diisopropylphenyl)imidazol)-2-ylidene; SIPr = (*N*,*N'*-bis(2,6-diisopropylphenyl)-4,5-dihydroimidazol)-2-ylidene) (Figure 14.1).

14.2.1 SYNTHESIS

The first synthetic procedure reported was straightforward and involved the simple fragmentation of the commercially available dimer, [Pd(allyl)Cl]$_2$, by NHC in THF.[9a] The NHC needed to be previously generated and isolated from the corresponding imidazolium salt (also commercially available) by reacting it with a strong base such as KOBu^t or KH. This step was eliminated for the synthesis of **1** by following a one-pot protocol in dry THF that yielded a nearly quantitative amount

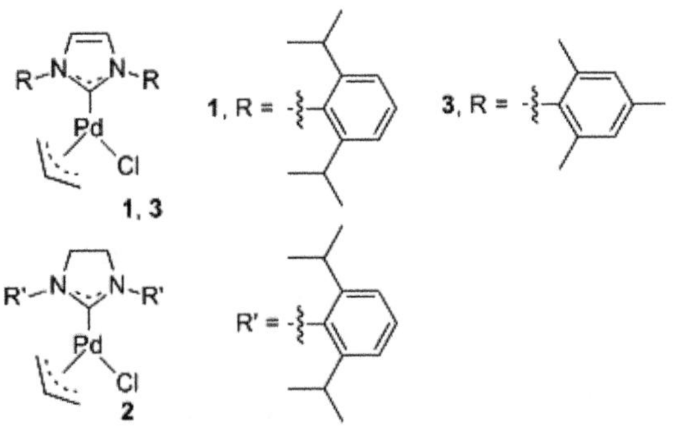

FIGURE 14.1 (IPr)Pd(allyl)Cl (**1**), (SIPr)Pd(allyl)Cl (**2**), and (IMes)Pd(allyl)Cl (**3**).

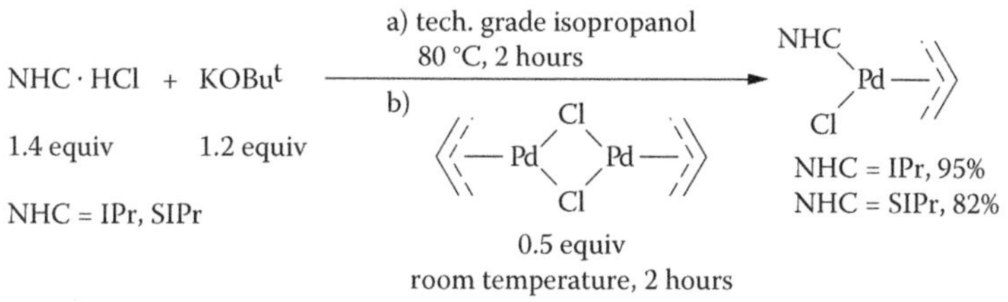

SCHEME 14.1 One-pot synthesis of **1** and **2**.

of the desired complex (on a 22-gram scale) without the need for prior isolation of the carbene.[9b] A similar protocol was described by Jensen and Sigman and carried out on a 0.5-mmol scale.[15] We recently reported a new one-pot protocol for the syntheses of **1** and **2** without the need for isolation or the carbene or the use of dry solvents (Scheme 14.1).[16] The deprotonation of the imidazolium salt is carried out in technical grade isopropanol, followed by the addition of the palladium dimer. When the reaction is complete, the complex is precipitated by addition of water in the original reaction vessel. A filtration in air affords the desired product in very high yields. In addition, the excess imidazolium salt can be easily recovered from the aqueous solution by a simple extraction. This straightforward procedure afforded on larger scale 10.8 g and 14.1 g of **1** and **2**, respectively.[17] The same procedure can be applied for the synthesis of (IMes)Pd(allyl)Cl (**3**) (IMes = (N,N'-bis(2,4,6-trimethylphenyl)imidazol)-2-ylidene), although in lower yield (42%) to afford 4.1 g of the desired complex.[17] A similar procedure is practiced by Umicore AG to synthesize **1** and **2** on larger (300-g) scale.

14.2.2 C–C Bond-Forming Reactions

14.2.2.1 Suzuki–Miyaura Cross-Coupling Reactions

Among the cross-coupling reactions,[18] the Suzuki–Miyaura reaction,[19,20] involving the coupling of an aryl halide or pseudo-halide with an organoboron reagent (usually boronic acids or esters) via transmetallation in the presence of a base, has emerged as one of the most common in academic and industrial laboratories because organoboron reagents exhibit many advantages[21]; for example, they are readily available by hydroboration and transmetalation, inert to water and related solvents as well as oxygen, generally thermally stable and tolerant toward various functional groups, and organoboron starting reagents and by-products exhibit low toxicity. We reported in several articles on the use of (IPr)Pd(allyl)Cl for this important reaction using very mild reaction conditions and short reaction times.[9,22] Large-scale reactions can be carried out with no decrease in activity and in similar conditions (Scheme 14.2).[23] Interestingly, the use of microwave heating was also found compatible with this precatalyst.[22a]

We carried out studies using technical-grade isopropanol as solvent with excellent results.[22b,24] Illustrated in Table 14.1 is the coupling of a series of aryl bromides and chlorides with a variety of arylboronic acids at low catalyst loadings. Aryl chlorides are very attractive substrates as they

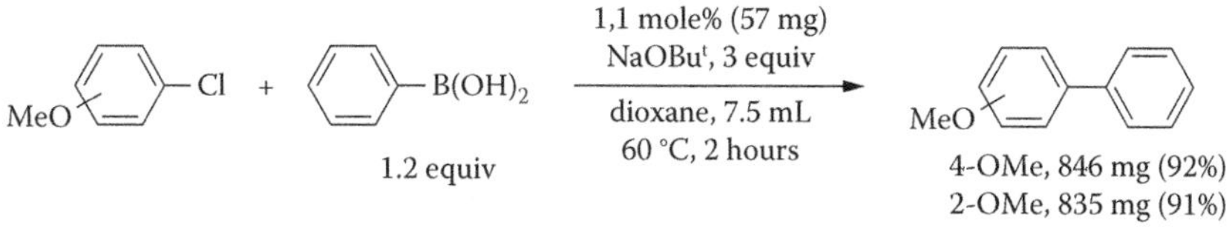

SCHEME 14.2 Large-scale Suzuki–Miyaura reaction catalyzed by **1**.

TABLE 14.1
Suzuki–Miyaura Cross-Coupling of Aryl Bromides and Chlorides with Boronic Acids at Low Catalyst Loadings

entry	Ar-X	boronic acid	product	time (h)	yield (%)[a]
1				3	92
2				2	93
3				3	83
4				3	92
5				3	91
6				1	85
7				1.5	91

[a]Isolated yields, average of two runs.

are available at low cost and in wide diversity, although their use in cross-coupling reactions has been somewhat limited due to their poor reactivity, attributed to the strength of the C–Cl bond.[25] When the temperature is increased in these reactions to 80°C, the catalyst loading can be reduced to 50 ppm with no loss of yield, and reactions reach completion in very short times. From an economical and industrial point of view, these reaction conditions are very appealing, especially regarding the use of an inexpensive and environmentally friendly solvent without predrying or purification. It is noteworthy that solutions of **2** in technical grade isopropanol decompose slightly over days, and the same solutions heated at 40°C for 48 h in air show little degradation of the precatalyst (5% by 1HNMR).[9b]

14.2.2.2 α-Ketone Arylation Reactions

The coupling of enolisable ketones and aryl halides, despite its great synthetic importance, has been less explored.[26] We recently reported on a very efficient system for the arylation of ketones with aryl halides and triflates using **2** as precatalyst.[27] Aryl triflates are a very attractive alternative to aryl halides, as they can be easily synthesized from readily available phenols.[28] Some examples of suitable substrates for this reaction are presented in Table 14.2. The use of **1** for this reaction has also been performed in large-scale reactions with very good results (Table 14.2, entries 6 and 7).[23]

14.2.2.3 Telomerization Reactions

Telomerization reactions, the formation of short oligomers from dienes, represent a very efficient organic transformation with an overall atom economy of 100%, and they have been the subject of intensive research in both academic and industrial laboratories.[29] Complexes of palladium are known to catalyze the reaction of dienes with a variety of nucleophiles.[2a] Mechanistically, the reactions are thought to proceed by allyl coordination of two butadiene molecules to a palladium(0) center followed by the formation of a C–C bond. The eight-carbon chain is then attacked by a nucleophile at the terminal or at the 3 position. The reaction usually leads to a mixture of cis/trans isomers and *n*- and *iso*-products. When the nucleophile is methanol, 1-methoxyocta-2,7-diene 1 (*n*-product) is generally the major product, which is a useful precursor for plasticizer alcohols (octanols), solvents, corrosion inhibitors, and monomers for polymerization.[30]

Recently, Beller and coworkers reported on the use of complexes **1** and **3**, among other (NHC)Pd complexes, for telomerization reactions of 1,3-butadiene with methanol.[29] Although **1** afforded the desired linear product in acceptable yields and selectivity, **3** proved to be an excellent precatalyst for this reaction at very low loadings (1,3-butadiene:catalyst 100 000:1), and even lower loadings can be used with excellent results when an excess of the imidazolium salt IMes·HCl is added (Table 14.3).

Our group also reported on the use of **1** for telomerization reaction of 1,3-butadiene with primary and secondary amines.[31] The use of this precatalyst in combination with a salt containing a noncoordinating anion (NaPF$_6$), to generate *in situ* a cationic species, proved to be very efficient in short reaction times and mild reaction conditions (Table 14.4). All reactions were carried out starting with 5 mmol of amine, affording the desired products in amounts ranging from 553 to 1136 mg. Scaling up this reaction does not appear to be problematic.

14.2.3 C–N Bond Formation Reactions

The Buchwald–Hartwig amination reaction has become a most efficient method for forming a new C–N bond, and it is mediated by a Pd catalyst.[3,32] Our group reported on the use of **1** to catalyze amination reactions of aryl triflates with very good results (Table 14.5).[22a]

Aryl chlorides can also be used as substrates even on a large scale using this precatalyst (Table 14.6).[23] We also reported on the use of **2** as an effective precatalyst in amination reactions (Table 14.7).[9a] In general, shorter reaction times are required using this complex as precatalyst. Complex

TABLE 14.2
α-Ketone Arylation Reactions with Aryl Halides and Triflates

entry	ketone	aryl halide	product	temp (°C)	time (h)	yield (%)[a]
1				60	1	80
2				60	1	90
3				70	4	71
4				60	0.5	68
5				60	1	87
6				70	5	85[b]
7				70	5	85[c]

[a]Isolated yields. average of two runs. [b]4.75 mmol scale. [c]6.10 mmol scale.

TABLE 14.3
Telomerization Reactions of 1,3-Butadiene and Methanol Catalyzed by 1 and 3

entry	catalyst	loading[a]	yield (%)[d]	n:iso	chemoselec. (%)[e]	TOF (n+iso)[f]	TOF (n+iso)[f]
1	1	100,000:1[b]	46	92:8	96	46,000	2875
2	3	100,000:1[b]	94	98:2	99	94,000	5875
3	3	1000,000:1[c]	19	98:2	90	190,000	11875
4	3	1000,000:1[cg]	89	98:2	98	890,000	55625

[a]1,3-butadiene: catalyst. [b]General conditions: 16 h, 70 °C, 1 mol% NaOMe, MeOH/1.3-butadiene 2:1.
[c]90 °C, [d]Yield of n+iso. [e](n+iso)/(n+iso+byproducts). [f]Calculated with respect to 1,3-butadiene.
[g]IMes HCl added (40 equiv/equiv catalyst).

3 proved to be very efficient for intramolecular amination reactions leading to the synthesis of *Cryptocaria* alkaloids *rac*-cyptaustoline (4) and *rac*-cryptowoline (5) (Scheme 14.3).[33]

14.2.4 CATALYTIC DEHALOGENATION REACTIONS

The dehalogenation of aryl halides,[36] and especially aryl chlorides, represents an important chemical transformation in organic synthesis,[35] and due to the high toxicity of polychlorinated arenes (i.e., PCBs), it is also of importance to environmental remediation.[36] Our group recently reported on a very efficient system for the catalytic dehalogenation of aryl chlorides using **1** as catalyst, using NaOBu[t] as base, and technical grade isopropanol as solvent at 60°C.[22] When the reactions were carried out using microwave heating, the catalyst loading could be reduced to 0.025 mol% with no significant decrease in the yields. Some examples with conventional heating are provided in Table 14.8.

14.3 (NHC)Pd(acac)$_n$ (n = 1 OR 2) COMPLEXES

We recently reported on a series of novel (NHC)Pd (II) complexes bearing 1 or 2 acetylacetonato (acac) ligands.[10] The complexes were prepared using Pd(acac)$_2$ as the Pd precursor. Their activity in C–C and C–N bond-forming reaction was also tested (Figure 14.2).

14.3.1 SYNTHESIS

Direct reaction of the free carbene IPr with Pd(acac)$_2$ at room temperature in anhydrous toluene yielded (IPr)Pd(acac)$_2$ (**6**) as a yellow powder (Scheme 14.4). Reaction of **6** with an equimolecular amount of HCl at room temperature produced the new species (IPr)Pd(acac)Cl (**7**) in nearly quantitative yield (Scheme 14.5).

Because preliminary tests of **6** and **7** in catalytic reactions showed the superior activity of **7**, we realized the convenience of synthesizing **7** without the need to isolate the (IPr)Pd(acac)$_2$ intermediate. A one-pot, multigram synthesis of **7** is summarized in Scheme 14.6. Reaction of the free carbene IPr with Pd(acac)$_2$ in anhydrous 1,4-dioxane at room temperature, followed by the addition of an equimolecular amount of HCl, leads to the formation of the desired product. An improved synthesis of **7** was recently developed.[37] With this protocol, isolation of the carbene is no longer required, and the complex can be directly prepared from commercial sources in excellent yields (Scheme 14.7).

TABLE 14.4
Telomerization Reactions of 1,3-Butadiene and Amines Catalyzed by 1

entry	amine	product	time (h)	yield (%)[a]
1			0.25 24 (0.02 mol% cat) 4 (rt)	94 91 91
2			0.5	98
3			6	95
4			0.5	89
5			2	54
6			1	95
7	H_2N-Me		3	92
8			1	70

[a] Isolated yields, average of two runs.

TABLE 14.5
Buchwald–Hartwig Amination Reactions with Aryl Triflates Using 1

entry	triflate	amine	product	time (h)	yield (%)[a]
1				3	80
2				3	71
3				3	90
4				6	85
5				3	90
6				3	90
7				3	88
8				3	90

[a] Isolated yields, average of two runs.

TABLE 14.6
Large-Scale Buchwald–Hartwig Amination Reactions Using 1

entry	chloride	amine	product	yield (%)[a]
1				97[b]
2				97[c]

[a]Isolated yields. [b]Chloride 5.70 mmol; amine 5.70 mmol; NaOBu^t, 5.29 mmol; THF 12 mL. [c]Chloride, 4.6 mmol; amine, 4.6 mmol; NaOBu^t, 5 mmol; D ME 11 mL.

TABLE 14.7
Buchwald–Hartwig Amination Reactions with Aryl Triflates Using 2

entry	chloride	amine	product	temperature	time (hours)	yield (%)[a]
1				rt	24	88
2				rt	14	96
3				rt	1.3	93
4				50	1	95

[a]Isolated yields, average of two runs.

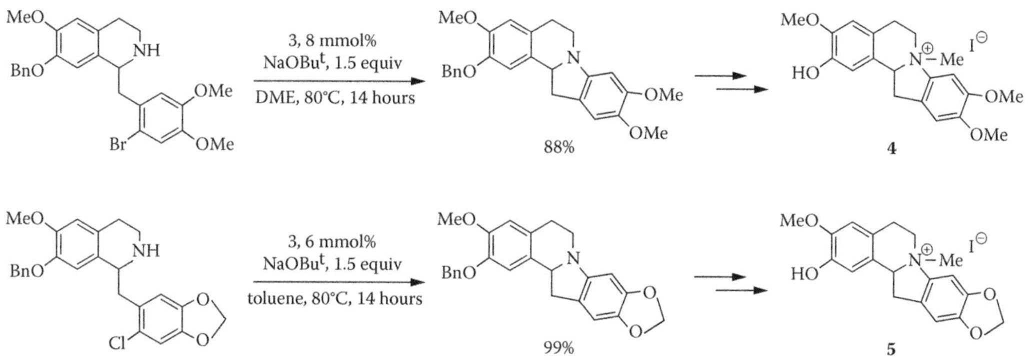

SCHEME 14.3 Synthesis of *Cryptocaria* alkaloids.

14.3.2 C–C AND C–N BOND-FORMING REACTIONS

Complex **7** was tested for α-ketone arylation reactions (Table 14.9) and Buchwald–Hartwig amination reactions with activated and unactivated aryl chlorides (Table 14.10). In both cases, very good results were obtained using mild temperatures in short reaction times.

14.4 SUMMARY

We presented a series of (NHC)Pd complexes that display very high activity and versatility as precatalysts for selected C–C and C–N bond-forming reactions. The generality of their activity in related C–C and C–X (X = N, O, S) reactions is an area of continuing investigation in our laboratories. The syntheses of these complexes are straightforward and can be carried out on multigram scale from commercially available reagents in very high yields. As an added advantage, all of these precatalysts are air- and moisture-stable complexes that display very high long-term stability.[38]

TABLE 14.8
Catalytic Dehalogenation of Aryl Chlorides Using 1

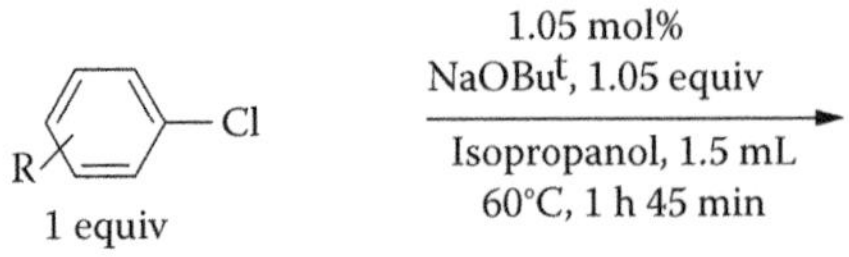

entry	aryl chloride	product	yield (%)[a]
1	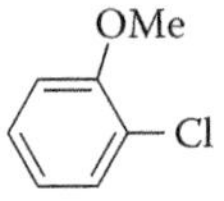	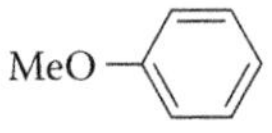	95
2	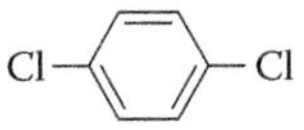		97[b]
3	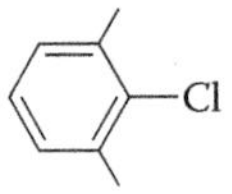	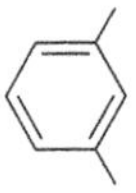	91
4	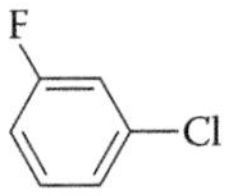	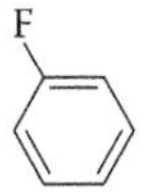	98
5	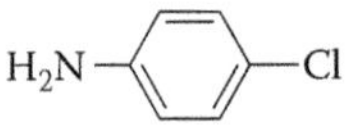	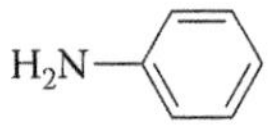	95
6	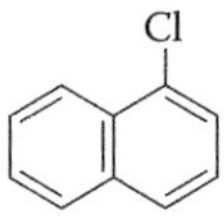	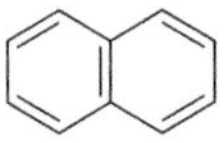	100
7	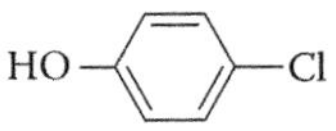	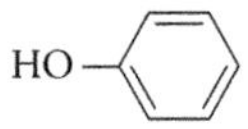	100

[a]GC yields, average of two runs. [b]1 mol% 1, 2.1 equiv of base.

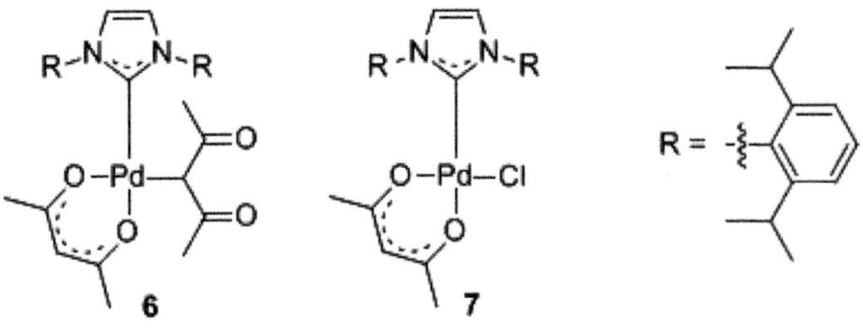

FIGURE 14.2 (IPr)Pd(acac)$_2$ (**6**) and (IPr)Pd(acac)Cl (**7**).

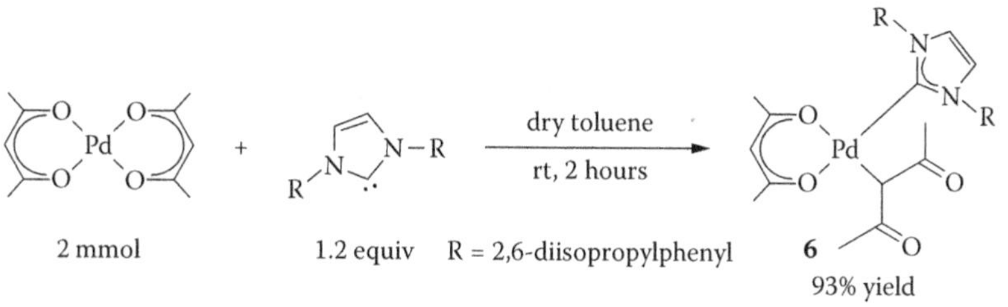

SCHEME 14.4 Synthesis of (IPr)Pd(acac)$_2$.

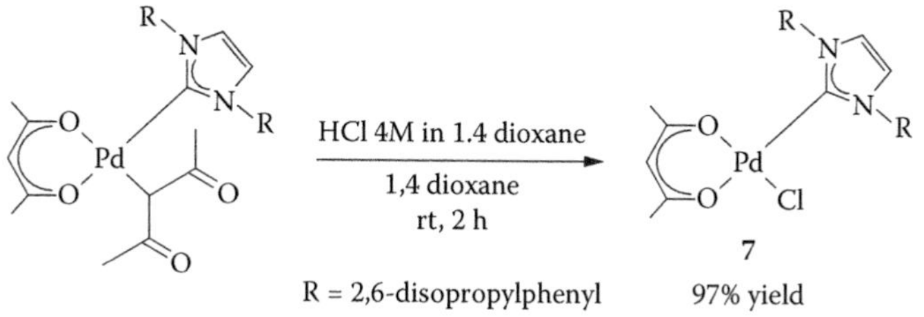

SCHEME 14.5 Synthesis of (IPr)Pd(acac)Cl.

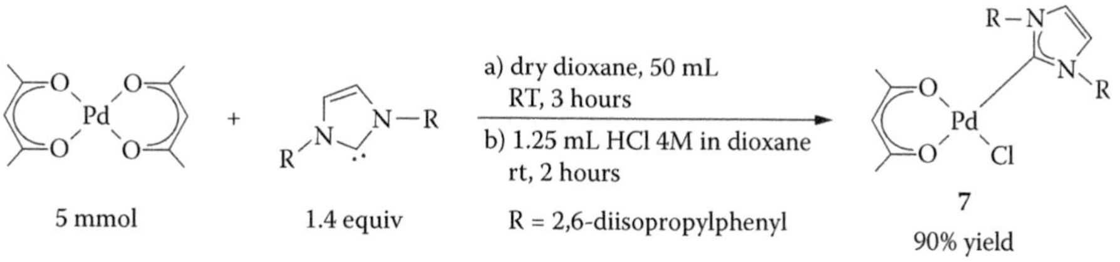

SCHEME 14.6 One-pot protocol for the synthesis of **7**.

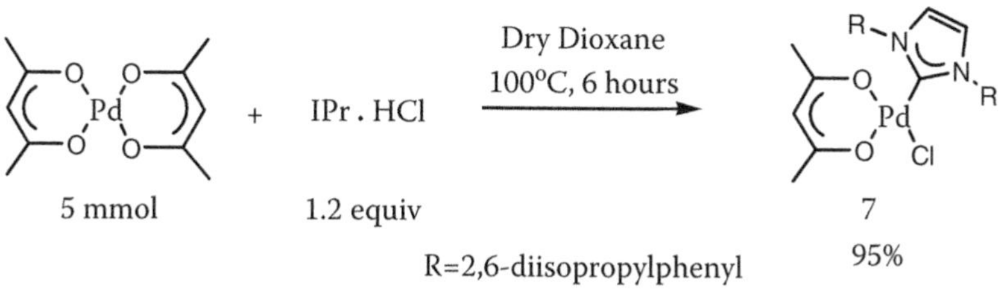

SCHEME 14.7 Improved one-pot protocol for the synthesis of **7**.

TABLE 14.9
α-Ketone Arylation Reactions with Aryl Chlorides Using 7

| | 1 mmol | 1.1 mmol | | | |

entry	aryl chloride	ketone	product	time (h)	yield (%)[2]
1				1	97
2				10	70
3				2	86
4				1	95
5				1.5	92
6				2	89

[2]Isolated yields, average of two runs.

TABLE 14.10
Buchwald–Hartwig Amination Reactions with Aryl Chlorides Using 7

$$R\text{—}Cl \ + \ R'_3\,NH \xrightarrow[\substack{D\ ME,\ 1\ mL \\ 50\ ^\circ C}]{\substack{7,\ 1\ mol\% \\ KOBu^t,\ 1.1\ equiv}} R\text{—}NR'_3$$

1 mmol   1.1 mmol

entry	aryl chloride	amine	product	time (h)	yield (%)[a]
1				0.5	97
2				0.5	98
3				1.5	90
4				4	99
5		Bu_2NH		6	95
6				10	93[b]

[a]Isolated yields, average of two runs. [b]2.1 equivalents of aryl chloride used.

REFERENCES AND NOTES

1. Tsuji, J. *Synthesis* 1990, 739–749.
2. (a) Tsuji, J. *Palladium Reagents and Catalysis*; John Wiley: New York, 1998. (b) Negishi, E. *J. Organomet. Chem.* 2002, *653*, 34–40.
3. (a) Wolfe, J.P.; Wagaw, S.; Marcoux, J.-F.; Buchwald, S.L. *Acc. Chem. Res.* 1998, *31*, 805–818. (b) Hartwig, J.F. *Acc. Chem. Res.* 1998, *31*, 852–860.
4. Applications of phosphine ligands in homogeneous catalysis: (a) Parshall, G.W.; Ittel, S. *Homogeneous Catalysis*; J. Wiley & Sons: New York, 1992. (b) Pignolet, L.H., Ed. *Homogeneous Catalysis with Metal Phosphine Complexes*; Plenum: New York, 1983.
5. Collman, J.P.; Hedegus, L.S.; Norton, J.R. Finke, R.G. *Principles and Applications of Organotransition Metal Chemistry*, 2nd ed.; University Science: Mill Valley, CA, 1987.
6. (a) Arduengo, A.J., III; Rasika Dias, H.V.; Harlow, R.L.; Kine, M. *J. Am. Chem. Soc.* 1992, *114*, 5530–5534.
7. (a) Hillier, A.C.; Nolan, S.P. *Chimica Oggi/Chemistry Today* 2001, *7/8*, 10–16. (b) Huang, J.; Grasa, A.; Nolan, S.P. *Org. Lett.* 1999, *1*, 2053–2055. (c) Huang, J.; Nolan, S.P. *J. Am. Chem. Soc.* 1999, *121*, 9889–9890. (d) Zhang, C.; Huang, J.; Trudell, M.L.; Nolan, S.P. *J. Org. Chem.* 1999, *64*, 3804–3805. (e) Lee, H.M.; Nolan, S.P. *Org. Lett.* 2000, *2*, 1307–1309.
8. (a) Weskamp, T.; Bohm, V.P.W.; Herrmann, W.A. *J. Organomet. Chem.* 2000, *600*, 12–22. (b) McGuiness, D.S.; Cavell, K.J.; Skelton, B.W.; White, A.H. *Organometallics* 1999, *18*, 1596–1605. (c) Stauffer, S.R.; Lee, S.; Stambuli, J.P.; Hauck, S.I.; Hartwig, J.F. *Org. Lett.* 2000, *2*, 1423–1426. (d) Bohm, V.P.W.; Gstottmayr, C.W.K.; Weskamp, T.; Herrmann, W.A. *J. Organomet. Chem.* 2000, *595*, 186–190. (e) For a recent review of monoligated palladium species as catalysts in cross-coupling reactions: Christmann, U.; Vilar, R. *Angew. Chem., Int. Ed.* 2005, *44*, 366–374.
9. (a) Viciu, M.S.; Germaneau, R.F.; Navarro-Fernandez, O.; Stevens, E.D.; Nolan, S.P. *Organometallics* 2002, *21*, 5470–5472. (b) Viciu, M.S.; Navarro, O.; Germaneau, R.F.; Kelly, R.A., III; Sommer, W.; Marion, N.; Stevens, E.D.; Cavallo, L.; Nolan, S.P. *Organometallics* 2004, *23*, 1629–1635.
10. Navarro, O.; Marion, N.; Stevens, E.D.; Scott, N.M.; González, J.; Amoroso, D.; Bell, A.; Nolan, S.P. *Tetrahedron* 2005, *61*, 9716–9722.
11. Viciu, M.S.; Stevens, E.D.; Petersen, J.L.; Nolan, S.P. *Organometallics* 2004, *23*, 3752–3755.
12. (a) Viciu, M.S.; Kelly, R.A., III; Stevens, E.D.; Naud, F.; Studer, M.; Nolan, S.P. *Org. Lett.* 2003, *5*, 1479–1482. (b) Viciu, M.S.; Kissling, R.M.; Stevens, E.D.; Nolan, S.P. *Org. Lett.* 2002, *4*, 2229–2231.
13. (IPr)Pd(allyl)Cl and (SIPr)Pd(allyl)Cl are commercially available from Strem Chemicals (Newbury-port, MA) for small quantities (hundreds of mg) and from Umicore AG (Brussels, Belgium) for larger quantities.
14. For a short review of uses of (IPr)Pd(allyl)Cl, see Scott, N.M.; Navarro, O.; Briel, O.; Nolan, S.P. *Chimica Oggi/Chemistry Today* 2005, *23*, 25–29.
15. Jensen, D.R.; Sigman, M.S. *Org. Lett.* 2003, *5*, 63–65.
16. Navarro, O.; Nolan, S.P. *Synthesis* 2006, 366–367.
17. From 10 mmol (3.66 g), 15 mmol (5.49 g), and 10 mmol (3.66 g) of [Pd(allyl)Cl]$_2$, respectively.
18. (a) Heck, R.F. *Palladium Reagents in Organic Synthesis*; Academic Press: New York, 1985. (b) Diederich, F., Stang, P.J., Eds. *Metal-Catalyzed Cross-Coupling Reactions*, 2nd ed; Wiley-VCH: Weinheim, 2004.
19. For reviews, see (a) Miyaura, N. *Top. Curr. Chem.* 2002, *219*, 11–59. (b) Suzuki, A. *J. Organomet. Chem.* 1999, *576*, 147–168. (c) Bellina, F.; Carpita, A.; Rossi, R. *Synthesis* 2004, *15*, 2419–2440.
20. Herrmann, W.A.; Reisinger, C.-P.; Spiegler, M. *J. Organomet. Chem.* 1998, *557*, 93–96.
21. Suzuki, A. *J. Organomet. Chem.* 2002, *653*, 83–90.
22. (a) Navarro, O.; Kaur, H.; Mahjoor, P.; Nolan, S.P. *J. Org. Chem.* 2004, *69*, 3173–3180. (b) Navarro, O.; Oonishi, Y.; Kelly, R.A.; Stevens, E.D.; Briel, O.; Nolan, S.P. *J. Organomet. Chem.* 2004, *689*, 3722–3727.
23. (a) Marion, N.; Navarro, O.; Kelly, R.A. III; Nolan, S.P. *Synthesis* 2003, *16*, 2590–2592. (b) Marion, N.; Navarro, O.; Kelly, R.A., III; Nolan, S.P. in *Catalysts for Fine Chemical Synthesis*; Roberts, S.M., Xiao, J., Whittall, J., Pickett, T.E., Eds.; Wiley: West Sussex, 2004; Vol. 3, chapter 4.8.
24. Marion, N., Navarro, O., Mei, J., Stevens, E.D., Scott, N.M., Nolan, S.P. *J. Am. Chem. Soc.* 2006, *128*, 4101–4111.

25. For a review in palladium-catalyzed coupling reactions of aryl chlorides: Littke, A.F.; Fu, G.C. *Angew. Chem. Int. Ed.* 2002, *41*, 4176–4211.

26. Culkin, D.A.; Hartwig, J.F. *Acc. Chem. Res.* 2003, *36*, 234–245.

27. Viciu, M.S.; Germaneau, R.F.; Nolan, S.P. *Org. Lett.* 2002, *4*, 4053–4056.

28. Echavarren, A.M.; Stille, J.K. *J. Am. Chem. Soc.* 1987, *109*, 5478–5486.

29. Jackstell, R.; Harkal, S.; Jiao, Ha.; Spannenberg, A.; Borgmann, C.; Roettger, D.; Nierlich, F.; Elliot, M.; Niven, S.; Cavell, K.; Navarro, O.; Viciu, M.S.; Nolan, S.P.; Beller, M. *Chemistry—A European Journal* 2004, *10*, 3891–3900, and references therein.

30. Yoshimura, N.; Tamura, M. (Kuraray Company, Ltd.), US 4356333, 1981 [*Chem. Abstr.* 1982, *96*, 103630s].

31. Viciu, M.S.; Zinn, F.K.; Stevens, E.D.; Nolan, S.P. *Organometallics* 2003, *22*, 3175–3177.

32. Hartwig, J.F. in *Modern Amination Methods*; Ricci, A., Ed.; Wiley-VCH: Weinheim, 2000.

33. Cämmerer, S.S.; Viciu, M.S.; Stevens, E.D.; Nolan, S.P. *Synlett* 2003, 1871–1873.

34. For a review, see Alonso, F.; Beletskaya, I.P.; Yus, M. *Chem. Rev.* 2002, *102*, 4009–4092.

35. (a) Terstiege, I.; Maleczka, R.E., Jr. *J. Org. Chem.* 1999, *64*, 342–343. (b) Dorman, G.; Otszewski, J.D.; Prestwich, G.D. *J. Org. Chem.* 1995, *60*, 2292–2297.

36. (a) Hutzinger, O.; Safe, S.; Zitko, V. *The Chemistry of PCBs*; CRC Press: Cleveland, 1974. (b) Mincher, B.J.; Randy, D.; Clevenger, T.E.; Golden, J. U.S. Patent 6132561, 2000. (c) McNab, W.W. Jr.; Ruiz, R.; Pico, T.M. U.S. Patent 6214202, 2001. (d) Morra, M.J.; Borek, V.; Koolpe, J. *J. Environ. Qual.* 2000, *29*, 706–715.

37. Marion, N., Ecarnot, E.C., Navarro, O., Amoroso, D., Bell, A., Nolan, S.P. *J. Org. Chem.* 2006, 71, 3816–3821.

38. To highlight the stability of these complexes, subjecting them to 90°F and nearly 100% humidity for 2 months resulted in no decomposition of these precatalysts. Furthermore, their activity was unaffected after this treatment. We can thank *Hurricane Katrina* for this ultimate stability test.

15 Toward Truly Efficient Organic Reactions in Water

Chikako Ogawa and Shū Kobayashi

CONTENTS

15.1 INTRODUCTION

Organic reactions in aqueous media are of current interest because water plays a key role as a solvent for green chemistry.[1] Indeed, water is a safe, harmless, and environmentally benign solvent. In addition, from practical and synthetic standpoints, the benefits of using water are immediately obvious, as it is unnecessary to dry solvents and substrates for reactions in aqueous media, and aqueous solutions of substrates or hydrated substrates can be used directly without further drying. Moreover, water has unique physical and chemical properties such as high dielectric constant and high cohesive energy density compared with most organic solvents. These properties allow highly efficient and selective enzymatic reactions to be conducted in living systems under mild conditions. Notably, the medium of enzymatic reactions is water, which plays major roles in the reactions. If the unique nature of water could be exploited *in vitro* as it is *in vivo*, it would be possible to develop interesting reactions with unique reactivity and selectivity that cannot be attained with organic solvents.

Conversely, Lewis acid catalysis has attracted much attention in organic synthesis.[2] Unique reactivity and selectivity are often observed under mild conditions in Lewis acid-catalyzed reactions. Although various kinds of Lewis acids have been developed and many have been applied in industry, they must generally be used under strictly anhydrous conditions. The presence of even a small amount of water stops the reactions, because most Lewis acids immediately react with water rather than with substrates. In addition, recovery and reuse of conventional Lewis acids are difficult, and these disadvantages have restricted the use of Lewis acids in organic synthesis.

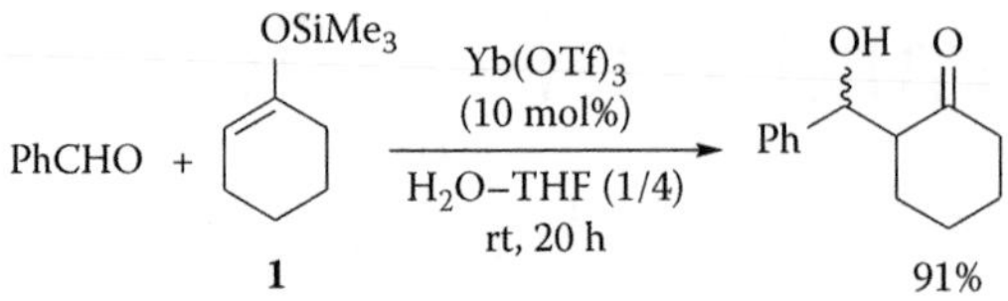

SCHEME 15.1

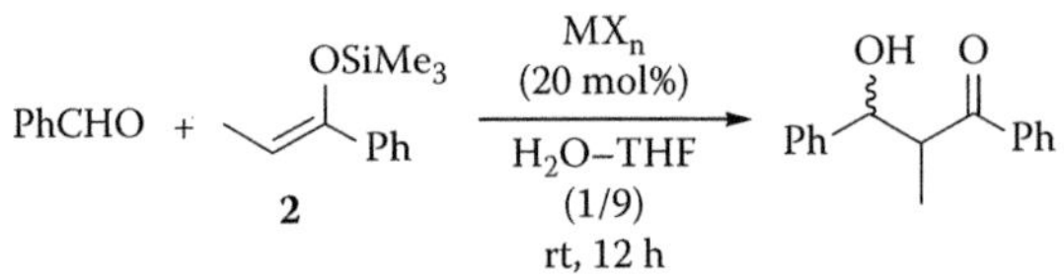

SCHEME 15.2

15.2 WATER-COMPATIBLE LEWIS ACIDS

Although most Lewis acids decompose in water, rare earth triflates (e.g., Sc(OTf)$_3$, Yb(OTf)$_3$) can be used as Lewis acid catalysts in water or water-containing solvents (water-compatible Lewis acids).[3] For example, the Mukaiyama aldol reaction of benzaldehyde with silyl enol ether **1** was catalyzed by Yb(OTf)$_3$ in water/THF (1/4) to give the corresponding aldol adduct in high yield (Scheme 15.1).[4] Interestingly, when this reaction was carried out in dry THF (without water), the yield of the aldol adduct was very low (ca. 10%). Thus, this catalyst is not only compatible with water but also activated by water, probably owing to dissociation of the counteranions from the Lewis acidic metal. Furthermore, the catalyst in this example can be easily recovered and reused.

Metal salts other than those derived from rare earth elements are also water-compatible Lewis acids. To find other Lewis acids that can be used in aqueous solvents and to determine the criteria for water-compatible Lewis acids, group 1–15 metal chlorides, perchlorates, and triflates were screened in the aldol reaction of benzaldehyde with silyl enol ether **2** in water/THF (1/9) (Scheme 15.2).[5] This screening revealed that not only Sc(III), Y(III), and Ln(III), but also Fe(II), Cu(II), Zn(II), Cd(II), and Pb(II) worked as Lewis acids in this medium to afford the desired aldol adduct in high yields.

These results reveal a correlation between the catalytic activity of the metal cations and two kinds of constants for the metal cations: hydrolysis constants (Kh) and exchange-rate constants for substitution of inner-sphere water ligands (water exchange-rate constants [WERC]).[6] Shown in Figure 15.1 are these constants for each metal cation, and metals that exhibited good catalytic activity in the screening (>50% yield) were found to have pK_h values that ranged from about 4 (4.3 for Sc(III)) to 10 (10.08 for Cd(II)) and WERC values greater than 3.2×10^6 M^{-1}s^{-1}. Cations having large pK_h values do not generally undergo efficient hydrolysis. When pK_h values are less than 4, cations are readily hydrolyzed to produce protons in sufficient number to cause rapid decomposition of the silyl enol ether. Conversely, when pK_h values are greater than 10, the Lewis acidities of the cations concerned are too low to catalyze the aldol reaction. Large WERC values may be required if a catalyst is to provide sufficiently fast exchange between the water molecules coordinated to the metal and the aldehyde substrate. "Borderline" species such as Mn(II), Ag(I), and In(III), which have pK_h and WERC values close to criteria limits, gave the aldol adduct in moderate yields. Whereas the precise activity of Lewis acids in aqueous media cannot be quantitatively predicted by pK_h and WERC values, their use has led to the identification of promising metal compounds as water-compatible Lewis acid catalysts[7] and also provided mechanistic insights into Lewis acid catalysis in aqueous media.

Li $^{+1}$ 13.64 4.7×10^7	Be — —					M $^{+n}$ pK_h a WERC b						B $^{+3}$ — —	C — —	N — —
Na $^{+1}$ 14.18 1.9×10^8	Mg $^{+2}$ 11.44 5.3×10^5											Al $^{+3}$ 4.97 1.6×10^0	Si $^{+4}$ — —	P $^{+5}$ — —
K $^{+1}$ 14.46 1.5×10^8	Ca $^{+2}$ 12.85 5×10^7	Sc $^{+3}$ 4.3 4.8×10^7	Ti $^{+4}$ ≤ 2.3 —	V $^{+3}$ 2.26 1×10^3	Cr $^{+3}$ 4.0 5.8×10^{-7}	Mn $^{+2}$ 10.59 3.1×10^7	Fe $^{+2}$ 9.5 3.2×10^6	Co $^{+2}$ 9.65 2×10^5	Ni $^{+2}$ 9.86 2.7×10^4	Cu $^{+2}$ 7.53 2×10^8	Zn $^{+2}$ 8.96 5×10^8	Ga $^{+3}$ 2.6 7.6×10^2	Ge $^{+4}$ — —	As — —
Rb — —	Sr — —	Y $^{+3}$ 7.7 1.3×10^7	Zr $^{+4}$ 0.22 —	Nb $^{+5}$ (0.6) —	Mo $^{+5}$ — —	Tc — —	Ru $^{+3}$ — —	Rh $^{+3}$ 3.4 3×10^{-8}	Pd $^{+2}$ 2.3 —	Ag $^{+1}$ 12 $>5 \times 10^6$	Cd $^{+2}$ 10.08 $>1 \times 10^8$	In $^{+3}$ 4.00 4.0×10^4	Sn $^{+4}$ — —	Sb $^{+5}$ — —
Cs — —	Ba $^{+2}$ 13.47 $>6 \times 10^7$	Ln $^{+3}$ 7.6 – 8.5 $10^6 – 10^8$	Hf $^{+4}$ 0.25 —	Ta $^{+5}$ (–1) —	W $^{+6}$ — —	Re $^{+5}$ — —	Os $^{+3}$ — —	Ir $^{+3}$ — —	Pt $^{+2}$ 4.8 —	Au $^{+1}$ — —	Hg $^{+2}$ 3.40 2×10^9	Tl $^{+3}$ 0.62 7×10^5	Pb $^{+2}$ 7.71 7.5×10^9	Bi $^{+3}$ 1.09 —

La $^{+3}$ 8.5 2.1×10^8	Ce $^{+3}$ 8.3 2.7×10^8	Pr $^{+3}$ 8.1 3.1×10^8	Nd $^{+3}$ 8.0 3.9×10^8	Pm — —	Sm $^{+3}$ 7.9 5.9×10^8	Eu $^{+3}$ 7.8 6.5×10^8	Gd $^{+3}$ 8.0 6.3×10^7	Tb $^{+3}$ 7.9 7.8×10^7	Dy $^{+3}$ 8.0 6.3×10^7	Ho $^{+3}$ 8.0 6.1×10^7	Er $^{+3}$ 7.9 1.4×10^8	Tm $^{+3}$ 7.7 6.4×10^6	Yb $^{+3}$ 7.7 8×10^7	Lu $^{+3}$ 7.6 6×10^7

[a] $pK_h = -\log K_h$. Reference 6a,b. [b] Exchange rate constants for substitution of inner-sphere water ligands. Reference 6c.

FIGURE 15.1 (See color insert following page 40). Hydrolysis constants (Kh) and water exchange rate constants for substitution of inner-sphere water ligands for metal cations.

15.3 LEWIS ACID–SURFACTANT COMBINED CATALYSTS

To achieve truly environmentally benign chemistry, water should be used as a sole solvent. If that is to be the case, then one big issue remains to be solved in this chemistry. How can you dissolve organic materials in water? In general, most organic compounds are insoluble in water, and therefore reaction media were required.

Lewis acid–surfactant combined catalysts (LASCs) are possible solutions to address this issue. LASCs designed from water-compatible Lewis acids are expected to act as surfactants as well as Lewis acids in water. One example of LASCs that can be readily prepared from scandium chloride and sodium trisdodecylsulfate in water is scandium trisdodecylsulfate ($[Sc(DS)_3]$), shown in Figure 15.2.

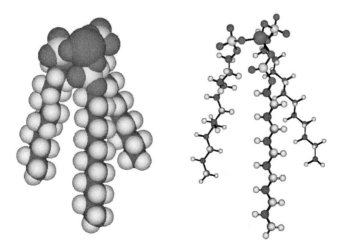

FIGURE 15.2 (See color insert following page 40). $Sc(O_3SOC_{12}H_{25})_3$.

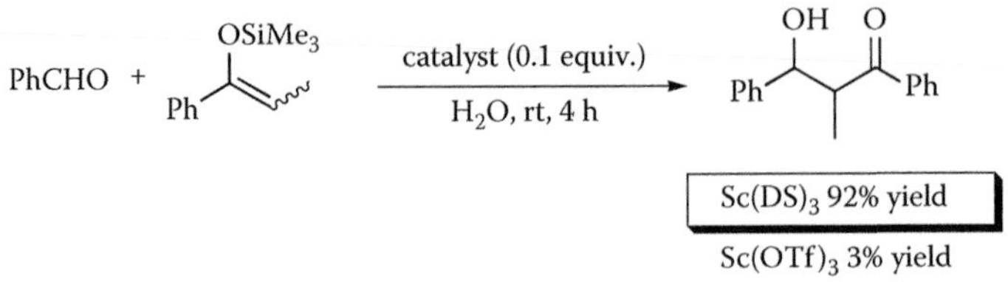

PhCHO + [OSiMe₃ / Ph structure] → catalyst (0.1 equiv.) / H₂O, rt, 4 h → [OH O / Ph — Ph structure]

Sc(DS)₃ 92% yield

Sc(OTf)₃ 3% yield

SCHEME 15.3

TABLE 15.1
Effect of Solvents

Solvent	Yield (%)	Solvent	Yield (%)
H₂O	92	THF	Trace
MeOH	4	Et₂O	Trace
Dimethylformamide	14	Toluene	Trace
Dimethyl sulfoxide	9	Hexane	4
MeCN	3	(neat)	31
CH₂Cl₂	3		

In fact, $Sc(DS)_3$ worked very efficiently in the aldol reaction of benzaldehyde with the silyl enol ether derived from propiophenone in water, whereas the reaction proceeded sluggishly when $Sc(OTf)_3$ was the catalyst (Scheme 15.3).[8]

A kinetic study on an initial rate of this reaction revealed that the reaction in water was nearly 100 times faster than that in dichloromethane. By comparison, under neat conditions, the reaction proceeded much slower, giving a low yield (Table 15.1).

A key to the success of this system was assumed to be the formation of stable emulsions. The size and the shape of emulsion droplets were examined by using transmission electron microscopy, which showed that a mere 0.08 mol% of $Sc(DS)_3$ was sufficient to form monolayers (Figure 15.3).

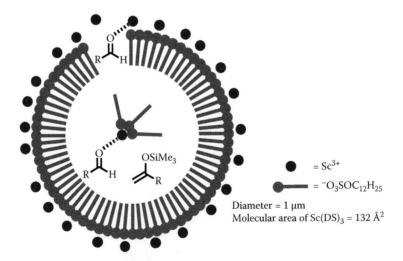

= Sc³⁺

= ⁻O₃SOC₁₂H₂₅

Diameter = 1 μm
Molecular area of Sc(DS)₃ = 132 Å²

FIGURE 15.3 (See color insert following page 40). A hydrophobic particle.

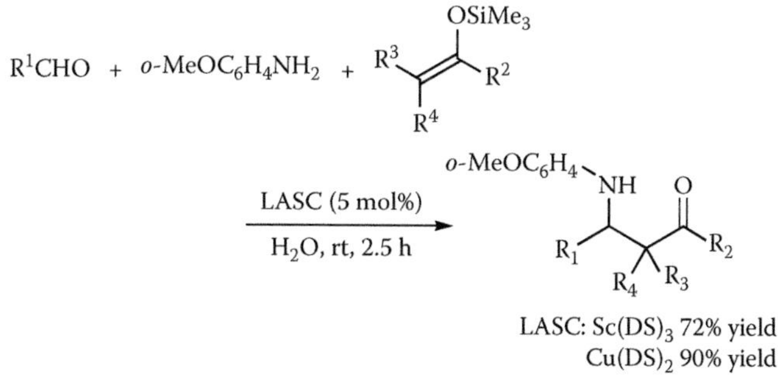

$$R^1CHO \ + \ o\text{-MeOC}_6H_4NH_2 \ + \ \overset{OSiMe_3}{\underset{R^4}{R^3}\diagup\diagdown R^2}$$

$$\xrightarrow[\text{H}_2\text{O, rt, 2.5 h}]{\text{LASC (5 mol\%)}}$$

(product structure with o-MeOC$_6$H$_4$—NH, R$_1$, R$_2$, R$_3$, R$_4$, O)

LASC: Sc(DS)$_3$ 72% yield
Cu(DS)$_2$ 90% yield

SCHEME 15.4

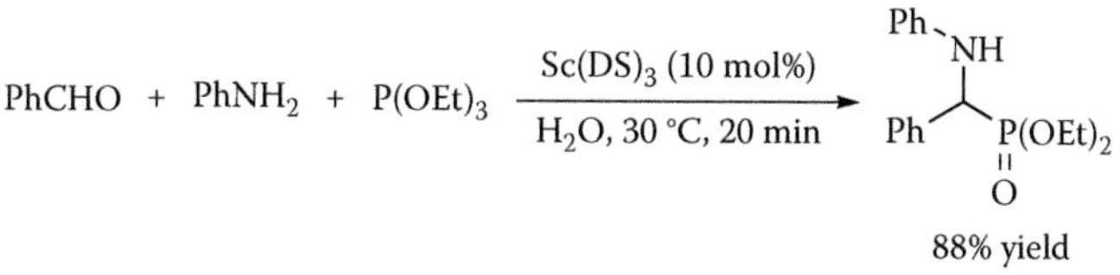

$$PhCHO \ + \ PhNH_2 \ + \ P(OEt)_3 \ \xrightarrow[\text{H}_2\text{O, 30 °C, 20 min}]{\text{Sc(DS)}_3 \ (10 \ \text{mol\%})} \ $$

(product: Ph—NH, Ph, P(OEt)$_2$, O)

88% yield

SCHEME 15.5

Based on these observations, the advantageous effect of water is attributed to the following factors:

1. Concentration of substrates and catalysts
2. Defense of hydrolysis of silyl enol ethers in hydrophobic environments formed by water and LASCs
3. High catalytic turnover caused by hydrolysis of scandium aldolates

Several LASC-catalyzed reactions in water have been developed based on these concepts. Several examples are summarized in Scheme 15.4 through Scheme 15.7:

1. Mannich-type reactions[9] (Scheme 15.4)
2. α-Aminophosphate synthesis[10] (Scheme 15.5)
3. Michael reaction[11] (Scheme 15.6)
4. Friedel–Crafts-type reaction[12] (Scheme 15.7)

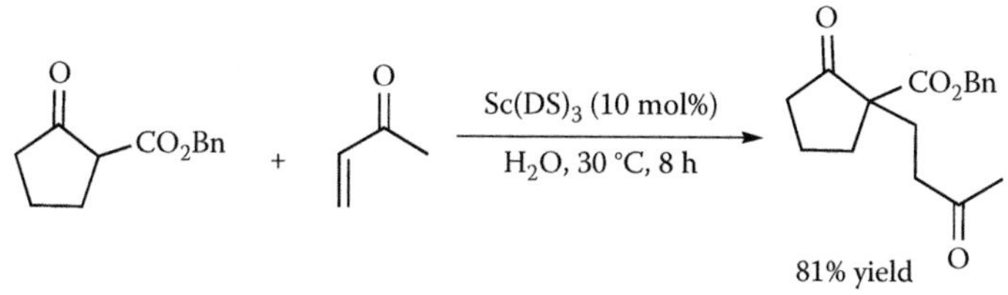

(reaction scheme: cyclopentanone with CO$_2$Bn + methyl vinyl ketone)

$$\xrightarrow[\text{H}_2\text{O, 30 °C, 8 h}]{\text{Sc(DS)}_3 \ (10 \ \text{mol\%})}$$

81% yield

SCHEME 15.6

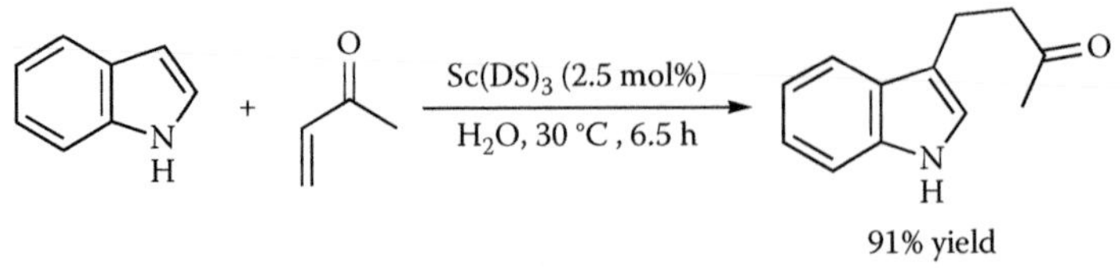

91% yield

SCHEME 15.7

$$CH_3(CH_2)_{10}CO_2H \ + \ CH_3(CH_2)_{13}OH$$

$$\xrightarrow[\text{H}_2\text{O, 40 °C, 48 h}]{\text{DBSA (10 mol\%)}} \ CH_3(CH_2)_{10}CO_2(CH_2)_{13}CH_3$$

>99% yield

SCHEME 15.8

15.4 BRØNSTED ACID–SURFACTANT COMBINED CATALYSTS

Dehydration is one of the most fundamental and still very important reactions in organic synthesis as well as in biology. A representative dehydration is an esterification reaction. Esterification of carboxylic acids is a very simple and very useful reaction in synthetic organic chemistry.[13] Generally, direct esterification of carboxylic acids with alcohols is carried out in organic solvents and requires one of two methods to shift the equilibrium between reactants and products. One method is the removal of water (azeotropically or using dehydrating agents) that is generated as the reaction proceeds, and the other is the use of a large excess of one of the reactants. Conversely, a new approach allows the esterification to be carried out, even in water, without using a large excess of reactants (Scheme 15.8). The direct esterification is a dehydration step, and it is remarkable that dehydration reactions in water proceeded smoothly in the presence of a catalytic amount of Brønsted acid such as dodecylbenzenesulfonic acid (DBSA).

The esterification is successful primarily because the surfactant-type catalysts and organic substrates (carboxylic acids and alcohols) in water form droplets whose interiors is hydrophobic. The surfactants concentrate a catalytic species such as a proton onto the droplet surface, where the reaction takes place.

When a 1:1 mixture of lauric acid and acetic acid was esterified with dodecanol in the presence of DBSA in water, the laurate ester was predominantly obtained in 81% yield, whereas yield in neat conditions was 63% (Scheme 15.9).

This unique selectivity is attributed to the hydrophobic nature of lauric acid and to the high hydrophilicity of acetic acid, and this result led to investigations of trans esterification (Scheme 15.10 and Scheme 15.11), etherification (Scheme 15.12), and thioacetalization (Scheme 15.13) in water.[14] Note again that dehydration proceeded smoothly in water in these reactions.

$$CH_3(CH_2)_{10}CO_2H \quad + \quad CH_3CO_2H \quad + \quad HOCH_3$$

$$(1:1:1)$$

$$\xrightarrow[\text{H}_2\text{O, 40 °C, 48 h}]{\text{DBSA (10 mol\%)}} \quad \begin{array}{l} CH_3(CH_2)_{10}CO_2CH_3 \ 81\% \ (63\%) \\ CH_3CO_2(CH_2)_{10}CH_3 \ 4\% \ (35\%) \end{array}$$

( ): neat conditions

SCHEME 15.9

$$CH_3(CH_2)_{10}CO_2CH_3 \; + \; CH_3(CH_2)_{11}OH$$

$$\xrightarrow[\text{H}_2\text{O, 40 °C, 48 h}]{\text{DBSA (10 mol\%)}} CH_3(CH_2)_{10}CO_2(CH_2)_{11}CH_3$$

>90% yield

SCHEME 15.10

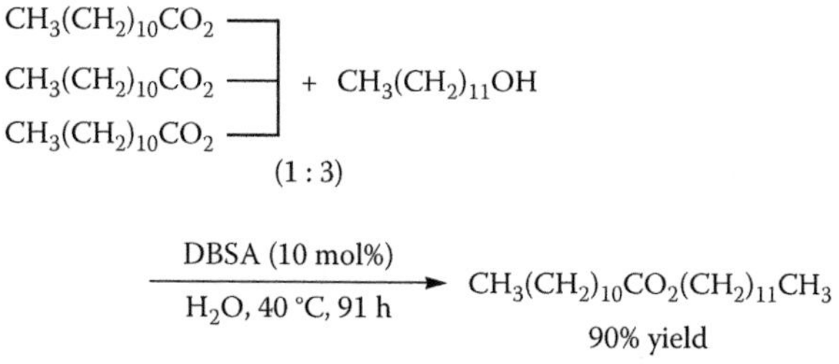

$$\left.\begin{array}{l} CH_3(CH_2)_{10}CO_2 \text{—} \\ CH_3(CH_2)_{10}CO_2 \text{—} \\ CH_3(CH_2)_{10}CO_2 \text{—} \end{array}\right] \; + \; CH_3(CH_2)_{11}OH$$

(1 : 3)

$$\xrightarrow[\text{H}_2\text{O, 40 °C, 91 h}]{\text{DBSA (10 mol\%)}} CH_3(CH_2)_{10}CO_2(CH_2)_{11}CH_3$$

90% yield

SCHEME 15.11

$$R^1OH \; + \; R^2OH \; \xrightarrow[\text{H}_2\text{O}]{\text{DBSA}} \; R^1OR^2$$

SCHEME 15.12

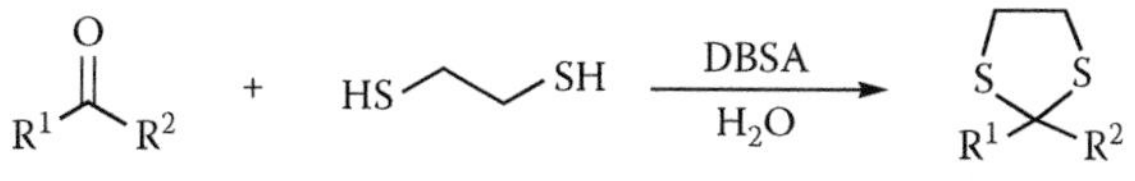

SCHEME 15.13

Several Brønsted acids as catalysts were tested in a model Mannich-type reaction in water, and DBSA afforded the desired product in high yield (Table 15.2, Scheme 15.14). Note that *p*-toluenesulfonic acid, which has a shorter alkyl chain than DBSA, gave only a trace amount of the product. This result suggests that (1) the long alkyl chain of the acid was indispensable for efficient catalysis, probably owing to the formation of hydrophobic reaction environments, and that (2) the strong acidity of DBSA was essential for the catalysis because a carboxylic acid having a long alkyl chain, lauric acid, was much less effective than DBSA.

TABLE 15.2
Effect of Brønsted Acids

Brønsted Acid	Yield (%)
Dodecylbenzenesulfonic acid	83
TsOH	Trace
$C_{11}H_{25}COOH$	6

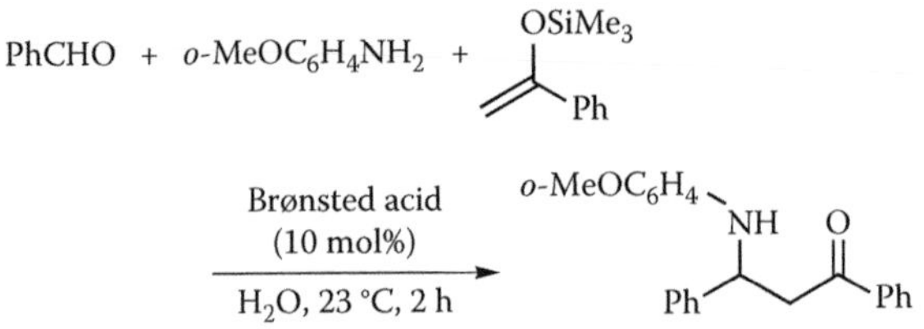

PhCHO + o-MeOC$_6$H$_4$NH$_2$ + [structure: OSiMe$_3$, Ph]

$$\xrightarrow[\text{H}_2\text{O, 23 °C, 2 h}]{\substack{\text{Brønsted acid} \\ \text{(10 mol%)}}}$$

[structure: o-MeOC$_6$H$_4$—NH, O, Ph, Ph]

SCHEME 15.14

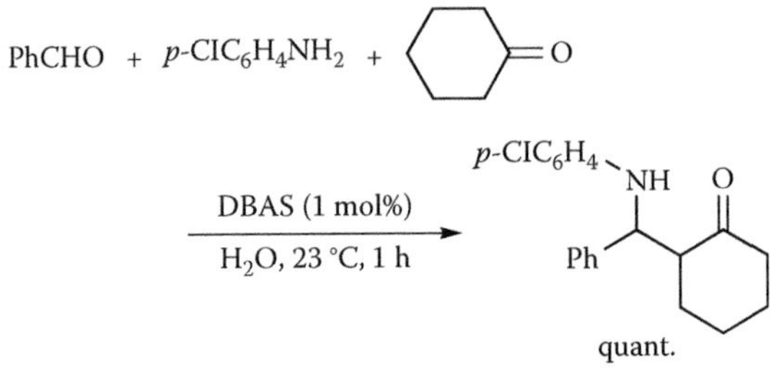

PhCHO + p-ClC$_6$H$_4$NH$_2$ + [cyclohexanone structure =O]

$$\xrightarrow[\text{H}_2\text{O, 23 °C, 1 h}]{\text{DBAS (1 mol%)}}$$

[structure: p-ClC$_6$H$_4$—NH, O, Ph, cyclohexyl]

quant.

SCHEME 15.15

From atom economy and environmental points of view, development of a new, efficient system for Mannich-type reactions in which the parent carbonyl compounds are directly used is desirable.[15] Remarkably, DBSA catalyzes Mannich-type reactions in a colloidal dispersion system using ketones as nucleophilic components (Scheme 15.15).[16]

15.5 ASYMMETRIC CATALYSIS

15.5.1 HYDROXYMETHYLATION

Due to increasing demands for optically active compounds, many catalytic asymmetric reactions have been investigated in this decade. Asymmetric catalysis in water or water/organic solvent systems is difficult, however, because many chiral catalysts are unstable in the presence of water.[17] In particular, chiral Lewis acid catalysis in aqueous media is extremely difficult because most chiral Lewis acids decompose rapidly in the presence of water.[18] To address this issue, catalytic asymmetric reactions using water-compatible Lewis acids with chiral ligands were developed (Figure 15.4).[19]

Formaldehyde is one of the most important C1 electrophiles in organic synthesis. Although hydroxymethylation of enolates with formaldehyde provides an efficient method to introduce a C1 functional group at the α-position of carbonyl groups, there have been few successful examples of catalytic asymmetric hydroxymethylation, which satisfies synthetic utility in terms of both yield and selectivity for a wide range of substrates.[20] To achieve such reactions, Lewis acid-catalyzed hydroxymethylation of silicon enolates[21] is promising. The reactions proceed regioselectively, and excellent substrate generality and synthetic efficiency can be expected.[22] As for the source of formaldehyde, use of a commercial aqueous solution of formaldehyde is the most convenient and avoids tedious and harmful procedures necessary to generate formaldehyde monomer from form-aldehyde oligomers, such as paraformaldehyde and trioxane.[23] Although it was previously reported that an aqueous solution of formaldehyde was successfully used for hydroxymethylation of silicon enolates **4**,[24] it is still difficult to realize catalytic asymmetric versions of this reaction. Quite recently, the catalytic asymmetric hydroxymethylation of silicon enolates in aqueous solvents was achieved.[25,26] In both cases, however, the enantioselectivities were moderate, and several issues

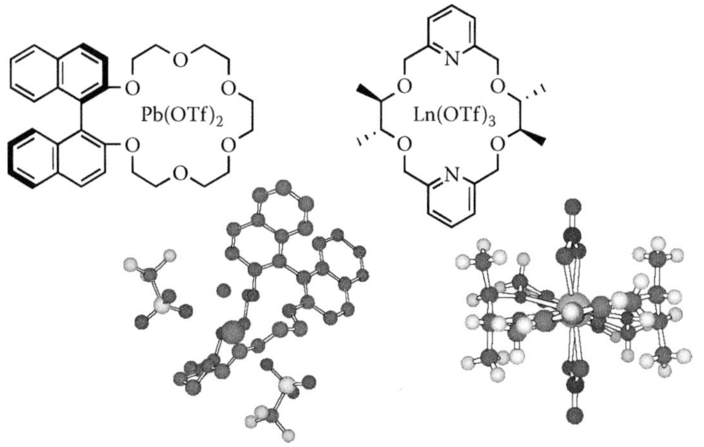

FIGURE 15.4 (See color insert following page 40). Chiral Pb and Ln catalysts for asymmetric Mukaiyama aldol reactions.

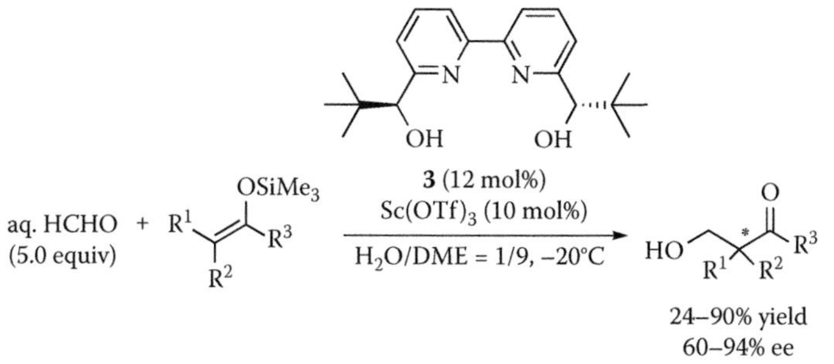

SCHEME 15.16

remained to be resolved. Achieving higher yields and selectivity in this reaction requires the development of a new catalytic system.

Catalytic asymmetric hydroxymethylation of silicon enolates using a Sc(OTf)$_3$·**3** complex as the catalyst was achieved (Scheme 15.16).[27a] In this reaction, a commercial aqueous solution of formaldehyde can be used, and as a result, this process can be conducted very easily and safely. This new catalytic system provides not only a useful method to synthesize optically active β-hydroxymethylated carbonyl compounds but also a guide to various kinds of catalytic asymmetric C–C bond-forming reactions in aqueous media.

Single crystals suitable for X-ray analysis were obtained from a ScBr$_3$·**3** complex (Figure 15.5).[27b] The complex adopts a pentagonal bipyramidal structure[28] in which the hydroxy groups of chiral bipyridine **3** coordinate to Sc^{3+} in a tetradentate manner. Formation of this type of structure may be a key for obtaining high enantioselectivity. In addition, on considering the absolute configurations of some of the hydroxymethylated products,[16,29] it is clear that formaldehyde tends to react with the same face of the silicon enolates in no relation to the substituents at the -position. Therefore, although the details are still unclear, it was assumed that the reactions proceed not via extended, acyclic transition states but via transition states in which the oxygen of a silicon enolate interacts with the chiral complex.

As an extension of this work, other metal salts (10 mol%) and chiral bipyridine **3** (12 mol%) were tested in the reaction of silicon enolate **2** with an aqueous formaldehyde solution, and remarkably it was found that Bi(OTf)$_3$[30] gave promising results. The result was also unexpected:

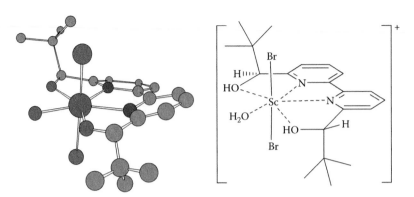

FIGURE 15.5 (See color insert following page 40). $[3 \cdot ScBr_2 \cdot H_2O]^+$ moiety in the X-ray structure of $[3 \cdot ScBr_2 \cdot H_2O] \cdot Br \cdot H_2O$. Hydrogen atoms are omitted for clarity.

there is a big difference in the ionic diameters between bismuth (2.34 for 8-coordination) and scandium (1.74 for 8-coordination), and because $Bi(OTf)_3$ is known to be hydrolyzed in the presence of water.[31] Indeed, only a trace amount of the hydroxymethylated adduct was obtained using $Bi(OTf)_3$ in the absence of the chiral bipyridine,[32] because it is known that silicon enolates such as **2** are rapidly decomposed by TfOH, which was easily generated from $Bi(OTf)_3$ in water. Conversely, decomposition of silicon enolate **2** was slow, and the desired hydroxymethylation proceeded in the presence of $Bi(OTf)_3$ and **3**. These results indicate that $Bi(OTf)_3$ was stabilized by chiral bipyridine **3** in water. Using $Bi(OTf)_3$ as a chiral bismuth catalyst, it was revealed that the desired product was obtained in 94% yield with 91% enantiomeric excess (using 1 mol% $Bi(OTf)_3$ and 3 mol% **3** in the presence of 5 mol% of 2,2-bipyridine). Several other substrates were applicable to this catalyst system.[33] The hydroxymethylation proceeded smoothly using an aqueous formaldehyde solution to afford the desired adducts in high yields with high enantioselectivities (Scheme 15.17). It is noteworthy that asymmetric quaternary carbons were constructed with high selectivities.

Data from several experiments revealed that the active catalyst was formed from an equimolar mixture of $Bi(OTf)_3$ and **3**. An X-ray crystal structure of the $BiBr_3 \cdot$**3** complex is shown in Figure 15.6. The complex adopts a pentagonal, bipyramidal structure in which the tetradentate ligand occupies four of the equatorial sites. The structure of the $Bi(III)Br_3$ complex of **3** is closely related to that of the corresponding $Sc(III)Br_3$ complex.[34] Nuclear magnetic resonance (NMR) analysis indicates that an active complex consisting of 1 equiv of $Bi(OTf)_3$ and 1 equiv of **3** was generated when an excess amount of **3** was added.

This work provides a new entry to water-compatible Lewis acids. For a long time, Lewis acids were believed to decompose in the presence of water. Contrary to this, we found that rare earth and other metal complexes are water compatible. In addition, we added $Bi(OTf)_3 \cdot$**3** complex as a water-compatible Lewis acid. It is noteworthy that $Bi(OTf)_3$ is unstable in the presence of water

aq. HCHO (5.0 equiv) + [OSiMe₃ enol ether with R¹, R², R³] $\xrightarrow[\text{H}_2\text{O/DME} = 1/4, \, 0°\text{C}]{\substack{\textbf{3} \, (3 \, \text{mol}\%) \\ Bi(OTf)_3 \, (1 \, \text{mol}\%) \\ Bipy \, (5 \, \text{mol}\%)}}$ [product HO–R¹R²–C*–C(=O)R³]

59–93% yield
77–95% ee

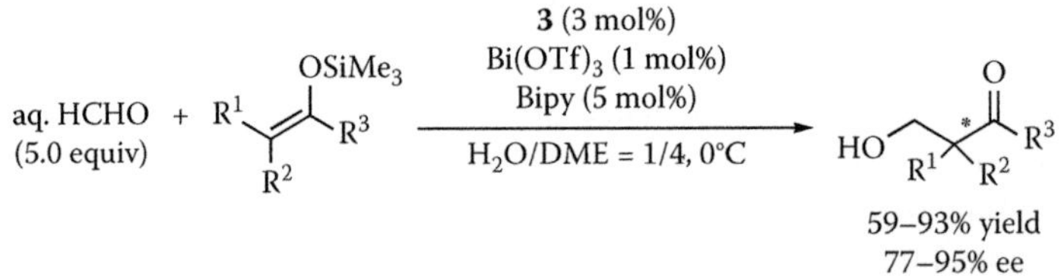

SCHEME 15.17

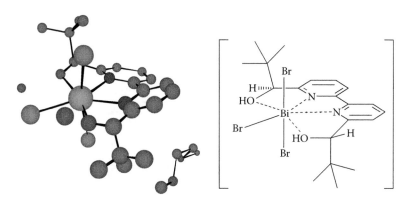

FIGURE 15.6 (See color insert following page 40). The X-ray crystal structure of [BiBr₃·1]·(H₂O)₂·dimethyl ether. Dimethyl ether is omitted for clarity.

but is stabilized by the basic ligand. Although discovery of water-compatible Lewis acids has greatly expanded their use in organic synthesis in aqueous media, conventional Lewis acids such as Al(III), Ti(IV), Sn(IV), still cannot be used in aqueous media under standard conditions. This restriction has been partially resolved by recent works.

15.5.2 Epoxide Ring Opening

Chiral β-amino alcohol units are found in many biologically active compounds and chiral auxiliaries/ligands used in asymmetric reactions.[35] Catalytic enantioselective synthesis of these chiral building blocks mainly relies on asymmetric ring opening of *meso*-epoxides. Several examples or reactions using a chiral catalyst (typically a chiral Lewis acid) are reported in the literature[36]; however, all of these reactions proceeded in organic solvents. Epoxides likely readily decompose under acidic conditions in water. Although Sc(OTf)₃·3 complex works as a chiral Lewis acid, as shown in Section 15.5.1), to extend the use of this novel chiral scandium complex to other reactions in water, the asymmetric ring opening of *cis*-stilbene oxide with aniline in water was investigated. The reaction proceeded smoothly in high yield with high enantioselectivity using 1 mol% of Sc(DS)₃ as a Lewis acid–surfactant combined catalyst[37] and 1.2 mol% of **3** in water (Scheme 15.18). Note that the ring-opening reaction proceeded smoothly in water and that no diol formation was observed. In general, even a trace amount of water has a detrimental effect on yield and enantioselectivity, and only a few examples of enantioselective Lewis acid-catalyzed reactions in pure water have been reported.[38] To the best of our knowledge, this example is the first to date of an asymmetric epoxide ring opening in pure water.[39–41]

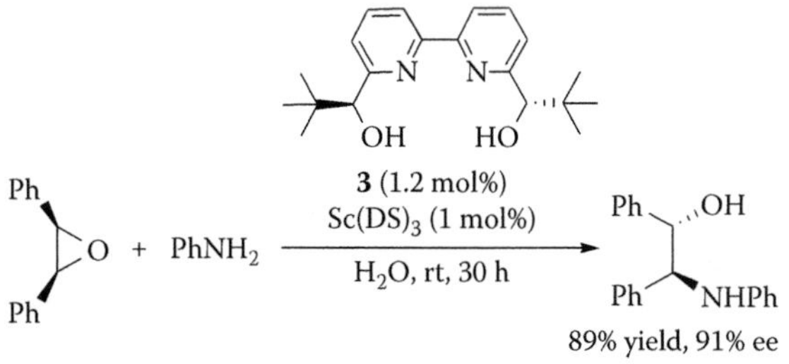

SCHEME 15.18

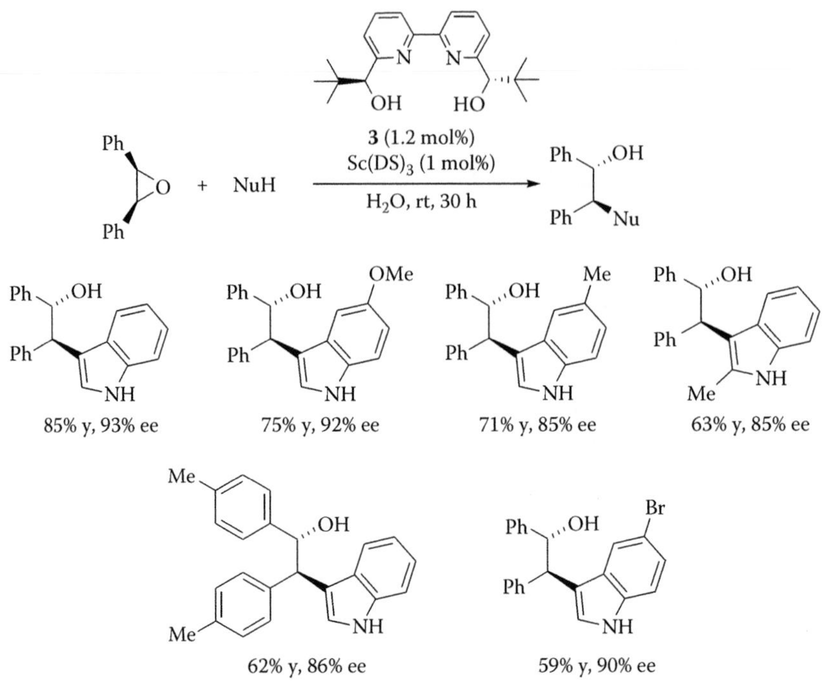

SCHEME 15.19

Moreover, catalytic asymmetric ring-opening reactions of *meso*-epoxides with indoles, alcohols, and thiols proceeded smoothly in the presence of catalytic amounts of Sc(DS)$_3$ and chiral bipyridine ligand **3** in water to afford β-amino alcohols in high yield and enantioselectivity (Scheme 15.19 and Scheme 15.20).[42] Note that an excellent hydrophobic, asymmetric environment has been created in water.

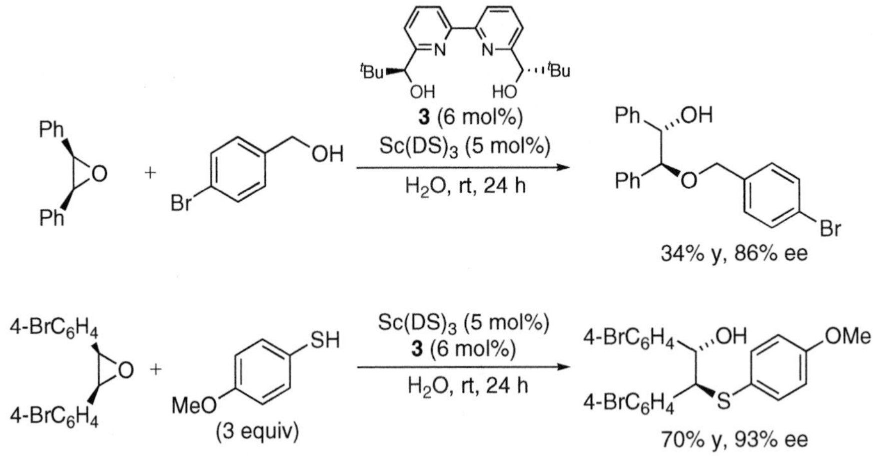

SCHEME 15.20

Catalytic asymmetric ring-opening reactions of *meso*-epoxides with aromatic amines also proceeded in the presence of a catalytic amount of bismuth triflate (Bi(OTf)₃) **3** and SDBS in pure water to give the corresponding β-amino alcohols in good yield with high enantioselectivity.[43]

15.5.3 MANNICH-TYPE REACTION

Asymmetric Mannich reactions provide useful routes for the synthesis of optically active β-amino ketones and esters, which are versatile chiral building blocks for the preparation of many nitrogen-containing biologically important compounds.[44] In recent years, various enantioselective Mannich reactions have been developed. Among them, catalytic enantioselective additions of silicon enolates to imines have been elaborated into one of the most powerful and efficient asymmetric Mannich-type reactions, primarily because silicon enolates can be prepared regio- and stereoselectively from various carbonyl compounds.[21]

Several examples of catalytic, asymmetric C–C bond-forming reactions catalyzed by water-compatible Lewis acids in aqueous media were reported. Realizing catalytic asymmetric Mannich-type reactions in aqueous media has been difficult, but in 2002 the first catalytic asymmetric Mannich-type reactions of an -hydrazono ester with silicon enolates in H_2O/THF by combining a stoichiometric amount of zinc fluoride and a catalytic amount of a chiral diamine and trifluoromethanesulfonic acid (TfOH) were revealed.[45,46] Furthermore, it was also found that in water without any organic cosolvents, these Mannich-type reactions proceeded in high yield with high stereoselectivity with ZnF_2, a cationic surfactant, and a chiral diamine having MeO groups on its aromatic rings.[47]

Imines are usually used as electrophiles in Mannich reactions.[48] Although some imines are easily prepared from the corresponding carbonyl compounds and amines, most necessitate dehydrative preparation using azeotropic distillation or dehydrating agents. In addition, imines are generally difficult to purify using distillation or column chromatography and unstable when stored for long periods. By contrast, N-acylhydrazones[49] are readily prepared from aldehydes and N-acylhydrazines and often isolated as much more stable crystals than the corresponding imines. Recently, we found that such electrophiles reacted smoothly with several nucleophiles in the presence of a catalytic amount of a Lewis acid.[50,51] Note that hydrazines such as the products of the Mannich reaction or allylation are interesting compounds not only because hydrazines can themselves be used as unique building blocks,[52] but also because N–N bond cleavage would lead to amine products. Furthermore, N-acylhydrazones have been successfully used in Sc(OTf)₃-catalyzed allylation in aqueous THF,[53] indicating that they can be regarded as imine surrogates stable even in aqueous media. Therefore, we decided to examine the catalytic asymmetric Mannich-type reactions of N-acylhydrazone with silicon enolates in aqueous media.

Diastereo- and enantioselective Mannich-type reactions of α-hydrazono ester 4 with silicon enolates in aqueous media have been achieved with a ZnF_2–chiral diamine complex (Scheme 15.21). This reaction seems to proceed with double activation in which Zn^{2+} acts as a Lewis acid to activate **4**, and fluoride anion acts as a Lewis base to activate silicon enolates. Both Zn^{2+} and fluoride anion were needed to obtain high yields and high enantioselectivities. The effect of the diamines **5a** having MeO groups on their aromatic rings is noteworthy, and these diamines provide the following advantages:

1. The reactions in aqueous THF were remarkably accelerated.
2. The reactions with some silicon enolates proceeded in high yields even in the absence of TfOH or NaOTf, which was needed in the reactions using **5b**.
3. The ZnF_2 loading could be reduced to 10 to 20 mol% without loss of yield and enantioselectivity, whereas more than 50 mol% of ZnF_2 was required in the reactions using **1a**.
4. The reactions in water without any organic cosolvents proceeded smoothly to give high yields and high stereoselectivities, and in contrast to most asymmetric Mannich-type

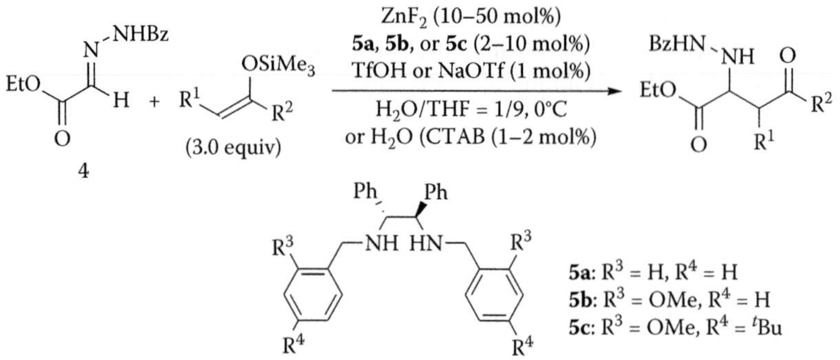

SCHEME 15.21

reactions, either syn or anti adducts were stereospecifically obtained from *(E)*- or *(Z)*-silicon enolates in the present reaction.

As for the reaction mechanism, some experimental evidence suggests that the ZnF_2–chiral diamine complex is the real catalytic active species and that it is regenerated from Me_3SiF that forms as the reaction progresses (Scheme 15.22, fluoride-catalyzed mechanism).

Finally, the present reaction proceeds smoothly only in water, without using any organic solvents, and in the presence or absence of a small amount of a cationic surfactant, cetyltrimethyl-ammonium bromide (CTAB).[54]

15.4 CONCLUSION

The use of water as a solvent in organic synthesis will play key roles in green chemistry. Despite the importance of Lewis acid-catalyzed reactions in laboratories as well as in industry, however, such reactions have not been carried out in aqueous media, because Lewis acids were believed to hydrolyze rapidly in the presence of water. Contrary to this belief, we found that rare earth and

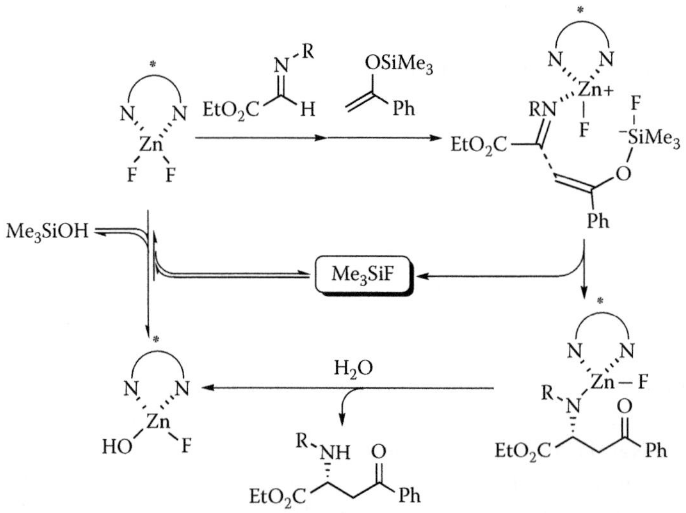

SCHEME 15.22

other metal complexes are water compatible. Moreover, chiral Lewis acid catalysis in aqueous media—that is known to be difficult owing to the instability of both Lewis acids and chiral Lewis acids even in the presence of a small amount of water—has been attained. Hydroxymethylation of silicon enolates with an aqueous solution of formaldehyde in the presence of $Sc(OTf)_3$–chiral bipyridine ligand or $Bi(OTf)_3$–chiral bipyridine ligand, Sc or Bi-catalyzed asymmetric *meso*-epoxide ring-opening reactions with amines, and asymmetric Mannich-type reactions of silicon enolates with N-acylhydrazones in the presence of a chiral Zn catalyst have all been developed. In each case, water plays key roles in these asymmetric reactions. In addition to demonstrating the synthetic utility of these enantioselective reactions, the studies provide a useful guide to the development of catalytic, asymmetric C–C bond-forming reactions in water.

REFERENCES AND NOTES

1. (a) Li, C.-J. and Chan, T.-H. *Organic Reactions in Aqueous Media*, Wiley & Sons, New York, 1997. (b) Grieco, P.A., Ed., *Organic Synthesis in Water*, Blackie Academic and Professional: London, 1998. (c) Lindstroem, U.M. *Chem. Rev.*, 2002, 102, 2751. (d) Sinou, D. *Adv. Synth. Catal.*, 2002, 344, 237. (e) Li, C-J. *Chem. Rev.*, 2005, 105, 3095.

2. (a) Schinzer, D., Ed., *Selectivities in Lewis Acid Promoted Reactions*, Kluwer Academic: Boston, 1989. (b) Yamamoto, H., Ed., *Lewis Acids in Organic Synthesis*, Wiley-VCH: New York, 2000. For an early review on reactions mediated by Lewis acidic metal cations in water, see (c) Hay, R.W. Lewis acid catalysis and the reactions of coordinated ligands, in *Comprehensive Coordination Chemistry*, Wilkinson, G.R., Gillard, D., and McCleverty, J.A., Eds., Pergamon Press: Elmsford, NY, Vol. 6, p. 411.

3. (a) Kobayashi, S. Lanthanide triflate-catalyzed carbon–carbon bond-forming reactions in organic synthesis, in *Lanthanides: Chemistry and Use in Organic Synthesis*, Kobayashi, S., Ed., Springer Publishing: New York, 1999, p. 63. (b) Kobayashi, S. *Eur. J. Org. Chem.*, 1999, 15. (c) Kobayashi, S. *Synlett*, 1994, 689. (d) Kobayashi, S., Sugiura, M., Kitagawa, H., and Lam, W.W.-L. *Chem. Rev.*, 2002, 102, 2227.

4. (a) Kobayashi, S. *Chem. Lett.*, 1991, 2187. (b) Kobayashi, S. and Hachiya, I. *J. Org. Chem.*, 1994, 59, 3590.

5. Kobayashi, S., Nagayama, S., and Busujima, T. *J. Am. Chem. Soc.*, 1998, 120, 8287.

6. (a) Baes, C.F., Jr. and Mesmer, R.E. *The Hydrolysis of Cations*, Wiley & Sons: New York, 1976. (b) Yatsimirksii, K.B. and Vasil'ev, V.P. *Instability Constants of Complex Compounds*, Pergamon: New York, 1960. (c) Martell, A.E., Ed., *Coordination Chemistry ACS Monograph 174*, Vol. 2, American Chemical Society: Washington, DC, 1978.

7. Fringuelli and coworkers reported use of Al(III), Ti(IV), and Sn(IV) as Lewis acids for epoxide opening reactions in acidic water. The pH was adjusted by adding H_2SO_4. Fringuelli, F., Pizzo, F., and Vaccaro, L. *J. Org. Chem.*, 2001, 66, 3554.

8. Manabe, K., et al. *J. Am. Chem. Soc.*, 2000, 122, 7202.

9. Kobayashi, S., Busujima, T., and Nagayama, S. *J. Chem. Soc., Chem. Commun.*, 1998, 19. See also Reference 8.

10. Manabe, K. and Kobayashi, S. *Chem. Commun.* 2000, 669.

11. Mori, Y., Kakumoto, K., Manabe, K., and Kobayashi, S. *Tetrahedron Lett.*, 2000, 41, 3107.

12. Manabe, K., Aoyama, N., and Kobayashi, S. *Adv. Synth. Catal.*, 2001, 343, 174.

13. (a) Manabe, K., Sun, X.-M., and Kobayashi, S. *J. Am. Chem. Soc.*, 2001, 123, 10101. (b) Manabe, K., et al. *J. Am. Chem. Soc.*, 2002, 124, 11971, and references therein.

14. Kobayashi, S., Iimura, S., and Manabe, K. *Chem. Lett.*, 2002, 10.

15. (a) Blatt, A.H. and Gross, N.J. *Org. Chem.* 1964, 29, 3306. (b) Yi, L., Zou, J., Lei, H., Lin, X., and Zhang, M. *Org. Prep. Proceed. Int.* 1991, 23, 673.

16. Manabe, K. and Kobayashi, S. *Org. Lett.*, 1999, 1, 1965.

17. Cornils, B. and Herrmann, W.A., Eds.. *Aqueous-Phase Organometallic Catalysis*, 2nd ed., Wiley-VCH: New York, 2004.

18. Manabe, K. and Kobayashi, S. *Chem. Eur. J.* 2002, 8, 4094.

19. For example, (a) Kobayashi, S., Nagayama, S., and Busujima, T. *Chem. Lett.*, 1999, 71. (b) Nagayama, S. and Kobayashi, S. *J. Am. Chem. Soc.*, 2000, 122, 11531. (c) Hamada, T., Manabe, K., and Kobayashi, S. *Angew. Chem., Int. Ed. Engl.* 2003, 42, 3927.

20. Catalytic asymmetric hydroxymethylation without using silicon enolates: (a) Fujii, M., et al. *Chem. Express*, 1992, 7, 309. (b) Kuwano, R., Miyazaki, H., and Ito, Y. *Chem. Commun.*, 1998, 71. (c) Torii, H., et al. *Angew. Chem., Int. Ed. Engl.*, 2004, 43, 1983.

21. For a review on silicon enolates, see Kobayashi, S., et al. In *Science of Synthesis, Houben-Weyl Methods of Molecular Transformation*, Vol. 4, Bellus, D., Ley, S.V., Noyori, R., Regitz, M., Schaumann, E., Shinkai, I., Thomas, E.J., and Trost, B.M., Eds., Thieme: New York, 2002, p. 317.

22. For reviews on asymmetric aldol reactions, see (a) Machajewski, T.D. and Wong, C.-H. *Angew. Chem., Int. Ed. Engl.*, 2000, 39, 1352. (b) Carreira, E.M. In *Comprehensive Asymmetric Catalysis*, Vol. 3, Jacobsen, E.N., Pflatz, A., and Yamamoto, H., Eds., Springer: New York, 1999, p. 998. (c) Nelson, S.G. *Tetrahedron: Asymmetry*, 1998, 9, 357.

23. Trioxane was used as a formaldehyde surrogate: Mukaiyama, T., Banno, K., and Narasaka, K. *J. Am. Chem. Soc.*, 1974, 96, 7503.

24. For hydroxymethylation of a silicon enolate in aqueous media without any catalysts, see Lubineau, A. and Meyer, E. *Tetrahedron*, 1988, 44, 6065.

25. Manabe, K., et al. *Tetrahedron*, 2003, 59, 10439.

26. Ozawa, N., et al. *Synlett*, 2003, 2219.

27. (a) Ishikawa, S.T., Hamada, T., Manabe,K., Kobayahi, S. *J. Am. Chem. Soc.*, 2004, 126, 12236. We performed the hydroxymethylation of **2** using 20 mol% of an Sc^{3+} source and 24 mol% of 3 in H_2O/1,4-dioxane at 0°C. As a result, $Sc(OTf)_3$ and $ScBr_3$ afforded almost the same results ($Sc(OTf)_3$: (b) 15 h, 86% yield, 84% enantiomeric excess; $ScBr_3$: 22 h, 75% yield, 83% enantiomeric excess).

28. Pentagonal bipyramidal structures have also been reported for $Sc(OTf)_3$ complexes with chiral ligands: (a) Evans, D.A., et al. *J. Am. Chem. Soc.*, 2001, 123, 12095. (b) Evans, D.A., et al. *J. Am. Chem. Soc.*, 2003, 125, 10780.

29. (a) Baliri, P.L., et al. *Enantiomer*, 1998, 3, 357. (b) Miyaoka, H., et al. *Tetrahedron: Asymmetry*, 1999, 10, 3189.

30. Gaspard-Iloughmane, H. and Le Roux, C. *Eur. J. Org. Chem.*, 2004, 2517.

31. Répichet, S., et al *Tetrahedron Lett.*, 2002, 43, 993.

32. $Sc(OTf)_3$ is a water-compatible Lewis acid, and it works well for hydroxymethylation even in the absence of a basic ligand. Kobayashi, S., et al. *Synlett*, 1993, 472.

33. Kobayashi, S., et al. *Org. Lett.*, 2005, 7, 4729.

34. The angle of O–Bi–O is 165°, whereas that of O–Sc–O is 151°. The torsional angle of two pyridines in the Bi complex is 27.0°, and that in the Sc complex is 19.4°. For the Sc complex, see Reference 27(a).

35. For reviews on the asymmetric synthesis and use of vicinal amino alcohols, see (a) Ager, D.J., Prakas, H.I., and Schaad, D.R. *Chem. Rev.*, 1996, 96, 835. (b) Bergmeier, S.C. *Tetrahedron*, 2000, 56, 2561. (c) Yamashita, M., Yamada, K., and Tomioka, K. *Org. Lett.*, 2005, 7, 2369. (d) Kolb, H.C. and Sharpless, K.B. In *Transition Metals for Organic Synthesis*, Beller, M., Bolm, C., Eds., Wiley-VCH: New York, 1998, p. 243.

36. (a) Hou, X.L., et al. *Tetrahedron: Asymmetry*, 1998, 9, 1747. (b) Sagawa, S., et al. *Org. Chem.*, 1999, 64, 4962. (c) Sekine, A., Ohshima, T., and Shibasaki, M. *Tetrahedron*, 2002, 58, 75. (d) Schneider, C., Sreekanth, A.R., and Mai, E. *Angew. Chem., Int. Ed. Engl.*, 2004, 43, 5691. (e) Carrée, F., Gil, R., and Collin, J. *Org. Lett.*, 2005, 7, 1023.

37. Kobayashi, S. and Wakabayashi, T. *Tetrahedron Lett.*, 1998, 39, 5389. See also Reference 8.

38. (a) Otto, S. and Engberts, J.B.F.N. *J. Am. Chem. Soc.*, 1999, 121, 6798. (b) Sinou, D., Rabeyrin, C., and Nguefack, C. *Adv. Synth. Catal.*, 2003, 345, 357. See also Ref. 18.

39. Azoulay, S., Manabe, K., and Kobayashi, S. *Org. Lett.*, 2005, 7, 4593.

40. For examples of racemic epoxide ring opening in water, see (a) Iranpoor, N., Firouzabadi, H., and Shekarize, M. *Org. Biomol. Chem.*, 2003, 1, 724. (b) Fan, R.H. and Hou, X.L. *J. Org. Chem.*, 2003, 68, 726. (c) Ollevier, T. and Lavie-Compain, G. *Tetrahedron Lett.*, 2004, 45, 49.

41. Schneider et al. reported the same reactions in an organic solvent. See Reference 36(d).

42. Boudou, M., Ogawa, C., and Kobayashi, S. *Adv. Synth. Catal.*, 2006, 348, 2585.

43. Ogawa, C., Azoulay, S., and Kobayashi, S. *Heterocycles* 2005, 66, 201.

44. For reviews on asymmetric Mannich reactions, see (a) Kobayashi, S. and Ishitani, H. *Chem. Rev.*, 1999, 99, 1069. (b) Taggi, A.E., Hafez, A.M., and Lectka, T. *Acc. Chem. Res.*, 2003, 36, 10. (c) Kobayashi, S. and Ueno, M. In *Comprehensive Asymmetric Catalysis*, Jacobsen, E.N., Pfaltz, A., Yamamoto, H., Eds., Springer: New York, 2004, supplement 1, chap. 29.5, 143–159.

45. After our first report (Reference 46), other reports on catalytic asymmetric Mannich reactions in aqueous media appeared: (a) Wolfgang, N., et al. *J. Org. Chem.*, 2003, 68, 9624. (b) Ibrahem, I., Casas, J., and Córdova, A., *Angew. Chem.*, 2004, 116, 6690; *Angew. Chem., Int. Ed. Engl.* 2004, 43, 6528.

46. Kobayashi, S., Hamada, T., and Manabe, K. *J. Am. Chem. Soc.* 2002, 124, 5640.

47. Hamada, T., Manabe, K., and Kobayashi, S. *J. Am. Chem. Soc.*, 2004, 126, 7768.

48. For review on Mannich reactions, see Arend, M., Westermann, B., and Risch, N. *Angew. Chem.*, 1998, 110, 1096; *Angew. Chem., Int. Ed. Engl.*, 1998, 37, 1044.

49. Burk reported catalytic asymmetric hydrogenations of N-acylhydrazones: Burk, M.J. and Feaster, J.E. *J. Am. Chem. Soc.*, 1992, 114, 6266.

50. (a) Oyamada, H. and Kobayashi, S. *Synlett*, 1998, 249. (b) Kobayashi, S., Sugita, K., and Oyamada, H. *Synlett*, 1999, 138. (c) Manabe, K., et al. *J. Org. Chem.*, 1999, 64, 8054.

51. Kobayashi, S., Hasegawa, Y., and Ishitani, H. *Chem. Lett.*, 1998, 1131.

52. Hydrazine and its derivatives, in *Kirk-Othmer Encyclopedia of Chemical Technology*, 4th ed., Wiley & Sons: New York, 1995, vol. 13.

53. Kobayashi, S., Hamada, T., and Manabe, K. *Synlett* 2001, 1140.

54. (a) Hamada, T., et al. *J. Am. Chem. Soc.*, 2003, 125, 2989. (b) Hamada, T., Manabe, K., and Kobayashi, S., *Chem. Eur. J.*, 2006, 12, 1205.

16 The Chemical Development of the Commercial Route to Sildenafil Citrate

Peter J. Dunn

CONTENTS

16.1 INTRODUCTION AND BIOLOGY

Sildenafil citrate (the active ingredient in Viagra™) is a selective inhibitor of phosphodiesterase-5 (PDE5) and was the first agent with this mode of action for the treatment of male erectile dysfunction. The drug was filed for approval in September 1997, received priority review from the U.S. Food and Drug Administration (FDA) and was approved within 6 months. Viagra was launched in March 1998 and generated revenues of $1.9 billion in 2003. The mechanism of action for sildenafil citrate is shown in Figure 16.1.[1–3]

Sexual stimulation leads to the release of nitric oxide. This permeates through the membrane of the corpus cavernosum and stimulates the enzyme guanylate cyclase to increase cyclic guanosine monophophate (cGMP) levels in the corpus cavernosum. cGMP is the secondary messenger and acts on the smooth muscle, causing relaxation and this increases blood flow to the penis leading to an erection. There are high levels of PDE5 in the corpus cavernosum, and this enzyme hydrolyzes cGMP to give inactive GMP. Patients suffering from erectile dysfunction have lower levels of cGMP in the corpus cavernosum, and these levels are quickly hydrolyzed by PDE5. Sildenafil binds to PDE5, inhibiting its action, allowing higher levels of cGMP to build up in the corpus cavernosum, leading to an erection. Information regarding the way sildenafil binds to PDE5 is now available; the crystal structure of sildenafil, in the catalytic domain of human PDE5, has been solved (Scheme 16.1).[4]

Base-catalyzed condensation of 2-pentanone and diethyl oxalate gave the diketoester **1**. Cyclization of **1** with hydrazine gave the pyrazole **2**, which could be methylated with good selectivity to give the pyrazole **3**. Subsequent hydrolysis reaction gave the acid **4**. Classical nitration and conversion of the acid to the amide via the acid chloride gave the nitropyrazole **5**. This intermediate is common to all of the synthetic routes discussed in this chapter. The nitropyrazole **5** was reduced

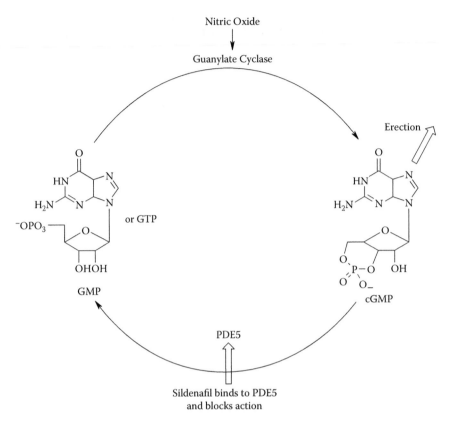

FIGURE 16.1 The mechanism of action for sildenafil. (See also, Boolell, M. et al., *Int. J. Impot. Res.* 8, 47, 1996; Campbell, S.F., *Clinical Science*, 99, 255, 2000; and Brock, G., *Drugs of Today*, 36, 125, 2000.)

with tin (II) chloride dihydrate in refluxing ethanol to give the requisite amine **6** that was acylated with 2-ethoxybenzoyl chloride to give the diamide **7**. Cyclization of **7** using aqueous sodium hydroxide and hydrogen peroxide gave the pyrazolo[4,3-d]pyrimidin-7-one **8**. Subsequent selective chlorosulfonation and reaction with *N*-methylpiperazine gave sildenafil which was converted to its citrate salt.

The source of the regiochemistry for the methylation reaction to give pyrazole **3** was intriguing. Subsequent work in chemical research and development showed that under basic conditions the main product was the *N*-2 regioisomer. Literature reports[6] have shown that 5-alkylpyrazole-3-carboxylates such as **2** exist in the tautomer like **2b**, both in the solid state and solution. Hence, it was proposed[7] that in solution the majority of the pyrazole exists as the *N*-2 isomer **2b** and under acidic or neutral conditions the lone pair of electrons on the *N*-2 isomer is delocalized into the pyrazole ring and reaction takes place on the "pyridine-like" nitrogen to give the *N*-1 alkylated isomer **3** as the main product. Under basic conditions, there is formation of an anion which leads to the more stable *N*-2 isomer (Scheme 16.2).

The medicinal chemistry route was completely linear and produced sildenafil citrate in 3.8% overall yield from 2-pentanone. Although these facts made it suboptimal for commercialization, the route was fully suitable for early project supply. Of course all steps were modified to make them suitable for scale-up. Some of the most important modifications were the following:

- Replacing the tin (II) chloride reduction with a catalytic hydrogenation. It was found that controlling sulfur impurities from the previous step (which involved thionyl chloride)

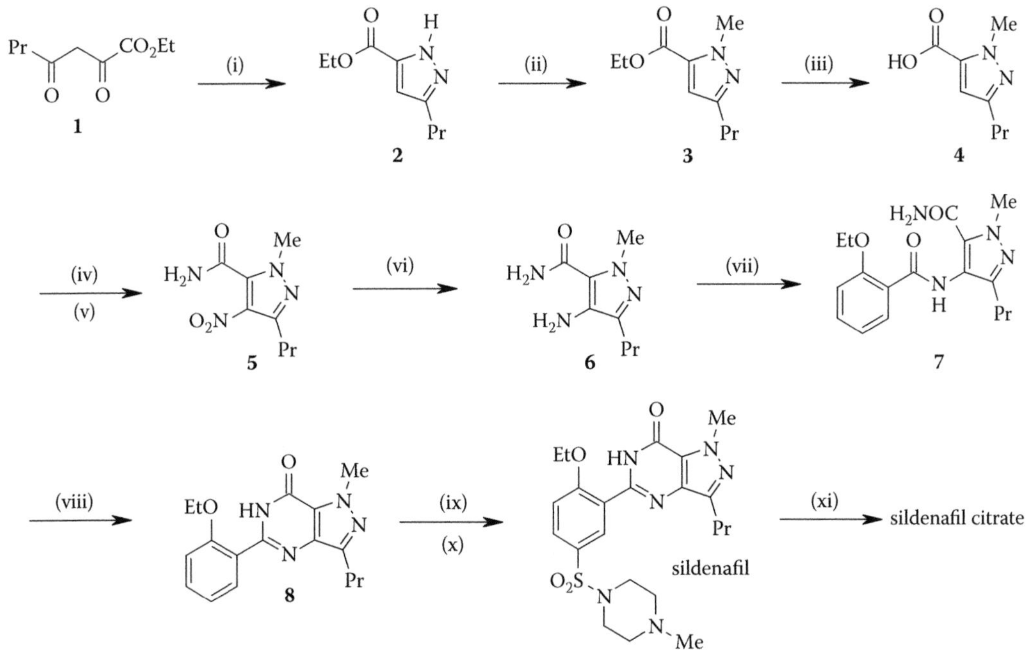

(i) NH_2NH_2, H_2O, 62% (ii) Me_2SO_4, 90 °C, 79% (iii) NaOH, H_2O, 80 °C, 71% (iv) HNO_3, H_2SO_4, 60 °C, 75% (v) $SOCl_3$, reflux, NH_3 (aq), 0 °C, 78% (vi) $SnCl_2$, EtOH, reflux, 94% (vii) 2-ethoxybenzoyl chloride, Et_3N, DMAP, CH_2Cl_2, 0 °C to RT, 40% (viii) NaOH, H_2O_2, H_2O, EtOH reflux, 72% (ix) $ClSO_3H$, 0 °C, 97% (x) N-methylpiperazine, EtOH, 88% (xi) citric acid, acetone (91%), then aqueous acetone recrystallisation (90%).

SCHEME 16.1 The medicinal chemistry route to sildenafil citrate.

was key to making the catalytic hydrogenation robust and reliable. Fortunately, we never had to use the tin (II) chloride reduction method during scale-up.

- Removing the hydrogen peroxide-based cyclization method to convert **7** to **8** and replacing with a $KOBu^t/^tBuOH$ cyclization that proceeded to 100% yield.
- Introducing a solvent into the exothermic methylation reaction to form pyrazole **3**. In the medicinal chemistry procedure, this reaction was performed neat.
- Toluene was introduced as a solvent for the preparation of **5**, whereas medicinal chemistry had made the acid chloride in neat thionyl chloride. This allowed the thionyl chloride level to be reduced from 1.6 to 1.2 equivalents.

With these and other modifications, the medicinal chemistry route was used to maintain supply for the project for 4 years, and 50 kg was prepared with this route. Further information on modifications to the medicinal chemistry route can be found in the literature.[8]

16.2 DEVELOPMENT OF COMMERCIAL MANUFACTURING ROUTE

Sildenafil citrate originally entered development as a candidate for the treatment of angina, but the early clinical results were disappointing. In 1994 Pfizer ran a small phase IIa trial with 12 patients suffering from male erectile dysfunction. Ten of those patients showed major improvements in their erections. Suddenly the project changed overnight. Sildenafil citrate had been the lowest-priority project in the Pfizer development portfolio, and now it became one of the highest. The medicinal

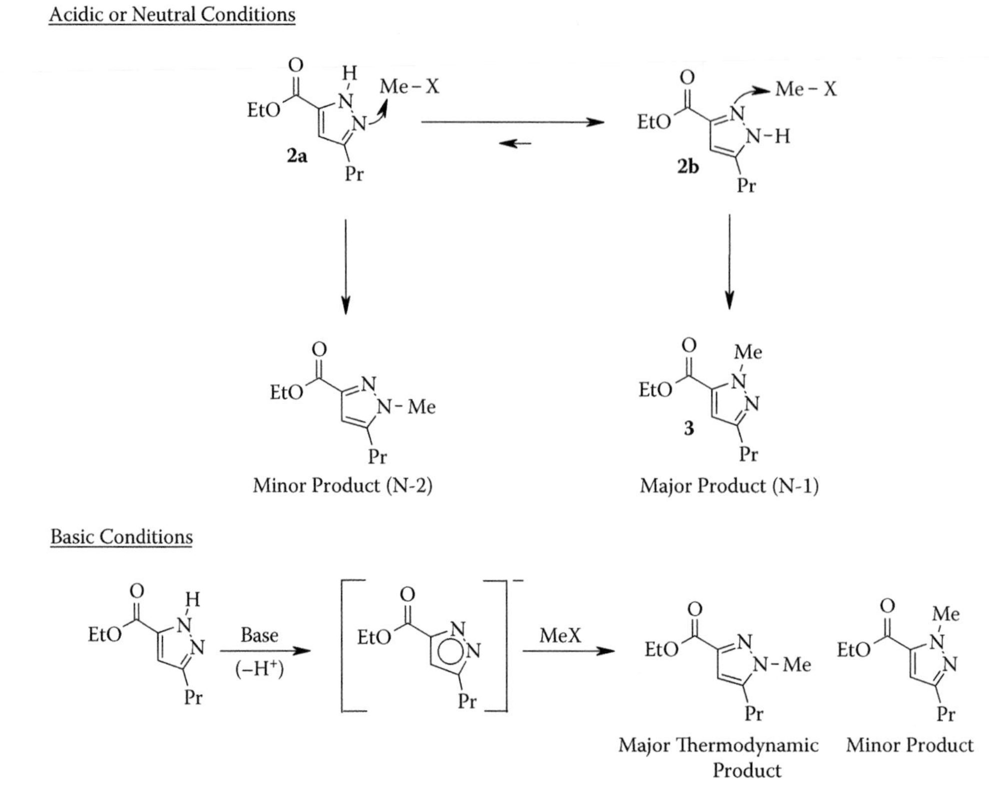
Acidic or Neutral Conditions
EtO
O
H
N
N
Me-X
2a
Pr
EtO
O
N
N-H
Me-X
2b
Pr
EtO
O
N
N-Me
Pr
Minor Product (N-2)
EtO
O
Me
N
N
3
Pr
Major Product (N-1)
Basic Conditions
Base
(−H+)
MeX
Major Thermodynamic Product
Minor Product

SCHEME 16.2 Proposed mechanism for pyrazole alkylation.

chemistry route could not cope with supplying material for a rapidly expanding clinical program. A new route of synthesis was urgently required.[7–9]

Many routes of synthesis proceeded through pyrazole **5**. Hence, a pragmatic approach was taken to nominate this material as the proposed regulatory starting material and focus on chemistry downstream of pyrazole **5**. The commercial route is shown in Scheme 16.3.

Chlorosulfonation of 2-ethoxybenzoic acid was found to be very regioselective; a mole of thionyl chloride needs to be added to ensure the intermediate sulfonic acid is converted to the sulfonyl chloride (see Scheme 16.3). The sulfonyl chloride was converted to sulfonamide **9** by reaction with *N*-methylpiperazine. Initially triethylamine was used as a base, and compound **9** could only be isolated as its hydrochloride-triethylamine double salt, which was very insoluble and difficult to use in subsequent steps. A key breakthrough was obtained when it was discovered that sulfonamide **9** also could be isolated as its highly crystalline zwitterion.

The process was quickly redesigned to make use of this new finding. After completion of the chlorosulfonation, the reaction was quenched to give the sulfonyl chloride, which was collected by filtration. The water-wet sulfonyl chloride was resuspended in water and converted to the sulfonamide by reaction with *N*-methylpiperazine in water. At the end of the reaction, the pH was adjusted to the isoelectric point by the addition of aqueous sodium hydroxide and compound **9** collected by filtration. Hence, the only solvent used to convert 2-ethoxybenzoic acid to the sulfonamide **9** is water.

As previously mentioned, the tin(II) chloride reduction used by medicinal chemistry was replaced by a palladium catalyzed hydrogenation reaction. The amine **6** could be isolated as a crystalline solid or salt, but it was more efficient to use the ethyl acetate solution of **6** directly in the next stage. The activation of the acid **9** was studied using thionyl chloride, oxalyl chloride, and

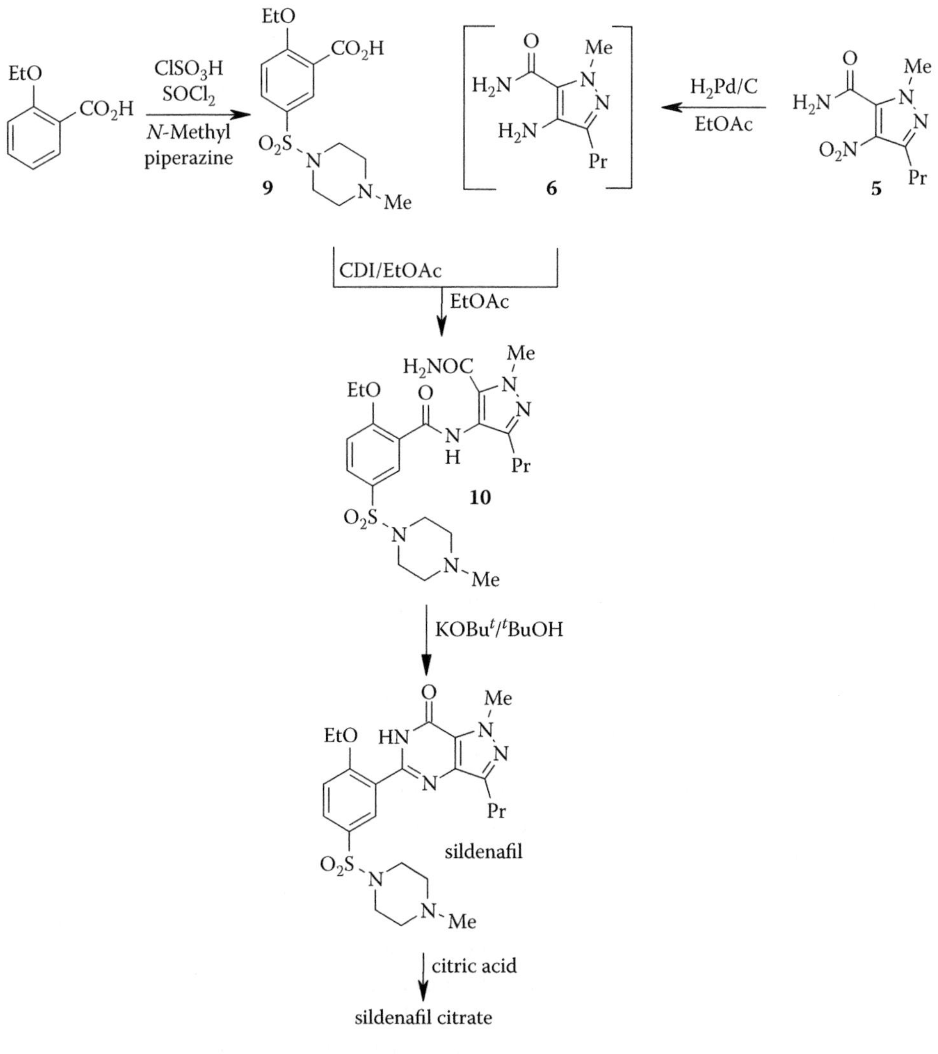

SCHEME 16.3 The commercial route to sildenafil citrate.

N,N'-carbonyl diimidazole (CDI). We wanted to use the same solvent for the activation as for the reduction as this would allow simple processing and efficient solvent recovery. Eventually we selected CDI on the basis that the very high chemical yield using that reagent, 96% yield over three chemical reactions, outweighed the small additional reagent cost. CDI costs around $8/mol. The amine **6** and the imidazolide are both highly soluble in ethyl acetate as is the main by-product imidazole. In contrast, the desired amide **10** had very low solubility. This led to a very simple process where the streams are mixed, reacted, and the product collected by filtration. There are no aqueous workups involved. Subsequently, a name has been coined for this type of efficient process—direct-drop or direct-isolation process.[10] The preparation of compound **9** is another example of a direct-drop process.

The cyclization reaction to convert the amide **10** into sildenafil was carried out by heating **10** for several hours with 1.2 equivalents of potassium t-butoxide in a concentrated solution in t-butanol. On completion of the cyclization, the reaction mixture was diluted with water and acidified with 4 M HCl to the isoelectric point (pH 7.5) to give a 95% yield of very high-quality sildenafil.

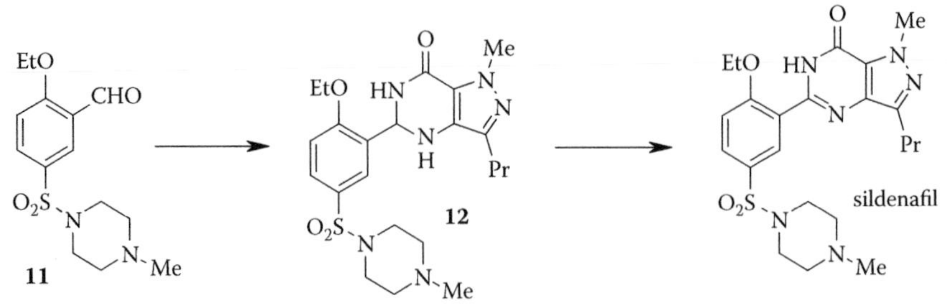

SCHEME 16.4 Alternative route to sildenafil via the aldehyde **11**.

This sildenafil was converted to sildenafil citrate using citric acid in 2-butanone. The process was carefully optimized using a statistical design approach to give a yield of 99 to 100% of sildenafil citrate.

16.3 ALTERNATIVE ROUTES TO SILDENAFIL

More than 15 different routes to sildenafil have been reported in the chemical or patent literature.[7] During the development program, the two main alternatives examined by Pfizer were synthesis via the aldehyde **11** to give dihydrosildenafil **12** followed by subsequent oxidation (Scheme 16.4) or synthesis via the halo derivatives such as **13** or **14** (Scheme 16.5).

Condensation of the aldehyde **11** with the amine **6** yields 52% dihydrosildenafil.[11] Other workers have shown that the yield can be improved to 95% by using an azeotropic distillation to remove the water by-product.[12] Dihydrosildenafil can be oxidized using Pd/C and a small quantity of trifluoroacetic acid at high temperatures or using sodium hydrogen sulfate—either method gives a good yield of sildenafil.[11,12]

Another potential synthesis combines the cyclization reaction of a compound like **13** or **14** with a nucleophilic displacement reaction, preferably using ethanol as solvent and either ethoxide or a hindered alkoxide as a base.[13] The reaction works for both compounds but for the fluoroderivative **14** the yield for the combined cyclization/displacement was 100%.

Despite the very high yield with the fluoro series, the approach was particularly interesting for the chloro series via compound **13** as 2-chlorobenzoic acid is cheaper than 2-ethoxybenzoic acid. The synthesis of compound **13** is shown in Scheme 16.6.[13]

In the end due to the time pressures of the development program and because of the high efficiency of the commercial route, all of these alternative approaches were put on hold.

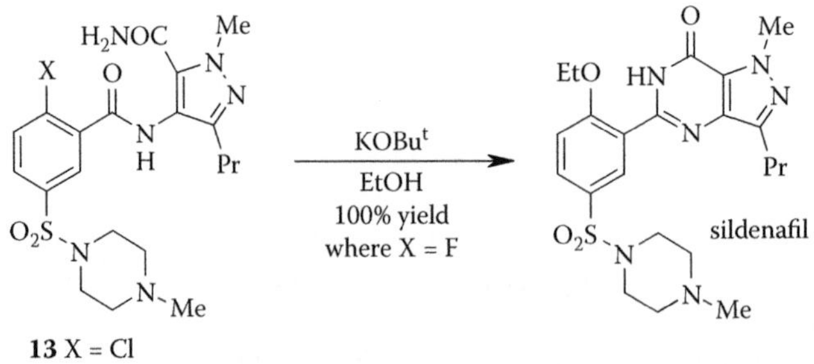

SCHEME 16.5 Alternative routes to sildenafil via the halo derivatives **13** and **14**.

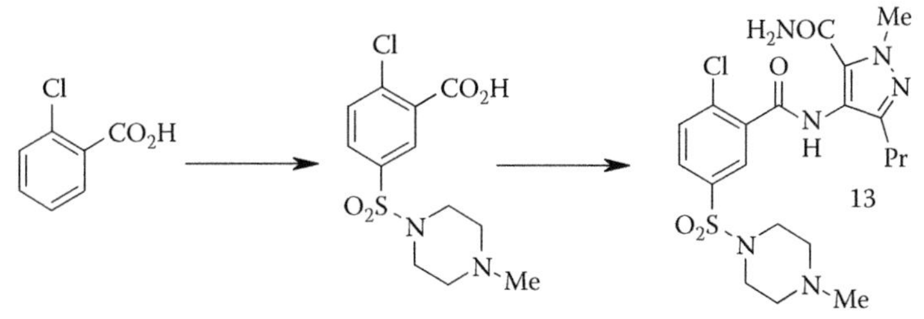

(i) ClSO$_3$H (4g/g) 95 °C for 6h, cool quench into H$_2$O, collect by filtration and dry (85%)
(ii) *N*-methylpiperazine (1.25 eq), H$_2$O, 25 °C, collect by filtration (82%) (iii) compound
6 (1eq), Et$_3$N (leq), WSCDI (1eq), HOBT (1eq) CH$_2$Cl$_2$, 25 °C, 48h, work up to give **13** (81%).

SCHEME 16.6 The synthesis of compound **13**. (See also, Dunn, P.J. and Levett, P.C. European Patent EP 0994 115.)

16.4 ALTERNATIVE SYNTHESIS OF PYRAZOLE 5

As previously mentioned, nitropyrazole **5** is one of the regulatory starting materials for sildenafil citrate, so not surprisingly the compound has attracted the attention of fine chemical companies and within Pfizer. The optimization of the condensation reaction to make intermediate **1** has been previously reported.[7] In the medicinal chemistry synthesis, conversion of **1** to compound **3** was accomplished via reaction with hydrazine followed by a methylation reaction. Clearly a more efficient way of making this transformation is to use a regioselective condensation reaction with methylhydrazine.[14–16]

This reaction is very interesting; if compound **1** is added to methylhydrazine in ethanol, then the desired pyrazole **3** was formed as a 10:1 mixture with the minor regioisomer **15**. If the addition is reversed and methylhydrazine is added to **1** in the same solvent (ethanol) and at the same temperature (5 to 10°C), then the ratio of **3** to **15** was reversed to 4:5. The 10:1 mixture of regioisomers formed under the best conditions was readily separated by distillation with the desired isomer boiling at 125 to 128°C/13 mm which is 40°C lower than the undesired regioisomer.[7]

To complete the synthesis, a nitration reaction is required followed an ammonolyis reaction. Workers at Bayer AG[16] have reported these transformations as highly efficient processes (Scheme 16.7). The pyrazole **3** was nitrated with anhydrous nitric acid using anhydrous sulfuric acid as solvent. After quenching, extraction, and evaporation, the resulting nitro ester was obtained, as an oil, in 97% yield. The nitro ester **16** was then converted through to the amide **5** using methanolic ammonia at 50°C. The yield for the ammonolysis is 98%; hence, the overall yield from methylhydrazine to pyrazole **5** is 74.7% [7,14–16] and the overall yield from methylhydrazine to sildenafil citrate is 67%. The electron-withdrawing power of the nitro group is key to activating the ester in compound **16** to ammonolysis. The alternative reaction sequence via compound **17** was unsuccessful.[7]

Workers at the India Orchid company[17] have shown that the condensation of 2-pentanone with diethyl oxalate may be catalyzed by sodium methoxide which is cheaper than sodium ethoxide. After further condensation with hydrazine hydrate, the pyrazole **18** was obtained as a mixture of methyl and ethyl esters. Methylation with dimethyl sulfate was performed neat, as in the Pfizer medicinal chemistry synthesis.[5] The mixture of the methylated pyrazoles **19** was then nitrated and subjected to ammonolysis to give the desired pyrazole intermediate **5** (Scheme 16.8).

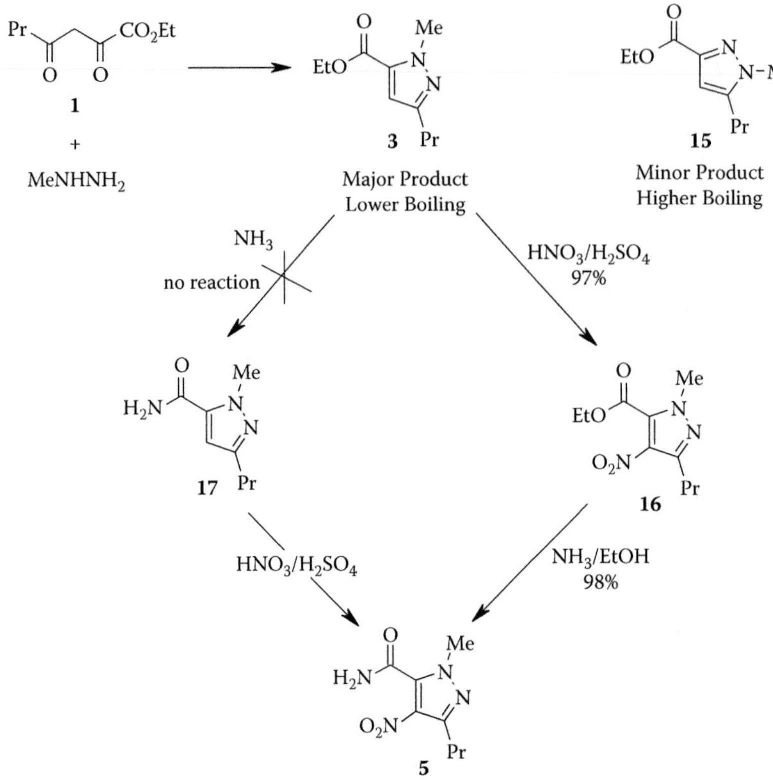

SCHEME 16.7 An alternative synthesis of pyrazole **5**.

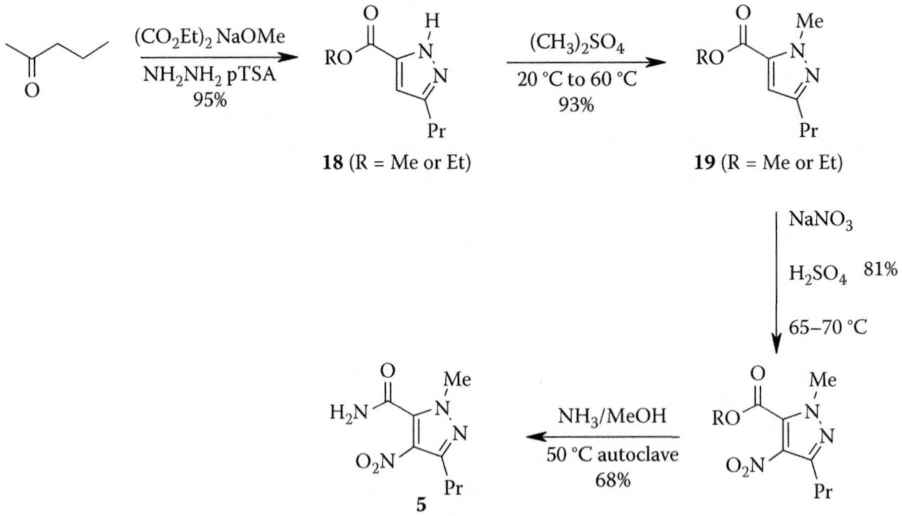

SCHEME 16.8 Another alternative synthesis of pyrazole **5**.

16.5 ENVIRONMENTAL PERFORMANCE OF COMMERCIAL ROUTE

The sildenafil citrate synthesis has an exceptionally low environmental profile having an E factor of 6 compared with an industry average of 25 to 100 for pharmaceutical products.[18] Green chemistry metrics such as E factor, reaction mass efficiency, organic waste, aqueous waste, and vapor emissions were measured or calculated at various time points for the sildenafil citrate synthesis and have been reported in detail in the green chemistry literature.[8] Perhaps the most striking of these metrics is the organic waste metric shown in Figure 16.2.

The medicinal chemistry synthesis produced 1300 L of waste per kilo of product, the majority of which was methylene chloride. After 4 year of chemical development, this had been reduced to 100 L/kg including a substantial reduction in methylene chloride use. However, a major step forward in implementing an environmentally friendly process was the introduction of a commercial route, which not only dramatically improved the yield but also set up the process for solvent recovery. In the manufacture of pharmaceuticals, diligent solvent recovery is very often required to fully optimize the environmental performance of a synthesis. The recovery of ethyl acetate and toluene was introduced in 1998, the year Viagra was launched, and this was followed by the recovery of 2-butanone which brought the solvent waste figure to 6.3 L of solvent per kilo, which is where the process currently stands. One problem with the current synthesis is that *t*-butanol, which is used as a solvent in the final cyclization step, is soluble in water in all proportions and is difficult to recover. At Pfizer there is a continued drive for improved environmental performance. A new process developed by production colleagues in Pfizer's plant in Ringaskiddy, Ireland, uses a different solvent for the cyclization step, and this new solvent can be recovered. The new process has been successfully demonstrated in production plant and when fully implemented would give a final optimized solvent usage of 4 L/kg. The data shown in Figure 16.2 are calculated for the preparation of sildenafil citrate from pyrazole **4** and 2-ethoxybenzoic acid. So the data are calculated on a like-by-like basis. It should be said that chemistry via the methylhydrazine route (Scheme 16.7) also has a clean environmental profile, using the atom economy principle.[19] The only by-products in the preparation of compound **5** are water and ethanol.

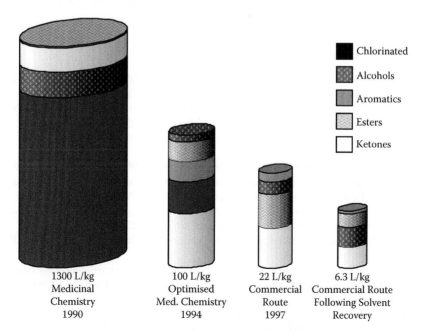

FIGURE 16.2 The amount of organic waste produced by the sildenafil citrate processes at various time points.

In 2003 Pfizer was awarded the Crystal Faraday Award for green chemical technology by the Institute of Chemical Engineers in recognition of the environmentally friendly manufacturing process to make sildenafil citrate.

16.6 CONCLUSIONS

The commercial route to sildenafil citrate has all the desired attributes required in chemical development:

- A safe, robust route with high throughput in production plant
- Reliable production of a very high-quality material
- A high yielding process, 67% overall yield compared with the discovery yield of 3.8%
- An exceptionally clean environmental performance

Often with the passage of time, the synthetic route used for the initial commercialization is surpassed by better chemistry. Sildenafil citrate is a high-profile molecule and more than 15 different syntheses of sildenafil citrate have been reported; however, the Pfizer commercial route remains the highest yielding and probably the most efficient way of making the molecule.

On December 3, 2004, Pfizer filed for approval to market a sildenafil citrate formulation under the trade name Revatio™ for the treatment of pulmonary hypertension. The FDA granted priority review and again the drug was approved within 6 months on June 3, 2005, adding yet another twist to the sildenafil citrate story.

ACKNOWLEDGMENTS

Many Pfizer colleagues were involved in the development of sildenafil citrate including Steve Belsey, Brian Corby, Gina Coghlan, Clare Crook, David Dale, Jeff Duke, Stephen Galvin, Darren Gore, Mike Hughes, Andy Pearce, Rosemary Prior, Pat Searle, and Albert Wood.

REFERENCES

1. Boolell, M. et al., Sildenafil: An orally active type 5 cyclic GMP-specific phosphodiesterase inhibitor for the treatment of penile erectile dysfunction, *Int. J. Impot. Res.* 8, 47, 1996.
2. Campbell, S.F., Science art and drug discovery: a personal perspective, *Clinical Science*, 99, 255, 2000.
3. Brock, G., Sildenafil citrate (Viagra™), *Drugs of Today*, 36, 125, 2000.
4. Lee, J.L. et al., Structure of the catalytic domain of human phosphodiesterase 5 with bound drug molecules, *Nature,* 425, 98, 2003.
5. Bell, A.S., Brown, A.S., and Terrett, N.K. European Patent, EP 0463 756,1991. Terrett, N.K. et al., Sildenafil (Viagra™), a potent and selective inhibitor of type-5 cGMP phosphodiesterase with high utility for the treatment of male erectile dysfunction, *Bioorg. Med. Chem. Lett.,* 6, 1819, 1996.
6. Elguero, J. et al., Packing modes in eight 3-ethoxycarbonylpyrazole derivatives. Influence on the crystal structure and annular tautomerism, *Heterocycles,* 50, 227, 1999.
7. Dunn, P.J., The synthesis of commercial phophodiesterase(V) inhibitors, *Org. Process Res. Dev.,* 9, 88, 2005.
8. Dunn, P.J., Galvin, S., and Hettenbach, K., The development of an environmentally benign synthesis of sildenafil citrate (Viagra™) and its assessment by Green Chemistry metrics *Green Chemistry,* 6, 43, 2004.
9. Dunn, P.J. et al., The chemical development of the commercial route to sildenafil: a case history, *Org. Process Res. Dev.,* 4, 17, 2000. See also Dunn, P.J. and Wood, A.S. European Patent, EP 0812 845.

10. Chen, C.-K. and Singh, A.K., A "bottom up" approach to process development: application of physiochemical properties towards the development of direct-drop processes, *Org. Process Res. Dev.*, 5, 508, 2001; Anderson, N.G., Assessing the benefits of direct isolation processes, *Org. Process Res. Dev.*, 8, 260, 2004.
11. Bunnage, M.E., Levett, P.C., and Thomson, N.M. World Patent, WO 01/98303.
12. Achmatowicz, et al, World Patent, WO 01/22918.
13. Dunn, P.J. and Levett, P.C. European Patent, EP 0994 115.
14. Muller, N. and Matzke, M., U.S. Patent 6 444 828. (Note that under certain conditions the regiochemistry for the methylhydrazine reaction can be 13:1.)
15. Muller, N. and Matzke, M., U.S. Patent 6 297 386.
16. Heuer, L., Muller, N., and Steffan, G. European Patent, EP 0 819 678, in German, same authors US Patent 5 969 152, 1999 and U.S. Patent 6 025 499, in English.
17. Chaudhari, D.T., Deshpande, P.B., and Khan, R.A.R. European Patent, EP 1 077 214.
18. Sheldon, R.A., Organic synthesis—past, present and future, *Chem. Ind.* 903, 1992; Sheldon, R.A., Catalysis and pollution prevention, *Chem. Ind.* 12, 1997.
19. Trost, B.M., The atom economy—a search for synthetic efficiency, *Science*, 254, 1471, 1991.

17 Stereoselective Enzymatic Synthesis of Intermediates Used for Antihypertensive, Antiinfective, and Anticancer Compounds

Ronald L. Hanson

CONTENTS

17.1 INTRODUCTION

The Process Research and Development Department at Bristol-Myers Squibb includes a small enzyme technology group. Although the synthetic routes and processes developed for production of drug candidate compounds rely mainly on conventional chemistry, there are certain cases, particularly involving the production of chiral intermediates, where an enzymatic process may be advantageous. The application of enzymes in process development takes advantage of their enantioselectivity, regioselectivity, ease of scale-up, and reactivity near ambient temperature and atmospheric pressure. In this chapter, the development of processes for the production of chiral amino acid intermediates for an antihypertensive compound will be discussed in detail and an overview of some enzyme applications in the synthetic routes to antibacterial, antiviral, and anticancer compounds will be given.

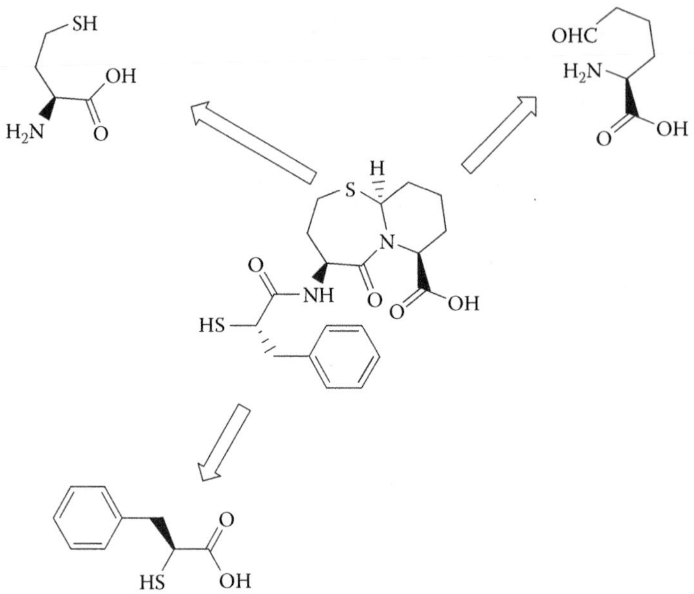

SCHEME 17.1 Vanlev synthetic strategy.

17.2 Vanlev®

Enzymatic processes were used in several versions of the synthetic routes to chiral amino acid intermediates needed for the synthesis of Vanlev (Omapatrilat, BMS-186716),[1] a vasopeptidase inhibitor intended for the treatment of hypertension. Vanlev inhibits angiotensin-converting enzyme, a zinc metalloprotease responsible for converting angiotensin I to angiotensin II, which elevates blood pressure. Vanlev also inhibits neutral endopeptidase which degrades atrial naturietic polypeptide, a hormone produced by the heart that decreases blood pressure. The overall synthetic strategy for Vanlev is shown in Scheme 17.1. Of the three precursors, homocysteine, (*S*)-2-thio-3-benzenepropanoic acid, and (*S*)-2-amino-6-oxohexanoic acid, most of the enzymatic efforts were directed toward preparation of the latter.

17.2.1 RESOLUTION WITH ACYLASE

In the initial discovery synthesis, resolution of racemic N-acetyl-6-hydroxynorleucine to L-6-hydroxynorleucine was carried out using 1 mg of L-amino acid acylase from pig kidney per gram of racemic substrate.[1] Although this process worked well with the more purified laboratory batches of racemic N-acetyl-6-hydroxynorleucine, some of the pilot-plant batches apparently contained impurities that inhibited the pig kidney acylase and prevented the reaction from going to completion. A more rugged acylase from *Aspergillus* (Amano acylase 30000) was suitable for carrying out the resolutions of all batches but was a crude preparation requiring 100 mg of enzyme per gram of substrate, resulting in difficulties in the isolation of the amino acid product by direct crystallization. To solve this problem, a process was developed (Scheme 17.2) that entailed first incubating 10% racemic N-acetyl-6-hydroxynorleucine in water with 5 mg of Amano acylase 30000 per gram of substrate for 1 h, followed by addition of 1.5 mg/g of Sigma hog kidney acylase for 24 to 28 h. This process worked well for all batches, and apparently whatever impurities were inhibiting the pig kidney acylase were removed by the initial charge of *Aspergillus* acylase. By this procedure, a total of 40.8 kg of L-6-hydroxynorleucine was prepared in close to 50% high-performance liquid chromatography (HPLC) yield with 97 to 99% ee.

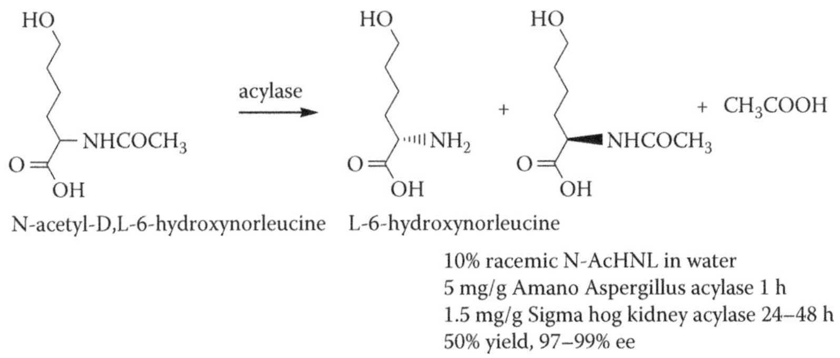

N-acetyl-D,L-6-hydroxynorleucine   L-6-hydroxynorleucine

10% racemic N-AcHNL in water
5 mg/g Amano Aspergillus acylase 1 h
1.5 mg/g Sigma hog kidney acylase 24–48 h
50% yield, 97–99% ee

SCHEME 17.2 Enantioselective hydrolysis of racemic N-acetyl-6-hydroxynorleucine.

17.2.2 PREPARATION OF L-6-HYDROXYNORLEUCINE BY REDUCTIVE AMINATION USING AMINO ACID DEHYDROGENASE

Because the resolution with acylase gave a theoretical maximum yield of only 50% and required separation of the desired product from the unreacted enantiomer at the end of the reaction, we next tried to prepare the amino acid by reductive amination of the corresponding ketoacid, a process with a theoretical maximum yield of 100%. A variety of ketoacids can be converted to L-amino acids by treatment with ammonia, reduced nicotinamide adenine dinucleotide (NADH), and a suitable amino acid dehydrogenase.[2] 2-Keto-6-hydroxyhexanoic acid (in equilibrium with its cyclic hemiketal form) was prepared by chemical synthesis starting from 4-chloro-1-butanol, which was O-protected, then converted to a Grignard reagent which was added to diethyl oxalate, followed by hydrolysis of the ester and deprotection of the hydroxyl group. Initial screening, with formate dehydrogenase for regeneration of NADH, showed that phenylalanine dehydrogenase from *Sporosarcina* sp. and beef liver glutamate dehydrogenase converted 0.1 *M* 2-keto-6-hydroxyhexanoic acid completely to L-6-hydroxynorleucine. Leucine dehydrogenase partially purified from *Bacillus sphaericus* ATCC 4525[3] and alanine dehydrogenase from *Bacillus subtilis* were not active. Additional screening with spectrophotometric enzyme assays (i.e., monitoring the rate of NADH oxidation in the reaction) of commercially available amino acid dehydrogenases and extracts of 132 cultures from our collection identified *Thermoactinomyces intermedius* ATCC 33205 as containing the most active enzyme. This strain has been shown to be a source of thermostable phenylalanine dehydrogenase[4] as well as leucine dehydrogenase.[5]

Beef liver glutamate dehydrogenase was used for preparative reactions at 10% total substrate concentration. As depicted in Scheme 17.3, 2-keto-6-hydroxyhexanoic acid, sodium salt, (in equilibrium with 2-hydroxytetrahydropyran-2-carboxylic acid, sodium salt), was converted to L-6-hydroxynorleucine; the reaction required ammonia and NADH. Nicotinamide adenine dinucleotide (NAD) produced during the reaction was recycled to NADH by the oxidation of glucose to gluconic acid using glucose dehydrogenase from *Bacillus megaterium*. The optimum pH for glutamate dehydrogenase with this substrate was determined to be about 8.8, and glucose dehydrogenase had a broad pH optimum centered at about 8.5. The reaction was complete in about 3 h with reaction yields of 89 to 92% and with ee >99%.

Chemical synthesis and isolation of 2-keto-6-hydroxyhexanoic acid required several steps. In a second, more convenient process (shown in Scheme 17.4), the ketoacid was prepared by treatment of racemic 6-hydroxynorleucine (produced by hydrolysis of commercially available 5-(4-hydroxybutyl) hydantoin) with D-amino acid oxidase and catalase. After the ee of the remaining L-6-hydroxynorleucine had risen to >99%, the reductive amination procedure was used to convert the mixture containing 2-keto-6-hydroxyhexanoic acid and L-6-hydroxynorleucine entirely to

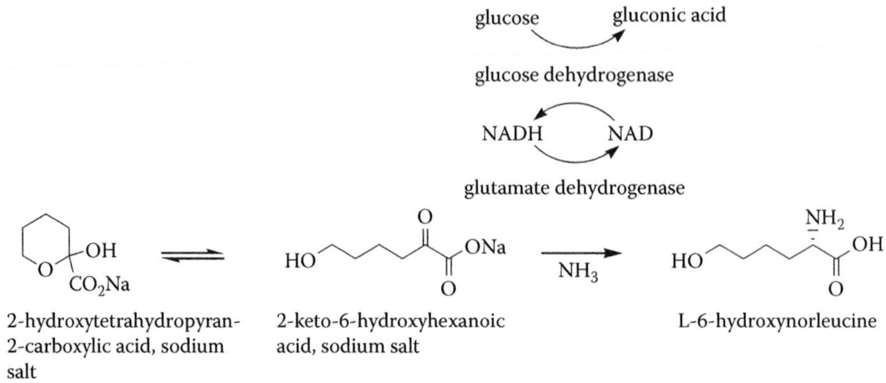

SCHEME 17.3 Conversion of 2-keto-6-hydroxyhexanoic acid to L-6-hydroxynorleucine.[2]

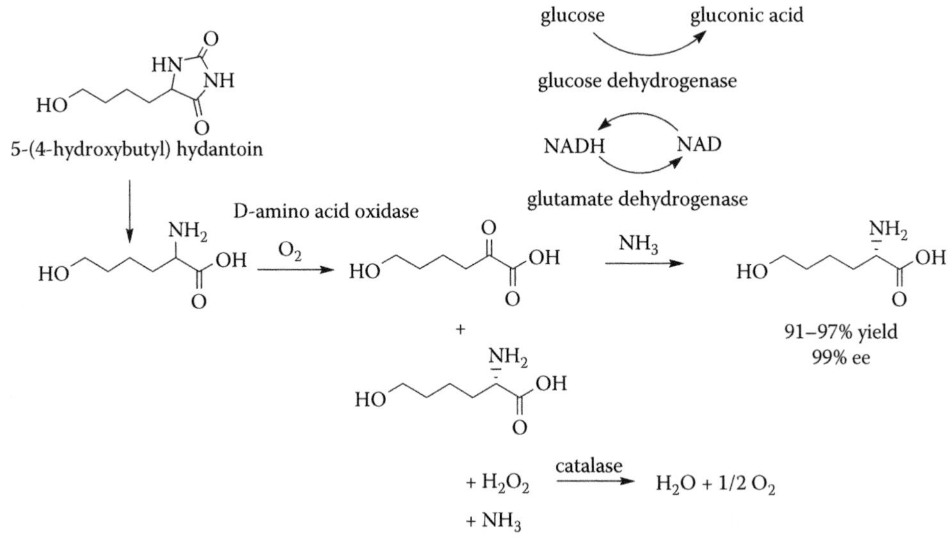

SCHEME 17.4 Conversion of racemic 6-hydroxynorleucine to L-6-hydroxynorleucine.[2]

L-6-hydroxynorleucine with yields of 91 to 97% and ee > 99%. Sigma porcine kidney D-amino acid oxidase and beef liver catalase or *Trigonopsis variabilis* whole cells (source of oxidase and catalase)[6] were used successfully for this transformation.

17.2.3 Preparation of Allysine Ethylene Acetal by Reductive Amination Using Phenylalanine Dehydrogenase

In the original route to BMS-186716, L-6-hydroxynorleucine was coupled with S-acetyl-N-Cbz-L-homocysteine, then oxidized to the aldehyde.[1] In the final route used for the synthesis of Vanlev, a process for the production of allysine ethylene acetal by enzymatic reductive amination of the corresponding keto acid was developed and scaled up to avoid this oxidation step (Scheme 17.5). For this process formate dehydrogenase (FDH) was used for regeneration of NADH. Although both the glucose and formate dehydrogenase reactions are equally effective for regeneration of NADH, the easy removal of CO_2 compared to gluconic acid makes this a preferable method for isolation

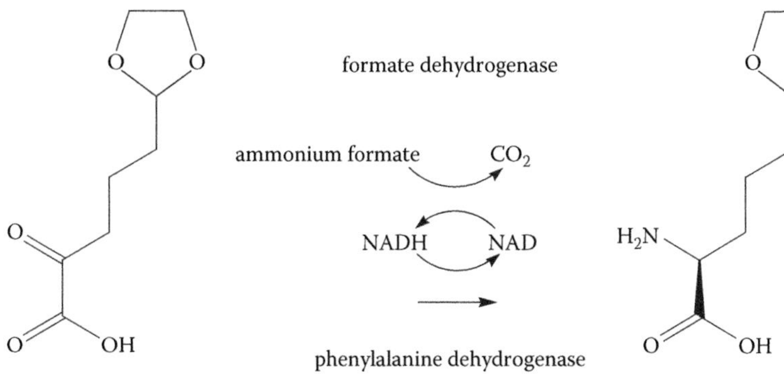

SCHEME 17.5 Reductive amination of keto acid acetal.

of the amino acid product. Screening of commercially available amino acid dehydrogenases as well as some strains from our culture collection showed that glutamate, alanine, leucine, and phenylalanine dehydrogenases (listed in order of increasing effectiveness) gave some of the desired product.[7] An extract of *Thermoactinomyces intermedius* ATCC 33205 was an effective source of phenylalanine dehydrogenase (PDH) for the reaction.

T. intermedius IFO14230 (ATCC 33205) was first identified as a source of PDH by Ohshima et al.[4] The enzyme was purified and characterized[4] and then cloned and expressed in *E. coli* by the same workers.[8] The enzyme was reported to be moderately specific for deamination of phenylalanine and to carry out the amination of some keto acids at a much lower rate than amination of phenylpyruvate.[4] In our screening, the enzyme was the most effective amino acid dehydrogenase identified for the reductive amination of the keto acid acetal.

FDH from *C. boidinii* was introduced by Whitesides and Shaked,[9] and by Kula, Wandrey, and coworkers[10] for regeneration of NADH. The advantages of this enzyme reaction are that the product CO_2 is easy to remove, and the negative reduction potential ($E'_o = -0.42$ v) for the FDH reaction drives the reductive amination to completion.

The optimum pH for the reductive amination of **1** by PDH from *T. intermedius* ATCC 33205 was about 8.7. Because **1** was much more stable under the reaction conditions at pH 8.0 than at 8.7, the reaction was carried out at pH 8.0. This pH is also well suited to the use of FDH from *C. boidinii,* which was reported to have a broad pH optimum of 7.5 to 8.5.[11] FDH from *P. pastoris* was reported to have a pH optimum of 6.5 to 7.5,[12] or 7.5,[13] with more than 80% of maximal activity at pH 8.[12,13]

Wet cells, heat-dried cells, extracts, and immobilized enzymes were all useful for the reaction, but heat-dried cell preparations were the simplest and most convenient enzyme source to use. Heat-dried cells were produced by drying the cells under vacuum at 54°C, then milling to <10 mesh. Our initial procedure for the conversion of **1** to **2** used dried *T. intermedius* from a 250-liter fermentor as a source of PDH and heat-dried *C. boidinii* as a source of FDH. When trying to scale up the procedure, recovery of the *T. intermedius* cells from larger fermentations was a problem because the cells lysed at the end of the growth period. The larger tanks take considerably longer to cool down and harvest and little cell paste could be recovered. The problem was solved by using heat-dried recombinant *E. coli* containing cloned *T. intermedius* PDH inducible with isopropylthiogalactoside and again using heat-dried *C. boidinii* as a source of FDH. Although the kinetics of the reaction were about the same, when using the same number of units of PDH, the higher specific activity of the recombinant *E. coli* compared to the *Thermoactinomyces intermedius* decreased the amount of dried cells required and made isolation of the product easier.

A third-generation procedure, using heat-dried recombinant *P. pastoris* containing *T. interme-dius* PDH and endogenous FDH, both inducible with methanol in this strain, allowed both enzymes to be produced during a single fermentation, and they were conveniently produced in about the right ratio that was used for the reaction. To carry out this fermentation, the culture was initially grown on 1% glycerol for about 16 h until glycerol was exhausted from the medium, then a methanol feed was begun to maintain the methanol concentration at about 0.1% for 48 h to induce production of both enzymes. The *Pichia* reaction procedure had the following modifications relative to the *E. coli/C. boidinii* procedure: twice the concentration of **1** and one-quarter the amount of NAD were used, and dithiothreitol was omitted. Decreasing the ratio of NAD to **1** and omitting dithio-threitol considerably decreased the materials cost of the process. Although FDH from *P. pastoris* is reported to be sensitive to sulfhydryl reagents, and mercaptoethanol or dithiothreitol was used during the purification of this enzyme[12,13] as well as PDH,[4] for a single use of the cells it was not necessary to add any dithiothreitol.

The procedure using heat-dried cells of *E. coli* containing cloned PDH and heat-dried *C. boidinii* was scaled up without any problems. A total of 197 kg of **2** was produced in three 1600-L batches using a 5% concentration of **1** with an average yield of 91.1 *M%* with ee greater than 98%. The procedure with *P. pastoris* was also scaled up to produce 15.5 kg of **2** in 97 M % yield with ee greater than 98% in a 180-L batch using 10% keto acid concentration.

The route using the amino acid acetal for preparation of the bicyclic intermediate for Vanlev is summarized in Scheme 17.6. The amino acid acetal is converted to the dimethyl acetal methyl ester, then coupled with N-protected homocystine to give a dipeptide dimer. The dimer is converted to the monomer with dithiothreitol or tributylphosphine. Acidification of the monomer gives the aldehyde that cyclizes to the bicyclic intermediate with concomitant hydrolysis of the ester.

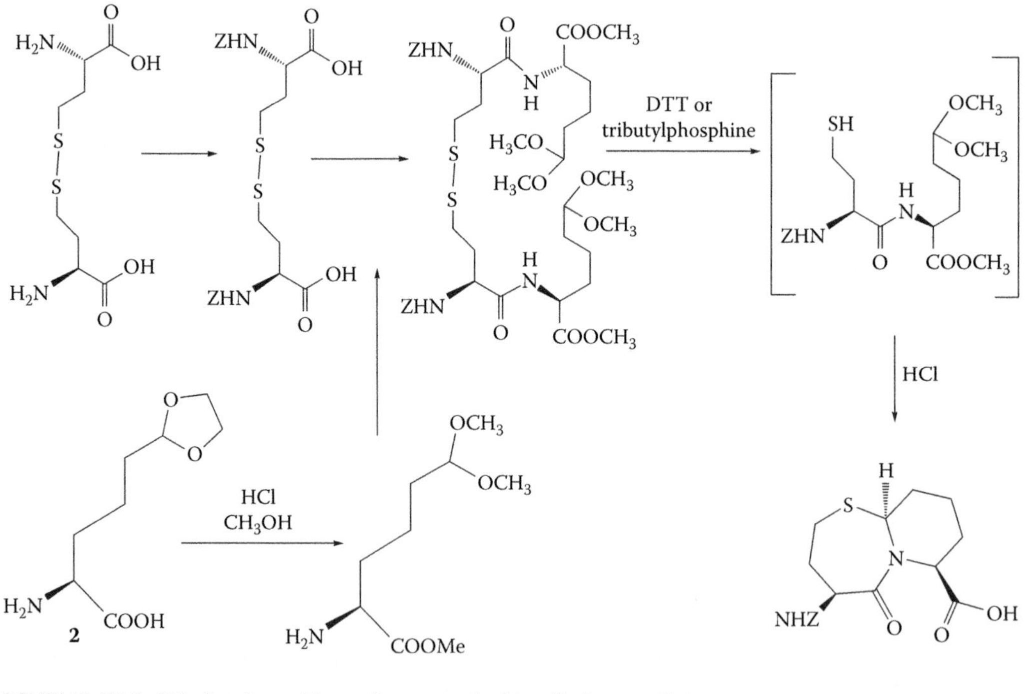

SCHEME 17.6 Chiral amino acid acetal route to the bicyclic intermediate.

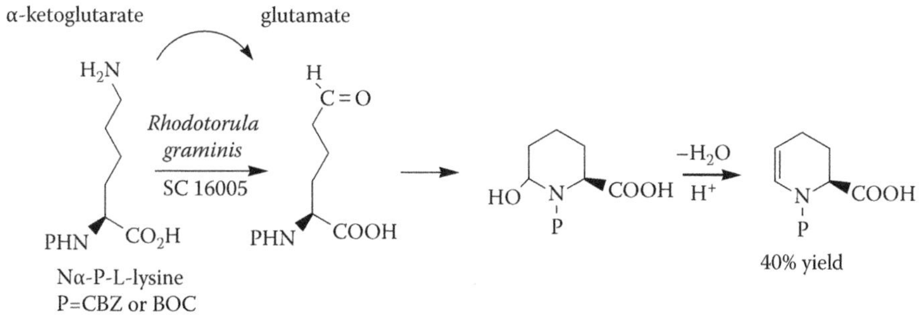

SCHEME 17.7 Transamination of α-Cbz- or α-BOC-lysine.

The preparation of **2** by enzymatic reductive amination provides the single enantiomer of the amino acid acetal by a shorter route than the previously published eight-step synthesis of racemic allysine ethylene acetal.[14] Unlike the previously described route, the enzymatic route does not require addition and removal of protecting groups, and therefore gives better atom economy. The synthesis of keto acid **1** and enzymatic reductive amination to **2** as described proved to be suitable for the preparation of the large quantities of the vasopeptidase inhibitor needed for clinical trials.

17.2.4 Use of Lysine-ε-aminotransferase

L-lysine was also considered as an inexpensive starting material for preparation of (*S*)-2-amino-6-oxo-hexanoic acid. Although L-lysine can be oxidized to α-amino adipic acid-δ-semialdehyde using lysine ε-dehydrogenase from *Agrobacterium tumefaciens* or by using a transaminase found in several strains, the product is not a stable isolable compound that can be used as an intermediate in a synthetic route. We found that α-Z- or α-Boc-lysine could be converted by transamination with α-ketoglutarate using cell suspensions of *Rhodotorula graminis* SC16005 to (S)-1-(Z- or Boc)-1,2,3,4-tetrahydropyridine-2-carboxylic acid (Scheme 17.7) after cyclization and acid-catalyzed dehydration.[15] Although these products could be isolated in about 40% yield, they were unreactive in the coupling reaction for synthesis of the bicyclic intermediate.

L-lysine, however, could be used in the synthesis as a dipeptide. A novel dipeptide lysine-ε-aminotransferase was isolated which allowed Z-homocystine lysine dipeptide to be used. A strain containing the transaminase was isolated from a soil sample by selection for growth with α-Z-lysine as the sole nitrogen source. The reasoning was that the α-N-protection would allow growth of only those strains that could utilize the ε-amino as a nitrogen source. The organism was then identified as a strain of *Sphingomonas paucimobilis* and the transaminase was purified, cloned, and expressed in *E. coli*.[16] The reactions for use of this transaminase are outlined in Scheme 17.8. Z-Homocystine lysine dipeptide dimer was converted to the dipeptide monomer with dithiothreitol or tributylphosphine. The dipeptide monomer was treated with α-ketoglutarate and the ε-transaminase to convert it to the aldehyde, which cyclizes in the presence of acid to give the bicyclic intermediate for Vanlev in about 65 to 70% yield. The glutamate produced by the transamination can be recyled to α-ketoglutarate using a glutamate oxidase that has been purified from *Streptomyces noursei* SC6007 and cloned and expressed in *Streptomyces lividans*. Although this reaction was demonstrated on a 20-g scale, it was not scaled up further owing to the difficulty of synthesis of the dipeptide dimer. The amino acid acetal route was used to supply the bulk of the API for clinical testing.

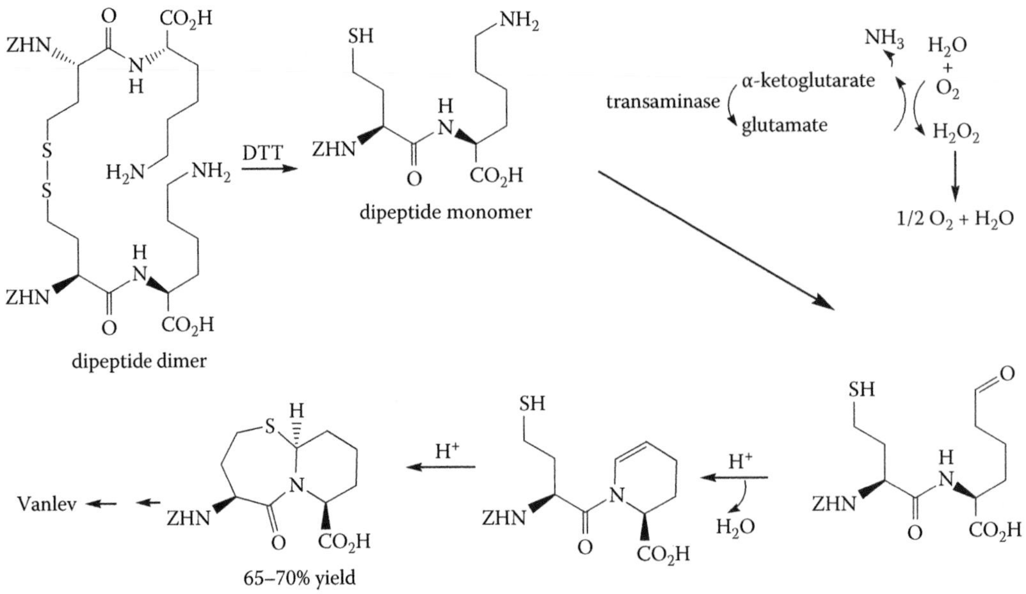

SCHEME 17.8 Use of lysine ε-aminotransferase from *Sphingomonas paucimobilis*.

17.3 HYDROXYLATION OF MUTILIN

Pleuromutilin (Scheme 17.9) is an antibiotic from *Pleurotus* or *Clitopilus* basidiomycete strains which kills mainly Gram-positive bacteria and mycoplasms. Metabolism of pleuromutilin and derivatives results in hydroxylation by microsomal cytochrome P-450 at the 2- or 8-position and inactivates the antibiotics. Modification of the 8-position of pleuromutilin and analogs was of interest as a means of preventing the metabolic hydroxylation. Microbial hydroxylation at the 8-position of pleuromutilin or mutilin would provide a functional group at this position to allow further chemical modification at this site to block metabolic hydroxylation when the compounds are administered to humans or animals.[17] The target analogs would maintain the biological activity of the parent compounds, yet not be susceptible to metabolic inactivation.

Biotransformation of mutilin and pleuromutilin by microbial cultures was investigated to provide a source of 8-hydroxymutilin or 8-hydroxypleuromutilin. Our approach here was to screen organisms from our culture collection which were known to carry out hydroxylations of other compounds. Growth of cultures in a medium containing soybean flour is known to be favorable for induction of hydroxylation enzymes. After an initial 24-h growth period, mutilin or pleuromutilin was added to the growth medium and incubation was continued for 2 to 5 days. Liquid chromatography/mass spectrometry (LC/MS) analysis of culture broths showed that several strains gave M+16 products from mutilin and one culture gave an M+16 product from pleuromutilin, suggesting addition of oxygen. Biotransformation products were extracted from culture broths with ethyl acetate, dried, and purified by chromatography on silica gel. *Streptomyces griseus* strains SC 1754 and SC 13971 (ATCC 13273) converted mutilin to (8S)-, (7S)-, and (2S)-hydroxymutilin (Scheme 17.9). *Cunninghamella echinulata* SC 16162 (NRRL 3655) gave (2S)-hydroxymutilin or (2R)-hydroxypleuromutilin from biotransformation of mutilin or pleuromutilin, respectively (Scheme 17.9). The biotransformation of mutilin by *Streptomyces griseus* strain SC 1754 was scaled up in 15-L, 60-L, and 100-L fermentations to produce a total of 49 g (8S)-hydroxymutilin (BMS-303786), 17 g (7S)-hydroxymutilin (BMS-303789), and 13 g (2S)-hydroxymutilin (BMS-303782) from 162 g of mutilin.

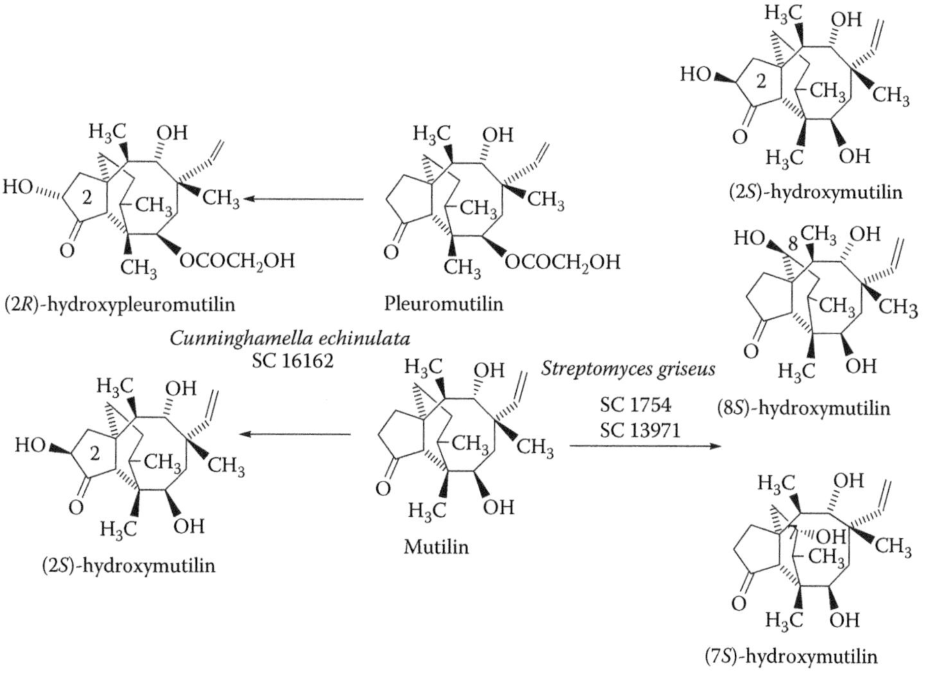

SCHEME 17.9 Hydroxylation of mutilin by *Streptomyces griseus* and *Cunninghamella echinulata*.[17]

This approach has also been used in our laboratories and others to provide a convenient source of many other drug metabolites, because microbial enzymes often produce the same metabolites as mammalian enzymes of drug metabolism. The metabolites produced by this method have been used for structure determinations, as analytical standards, to test for pharmacological activity[18] or toxicity, and as starting materials for further chemical modification as in the case of mutilin.[19]

17.4 REGIOSELECTIVE AMINOACYLATION OF LOBUCAVIR

Lobucavir (BMS 180194, Scheme 17.11) is a cyclobutyl guanine nucleoside analog recently under development as an antiviral agent for the treatment of herpes viruses and hepatitis B. A prodrug form in which one of the two hydroxyls is coupled to valine, BMS 233866 (Scheme 17.11), was also considered for development. Regioselective aminoacylation by valine is difficult to achieve by chemical procedures but appeared to be possible via an enzymatic approach. Regioselective acylation of only one of the two hydroxyls of lobucavir with N-protected L-valine is a problem that we approached by chemical aminoacylation of both hydroxyl groups followed by selective enzymatic hydrolysis and also by selective enzymatic aminoacylation.

Either hydroxyl group of lobucavir could be selectively aminoacylated with valine by using enzymatic reactions.[20] Monoester **5** (82.5% yield) was obtained by selective hydrolysis of **3** with lipase M from *Mucor javanicus*, and **6** (87% yield) was obtained by hydrolysis of **4** with lipase from *Candida cylindracea* (Scheme 17.10). However, these two products were not the desired intermediates for the lobucavir prodrug. ChiroCLEC™ BL (cross-linked crystals of subtilisin) catalyzed a regioselective aminoacylation of lobucavir (7 g) using Z-L-valine *p*-nitrophenyl ester as acyl donor to give **7** (8 g, 61% yield) (Scheme 17.11). Undesired side products were **5** (7.8%) and **3** (13%). The reaction was scaled up to give 607 g of **7** (53.8 M%) from a 600-g input of lobucavir and 4.5 kg of **7** (54.8 M%) from a 4.4-kg input of lobucavir.

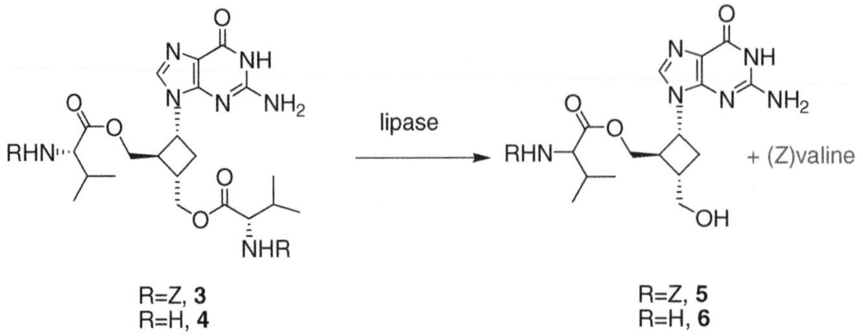

SCHEME 17.10 Regioselective hydrolysis of diester.[20]

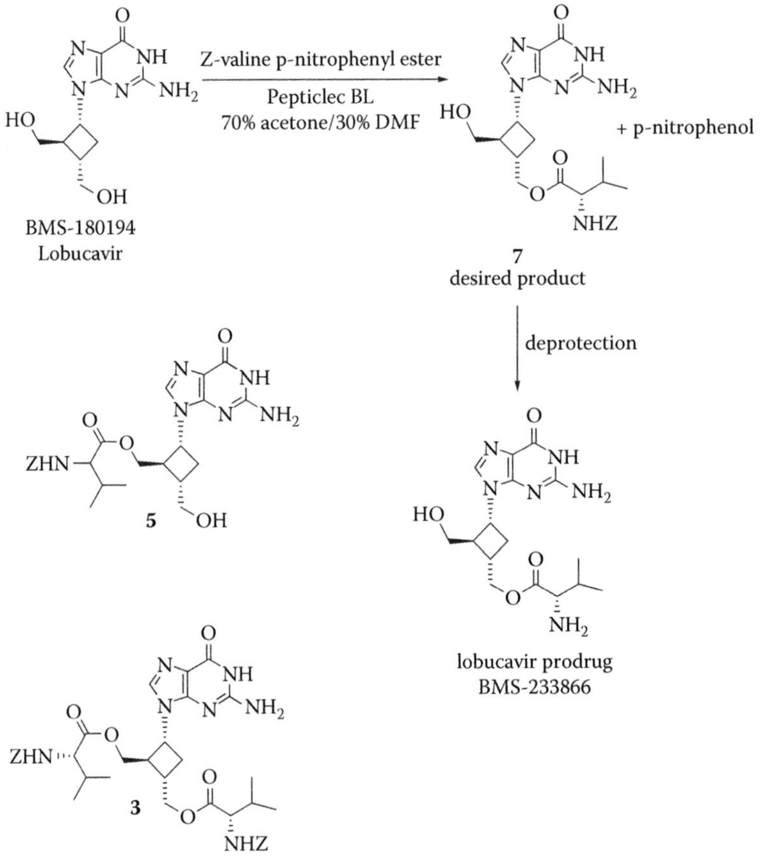

SCHEME 17.11 Transesterification using Pepticlec BL.[21]

The rate, yield, and regioselectivity of the transesterification of Z-valine p-nitrophenyl ester with lobucavir using ChiroCLEC™ BL were strongly affected by solvent. A 70% acetone/30% DMF mixture gave the highest rate and yield of **7** of any of the solvents tested. The rate and yield in the aminoacylation of lobucavir by ChiroCLEC™ BL were also strongly affected by the leaving group and N-protecting groups on the valine ester. When several Z-L-valine esters were compared, p-nitrophenyl = trifluoroethyl > N-hydroxysuccinimide >> methyl ester in rate and yield for the reaction. When Boc-L-valine *p*-nitrophenyl ester was used, followed by deprotection, a 76% yield of BMS 233866 was achieved using half the amount of ChiroCLEC™ BL required for the reaction with the Z-valine *p*-nitrophenyl ester. Undesired side products were **6** (8.3%) and **4** (11.5%).

ChiroCLEC™ PC (cross-linked crystals of *Pseudomonas cepacia* lipase) or Amano lipase PS30 (also *Pseudomonas cepacia* lipase) immobilized on Accurel polypropylene gave the highest yield and regioselectivity. Z-valine trifluoroethyl ester was transesterified with lobucavir to give **7** in 84% yield, with no **5** and only 0.4% **3**. With this lipase, Z-valine *p*-nitrophenyl ester was not an effective substrate. Moris and Gotor also noted that *p*-nitrophenyl esters of N-protected amino acids were not effective aminoacylating agents when using *Pseudomonas cepacia* lipase and other lipases.[21]

17.5 SEMISYNTHESIS OF PACLITAXEL

Paclitaxel (Figure 17.14), an anticancer compound first isolated by Wani et al., has been approved as a drug (Taxol®) for treatment of ovarian cancer, breast cancer, non-small-cell lung cancer, and AIDS-related Kaposi's sarcoma. It has generated additional interest because of its novel mechanism of action as an inhibitor of depolymerization of microtubules. Paclitaxel was initially purified from extracts of the bark of the Pacific yew, *Taxus brevifolia*, which raised concerns of damage to old-growth forests. Semisynthesis from 10-deacetylbaccatin-III (Figure 17.14), which can be obtained from needles and twigs, was then developed to provide a more renewable source of paclitaxel and analogs.[22] Total synthesis of paclitaxel, biosynthesis by cultured yew cells, and production by a fungus have also been reported, with only the cultured yew cells providing another renewable practical source for production of large amounts of material at the present time.

Enzymes were used for the chiral synthesis or resolution of side chains that can be coupled to 10-deacetylbaccatin III in the semisynthesis of paclitaxel and analogs. Racemic acetate **8** (Scheme 17.12) was converted to the undesired (3*S*)-alcohol **9a** by enantioselective hydrolysis using either Amano lipase PS-30 or lipase prepared in-house by fermentation of *Pseudomonas* sp. SC 13856.

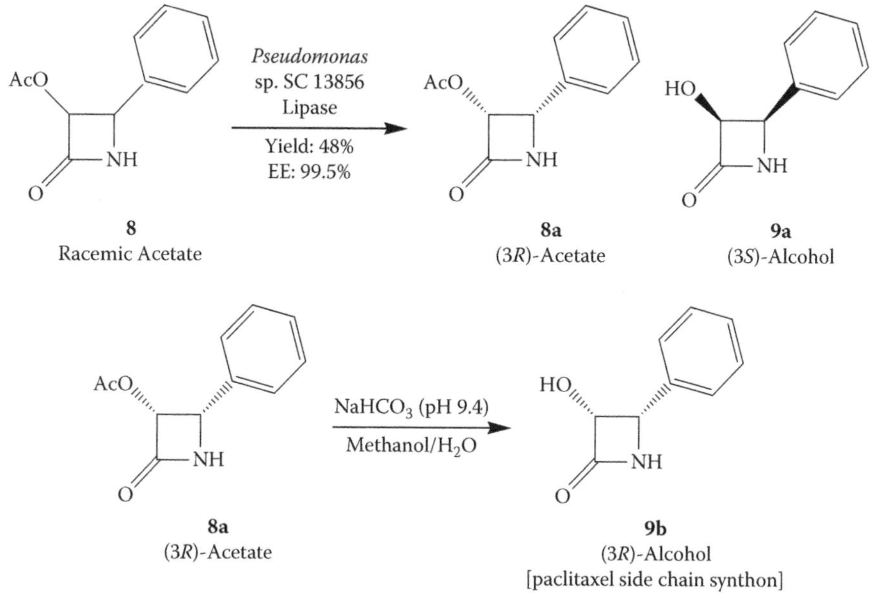

SCHEME 17.12 Resolution of paclitaxel C13 side chain synthon.

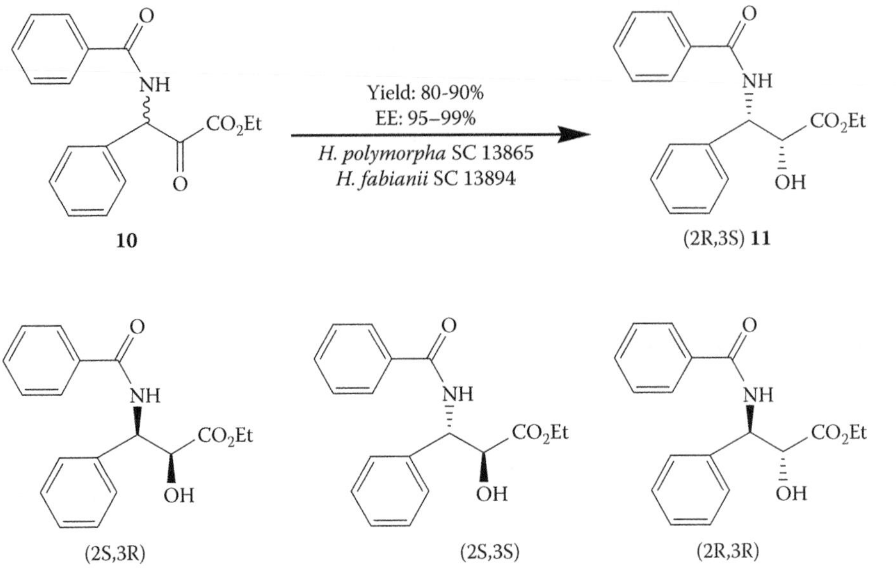

SCHEME 17.13 Taxol® side chain synthon from enantioselective reduction.

The unreacted (3R)-acetate **8a** was obtained in 48% yield (50% theoretical maximum) with 99% ee. Hydrolysis of the isolated acetate with $NaHCO_3$ in methanol gave the desired (3R)-alcohol **9b** that was coupled with baccatin to give paclitaxel.[23]

Another enzymatic approach to obtain the desired stereochemistry in the side chain is shown in Scheme 17.13.[24] The keto ester precursor **10** of the side chain ethyl ester can be reduced to the hydroxy (2R,3S) ester **11** using either of the yeasts *Hansenula polymorpha* SC 13865 or *Hansenula fabianii* SC 13894. Screening a variety of strains from our culture collection revealed many other strains that could carry out the reduction reaction, but the best yields and ee's using whole cells were obtained with the two strains of *Hansenula*. Of four possible reduction products, the desired product **11** is obtained with 95 to 99% ee and 80 to 90% yield. Because of rapid ketone/enol tautomerism, the enzymatic reduction can work as a dynamic resolution and fix the stereochemistry at both the 2- and 3-positions.

Paclitaxel and related compounds have also been found in various *Taxus* species in addition to the Pacific yew, occurring in roots, stems, wood, and needles as well as bark. Yew extracts contain a complex mixture of taxanes, with paclitaxel usually constituting less than 20% of the total taxanes. Isolation of paclitaxel from these mixtures is a difficult purification problem and contributed to the slow development of this compound as a drug. The most valuable material in this mixture for semisynthesis is 10-deacetylbaccatin-III. Microbial strains were isolated from soil samples containing C-13 deacylase and C-10-deacetylase enzyme activities that are able to convert mixtures of taxanes to 10-deacetylbaccatin-III, thereby increasing the amount and ease of isolation of this precursor for semisynthesis (Scheme 17.14).[22] Treatment of ethanol extracts, prepared either from whole plants of a variety of renewable yew cultivars or from material derived from the bark of *T. brevifolia*, with the two enzymes converted a complex mixture of taxanes primarily to 10-deacetylbaccatin-III, increasing the amount of this key precursor by 5.5- to 24-fold.

7-Xylosyltaxanes, including 7-xylosylpaclitaxel, 7-xylosyl-10-deacetylpaclitaxel, 7-xylosyl-cephalomannine, and 7-xylosyl-10-deacetylcephalomannine have been isolated from yew bark with 7-xylosyl-10-deacetylpaclitaxel being the most abundant taxane found in bark from *Taxus baccata* and *Taxus brevifolia*.[25] Treatment of taxane xylosides with periodate, lead tetraacetate, or other oxidizing agents, and hydrolysis of the product with a substituted hydrazine or acid have been

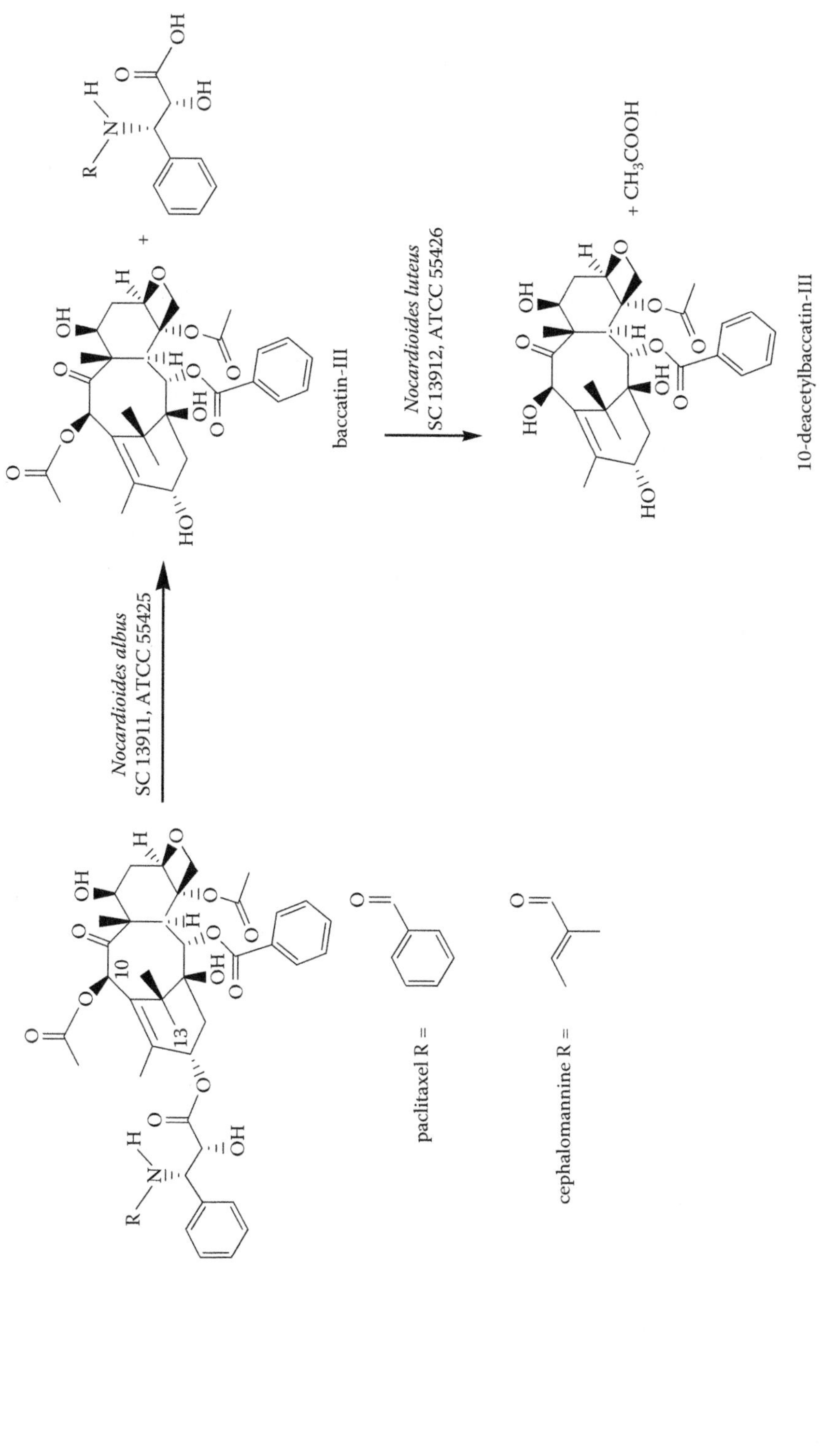

SCHEME 17.14 Reactions catalyzed by C13-deacylase and C10-deacetylase.

reported as chemical methods for removal of the sugar. Hydrolysis by a suitable xylosidase, in combination with hydrolysis by the C10 and C13 esterases described above, would provide a simple enzymatic method of recovering the 10-deacetylbaccatin-III moiety of 7-xylosyltaxanes for use in the semisynthesis of paclitaxel.

Screening of commercially available xylosidases, xylanases, and other glycosidases did not reveal any that removed xylose from 7-xylosylpaclitaxel or 7-xylosyl-10-deacetylpaclitaxel. Nine ATCC cultures reported to have xylanase activity were also inactive for hydrolysis of 7-xylosyl-10-deacetylpaclitaxel. Therefore, a screening procedure was devised to identify microorganisms from environmental samples able to remove xylose from these substrates. A total of 125 strains with xylosidase activity (as indicated by hydrolysis of 4-methyl-umbelliferyl-β-D-xyloside to the fluorescent 4-methylumbelliferone) were isolated from soil samples and a wood sample. These strains were abundant and easy to isolate. Of 86 isolates screened for their effectiveness with 7-xylosyl-10-deacetylpaclitaxel, nine of the isolates produced 10-deacetylpaclitaxel. Four of the isolated strains were identified and deposited in the ATCC repository: *Moraxella* sp. ATCC 55475, *Bacillus macerans* ATCC 55476, *Bacillus circulans* ATCC 55477, and *Micrococcus* sp. ATCC 55478. The *Moraxella* strain gave the highest yield of 10-deacetylpaclitaxel from 7-xylosyl-10-deacetylpaclitaxel and was further characterized. The strain was able to remove the xylosyl group from 7-xylosylpaclitaxel, 7-xylosyl-10-deacetylpaclitaxel, 7-xylosylbaccatin III, and 7-xylosyl-10-deacetylbaccatin III, thereby making the xylosyltaxanes available as sources of 10-deacetylbaccatin III for the semisynthesis of paclitaxel. (Scheme 17.15).[25]

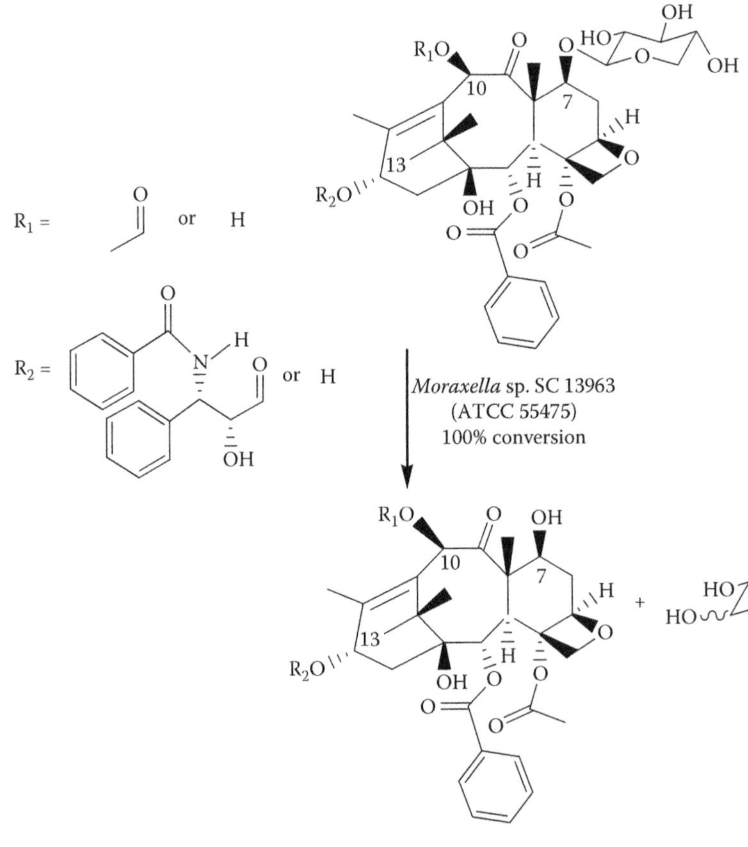

SCHEME 17.15 Reactions catalyzed by C7-xylosidase.

17.6 SUMMARY AND CONCLUSIONS

Unnatural amino acids are useful synthons for preparing many compounds with pharmacological activity. Resolution with L-amino acid acylase or D-amino acid oxidase, reductive amination with amino acid dehydrogenases, and aminotransferase were applied to the production of amino acid intermediates for Vanlev. We found reductive amination of α-keto acids using amino acid dehydrogenases to be a technology that has been widely applicable in various synthetic routes.[2,3,7] The reactions proceed at high substrate concentrations with a variety of structures and give exclusively the L-amino acid.

Monooxygenases were applied to prepare hydroxylated metabolites of the antibiotic pleuromutilin. This type of hydroxylation is often difficult to achieve by chemical methods and is useful for preparing metabolites of many drugs, because microbial strains are often found to give the same metabolites as mammalian systems.

Protease or lipase enzymes were useful for the regioselective aminoacylation of lobucavir. Lipase was also used for resolution of a synthon for the paclitaxel side chain. The paclitaxel side-chain ester was also prepared by reduction of a keto ester precursor. Enzymatic reduction of ketones to chiral alcohols is another reaction that has been widely applicable. C14-deacylase, C10-deacetylase, and C7-xylosidase were identified from microorganisms isolated from soil samples and were useful for converting complex mixtures of taxanes found in yew extracts primarily to 10-deacetyl baccatin III, a precursor for the semisynthesis of paclitaxel and analogs.

Enzymes used for biocatalytic processes may be commercially available, identified by screening culture collections, or available from a toolbox of enzymes previously cloned for other projects, or, if necessary, can be identified by enrichment, selection, and screening of strains from soil samples.

ACKNOWLEDGMENTS

The author thanks the following colleagues who contributed to this work: Amit Banerjee, David Brzozowski, Dana Cazzulino, Mary Jo Donovan, Ronald Eiring, Steven Goldberg, Jeffrey Howell, Robert Johnston, Rafael Ko, Thomas LaPorte, Michael Montana, Richard Mueller, Venkata Nanduri, Ramesh Patel, Laszlo Szarka, Thomas Tully, and Valerie Zanella. Other collaborators who contributed to chemical syntheses and product isolation are listed in the references.

REFERENCES

1. Robl, J.A.; Sun, C; Stevenson, J;, Ryono, D.E.; Simpkins, L.M.; Cimarusti, M.P.; Dejneka, T; Slusarchyk, W.A.; Chao, S; Stratton, L.; Misra, R.N. ; Bednarz, M.S.; Asaad, M.M.; Cheung, H.S.; Aboa-Offei, B.E.; Smith, P.L.; Mathers, P.D.; Fox, M.; Schaeffer, T.R.; Seymour, A.A.; Trippodo, N.C. *J. Med. Chem.* 1997, *40*, 1570–1577.
2. Hanson, R.L.; Schwinden, M.D.; Banerjee, A.; Brzozowski, D.B.; Chen, B.-C.; Patel, B.P.; McNamee, C.G.; Kodersha, G.A.; Kronenthal, D.R.; Patel, R.N.; Szarka, L.J. *Bioorg. Med. Chem.* 1999, *7*, 2247–2252.
3. Hanson, R.L.; Singh, J.; Kissick, T.P.; Patel, R.N.; Szarka, L.J. and Mueller, R.H. *Bioorg. Chemistry* 1990, *18*, 116–130.
4. Oshima, T.; Takada, H.; Yoshimura, T.; Esaki, N.; Soda, K. *J. Bacteriol.* 1991, 173, 3943–3948.
5. Oshima, T.; Nishida N.; Bakthavatsalam, S.; Kataoka, K.; Takada, H.; Yoshimura, T.; Esaki, N.; Soda, K. *Eur. J. Biochem.* 1994, 222, 305–312.
6. Wei, C.; Huang, J.; Tsai, Y. *Biotechnol. Bioeng.* 1989, 34, 570–574.

7. Hanson, R.L.; Howell, J.M.; LaPorte, T.L.; Donovan, M.J.; Cazzulino, D.L.; Zannella, V.; Montana, M.A.; Nanduri, V.B.; Schwarz, S.R.; Eiring, R.F.; Durand, S.C.; Wasylyk, J.M.; Parker, W.L.; Liu, M.S.; Okuniewicz, F.J.; Chen, B.-C.; Harris, J.C.; Natalie, K.J.; Ramig, K.; Swaminathan, S.; Rosso, V.W.; Pack, S.K.; Lotz, B.T.; Bernot, P.J.; Rusowicz, A.; Lust, D.A.; Tse, K.S.; Venit, J.J.; Szarka, L.J.; Patel, R.N. *Enzyme Microb. Technol.* 2000, *26*, 348–358.

8. Takada, H.; Yoshimura, T.; Ohshima, T.; Esaki, N.; and Soda, K. *J. Biochem.* 1991, *109*, 371–376.

9. Shaked, Z.; Whitesides, G.M. *J. Am. Chem. Soc.* 1980, *102*, 7104–7105.

10. Kula, M.R.; Wandrey, C. *Methods in Enzymology* 1987, *136*, 9–21.

11. Schütte, H.; Flossdorf, J.; Sahm, H.; Kula, M.-R. *Eur. J. Biochem.* 1976, *62*, 151–160.

12. Hou, C.T.; Patel, R.N.; Laskin, A.I.; Barnabe, N. *Arch. Biochem. Biophys.* 1982, *216*, 296–305.

13. Allais, J.J.; Louktibi, A.; Baratti, J. *Agric. Biol. Chem.* 1983, *47*, 2547–2554.

14. Rumbero, A.; Martin, J.C.; Lumbreras, M.A.; Liras, P.; Esmahan, C. *Bioorg. Med. Chem.* 1995, *3*, 1237–1240.

15. Patel, R.N.; Banerjee, A.; Hanson, R.L.; Brzozowski, D.B.; Parker, L.W.; Szarka, L.J. *Tetrahedron: Asymmetry* 1999, *10*, 31–36.

16. Patel, R.N.; Banerjee, A.; Nanduri, V.B.; Goldberg, S.L.; Johnston, R.M.; Hanson, R.L.; McNamee, C.G.; Brzozowski, D.B.; Tully, T.P.; Ko, R.Y.; LaPorte, T.L.; Cazzulino, D.L.; Swaminathan, S.; Chen, C.-K.; Parker, L.W.; Venit, J.J. *Enzyme Microb. Technol.* 2000, *27*, 376–389.

17. Hanson, R.L.; Matson, J.A.; Brzozowski, D.B.; LaPorte, T.L.; Springer, D.M.; Patel, R.N *Org. Proc. Res. & Dev.* 2002, *6*, 482–487, and references therein.

18. Fura, A; Shu, Y.-Z.; Zhu, M.; Hanson, R.L.; Roongta, V.; Humphreys, W.G. *J. Med. Chem.* 2004, *47*, 4339–4351.

19. Springer, D.M.; Sorenson, M.E.; Huang, S.; Connolly, T.P.; Bronson, J.J.; Matson, J.A.; Hanson, R.L.; Brzozowski, D.B.; LaPorte, T.L.; Patel, R.N *Bioorg. Med. Chem. Lett.* 2003, *13*, 1751–1753.

20. Hanson, R.L.; Shi, Z.; Brzozowski, D.B.; Banerjee, A.; Kissick, T.P.; Singh, J.; Pullockaran, A.J.; North, J.T.; Fan, J.; Howell, J.; Durand, S.C.; Montana, M.A.; Kronenthal, D.R.; Mueller, R.H.; Patel, R.N., *Bioorg. Med. Chem.* 2000, *8*, 2681–2687, and references therein.

21. Moris, F.; Gotor, V. *Tetrahedron* 1994, *50*, 6927–6934.

22. Hanson, R.L.; Wasylyk, J.M.; Nanduri, V.B.; Cazzulino, D.L.; Patel, R.N.; Szarka, L.J. *J. Biol. Chem.* 1994, *269*, 22145–22149, and references therein.

23. Patel, R.N.; Banerjee, A.; Ko, R.Y.; Howell, J.M.; Li, W.-S.; Comezoglu, F.T.; Partyka, R.A.; Szarka, L. *Biotechnol. Appl. Biochem.* 1994, *20*, 23–33.

24. Patel, R.N.; Banerjee, A.; Howell, J.M.; McNamee, C.G.; Brzozowski, D.; Mirfakhrae, D.; Nanduri, V.; Thottathil, J.K.; Szarka, L.J. *Tetrahedron: Asymmetry* 1993, *9*, 2069–2084.

25. Hanson, R.L.; Howell, J.M.; Brzozowski, D.B.; Sullivan, S.A.; Patel, R.N.; Szarka, L.J. *Biotechnol. Appl. Biochem.* 1997, *26*, 153–158, and references therein.

18 Designing Robust Crystallization Processes for Active Pharmaceutical Ingredients—From Art to Science

Dierk Wieckhusen

CONTENTS

18.1 BACKGROUND

During the early 1990s, crystallizations were considered more of an art than a science. During that time, process chemists would design a process that delivered an active pharmaceutical ingredient (API) with a certain particle size, and formulation chemists would use it to design a tablet or a capsule suitable for clinical trials or the marketplace. Since then, several events have changed that business model:

1. Concerns over the safety of tiny amounts of potential genotoxic impurities in APIs were driven by major advances in analytical instrumentation that allow detection of these impurities at parts per million levels.
2. New "intelligent formulations" that are more suitable for certain markets and that can enhance the life cycle of the drug have required more specialized and tailor-made particle-size APIs.
3. Internal pressure has increased to deliver drugs to the market more quickly, thus compressing development timelines to reach new drug applications (NDAs) in less time and potentially with less knowledge.

In response to these changes, process chemists turned to crystal engineering science, which emerged as a strong new discipline that assists in understanding the robustness of the process and the physical properties of an API and provides formulators the range of particle sizes and other powder properties they need to develop optimum formulations.

Along with the science, the tools to respond to these challenges have also changed. Molecular modeling has moved from the hands of specialists to become a widely used visualization tool of how molecules pack into the crystal lattice and the impact of that packing on the physical properties of APIs. Process analytical technology (PAT) has also advanced to allow us to observe crystals growing in real time and understand the impact of seeding and mixing on the morphology of particles. New tools are becoming available every day that will continue to change the art of designing and growing crystals into the science it should be.

18.2 LARGE, COMPLEX, ORGANIC MOLECULES

The majority of the crystallization literature deals with inorganic crystals, and the organic chemistry literature mainly covers small organic molecules with few degrees of freedom such as glutamic acid, phenytoin, or paracetamol. The reality in the pharmaceutical industry is often much more complicated—molecular weights above 1000 g/mole are common, and complex molecular structures have an effect on nucleation and growth kinetics as well as on the likelihood of polymorph formation.

Studies of crystals of compounds of pharmaceutical interest have not yet become fashionable in the academic arena. Some academic colleagues are reluctant to switch to more complicated pharmaceutical molecules—be it because of the difficulty in obtaining supplies of these compounds or the potential toxicity that often requires special handling. I do not want to go too deeply into this discussion; however, I believe that pharmaceutical crystallization is not a fashion that will soon disappear. People will always need medications, and because every drug has its own crystallization characteristics, there will be plenty of work to do in this rewarding field.

Why are complicated organic molecules so special?

- They can possess complicated polymorphic/pseudomorphic behavior.
- Different salt forms usually behave differently.
- They may contain certain potent by-products or impurities that are inherent in their chemical synthesis.
- They may exhibit multiple conformational degrees of freedom, thus showing slow crystallization kinetics.
- Their chemical stability is not guaranteed but can be influenced by the crystal form.
- Various faces of a crystal may exhibit different properties.

18.3   FIXING TARGETS

Before designing an API crystallization process, the targets have to be fixed on a case-by-case scenario. First, it is important to know the anticipated market form for the drug because it determines to a great extent targeted API properties such as particle size, shape, bulk density, specific surface area, polymorph, yield (price), and dissolution characteristics (e.g., sustained release, fast release). Specifications usually should be determined in an iterative manner: during development, various drug substance qualities should be crystallized and tested by the formulators. This process is the only way to find the real limits for the drug product process and to set rational and sound specifications. Pharmacists often urge chemical development to stay within certain specifications during development, although there is no real need to do so. Their request is usually due to the fear that they may not be able to formulate material that is different from that they previously obtained. Setting these limits too early during development hinders the freedom needed later for scale-up and the selection of the API manufacturing site and the equipment needed and, most important, could inhibit creativity. This statement is not one opposing drug substance specifications, but setting these specifications too narrowly will generate unnecessary limitations during the different phases of development. The smartest procedure would be to produce and test a wide range of drug substance properties during development and in this way get a sense of what kinds of materials the pharmaceutical process can tolerate. The powder properties that impact quality and robustness can then be fixed. These features are worth in-depth consideration because they can have a tremendous impact on the manufacturability and, potentially, the profitability of the chemical process.

Coming back to the setting of the targets, there are three perspectives to consider when designing an API crystallization process. From the perspective of the development chemist, the process should reach a certain yield and purity, the polymorph or solvate has to be the desired one, and the process must be safe, easy, and reproducible. Colleagues in chemical production have different demands: freedom in the choice of manufacturing equipment; quick filtrations, fast drying times, high throughput, and robustness; and, most importantly, minimal costs. The pharmacist customer, who has to formulate the drug substance and therefore has the right to raise concerns, prefers a drug substance that is processible and provides consistent quality attributes, especially in terms of dissolution rate and bioavailability. To ensure these qualities, consistent powder properties, such as particle size distribution (PSD), shape, flowability, tablettability (if applicable), bulk density, and stability, are required, all of which depend on a robust crystallization process.

18.3.1   PARTICLE SIZE

For the process chemist, particle size affects the filtration, isolation, and drying kinetics of the API, and PSD is one of the most critical quality attributes that formulation scientists require. Particle size affects the manufacturability of the drug product in terms of flow characteristics, interactions with excipients, and bioavailability.

Certain products, such as inhaled candidates, require extremely fine particles on the order of 1 to 5 microns. In this case, the milling of the API is absolutely critical to achieve the proper delivery profiles of the drug. Other products may require coarse particles that are more suitable for capsules.

How can particle size be controlled? The answer is clear: controlling primary nucleation and avoiding secondary nucleation and attrition can potentially enable chemists to achieve the same crystal size distribution with each batch. A later discussion outlines some of the inherent difficulties in these strategies and how best to overcome them.

18.3.2 Yield

Yield in the API crystallization process is an important target because it determines to a great extent the profitability of the process. Both solvent and process should be chosen such that a high yield with the desired purity can be obtained. Process variables can have a negative impact on purity—if solubility is decreased too much, by-products can be driven out of solution—therefore, processes are not typically pushed to yields greater than 95%. An important fact here is that in many cases equilibrium yields are not reached for kinetic reasons. Crystallizations at low temperatures require time and patience, especially in the case of complex organic molecules with many degrees of freedom. Therefore, it is worthwhile to compare solubility at the final temperature with the yield obtained. In the case of cooling crystallizations, if the calculated yield is higher than that obtained, cooling rates should be decreased, or holding time at final temperature should be increased in order to allow for total desupersaturation and, hence, maximal yield. In the case of drowning out crystallizations, similar considerations are required. An important effect to remember is that although adding antisolvent reduces solubility on one hand, it increases the total amount of solvent mixture, so the gain in yield may not be as high as expected, and cooling crystallization, if well designed, could be a valuable alternative for many antisolvent addition processes.

18.3.3 Purity

Purity is determined by the amount of by-products present in the final crystallization step, the solubility of these by-products in the mother liquor before filtration, and the crystallization rate. One has to ensure that all by-products are soluble in the mother liquor, otherwise they may crystallize or oil out with the API and will be difficult to remove by the wash, or they will remain in solution (i.e., the mother liquor will be supersaturated with these by-products, which, in time, will nucleate, crystallize, and potentially create future purity problems).

It is extremely important to understand where by-products are located. They are mainly:

1. Attached to the surfaces of the API crystals: Attachment can occur through molecular adsorption onto the crystal surface, oiling out, and adherence to the crystals. In this case, a washing process may help.
2. Crystallized separately: If by-products crystallize as well, it will be difficult to eliminate them. The best way is to find a new solvent that has higher solubility for these substances. In rare cases, one might even consider crystallizing the by-product first and then removing it.[1]
3. Equally incorporated into the crystal lattice: If the by-product has a structure similar to that of the drug substance, rejection of the by-product by the crystal is limited. The only way to get high purity then may be through very slow crystallization, which improves this selection.
4. Enriched in the crystal core: This occurs if crystallization begins too rapidly and the by-product is incorporated earlier in the crystallization process rather than toward the end where supersaturation is lower and crystal growth rates are smaller.

5. Enriched in the outer part of the crystal: The presence of by-products in the outer part of the crystal indicates that the supersaturation for by-products is reached during the crystallization process. Further increase in supersaturation forces incorporation into the crystal.

Another cause of impurity is incorporation of mother liquor into the crystal lattice (inclusion); however, this rarely happens if crystallization is slow and precipitation is avoided.

It is easy to find out which of these mechanisms is dominating by performing washing trials. One should check high-performance liquid chromatography purity of the wet filter cake, washed filter cake, mother liquor, and extremely washed filter cake with about 30% and then 70% of the API washed off. Data from these experiments will provide a clear picture of the mechanism of incorporation, which can suggest corrective measures.

18.3.4 Polymorphism and Solvate Formation

Delivering the right polymorphic form or solvate is crucial because these forms usually differ in their solubility and dissolution rate and because this area usually generates much regulatory scrutiny when a new drug application (NDA) is filed (International Conference on Harmonisation [ICH] Guideline Q6B, see Web site of ICH: www.ich.org).

18.3.5 Morphology

To a great extent, morphology determines the flow properties and other physical characteristics of an API powder. There are kinetic and thermodynamic reasons why a certain morphology is established. Computer models are available to calculate the thermodynamic morphology based on attachment energy models, for example. These models assume that incorporation of a molecule is much more likely on a crystal face with high attachment energy. In the case of API salts, needle-like structures are often observed, originating most frequently in the formation of ionic bonds along one axis of the crystal. A similar explanation can be used for nonionic molecules forming hydrogen bonds along one axis. If shape is a result of preferred attachment to certain crystal faces, it is obvious that morphology can be modified by additives or solvent molecules that exhibit preferred adsorption—for example, onto the fast-growing face—and, hence, inhibit the incorporation of API molecules. Using solvents as additives is by far the most elegant way to manipulate crystal shape, but one can also use additives, which is less common in pharmaceutical crystallization because the addition of any product other than the API has to be declared and explained.

Coarse crystallization and consecutive milling can also modify shape. In milling it is more likely that the crystal breaks along its longest axis, and therefore aspect ratio changes toward more compact morphology. The same principles apply for platelets, for which filtration can be especially critical.

18.4 THERMODYNAMIC BACKGROUND

18.4.1 Determining Solubility Curve

Before the development of an API crystallization process can begin, it is imperative that the solubility curves in various solvents be measured—that is, measurement of solubility as a function of temperature (Figure 18.1).

These curves are essential to design the process because they provide information about how much the drug has to be diluted and whether cooling can be used to generate supersaturation and result in adequate yield. Complex solubility curves are a strong indicator of polymorphs with enantiotropic behavior or solvates.

Equation 18.1 calculates yield loss Y [%] at a given solubility S [%] at a temperature T and a dilution D (factor of solvent used compared to API mass):

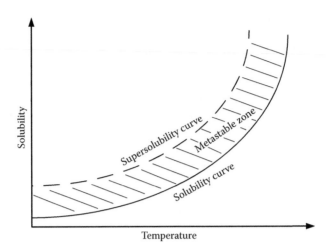

FIGURE 18.1 Solubility and supersolubility curve of an API agent.

$$Y = S_T * D \qquad (18.1)$$

Using this formula for a measured solubility at low temperatures immediately gives an idea about potential yield and about the feasibility of a cooling crystallization process. In the metastable zone no nucleation is observed even though the solution is supersaturated. Spontaneous nucleation will only occur when the supersolubility curve is reached.

18.4.2 POLYMORPHISM

Pharmaceutical companies make substantial efforts to search for new polymorphs and pseudopolymorphs for several reasons:

- Different forms have different physical behavior, especially different solubility and dissolution kinetics.
- Regulatory authorities demand intensive investigations to find potential forms and especially to characterize them, rank them with regard to their stability, and control the registered form.
- It is vitally important that companies control the desired polymorph during many years of production; losing this control can cause severe problems (see below).
- Knowing as many polymorphs as possible enables selection of the appropriate one for development and sometimes ensures freedom to operate and perhaps prolong the patent life of a drug.

Some basic theories on polymorphism appear below. For deeper studies there are excellent textbooks and papers available.[2-5] For a study of disappearing polymorphs, please read the article by Dunnitz and Bernstein.[6]

18.4.2.1 Monotropy

In the case of monotropism, one form is thermodynamically stable (and hence has the lowest solubility) over the whole temperature range (Figure 18.2).

In some cases, the solubility curves of two forms are very close together, making it difficult both to determine which form is the stable one and to control the desired form. Although selecting the metastable form is not at all recommended, the hatched area in Figure 18.2 indicates the region

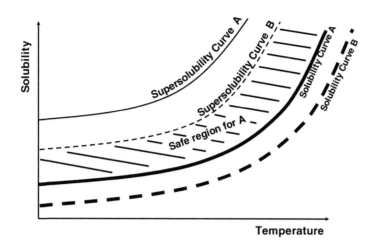

FIGURE 18.2 Solubility diagram for two monotropic polymorphs A and B.

where the metastable form could be produced with a certain safety. However, it has to be noted that supersolubility curves are not fixed, but can depend on the presence of dissolved impurities (that can hinder nucleation) and scratches on the walls of the reactor (that can promote nucleation). Therefore the recommendation applies only to manufacture the thermodynamically stable form.

18.4.2.2 Enantiotropy

In the case of enantiotropy, stability ranking changes at a certain temperature, the so-called transition temperature T_t. One form is thermodynamically stable and less soluble below this temperature; the other form is stable at temperatures above T_t.

Understanding the relationship between the polymorphs of a developing candidate and how they can transform from one to the other is critical; otherwise, it is difficult to design a robust crystallization process. Personal experience suggests choosing the most stable form for development. In some cases, chemists may be urged by the organization to select a metastable form because it shows better bioavailability, but in the long run it is very difficult to work against thermodynamics, especially because kinetics cannot always be controlled. What if the unwanted thermodynamically stable form is just hindered by some by-products that might disappear at some time (e.g., after optimization of the chemical synthesis)? It is worthwhile to experiment with the stable form before setting it aside for a form that has other appeals in the initial stages of development. A case in which the sudden occurrence of a new, more stable polymorph caused severe problems for a pharmaceutical company was the development of Norvir® (Abbott Pharmaceuticals, Abbott Park, Illinois).[7]

In the case of enantiotropy, the API production process should be performed at a temperature considerably below the transition point (T_t) to avoid nucleation of the high-temperature form.[8] If it is for the sake of dilution necessary to start at temperature above T_t, consider keeping a second vessel at lower temperature (below T_t) together with seeds, and carefully transfer the hot solution into this vessel. It is unsafe to rely on the transformation of the high-temperature modification into the low-temperature modification below T_t, because the kinetics of this process may, again, be uncontrolled. Furthermore, it is hardly possible to control PSD during transformation, and the PSD—adjusted, for example, in a seeding process—above T_t may be significantly changed.

It is also important to know that the stability ranking of true polymorphs (i.e., no solvates) is independent of the solvent used.[9,10] For solvates, however, there is a dependency of stability on solvent composition and temperature. Therefore, if solvent effects are observed that seem to influence polymorph stability, they are of a kinetic nature, or the crystal form is not a true polymorph but a solvate.

18.4.3 FINDING STABLE POLYMORPH

The easiest, and probably safest, way to determine which of two forms is the stable one at a given temperature is to perform equilibration studies. Equal amounts of both forms are suspended in a solvent, which dissolves part of the API, and are equilibrated at the temperature for a few days or a week. The suspension is then filtered, and the wet filter cake is analyzed using x-ray powder diffractometry. If the powder pattern shows an increase of one form, that form is the stable form. The longer it takes for the full transformation to occur, the smaller the driving force (e.g., because the temperature is close to T_t or the solubility difference of both forms is minute). Be aware that "equilibration" of just one form is insufficient to prove stability—the stable form has to nucleate, and it may not do so readily.

18.4.4 SOLVATE FORMATION

Solvated forms have to be treated as different chemical entities, and therefore, the rules given for polymorphism cannot be applied to relate two different solvates even though polymorphism within a certain solvate is often observed (and then these rules can be applied). One distinguishing feature of solvated forms is that their relative stability usually depends on the solvent composition. Equilibrium studies are an effective means of investigating the stability of solvates in various solvent mixtures and at different temperatures.

18.4.5 AMORPHOUS ACTIVE PHARMACEUTICAL INGREDIENTS/OILING OUT

In rare cases, it is impossible to find a crystalline form, and the API has to be manufactured as an amorphous material or oil. In these cases, the purification of the final step usually has to take place using chromatography or extraction, a clear disadvantage compared to purification through crystallization processes. A second disadvantage of amorphous APIs is the lower chemical and physical stability and the latent danger that a crystalline form might occur some day, making it difficult to reproduce the amorphous form. Most common drug substances have enough possibilities for strong bonds to build up crystal lattices, so the absence of crystals may be triggered simply by the hindrance of nucleation—for example, due to by-products present in the mother liquor. It is important to remember that an API solution contains a cocktail of specific additives available in the form of by-products that have formed during the various chemical steps. These by-products can be similar to the API or to parts of the API molecule and thus can be expected to be able to attach to the faces of the crystal. This attachment can hinder crystal growth at the face to which they are attached, modifying shape or even suppressing nucleation.

Another source for amorphous material is milling, especially high-energy milling as air-jet milling and so-called oiling out, which is the formation of amorphous material through too sharp an increase in supersaturation so that a miscibility gap is reached and the system is forced to decompose (spinodal decomposition). Supersaturation can be achieved by adding an antisolvent or through rapid cooling of the mixture. In this case, crystallization kinetics may be too slow, and a phase separation of the supersaturated system may occur. For more background on amorphous drug substances, see work by Hancock and Zografi.[11]

18.5 ROLE OF KINETICS

In the majority of cases, APIs are comparatively large organic molecules and thus have a more pronounced response to kinetic effects than do small molecules. The formation of metastable forms is induced by kinetic factors, and there are two steps in API crystallization that can be affected by such mechanisms: the formation of clusters of a critical size with an energy barrier which is responsible for the metastable zone width and their growth (i.e., the incorporation of molecules in the right conformational arrangement to fit into the crystal lattice, which is associated with the release of the heat of crystallization). Recall that a variety of potent additives may be present in the mother liquor in the form of by-products. These by-products are by nature similar to the API and may therefore attach to clusters and hinder nucleation, or attach to certain crystal faces and thus slow crystal growth or modify the crystal habit.

Supersaturation as the driving force in crystallization must be adapted to the crystallization kinetics to avoid the formation of metastable forms (which are kinetically favored if observed at all) and to prevent the incorporation of by-products into the crystal lattice. In the case of spontaneous nucleation, nucleation kinetics determines the number of nuclei formed and thus, to a great extent, the average particle size and final PSD. It is obvious that in the case of varying impurity profiles and by-product patterns, which are typical for early API development, substantial risk may be associated with processes that rely on spontaneous nucleation—that is, the desired PSD or polymorphic form may no longer be accessible owing to impurities. To minimize this risk, seeding is recommended for initiating API crystallization processes. Nucleation and growth kinetics have to be thoroughly investigated and duly understood to ensure the definition and specification of robust process parameters. Generally, it is advisable to select the crystal form that is thermodynamically stable at the conditions envisaged for production and storage. Although it may be tempting and even appear kinetically controlled in some cases, the selection of a metastable form for API development should be avoided at all cost. As outlined earlier, kinetic conditions during a process may vary substantially (e.g., due to the by-product pattern), and unforeseen effects such as the nucleation of an unwanted form may then occur. The only reliable factor that should be considered in process design is thermodynamics.

18.6 SALT AND SOLVENT SELECTION

Salt selection is a wide field that is not covered in detail here. Worth mentioning, however, is the fact that each salt should be treated as a new chemical entity; the knowledge collected for one form cannot usually be transferred to a new salt form. This approach increases workload on the one hand but presents opportunities to find improved physicochemical properties or processing behavior with a new salt, on the other hand. Details on pharmaceutical salt selection are provided elsewhere.[12]

Several criteria have to be fulfilled in the selection of a solvent or solvent mixture: the solvent should not react with the drug substance, and it should preferably be cheap and nontoxic. The U.S. Food and Drug Administration and ICH guideline Q3C classify standard organic solvents according to their toxicity and suggest limits for residual solvent content. These limits have to be met in industrial practice, and low-toxicity class 1 solvents such as alcohols or esters are generally first choices. Another aspect that must be considered is the solubility of the API in the chosen solvent or solvent mixture as a function of temperature. To facilitate the design of a robust, high-yield cooling crystallization procedure, API solubility should be high at elevated temperatures and particularly low at the typical lower end of the industrially accessible temperature range. These features avoid excessive dilution and produce acceptable yields. It is worth pointing out that too low a solubility might cause problems in the depletion of by-products. Furthermore, the tendency to form solvates with certain classes of solvents has to be considered. Any solvent that shows a tendency to form solvates should be avoided to minimize the risk of unwanted surprises during process development and later process transfer to chemical production. Solvent removal during drying should also be considered, especially for toxic solvents that have specified residual solvent limits in the low parts per million range.

Finally, solvent recovery should be considered early during process selection and development, especially if large amounts of solvent are used in the processing of high-volume APIs or if solvent mixtures have to be separated and purified. If possible, solvent mixtures should generally be avoided, not only to enhance solvent recovery but also because multiple solvents may cause multiple chances for solvate surprises.

Solvent screening is often performed to find as many polymorphic or solvated forms as possible and to find the optimal form and secure patent protection of different API forms. Apart from the case of solvates, the role of the solvent is kinetic only; therefore, one need not stay with solvents that promote the formation of the desired, stable form—if seeding is applied, any other solvent chosen according to the criteria above will also work.

18.7 NUCLEATION AND SEEDING

Industrial crystallization processes are typically initiated either by increasing supersaturation to a level above which primary homogenous nucleation is triggered (e.g., by cooling, antisolvent addition, or solvent evaporation) or by adding seeds after a desired value of supersaturation has been attained. In the case of primary homogeneous or heterogeneous nucleation, the nucleation rate strongly depends on supersaturation and the interfacial tension between the crystal and the solvent phases. By-products and the presence (or absence) of heterogeneous nuclei may very strongly affect the starting point of any industrial crystallization process, thus rendering robust processes and product quality control difficult if not impossible, at least in a number of cases. It is thus much more convenient to add seed crystals at a relatively low level of supersaturation, thereby avoiding spontaneous nucleation, and then crystallize the API with high purity at low to moderate supersaturation. Ideally, the final PSD is determined by the size and amount of seed crystals added, assuming that secondary nucleation is efficiently prevented. Under the above assumptions, the average particle size of the final product can be calculated from Equation 18.2, using only total seed mass and average seed size, which are both typically available in industrial practice[13]:

$$L_{final} = L_{seed} * (M_{final} / M_{seed})^{1/3} \qquad (18.2)$$

Examples of the successful implementation of the seeding concept are illustrated in Figure 18.3. In these cases, vastly differing particle sizes could be obtained by varying the seeds used.

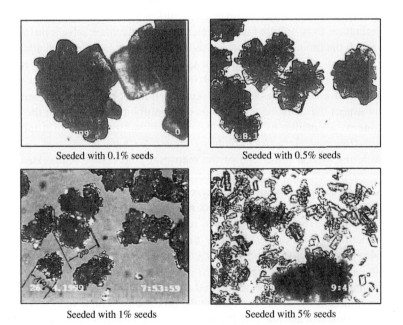

Seeded with 0.1% seeds	Seeded with 0.5% seeds
Seeded with 1% seeds	Seeded with 5% seeds

FIGURE 18.3 (See color insert following page 40). API particle size as a function of the amount of seeds added.

The samples in Figure 18.3 were manufactured during the development of an actual API crystallization process. In this particular case, the objective was to assess the effect of product particle size and PSD on the downstream drug product formulation process. An understanding of these effects is useful for setting reasonable specifications for particle size and PSD for the drug substance manufacturing procedure.

It is worth keeping in mind, however, that the simplified Equation 18.2 assumes that there are no additional nucleation events. Nucleation can only be prevented by robustly and efficiently controlling supersaturation in a low to medium range throughout the crystallization vessel and by avoiding the generation of large amounts of fresh nuclei by attrition (i.e., by particle–particle, particle–wall/baffle or particle–stirrer collisions). Both effects can be prevented by building supersaturation in a slow and controlled manner and by maintaining it in a low to medium range throughout the process. Generally, introducing a defined holding time after the addition of the seed material is recommended to equilibrate the system and to allow the seed material to deplete some of the available supersaturation in the system before process supersaturation is generated by cooling or antisolvent addition. Stirring should be kept at a level just sufficient to keep the crystals suspended. This detail is particularly critical for coarse particles and strongly depends on the stirrer type as well. For large crystals, stirrers with high axial flow are recommended (e.g., pitched-blade stirrers); in extreme cases, air bubbles may be the only way to keep large crystals suspended without destroying them by abrasion.[14]

Seeding is an efficient and underestimated tool for the development of robust API crystallization processes,[13] often because process chemists are reluctant to open the crystallizer during the process and add "impurity." This concern may be legitimate, but seeding is still a critical success factor for an API crystallization process because it frequently delivers superior crystallization and is, in the long run in many cases, the only way to reproducibly control the onset of a crystallization process. Seeding is not just the addition of a few crystals to initiate uncontrolled bulk crystallization but instead should be used to tailor the final product properties by adding a defined number of particles with a defined size distribution. Appropriate ways of efficiently dispersing the seed material in the process solvent or solvent mixture are thus of key importance.

In industrial practice, high-shear equipment is frequently used to deaggregate larger chunks and achieve sufficient dispersion. The addition of dry seed material may appear to be an attractive alternative, but it is often associated with operational issues. In particular, the dispersion in a nitrogen flow can be rather demanding from a safety and operational hygiene perspective, and agglomeration or aggregation can never be excluded, particularly when the material has previously been stored over a long period of time.

Another major advantage of seeded crystallization processes is the opportunity to add the desired polymorph at low to moderate supersaturation levels, thus successfully suppressing unexpected thermodynamic effects. This advantage can be explained by the energy barrier that a new, more stable form would have to overcome before the first nucleus is formed; in other words, the supersolubility curve of this new form would have to be reached to trigger nucleation of the new form. This effect for the case of monotropism is illustrated in Figure 18.2. If supersaturation is kept at a value below the metastable zone limit of the stable form (i.e., within the hatched area), then there is no danger that the new, more stable form actually occurs. The same advantageous effect can be exploited for solvents: if the desired solvated form is added as a seed, other forms are less likely to result. Notably, supersolubility curves or metastable zone limits, unlike solubility curves, are not determined by thermodynamics. They describe kinetic processes—that is, the formation of nuclei under conditions such as specific cooling rates, equipment, and by-product spectra, among others. A change in one or all of these parameters also changes supersolubility curves. Such a change may occur if, for example, a switch to an older reactor with scratched vessel walls becomes necessary for capacity reasons. In that case, nucleation may start at a substantially earlier time. Early nucleation may also occur if certain by-products that have hindered the onset of nucleation suddenly disappear.

The general recommendation is to keep supersaturation levels in the low to medium range to prevent additional nucleation events. This strategy has an additional beneficial effect on purity: owing to slow crystal growth, the purity of the final crystals is usually higher compared to that in the case of fast growth rates. Finally, the addition of seeds efficiently avoids the formation of amorphous material, which can later recrystallize in an uncontrollable fashion.

18.8 GENERATION OF SUPERSATURATION

After clear filtration,* which is typically performed with unsaturated stable solution to avoid solids formation and subsequent clogging of equipment piping, the API solution is transferred into the crystallizer. Before the actual crystallization process is initiated, the solution is supersaturated by means of cooling, antisolvent addition, or evaporation.

18.8.1 COOLING

Cooling is the recommended way to generate supersaturation for a number of reasons with operational importance:

1. Cooling can be applied very gently and in a very controlled manner by making use of a large reactor surface area, thereby substantially reducing the risk of uncontrolled, spontaneous nucleation.
2. In the cooling mode, supersaturation can be efficiently controlled using readily available solubility data as a function of temperature.
3. Scale-up is typically uncritical if excessive cooling rates are avoided. Large temperature gradients can carry the risk of spontaneous solids formation or encrustation on larger scale, when the surface-to-volume ratio is smaller.
4. Keeping the level of suspension in the crystallization vessel constant also prevents crust formation against the vessel walls.

Three typical and frequently used cooling profiles are illustrated in Figure 18.4.[15]

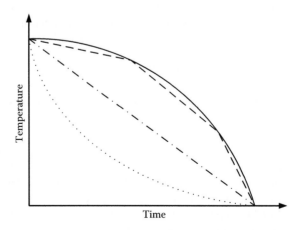

FIGURE 18.4 Different cooling modes; dotted line = natural cooling; dashed line = programmed cooling; middle line = linear cooling.

---

* In production, the API solution is usually clear-filtered prior to the initiation of crystallization in order to remove any solid dust particles or abraded particles from seals, and so forth, which might otherwise contaminate the API.

18.8.1.1 Linear Cooling

Linear cooling is widely used in industrial practice and can be recommended if cooling rates are not too high. When time allows in the given application (e.g., if only one batch per day is performed), the period for cooling can be chosen to maintain supersaturation at a sufficiently low level, effectively preventing any primary and secondary nucleation events.

18.8.1.2 Programmed Cooling

In programmed cooling, the temperature gradient varies throughout the crystallization procedure. Cooling is very slow at the beginning of the process and gradually increases as more crystal surface becomes available to consume increasing levels of supersaturation with time. In this way, the growth rate (expressed in microns per minute) is kept more or less constant, and thus high purities are typically obtained. For optimal control of supersaturation in an industrial crystallization process, process analytical technologies (PATs) can be applied to take into consideration process parameters such as a varying by-product content, which may adversely affect crystal growth on-line and *in situ*. Attenuated total reflection Fourier transform IR spectroscopy is a commonly used PAT, among many other commercially available on-line sensors and actuators. If programmed cooling is impossible in an existing control unit of a pilot- or production-scale manufacturing plant, it can easily be mimicked and approximated by sequential steps of linear cooling with increasing temperature gradients (dashed line in Figure 18.4).

18.8.1.3 Natural Cooling

Natural cooling combines high cooling rates at the very beginning of the crystallization process with high-solubility gradients at elevated process temperatures. Very high supersaturations are thus generated early in the process, when only a limited number of seeds and thus surface area are available in the system to consume excess supersaturation. This cooling mode is generally not recommended for API crystallizations.

18.8.2 Addition of Antisolvent

Adding an antisolvent to an API solution is a widely used approach to generate supersaturation in industrial solids formation processes. It is frequently applied when good yields are mandatory, but many process chemists may not be entirely aware of the risks associated with this procedure. The solvent composition in the process changes; hence, thermodynamic stability of pseudopolymorphs may also change. If these fluctuations are ignored, severe problems may occur during scale-up or later in production. Furthermore, excessive supersaturation levels may be generated at the locus of antisolvent addition in the reactor, often entailing spontaneous solids formation by precipitation in this region. This outcome can depend on the addition rate as well as on the way the solvent is introduced and distributed in the reactor. Spray balls may be used to decrease local supersaturation, but scale-up is difficult nonetheless.

Usually, antisolvent addition is used in precipitation processes—for example, when particles are to be formed spontaneously at high nucleation rates. This procedure typically yields very fine product and should be applied only if the API cannot be produced in a different way. Seeding may potentially be considered to better control the process, but supersaturation at the addition point is usually so large that primary or secondary nucleation can occur additionally, and fines* are generated. If antisolvent addition cannot be avoided for reason of process yield, it is recommended that the antisolvent be added before adding seeds and generating further supersaturation—that is,

---

* Fines is a fine fraction which in this case is generated by nucleation (i.e., the particles formed due to this event are finer than the already growing seed crystals). In this way, bimodal distributions may be obtained.

in the initial API solution after clear filtration or at the end of the crystallization process, after most of the product has already crystallized and a high crystal surface area is available that can consume the process supersaturation, thus avoiding accumulation and reducing the risk of spontaneous nucleation.

18.8.3 DISTILLATION

Distillation is also not an ideal approach to generating supersaturation—solvent composition changes in the case of solvent mixtures, which may trigger unfavorable solvate formation. Distillation poses an even greater risk for crust formation on the vessel wall or on the stirrer, both of which are in contact with a wide variation of solvent composition up to dryness throughout the entire process. If unwanted solvates or polymorphs are generated in such a manner, there is a risk of inoculating the entire batch with an unwanted crystal modification or solvate. Another drawback is the difficulty to monitor the solvent composition and determine the ideal point for seed addition at moderate supersaturation. This issue can be tackled by making use of suitable PAT. Furthermore, crusting often occurs in evaporative crystallization processes owing to the decreasing liquid level in the reactor, again entailing product loss.

18.8.4 ADDING REACTION PARTNER OR REACTION PRECIPITATION

In cases in which the product of a chemical reaction or salt formation is insoluble, nucleation events and precipitation may start instantaneously. These events can be difficult to control, but nevertheless one should investigate the possibility of designing an addition-controlled process—for example, by adding the reaction partner slowly—which may be coupled with appropriate seeding once supersaturation is reached. Scale-up in this area is generally difficult, and the addition mode (e.g., using spray balls or other devices to optimize the distribution of the reactant in the crystallization vessel) plays an important role. For details on precipitation, refer to the work of Söhnel and Garside.[16]

18.9 DOWNSTREAM PROCESSING

Downstream processing of APIs using filtration and drying is often underestimated and can frequently be a bottleneck in API manufacturing processes. It is therefore vital to keep these downstream unit operations in mind when a salt form or API candidate is selected or when the crystallization process is designed.

18.9.1 FILTRATION

Filtration on a laboratory scale is usually performed with relatively small amounts of API and thus relatively small filter cake heights. Because the time for cake formation scales quadratically with cake height, potential filtration issues on a large scale are often unforeseen in early phase process development. The filtration time of a suspension on plant scale, t_P, as a function of filtration time, t_L, and cake height, h_L, on a laboratory scale is determined by Equation 18.3:

$$t_p = t_L \cdot \left(\frac{h_p}{h_L}\right)^2 \tag{18.3}$$

For washing with a constant cake height, Equation 18.4 is used in which filtration time scales linearly with cake height:

$$\dot{V} = \frac{1}{K} \cdot \frac{\varepsilon}{(1-\varepsilon)^2} \cdot \frac{\Delta p \cdot dp^2}{\eta \cdot h} \cdot A \qquad (18.4)$$

where $\dot{V}$ is the flow rate; K is a constant; ε is porosity, void fraction; Δp is pressure difference; dp is particle size; A is filter area; η is viscosity; and h is cake height.

In Equation 18.4, it is shown that filtrate flow rate strongly depends on particle size, which additionally influences the porosity of the filter cake. One can readily conclude that controlling both the particle size and PSD and avoiding the formation of fines[2] (which may be triggered by unwanted nucleation events in an industrial crystallization process, for example) are vitally important.

If centrifuges are used, flow can further be decreased if the cake exhibits some compressibility after exposure to a centrifugal force field. Then the porosity typically decreases at increased filtration times. Peeling should be performed with low centrifuge speed and high knife speed to minimize the mechanical stress on the API crystals.

18.9.2 Drying

Drying on a plant scale is usually performed in agitated dryers (e.g., paddle dryers or double cone dryers). These can feature drying characteristics that differ significantly from those of commonly used laboratory-scale equipment. These features must be anticipated to avoid surprises such as lumping, abrasion, or larger drying times during scale-up. Most important, the vacuum in the plant usually is not as strong as that on a laboratory scale, where values below 1 mbar can be achieved rather easily. It is therefore crucial to know the conditions for solvent release, especially in the case of solvates. For this purpose, vacuum thermogravimetry systems (available commercially from various suppliers) can be used. These systems deliver information about solvent release as a function of pressure at a given temperature and thus allow for a better prediction of drying on plant scale.

A second aspect to consider is the mechanical stress acting on API particles in large-scale dryers, especially paddle dryers, which can lead to attrition and particle breakage (Figure 18.5).

This aspect is rather difficult to assess, because there is no readily available equipment for small-scale testing. For this purpose, we built two paddle dryers with volumes of 1 and 10 l in our workshop. Although it is obvious that the shear forces generated by the stirrer in a large powder bed are much higher than those in small-scale equipment, suitable scale-up criteria have yet to be defined and investigated thoroughly. If tailor-made bench dryers are inaccessible, it is recommended that mechanical stress be minimized to the level absolutely required to get rid of the solvent. A first hint of suitable operating conditions can be obtained from the drying characteristics that are assessed using thermogravimetric analysis. If the release of the solvent is uncritical and takes place at relatively high pressure (e.g., between 50 and 100 mbar), mechanical stirring can probably be limited if not stopped for certain intervals. The objective of stirring is then mainly reduced to improving heat transfer to the wet filter cake. For wet and sticky filter cakes, it is recommended that the stirrer not be stopped before the second drying phase is reached and flow improves, as it may be impossible to restart the stirrer, or the stirrer may even break owing to the high torque required. Sometimes, mechanical stress also helps to improve drying by breaking up aggregates and releasing the solvent entrapped therein. In other cases, the movement of the wet filter cake can lead to severe lumping by aggregation.

18.10 TO MILL OR NOT TO MILL?

Although some recommend that milling be avoided to save money, the author suggests that the benefits of milling usually outweigh its costs by evening out varying PSDs in chemical production. Experience shows that due to changes in starting material quality (purity), improvements in the synthesis, modifications in crystallization processes, and changes in production equipment or site

(A)

(B)

FIGURE 18.5 Influence of mechanical stress on drying; (A) API after filtration; (B) API after drying in a paddle dryer.

may make it difficult to keep API PSD constant over the years. It is therefore advantageous to have a final product-shaping step before the drug substance is delivered to the formulators. Bear in mind, however, that PSDs generated directly by crystallization and PSDs generated by milling do not necessarily exhibit the same powder properties. It is obvious that a broken surface has a different energy from a grown surface. Hence, there can be significant differences in the physical behavior of these materials. In the case of micronization in high-energy mills, it can be assumed that part of the crystal surface is amorphicized by mechanical impact or high temperatures. This amorphous content can take up moisture and recrystallize, thereby causing undesired effects. Generally speaking, the best strategy is to crystallize the drug substance a little coarser than is absolutely required and then use milling as a final shaping tool rather than try to provide the right particle size by direct crystallization.

18.11 SCALE-UP OF INDUSTRIAL CRYSTALLIZATION PROCESSES—MISSION IMPOSSIBLE?

Literally speaking, the scale-up of crystallization processes may frequently appear to be a "mission impossible," or at least a demanding endeavor, as changing the size of a crystallizer drastically affects numerous important parameters governing the crystallization process.[17,18]

An obvious example is the surface-to-volume ratio, the value of which decreases with scale-up. Subsequently, high cooling rates can only be maintained in large-scale industrial practice if the jacket temperature is decreased substantially. This decrease may critically affect the nucleation rate; however, as solution temperatures at and close to the crystallizer wall are also lowered, potentially enhancing primary or secondary nucleation in this region, the decrease may then cause crusting on the vessel walls.

Another example is considering stirrer speed and stirring efficiency. It is impossible to keep both tip speed—and thus the key parameter governing crystal abrasion—and power input by unit volume constant. In other words, tip speed has to be increased to keep the specific energy input upon scale-up constant. This may be critical in the case of large, brittle crystals that exhibit a pronounced tendency to settle and tend to break at higher stirrer speed. The only solution may then be to change the stirrer to a type with a higher axial flow component (e.g., from impeller to pitched-blade stirrer, which features more pronounced axial flow than the impeller—that is, it can be operated at lower tip speeds and still keep relatively large crystals in suspension, successfully avoiding particle sedimentation). Although changing stirrer type can be used to scale-up critical crystallization processes, this solution may not necessarily be viable in other cases. For example, if nucleation relies on fast and efficient micromixing, changing the stirrer type is infeasible.

Physical barriers to crystallization may be impossible to overcome during scale-up, and governing operating parameters in crystallization interact rather heavily, particularly on large scales. It is therefore indispensable to work out the factors and critical process parameters defining any API crystallization process and then derive appropriate scale-up criteria compatible with these major parameters only. In the case of industrial precipitation processes, such planning may, for example, mean that mixing needs to be analyzed and characterized thoroughly to facilitate process design and transfer to a larger scale. Other parameters that may need to be addressed include stirrer type, power input, and antisolvent addition locus, rate, and dispersion into the API solution. In conventional cooling crystallization, seeding is a powerful tool for eliminating many unknowns about mixing and the effect of solution homogeneity on nucleation, and thus comes highly recommended for the design of robust crystallization processes. Once the start of formation is efficiently, robustly, and reproducibly controlled through seeding, process control can focus on supersaturation generation and stirrer speed, which can be kept in an uncritical range.

18.12 RECOMMENDATIONS FOR DESIGN OF ROBUST CRYSTALLIZATION PROCESSES

The examples in this chapter reveal potential pitfalls in the design and scale-up of industrial crystallization processes. These may successfully be circumvented to yield robust and efficient processes by considering a few major points in the design phase. Most important, increasing process robustness reduces the number of degrees of freedom in the given process. Consequently, nucleation—which is generally difficult to control—should be suppressed using seed crystals with sufficient surface area, supersaturation should be generated in a smooth and controlled manner, and polymorphism and solvate formation have to be understood and successfully controlled (by means varying temperatures and/or solvent selection, for example).

With all that in mind, a certain level of uncertainty remains in all applications, as also outlined above. Solutions in one case may not help at all in another, and even small alterations in the molecule usually have dramatic effects on the crystal lattice and hence on physical properties, crystal shape, solubility, sensitivity toward by-product effects, and so on. It is again worth noting that ideally, the physicochemical characteristics of the API should be well known and understood, including special features such as solvate formation or polymorphism.

There are nine important questions that should be raised during the design of any crystallization process. These questions should also be asked when a process is transferred from the laboratory to the pilot and, eventually, to production scales. These questions are summarized as follows:

1. What are the desired drug substance properties for drug product formulation, and how are they taken care of in the process?
2. How is supersaturation generated and why? Is the use of an antisolvent really necessary?
3. How is nucleation controlled? Is seeding possible?
4. Polymorphism: are other polymorphs known? How are they related thermodynamically?
5. Are solvates known? Is the use of solvent mixtures really necessary?
6. Is the process controlled thermodynamically or kinetically?
7. Does the solubility at the final process temperature correlate with the yield?
8. What is the expected filtration time on plant scale?
9. Is the drying behavior characterized? Is abrasion in agitated driers a potential problem?

Process design and scale-up always carry risk, and the reasons for failure are often tedious to work out. Although it is clear that answering the above questions may not overcome all problems and risks in industrial practice, doing so with a solid understanding of the characteristics of the drug substance to be crystallized should help to minimize these risks.

REFERENCES

1. Küsters, E., Heuer, Ch., and Wieckhusen, D. *Journal of Chromatography A*, 2000, 874, 155–165.
2. McCrone, W.C. In *Physics and Chemistry of the Organic Solid State*; Fox, D., Labes, M.M., and Weissberger A., Eds.; Interscience: New York; 1965, 726–767.
3. Bernstein, J. *Polymorphism in Molecular Crystals*; Clarendon Press: Oxford, New York; 2002.
4. Burger, A. and Ramberger, R. *Mikrochim. Acta* 1979, II, 259–271.
5. *FDC Quality Control Records*, 1996, 30(6), 1–5.
6. Dunitz, J.D. and Bernstein, J. *J. Acc. Chem. Res.* 1995, 28, 193–200.
7. Chemburkar, S.R. et al. *Org. Process Res. Dev.*, 2000, 4(5), 413–417.
8. Müller, M. et al. *Crystal Growth & Design*, 2006, 6 (4), 946–954.
9. Khoshkhoo, S. and Anwar, J. *J. Phys. D: Appl. Phys.* 1993, 26, B90–B93.
10. Threlfall, T.L. *Analyst*, 1995, 120, 2435–2460.
11. Hancock, B.C. and Zografi, G. *J. Pharm. Sci.* 1997, 86, 8.
12. Stahl, P.H. and Wermuth, C.G. *Handbook of Pharmaceutical Salts, Properties, Selection and Use*; Wiley-VCH: New York; 2002.
13. Heffels, S.K. and Kind, M. *Proceedings of the 14th International Symposium on Industrial Crystallization in Cambridge* 1999, 14, 2234–2246.
14. Braun, B., Grön, H., and Tschernjaew, J. *J. Cryst. Growth Des.* 2004, 4, 915–920.
15. Nyvlt, J. and Mullin, J.W. *Chem. Eng. Sci.* 1971, 26, 369–377.
16. Söhnel, O. and Garside, J. *Precipitation: Basic Principles and Industrial Applications*; Butterworth-Heinemann: Boston; 1992.
17. Oldshue, J.Y. In *Handbook of Industrial Crystallization*; Myerson, A.S., Ed.; Butterworth-Heinemann: Boston, 1993.
18. Genck, W.J. *Chem. Eng. Prog.* 2003, 99, 36–44.

19 *In Situ* Mid-Infrared Spectroscopy for Process Development

David A. Conlon, Bill Izzo, J. Christopher McWilliams, Robert A. Reamer, Feng Xu, and Paul Collins

CONTENTS

19.1 APPLICATIONS AND ADVANTAGES

There are numerous analytical methods that can be applied to process monitoring, and the optimal technique depends upon the physical properties most relevant to the transformation in question. For example, while gas chromatography (GC) or high-performance liquid chromatography (HPLC) would be applicable to monitoring the progress of a Diels–Alder reaction, acid titration would be more appropriate for a step in which an HCl salt is formed. In the past, analytical methods for process monitoring in the pharmaceutical industry have been predominantly indirect or "off-line" in nature. That is, they are not real-time, direct measurements of processes. Rather, they are measurements of samples taken from the process, and typically transformed into a new mixture. For example, HPLC analysis usually involves sampling and dilution prior to analysis. Although often an accurate measurement of the reaction mixture, it is not a direct analysis of the actual reaction medium. In contrast, *in situ* or real-time monitoring involves analysis of the reaction mixture directly through incorporation of the analytical measurement device into the process vessel or stream.

There are several advantages of real-time monitoring. Unstable intermediates can be detected and monitored for conversion and stability, often a key factor in multistep transformations such as the preparation and subsequent reaction of main group organometallics. In addition, intermediates not observable by indirect analysis can provide mechanistic insight. Uncertainties and errors that arise from quenching conditions are eliminated with real-time monitoring. For example, attempting to quench low-temperature samples can result in errors arising from the necessity to transfer a sample from the reaction to a quenching medium, usually requiring transport through a heated medium (e.g., subsurface to surface line). Errors can also occur if the quench does not result in clean or complete conversion to a calibrated intermediate. There are also distinct disadvantages of real-time monitoring. In most cases, the technology and techniques require careful calibration to achieve the same accuracy usually obtainable by indirect analysis. An indirect method such as HPLC, GC, or nuclear magnetic resonance (NMR) is still required to maintain calibration of the instrument. Calibration, maintenance, and troubleshooting of the instruments applied to real-time monitoring typically require qualified personnel with specialized training. Thus, there is a balance of benefit and costs that has to be considered when applying real-time monitoring to manufacturing processes.

The nature of the pharmaceutical business necessitates that process chemistry place as much emphasis on product quality as it does on developing efficient synthetic routes to final API (active pharmaceutical ingredient). This is not only an ethical imperative, but one that is increasingly regulated by the U.S. Food and Drug Administration (FDA). Consequently, once an efficient process that produces high-quality API has been achieved in the laboratory, it is critical to design means for consistently reproducing those results in a manufacturing arena. This, in turn, usually translates to developing analytical techniques that aid in understanding and monitoring the individual processing steps that lead to final API. Process analytics technologies (PATs) has emerged as a prominent and powerful tool in the development and manufacture of pharmaceutical products and pharmaceutical ingredients to address these issues. This chapter serves as an introduction to the area of PAT, an emerging field with respect to discovery, development, and commercialization within the pharmaceutical industry. PAT is defined by the FDA as "a system for designing, analyzing, and controlling manufacturing through timely measurements (i.e., during processing) of critical quality and performance attributes of raw and in-process materials and processes with the goal of ensuring final product quality."

The analytical portion of this system covers the areas of chemical and physical attributes of a product or the attributes of the process through which a product is generated. The quality of the final product is normally closely linked to the attributes of the raw materials and the process performance. At the laboratory scale, these techniques are traditionally referred to as *in situ* or real-time monitoring.[1] The PAT designation expands the traditional role to incorporate modeling, feedback control, and real-time release of a product within a regulated industry. This definition includes traditional sensors such as temperature, pressure, density, and conductivity; however, the more recent focus is on advanced spectroscopic techniques such as mid-infrared, near-infrared, Raman, and laser-light scattering. These techniques are applied at both the laboratory and manufacturing scales in qualitative and quantitative modes to assess the identity and quantity of stable and transient chemical species as well as particle size and crystal morphology.

The chemical and petroleum industries have leveraged the power PAT for decades to reduce cost and improve quality. The pharmaceutical industry, however, has not fully embraced the technology in a manufacturing arena.[2] The reasons for the disparity can be partially traced to the increased regulatory burden associated with commercialization of a new chemical entity. In addition, manufacturing of active pharmaceutical ingredients (APIs) or drug substances within the pharmaceutical industry has historically been dominated by batch processing due to the annual production volumes, inventory management strategy, and the need to meet the required chemical purity and physical characteristics of the API. Continuous processing is neither necessary nor advantageous in most production routes for new chemical entities. In contrast to API manufacturing, drug product manufacturing (tablets, capsules, parenteral preparations, etc.) is more amenable to continuous processing, thus it is not surprising that implementation of PAT in this arena far exceeds that of active pharmaceutical ingredients.[3–5]

In 2004, the FDA launched a broad-reaching Quality by Design Initiative focused on improving the quality and safety of drugs while simultaneously reducing cost to consumers.[6] The cornerstone of the initiative is the acceptance and implementation of PAT across all aspects of the pharmaceutical industry. The industry is currently expanding the role of PAT throughout discovery, development, and commercialization[7–9] In the area of chemical development, PAT has substantially increased the scientific understanding for manufacturing of drug substance and drug product. Within the manufacturing arena, PAT can increase efficiency through improved sample integrity, reduction in cycle time, feedback control, and minimization of off-specification product.[10,11] In the development area, real-time monitoring of the concentration of reagents, intermediates, and products at extreme conditions of temperature and pressure rapidly provides fundamental understanding of the underlying mechanisms and kinetics for organic and inorganic reactions.[12] This understanding, in turn, facilitates the design and optimization of economically viable and safe manufacturing processes in the pharmaceutical industry. In addition, safety and industrial hygiene risks as well as process safety risks can be substantially mitigated through implementation of PAT.[13-15]

PAT encompasses a diverse range of measurement techniques from conductivity to Raman spectroscopy. With regards to API process development, spectroscopic techniques (near-infrared [NIR], mid-infrared [mid-IR], and Raman, etc.) have all demonstrated enormous utility within the laboratory, pilot-plant, and manufacturing areas.[16] One of the most versatile technologies in this area is mid-IR owing to its specificity and adaptability to manufacturing environments.[17] Mid-IR, has become a standard tool for investigating and developing complex reactions owing to its molecular specificity. This technique can be used to identify a reaction product or transient intermediate from the characteristic mid-IR spectrum of the molecule and can also be used to measure the concentration of a compound in a complex mixture.

The development of the attenuated total reflectance (ATR) technology with its short path length (10 to 20 μm) allows the use of mid-IR directly on process streams including slurries.[18] The use of chemically inert and hard diamond as the ATR optical element provides an excellent interface with the process stream, with available technology covering a broad range of temperatures (80 to 250°C), pressure (0 to 15,000 psi), and pH (0 to 14). Other materials such as silicon and zirconia provide greater sensitivity in certain spectral ranges; however, these materials have limitations with respect to pH (<10 for silicon) and optical range (limited to >1500 cm[1] for zirconia). Although mid-IR ATR provides a versatile and valuable *in situ* tool for laboratory development and scientific investigation, many disadvantages exist for the implementation of this technology pilot plants and manufacturing facilities. The main disadvantage of IR compared to NIR and Raman is the required proximity (3 to 5 feet) of the analyzer to the ATR probe. Standard instruments use rigid conduit containing mirrors to direct incident and reflected radiation to and from the process stream. Newer technology involving silica fibers in place of the conduit and mirrors is available; however, length is still restricted to approximately 1 meter, and the silica reduces spectral range and throughput of energy. Some of the proximity issues can be addressed by using a recycle loop to bring the process stream to the analyzer. This approach can, however, add additional process risk if the operating temperature deviates significantly from ambient conditions. Another major challenge for the operation of process analyzers is electrical classification in a process area. For nonaqueous processing, explosion-proof purging is required. This increases the size of the analyzer and introduces additional utility requirements to maintain purging.

The next three chapters are devoted to three case studies in the application of mid-IR spectroscopy to API process development. The examples provided cover the spectrum of development—laboratory, preparatory, and pilot-plant scales. The first two examples demonstrate the utility of mid-IR spectroscopy in identifying reaction intermediates to propose or confirm underlying reaction mechanisms. The first example involves reaction at cryogenic conditions and provides a contrast to the more traditional investigational tool of nuclear magnetic resonance (NMR). The second example expands upon the theme of reaction mechanism by demonstrating the collection of kinetic data by mid-IR, and the subsequent verification of a reaction mechanism by kinetic modeling. The last example focuses predominately on the quantitative method development with mid-IR as an alternate approach to traditional methods such as high-performance liquid chromatography (HPLC), Karl Fisher (KF), and NMR. In addition, these chapters will provide some perspective on the cost, drivers, and regulatory aspects of implementation at pilot-plant and manufacturing scales.

REFERENCES

1. Hassel, D.C.; Bowman, E.M. *Applied Spectroscopy*, 1998 *52*, 18A.
2. Abboud, L., Hensley, S., *Wall Street Journal*, September 3, 2003.

3. Ge, Z., Buchanan, B., Timmermans, J., DeTora, D., Ellison, D., Wyvratt, J. *Process Control and Quality*, 1999, *11*, 277.
4. Higgns, J.P., Arrivo, S., Reed, R.A, *Journal of Pharmaceutical Sciences*, 2003, *92*, 2302.
5. Wikstrom, H., Marsac, P.J., Taylor, L.S. *Journal of Pharmaceutical Sciences*, 2004, *94*, 209.
6. Watts, D.C. Process Analytical Technology (PAT): What's in a name?, Science Seminar Series for the Office of Commissioner, Center for Drug Evaluation and Research, Food and Drug Administration, April 9, 2004.
7. Wartwig, S., Neibert, R.H.H., *Advanced Drug Delivery Reviews*, 2005, *57*, 1144.
8. Gurden, S., Westerhuis, J.A., Smilde, A.K. *AIChE Journal*, 2002, *48*, 2283.
9. Yan, B., Fang, L., Zhao, J., Irving, M.M. *Analytical Sciences*, 2001, *17*, i487.
10. McKelvy, M.L., Britt, T.R., Davis, B.L., Gillie, K., Graves, F.B., Lentz, L.A. *Analytical Chemistry*, 1998, *70*, 119R.
11. Reference 3.
12. Palucki, M., Lin, Z., Sun, Y. *Org. Process Res. Dev. 2005,* 9 (2), 141.
13. An Ende, D., Clifford, P., DeAntonis, D., SantaMaria, C., Brenek, S. *Org. Process Res. Dev.*,1999, *3*, 319.
14. Espinoza, L., Lucas, D., Littlejohn, D. *Applied Spectroscopy*, 1999, *63*, 97.
15. Espinoza, L., Lucas, D., Littlejohn, D. *Applied Spectroscopy*, 1999, *63*, 103.
16. Wartwig, S., Neibert, R.H.H. *Advanced Drug Delivery Reviews*, 2005, *57*, 1144.
17. Reference 3.
18. Harrick, N.J. *Internal Reflection Spectroscopy*, Harrick Scientific Corporation, Ossining, New York, 1979.

20 Optimizing an Asymmetric Homologation in a Tandem Asymmetric Homologation–Homoaldol Process

J. Christopher McWilliams and Robert A. Reamer

CONTENTS

20.1 INTRODUCTION

The Merck labs described a process for coupling an asymmetric homologation to an asymmetric homoaldol reaction in a previous communication.[1] In this tandem asymmetric transformation, chiral amide enolates derived from aminoindanol amides (**I**) were homologated to the zinc homoenolates (**II**), which were subsequently treated with aldehydes following transmetallation to titanium, yielding the homoaldol products (**III**) with two new stereocenters (Scheme 20.1). This case study details the successful development of this complex process as a direct consequence of insights provided by *in situ* infrared (IR) monitoring. In addition, *in situ* IR enabled a more in-depth understanding of the intermediate zincate enolate structures, which was not available via alternative methods, such as low-temperature nuclear magnetic resonance (NMR). Although we used NMR as a key complement to the IR observations, attempts to use NMR exclusively were not tenable. Many of the intermediates in the process were water, air, and temperature sensitive. In addition, the reaction system consisted of binary and ternary solvent mixtures, further complicating the NMR spectra. When combined with the requirement to conduct the chemistry on very small scale to accommodate NMR tubes, obtaining NMR data proved a very involved process.[2] In contrast, by simply conducting the reaction in a vessel

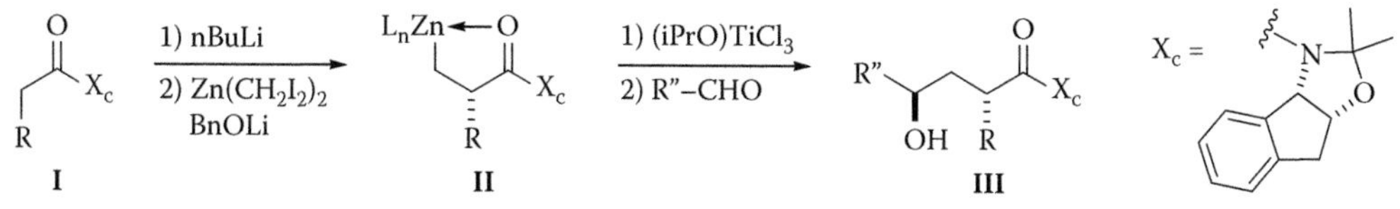

SCHEME 20.1 Adapted with permission from the Journal of the American Chemical Society, Vol. 118, No. 47, 11970–11971. Copyright 1996 American Chemical Society.

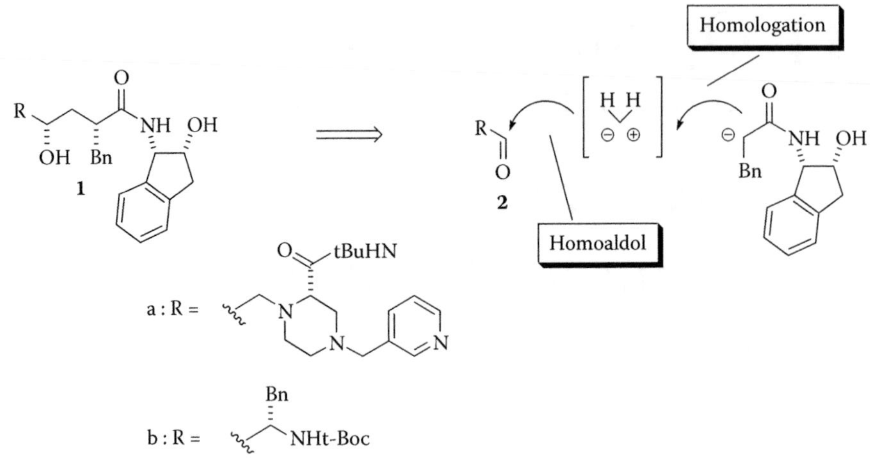

SCHEME 20.2

fitted with the IR probe, we were able to obtain *in situ* IR data. All other issues, such as scale and temperature control, were unaffected by the choice to employ *in situ* IR.

The motivation for undertaking this study arose from the need to develop efficient synthetic routes to HIV protease inhibitors that were in early development at the time, such as indinavir (**1a**) and **1b** (Scheme 20.2). One proposed route consisted of setting both stereocenters via a one-pot, tandem asymmetric homologation coupled to an asymmetric aldol. The precedent for this transformation consisted of the known high stereoselectivity in the alkylation of lithium enolates derived from **1** and reports describing stereoselective homoaldol reactions between ester and amide homoenolates and α-amino aldehydes.[3] A report describing homologations of ketone and aldehyde enolates with zinc carbenoids appeared an attractive candidate for the bipolar methylene component.[4] Although promising, nothing was known about the behavior of these reagents toward other types of enolates, such as esters or amide enolates. With this as background, a single experiment in which the lithium enolate derived from amide **4** was sequentially treated with a solution containing bis(iodomethyl)zinc (*BIZ*), a mixed titanium species, and aldehyde **2b** provided the tandem homologation–homoaldol product **6** with high diastereoselectivity, albeit in low yield (<30%) (Scheme 20.3).[5]

20.2 PROCEDURES

Having established proof of concept, the objective became optimizing the potential of the reaction by improving the overall yield. Due to the complexity of the reaction, and the distinct steps involved,

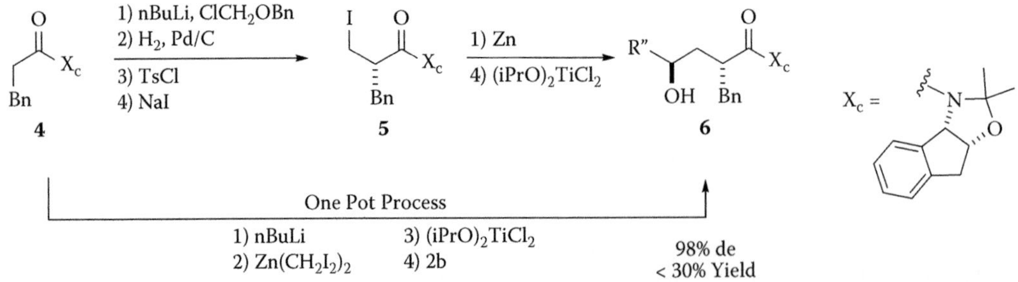

SCHEME 20.3 Adapted with permission from the Journal of the American Chemical Society, Vol. 118, No. 47, 11970–11971. Copyright 1996 American Chemical Society.

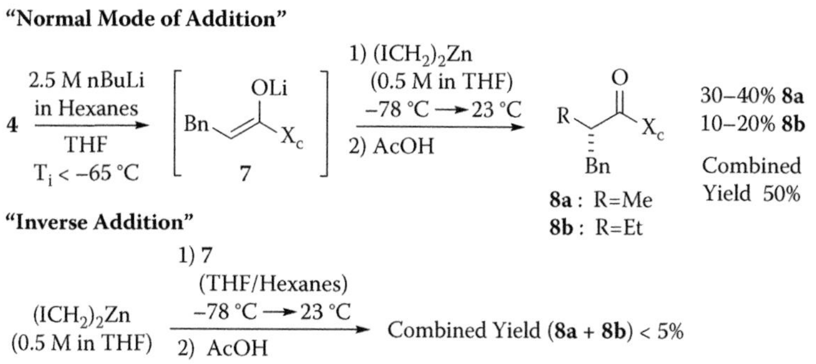

"Normal Mode of Addition"

$$4 \xrightarrow[\substack{\text{THF} \\ T_i < -65\,°C}]{\substack{2.5\ M\ nBuLi \\ \text{in Hexanes}}} \left[\substack{\text{Bn} \diagdown \diagup \substack{OLi \\ X_c} \\ \textbf{7}} \right] \xrightarrow[\substack{\text{2) AcOH}}]{\substack{1)\ (ICH_2)_2Zn \\ (0.5\ M\ in\ THF) \\ -78\,°C \rightarrow 23\,°C}} \substack{R \diagdown \diagup \substack{O \\ X_c} \\ \text{Bn}}$$

30–40% **8a**
10–20% **8b**

Combined
Yield 50%

8a : R=Me
8b : R=Et

"Inverse Addition"

$$\substack{(ICH_2)_2Zn \\ (0.5\ M\ in\ THF)} \xrightarrow[\substack{2)\ AcOH}]{\substack{1)\ \textbf{7} \\ (THF/Hexanes) \\ -78\,°C \rightarrow 23\,°C}} \text{Combined Yield (\textbf{8a} + \textbf{8b}) < 5\%}$$

SCHEME 20.4 Adapted with permission from the Journal of the American Chemical Society, Vol. 118, No. 47, 11970–11971. Copyright 1996 American Chemical Society.

evaluating each step individually was the obvious strategy. Prior results indicated the reaction between the titanium homoenolate derived from **5** and **2a** provided reasonable yields of **6**. Thus, we realized the key issues likely resided in the steps prior to the homoaldol reaction.

The first step in the sequence was lithium enolate (**7**) formation. Previous studies with this substrate indicated that **4** was prone to partial decomposition when subjected to n-BuLi.[6] Our observations were consistent with this, typically observing 8 to 10% loss of **4** via decomposition pathways. Several alternative strong bases were examined, including tert-BuLi, sec-BuLi, and LiTMP. Unfortunately, none of these alternatives resulted in improved results. Weaker bases, such as LiHMDS, LDA, and the zincate derived from mixing n-BuLi with Et$_2$Zn, were not basic enough to sufficiently deprotonate **4**. Although alternative strategies were available, we chose to accept the loss in the short term until further optimization revealed the potential of subsequent steps.[7] The amount of decomposition was minimized by maintaining low internal temperature (−65°C) during the course of deprotonation.

The second step in the sequence consisted of homologation of **7** to the presumed homoenolate of type **II**. We evaluated the efficiency of this reaction by quenching the reaction mixture with acid, then assaying for the amount of methylated product **8a** (Scheme 20.4). The first observation made was the presence of the ethylated side product **8b**. This is a likely consequence of alkylation of the ethyl iodide formed as a by-product in excess when *BIZ* is prepared by metathesis between diiodomethane and diethylzinc (Equation 20.1).[8]

$$Et_2Zn + CH_2I_2 \xrightarrow[\text{or THF}]{\text{DCM}} (ICH_2)_2Zn + 2\ EtI \qquad (20.1)$$

In addition, we made other critical observations. The overall conversion to **8a** and **8b** was always around 50%, regardless of the varying distribution of the two products. When the order of mixing was reversed such that the solution of **7** was added via cannula to a precooled solution of *BIZ*, very little of either product was observed. Finally, the bright yellow color indicative of the solution containing **7** tracked with enolate reactivity. In the normal mode of addition, the yellow color had dissipated once 50 mole% of the *BIZ* solution had been added. In the inverse mode of addition, the yellow color dissipated immediately upon making contact with the solution containing excess *BIZ*. From these observations, we proposed that a new, less-reactive species was forming when the *BIZ* solution and the lithium enolate were mixed. In the presence of excess reagent, this unreactive species formed more rapidly than homologation or ethyl iodide alkylation.

Thus, we set out to determine whether we could directly observe and identify the unreactive species, and this is when we turned to *in situ* IR for insights. It seemed reasonable that a lithium-to-zinc transmetallation was a likely cause of the observed attenuated reactivity. It is well established that zinc enolates are much less reactive than the corresponding lithium enolates.[9] The addition of

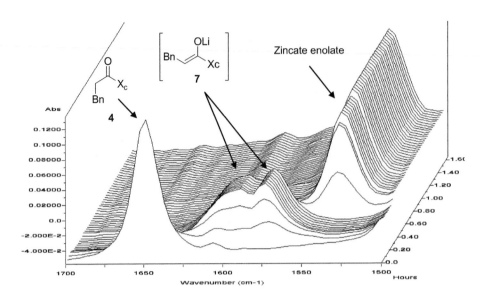

FIGURE 20.1 The addition of nBuLi, followed by Et$_2$Zn to **4**.

dialkylzinc reagents to lithium enolates has been utilized as a protocol for intentionally reducing their reactivity.

Due to the potential complications of evaluating mixtures containing **7** and *BIZ*, we began by looking at the reaction of the lithium enolate **7** with the unreactive dialkylzinc, diethylzinc. Figure 20.1 shows our first *in situ* IR experiment.[10] The initial spectrum showed an amide stretch at 1648 cm^{-1} representing **4** in THF solution. Upon addition of nBuLi, the amide dissipates, and three new absorbances arising from the lithium enolate **7** are observed at lower wave numbers (1575, 1590, and 1603 cm^{-1}). That there are several absorbances representing **7** indicates the presence of isomers, aggregates, or mixtures thereof. The NMR of **7** supports the presence of C–N rotamers (*vide infra*), which is typical of most amides derived from this aminoindanol chiral auxiliary.[11] Addition of diethylzinc to **7** resulted in the rapid and complete formation of a new species with an absorbance at 1536 cm^{-1}.

In a complementary experiment, the lithium enolate **3** was added in an inverse mode to a solution of *BIZ* in THF. Similarly, a new species represented by the absorbance at 1559 cm^{-1} formed rapidly and completely (Figure 20.2). No apparent free lithium enolate **7** was observed in either experiment. Furthermore, one can see directly from the IR spectra that both species were indefinitely stable and unreactive at –70°C.

The infrared experiments revealed to us the formation of a new species, which was apparently an amide zincate enolate, the first observation of its kind.[12] The structure of this enolate was not clear from the data, however. The absorbances at 1536 cm^{-1} and 1559 cm^{-1} appeared too low to assign to either an oxygen-metallated enolate carbon–carbon double bond structure as in **9**, or to a carbon-metallated enolate with a free amide carbon–oxygen double-bond stretch as in **10** (Figure 20.3). Rather, a carbon-metallated structure retaining a strong metal–oxygen bond, as in **11**, seemed a more accurate description. The dative bond between the carbonyl oxygen and metal center results in a lower bond order for the carbonyl, which is reflected in a carbonyl stretch at substantially lower wave numbers.

To provide support for this assignment, we examined the IR and NMR spectra of several species derived from the lithium enolate **4** (Scheme 20.5, Table 20.1). The ^{13}C chemical shift at C$_2$, along with the one-bond C$_2$ carbon–hydrogen coupling constant (J$_{CH}$), provides a distinction between the oxygen-metallated structure and the two carbon-metallated structures, whereas the ^{13}C chemical

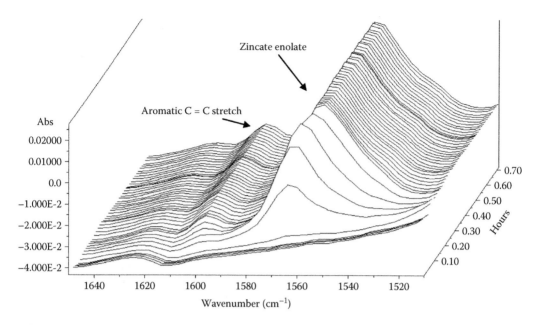

FIGURE 20.2 The inverse addition of **4** to a solution of *BIZ*.

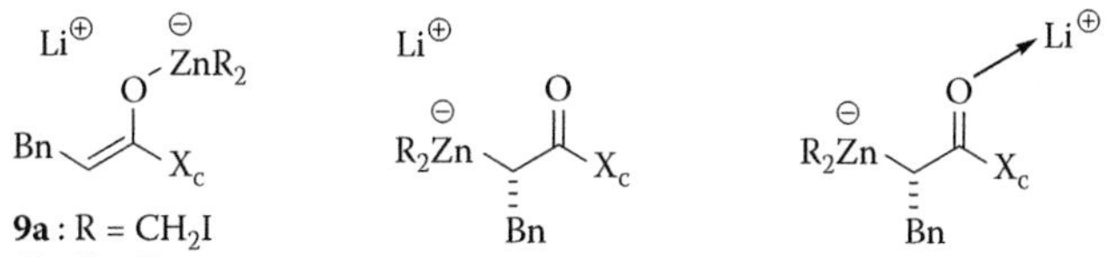

9a : R = CH₂I
9b : R = Et₂

10a : R = CH₂I
10b : R = Et₂

11a : R = CH₂I
11b : R = Et₂

FIGURE 20.3 Possible structures for the zincate enolate.

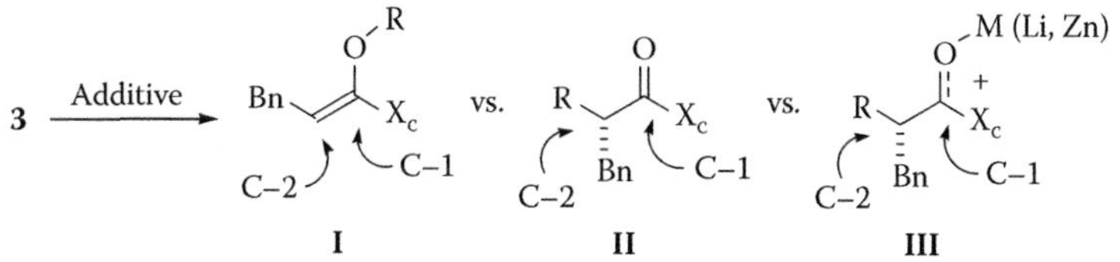

SCHEME 20.5

shift at C_1 and the infrared stretching frequency can be used to distinguish between the two carbon-metallated structures (**II** and **III**). The starting amide **4** provides representative data for an sp^3-hybridized carbon at C_2, and no oxygen–metal bonding. In contrast, the lithium enolate **7**, with a downfield C_2 chemical shift at 73 ppm, and a large $^1J_{CH}$ coupling constant of 150 Hz, is clearly an sp^2-hybridized carbon (i.e., an oxygen-metallated enolate). When the lithium enolate **7** is quenched with chlorotrimethylsilane, a new species is formed which, by its high field C_2 chemical shift (42 ppm) and small $^1J_{CH}$ coupling constant (123 Hz), is clearly an sp^3-hybridized species. The similarity

TABLE 20.1
Physical Data of Representative Enolates

Additive	R	C-1[a]	C-2[a]	J_{C2-H}	IR (cm^{-1})	Structure
H$^+$	H[b]	169	39	130	1648	**II**
None	Li	160	73	150	(1597, 1575)	**I**
TMSCl	TMS	170	42	123	1629	**II**
ZnI$_2$	ZnI	178	45	127	1567	**III**
(ICH$_2$)$_2$Zn	(ICH$_2$)$_2$Zn	179	45		1559	**III (11a)**
Et$_2$Zn	Et$_2$Zn	177	52	136	1536	**III (11b)**

[a] PPM (^{13}C NMR).
[b] Starting benzylacetamide **4**.

in the C$_2$ chemical shift and IR stretching frequency when compared to starting amide **4** indicates it is a carbon-metallated enolate with no metal–oxygen interaction. This result was somewhat surprising in light of the penchant for silicon to form oxygen-metallated enolates. Thus, it is apparent that the enolate derived from amide **7** has a preference to form carbon-metallated structures, even with electrophilic reagents that typically react at oxygen.

We then examined the structures of the three zinc enolate species derived from ZnI$_2$, *BIZ*, and Et$_2$Zn. It is clear from the data in Table 20.1 that all of these species are similar in structure.[13] The C$_1$ chemical shifts range from 45 to 52 ppm with $^1J_{CH}$ coupling constants ranging from 127 to 136 Hz. Thus, all three enolates are sp^3-hybridized at C$_1$. The C$_1$ chemical shift has shifted 8 ppm downfield from the starting amide **7**, implying less electron density at this carbon. In addition, the infrared absorbances for all species are below 1567 cm^{-1}. These latter two pieces of data indicate significant metal-oxygen bonding.[14] Interestingly, as one progresses to more electropositive zinc species (Et$_2$Zn→*BIZ* →ZnI$_2$), the infrared absorbance frequency shifts to higher wave numbers. This is likely a result of electron-pull toward the increasingly electropositive zinc atom bonded to carbon, and away from the metal atom bonded to oxygen, increasing the strength of the carbonyl bond.

We were also curious as to the identity of the metal bound to the carbonyl oxygen, which could be either zinc or lithium. One would expect zinc in the case of the enolate derived from ZnI$_2$, which is not actually a zincate, but the more well-known zinc enolate Reformatsky reagent. The structures of these enolates have been established to be dimeric, with zinc acting as the bridging metal via bonding to both carbon and oxygen. A similar structure for the zincate species is possible, although no precedent existed for this. We decided to probe this question by adding a complexing agent, 12-crown-4. Accordingly, we subjected the zincate enolate **11b** derived from **7** and Et$_2$Zn to an equivalent of 12-crown-4. What we observed was a distinct 12 cm^{-1} shift to higher wave numbers (Figure 20.4). The addition of more than one equivalent of 12-crown-4 had no further effect on the IR spectrum. This indicates a clear interaction between the zincate enolate and 12-crown-4. The shift to higher wave numbers is consistent with binding of 12-crown-4 to metal bound to oxygen, which results in a weakening of the carbonyl–metal bond and a corresponding increase in the carbonyl carbon–oxygen bond order.[15] We currently ascribe these results to complex formation with lithium, as opposed to zinc (Equation 20.2).[16] Although complexes of larger metals and 12-crown-4 are known, those for zinc are limited to highly unsaturated metal complexes.[17]

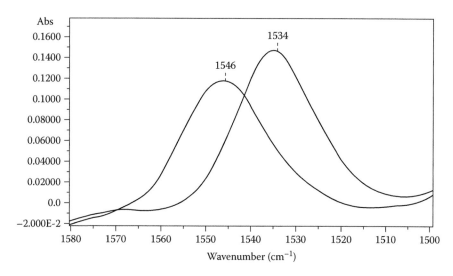

FIGURE 20.4 (See color insert following page 40). Infrared absorbance shift upon addition of 12-crown-4 to **11b**.

$$(20.2)$$

Having established the nature of the unreactive zincate enolate as **11a**, we now needed to discover a means of "activating" the zincate toward homologation. From our original observations, we noted that homologation occurred to some extent during the early phases of the normal addition mode of *BIZ* to **4**. Although one explanation is that free lithium enolate was solely responsible for attack on *BIZ* and EtI, this was not consistent with the presumed mechanism of carbenoid homologations, which are assumed to proceed through 1,2-migrations. In contrast, **11a** is structurally predisposed for a 1,2-migration from carbon. Thus, another possibility was that **11a** becomes activated in the presence of excess **7**. The activator, in this case, could be a second molecule of **7**. This would also be consistent with the observed 50% conversions.

Once again, direct observation of the effects of a second equivalent of **4** added to **11a** could be easily obtained by *in situ* IR. Accordingly, a solution of **11a**, prepared by inverse mode addition of **4** to *BIZ*, was treated with a second equivalent of **4** at low temperature (Figure 20.5). What was observed was the rapid formation of a homoenolate species represented by the carbonyl stretch at 1602 cm^{-1}, and an 8 cm^{-1} shift in the zincate enolate absorbance to 1551 cm^{-1}. The homologation occurred too rapidly to observe either free lithium enolate **7** or any new high-energy intermediate that may have formed prior to homologation. Thus, the species represented by the absorbances at 1602 cm^{-1} and 1551 cm^{-1} is consistent with the mixed zincate **13** (Equation 20.3). The absorbance at 1602 cm^{-1} representing the homoenolate amide carbonyl is consistent with dative bonding to the zinc metal, forming a five-membered metallocycle. The unreacted enolate exists as a carbon-bound zincate species, although the carbonyl stretch has shifted slightly to 1551 cm^{-1}. High-performance liquid chromatography (HPLC) analysis of this mixture confirmed that it was an approximate 50:50

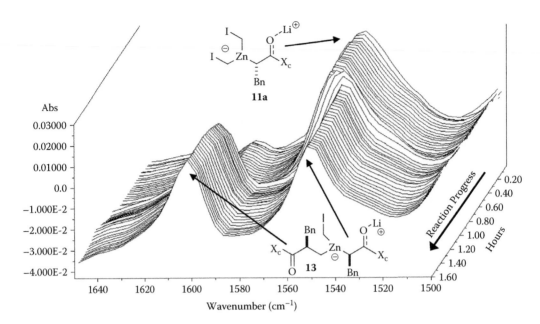

FIGURE 20.5 Addition of enolate **7** to zincate enolate **11a**.

mixture of enolate and homoenolate.[18] The shift of the carbon-metallated enolate carbonyl absorbance to lower wave numbers is consistent with an increase in electron density on the zinc metal
center, resulting in an electron push toward the counterion. This experiment provided direct evidence
in support of a hypothesis that suggested the second equivalent of enolate accounts for the observed
reactivity.

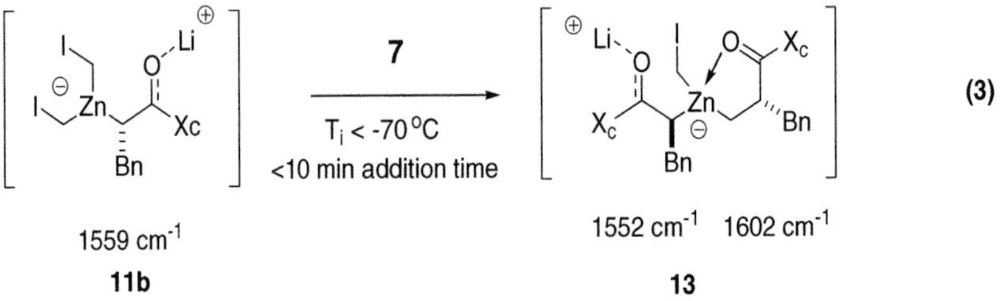

$$ (3) $$

$$ (20.3) $$

$1559\ \text{cm}^{-1}$

11b   **13**

The *in situ* IR observations shown in Figure 20.5 provided the key result that led us to propose
that an activator was needed to drive the homologation of zincate enolate **11a**, and that the activator
in this case was a second equivalent of **7**. Our observations appeared similar to those described by
Harada and Oku.[19] They had observed certain α-halo diorganozinc compounds could be driven to
undergo 1,2-migration if an excess of alkoxide was added (Scheme 20.6). Because higher conver-

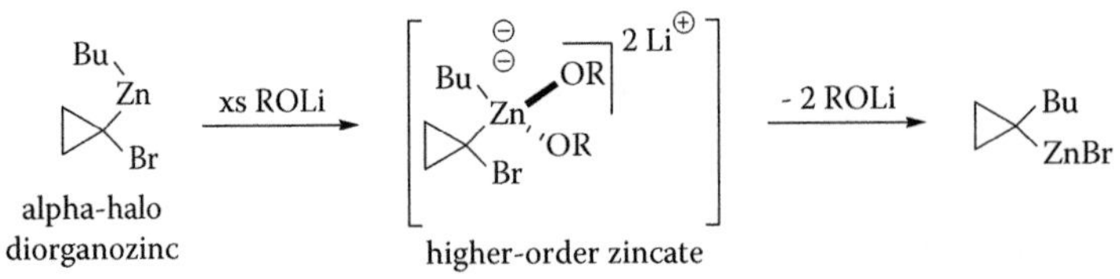

SCHEME 20.6

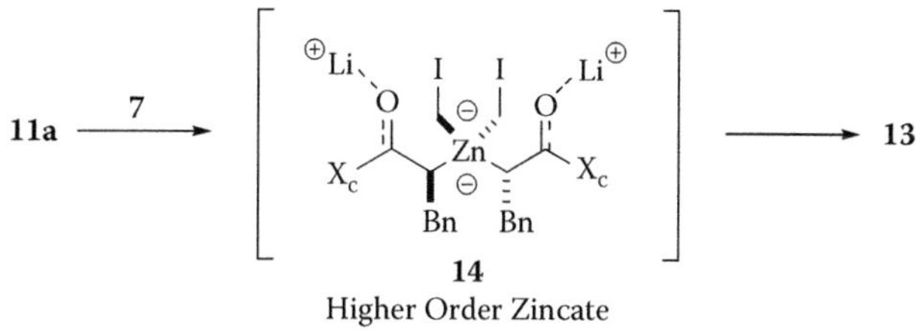

$$11a \xrightarrow{7} \left[\begin{array}{c} \overset{\oplus}{Li} \diagdown O \quad \overset{I}{\underset{\ominus}{|}} \quad \overset{I}{\underset{|}{|}} \quad O \diagup \overset{\oplus}{Li} \\ X_c \overset{\|}{\diagup} \quad \underset{\ominus}{Zn} \quad \diagdown X_c \\ \underset{Bn \quad Bn}{} \end{array} \right] \longrightarrow 13$$

14
Higher Order Zincate

SCHEME 20.7

sions were observed when >1 equivalent of alkoxide was used (e.g., chelating dialkoxides), they postulated that a higher-order zincate was responsible for providing the energy required to drive the migration step. By analogy, we suspected the higher-order zincate **14** was the driving force behind the 1,2-migration (Scheme 20.7). Only **14** contained sufficient electron density around the zinc metal center to drive the 1,2-migration, leading back to a lower-order zincate **13**. Like the zincate enolate **11a**, **13** lacked the driving force required to undergo a second migration.

Working from this hypothesis, the solution to obtaining higher yields in the migration step was reduced to finding a surrogate for **7** to act as ligand to generate a higher-order zincate. Using the precedent of Harada and Oku, we evaluated several lithium alkoxides (Table 20.2).[20] In contrast to the results observed by Harada and Oku, dialkoxides were not effective, nor were lower-chain alcohols. The reason for this was likely the low solubility of these alkoxides in hexane/THF mixtures at low temperature.[21] The more soluble lithium n-propoxide and benzyloxide provided the best results.

Naturally, we followed the reaction by *in situ* IR (Figure 20.6). We learned from our inverse addition experiments that "quenching" **7** as the zincate enolate **11a** had the benefit of minimizing

TABLE 20.2
Conversion to Homoenolate
upon Addition of Alkoxides

Alkoxide	Homoenolate (LCAP)[a]
EtOLi	35, 65[b]
nPrOLi	74–82[c]
BnOLi	70–82[c] (78%)[d]
LiO(CH$_2$)$_2$OLi	31
LiO(CH$_2$)$_3$OLi	44
Me$_2$N(CH$_2$)$_2$OLi	22, 38[b]
TMEDA	28

[a] Results are based upon the "normal mode" of addition. Consequently, significant amounts of **8b** were also observed.
[b] After warming to 23°C.
[c] >99A% conversion of **4** to **8a** + **8b** observed.
[d] Isolated yield.

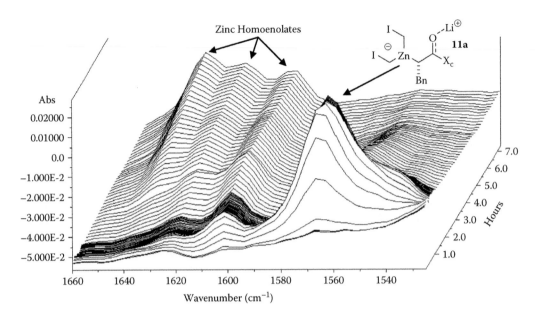

FIGURE 20.6 Homologation with lithium benzylalkoxide.

Diorganozinc homoenolate **15** : R = CH$_2$I

Zincate homoenolate **16** : R = CH$_2$I

Higher order zincate homoenolate **17** : R = CH$_2$I

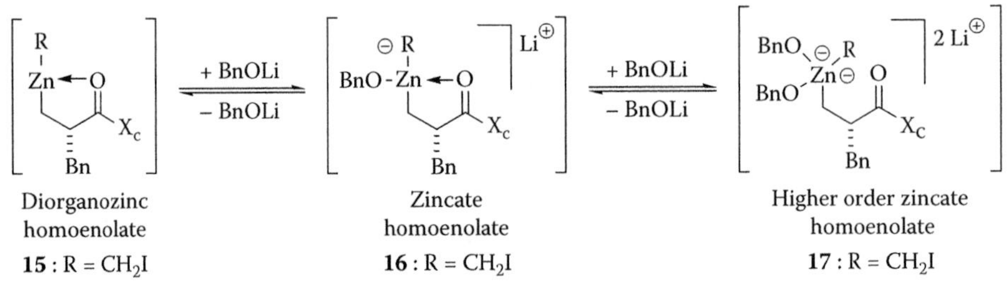

SCHEME 20.8

ethyl iodide alkylation. Accordingly, we conducted subsequent homologations in this manner to reduce the formation of this undesired side product.[22] If our hypothesis held true, we should be able to drive the conversion of all of **11a** to homoenolate product. We found that upon adding an excess of lithium benzylalkoxide, a gradual conversion of **11a** to homoenolate was observed.[23] In these early studies, three equivalents of alkoxide were required to achieve reasonable reaction rates. Later, we found that less alkoxide was needed upon further understanding of the reaction and additional optimization (*vide infra*). Under optimized conditions, typical conversions were >95%.

Although very pleased with the outcome, we were curious about the presence of three absorbencies, all of which presumably represented homoenolate. This observation contrasted with the single absorbance representing the homoenolate **13**. We speculated that the additional absorbances arose due to the equilibria shown in Scheme 20.8, wherein the presence of excess alkoxide results in equilibrium between the diorganozinc homoenolate **15**, mixed alkoxy zincate **16**, and higher-order zincate **17**. Once again, we employed an *in situ* IR experiment to clarify the situation. Accordingly, we subjected the homoenolate solution to aliquots of acetyl chloride (Figure 20.7). Each addition of acetyl chloride (25 mole% relative to added BnOLi) resulted in disappearance of the acetyl chloride absorbance at 1810 cm^{-1}, increase in the absorbance at 1741 cm^{-1} (benzyl

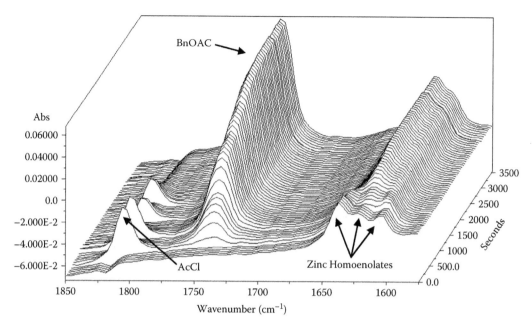

FIGURE 20.7 Addition of acetyl chloride to homoenolate solution.

acetate), and consumption of the homoenolate absorbance at 1637 cm^{-1} with concomitant increases in the homoenolate absorbances at 1621 cm^{-1} and 1602 cm^{-1}. Thus, it appeared that the absorbance at 1637 cm^{-1} corresponded to an alkoxide-complexed homoenolate, presumably the zincate enolate **16**, while two absorbances representing the diorganozinc homoenolate remained after all the alkoxide had reacted to form benzyl acetate. One explanation for the presence of two absorbances for **17** could be the existence of homo- and hetero-diorganozinc species (e.g., the bishomoenolate, wherein R = homoenolate) resulting from ligand exchange processes.

To further optimize the homologation step, we reasoned that only 50 mole% of the *BIZ* reagent should be required to complete the homologation of **7**, as there are two iodomethyl equivalents per molecule of *BIZ*. In support of this hypothesis, we had already shown that a second equivalent of **7** added to the zincate **11a** resulted in 50% conversion to **13**. Thus, one would expect added alkoxide to **13** should be able to drive the remaining 50% to completion. However, initial attempts to use 50% of *BIZ* resulted in incomplete conversions to product. We suspected the preparation of *BIZ* was not consistently yielding a homogeneous solution of desired reagent. Thus, we examined the product mixture arising from the preparation described in the literature.[24] According to the literature procedure, diiodomethane is added to diethylzinc at low temperature (–60°C), then the mixture is warmed to 0°C and stirred for 30 min. When we followed the reaction progress by NMR, we observed that a substantial amount of the reagent was decomposing to yield iodomethylzinc iodide and ethylene when the solution was warmed to 0°C.[25] Previous attempts to homologate **7** with iodomethylzinc iodide resulted in very low yields of **8a**. Furthermore, the addition of alkoxides to solutions of the iodomethylzinc enolate derived from mixing **7** and iodomethylzinc iodide did not proceed to product **8a**. Thus, it appeared that a better preparation of *BIZ* was needed to maximize the efficiency of the homologation. Once again, we turned to *in situ* IR as a means of monitoring the metathesis reaction. What we observed was that the first metathesis step between diethylzinc and diiodomethane was rapid at –60°C, yielding 1 equivalent of iodomethyl ethyl zinc and ethyl iodide (Scheme 20.9). However, the second metathesis was very slow at this temperature, and required heating to –40°C to achieve a rate that would allow complete conversion within 1.5 h.[26] The completion of the reaction could be monitored by following the disappearance of the CH$_2$ wag

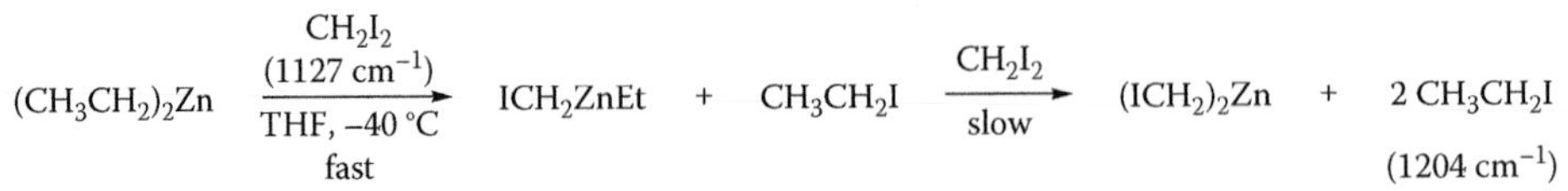

$$(CH_3CH_2)_2Zn \xrightarrow[\substack{THF, -40\ ^\circ C \\ fast}]{\substack{CH_2I_2 \\ (1127\ cm^{-1})}} ICH_2ZnEt + CH_3CH_2I \xrightarrow[slow]{CH_2I_2} (ICH_2)_2Zn + 2\ CH_3CH_2I$$

$$(1204\ cm^{-1})$$

SCHEME 20.9

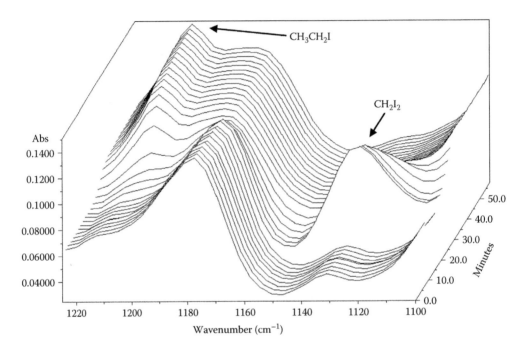

FIGURE 20.8 Monitoring the formation of *BIZ*.

of diiodomethane at 1127 cm^{-1} and the appearance of the CH$_2$ wag of ethyl iodide at 1204 cm^{-1} (Figure 20.8). Thereby, we were able to control and monitor the formation of *BIZ* at a temperature at which decomposition to iodomethylzinc iodide was minimized. Using this procedure, we were able to induce complete conversion of **7** to homoenolate using only 0.55 equivalents of *BIZ*. In addition, the rate of homologation was observed to increase, and as few as 1.1 equivalents of lithium benzylalkoxide were required.

We now felt that the first step in the sequence was understood well enough, and optimized to achieve consistently high conversion using a minimum of reagents. The next step consisted of transmetallation to form the titanium enolate. Monitoring by *in situ* IR, we found that the addition of enough TiCl$_4$ to react with all the lithium benzylalkoxide had the same effect as the addition of acetyl chloride (Figure 20.9). Upon addition of another equivalent of TiCl$_4$, all of the absorbances representing homoenolate dissipated with concomitant formation of a broad absorbance at 1559 cm^{-1}. This absorbance is consistent with prior reports of titanium ester homoenolates, and represents the titanium enolate **18** wherein the carbonyl oxygen completes a five-membered metallocycle via dative bonding to the titanium metal center (Scheme 20.10).[27]

Optimization of the titanium homoenolate formation consisted of selecting the most appropriate titanium ligands (i.e., the ratio of chloride to alkoxide, as well as alkoxide structure). In our original communication, we described the use of ClTi(OiPr)$_3$ as the transmetallation reagent. This selection was made partly based upon prior precedent, but also upon an observation made when employing

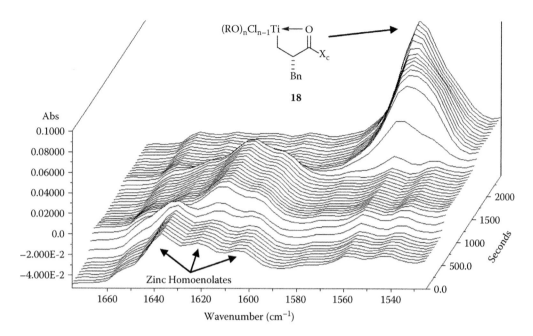

FIGURE 20.9 The addition of $TiCl_4$ to zinc homoenolate mixture.

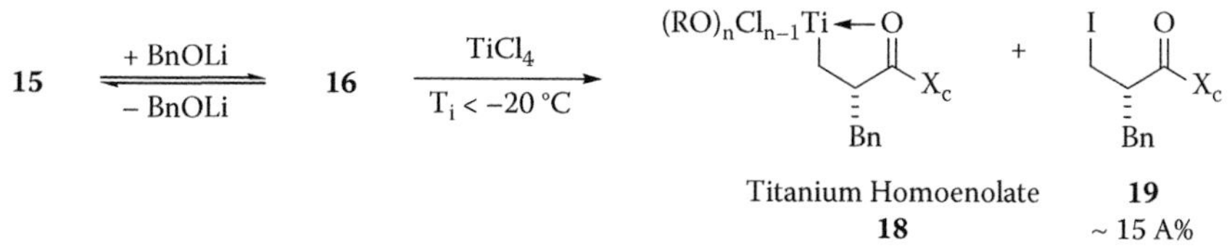

SCHEME 20.10

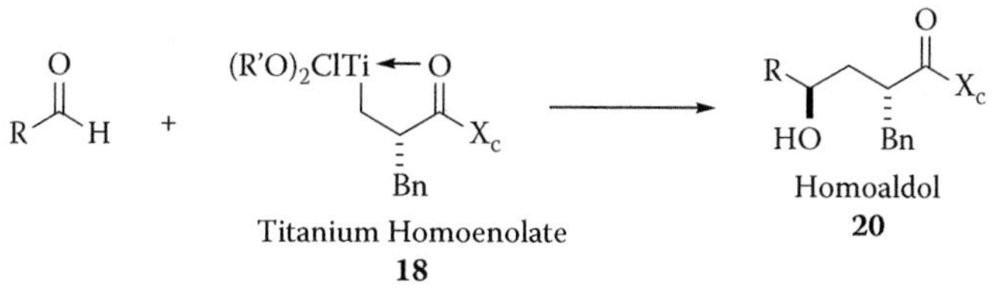

SCHEME 20.11

$TiCl_4$. Upon addition of $TiCl_4$ to the homoenolate solution, we observed the formation of a purple color, and the product mixture contained up to 15% of a new component that was identified as the iodide **19** (Scheme 20.10). The mechanism of this transformation was not determined but appears to involve a net redox reaction converting iodide to iodine, initiated by the addition of $TiCl_4$ to the reaction mixture. In contrast, the purple color and iodide product **19** were not observed when $ClTi(OiPr)_4$ was employed.

After making these adjustments, we were left to the final step in the sequence, the homoaldol with aldehydes (Scheme 20.11). Optimization of this step focused on achieving maximum diastereoselectivity while maintaining reasonable reactivity (Table 20.3). In the case of the L-Boc-

TABLE 20.3
Tandem Asymmetric 1,2-Migration/Homoaldol Reactions

Entry	Aldehyde (R')	Product	T (°C)	de (%)[b]	Yield (%)[c]
1	BocNH, Bn	6a	-20	≥ 99	59[d]
2	BocNH	6b	-20	≥ 99	53[d]
3	phenyl	6c	-40	82	50 (76)[e]
4	2-propenyl	6d	-40	76	55[f]
5	cyclohexyl	6e	-40	86	55 (73)[g]
6	iso-propyl	6f	-50	76	53 (74)
7	n-butyl	6g	-20	64	53 (68)

[a] Reactions were run with 5 equivalents of aldehyde, except for entries 1-2, in which 0.5 equivalents of the corresponding aldehydes was employed. [b] Refers to stereoselectivity of homoaldol reaction as determined by GLC after silylation of the crude mixtures, or by HPLC. [c] Isolated yield. Yields in parentheses are based upon homoenolate, the molarity of which was determined by HPLC, correcting for response factors. [d] Yield based upon aldehyde.[2] [e]Transmetallation with $TiCl_4$. [f]Based upon recovered aldehyde. [g]Yield of major isomer only.

(Adapted with permission from the *J. Am. Chem. Soc.*, Vol. 118, No. 47,11970–11971. Copyright 1996 American Chemical Society.)

aminoldehydes, the diastereoselectivity was very high (entries 1 and 2). The homoaldol with prochiral aldehydes decreased as the size of the aldehyde substituent decreased (entries 3 through 7). Diastereoselectivity was increased by conducting the reaction at lower temperatures, but required using excess aldehyde to achieve reasonable conversions.

We also noted that the homoaldol diastereoselectivity was dependent upon the alkoxide structure (Table 20.4). Smaller alkoxide ligands yielded the homoaldol products **20** with higher diastereoselectivity. Thus, we were able to improve upon the diastereoselectivity simply by adjusting the alkoxide source used in the homologation and transmetallation steps.

TABLE 20.4
The Effect of Alkoxide Structure on Homoaldol Diastereoselectivity

Aldehyde (R)	Temp (° C)	Alkoxide (R')	de
Ph	-20	nPr	86
Ph	-40	Bn	82
Ph	-40	3:1 Bn/iPr	76
iPr	-40	nPr	88
iPr	-40	3:1 Bn/iPr	76

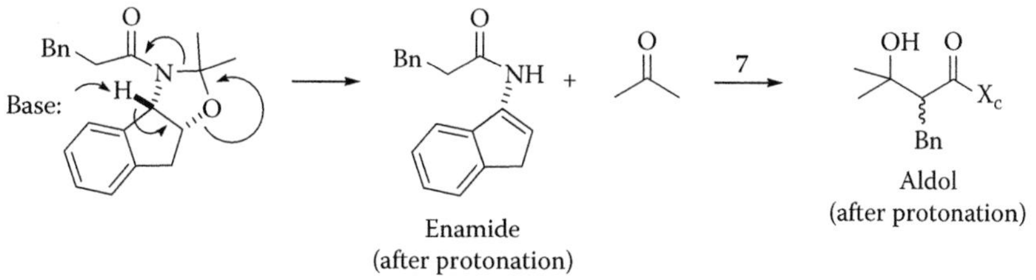

SCHEME 20.12

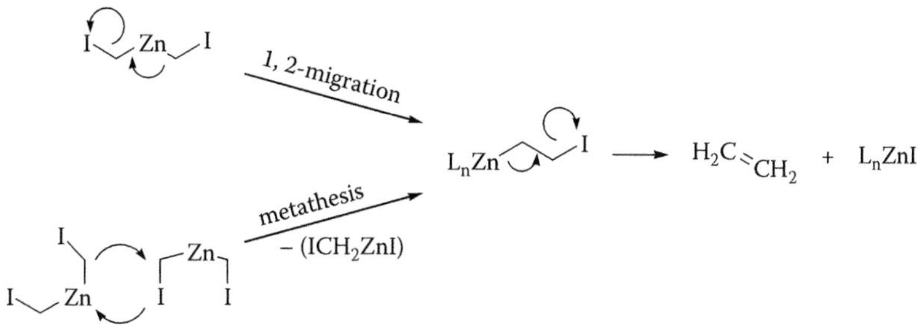

SCHEME 20.13

Our initial objective was to optimize the yield of the tandem asymmetric homologation–homoaldol for the formation of hydroxyamide **6a**. In the end, we were able to double the yield from <30 to 59%. Although we cannot go so far as to say that we have sufficient data to definitively support the proposed structural assignments of the observed intermediates, the improvements in reaction efficiency can be attributed to a better understanding of the process as a consequence of the observations made with *in situ* IR. Direct observation of the unreactive zincate enolate helped us move toward a working hypothesis. The direct observation of the formation and existence of the presumed mixed homoenolate zincate enolate solidified our working hypothesis and enabled a new approach wherein we could control the reactivity by using enolate surrogates. Additionally, we used our knowledge of the zincate reactivity to minimize ethyl iodide alkylation by using a mode of addition that would attenuate the reactivity of the lithium enolate. Following our initial NMR observations of *BIZ* degradation at elevated temperatures, we found *in situ* IR to be a convenient means of gaining qualitative rate data to optimize *BIZ* formation at temperatures at which degradation was minimized. Final changes in the identity of the titanium transmetallation reagent and alkoxides employed minimized side-product formation and improved diastereoselectivity in the homoaldol step. Although further improvements could have been pursued, we felt we had successfully made the transition from a proof of concept to a practical synthetic transformation.

REFERENCES AND NOTES

1. McWilliams, J.C.; Armstrong, III, J.D.; Zheng, N.; Bhupathy, M.; Volante, R.P. Reider, P.J. *J. Am. Chem. Soc.* 1996, *118*, 11970.
2. Transferring solutions from larger vessels into NMR tubes via cannulas was somewhat effective, but warming of the reaction mixtures was unavoidable.

3. (a) Armstrong III, J.D.; Hartner, Jr., F.W.; DeCamp, A.E.; Volante, R.P.; Shinkai, I. *Tetrahedron Lett.* 1992, *33*, 6599 (b) DeCamp, A.E.; Kawaguchi, A.T.; Volante, R.P.; Shinkai, I. *Tetrahedron Lett.* 1991, *32*, 1867. (c) Askin, D.; Wallace, M.A.; Vacca, J.P.; Reamer, R.A.; Volante, R.P.; Shinkai, I. *J. Org. Chem.* 1992, *57*, 2771.

4. (a) Sidduri, A.; Rozema, M.J.; Knochel, P. *J. Org. Chem.* 1993, *58*, 2694, and references cited therein. (b) Whitlick, Jr., H.W.; Overman, L.E. *J. Org. Chem.* 1969, *34*, 1962.

5. Although the hydroxyl functionality in **6** is epimeric to **2a** and **2b**, this route was still considered of interest due to available methods for stereochemical inversion, such as the Mitsunobu reaction. Regardless, the inherent potential efficiency of the transformation warranted additional studies.

6. Previous studies from the Merck labs have shown that decomposition occurs via deprotonation of the benzylic proton in the chiral auxiliary, followed by extrusion of acetone and trapping by enolate **7** as the aldol product. Consequently, two molecules of **4** are converted to a molecule of enamide and a molecule of aldol product (Scheme 20.12).

7. One could opt to change the amide auxiliary. However, supplies of **4** were plentiful, and the difficulty in deprotonation of **4** also appeared to correlate with the high reactivity observed in the corresponding enolate.

8. See Reference a and Furukawa, J.; Kawabata, N.; Nishimura, J. *Tetrahedron Lett.* 1966, 3353.

9. Morita, Y.; Suzuki, M.; Noyori, R. *J. Am. Chem. Soc.* 1989, *54*, 1785.

10. Infrared experiments were generally run on a scale of 6 to 25 mmol in **4**, and at a starting concentration of 0.4 to 0.5 M in THF. See Reference 1 for further details.

11. Attempts to detect an equilibrium process through dilution and variable temperature experiments were unfruitful.

12. To our knowledge, there are no current reports of observed ester or amide zincate enolates.

13. The NMR of the mixture of **7** and *BIZ* yielded a mixture of species. However, each of the species had a similar structure with carbon shifts in the region described in Table 20.1.

14. A recent report on the first structurally characterized organozinc amide enolate has appeared, and the observations are consistent with ours. Hlavinka, M.L.; Hagadorn, J.R. *Organometallics* 2005, *24*, 4116.

15. Although drawn as monovalent in Equation 20.2 for clarity, it is assumed that lithium is tetracoordinate, and 12-crown-4 is displacing other ligands bound to lithium, such as THF.

16. We also observed a similar shift when dimethyl sulfoxide (DMSO) was added to **11a**. Interestingly, the addition of (TMEDA N,N,N′,N′-tetramethylethylenediamine) resulted in reformation of the lithium enolate, as determined by *in situ* IR and comparison of the kinetics in the alkylation of 1-iodopropane. Apparently, the formation of the lithium enolate and TMEDA complexed Et_2Zn is thermodynamically favored over complexation of TMEDA to the lithium cation in **11a**.

17. Junk, P.C.; Smith, M.K.; Steed, J.W. *Polyhedron*, 2001, *20*, 2979.

18. Following an acetic acid quench, HPLC analysis indicated a 1:1 mixture of **4** and **8a**.

19. Harada, T.; Katsuhira, T.; Hattori, K.; Oku, A. *J. Org. Chem.* 1993, *58*, 2958.

20. Note that these results represent the addition of lithium alkoxide to a reaction mixture in which *BIZ* had been added to **7** (i.e., "normal addition mode"). Thus, 20 to 30% of the conversion was complete prior to the addition of alkoxide, and 10 to 20% of the mixture consisted of **8b**. This approach was pursued as a matter of convenience.

21. We observed solids when the lithium alkoxides derived from ethanol, ethylene glycol, and propylene glycol were cooled to reaction temperature.

22. We still observed ~3 A% **8b** under these conditions.

23. Corroborating HPLC data, following acetic acid quench, were also obtained.

24. Reference 4a.

25. An authentic sample of iodomethylzinc iodide was prepared from zinc metal and diiodomethane. The formation of ethylene is interesting. It likely arises from an E1cb reaction from an α-iodoethylzinc compound. This, in turn, could arise from a 1,2-migration or metathesis pathway (Scheme 20.13).

26. Adding diiodomethane to diethylzinc at –60°C resulted in an initial exotherm. We used the energy of this exotherm to warm the internal temperature to –40°C.

27. (a) Nakamura, E.; Oshino, H.; Kuwajima, I. *J. Am. Chem. Soc.* 1986, *108*, 3745. (b) Nakamura, E.; Kuwajima, I. *J. Am. Chem. Soc.* 1983, *105*, 651.

21 Development of Efficient One-Pot Process in the Synthesis of Sitagliptin: Application of Online-Infrared for Kinetic Studies to Probe the Reaction Mechanism

Feng Xu

CONTENTS

21.1 INTRODUCTION

Sitagliptin (**1**), a potent DPP-IV inhibitor, is currently undergoing development for the treatment of type II diabetes. The current manufacturing process, outlined in Scheme 21.1, features an extremely efficient one-pot through process[1] (**2** → **6**). The ketoamide **5** is directly prepared by mixing pivaloyl chloride with **2** and Meldrum's acid in the presence of i-Pr$_2$NEt in acetonitrile followed by the treatment with triazol **4**[2] and a catalytic amount of trifluoroacetic acid. Without isolation of **5** or any workup, NH$_4$OAc and MeOH are added to allow the formation of the enamine **6**, which is directly crystallized and isolated in 82% overall yield and 99.6 wt% purity through a simple filtration from the one-pot, three-step reaction mixture. Then, **6** is converted to sitagliptin through highly enantioselective Rh(I) catalyzed hydrogenation.[3] The case study here focuses on the design and development of the one-pot through process with detailed mechanistic understanding about the amide formation from Meldrum's adduct **3** by applying online-infrared (IR) analysis. The initial observations made during the development of this process provided the basis for an in-depth examination of the mechanistic aspects of the reaction. The combination of online IR monitoring and chemometric analysis made it possible to profile the concentration of both the

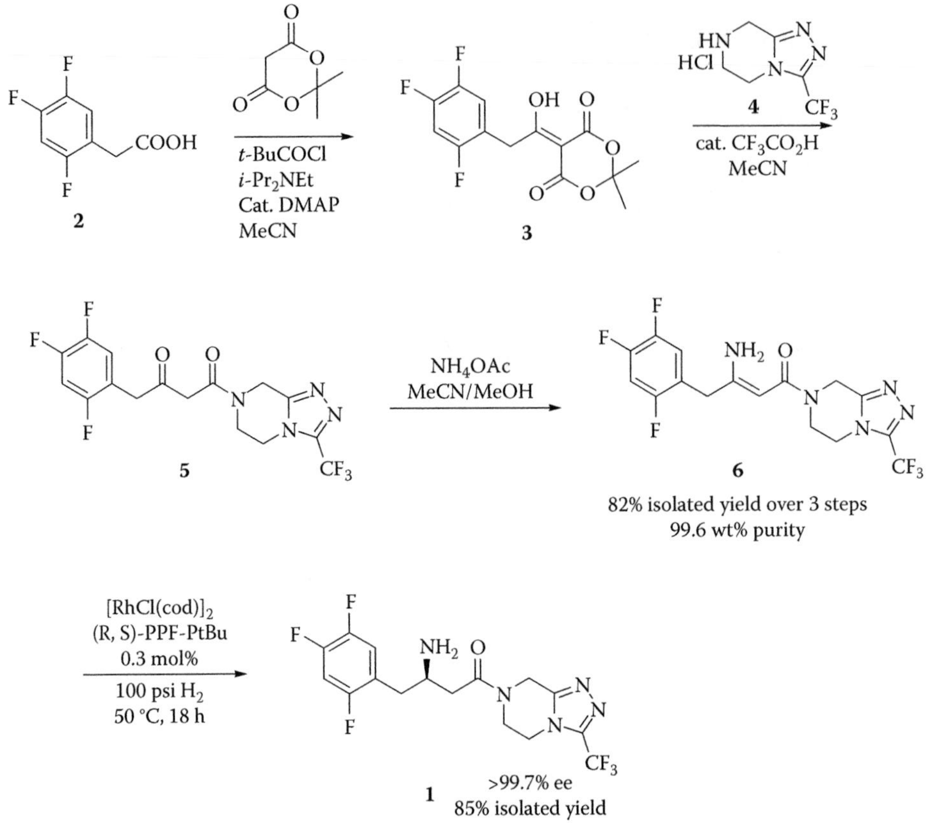

SCHEME 21.1 The ultimate manufacturing route to sitagliptin.

anionic and free acid forms of the Meldrum's adduct **3**, respectively, in real time for kinetic studies, which provided a crucial tool for delineating the mechanism of this complicated reaction sequence.

β-Keto esters and amides are versatile intermediates in organic synthesis and often are prepared from acyl Meldrum's acid adducts.[4] This method involves formation of acyl Meldrum's adducts by reaction of Meldrum's acid with activated carboxylic acids followed by decarboxylation in the presence of nucleophiles such as alcohols or amines (Scheme 21.2). The ability of readily available acyl Meldrum's acids to react with various nucleophiles allows quick access to a variety of functionalized compounds. Application of this methodology in synthetic chemistry has been widely exploited. For example, acyl Meldrum's adducts can react with imines to prepare pyridinones[5] or 1,3-oxazinones.[6] However, the conversion of acyl Meldrum's acids to the corresponding β-keto esters/amides or reactions involving decarboxylation of acyl Meldrum's acid is not well understood.[4,7] Four tentative reaction pathways were proposed: (1) nucleophilic addition–elimination pathway via intermediate **7**,[8] (2) formation of α-oxoketene **8**,[6,8b,9–11] (3) formation of protonated α-oxoketene **9**,[5] and (4) reaction via intermediate **10**.[9a,12]

21.2   ONE-POT THREE-COMPONENT COUPLING

The synthetic process to prepare keto amide **5** required coupling the carboxylic acid **2**, Meldrum's acid, and the triazole **4** and consists of three key steps: carboxylic acid activation, reaction of the activated acid with Meldrum's acid, followed by reaction with the amine and decarboxylation. As

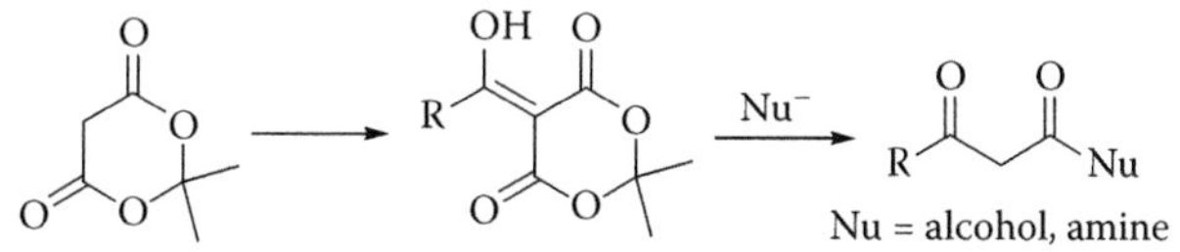

SCHEME 21.2 Preparation of β-keto esters and amides from acyl meldrum's acids β.

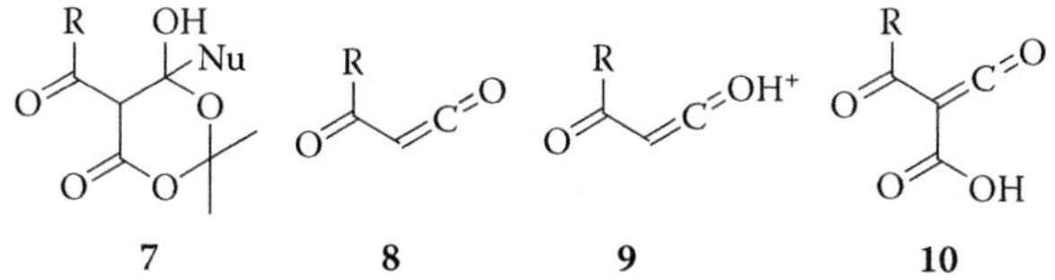

FIGURE 21.1 Possible reaction intermediates proposed in the literature.

outlined below, a process was defined that allowed all three steps to be carried out in an efficient, one-pot through process.

Carboxylic acids can be activated in several ways, such as conversion to the corresponding acid chlorides, imidazole amides, anhydrides, or by using coupling reagents such as DCC.[13] We chose pivaloyl chloride for initial development, because the hindrance of the pivaloyl group should favor the desired selectivity with the corresponding mixed anhydride **11**.[14,15]

After preliminary optimization, **3** was prepared by directly adding 1.1 to 1.2 equiv of pivaloyl chloride to a solution of Meldrum's acid, the carboxylic acid **2**, catalytic DMAP, and i-Pr$_2$NEt at ambient temperature followed by warming to 55°C. The use of 5 to 10 mol% of DMAP was necessary to achieve >97% conversion in less than 3 h. In order to maintain the high process productivity and the reaction homogeneity, the use of the combinations of i-Pr$_2$NEt with Me$_2$NAc or MeCN as solvent was discovered to be the best choice.

Preliminary studies showed that **3** was unstable in acidic media at ambient temperature[16] but was stable in basic solution or as a crystalline solid. In order to overcome the instability of the free acid of **3** which would occur upon attempted isolation, a through process to prepare **5** was required. However, when **3** was formed in the presence of 3 equiv of i-Pr$_2$NEt in 95% yield, and treated directly with 1.0 equiv of triazole HCl salt **4**, the decarboxylation/aminolysis reaction was slow and resulted in incomplete reaction (even at 90°C for 24 h). The reaction stalled at about 80% conversion.

This was rationalized and remedied as follows. When Meldrum's acid (pK_a = 4.97 in water[17]) and **2** are treated with pivaloyl chloride to prepare **3** (pK_a = 3.1 in water[18]), 1 equiv of HCl and 1 equiv of t-BuCO$_2$H are generated. The use of i-Pr$_2$NEt neutralizes the acids generated during the reaction, helps to achieve full conversion, and keeps the product **3** (which is also an acid) as its stable anionic form in the reaction solution. Because the pK_a of t-BuCO$_2$H is 5.01,[18] which is about two units above **3**, we decided to reduce the i-Pr$_2$NEt charge to 2.1 equivalent to quench only the free acid **3** and HCl generated during the reaction. Under these conditions, we found that formation of **3** still performed well (95% yield, 97% conversion), and the decarboxylation/aminolysis now proceeded smoothly at 70°C to achieve 98% conversion within 6 h, after 1 equiv of triazole HCl salt **4** was added to the through process solution. This dramatic improvement was made simply by reducing the base charge to accelerate the reaction rate and to drive the reaction to completion.

TABLE 21.1
Acidity Effects on Decarboxylation/Aminolysis

Acids	pKa	Conversion at 50–55°C, 6h[a]
Conc. HCl	–2.2	>99%
MeSO$_3$H	–0.6[18b]	>99%
CF$_3$CO$_2$H	0.23[18b]	>99%
o-NO$_2$PhCO$_2$H	2.17	99%
3	3.1[18a]	
m-NO$_2$PhCO$_2$H	3.45[18b]	95%
HOAc	4.76[18b]	61%
t-BuCO$_2$H	5.01[18b]	20%

[a] 0.3–0.5 equiv of acids were added after **4** was added to a through-process stock solution of **3** in MeCN.

However, the overall reaction process is sensitive to the initial i-Pr$_2$NEt charge as the base has opposing effects on the two steps. The rate of decarboxylation decreases as the initial base charge increases. If the base charge is <1.9 equiv, formation of **3** suffers from conversion and yield, while the decarboxylation is very fast and reaches completion within 2 h at 70°C and the overall yield for the two steps is reduced to <85%. If the base charge is >2.2 equiv, >98% conversion for formation of **3** can be achieved, while the decarboxylation is slow and cannot reach full conversion at 70°C. Thus, the amount of base must be carefully balanced (2.0 to 2.1 equiv) to achieve maximum conversion and yield.

21.3 ACID EFFECTS IN ONE-POT PROCESS

The observations outlined above suggested the decarboxylation/amination reaction required a somewhat acidic medium for optimum results. Likewise, a careful review of literature reports for the preparation of β-keto amides revealed that the amine component was always undercharged relative to the acyl Meldrum's acid intermediate, in those cases where isolated acyl Meldrum's acid intermediate was used, again suggesting that slightly acidic conditions might be beneficial.[8] Therefore, the effects of acid on the decarboxylation step were further studied. Several acids of varying pK_a were screened to determine the role of acidity on decarboxylation (Table 21.1). Examination of Table 21.1 indicates the pK_a of the acyl Meldrum's adduct **3** is the key to determining whether an acid is strong enough to assist in the completion of the decarboxylation reaction. If an acid is stronger than **3** and converts the anion form of **3** to its free acid form, the reaction performs well in terms of the reaction rate and conversion. Otherwise, the reaction is sluggish if an acid used is weaker than **3**. Therefore, weak acids such as HOAc and t-BuCO$_2$H are ineffective in driving the multi-acid–base equilibria in the reaction mixture toward the free acid form of **3**, and incomplete conversion is observed.

Building on these findings, the reaction was further optimized with CF$_3$CO$_2$H (TFA) chosen for its appropriate acidity. The optimized protocol involves formation of **3** with pivaloyl chloride and Meldrum's acid, the addition of **4**, followed by addition of TFA, typically 0.3 equiv. In this way, the significant aminolysis rate variation caused by the slight changes in the i-Pr$_2$NEt base charge is minimized after the TFA "pH adjustment." Finally, the decarboxylation/aminolysis reaction is carried out at lower temperature (50°C versus the previous 70°C).[19]

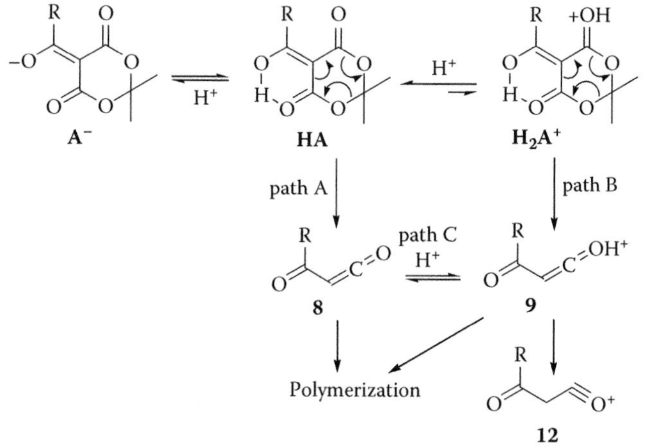

SCHEME 21.3 Possible pathways for self-decomposition. See References 1 and 20. (Reprinted with permission from the *J. Am. Chem. Soc.*, 126, 40, 13002–13009. Copyright 2004 American Chemical Society.)

21.4 KINETIC STUDIES ON SELF-DECOMPOSITION OF ACYL MELDRUM'S ADDUCT

These noteworthy observations made during process development prompted us to study the reaction in more detail. Our kinetic studies began with an investigation of the decomposition of **3** in the absence of an added nucleophile, which can precede either via the neutral compound (path A, Scheme 21.3) or the protonated species (path B, Scheme 21.3). As discussed below, these pathways can be distinguished based on the response of the kinetics to added acid or base.[20] For pathway A, the reaction rate constants will be unaffected by added acid or base, as shown in Equation 21.1 and Equation 21.2.

$$HA \xrightarrow[k]{slow} 8 + Me_2CO + CO_2 \tag{21.1}$$

$$rate = k[HA] \tag{21.2}$$

For pathway B, Equation 21.3 and Equation 21.4 apply.

$$2HA \underset{}{\overset{k_1}{\rightleftharpoons}} H_2A^+ + A^- \tag{21.3}$$

$$H_2A^+ \xrightarrow[k]{slow} 9 + Me_2CO + CO_2 \tag{21.4}$$

$$[H_2A^+] = \frac{K_1[HA]^2}{[A^-]}$$

$$rate = k[H_2A^+] \tag{21.5}$$

$$= \frac{kK_1[HA]^2}{[A^-]}$$

In the absence of acid or base, $[H_2A^+] = [A]$. Thus,

$$\text{rate} = k\left[H_2A^+\right] = k\sqrt{K_1}\left[HA\right] \tag{21.6}$$

If <1 equiv of base B is added to the system, A would increase and lead to a reduced rate (Equation 21.5).

In the presence of a strong acid, the equilibrium (Equation 21.3) is shifted, and $[H_2A^+]$ is increased. Therefore, the reaction rate would increase.

$$HA + B \rightleftharpoons A^- + HB^+ \tag{21.7}$$

In addition, protonation of ketenes has been studied both theoretically and experimentally.[21,22] The preferred protonation of ketene is at C_β to afford the acylium ion **12** instead of **9**. In the gas phase, **12** is lower in energy than the α-protonated ketene **9**.[22,23,24,25] It is unlikely that **9** can be formed through direct protonation of **8** via pathway C, because this process has a high energy barrier and **9** is disfavored thermodynamically.[24]

At this point it became clear that the capability to obtain the kinetic profile of the free acid **3** as well as its anionic form was essential to understanding the reaction pathway. Therefore, attempts to quantify the free acid and its anionic form in solution by applying online IR technique were initiated. Titration with i-Pr$_2$NEt showed that the free acid **3** was stoichiometrically converted to its anionic form as expected based on the several unit differences of pK_a between **3** and i-Pr$_2$NEt. Although the IR spectra of the free acid **3** and its anion are substantially different, the overlapping peaks in the online IR spectra obtained during a reaction or titration are subtle and make obtaining the kinetic profiles of the free acid **3** and its anion quite difficult. To fully distinguish each component in the complex mixture, a more sophisticated data analysis method had to be applied in order to resolve each component over the entire reaction period.

The method we chose was principal component analysis, an algorithm that has been used to extract chemical and process information from reaction spectra.[25,26] To extract the spectra of both free acid **3** and its anionic form, principal component analyses were first performed on the titration spectra of free acid **3** with i-Pr$_2$NEt, using ConCIRt (version 2, from Mettler Toledo[27]). After several attempts, optimal spectral regions (1800 to 1680, 985 to 875, and 725 to 675 cm^1) containing major spectral features of both acid **3** and its anion, and having the least interferences by other coexisting species, were identified. The component spectra of the free acid **3** and its anion within these three optimal regions were obtained and are illustrated in Figure 21.2. Peaks at 704.3, 922.3, and 1739.4 cm^1 represent the free acid **3**, and peaks at 702.3, 947.7, and 1717.6 cm^1 correspond to the anionic form of **3**.[28] Thus, quantitative monitoring of the real-time concentration change of HA and A in the reaction system became possible by feeding the above extracted spectra (Figure 21.2).

With an analytical method in hand, kinetic studies were initiated. The self-decomposition kinetic studies were studied first by using isolated adduct **3** in a homogenous Me$_2$NAc solution, which provided a simplified system by eliminating all the acid–base equilibria that exist in the one-pot process solution. As mentioned above, preparation of **5** is equally effective in either acetonitrile or Me$_2$NAc. As shown in Figure 21.3, a first-order dependence in the acid form **3** (HA) was clearly observed, with a k_{obs} of 4.09×10^4 s^{-1}.

In addition, the kinetic profiles for the self-decomposition of **3** in the absence/presence of TFA or i-Pr$_2$NEt (Figure 21.3; Table 21.2, entries 4 to 6) indicate that an acid or base has almost no effect on the rate constants. The observed first-order kinetics in HA only, along with the lack of catalysis by TFA, allowed us to rule out the pathway via protonated ketene **9** for the decomposition of acyl Meldrum's adducts.[29]

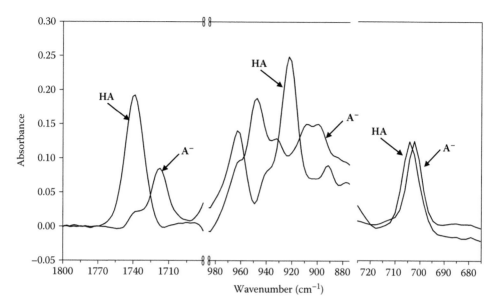

FIGURE 21.2 Extracted spectra of the free acid (HA) and anion (A⁻, i-Pr$_2$NEt salt) of **3** obtained in Me$_2$NAc. (Reprinted with permission from *J. Am. Chem. Soc.*, 126, 40, 130002–13009. Copyright 2004 American Chemical Society.)

21.5 KINETIC EVIDENCE AGAINST NUCLEOPHILIC ADDITION–ELIMINATION MECHANISM

Based on the kinetics described above, the self-decomposition of **3** is consistent with reaction via an α-oxoketene intermediate. However, for reactions in the presence of a nucleophile, the nucleophilic addition–elimination through intermediates such as D1 or D2, formed from HA or H$_2$A$^+$, could become the preferred pathway. Given that proton exchange steps are fast, the addition–elimination step or possibly the following fragmentation of D1 or D2 would likely be the rate-determining step. Reaction rates or k_{obs} should be different if nucleophilic addition was involved in the reaction, or if fragmentation of an intermediate already incorporating the nucleophile was the rate-limiting step.

As outlined above, isolated acyl Meldrum's adducts react with amines to afford the corresponding β-ketoamides. More interestingly, nonbasic amine derivatives, such as carbamates,[30] amides,[31] and hydroxylamines,[32] as well as alcohols, can also react with these adducts to effectively provide the corresponding 1,3-dicarbonyl compounds. To investigate the effect of the nature of the nucleophile, we initiated these studies with BocNHOBoc, because it is nonbasic and is a much weaker bulky nucleophile than the triazole **4**. Comparison of the k_{obs} with other weak or strong nucleophiles would reveal whether nucleophilic addition or the following elimination is involved in the rate-determining step.

Kinetic studies were carried out by online IR. First-order kinetics in the free acid **3** and zero-order kinetics in BocNHOBoc were clearly observed. The yield and rate were almost identical when 1 or 2 equiv of BocNHOBoc were used. The k_{obs} values (Table 21.2, entries 1 to 2) obtained by online IR are very close to those measured for the self-decomposition of **3**, in which the rate-determining step is likely the formation of α-oxoketene **2**.[33] To conclude, the k_{obs} values and the observed first-order kinetics in HA do not support the nucleophilic addition–elimination mechanism and are consistent with a pathway via the oxoketene **8**.

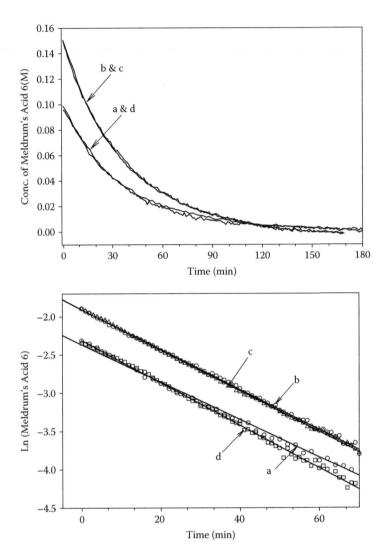

FIGURE 21.3 Self-decomposition of the free acid **3** in Me$_2$NAc. Top: Plots of [HA] versus time; bottom: plots of Ln[HA] versus time. (a) **3** at 51.5°C; (b) in the presence of 0.5 equiv TFA at 48.3°C; (c) in the presence of 1.0 equiv TFA at 48.8°C; (d) In the presence of 0.5 equiv i-Pr$_2$NEt at 52.5°C. Mechanistic Evidence for an α-Oxoketene Pathway in the Formation of β-Ketoamides/esters via Meldrum's Acid Adducts.

These studies were extended with triazole free base **4** as nucleophiles. Introduction of an undercharge of a basic nucleophile such as an amine introduces a new fast acid–base equilibrium (Equation 21.8).

$$HA + Nu \rightleftharpoons A^- + HNu^+ \tag{21.8}$$

In this case, **HA** is constant before most of Nu is consumed (*vide infra*). Therefore, Equation 21.2 and Equation 21.5 would result in different kinetic profiles. As outlined above, the anionic form **A⁻** is very stable and will not react with amines. Negative first-order kinetics in **A⁻** would be observed if a protonated α-oxoketene is involved in the main reaction pathway. Otherwise, given

TABLE 21.2
Measured k_{obs} in Me_2NAc with Isolated Meldrum's Adduct

Entry	Substrates	Additive	$k_{obs} \times 10^{-4}$ (s^{-1})	Temp (°C)	Isolated yield
1	1 equiv BocNHOBoc		3.93	50.8	78%, 13[a]
2	2 equiv BocNHOBoc		3.35	50.1	78%, 13
3	4 free base		4.81	51.4	91%, 5
4	None	None	4.09	51.5	
5	None	1.0 equiv TFA	4.37	48.8	
6	None	0.5 equiv i-Pr$_2$NEt	4.68	52.5	

(a) **13** =

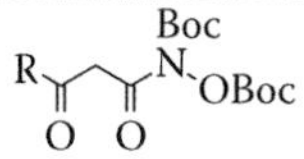

R = 2,4,5-trifluorobenzyl.

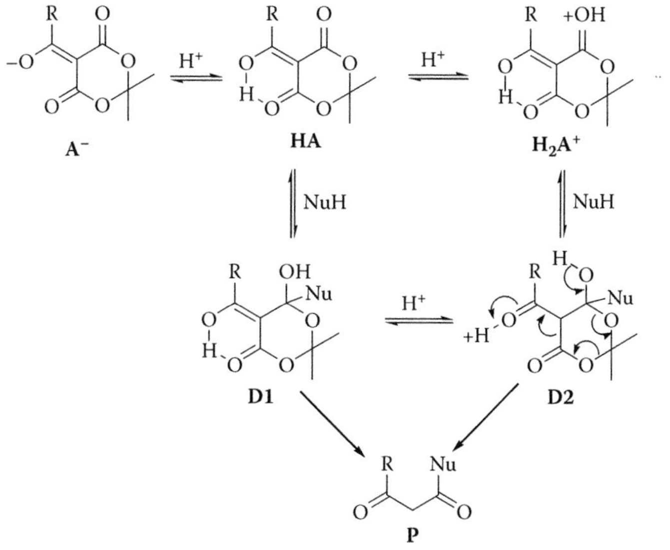

SCHEME 21.4 Possible nucleophilic addition-elimination pathway.[21] (Reprinted with permission from *J. Am. Chem. Soc.*, 126, 40, 13002–13009. Copyright 2004 American Chemical Society.)

that the acid–base equilibrium is rapid, pseudo-zero-order kinetics in **A**⁻ would be expected. Once again we turned to online IR to probe the reaction kinetic profiles to differentiate the concentration changes of the anionic form **A**⁻ and the free acid form **HA** during the reaction.

The use of the triazole **4** resulted in pseudo-zero-order kinetics in [**A**⁻]. As shown in Figure 21.4, the observed concentration profile of the free acid form **HA** is almost unchanged until the majority of the anionic form **A**⁻ is consumed, due to the rapid acid–base proton exchange (Equation 21.8). The kinetic profiles of **HA** and **A**⁻ obtained/extracted by using online IR exactly matched the predicted patterns as described above. The triazole **4** and BocNHOBoc all gave similar k_{obs} values, which closely matched those measured for the self-decomposition of **3**. The k_{obs} values (Table 21.2) further confirmed that nucleophiles do not react directly with **HA** or **H$_2$A⁺**. The fact

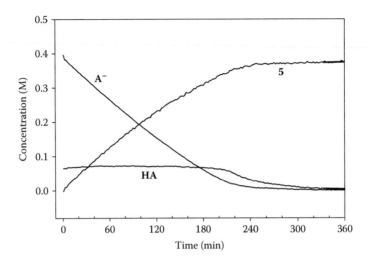

FIGURE 21.4 Plots of concentration of the free acid and anion forms of **3** and product **5** in DMAc solution versus time at 51.4°C.

that the reaction rate is zero order in **A⁻** in the presence of amine nucleophile **4** also implies that the main reaction pathway is through an α-oxoketene rather than a protonated α-oxoketene intermediate.

21.6 KINETIC PROFILES IN PROCESS SOLUTION

Finally, the kinetic profiles in the "real" process solution were examined. In these experiments, amine HCl salts were used. As mentioned above, 1 equiv of *t*-BuCO$_2$H and 1 equiv of HCl are formed during the formation of **3**. The 2 equiv of *i*-Pr$_2$NEt charged at the beginning of the reaction is thus used to neutralize the 1 equiv of HCl and 1 equiv of free acid **3** to their corresponding salts. Maintaining **3** as its anion form is one of the keys to making this process robust by minimizing its decomposition, because **3** remains as its anionic form in the solution until a catalytic amount of TFA is added. TFA provides the proton source to partially convert the anionic form of **3** to its free acid, while TFA in turn is converted to its carboxylate salt. As the decarboxylation of **3** proceeds, the triazole HCl salt **4** is converted to amide **5**, and the liberated H⁺ converts the anionic form of **3** (**A⁻**) to its free acid form (**HA**). With this overall acid–base–salts turnover cycle (Scheme 21.5) in mind, the outcome of the kinetic results for the real process can be readily understood.

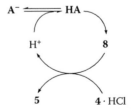

SCHEME 21.5 Acid-base-salt turning cycle in the one-pot process.

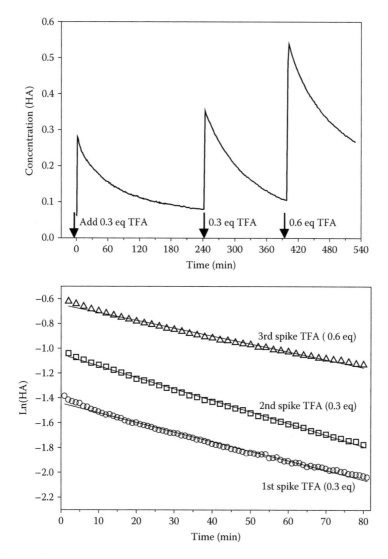

FIGURE 21.5 Top: Plot of concentration of the acid form of **3** versus time for self-decomposition of **3** in acetonitrile process solution at 49.5°C. Bottom: Plots of Ln **3** versus time.

The requirement of one full equivalent of TFA to complete the self-decomposition of **3** in the process solution confirms the acid–base turnover cycle. Addition of TFA converts the anion form of **3** into its free acid which initiates decomposition. NMR analyses of the decomposed products indicated that polymers or oligomers were formed. The reaction profile by online IR clearly showed that **3** decomposed after aliquots of TFA were introduced sequentially (Figure 21.5). Kinetic analysis indicated first order in HA and zero order in A^-. The k_{obs} values measured at each of the three TFA charge stages are listed in Table 21.3 (entries 3 to 5).

Figure 21.6 shows the kinetic profile under the actual process conditions. The reaction profile (combination of the anion and free acid of **3** versus time) obtained by high-performance liquid chromatography (HPLC) analysis also matched the online IR data. Again, the use of online IR coupled with principal component analysis provided the means to profile HA and A^- during the reaction. Formation of the free acid form **3** (HA) was immediately observed upon addition of a catalytic amount of TFA. This clearly shows that constant liberation of the free acid form HA

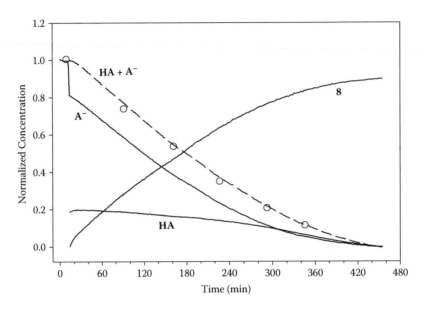

FIGURE 21.6 Plots of concentration of the free acid and anion forms of **3**, and **5** in acetonitrile process solution versus time. Reaction conditions: 0.3 equiv of TFA and 1.0 equiv of triazole HCl salt **4** at 49.5°C. [HA] plus [A⁻]: dashed line is based on online IR data. Circles are obtained by HPLC analyses. For reaction sequence, see Scheme 21.5. (Reprinted with permission from *J. Am. Chem. Soc.*, 126, 40, 13003–13009. Copyright 2004. American Chemical Society.)

through a fast acid–base proton exchange during the reaction is the key to achieving high conversion (>99%). This is also consistent with the observed variation in decarboxylation rate and conversion if TFA is not introduced into the reaction system, because any excess base would prevent complete conversion to the free acid form.

Under process conditions, the decarboxylation is pseudo zero order in the anionic form **3**, which is the same as when isolated **3** is treated with free amine nucleophiles in Me₂NAc. The reaction rate increases as more TFA is charged, because the initial concentration of HA is increased. However, the k_{obs} values are almost the same (Table 21.3, entries 1 and 2). The data are consistent

TABLE 21.3
Measured k_{obs} in Acetonitrile Process Solution[20]

Entry	Substrates	k_{obs} x 10^{-4} (s⁻¹)	Temp (°C)	Yield[a]
1	**4** HCl salt	2.74	49.5	91%, **5**
2	**4** HCl salt[b]	3.30	49.6	90%, **5**
3	None[c]	2.13	49.5	
4	None[c]	2.16	49.5	
5	None[c]	1.48	49.5	

[a] Assay yield calculated by HPLC using an external reference standard.

[b] 0.5 equiv of TFA was charged.

[c] Self-decomposition of **3** in the process MeCN solution by stepwise addition of TFA. Entry 3, with first 0.3 equiv of TFA addition; Entry 4, with second 0.3 of equiv TFA addition; Entry 5, with last 0.6 equiv of TFA addition.

Source: Adapted with permission from the *Journal of the American Chemical Society*, 126, 40, 13002–13009. Copyright 2004 American Chemical Society.

with the rate-determining step being the formation of the α-oxoketene intermediate **8** under the one-pot conditions developed for synthetic applications.

21.7 CONCLUSION

A practical one-pot process for preparation of β-keto amides/enamine has been developed. The use of online IR and subsequent principal component analysis for kinetic studies were critical as tools for profiling the concentration changes of the anion and free acid forms, as well as the product, throughout the course of the reaction. The mechanistic understanding about the reaction provided an in-depth understanding to design and develop the efficient, one-pot process.

ACKNOWLEDGMENTS

This chapter represents the outstanding contributions of many scientists within Merck Research Laboratories. The author hereby expresses his sincere gratitude to the many colleagues whose hard work and dedication led to the success of this program. Their names appear in References 1 through 3.

REFERENCES AND NOTES

1. Xu, F., Armstrong, J.D., III; Zhou, G.X.; Simmons, B.; Hughes, D.; Ge, Z.; Grabowski, E.J.J. *J. Am. Chem. Soc.* 2004, *126*, 13002.
2. Balsells, J.; DiMichele, L.; Liu, J.; Kubryk, M.; Hansen, K.; Armstrong, J.D., III. *Org. Lett.* 2005, *7*, 1039.
3. Hsiao, Y.; Rivera, N.R.; Rosner, T.; Krska, S.W.; Njolito, E.; Wang, F.; Sun, Y.; Armstrong, J.D., III; Grabowski, E.J.J.; Tillyer, R.D.; Spindler, F.; Malan, C. *J. Am. Chem. Soc.* 2004, *126*, 9918.
4. For reviews, see (a) Far, A.D. *Angew. Chem., Int. Ed. Engl.* 2003, *42*, 2340. (b) Gaber, A.E.M.; McNab, H. *Synthesis* 2001, 2059. (c) Chen, B.C. *Heterocycles* 1991, *32*, 529. (d) Huang, X. *Youji Huaxue* 1986, 329.
5. (a) Pemberton, N.; Emtenas, H.; Bostrom, D.; Greenberg, W.A.; Levin, M.D.; Zhu, Z.; Almqvist, F. *Org. Lett.* 2005, *6*, 1019. (b) Emtenas, H.; Alderin, L.; Almqvist, F. *J. Org. Chem.* 2001, *66*, 6756.
6. Emtenas, H.; Soto, G.; Hultgren, S.J.; Marshall, G.R.; Almqvist, F. *Org. Lett.* 2000, *2*, 2065.
7. The mechanism involving acyl Meldrum's acids in solution was never clarified. Several proposed reaction pathways are often found in the same publication.
8. For examples, see (a) Pak, C.S.; Yang, H.C.; Choi, E.B. *Synthesis* 1992, 1213. (b) Svetlik, J.; Goljer, I.; Turecek, F. *J. Chem. Soc. Perkin Trans. I* 1990, 1315. (c) Sato, M.; Yoneda, N.; Katagiri, N.; Watanabe, H.; Kaneko, C. *Synthesis* 1986, 672. (d) Oikawa, Y.; Sugano, K.; Yonemitsu, O. *J. Org. Chem.* 1978, *43*, 2087.
9. For additional examples, see (a) Yamamoto, Y.; Watanabe, Y.; Ohnishi, S. *Chem. Pharm. Bull.* 1987, *35*, 1860. (b) Sato, M.; Ogsawara, H.; Yoshizumi, E.; Kato, T. *Chem. Pharm. Bull.* 1983, *31*, 1902. (c) Sato, M.; Ogsawara, H.; Yoshizumi, E.; Kato, T. *Heterocycles* 1982, *17*, 297.
10. Zawacki, F.J.; Crimmins, M.T. *Tetrahedron Lett.* 1996, *37*, 6499.
11. For recent reviews about α-oxoketenes **8**, see (a) Tidwell, T.T. *Ketenes*; John Wiley & Sons: New York, 1995. (b) Wentrup, C.; Heilmayer, W.; Kollenz, G. *Synthesis* 1994, 1219. (c) Tidwell, T.T. *Acc. Chem. Res.* 1990, *23*, 273.
12. (a) Hamilakis, S.; Kontonassios, D.; Sandris, C. *J. Heterocycl. Chem.* 1996, *33*, 825. (b) Sato, M.; Takayama, K.; Abe, Y.; Furuya, T.; Inukai, N.; Kaneko, C. *Chem, Pharm. Bull.* 1990, *38*, 336. (c) Yamamoto, Y.; Watanabe, Y. *Chem. Pharm. Bull.* 1987, *35*, 1871.

13. (a) Sorensen, U.S.; Falch, E.; Krogsgaard-Larsen, P. *J. Org. Chem.* 2000, *65*, 1003. (b) Hamada, Y.; Kondo, Y.; Shioiri, T. *J. Am. Chem. Soc.* 1989, *111*, 669. (c) Alker, D.; Campbell, S.F.; Cross, P.E.;Burges, R.A.; Carter, A.J.; Gardiner, D.G. *J. Med. Chem.* 1989, *32*, 2381. (d) Maibaum, J.; Rich, D.H. *J. Med. Chem.* 1989, *32*, 1571. (e) Shinkai, I.; Liu, T.; Reamer, R.A.; Sletzinger, M. *Tetrahedron Lett.* 1982, *23*, 4899. (f) Mohri, K.; Oikawa, Y.; Hirao, K.-I.; Yonemitsu, O. *Chem. Pharm. Bull.* 1982, *30*, 3097. (g) Houghton, R.P.; Lapham, D.J. *Synthesis* 1982, 451.

14. Although **3** can be obtained in high yield by acylating Meldrum's acid with 2,4,5-trifluorophenylacetyl chloride in the presence of various amine bases, without aqueous workup, the next through process step to the ketoamide **5** does not perform well in terms of impurity profile and conversion.

15. ^{1}H NMR studies showed that the background reaction between pivaloyl chloride and Meldrum's acid in CD_3CN in the presence of *i*-Pr_2NEt and 10 mol% DMAP at 0°C to ambient temperature is slow. Under these reaction conditions, formation of **14** is negligible in comparison to the reaction rate for formation of **3**.

16. It is known that acyl Meldrum's acids are not always stable. For example, see Reference 8b.

17. Pihlaja, K.; Seilo, M. *Acta Chem. Scand.* 1969, *23*, 303.

18. (a) Measured by titration in water at ambient temperature. (b) Refs. to pka in H2O: Sober et al. Hbk. Biochem. Selected Data for Mol. Biol., 2nd ed., CRC Press, 1970, J-89, J-191, J194.

19. This process has been successfully and reproducibly carried out in 300-kg scales.

20. For a less-complicated consideration, all the tautomers, rotamers, as well as resonance structures are not specifically considered here. For example, although these equilibriums can be included in the kinetic analyses, the outcome reaction rate results in the same kinetic effects on [HA] or [A] as described in the text. In order to have more focused discussion in the text, detailed kinetic analyses of all other possible mechanisms, which can be done as described in the text, are not listed here.

21. For recent reviews, see (a) Tidwell, T.T. *Ketenes*; John Wiley & Sons: New York, 1995. (b) Wentrup, C.; Heilmayer, W.; Kollenz, G. *Synthesis* 1994, 1219. (c) Tidwell, T.T. *Acc. Chem. Res.* 1990, *23*, 273.

22. For leading references, see (a) Gong, L.; McAllister, M.A.; Tidwell, T.T. *J. Am. Chem. Soc.* 1991, *113*, 6021. (b) Lien, M.H.; Hopkinson, A.C. *J. Org. Chem.* 1988, *53*, 2150. (c) Armitage, M.A.; Higgins, M.J.; Lewars, E.G.; March, R.E. *J. Am. Chem. Soc.* 1980, *102*, 5064.

23. (a) Birney, D.M. *J. Org. Chem.* 1994, *59*, 2557. (b) Tortajada, J.; Berthomieu, D.; Morizur, J.-P.; Audier, H.-E. *J. Am. Chem. Soc.* 1992, *114*, 10874. (c) Leung-Toung, R.; Peterson, M.R.; Tidwell, T.T.; Csizmadia, I.G. *J. Mol. Struct.* 1989, *183*, 319. (d) Bouchoux, G.; Hppilliard, Y. *J. Phys. Chem.* 1988, *92*, 5869.

24. As reported, a 43 kcal/mol gap exists between CH_3CO^+ and $CH_2=C=O^+H$. See (a) Nobes, R.H.; Bouma, W.J.; Radom, L. *J. Am. Chem. Soc.* 1983, *105*, 309. (b) Vogt, J.; Williamson, A.D.; Beauchamp, J.L. *J. Am. Chem. Soc.* 1978, *100*, 3478.

25. Malinowski, E.H.; Howery, D.G. *Factor Analysis in Chemistry*; John Wiley & Sons: New York, 1980.

26. Cameron, M.; Zhou, G.X; Hicks, M.B.; Antonucci, V.; Ge, Z.; Lieberman, D.R.; Lynch, J.E.; Shi, Y.-J. *J. Pharm. Biomed. Anal.* 2002, *28*, 137.

27. Previously known as ASI Applied Systems, Millersville, Maryland.

28. The obtained online IR kinetic profiles of the combination of anion and free acid form of **3** as well as the formation of the product matched very well with the HPLC kinetic profile as shown later in Figure 21.6.

29. A stepwise formation of the oxoketene by loss of acetone to form intermediates such as **10**, followed by decarboxylation, cannot be ruled out. The fact that the reaction rate is unaffected by the increasing concentration of acetone formed during the reaction provides some evidence against the pathway via intermediates such as **10**, if reversible formation of these intermediates is the rate-determining step.

30. Sorensen, U.S.; Falch, E.; Krogsgaard-Larsen, P. *J. Org. Chem.* 2000, *65*, 1003.

31. Yamamoto, Y.; Ohnishi, S.; Azuma, Y. *Chem. Pharm. Bull.* 1982, *30*, 3505.
32. Mohri, K.; Oikawa, Y.; Hirao, K.-I.; Yonemitsu, O. *Chem. Pharm. Bull.* 1982, *30*, 3097; *Heterocycles* 1982, *19*, 515.
33. Shelkov, R.; Nahmany, M.; Melman, A. *J. Org. Chem.* 2002, *67*, 8975.

22 Mid-Infrared Monitoring Applications during Development of the Vinyl Ether Formation Step in the Preparation of Aprepitant (Emend®)

David A. Conlon, Bill Izzo, and Paul Collins

CONTENTS

22.1 INTRODUCTION

Process chemists are responsible for preparing multikilogram quantities of drug candidates to support toxicological studies, formulation development, and clinical trials and to design practical, efficient, environmentally benign, and economically viable commercial processes for approved products. Acceleration of the development time line for the discovery and development of new pharmaceutical agents and the requirement for reproducible and validated manufacturing processes in the pharmaceutical industry have led to an increased application of process analytics for real-time monitoring of chemical processes. One of the most versatile technologies in this area is mid-infrared (IR) owing to its specificity and adaptability to manufacturing environments. *In situ* mid-IR analysis of chemical processes has long been a prominent tool for investigating complex synthetic chemical reactions in the laboratory, offering a direct observation window into the chemical transformations occurring during processing. This, in turn, provides greater scientific understanding for development and optimization of efficient and safe chemical processes. The emergence of mid-IR monitoring as a routine part of laboratory process development has naturally advanced to larger-scale applications. Knowledge gained from laboratory experiments is easily incorporated into pilot-plant and commercial production of active pharmaceutical ingredients (APIs). In-line spectroscopic analysis offers several advantages over traditional "grab sampling" techniques. The potential advantages of in-line monitoring include improved sample integrity, reduction in cycle time, reduction of safety hazards, opportunity for feedback control and improved optimization, and reduction in operating costs.

This case study describes the process chemistry, the laboratory development, and the application of Mettler Toledo's[1] ReactIR™ technology for in-line monitoring of several quantitative mid-IR applications during the early process development for aprepitant (Emend®), approved for the treatment of chemotherapy-induced nausea and vomiting.[2] The most significant mid-IR application was the real-time monitoring of reaction progress and endpoint determination during the formation of a vinyl ether intermediate via a dimethyltitanocene olefination reaction. This application was fully developed and successfully demonstrated at the pilot-plant scale.[3] Two secondary mid-IR applications were initially investigated as replacements for traditional off-line sampling. These applications were demonstrated at the laboratory scale but not implemented during the initial preparation of API in the Merck pilot plant. All three applications led to increased process understanding and had the potential for substantial improvement in the quality of analytical information and reduction in operating costs. Outlined in this chapter will be the analytical and process methodology employed in the development of the off-line and in-line mid-IR monitoring techniques as well as the key knowledge acquired through the pilot-plant demonstration of the technology.

Aprepitant (Emend[R])

22.2 VINYL ETHER FORMATION

Several different synthetic approaches[4] to aprepitant were pursued; however, an optimized route originally reported by medicinal chemists at Merck[5] was used in the pilot plant to support clinical development. The medicinal chemists at Merck used an olefination reaction to introduce the methyl group on the benzylic carbon adjacent to the 3,5-bis(trifluoromethyl)phenyl moiety in aprepitant. Dimethyltitanocene (DMT) was selected for the olefination because the reagent is easy to prepare and does not generate acid by-products that are detrimental to the acid labile vinyl ether product. Formation of the vinyl ether (Scheme 22.1) was followed by catalytic hydrogenation that reduced the vinyl ether with concomitant cleavage of the *N*-benzyl group. The final step in the synthesis of aprepitant was the alkylation of the morpholine nitrogen with the triazolinone side chain.

One of the main challenges in developing the olefination route was subsequent reaction of the vinyl ether with excess DMT to form an ethyl impurity.[6] Therefore, initial development efforts on the olefination reaction focused on in-line monitoring by mid-IR to determine the reaction endpoint and limit the overreaction and formation of the ethyl impurity. Typically, online monitoring can reduce endpoint determination by 1 to 2 h versus traditional off-line assays. Laboratory runs clearly

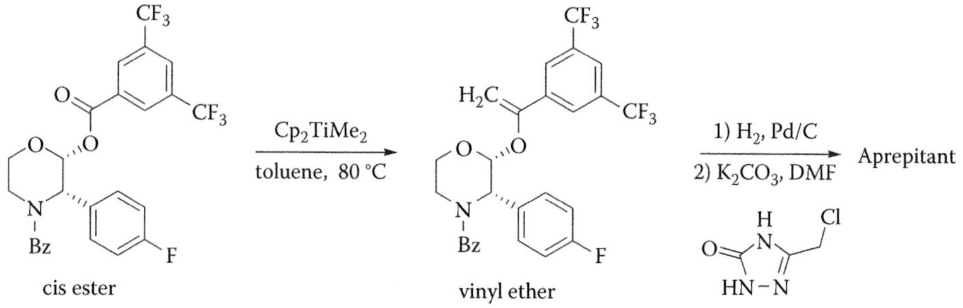

SCHEME 22.1 Conversion of the cis-ester to the vinyl ether.

demonstrated the feasibility of the approach. The reaction progress could be monitored by the disappearance of the relevant IR bands for the cis-ester (1740 cm^{-1}) and DMT (1247 cm^{-1}) as well as the appearance of the band at 1625 cm^{-1} which was attributed to the formation of the vinyl ether (Scheme 22.1).

In parallel, a reaction engineering approach involving a scavenger reagent with intermediate reactivity with DMT was investigated. This scavenger compound would ideally be less reactive than the cis-ester, so it did not interfere with the olefination reaction but be more reactive than the vinyl ether. Screening experiments with a series of esters identified 1,1-dimethyl-2-phenylethyl acetate (DMPEA) as a compound that had the desired reactivity.[7]

22.3 GENERAL CONDITIONS FOR OLEFINATION REACTION

The reaction is run in toluene using 1 eq of the cis-ester, 2.5 eq of dimethyltitanocene (DMT), 0.062 eq of titanocene dichloride (TiDi), and 0.75 eq of the scavenger ester, DMPEA. The reaction is run in batch mode at 80 to 85°C for approximately 5 h. The reaction endpoint is determined by high-performance liquid chromatography (HPLC) assay (>99% conversion of the cis-ester). The DMT reagent is prepared in a separate step in tetrahydrofuran (THF) from titanocene dichloride and methylmagnesium chloride. The cis-ester is charged to the DMT solution, and the system is solvent switched to toluene by vacuum distillation. The reaction is fairly robust; however, the DMT is unstable in the presence of protic compounds such as water. As a result, verifying the removal of water following the solvent switch is a critical in-process assay.

Although online monitoring was an improvement from the off-line assay, internal capabilities, both hardware and experience, for implementation in the pilot plant did not exist. In addition, the scalability of the technique with regard to cool-down time cycles and overreaction during the cool down was a concern. As a result, the reaction engineering approach was developed as the primary solution to overreaction of the vinyl ether, providing a robust and immediate solution. To address the issue of internal mid-IR capabilities, the implementation of online monitoring of this reaction was also pursued.

22.4 MID-IR LABORATORY DEVELOPMENT

The mid-IR work in the laboratory was initially performed with Mettler Toledo's DiComp™ probe inserted directly into the reaction vessel. This sampling technology, while ideal for laboratory applications, is not compatible with most existing process equipment within a pharmaceutical pilot plant. As a result, an alternate sampling technology, Mettler Toledo's Streamline™ flow cell equipped with a DiComp Sentinel™ attenuated total reflectance (ATR) probe, was investigated for pilot-plant applications. This sampling technology is an off-the-shelf product that can be interfaced with pilot-plant vessels through a recycle loop. Inherent advantages and disadvantages exist with a recycle loop relative to the laboratory setup. A distinct advantage of the approach is the segregation of the process analytic technology from the process equipment which allows more facile cleaning and optimization of the spectrophotometer. It also provides a facile means of collecting off-line samples for method development. A distinct disadvantage of the approach is the added complexity of the equipment setup and operation. In addition, temperature control for processing which deviates substantially from ambient conditions can represent a serious engineering challenge and process risk. A second potential disadvantage of this technology is that each probe has different character-istics and a calibration performed with one cannot be transferred to another. This often necessitates calibration on the process analyzer to be used for the analysis.

22.5 PILOT-PLANT SETUP AND OPERATING PROCEDURES

Development work in the pilot-plant production of the vinyl ether intermediate was performed on a Mettler Toledo ReactIR MP™. This instrument was the first generation of explosion-proof manufacturing technology available from Mettler Toledo. Setup requirements for the incorporation of the ReactIR MP™ monitoring system into the vinyl ether process in the pilot plant were minor. Installation of a fiber optics cable and protective conduit from the pilot-plant operations area to the control room, allowing communication between the acquisition computer (contained within the ReactIR MP™) and the monitoring station, was the most extensive mechanical modification.[8]

The Streamline flow cell was connected to a glass-lined 500-gal vessel via a recycle loop as shown in Figure 22.1. A Yamada pump (1/2" pump with Kynar® internals) was used to circulate the reaction solution from the vessel through Mettler Toledo's Streamline® flow cell and back to the vessel via the above surface line. The process lines (3/8" JIC) composing the recycle loop were partially insulated, but not heat traced, resulting in a temperature drop between the vessel and the return line. Although the temperature at the probe was approximately 10°C lower than the batch temperature, the probe temperature did not fluctuate substantially once the temperature in the vessel reached a steady state. This temperature difference did not adversely affect the reaction yield, cycle time, or method development.

The ReactIR MP™ unit is designed from a safety and performance perspective for operations in manufacturing environments. One of the operational problems occasionally encountered with in-line spectroscopic techniques is fouling of optical components. To avoid fouling of the probe and to insure a clean spectroscopic window, the ReactIR MP™ is equipped with a 200-W ultrasonic

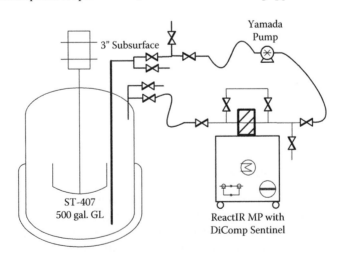

FIGURE 22.1 Pilot-plant setup for in-line mid-IR monitoring.

cleaner mounted perpendicular to the probe surface. This cleaning device can be activated at the instrument in 5-sec bursts. During processing, cleaning of the probe was initiated immediately following collection of grab samples for HPLC analysis. An alternate approach to probe cleaning consists of a bypass loop on the main process recycle with flushing of a pure solvent to waste. These approaches are not easily implemented with direct insertion of the spectroscopic probe into a process vessel.

The batch processing and mid-IR monitoring started after completion of all charges to the vessel. Flow through the recycle loop was initiated, and warm water (85 to 90°C) was placed on the jacket of the vessel. A reaction temperature of ~75°C (considered $t = 0$) was reached after 1 to 2 h of heating. The reaction kinetics were considerably slower at temperatures below 80°C; however, some conversion of the cis-ester occurred during the heating period. Samples for off-line HPLC analysis were collected from a sample tap just upstream of the IR probe.

22.6 PILOT-PLANT RESULTS

Five batches of the vinyl ether reaction were monitored by mid-IR using ReactIR MP™ (Mettler Toledo) during a pilot-plant campaign. Each batch yielded approximately 215 kg of vinyl ether for downstream production of aprepitant (Emend®) for phase II and III clinical trials.

Figure 22.2 and Figure 22.3 show the IR spectral regions for cis-ester and vinyl ether collected during one batch with reaction time on the z-axis. Spectra were collected every 4 min at 8 cm^{-1} resolution and 256 scans. The absorbance from 1755 to 1715 cm^{-1} is a combination of the carbonyl stretches for cis-ester and DMPEA. DMT reacts with DMPEA much more slowly than the cis-ester, and thus, temporal changes in the height and area of this peak are primarily due to reaction of the cis-ester, and a shift in peak maximum from 1741 cm^{-1} (cis-ester) to 1737 cm^{-1} (DMPEA) is readily apparent in the profile. The spectral region shown in Figure 22.3 was used to monitor vinyl ether formation. The absorbance at 1625 cm^{-1} was assigned to the methylene stretch of the vinyl ether. This absorbance overlaps with a much larger absorbance at 1606 cm^{-1}, which is a

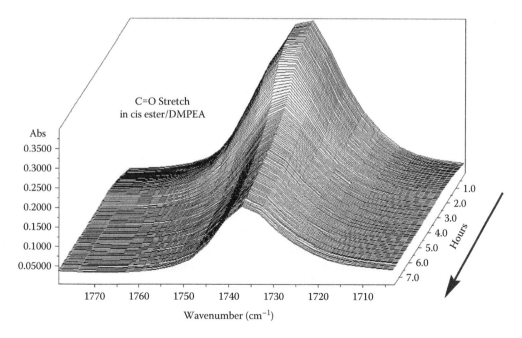

FIGURE 22.2 Spectral region for monitoring cis-ester and DMPEA.

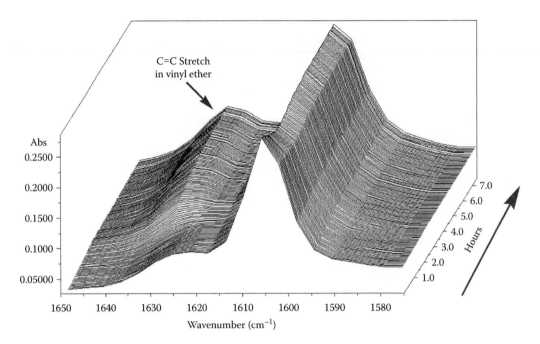

FIGURE 22.3 Spectral region for monitoring vinyl ether.

characteristic aromatic ring stretch. The presence of an additional vinyl stretch related to the conversion of DMPEA to its vinyl adduct is not readily apparent in Figure 22.3, although some conversion of the DMPEA is expected at late reaction times.

22.7 MID-IR QUANTITATIVE ASSAY DEVELOPMENT

During processing in the pilot plant, 13 samples from two separate batches were collected and assayed by HPLC for cis-ester and vinyl ether concentration. These assays were used as inputs into a partial least squares (PLS) algorithm supplied as part of Mettler Toledo's software in order to develop an IR calibration for the reaction. Both peak height and peak areas for the cis ester and vinyl ether were used in the calibration. The cis-ester region overlapped with DMPEA; however, the rate of disappearance of DMPEA is much slower than that of cis-ester. The area calculation coupled with the peak height for the cis-ester resulted in an excellent calibration at moderate to high concentrations of cis-ester (20 to 160 g/L). The region chosen for vinyl ether contained an additional solvent absorbance (C=C stretch of toluene) that was assumed to remain constant during the reaction.

The PLS calibration constructed from the off-line samples was used during subsequent batches to predict the concentrations of cis-ester and vinyl ether in real time. For the batch shown in Figure 22.4, four process samples were taken to assess the calibration performance and show the correlation between the mid-IR and HPLC assays. Table 22.1 and Figure 22.4 show that excellent agreement between HPLC and IR assays was observed. The average absolute error between the liquid chromatography (LC) and IR assays was 1.7 g/L for cis-ester and 1.6 g/L for vinyl ether. Termination of the reaction was based on the HPLC assay for the 4.75 h process sample which indicated that 99.4% of the cis-ester had reacted.

The mid-IR quantitative method for the vinyl ether step yielded excellent results considering the limited number of process samples used to construct this calibration curve. Expanding the number of calibration samples would undoubtedly improve the calibration, particularly at the low

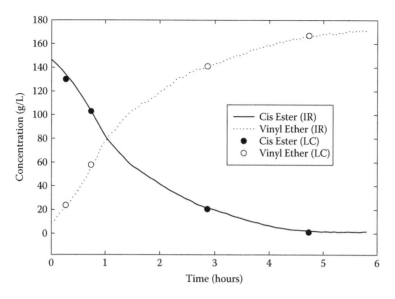

FIGURE 22.4 Comparison of infrared and liquid chromatography concentration profiles.

TABLE 22.1
Comparison of Liquid Chromatography and Infrared Assays

Time (h)	Cis-Ester[a]		Vinyl Ether[a]	
	HPLC	MID-IR	HPLC	MID-IR
0.25	130.0	133.2	23.7	23.6
0.75	103.0	103.3	57.8	55.5
2.75	20.5	21.5	141	140.9
4.75	1.1	3.3	167	163.2

[a] g/L.

concentrations of cis-ester. Further improvements could be realized by adding the concentrations for DMPEA and its vinyl ether product to the calibration. This would provide additional constraints on the PLS algorithm and allow deconvolution of the overlapping DMPEA and cis-ester absorbances. Full method development for manufacturing would likewise include variations in DMPEA charge, initial water content, and temperature.

22.8 PREPARATION OF DIMETHYLTITANOCENE

Dimethyltitanocene (DMT) was developed by Petasis for the conversion of the carbonyl moiety in esters, ketones, and amides to the corresponding olefin.[9] The Petasis reagent is very effective at converting esters to vinyl ethers, and isolated yields are higher with this reagent due to the neutral conditions used to generate the titanium carbene. The details of the reaction of DMT with esters was published by Hughes et al.,[10] and an improved procedure for the preparation of DMT from titanocene dichloride and methyl magnesium chloride was reported by Payack et al.[11] DMT is prepared by the addition of methylmagnesium chloride to a slurry of titanocene dichloride (TiDi) in THF and passes through the monomethylmonochloro intermediate (Scheme 22.2). Hydrogen

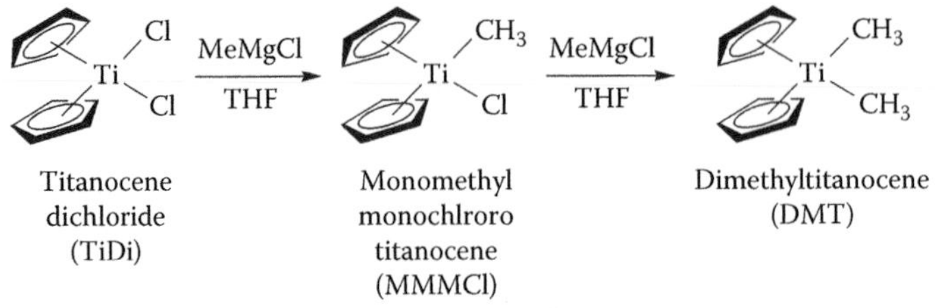

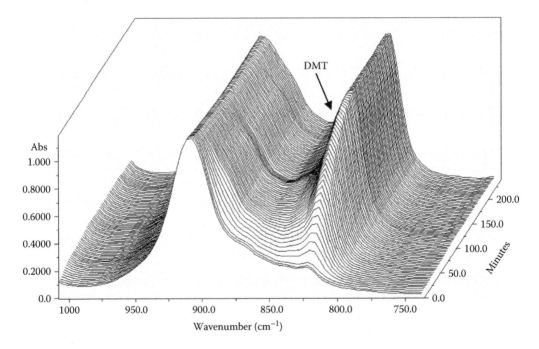

SCHEME 22.2 Preparation of dimethyltitanocene.

nuclear magnetic resonance (^{1}H-NMR) was initially used to analyze the formation of DMT. Subsequently, Vailaya et al. developed iodometric titration and gas chromatography (GC) methods to replace the quantitative NMR assay.[12] To give additional insight into the physical process of this critical reaction, a quantitative IR method was developed and demonstrated on the laboratory scale that used the NMR assay as a reference method. Figure 22.5 clearly shows the formation of an absorbance at 814 cm^{-1} that parallels the methylmagnesium chloride addition and is distinct from the absorbances in the Grignard solution.

Titanocene dichloride, methyl magnesium chloride, DMT, and the dimer decomposition product of DMT are all visible by mid-IR. The absorbances for titanocene dichloride and DMT overlap, and monomethylmonochloro titanocene, a transient intermediate observed by NMR and normal-phase HPLC, was initially not detected by mid-IR but was assumed to have similar absorbance.

The absorbance at 814 cm^{-1} includes all three of the titanocene compounds. Stand-alone peaks for TiDi, monomethylmonochloro titanocene, or DMT were not observed; however, the chemometrics package in the Quant-IR program was successfully used to extract concentration information for each component. ^{1}H NMR data and the IR spectra were used to construct a calibration curve

FIGURE 22.5 ReactIR™ (Mettler Toledo) data for formation of dimethyltitanocene.

TABLE 22.2
Comparison of Calibration Methods for Dimethyltitanocene (DMT) Formation

| | | Average Percent Error | |
Calibration Method	Reference Method	DMT Concentration	MMMCl Concentration
Inverse P-matrix with four frequencies	¹H NMR	8.8%	43.2%
Inverse P-matrix with automatic frequency selection	¹H NMR	0.6%	0.5%
Inverse P-matrix with automatic frequency selection	HPLC	7.9%	—

using an inverse least squares (P-matrix) calibration. Initially, four peaks were selected, and this gave a reasonable fit with an average percent error for DMT of 8.8% and 43.2% for monomethylmonochloro titanocene. Using the automatic frequency selection option within QuanIR® which selected five (different) frequencies and ¹H NMR data significantly improve the calibration with the average percent error for DMT of 0.6% and for monomethylmonochloro titanocene of 0.5%. Using HPLC data (g/L) for DMT and the automatic frequency selection option gave an average percent error of 7.9%. There was also more scatter in the data from HPLC than the data from NMR; however, the response factor for monomethylmonochloro titanocene was not available, so a calibration curve cannot be constructed using HPLC data. Results from the various calibration procedures investigated are summarized in Table 22.2.

To further investigate the DMT formation, an experiment was performed where the methyl-magnesium chloride addition was stopped at different intervals to observe the corresponding response in the reaction profile. An advantage of this analysis method is that concentration profiles can be obtained on heterogeneous samples that typically are difficult to assay quantitatively. The concentration of DMT and monomethylmonochloro titanocene versus the reaction time is shown in Figure 22.6. TiDi has low solubility in the reaction solvent, so the methylmagnesium chloride

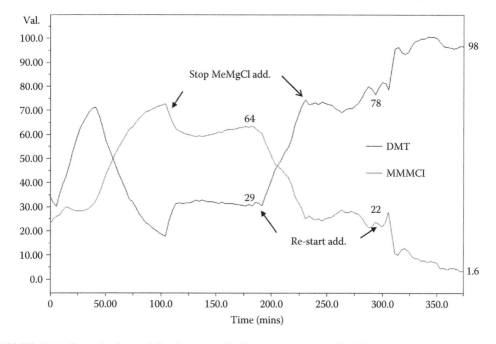

FIGURE 22.6 (See color insert following page 40). Dimethyltitanocene (DMT) and monomethylmonochloro titanocene concentration profiles from discontinuous addition of Grignard solution. The concentration of DMT and MMMCl determined by 1H NMR is shown for three samples.

initially added converts all the soluble TiDi to DMT. A disproportionation reaction between DMT and TiDi results in the formation of the monomethylmonochloro derivative. This creates some confusion when using traditional grab samples owing to the changing concentration of the species due to disproportionation as TiDi slowly dissolves. Continued addition of MeMgCl leads to the immediate conversion of monomethylmonochloro titanocene to DMT.

22.9 DETERMINATION OF WATER LEVELS

The presence of water or other protic solvents decreases the stability of dimethyltitanocene (DMT). Karl Fisher (KF) titrations were shown to be unreliable due to the reaction of DMT with iodine, a component of the KF reagent. A quantitative analysis for water using mid-IR was investigated to determine whether this technique could replace the traditional method of pulling samples and conducting KF titrations.[13] The DMT concentrate (DMT, cis-ester, and toluene) was dried overnight with molecular sieves and used as the 0 μg/mL sample. The water content was then incrementally increased by adding a 50% water solution in THF and spectra collected (Figure 22.7).

A QuantIR calibration was prepared (Figure 22.8) from this data and additional data points and used to assay samples of the DMT concentrate (Table 22.3).

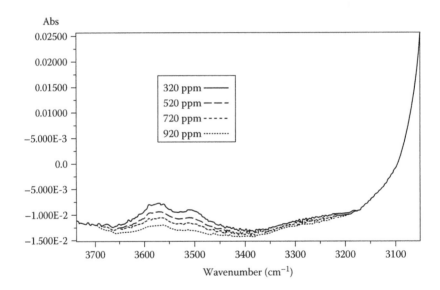

FIGURE 22.7 (See color insert following page 40). Water concentration in the DMT solution.

TABLE 22.3
Water Concentration
Determined by Fourier
Transform Infrared (FTIR)

Sample Number	FTIR (μg/mL)
1	285
2	380
3	336
4	275

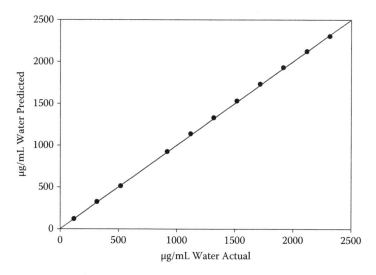

FIGURE 22.8 Plot of the actual water concentration versus the levels predicted by QuantIR.

The water concentration was then determined in four process samples from the pilot plant using this QuantIR method. Unfortunately, a reliable independent method for determining the actual water content in these samples was not available at the time this study was performed, thus determining the method accuracy and precision was not possible.[14]

This study illustrated a drawback to using IR for quantitative analyses because you must rely upon an independent assay for the calibration. The limited number of runs conducted did, however, demonstrate feasibility of online determination of water content. From a manufacturing perspective, this technology would greatly reduce the cycle time for the process by eliminating the need to suspend the vacuum concentration used to azeotropically remove water in order to obtain a sample for off-line analysis.

22.10 CONCLUSION

Three quantitative assays were developed using *in situ* mid-IR for the analysis of water concentration, DMT concentration, and the conversion of ester to a vinyl ether. One instrument could, therefore, potentially replace three different instruments. The advantage of this technique is the potential to monitor and analyze a process in real time, which can be critical for accurate quantification of unstable intermediates and useful when established analytical methods are unable to provide accurate results. Online monitoring provided valuable information for process optimization, and the mechanistic information regarding the formation of an intermediate was extracted from overlapping spectral data using the QuantIR program. However, this is only possible if an independent quantitative assay is available for the construction of the calibration table. Knowledge gained from laboratory experiments can be incorporated into pilot-plant campaigns that can then be incorporated into a potential commercial process. As the understanding of the process increases over the course of development, the need for online monitoring may decrease. Incorporation of online monitoring in a manufacturing process is contingent on an evaluation of the cost–benefit ratio and will ultimately be driven by business considerations. This project integrated the different phases of development at Merck Research Laboratories by involving the Process Research, Chemical Engineering R&D, and Analytical Research departments.

REFERENCES AND NOTES

1. Formerly ASI.
2. Navari, R.M.; Reinhardt, R.R.; Gralla, R.J.; Kris, M.G.; Hesketh, P.J.; Khojasteh, A.; Kindler, H.; Grote, T.H.; Pendergrass, K.; Grunberg, S.M.; Carides, A.D.; Gertz, B.J. *New. Engl. J. Med.* 1999, *340*, 190–195.
3. Payack, J.F.; Huffman, M.A.; Cai, D.; Hughes, D.L.; Collins, P.C.; Johnson, B.K.; Cottrell, I.F.; Tuma, L.D. *Org. Process Res. Dev.* 2004, *8*, 256–259.
4. (a) Brands, K.M.J.; Payack, J.F.; Rosen, J.D.; Nelson, T.D.; Candelario, A.; Huffman, M.A.; Zhao, M.M.; Li, J.; Craig, B.; Song, Z.J.; Tschaen, D.M.; Hansen, K.; Devine, P.N.; Pye, P.J.; Rossen, K.; Dormer, P.G.; Reamer, R.A.; Welch, C.J.; Mathre, D.J.; Tsou, N.N.; McNamara J.M.; Reider, P.J. *J. Am. Chem. Soc.* 2002, *125*, 2129–2135. (b) Zhao, M.M.; McNamara, J.M.; Ho, G.-J.; Emerson, K.M.; Song, Z.J.; Tschaen, D.M.; Brands, K.M.J.; Dolling, U.-H.; Grabowski, E.J.J.; Reider, P.J.; Cottrell, I.F.; Ashwood, M.S.; Bishop, B.C. *J. Org. Chem.* 2002, *67*, 6743–6747. (c) Pye, P.J.; Rossen, K.; Weissman, S.A.; Maliakal, A.; Reamer, R.A.; Ball, R.; Tsou, N.N.; Volante, R.P.; Reider, P.J. *Chem. Eur. J.* 2002, *8*, 1372–1376.
5. Hale, J.J.; Mills, S.G.; MacCoss, M.; Finke, P.E.; Cascieri, M.A.; Sadowski, S.; Ber, E.; Chicchi, G.G.; Kurtz, M.; Metzger, J.; Eiermann, G.; Tsou, N.N.; Tattersall, F.D.; Rupniak, N.M.J.; Williams, A.R.; Rycroft, W.; Hargreaves, R.; MacIntyre, D.E. *J. Med. Chem.* 1998, *41*, 4607–4614.
6. Ethyl impurity.

ethyl impurity

7. Huffman, M.A. U.S. Patent 6,255,550, 2001.
8. Additional service requirements for the MP unit included high-pressure instrument air (>80 psig) for purging the explosion-proof enclosure, low-pressure nitrogen (5 psig) for purging the optical conduit, and 20 to 25°C deionized water for internal temperature control of the explosion-proof enclosure. An attempt to use nitrogen to purge the MP was made; however, supply pressure deviations were experienced, resulting in discontinuous purging of the unit and loss of power. This necessitated the use of high-pressure instrument air for instrument purging. The main concern with using air versus nitrogen was the presence of pump oil which could potentially damage the instrument optics. To avoid such contamination, an oil trap was placed between the instrument air header and the MP unit.
9. (a) Petasis, N.A.; Bzowej, E.I. *J. Am. Chem. Soc.* 1990, *112*, 6392–6394. (b) Petasis, N.A.; Lu, S.-P. *Tetrahedron Lett.* 1995, *36*, 2393–2396.
10. Hughes, D.L.; Payack, J.J.; Cai, D.; Verhoeven, T.R.; Reider, P.J. *Organometallics* 1996, *15*, 663–667.
11. (a) Payack, J.F.; Hughes, D.L.; Cai, D.; Cottrell, I.; Verhoeven, T.R.; Reider, P.J. *Org. Prep. Proceed. Int.* 1995, *27*, 715–717. (b) Payack, J.F.; Hughes, D.L.; Cai, D.; Cottrell, I.F.; Verhoeven, T.R. *Org. Synth.* 2002, *79*, 19–26.
12. Vailaya, A.; Wang, T.; Chen, Y.; Huffman, M.A. *J. Pharm. Biomed. Anal.* 2001, *25*, 577–588.
13. am Ende, D.J.; Clifford, P.J.; DeAntonis, D.M.; SantaMaria, C.; Brenek, S.J. *Org. Process Res. Dev.* 1999, *3*, 319–329.
14. A reliable NIR method was developed and implemented in the pilot plant. Y. Chen, unpublished results.

23 Process Analytical Technology in the Manufacture of Bulk Active Pharmaceuticals— Promise, Practice, and Challenges

C.A. Mojica, L. St. Pierre-Berry, and F. Sistare

CONTENTS

23.1 PART I—CHALLENGES AND OLD PARADIGMS

The "vision" underlying the use of process analytical technologies (PATs) in the manufacture of pharmaceuticals has been articulated as "…a system for designing, analyzing, and controlling manufacturing through timely measurements (i.e., during processing) of critical quality and performance attributes of raw and in-process materials and processes with the goal of ensuring final product quality."[1] The examples that we cite and the opinions expressed herein are the result of our collective experience as implementers of PAT technologies in a bulk API manufacturing plant.

We have embraced this relatively modern definition of PAT for the purposes of this chapter. In particular, PAT in this context includes any analytical technique that provides data in real time to manufacturing processes involved in making bulk active pharmaceutical ingredients (APIs) with the primary intent to ensure process control and thus product quality.

In general, the pharmaceutical industry has been criticized for its lack of modernization in this area. It has been said that PAT as an enabling technology has been established practice among petrochemical, polymer, and food industries long before the pharmaceutical sector embraced it—a comment many times heard not only from the media, but from industry regulators as well. In a 2003 article by *The Wall Street Journal*, reporters commented that "The pharmaceutical industry

... even as it invents futuristic new drugs, its manufacturing techniques lag far behind those of potato-chip and laundry-soap makers."[2]

An examination of the reasons behind the apparent timidity on the part of otherwise resource-rich and innovative companies to implement PAT helps frame the industry's challenges in this area. New perspectives, coupled with those of regulators, have developed to the point where it appears that PAT are now earmarked to become a critical feature in modern API manufacturing.

The basic challenges and historical reasons for the apparent lack of progress in PAT implementation can be traced to two factors: the use of classical organic chemistry as the underlying foundation for API manufacturing and the influence of government regulations.

23.1.1 CLASSICAL ORGANIC CHEMISTRY AND TRADITIONAL TESTING

A large percentage of the manufacture of API pharmaceuticals is essentially the commercialization at large scale of the same reactions performed by organic chemists at laboratory scale. This applies to the analytical technologies as well as chemical technologies. In this model, there are natural stopping points along the synthetic route where detailed analysis of progress, both in terms of yield and quality, is practical. In terms of in-process controls, the traditional view has been that if particular reaction parameters are strictly controlled within acceptable ranges, then accurate in-process information in real time is of secondary importance, or perhaps even unnecessary. Only recently, and in great measure by efforts spearheaded by regulators coming to terms with the conundrum of process validation (something they themselves help conjure) has this view been modified.[3] However, inasmuch as manufacturing processes mimic one-step-at-a-time laboratory processes, analytical technologies in real time (i.e., PAT) may seem to some in the industry as superfluous. As processes move to a more continuous model, PAT becomes more relevant.

In general terms, petrochemical and food processes (i.e., making potato chips and laundry detergents) are shorter, continuous, or semicontinuous, and heavily engineered compared to bulk pharmaceutical manufacturing operations. In addition, given the short processes involved, regulatory scrutiny and controls are applied to the finished product, not to each step along the way; this allows a level of flexibility that the pharmaceutical industry does not presently enjoy and has never experienced. Thus, the unfavorable comparison between the two industries in the area of PAT is, in our view, not entirely justified.

Bulk pharmaceutical manufacturing processes, by which we mean organic chemical reactions, can benefit greatly from the application of online PAT; this in itself is not news. However, it took a while for the pharmaceutical industry to completely understand and assimilate the added benefits of the technology. In the early 1990s, when industry leaders started to have serious discussions around PAT for bulk manufacture, the primary driver for the investment was cost reduction resulting from the elimination of typical laboratory-based in-process testing. This was a direct extension of what had been achieved in the manufacture of some finished drug products, and in the food and petrochemical industries. There are several inherent inefficiencies surrounding laboratory in-process testing. For instance, the sampling requires stopping the operation and manual intervention by human operators. Manual intervention many times can raise safety and quality concerns. The stop-sample-and-wait mode can be slow and inefficient with turn-around times for analytical results lasting several hours not uncommon. Finally, the quality of the sample, or the sample actually changing en route, are variables that add additional complications. Thus, the industry had some strong justifications in examining this operational model vis-à-vis what PAT could offer.

An example of how time consuming and labor intensive the stop-sample-and-wait mode of testing can be is represented in Figure 23.1. In this simple example, we compare the typical steps involved in determining whether a particular product is dry (i.e., within specifications) by the conventional method versus an alternative real-time PAT drier application. The task of measuring product dryness, and assuring product quality, by conventional methods involves breaking vacuum, having trained operators suit-up in personal protective equipment (PPE), securing and labeling an

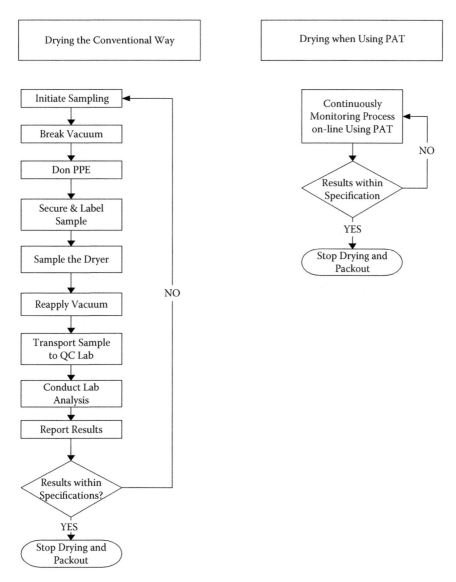

FIGURE 23.1 (See color insert following page 40). Process steps in a conventional drier monitoring operation versus a process analytical technology (PAT) drier application.

appropriate container for the sample, sampling the dryer, reapplying vacuum once a homogenous sample is secured, transporting the sample to the quality control laboratory, performing the laboratory analysis, verification, and finally the reporting of results. When PAT is used, the dryer can be continuously analyzed without having to break vacuum or involving any manual intervention.

It seems clear that the conventional formula for drier monitoring is very time consuming and involves many more steps than an alternate PAT application. In many cases, the time invested in the total drying operation can be reduced by at least half. This observation can prove important, especially if, as it is many times the case, drying is the rate-limiting step of the manufacturing process. In addition, PAT is able to monitor multiple parameters, such as water, residual solvents, and even particle size. In fact, a combination of parameters can be analyzed at the same time and, in many cases, with the same PAT instrument. The same principles and observations apply to almost every typical in-process analysis in API bulk manufacturing. It is in this particular area that during the past 15 years industries such as food, petrochemicals, and polymers have been using PAT.

However, the industry encountered significant challenges in adopting PAT to its fullest extent. For one thing, PAT applications offer the biggest paybacks when coupled with automated process control systems that allow for continuous feedback and process modification in real time. These vary from company to company and within each company's manufacturing sites. This situation, in turn, makes formulating a comprehensive multi-site strategy difficult and forces management to commit to long implementation timelines. The second and biggest hurdle was government regulation.

23.1.2 GOVERNMENT REGULATIONS

Recent comments by regulators notwithstanding, in our opinion, government regulation has been by far the most significant factor influencing PAT implementation in the pharmaceutical industry. By statutory mandate, the industry must disclose practically every detail of the manufacturing process to government agencies, and must comply with strict control, disclosure, and approval rules regarding changes to the manufacturing process once these are originally filed with the government. These rules apply to the analytical technologies and methodologies used to monitor the manufacturing process as well. In terms of the chemical and analytical technologies represented, these regulatory submissions are the best the companies could develop at the time of the filing, within cost and business constraints. Thus, they represent a static snapshot of the most pragmatic manufacturing process at a point in time. Moreover, the regulatory approval process to amend already filed processes is not designed to accommodate changes in manufacturing or testing rapidly or with ease. The end result is that progress toward implementation of new technologies is slow.

It is fair to say that this form of operation has served both the industry and the regulators well. Government agencies, mandated to safeguard the public safety, are seen as carefully regulating the manufacturing process, and thus the industry overall. For the industry, well-publicized compliance of government regulations coupled with positive plant inspections can be represented as enhanced product safety to customers, and as a competitive advantage to investors. However, the administrative and bureaucratic systems managing process change, both governmental and within the companies, are of such complexity that any change must have substantial justification to warrant the administrative effort and the potential regulatory exposure. Most postapproval filings in bulk manufacturing that are not simply clarifications or fixing errors are justified by a financial impact of some significance. The industry has been careful in weighing the advantages and disadvantages of very innovative technologies and not launching into implementation without a strong justification, particularly if financial gains are "soft," or as is the case with PAT, if there are already well-established and well-understood systems that deliver reliable results (i.e., typical laboratory-based testing). Thus, in terms of implementation of PAT, the trend has been to limit investment to relatively new processes and for special applications. Older or well-established processes have not seen the benefit of regulatory PAT applications. However, it seems clear that by maintaining this model, the industry is losing in efficiency, process understanding, and process capability, and is missing optimization opportunities.

The success of this industry/government relationship has meant complete institutionalization, acculturation, and enhancement of all the peripheral systems that support it, on both the industry and government sides. For example, one can argue that the analytical scrutiny at defined steps has naturally led in part to the adoption of more and more richly documented validation exercises of analytical methods. These validations are well liked by regulators (validation is indeed the law) because they provide a seemingly thorough technical examination of the ability of the analytical method to perform the task intended. In essence, it provides documented assurances that the analytical data are reliable up to the very limits of statistical certainty. Because the data are thoroughly proven to be reliable, the discussion with regulatory inspectors many times revolves around the rigor of these documented validations. In response to any potential exposure, the industry (by nature already very risk averse) thus augments its documentation and method suitability exercises (i.e., validation requirements) even more. If these methods were not filed, it is arguable whether the industry would invest as many resources on validation as it feels it needs to do at present. More to the point, the specific details surrounding method suitability and validation paradigms have had an adverse effect on the implementation of PAT methods because of the intrinsic differences between PAT and conventional analytical methods. Moreover, the institutionalization of this operational model has created a culture in both the industry and government regarding how these validation packages should look and what type of data should be represented, expectations that may prove difficult to change in the short term. Some have argued that the industry should take a stand and validate a method/process to the intended purpose, not to the extremes that the industry currently validates.

For example, validation of a typical laboratory high-performance liquid chromatography (HPLC) method involves altering samples to be well outside the limits where it is intended to operate (e.g., for an assay method intended to measure product close to a value of X, it would be typical to demonstrate suitability in the range of 80 to 120% of the target X assay value). In addition, the typical validation package would include parameters such as limit of detection, limit of quantitation, sample stability, method precision (analyzing the same sample multiple times), robustness (using two different analysts and instruments on different days), among many. Some of these parameters lose meaning when considering typical PAT application. For example, one cannot build a PAT model with a range of 80 to 120% from a target value in a real-time PAT application for reaction completion, because the data only goes from 0 to 100% (0% product at the beginning to 100% at the end). When trying to validate an in-line PAT near-infrared (NIR) spectroscopy drier method, for example, one can only build a reliable chemometric model with typical product manufactured within a tight range of operating limits. Working outside typical operational parameters compromises the chemometric model and forces the use of laboratory-generated samples which adds an unnecessary level of variation. Spiking the sample with known impurities or multiple repetitions of the same measurement is not only difficult because the process is not static, but it seems meaningless because it does not reflect reality nor take into account processing parameters, such as agitation, temperature, pressure, and so forth. Finally, testing for robustness, such as different instruments and operators on different days, is unnecessary, because the PAT applications are typically specific to equipment and process. Testing for operator variation obviously adds no value, because the PAT instrument/model is controlled by an automated system with no manual intervention from operators (for some well-articulated angst on the topic of PAT validation and handling PAT data, see Reference 4).

Resolving this incompatibility between PAT methods and the current expectations around method validations has proven to be perplexing to both the industry and regulators. A new clearly understood, well-publicized rule book needs to emerge and be tested by actual PAT filings and regulatory inspections.[5] In established processes, the preferred method to move to PAT applications from older methodologies appears to be filing comparability or equivalency protocols. In general terms, it remains to be seen how well received these protocols will be by filing agencies, and how well they will stand up to the test of actual inspections.

In addition to PAT methods validation, the industry has expressed some anxiety regarding how real-time in-process data would be viewed by regulatory agencies with respect to process validation. The problem stems from another institutional paradigm around what process validation means. Ironically, this concept of process validation may prove critical in finally integrating PAT into mainstream API manufacturing.

Process validation, formally articulated in 1987 and similar in theory to analytical method validation, is defined as "… establishing documented evidence which provides a high degree of assurance that a specific process will consistently produce a product meeting its pre-determined specifications and quality attributes."[6]

Almost by convention, a process is pronounced "validated" if the company can produce documentation on three identically run consecutive batches that result in product that meets specifications. The picture is one of permanent static control. With the introduction of PAT applications, which generate a continuous stream of real-time data that, in turn, allow for the possibility of real-time process modification, the picture of static control is shattered. Thus, a conundrum some in the industry feared that with PAT the company would be generating data that could be used by regulators to demonstrate that processes were really not "validated" after all.

The static three-batch validation exercise, the applied derivative of the definition above, has started to look antiquated, if not counterproductive, to regulators because it is seen as stifling innovation. Through the years, both the industry and regulators have worked hard to establish documented evidence that supports the picture of permanent static control. By the year 2000, neither side was happy with their creation. The industry felt burdened with administrative systems documenting processes that were, in fact, not under permanent static control, and everyone knew it (see Reference 3). Recent criticism leveled at the industry specifically from regulators pointed out that

> High residual uncertainty in an approved and validated manufacturing process can also impede a firm's ability to collect additional information (e.g., fear of jeopardizing the validation status) to develop an understanding of the sources of variability and to investigate root-cause of out-of-specifications. As a result, the current system leans towards testing to document quality and just in case manufacturing as means to overcome uncertainty in decisions related to quality of a batch and meeting market demand. In the current state, innovation and continuous improvement in pharmaceutical manufacturing is discouraged to a large extent by uncertainty.[7]

That the regulators helped delineate the expectations that led to this sad state of affairs, a process that continues, is a detail that seems to have been overlooked. In any case, regulators at both sides of the Atlantic and in Japan have recently defined a new "desired state"[8] that stresses the value of process understanding as a new basis for validation. This is a vision where variation, a fact of life in any industrial process, must be understood. Specifically, the vision of the "desired state" has been articulated as one where

1. Product quality and performance are achieved by design of effective and efficient manufacturing processes.
2. Product specifications are based on a mechanistic understanding of how formulation and process factors impact product performance.
3. Manufacturers are able to effect continuous improvement and continuous "real time" assurance of quality.

Full implementation of PAT can play a significant role in the realization of this new emerging paradigm. As presented previously, commercialization/inclusion of PAT in the manufacture of pharmaceuticals has had to deal directly, and with different levels of success, with the main characteristics of the current manufacturing model (classical organic chemistry and government regulations). Used to its full potential, PAT can only increase the industry's fundamental understanding of the processes they are running, which relate, in particular, to points 1 and 3 of the "desired state." This new vision opens the door to the utilization of PAT to populate databases on a multitude of process parameters, which then can lead to the "mechanistic understanding" of processes, which in turn should lead to "continuous 'real time' assurances of quality." The direction taken toward achieving the original goal of validated processes may thus have been shifted from an emphasis on documentation to one of demonstrating process understanding. PAT, in combination with multivariate statistical process control and other such tools, could become the cornerstone in this new paradigm. The pharmaceutical sector may now be ready to follow the food, polymer, and petrochemical sectors in duplicating their results in the form of the following[9]:

- Assured acceptable end product quality at the completion of the process
- Reduced and even eliminated some finished product release testing (achieved real-time release)

- Reduced manufacturing costs
- Reduced production cycle times
- Increased throughput
- Reduced rejects, scrap, and reprocessing
- Reduced raw material cost by precise formulation
- Increased automation, improved operator safety, and reduced human errors
- Improved efficiency and managed variability

23.2 PART II—PAT AND REALIZATION OF NEW "DESIRED STATE" PARADIGM

The concept of "process understanding" can be thought to consist of several levels. In the context of bulk API manufacturing, implementation of PAT applications can be broken down into unit operations, each reflecting different levels of process understanding. PAT applications for each of these different unit operations will pose slightly different challenges, and successful commercialization may depend on the sophistication of peripheral process control systems such as level of process automation and types of equipment available. In addition, several of the old regulatory questions, specifically around PAT method validation and equivalency protocols, will remain untested for a period of time.

Listed below are areas for PAT research that we believe align well with the emerging "desired state" paradigm. These are listed in order of delivering increasing levels of process understanding. There is a considerable element of subjectivity in formulating this order, but the steps also happen to follow a pattern of increasing technical complexity.

23.2.1 RAW MATERIAL APPLICATIONS

Many raw materials in API manufacturing come from unique and relatively small numbers of suppliers. Depending on the nature of the relationship with the supplier, PAT applications such as NIR can be developed where raw materials are tested and approved, eliminating the sample–test–release model commonly used at present. This PAT-base mode of raw material testing and approval aligns itself well with many other of the industry's improvement programs, such as 6-Sigma and Just-In-Time initiatives. The main advantages of implementing a program such as this include the minimization of the time spent in the approval process. NIR technology is particularly well suited for this task. A typical approach is to compare the incoming material with a specific reference library derived from past deliveries that have met specifications and performed well in the process, and then assess how the new incoming lots compare. Acceptability thresholds as narrow as needed in order to ensure quality can be programmed into the NIR instrument software.

Although the analytical technique and the overall application are relatively simple to implement at the ground level, the main drawback to implementation of this strategy is regulatory. Specifications for raw materials are also part of the new drug application (NDA), and thus, analytical methods are either part of the initial filing or any new ones need to be included later as a postfiling alternative method. Either approach carries with it regulatory concerns. Including this type of method in the original NDA may require inclusion of the standard libraries. This would force manufacturing sites to always use the same suppliers selected during the original process development. It would also limit the flexibility of adding new suppliers or new libraries, as this would require potentially time-consuming supplemental submission to the filings. Pursuing a strategy of filing alternative postapproval NIR methods, justified after successful equivalency protocols are completed, may prove administratively complex if the same raw material is used in many processes (common organic solvents, for example), thus requiring individual filing for each process. This also applies if the process is filed separately in many countries. The fact that this type of PAT project does not carry a significant economic return may make it ultimately unattractive for many in the industry. However,

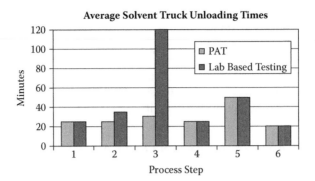

FIGURE 23.2 (See color insert following page 40). Average times involved in testing, release, and unloading a solvent tank truck (study conducted by Pfizer Inc., Groton, CT).

this type of PAT project can be successfully incorporated within a program of conditional approvals designed to expedite deliveries to the plant (i.e., proceed at minimal risk pending approval by conventional lab-based testing) or when dealing with transfer of intermediates between plants.

For example, the typical process for accepting bulk solvents for use in both processing and equipment cleaning can be very time consuming. Typically, when a solvent delivery truck arrives at a plant, it is escorted to the unloading area where it is inspected and all of the paperwork is reviewed to ensure that the material meets the agreed quality. If the truck passes the physical inspection and all of the necessary paperwork is in order, the material in the solvent truck is sampled and delivered to the testing laboratories to be analyzed following at least the testing regime specified in the particular filing. Typically these tests include appearance, identity by GC, or any of several spectroscopic methods as well as refractive index and specific gravity. In some instances, if by prior agreement all regulatory testing has been already done by the supplier following filed methodology, the receiving site just completes a review of the certificate of analysis that comes with the delivery. In any case, a great deal of time is spent in testing, paperwork review, and the administrative processes associated with "releasing" the solvent delivery (see Figure 23.2).

Process steps include the following:

1. Pick up/escort truck to unloading area
2. Secure truck, safety inspection, bring sample to lab
3. Quality operations (QO) labs testing/approval
4. Connect truck for off loading to storage tanks
5. Unload truck
6. Disconnect truck/escort off site.

The alternative PAT application is relatively simple. The operators involved in the physical inspection of the truck can also analyze the sample at-line using an NIR application. The sample is then compared to a standard library derived from past deliveries that met specifications. The NIR chemometric model is more than capable of determining whether the material is the solvent intended, and whether it is comparable to past deliveries. The process is practically automatic and carries a very low compliance risk. This PAT application demonstrated a close to 40% reduction in the time spent approving a solvent delivery. At the ground level, the truck would have been unloaded for use "at risk" and a sample sent for "official" testing and QO approval. The same strategy can be applied to other deliveries of relatively common chemicals, such as sodium carbonate, sodium chloride, permanganates, potassium *t*-butoxide, or transfer of advanced intermediates between manufacturing plants.

23.2.2 Solvent Recovery Applications

PAT applications in solvent recovery enjoy particular success because many times they realize financial paybacks associated with increased recovery efficiencies and productivity. Typically, either NIR or online GC applications can be used to monitor residual water or specific impurities (other solvents) in the recovery of a large number of expensive solvents. This form of quality assessment in real time allows the solvent recovery operators to modify the column conditions to suit their needs and thus optimize the recovery process. In the vast majority of cases, the recovery process is not part of the regulatory filing; thus, there is a considerable amount of flexibility allowed in the actual running of the operation. Depending on the specific PAT application, once the recovery operation is complete, a homogenous sample of the recovered batch is secured and tested/released as done for virgin material. If the stream is part of a closed recycled loop and specific quality assurance concerns have been addressed, many times the recovered solvent can be reused without further testing. Online gas chromatography (OL-GC) offers a slight advantage over NIR applications for solvent recovery, because GC is the analytical method of choice used in the typical regulatory testing pattern for solvents, and thus it enjoys certain credibility with quality assurance functions. It also shows higher sensitivity in many applications and does not require generation of solvent-specific libraries or chemometric models to calculate concentration of different impurities.

One example in which OL-GC has proven to be valuable is in recovery of methanol from process streams. In this example, methanol was being recovered primarily to remove water and small levels of isopropanol. Purification is achieved by fractional distillation based on the differences in boiling points of the components. Traditionally, the recovery was performed without the aid of real-time results. Based on past experience, operators would manually secure samples from the recovery column at predetermined times and run water analyses using standard Karl Fisher (KF) methodology. If results were found acceptable, samples would then be sent to the quality control in-process testing lab for GC assay. The actual time spent in this operation from securing the first sample to off-line KF through GC results was approximately 130 min. If the off-line GC assay was unsatisfactory, the whole process would start anew. During this time, the recovery column was running blind.

For a modest capital investment, the OL-GC provided accurate real-time results for both water and percent isopropanol, every 4 min, thus reducing the total sample time to practically nothing. The OL-GC was interfaced with the recovery towers automated control systems such that percent water and percent isopropanol results were immediately available to the operators at the control room console for process fine-tuning and optimization. The added control realized by the use of this PAT application resulted in the ability to rapidly respond to column upsets, and increased productivity. Moreover, all the samples eventually sent to QO for testing and approval met specifications, which meant no recovery failures or associated lab investigations (failing test results are typically investigated to verify the analytical method is performing as intended). Ultimately, when enough confidence in the system was gained, the QO results, although important for compliance reasons, were proven to be technically unnecessary.

Online GC results have since been reconfirmed using compendial methods and have been proven to be just as accurate. For example, analysis for water content by KF versus OL-GC results showed essentially no difference because, the PAT application differed from the QO lab results by 0.1 to 0.2 % (Table 23.1). This difference was attributed mainly to sampling technique.

23.2.3 Filter-Dryer Applications

PAT applications in driers, agitated filter-driers in particular, allow for the study and optimization of several important focus areas for API manufacture and can also offer substantial benefits to the plants in the form of increased productivity. In particular, many APIs are hydrates or solvates that typically require careful attention during the late stages of the drying operation in order to avoid

TABLE 23.1
Online Gas Chromatography versus Compendial Assay

Sample	Water OL-GC	KF	Difference (%)	% IPA OL-GC	QO Lab	Difference (%)
1	0.31	0.33	0.02	0.33	0.27	0.06
2	0.154	0.14	0.01	0.27	0.35	0.08
3	0.21	0.25	0.04	0.48	0.31	0.17
4	0.23	0.19	0.04	0.55	0.38	0.17
5	0.28	0.23	0.05	0.58	0.44	0.14

unintended removal of water or solvent. NIR applications are particularly useful in this regard, as product-specific libraries can yield accurate chemometric models based on typical/desired material. The PAT data thus allow plant operators to follow the drying process in real time and either modify conditions or stop the drying process when appropriate. A secondary application in the dryers involves the monitoring of particle size. NIR PAT applications are also useful in this regard, as NIR is responsive to changes in particle size distribution. Control of particle size is of crucial importance in many bulk API products. The required finished goods particle size distribution is usually controlled by a separate unit operation such as milling. However, changes in the particle size of the input material can have a drastic effect on the effectiveness of the milling operation and thus the final product. This is of singular importance in automated agitated filter-driers, where prolonged agitation can cause particle attrition and generation of fines. Therefore, reproducible and robust control of the particle size distribution of the material coming out of the drier is very important to ensure a robust and reproducible milling operation and, ultimately, is important to the physical characteristics of the final API.

Regulatory issues around PAT drying operations are usually uncomplicated, because in most cases, the regulatory specifications for solvent and water content apply only to the final composite sample of the milled material. Similar to the previous two applications, the PAT drier check data are thus usually treated "for information only." In essence, the PAT data do not carry any regulatory entanglements and can be thought of as a sophisticated alarm that lets the operator know when the batch is ready to be unloaded, or when to take the "official" in-process control (IPC) drier check sample to the lab. The particle size information referred to earlier can also be regarded in the same way.

The real-time data have other potential benefits, coupled with automated drying systems and 6-Sigma-type initiatives (such as control charting or process modeling), the information can be used to build an extremely reliable process enjoying a high level of robustness and predictability. This operational model consisting of process understanding based on actual plant data, PAT data in particular, is a good example of what regulators are referring to when speaking of the "desired state."

In terms of financial payback, drier applications offer direct benefits of increased productivity. In many cases, drying times are set to a standard number of hours based on past experience (i.e., the longest it has ever taken to dry the particular batch of that product). The time thus allowed for drying can be longer than necessary if these drying times are based on data from nontypical or outlier batches. With a PAT application, because there is no need to set a particular "sure bet" drying time, the usual result is that drying times are cut significantly. Reduction in drying times results in productivity increases, particularly in operations involving filter-driers, as these combine two unit operations into one. Other benefits, already mentioned, include a more consistent product in terms of particle size and more reliability resulting from conditions that will not over-dry the product.

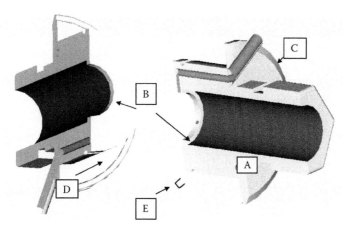

FIGURE 23.3 (See color insert following page 40). Typical process analytical technology (PAT) probe holder for automated filter-driers. A PAT probe sits in section (A). Nitrogen is supplied through the line going out through small holes (B). Plate (C) is the outside of the drier; plate (D) is flush with the inside wall of the drier. Product "samples" are deposited in the indentation marked (E) and are blown off by nitrogen through (B) (Pfizer, patent pending).

Depending on the specific application and exactly how the sample is presented to the PAT instrument, general concerns such as sample homogeneity and analytical specificity can be very successfully addressed. For instance, a common solution to the homogeneity concern involves the use of PAT in agitated filter-driers where the PAT probe sits in a sample well (see Figure 23.3), a sample is deposited in the well cavity, it is read by the PAT probe, and the sample is expelled by means of a nitrogen blast. Multiple readings per agitation period can be obtained and averaged; the sheer volume of data taken ensures that the data represent the batch overall. Concerns about method specificity, assurances that one is truly reading residual solvent and not some other parameter like particle size, color, and so forth, can be successfully addressed by using a suitably generated chemometric model of the product being dried.

One example of a PAT drier application used for both process understanding and to improve cycle times involves a typical product transfer between manufacturing sites. In this example, API X was being transferred from one manufacturing site (plant A) to another manufacturing site (plant B). Plant A used a tray drier, whereas plant B used an agitated Rosenmünd filter-drier. This change in equipment had the net effect of producing dried material with different particle size distributions, which resulted in rendering one material suitable for formulation (tray-drier product), whereas the other proved problematic. The original formulation, a compressed tablet consisting mostly of API, was developed using tray-dried product. In general, tray drying is particularly labor intensive and can present issues related to batch homogeneity and foreign matter contamination. The filter-drier operation is completely automated and involves a minimum of manual intervention, but if not monitored properly, prolonged agitation can cause particle attrition or overdrying.

A NIR PAT application was developed to study the Rosenmünd drying operation and to determine what factors affected the formation of fines. The NIR data were used to monitor the levels of residual solvent. In parallel with this, samples were obtained at different intervals for particle size analysis using Malvern. A correlation between the two data sets showed that particle attrition occurred primarily once the solvent had been removed down to a certain level (see Figure 23.4). The solvent in this case was acting as a lubricant preventing the particles from breaking. Ultimately, an operational PAT application was built where NIR would monitor the drying, and once a certain level of residual solvent was achieved, agitation inside the drier was changed to an intermittent mode. The PAT application proved critical because not all batches started the drying cycle with the same amount of residual solvent, and thus no single "standard" drying process would

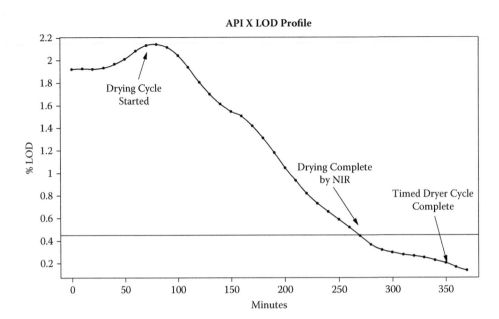

FIGURE 23.4 Percent loss of drying versus time in process analytical technology (PAT) near-infrared (NIR) application for active pharmaceutical ingredient (API) *X*.

have ensured that one batch was not going to be over- or underdried. The PAT-controlled drying process prevented particle attrition and assured that the product was not overdried; ultimately, formulation trials with this material proved successful.

23.2.4 NONREGULATORY IN-PROCESS ANALYSIS

PAT applications offer the most benefit in terms of process understanding and realization of the "desired state" when used to directly control the manufacturing process.

From a holistic perspective, in the current paradigm (see Figure 23.5), a sample is taken at the end of each process step, starting with the raw materials and ending at the finished product. This operational model, alluded to earlier when discussing classic organic chemistry, only gives us information about the end results—that is, what happened after completion of each step. It does not give us any information about what happened during the processing and thus does not contribute much to "process knowledge." Process knowledge in this sense must be inferred by the end results. The main feature of this operational model is that quality is secured by testing at the end of each step in order to reduce the risk of moving to the next stage. One of the disadvantages of this approach is the lack of a detailed real-time picture of the intricacies of the chemical process as run at the plant scale. Any mechanistic information must be, in the best of cases, obtained from independent laboratory experiments and translated to plant-scale process operations and timelines. Although an exact mechanistic pathway within each step was originally thought not to be critical (i.e., the static-control picture referred to previously), the new industry perspective is that in order to ensure true process robustness and control of variation, a higher level of understanding is crucial.

If one were to analyze for quality continuously or multiple times throughout the process using PAT applications, not only would one also achieve the goal of reducing the risk of moving to the next stage, but in addition, one would gain more knowledge about what is occurring during processing, which would allow for more informed responses to, if required, modify key process parameters. Using PAT helps to ensure that the same processing path is taken every time.

The distinction made in this section versus the next (Regulatory In-Process Analysis) is really around gaining process understanding on parameters not particularly specified in the NDA. Obviously,

Process Control Philosophy–Paradigm Shift

Conventional approach- lab based

End of phase testing of quality to reduce the risk in moving to
next stage

| Obtain Raw Materials | → | Run Reactions | → | Dry | → | Isolate & Pack Out |

P.A.T approach- process based, at-line or on-line

| Obtain Raw Materials | → | Run Reactions | → | Dry | → | Isolate & Pack Out |

Continuously or more frequently test quality during each phase,
to remove the risk in moving to the next stage

FIGURE 23.5 (See color insert following page 40). Scheme of conventional versus process analytical technology (PAT) process control paradigms.

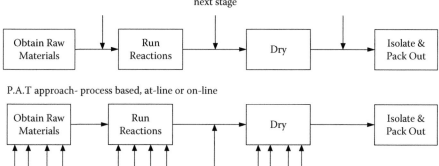

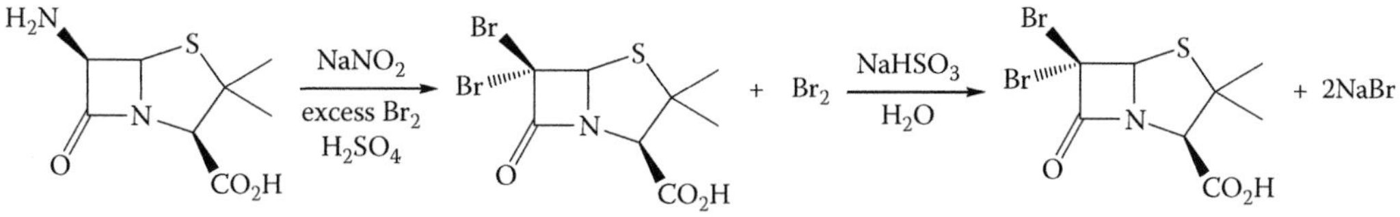

SCHEME 23.1 Synthesis of 6,6-dibromopenicillanic acid.

an argument can be made that the understanding of variables affecting important process parameters is of great interest to regulators, whether these are filed or not. However, there are different expectations at ground level regarding how any nonregulatory PAT application has to be validated, and because any such application is conducted in addition to all other regulatory IPCs, there are lower regulatory exposure risks involved. In any case, some of these applications may involve processes that, although important to the ease of production operations and maybe even some quality parameter, are somewhat peripheral to the main reaction. We offer two such examples.

The first involves the use of a redox probe in a bromine destruction step.[10] Oxidation/reduction reactions require species that have a propensity to give up or receive electrons; this exchange can be measured by differences in potential as the reaction progresses. The synthetic reaction involved in this application is shown in Scheme 23.1.

The chemical process before the installation of the redox probe involved sequential charges of a set amount of sodium bisulfite to quench excess bromine, followed by a series of off-line IPC testing to determine any excess bromine. At this point, the bromination reaction was deemed over, and this operation was deemed part of the workup. Under the charge-sample-and-test model, plant over- and undercharges of sodium bisulfite were not uncommon. Overcharges of bisulfite were particularly problematic as too much excess could compromise the quality of the desired product; thus, bromine had to be charged again and the process repeated from the beginning.

This particular redox reaction has electronic potentials for bromine = 1.066 E°/V and sodium bisulfite = +1.12 E°/V, which are easily measured online using a common redox probe. The corresponding millivolt response for bromine is approximately 750 mv, whereas sodium bisulfite response is around 350 millivolts (mv). Charting the millivolt response (see Figure 23.6) associated

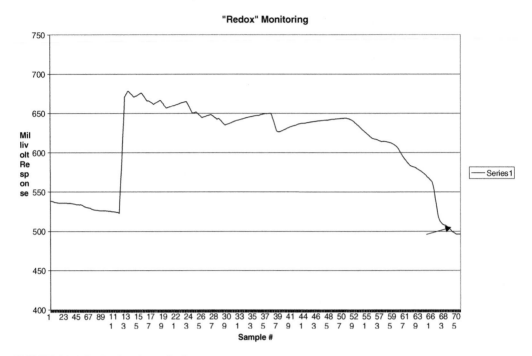

FIGURE 23.6 Redox heads-up display.

with the amount of reducing agent added can easily follow the bromine reduction. The reduction reaction endpoint does not go to completion and must be stopped at a point when a slight excess of bromine is present. The endpoint was previously verified by sampling and testing using the reaction mixture's ultraviolet (UV) absorbance, which was our traditional method; the data were previously gathered off-line.

The results proved relatively simple to interpret; when the millivolt response from the redox probe was ~550 mv (see Figure 23.6), the ultraviolet/visible (UV/Vis) absorbance was less than 0.100, and a slight excess of bromine remained.

The benefits of the application proved to be obvious to production. The optimization of the sodium bisulfite charge using online redox monitoring eliminated reprocessing due to overcharging the reducing agent, eliminated hazardous sampling and operator exposure, and eliminated IPC test time. Issues around reprocessing and overcharge of bisulfite were reduced by more than ten-fold after the implementation of this application.

Additional process information could be derived from continuous online redox reaction monitoring. Reaction data such as the amount of sodium bisulfite used from batch to batch would indicate varying amounts of unreacted bromine that provided an indication of incomplete reaction or the prevalence of other undesired side reactions. The amount and quality of the data derived from the online redox provided an excellent opportunity for process optimization, control, and real validation.

Our second example of in-process control testing of a "peripheral" nonregulatory application involves the use of online mass spectrometry to monitor reaction off-gases. In our manufacturing plant in Connecticut, environmental permitting related to the atmospheric release of volatile organic compounds (VOCs) is capped to a 24-h average of no more than 50 ppm. These readings, which are obtained every 60 sec, are taken at the gaseous outflow of all plant operations after a series of scrubbers remove VOCs by chemical decomposition or reaction and ultimately by condensation of the most volatile components. Analysis for VOCs is done by a commercial in-line gas chromatograph called CEMS (continuous emission monitoring system), which is automatically calibrated with methane.

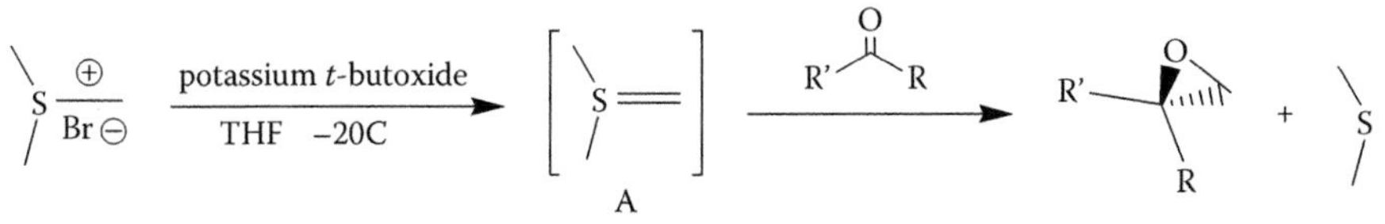

SCHEME 23.2 Sulfur ylide methylenation.

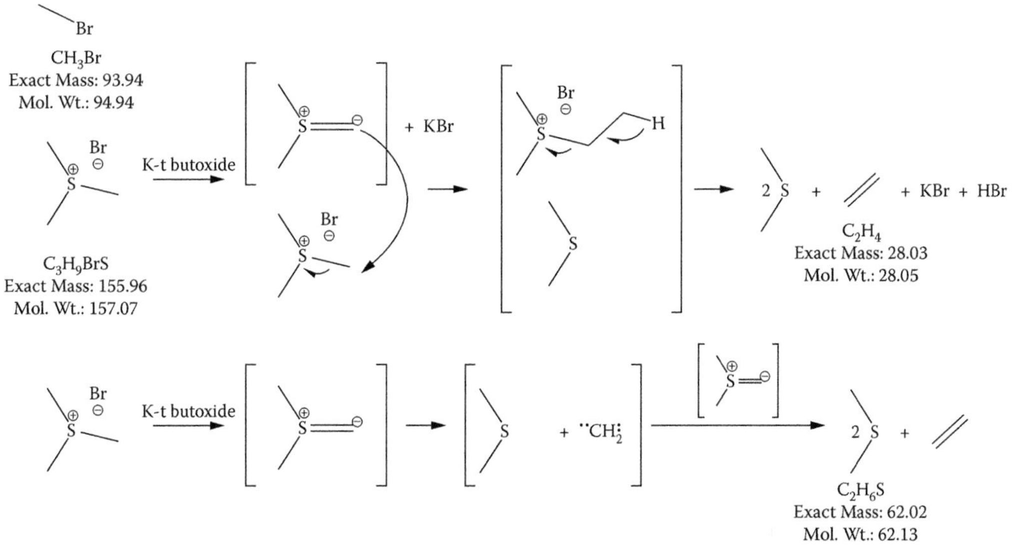

SCHEME 23.3 Potential culprits for high continuous emission monitoring system (CEMS) readings in ylide reaction.

During the manufacture of one key intermediate, the CEMS unit invariably registered spikes in VOC readings. In particular, the CEMS spikes appeared at specific unit operations related to a sulfur ylide reaction (Scheme 23.2).

The CEMS unit has proven itself to be more than appropriate for the intended purpose of measuring organic compounds escaping our scrubber system; however, it is powerless when it comes to offering information that could lead to identification of what is causing the high readings. Our very first suspect for high VOC emissions was the known dimethylsufide by-product. There were other suspects as well, two in particular enjoyed popular support from various members of the investigating team, one was bromo methane, presumably derived from disproportionation of trimethylsulfonium bromide, and the second was ethylene (see Scheme 23.3).

The suspect status of these two compounds was based on the fact that neither would have been affected to a significant extent by the two sequential potassium permanganate (later Oxone®) and other scrubbing systems we had installed to deal with dimethylsulfide. What was needed was a PAT application that could tell us exactly what was going through all the scrubbing systems and giving us high CEMS readings.

An online gas chromatograph/mass spectrometer (GC/MS) was thus installed just before the CEMS unit, and we watched, with the anticipation only other PAT geeks can understand, for the next CEMS spike. The GC/MS reading at the next CEMS spiked is shown in Figure 23.7.

Bromomethane and dimethylsulfide were eliminated as being responsible for the CEMS spikes. Evidence for ethylene (MW 28.05) was inconclusive from just the online GC/MC, but by subtracting background readings (nitrogen used routinely as an inerting gas has a MW of 28.01) an enhanced peak was shown with the right molecular weight present only during CEMS spikes (see Figure 23.8).

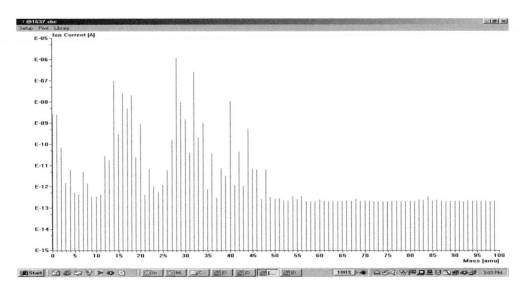

FIGURE 23.7 OmniStar™ mass spectrometry data. Amu range 0 to 100 [w].

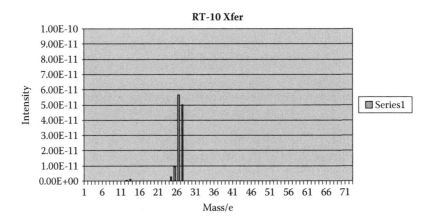

FIGURE 23.8 Gas chromatography/mass spectrometry OmniStar™ reading minus background readings.

Ultimately, formation of parts per million (ppm) levels of ethylene was corroborated by very carefully controlled laboratory experiments that duplicated plant timelines and realistic temperature swings. Eventually, the analytical data were utilized to justify the connection of a thermal oxidizer to deal with the ethylene emissions. Needless to say, if this issue had not been resolved successfully, the plant in Connecticut would have found it difficult to continue manufacturing this compound.

This same ylide reaction (Scheme 23.2) offers us a second example to demonstrate the use of PAT for in-process testing in real time. Operationally, formation of the ylide intermediate was deemed complete after addition of all specified reagents and following a set time frame. In practical terms, sampling the ylide (A in Scheme 23.2) at –20°C coupled with having to develop and implement a derivatization IPC method would have been difficult. Thus, when the process was filed, no IPC for this step was included. The reaction completion was assessed with a specific test for residual starting material in the isolated epoxide product. This type of strategy, which is not that uncommon, is affectionately referred to by PAT people as a "postmortem."

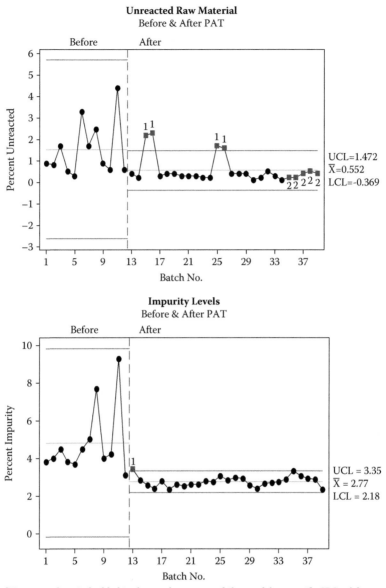

FIGURE 23.9 (See color insert following page 40). Control chart of percent unreacted before and after PAT (top) and impurity levels before and after implementing PAT (bottom).

Moreover, experience gathered during early process development suggested that the ylide formation reaction was not critical to product quality as long as the reaction was conducted below −10°C. At the initial manufacturing start-up campaign, great variability in product quality was observed—in particular, in the percent unreacted starting material and levels of one key impurity. As shown in Figure 23.9, the high variability hinted strongly that this process was not in control. To understand this process, an NIR reaction monitoring application was developed.

Several batches were studied by NIR, and it was determined that formation of ylide in high yield was crucial to both levels of unreacted starting materials and impurities. This may appear obvious, unless one remembers that the variable that was previously identified as crucial

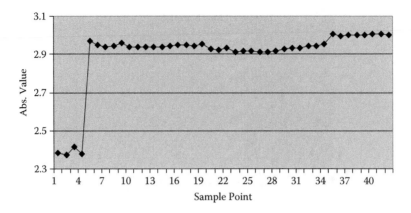

FIGURE 23.10 Net absorbance change for a successful ylide reaction.

(temperature), was in perfect control. In addition, many other variables were studied, such as addition rates, agitation rates, placement of the temperature probe, and so forth, as there never seems to be a limit to potential variables present in plant operations. What was found was that, for unknown reasons, some ylide reactions were proceeding to a higher degree of reaction completion than others. For example, we observed that occasionally the rate of ylide formation was slower, or in some cases not proceeding at all, as shown in Figure 23.8 and Figure 23.9. Focusing the NIR at a particular wave number, we observed that once the potassium t-butoxide was added, a favorable/complete ylide reaction exhibited a sizable jump in absorbance, whereas no such jump was seen upon potassium t-butoxide addition in an unfavorable/incomplete ylide reaction. In the first case (Figure 23.10), the ylide reaction went to completion very quickly and resulted in low levels of unreacted material and impurities. In the second case (Figure 23.11), the reaction did not go to completion, thus resulting in high levels of unreacted material and impurities.

Once this was understood, we were able to track down the critical process parameters affecting the ylide formation. Ultimately, we eliminated many operational suspects (agitation, reagent addition rates, hot spots, and others) and discovered that differences in the potassium t-butoxide were responsible for differences in ylide reaction rates. Operationally, the NIR is utilized to decide whether to proceed with a particular ylide batch or to discard it and make a new one. The net impact was much-improved robustness and product quality.

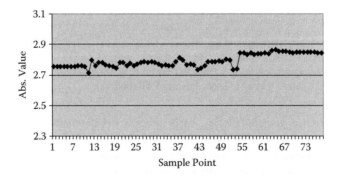

FIGURE 23.11 Net absorbance change for an incomplete ylide reaction.

23.2.5 REGULATORY IN-PROCESS TESTING

Using PAT applications for filed in-process testing offers the same basic advantages as for nonregulatory tests in terms of process understanding and robustness. The basic difference is the regulatory exposure associated with routine and exclusive use of these tests to make decisions in the manufacture and approval of API products. For processes that are just now being developed, PAT applications can be made part of the original NDA. On the operational side, one problem is that not all manufacturing plants have the same level of expertise or equipment capabilities to implement PAT applications; thus, the tendency has been to use PAT only during early development, with the objective of gaining crucial process knowledge early on, but to rely on standard laboratory test methodology at plant scale. On the strategic level, filing PAT applications in the original NDA, particularly NIR methods, means having to deal with the nagging issue of standards libraries. From the industry perspective, there is a balancing act between investing financial and human resources in early development to produce the necessary number of lots to provide manufacturing with a robust chemometric model versus fulfilling the minimum requirements expected from the regulatory agencies strictly for filing purposes.

Another strategy is to file a comparability protocol between typical methods and PAT methods in the original NDA. The number of batches produced in early process development, the size of the clinical trials, and other factors may determine whether manufacturing will be tasked with completing the comparability protocol. Recent experience seems to suggest that this will be the case. The FDA has defined comparability protocols as follows:

> A well-defined, detailed, written plan for assessing the effect of specific CMC changes on the identity, strength, quality, purity and potency of a specific drug product as they may relate to the safety and effectiveness of the product. A comparability protocol describes the changes that are covered under the protocol and specifies the tests and studies that will be performed, including the analytical procedures that will be used, and acceptance criteria that will be met to demonstrate that specified CMC changes do not adversely affect the product.[11]

In principle, if PAT methodology was not included in the original NDA, it can be filed later as a supplemental submission. The big analytical challenge revolves around establishing comparability when old methodology is being eliminated from the filing or being made an optional alternative. Based on published draft guidance,[11] it is recommended that the protocol include a detailed plan of how equivalency to the compendial method is to be determined, and to outline in detail how the calibration curve/chemometric model will be developed as well as the criteria for demonstrating equivalency. For the most part, API manufacturers have not yet subjected large numbers of PAT comparability protocols to the rigors of regulatory field inspections, so the industry has yet to test the waters on these expectations. Additionally, the administrative burden of registering new methods in different markets around the world can prove to be onerous and time consuming. Given all these obstacles, it is understandable that so far the industry has been cautious in filing PAT applications.

23.2.6 STRATEGIC PAT AND PROCESS VALIDATION

In the "desired state" vision that both regulators and pharmaceutical industry are currently pursuing, product quality and performance are achieved and assured by design of effective and efficient manufacturing processes:

> [The] description of the desired state aims to enhance the utility of product and manufacturing process design knowledge and understanding for quality assessment. When product and process knowledge is shared with the FDA, it can provide a basis to recognize the level of understanding achieved for risk-based regulatory decisions and to provide flexibility for continuous improvement and innovation for those companies that have demonstrated an ability to manage risk to quality.[7]

Many aspects of this emerging paradigm are still undefined and untested; however, it seems apparent that regulators are willing to contemplate some degree of flexibility in decision making, process modifications, or innovations if companies can demonstrate unequivocally they possess high levels of process understanding. Degrees of "process understanding" thus provide a new basis for "real" process validation, as defined previously. PATs are seen as key enabling tools in achieving these high levels of process knowledge. In this sense, investment in an overall manufacturing strategy that encompasses significant reliance in PAT can be regarded as a strong strategic advantage.

For the pharmaceutical industry to be successful in realizing the promise of increased flexibility and the ability to implement innovation with some rapidity, it will have to redesign many peripheral systems that support the current manufacturing model, and successfully test these with regulators. Regulators, on the other hand, will have to demonstrate an open mind and pragmatism when making judgments around what constitutes acceptable levels of "process understanding." If the industry feels that the expectations around process knowledge are unreasonable, or too burdensome or expensive to realize, the "desired state" idea will probably devolve back to the current business as usual. However, the most likely outcome, given that both the regulators and the industry want to move in this direction, is that a consensus around some of the regulatory issues raised herein will be developed, and change will slowly be realized.

"Desired state" or not, what seems apparent is that PAT can only help in moving forward process understanding at all levels and for practical business reasons. Much around how to deal with regulatory filings of PAT methods remains virgin territory, but there are other applications with good paybacks and little or no regulatory exposure that deserve attention. Moving in those directions now will develop the internal expertise and language plus operational and documentation systems to make the transition to regulatory applications much easier.

REFERENCES

1. Definition taken from the Web site of the Food and Drug Administration's Center for Drug Evaluation and Research–Office of Pharmaceutical Sciences (www.fda.gov/cder/OPS/PAT.htm#Introduction).
2. Leila Abboud and Scott Hensley; Factory shift: new perspectives for drug makers—update the plants after years of neglect, industry focuses on manufacturing FDA acts as a catalyst. *The Wall Street Journal*, CCXLII (45), September 3, 2003.
3. Laura Bush; The end of process validations as we know it. *Pharmaceutical Technology*, August, 36–41, 2005.
4. Thomas Layloff; An overview of process analytical technology in the pharmaceutical industry. *American Pharmaceutical Review*, 7 (3), May–June, 30–34, 2004.
5. The FDA has published a guideline for PAT, but it lacks many specifics in the area of method validation and comparability protocols. See *PAT—A Framework for Innovative Pharmaceutical Development, Manufacturing, and Quality Assurance*. FDA Rockville, MD, September 2004.
6. U.S. Food and Drug Administration; *Guideline on General Principles of Process Validation*, FDA, Rockville, MD, May 1987.
7. Ajaz S. Hussain (FDA CDER Director); Process analytical technology; a first step in a journey towards the desired state. *Journal of Process Analytical Technology*, 2 (I), January–February, 8–13, 2005.
8. *Innovation and Continuous Improvement in Pharmaceutical Manufacturing: The PAT Team and Manufacturing Science Working Group Report: A Summary of Learning, Contributions and Proposed Next Steps for Moving towards the "Desired State" of Pharmaceutical Manufacturing in the 21st Century* (www.fda.gov/cder/gmp/gpm2004/manufSciWP.dft).
9. Kourti, T.; Process analytical technology and multivariate statistical process control. Wellness index of process and product—part 1. *Process Analytical Technology*, 1 (1), 13–19, 2004.
10. Frank Sistare, Laurie St. Pierre Berry. and Carlos A. Mojica; *Journal of Organic Process Research and Development* 2005, 9 (3), 332.

11. U.S. Food and Drug Administration, *DRAFT Guideline for Industry: Comparability Protocols—Protein Drug Products and Biological Products—Chemistry, Manufacturing, and Controls Information,* September 3, 2003.

24 PEGylation of Biological Macromolecules

John J. Buckley, Rory F. Finn, Jianming Mo, Laura A. Bass, and Sa V. Ho

CONTENTS

24.1 INTRODUCTION

Proteins, peptides, and other bioactive macromolecules have great potential as therapeutics due to their high degree of selectivity and potency. The utilization of these molecules as biopharmaceuticals can be, however, very challenging due to their rather unfavorable physico-chemical characteristics

TABLE 24.1
Commercial PEGylated Biomolecules

Approved PEGylated Products	Company
PEG-GCSF (Neulasta)	Amgen
Pegvisomant (Somavert)	Pfizer
PEG Interferon-alpha-2a (PEGasys)	Roche
PEG Interferon-alpha-2b (PEG-Intron)	Schering-Plough
PEG-adenosine deaminase (Adagen)	Enzon
PEG-asparaginase (Oncospar)	Enzon
PEG-Aptanib (Macugen)	Eyetech/Pfizer

such as large size, relatively low solubility, instability, rapid clearance in plasma, and immunogenicity [1]. Over the past 10 to 15 years, a great deal of work has focused on methods for formulating, delivering, and extending the duration of action of proteins, peptides and oligonucleotides [2]. Increasing plasma half-life has become a focused target for pharmaceutical companies for both new protein therapeutics and second-generation versions of current protein drugs. As many of these molecules are administered parenterally, with half-lives that necessitate once daily or even more frequent dosing, extending the duration of action may have a pronounced effect on the utility of the drug simply by improving patient compliance through convenience.

Currently, the method of choice for increasing the plasma circulation times of protein and peptide therapeutics is PEGylation, the covalent attachment of poly (ethylene glycol), or PEG, to those compounds [3-6]. This is evidenced by the six approved PEGylated protein therapeutic drugs currently being marketed (Table 24.1). Also shown in the table is Macugen, the first approved PEGylated oligonucleotide drug. PEGylation of biomolecules can not only provide advantages in terms of clearance rates, but can also result in improved safety profiles, increased efficacy, decreased dosing frequency, improved solubility and stability, and reduced immunogenicity [7–16]. Typically, the molecular weight of the conjugated PEG polymer, which has the general chemical formula $HO-(CH_2CH_2O)_n-H$, can range from several hundred to 60,000 daltons. With their propensity to be highly hydrated in aqueous solutions (two or three water molecules associated with each ethylene glycol subunit) PEGs extend their large hydrodynamic volumes to their conjugates such that molecules of such sizes, which would normally pass through the renal ultra filtration system, are retained [7–10,17,18]. The large hydrodynamic volumes of PEGs can also shield the conjugated proteins, peptides or nucleotides from interactions with other molecules such as proteases and immune clearance mediators, leading to better plasma concentrations and extended half lives. PEGs have been shown to reduce immunogenicity and antigenicity indirectly through reduction of protein aggregation or directly through coverage of epitopes, the part of an antigen recognized by the immune system [13–16,19,20]. PEGs can also bestow physical attributes to bio-molecules that directly enhance their drug-ability through improved handling and processing, such as increased solubility, decreased aggregation, and increased stability [15,21,22].

Along with the many described advantages, PEGylation, however, has also brought new challenges in terms of processing, characterization and formulation. PEGylation typically adds steps to the process stream, resulting in higher production cost, and generally requires additional or specialized characterization tools. In this chapter we will focus on key strategies and considerations necessary for the development of PEGylated therapeutic biomolecules.

24.2 PEGYLATION CHEMISTRY

A number of excellent reviews of the chemistry of PEGylation have been written in recent years [23–28]. For a more highly detailed discussion, the reader may wish to refer to these references.

FIGURE 24.1 Polyethylene glycol structures.

The following is a summary of some of the major points of PEGylation chemistry of biological macromolecules.

24.2.1 CHEMISTRY AND PROPERTIES OF POLYETHYLENE GLYCOL (PEG)

PEG reagents used for PEGylation are typically purified to obtain a relatively narrow molecular weight distribution, determined by the polydispersity index. Polydispersity is the ratio of the weight average molecular weight (Mw) to the number average molecular weight (Mn), which indicates the distribution of individual molecular weights in a batch of polymers). Low molecular weight PEGs tend to have a lower polydispersity of ~ 1.01 whereas the polydispersity of higher molecular weight PEGs is about 1.1 or less [26]. PEGs are available in mono methoxy (m-PEG) or diol form (Figure 24.1) depending on whether methanol or water is used as the initiator in polymerization, respectively [26]. Compared to other water-soluble polymers, PEGs are readily available from multiple suppliers in purified form with a narrow molecular weight distribution. PEG's affinity for water contributes to many of its desirable properties when attached to biological molecules. PEGylated molecules show increased molecular size and solubility along with reduced rate of clearance by the kidneys, increased resistance to proteolytic degradation and reduced immunogenicity [33].

For attachment to macromolecules, PEG is activated via the free hydroxyl groups at the end of the polymer chain by adding a reactive functional group [28]. The mPEG form is used for the synthesis of mono functional reagents. Difunctional reagents may be prepared from the diol form. PEG has the unique characteristic of being soluble in both organic and aqueous solvents, allowing for flexibility in the selection of chemistries used for addition of reactive functional groups to the PEG chain [6].

24.2.2 EVOLUTION OF PEG STRUCTURE

First generation PEG reagents were characterized as having molecular weight <12 kDa with significant levels of difunctional PEG derivatives. The presence of this difunctional PEG at high levels in the starting polymer, a consequence of the synthesis and purification of the methoxy PEG, leads in turn to the possible formation of cross-linked products in the reaction mixture [24]. Removal of these reaction by-products was necessary as part of a synthesis scheme to produce PEG modified macromolecules. Improvements in polymer and conjugation reaction chemistry have now resulted in a second generation of PEG reagents [24,28]. These newer reagents have higher purity and reaction specificity due to lower levels of diol impurities. This has been achieved through the use of mono carboxylic acid intermediates that can be prepared at high levels of purity through ion exchange chromatography [28]. With improvements in polymer synthesis, PEGs with higher molecular weights and well-controlled molecular weight distributions are now available. This makes it possible to develop higher molecular weight PEG conjugates with potentially improved pharmacokinetic and stability properties. More complex reagent structures such as branched PEGs are also available.

TABLE 24.2
Pegylation Reagent Chemistries

Amine Selective	Abbrev.	Chemistry Class	Linkage	Reference
PEG tresylate	TS PEG	alkylating	amine	29, 30
PEG dichlorotriazine	DCT PEG	alkylating	amine	36
PEG benzotriazolyl carbonate	BTC PEG	acylating	carbamate	34
PEG succinimidyl carbonate	SC PEG	acylating	carbamate	31, 32
PEG p-nitrophenyl carbonate	PNPC PEG	acylating	carbamate	35
PEG trichlorophenyl carbonate	TCPC PEG	acylating	carbamate	37
PEG carbonylimidazole	CDI PEG	acylating	amide	38
PEG succinimidyl succinate	SS PEG	acylating	amide	46
PEG succinimidyl alkyl ester	MSPA, MSBA PEG	acylating	amide	40, 41
PEG propionaldehyde	PEG ALD	alkylating	amine	42, 43
Thiol Selective				
PEG maleimide	PEG MAL	alkylating	thioether	44,47
PEG vinyl sulfone	PEG VS	alkylating	thioether	45
PEG iodoacetamide	PEG IA	alkylating	thioether	48
PEG ortho pyridyl disulfide	PEG OPD	disulfide interchange or oxidation	disulfide	44
Miscellaneous Chemistries				
PEG epoxide	PEG epoxide	hydroxide/amine	ether	51
PEG amino	PEG NH2	carboxyl	ester	54
PEG hydrazide	PEG NNH2	carboxyl	ester	52
PEG isocyanate	PEG NCO	hydroxyl	urethane	55
PEG thiol	PEG SH	thiol	disulfide	53

24.2.3 PEG Reagent Linker Chemistry

24.2.3.1 Amine Selective Chemistry

Many of the early first generation PEG reagents primarily utilized amine attachment chemistry [28]. Included in this group are reagents such as SS PEG, SC PEG, and pNPC PEG [31,32,35] that are listed in Table 24.2. It should be noted that although these reagents react with primary amino groups in the target molecule, side reaction with other reactive functional groups such as serine or tyrosine (hydroxyl functional groups) has been observed [36]. The linkages resulting from these side reactions have been reported to be susceptible to hydrolysis [24]. The chemistry of amine selective reagents can be classified as acylating or alkylating. Note that the acylating chemistries including the carbonate and succinimidyl esters result in a net charge reduction on the target molecule through generation of an amide linkage. Alkylating chemistries (TS PEG, DCT PEG, and PEG ALD) preserve target molecule charge through generation of a secondary amine linkage and are not susceptible to hydrolysis. Most first generation PEG chemistries have been of the acylating reagent class. A list of the PEG functional group chemistries that are reactive with amines is found in Table 24.2. The PEG carbonate esters, while their reactivity is low, allow a degree of selectivity in reactions with amino groups. The succinimidyl succinate esters are more reactive to amines, but the linkage backbone contains an ester functionality that is susceptible to hydrolysis. Acylation reactions therefore can be slow and specific, or fast and less specific depending on the leaving group utilized.

An improvement in amine selective PEGylation chemistry came with the development of activated esters based on conversion of the mPEG terminal OH to a carboxyl group [39]. Esterification with N hydroxysuccinimide gave an activated ester that reacted readily with amines to give

mPEG propionic acid ester   NHS mPEG butanoic acid ester

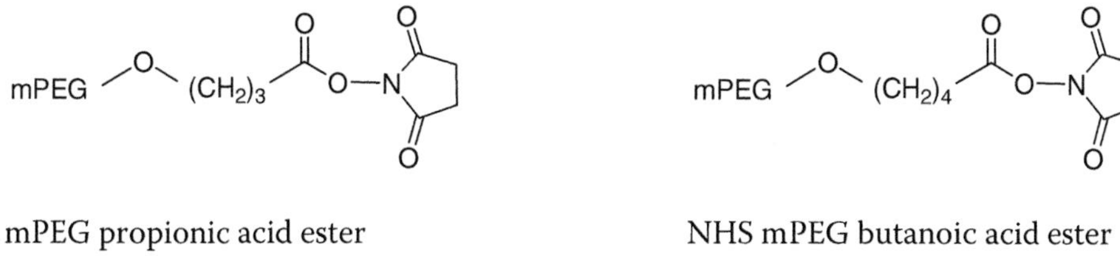

FIGURE 24.2 Activated NHS ester PEGylation reagent structures.

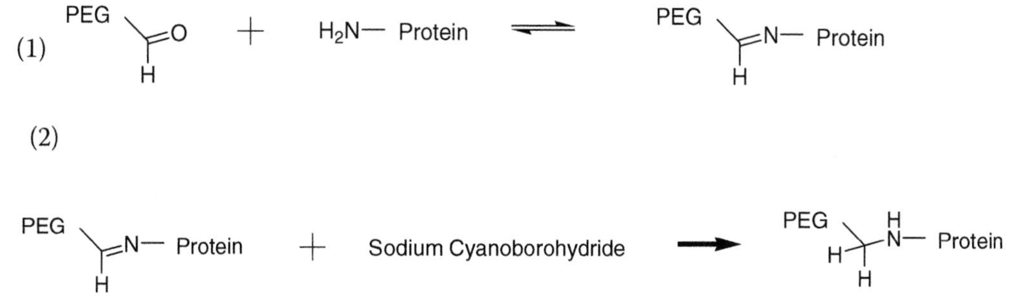

FIGURE 24.3 Reaction of aldehyde PEG reagents with proteins.

an amide with no hydrolysis susceptible linkages. The first of the activated ester compounds was prepared by reacting carboxymethyl PEG with N-hydroxysuccinimide (NHS) to form an activated NHS ester. However this compound is extremely reactive and decomposes rapidly via hydrolysis, making it difficult to use. Increasing the length of the alkyl chain increases stability of this class of PEG reagents. Thus, butanoic NHS ester is more stable than the propanoic acid ester [28]. Figure 24.2 shows the chemical structures for these two esters. This reduced tendency toward hydrolysis has resulted in many practical applications for this class of reagents [40,41].

An important PEG attachment chemistry for reaction with amino groups involves PEG aldehydes. In this chemistry, PEG aldehyde reacts with amino group on the target protein to form a reversible Schiff base linkage. The imine intermediate is then reduced with a suitable reductant such as sodium cyanoborohydride (Figure 24.3). The most notable feature of this reaction is that, when conducted at low pH, the reaction is specific for the amino terminus of the protein [42,43]. The reason for this selectivity is that, relative to other nucleophilic residues in the molecule, the amino terminus has a lower pKa. Therefore, this reaction scheme can be used to reduce the side reactions that form multipegylated products in other amine selective chemistries.

24.2.3.2 Thiol Selective Chemistry

PEG reagents reactive to thiol groups include vinyl sulfone PEG [45], maleimide PEG [44,45], iodoacetamide [48] and o-pyridyl disulfide PEG [43]. The first two of these react with a free thiol on the substrate of interest through a Michael addition reaction, iodoacetamide PEG through a nucleophilic substitution reaction, and o-pyridyl disulfide PEG through a disulfide interchange reaction. These reactions require the presence of a free thiol on the target molecule, which is rare in native proteins. The thiol functionality is, therefore, frequently genetically engineered into the molecule at an appropriate position.

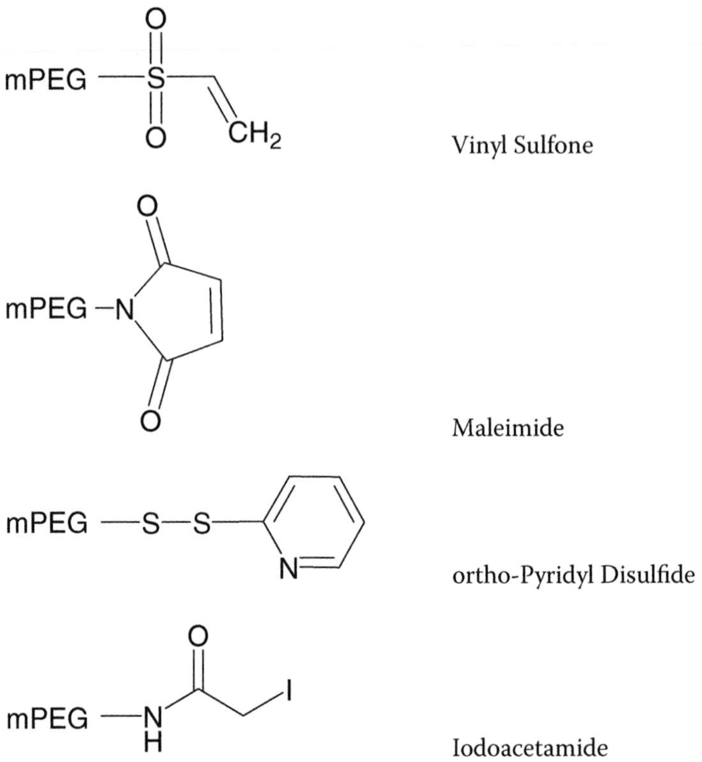

Vinyl Sulfone

Maleimide

ortho-Pyridyl Disulfide

Iodoacetamide

FIGURE 24.4 Structure of Thiol reactive PEG reagents.

24.2.3.3 Hetero Bifunctional PEG

Heterobifunctional PEGs are linkers used to join two separate entities together through a covalent linkage [48-50]. These reagents contain a variety of chemical functional group combinations selected for the specific targets to be attached. The bridging PEG chain provides a flexible and biocompatible linker. This chemistry allows the specific coupling of two different macromolecular entities or, the attachment of macromolecules to surfaces. Examples of heterobifunctional groups are amine/carboxylate (e.g. FMOC amine/NHS ester), sulfhydryl/amine (e.g. maleimide/NHS ester), and, for attachment to surfaces, silane/NHS ester and epoxy/NHS ester. Homo-bifunctional PEG's for cross-linking of macromolecules, e.g., bismaleimide PEG, are also available.

24.3 DISCOVERY PROCESS

24.3.1 TARGET IDENTIFICATION

As stated above, there are a plethora of reasons why one might choose to PEGylate proteins, peptides or nucleotide therapeutics. In designing PEGylated forms of these bioactive molecules one must balance the described advantages against certain disadvantages created by conjugation of PEG. Potentially, increasing the molecular weight and size of a molecule could sterically hinder receptor interactions, alter conformation, change binding properties, and/or alter bioactivities [8–9,18,20,60]. Therefore, PEG size, PEG conformation and sites of PEGylation all become very important considerations for developing PEGylated macromolecules with the desired pharmaceutical properties. Thus, even with the recent development of selective chemistries, achieving precise PEGylation at a site(s) with minimal negative effect on bioactivity is often a very challenging task.

Two strategies can be used to PEGylate a protein: 1) conjugation of PEG to appropriate native amino acid(s) within the given native protein or, alternatively, 2) insertion of a new residue or binding site into the protein to which a PEG can be site-selectively attached. As larger PEG molecules (>12,500 kDa) became available, mono PEGylation of proteins for increased duration of action has become the standard. If the molecule of interest has a free thiol one can use thiol-specific chemistry for attaching the PEG to a cysteine [10,63].

Aldehyde chemistry has been used to preferentially PEGylate the N-terminus of proteins [64]. These selective methods as well as nonselective PEGylation have been used successfully when native sequences and conformations are amenable to conjugation away from active sites.

In cases where N-terminal PEGylation may decrease the potency of a given protein and/or an unpaired cysteine residue may not exist within the native sequence, an alternative approach is to incorporate an amino acid into the protein to which a PEG group can be site-selectively attached. Several groups [47,66–67] have incorporated target cysteine residues into proteins, which can be PEGylated via thiol chemistry; others have added substrate residues for enzymatic PEGylation [69] or azido functionalized methionine analogs for PEGylation using Staudinger ligation [70]. Recently, technology has been developed to biosynthetically incorporate non-native amino acids (e.g., p-acetyl phenylalanine) for subsequent targeted PEG modification [71]. In most of these cases, sites of attachment are selected by a combination of experimentally derived structural information and predictive modeling. Along with the variety of linker chemistries that have been developed, PEG geometries other than linear should be considered. Branched, pendant and combed shaped reactive PEGs have become available and each might have a different effect on the protein's in vivo activity [6,8,62,73].

In cases where only limited information is available with regard to structure activity relationship and clearance mechanism, it may be necessary to generate a broad panel of conjugates, i.e. multiple PEG sizes, site selective chemistries and PEG conformations. Conversely, one can narrow synthesis options greatly if detailed structural information is available. Thus, a more targeted approach toward specific residues can be taken.

24.3.2 PEGylated Product Generation

Therapeutic entities that could potentially benefit from pegylation include proteins and peptides, nucleotides and phospholipids. Current PEGylated products on the market are mostly proteins, such as GCSF, human growth hormone antagonist and interferons (Table 24.1). Other types of biomolecules should gain similar benefits in PEGylated form. An example of a recent approved opthamology product is Macugen, which is an oligonucleotide of 28 units long PEGylated at the 5' end with a 40K branched NHS PEG reagents through an amino linker. Compared to proteins and peptides, which have many potentially reactive groups on the amino acid residues, oligonucleotides do not possess highly reactive groups and thus are not PEGylatable except through an activated linker. PEGylation of nucleotides is thus more efficient since the linker chemistry could be chosen to be both selective and facile.

To generate product candidates for screening, typically a number of PEGylated forms of the macromolecule of interest designed by the strategies described in target identification are synthesized on a relatively small scale (<100 mg) for screening. Following synthesis, the PEGylated molecule is typically purified from other reaction byproducts using column chromatography (ion exchange, size exclusion) [25].

24.3.3 SCREENING

In the selection of a lead PEGylated molecule there are many factors that must be considered, including synthesis and purification yields, stability, bioactivity, pharmacokinetics, biodistribution and immunogenicity. In general, high throughput in vitro bioassays can be used to screen the various PEG -macromolecule conjugates for rank order of bioactivity. However, in most cases, in vivo testing must be performed to truly assess efficacy from extended pharmacokinetics, improved bioavailability or decreased immunogenicity.

24.4 PROCESS TECHNOLOGY

24.4.1 PROCESS CONSIDERATION

PEGylation chemistry deals primarily with the chemical reactivity and specificity of the activated functional group on the PEG molecule in relationship with the targeted site(s) on the biomolecules of interest. Additionally, the size (molecular weight) and configuration (linear, branched, etc.) of the polymer as well as of the biomolecule will also affect the PEGylation reaction.

From a processing standpoint, PEGylation of macromolecules possesses several distinct characteristics. Partly due to the relative large sizes of both PEG reagents and biomolecules, the PEGylation reaction typically proceeds very slowly, taking from 1 to several days to complete. Unlike the inexpensive PEGs commonly used as a commodity, PEG reagents used for biomolecule therapeutics are highly purified, of narrow size distribution, and quite costly [26–28]. Nevertheless, typically a PEG/protein molar excess of at least 2, if not more, is used in the PEGylation step in order to drive the conversion of proteins to completion, since therapeutic proteins are invariably even more expensive than the PEG reagents. Optimization of the PEGylation reaction is thus a critical part of the overall process development. Typical reaction parameters include protein concentration, PEG to protein molar ratio, temperature, reaction time and solution buffers/pH. Some impurities present in the protein solution prior to conducting the PEGylation reaction can adversely affect the performance of this step due to competitive formation of mixed PEGylated products.

In addition to the above parametric optimization, innovative process approaches could involve manipulating the redox conditions of the reaction mixture, solvent properties, and PEG/biomolecule addition scheme, as well as blocking unintended reactive sites on the biomolecule. As shown by the example below in the kinetic modeling section, deactivation of the PEG reagent itself (loss of the reactive functionality) is also likely to occur during the PEGylation step and must be properly studied and kept under control. Potential causes for this deactivation are many and depend upon the chemical functionality of the PEG reagent, including hydrolysis, oxidation, and reaction with amine-containing buffers or impurities in the solution.

PEGylation of macromolecules is typically carried out in a batch reactor with adequate mixing, which fits well with biopharmaceutical manufacturing. Whether it is advantageous to add either the PEG reagent or the macromolecule stepwise (e.g. fed-batch) depends upon the reaction kinetics. Size exclusion chromatography has been used with PEGylation occurring in the mobile phase to exert some control on the degree of PEGylation and accomplish separation of the reaction species based on size [74]. Packed-bed or "on-column PEGylation" could be another option where the PEG reagent is anchored to a surface through a covalent linkage, and the macromolecule is free in solution. With this approach, once PEGylated, the PEGylated macromolecule is attached to the surface, which enables better control of the degree of PEGylation as well as facilitates the separation of the product from other components in the reaction mixture such as the unreacted macromolecule or PEG reagent [73,75–77]. However, due to the operational simplicity of batch reactors, more complex reactor types will need to demonstrate clear economic or product quality advantages to justify their implementation.

24.4.2 Protein PEGylation Kinetic Model

A sound reaction kinetic model is highly beneficial for rapid and effective optimization of the PEGylation step. As an example, a basic kinetic model was developed for the PEGylation of Growth Hormone Antagonist (GHA). This process utilizes mPEG-Succinimidyl Propionate 5K (mPEG-SPA), an electrophilic PEG reagent that reacts with lysine residues and terminal amines of proteins. The reaction between the activated ester of mPEG-SPA and amino groups forms a stable amide linkage. The model was utilized to optimize the performance of the PEGylation step with respect to overall protein yield, PEG usage, and the distribution of the desired PEGylated GHA species.

24.4.2.1 Kinetic Model

Figure 24.5 shows the schematic structure of GHA with the major pegylation sites indicated. There are nine potential reactive sites (8 lysine side chains and the amino terminal), with the amino terminal most reactive (100%). The PEG-SPA reagent can also be hydrolyzed in water. The kinetic model thus consists of nine sequential PEGylation reactions starting with 1 GHA to 1 PEG and ending with 1 GHA to 9 PEG molecules, the associated rate constants being k_1 through k_9. The hydrolysis or more generally deactivation of the PEG reagent also occurs concurrently and is modeled as a simple first-order decay. The inclusion of this inactivation reaction in the kinetic model has been found critical for the success of the model to accurately predict the performance of the PEGylation step. For simplicity the inactivation reaction is modeled as proportional to the concentration of the PEG reagent only. Clearly some factor(s) in the reaction mixture such as hydroxide ion (or pH) must be accountable for the inactivation of the PEG reagent; their effects are thus incorporated into the "apparent" rate constant k_d.

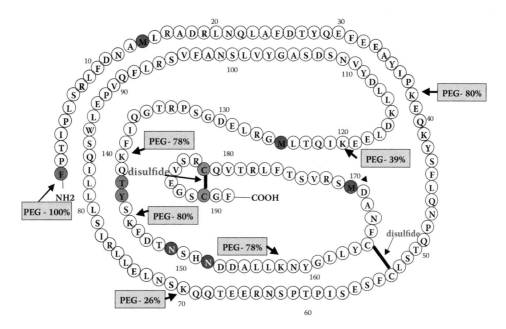

FIGURE 24.5 (See color insert following page 40). Potential PEGylation sites of human growth hormone antagonist (shown with their PEGylation reactivities relative to that of the N-terminus amine indicated in percent).

Modeled reactions:

$$PEG + GHA \xrightarrow{k1} PEG1.GHA$$

$$PEG + PEG.GHA \xrightarrow{k2} PEG2.GHA$$

$$PEG + PEG2.GHA \xrightarrow{k3} PEG3.GHA$$

$$PEG + PEG3.GHA.......$$

$$PEG + PEG8.GHA \xrightarrow{k9} PEG9.GHA$$

$$PEG \xrightarrow{kd} D.PEG$$

With the assumption of simple first order with respect to each reactant, given the initial levels in the reaction mixture of all the species involved (GHA, PEG, PEG1-GHA, etc.) and the values of all the rate constants (k_1 through k_9 and k_d), the above set of differential equations can be solved numerically to obtain the concentrations of all the species as a function of reaction time. The software package MicroMath® Scientist® for Windows™ version 2.01 was used for the numerical calculation in this study.

24.4.2.2   Rate Constants from Data Fit

With the experimental data of the main species (GHA, PEG1-GHA, PEG2-GHA ..., PEG9-GHA) as a function of time, the rate constants k_1 through k_9 can be obtained that best fit the data. While there are many different ways for this data fitting, one effective approach for this reaction system is to estimate k_1 first using the GHA concentration profile assuming a pseudo second order reaction consuming GHA. The exponential decline in the apparent value of k_1 as a function of reaction time is used to estimate the value of kd (PEG deactivation). The values for k_2 through k_9 can be estimated from concentration ratios of the main species at the end of the reaction. With these initial values for the rate constants, excellent fit of the data can typically be achieved with only minor adjustments. Alternatively, Gepasi, a different software package available on the internet, could be used to best fit the data for extracting the rate constants [79].

Due to the many rate constants involved, one set of data was not sufficient to accurately extract all the rate constants, primarily because not all the multi-PEGylated species are analytically detected at reliable levels. Under typical reaction conditions used in the manufacturing process, for instance, no PEG1-GHA through PEG3-GHA could be detected at any significant level. The experiments were therefore carried out specifically to extract a few rate constants from each set of data. For the initial PEGylation reactions, k_1 through k_3, the reactions were deliberately slowed down significantly to get good concentration time points for the corresponding PEGylated species. Once these rate constants were obtained with a good degree of confidence, they were used along with rate data under normal conditions for extracting the higher rate constants, say, k_4 through k_6. For the rate constants k_7 through k_9, the reactions were accelerated or run longer to get sufficient levels of the highly PEGylated species for data fitting.

The rate constants obtained with many sets of experimental data under wide ranges of GHA concentration and PEG/GHA molar ratio are shown in Table 24.3. The rate constants are fairly consistent under these conditions, with k_1 the highest and k_8 the lowest. This is expected since k_1 designates the reaction of PEG reagent to the most active site on GHA, i.e., the amino terminus, and k_2 through k_8 designate PEGylation at sites of decreasing reactivity.

TABLE 24.3
Fitted Rate Constants Obtained for PEGylation of GHA

GHA, g/L	PEG/GHA Molar Ratio	Average Values of Rate Constants for the PEGylation of GHA, L/mol-hr								PEG Deactivation Rate Constant, hr-1
		k_1	k_2	k_3	k_4	k_5	k_6	k_7	k_8	k_d
1–18	1–10	22,100	7,100	3,400	2,500	1,600	1100	650	N/A	1.3

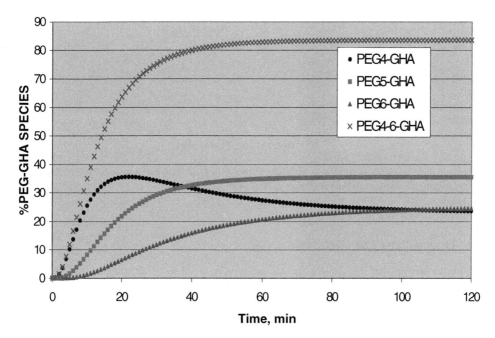

FIGURE 24.6 (See color insert following page 40). PEG-GHA 4-6 formation at 25°C, GHA = 14 mg/mL, (PEG/GHA) molar ratio = 8.5.

24.4.2.3 Model Prediction

Figure 24.6 shows the distribution of the PEG-GHA species 4 through 6. Note that PEG4-GHA level goes through a maximum around 20 minutes then declines as PEG5-GHA is building up with PEG6-GHA following, typical of sequential reactions.

Figure 24.7 shows an example of the use of the model prediction for optimizing the PEGylation step. With the rate constants shown, the PEG/GHA molar ratio was varied from 7 to 11 and the total amount of PEG-GHA(4-6) was computed as a function of time. As can be seen, the overall yield to the desired PEG-GHA species (in mole% of total) increases significantly from molar ratio of 7 to about 9 but not much after that. Also for the molar ratio of 9 and higher, the reaction is essentially complete in 90 minutes.

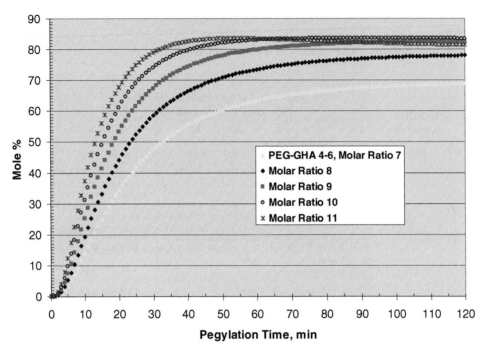

FIGURE 24.7 (See color insert following page 40). Effects of PEG/GHA molar ratio on yield of PEG-GHA(4 through 6). Reaction conditions: GHA = 10 mg/ml, PEG/GHA molar ratio = 7 to 11, T = 20°C. Rate constants: $k_1 = 23,000$, $k_2 = 9000$, $k_3 = 3300$, $k_4 = 2500$, $k_5 = 1750$, $k_6 = 1150$, $k_7 = 700$, $k_8 = 350$ in units of L/mol-hr; $k_d = 1.3$ hr^1.

24.4.3 Downstream Considerations

Potential downstream impacts, positively and negatively, need to be an integral part of the PEGylation evaluation. Both unreacted PEG and macromolecule, if any, will have to be separated from the PEGylated product, so do product-related impurities generated from side reactions such as PEGylated isomers, multi-PEGylated species and aggregates. PEGylation can thus be manipulated to not only maximize yield but also minimize impurities that are more difficult to remove by the subsequent downstream steps. Typically a chromatography step such as ion exchange will follow PEGylation to accomplish the required separation. Additionally, PEGylation could be combined with a separation step (e.g. resin adsorption, membranes, extraction) to enhanced yield and purity as well as facilitate recovery of unreacted polymer and/or biomolecules. A recent review by Fee and Van Alstine [57] provides a fairly comprehensive discussion on reaction engineering and separation issues associated with protein PEGylation.

An example of strong linkage between PEGylation and subsequent chromatographic separation is in the case of the Somavert process. The PEGylation of this human growth hormone antagonist was carried out using N-hydroxysuccinimide ester chemistry at a protein concentration of 10 g/L and a PEG reagent to protein weight ratio of 2:1. Since multiple reaction sites within the protein are PEGylated, distribution of the PEGylated species is quite broad, as shown in Figure 24.8. The reaction mixture was then subjected to hydrophobic interaction chromatography to remove unreacted PEGylation reagent and high molecular weight substances. Q Sepharose anion exchange chromatography was then used to separate differentially PEGylated species, enriching the desired PEGylated isomers that contain 4, 5 and 6 molecules of PEG per molecule of protein (denoted as PEG4, PEG5 and PEG6). Analysis of collected fractions can then be used to select for the desired PEGylated species. A typical pooled distribution of isomers in the final product is shown in Figure 24.9.

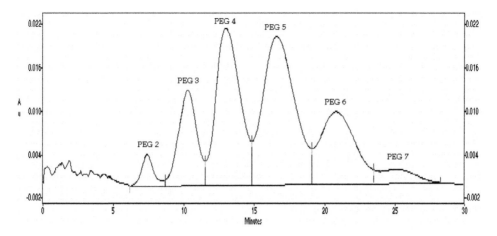

FIGURE 24.8 CE electropherogram showing the distribution of the various PEGylated species post-PEGylation reaction.

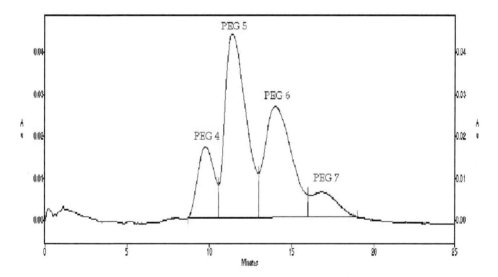

FIGURE 24.9 (See color insert following page 40). CE electropherogram of PEGvisomant formulated drug substance.

24.5 ANALYTICAL CONSIDERATIONS

Analytical consideration for PEGylated biological macromolecules generally consists of two aspects, characterization of the PEG reagents as raw materials and of the PEGylated macromolecules as active pharmaceutical ingredients (API) and drug products. Since PEG reagents have very different chemical and physical properties compared to biological molecules such as proteins, peptides and oligonucleotides, different analytical techniques have to be developed to ensure characterization and control of these biopharmaceuticals [26, 42, 58–59,61].

24.5.1 PEG REAGENTS ANALYSIS

Analytically, PEG reagents have very few properties that can be exploited to enable separation and detection of impurities and degradants. As discussed in the chemistry section, PEG reagents are made of repeating units of CH_2CH_2O with one end capped, usually with a methoxyl group, and the other end attached to a reactive linker. There are two major heterogeneities associated with PEG molecules; molecular weight distribution expressed as polydispersity and linker impurities. PEG reagents are synthesized as molecules with a molecular weight distribution centered at the desired value. Gel Permeation Chromatography (GPC) is commonly used to determine size polydispersity of PEG reagents. To measure heterogeneity of the linker for a PEG reagent, suitable analytical methods are developed based on the property of the linker. Since structurally linkers are quite different from the backbone in a PEG reagent, NMR can be used to confirm the identity and integrity of the linker. However, NMR usually is not sensitive enough to detect and quantify low levels of linker impurities. Additional techniques, such as Size Exclusion (SE) HPLC, are often required to monitor the extent to which the PEG reagent contains the appropriate linker (% substitution) and to ensure control of process related impurities (e.g., high molecular weight, low molecular weight, and linker variants).

24.5.2 PEGYLATED BIOMOLECULES ANALYSIS

Characterization and control of biological macromolecules requires a variety of analytical tools for detection of chemical and structural modifications. Attachment of a PEG moiety to the biomolecule results in an additional layer of complexity and often requires a unique set of analytical tools to ensure quality and control. In developing PEGylated biomolecules, three key attributes must be controlled to achieve the desired product: PEG reagent quality, PEGylation chemistry and stability of the PEG moiety post conjugation. Each of these is described in detail in the following sections.

24.5.2.1 Heterogeneity Caused by PEG Impurities

The PEG reagent can often contain variants that retain the reactive linker and thus conjugate to the biomolecule. Although these variants may differ in MW than the parental PEG, they may not be detected by the typical release assays due to the polydispersity of the PEG and the resulting lack of mass resolution. However, if the PEG impurities have properties that are significantly different in size and or charge than the parent, analytical methods can be employed for their detection. For example, conjugates formed with reactive PEG impurities of high or low molecular weight may produce species with sizes different from that of the desirable molecules, and size exclusion chromatography would be able to resolve them.

24.5.2.2 Heterogeneity Due to Sites and Extent of PEGylation

For some biomolecules, PEGylation can occur at multiple sites, usually with one dominating site and a few minor sites. Because potency is impacted by both the site and extent of PEGylation, these two attributes must be tightly controlled during manufacture and material release [62]. Analytical tools capable of detecting multiPEGylated species include HPLC (RP, SEC, IEX), SDS-PAGE and Capillary Electrophoresis (CE). In order to monitor PEGylation that occurs at sites other than the targeted site, (i.e. positional isomers) digestion of the protein is often required. This is then followed by a careful analysis of the obtained fragments, which can reveal the extent of PEGylation at each of the possible sites. For example, examination of a proteolytic map for a PEGylated protein with the N-terminus as the predominant PEGylation site showed that over 90% of the PEGylation occurred at the desired site, while a small percent of PEGylation did occur at side chains of lysine residues.

24.5.2.3 PEG Moiety Stability

There are several pathways by which the PEG moiety can degrade post-conjugation [28,65]. These include linker degradation, PEG chain truncation and fragmentation, and de-PEGylation. Multiple

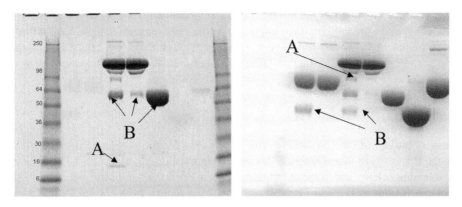

FIGURE 24.10 (See color insert following page 40). Detection of de-PEGylation and PEG chain truncation.

analytical methods can be employed to monitor these degradations. For example, if structural changes to the linker induce a change in charge, ion exchange HPLC may be a good option. For structural elucidation of this linker degradant, complete hydrolysis of the PEGylated biomolecule can be used to free the linker from both the biomolecule and the PEG, then LC-MS can be employed to detect any changes on the freed linker. For PEG truncation, fragmentation, and de-PEGylation, size exclusion chromatography coupled with different detections such as UV (for proteins) and RI (for PEG) can be implemented to resolve the degradation products from the parent and measure their relative abundance. Alternatively, gel electrophoresis combined with stains for both proteins and PEG can also detect and measure these types of degradation. Figure 24.10 shows SDS-PAGE images of a normal and stressed PEGylated protein using two separate stains, one for protein and the other for PEG. Degradation products due to de-PEGylation and PEG moiety truncation could be detected and their relative abundances measured.

24.6 MANUFACTURING CONSIDERATIONS

To date PEGylation of biological macromolecules in manufacturing primarily is carried out in simple batch reactors. Typical scaling issues associated with such an operation include mixing and temperature control if the reaction occurs at temperatures different from ambient. Obviously, the reaction should be carried out at the optimum biomolecule concentration and PEG: biomolecule molar ratio, which should have already been determined in laboratory and pilot studies. Cautions should be taken in preparing large volumes of PEG solution both in terms of physical safety (handling of fine, reactive powder) and reagent stability. Highly reactive PEG reagents such as NHS types are susceptible to hydrolysis in aqueous solutions and extended holding of such solutions should be avoided.

An important consideration in preparation for manufacturing is the source and availability of the required PEG reagent. While the supply situation is improving, large amounts of certain PEG reagents of consistent quality may not be readily available in a timely manner. Typical of synthetic, high molecular weight polymers, PEG reagents are not as well characterized and their properties are not as well controlled as one would expect with small molecule reagents. Key issues include PEGylation reactivity, PEG-related impurities and trace process impurities, and stability of both the PEG chain and the linker between PEG and the reactive functional group. A simple use test to monitor PEGylation performance between reagent batches is highly recommended. A good PEGylation kinetic model worked out during the process development stage would be beneficial in determining the variation of the reaction rate constants for PEG reagent comparability evaluation.

Consistent stability of the PEG chain and the linker under various conditions of handling and storage, which may be affected by trace impurities in the PEG reagent, is more difficult to address from batch to batch. However, this aspect, if not monitored properly, could cause significant problems in some API batches either in failing to meet product specifications and/or in reduced efficacy.

24.7 SUMMARY

PEGylation of biological macromolecules can greatly enhance the plasma half-life of a drug, thus reducing the frequency of administration. It can also improve the safety profile of the drug through reduced immunogenicity and most importantly can improve the stability and solubility of the biomolecule being dosed. We will likely see more NDA approvals in the future for this continually emerging area of biomolecule therapeutics. The complex characteristics of these molecules pose many challenges to process scientists with the unusual combination of synthetic chemistry constraints, specialized analytical methods, and potentially unstable starting materials, intermediates and drug substances. This chapter was not meant for the experts in the field but rather to inform the reader of the issues associated with this future trend in macromolecule therapeutics. We hope this chapter has conveyed the complexity of the issues and the challenges.

REFERENCES

1. Kompella, U.B. and Lee, V.H.L., Pharmacokinetics of peptide and protein drugs, in *Peptide and Protein Drug Delivery*, Lee, V.H.L., Ed., Marcel Dekker:New York, NY, 1991, 391–404.
2. Cleland, J.L., Daugherty, A., and Mrsny. R., Emerging protein delivery systems, *Current opinion in Biotechnology*, 12, 21, 2001.
3. Nucci, M. L., Shorr, R., and Aubuchowski, A. The therapeutic value of poly (ethylene glycol)-modified proteins, *Adv. Drug Delivery Rev.*, 6, 133, 1991.
4. Inada Y, Furukawa, M., Sasaki H., Kodera, Y., Hiroto, M., Nishimura, H., Matsusshima, A., Biological and Biotechnological applications of PEG and PM-modified proteins. *Tibtech.* 1995;13:86–90.
5. Harris, J.M. and Chess, R.B., Effect of PEGylation on pharmaceuticals, *Nat. Rev.Drug Discov.*, 2, 214, 2003.
6. Pasut, G., Guiotto, A., and Veronese, F. M., Protein, peptide and non-peptide drug PEGylation for therapeutic application, *Expert Opinion on Therapeutic Patents*, 14, 6, 859, 2004.
7. Youngster, S., Wang, Yu-Sen, Grace, M., Bausch, J., Bordens, R., Wyss, D. F. Structure, biology and therapeutic implications of PEGylated interferon Alpha-2b. *Current Pharmaceutical Design* 2002; 8:2139–2157.
8. Bailon, P., Palleroni, A., and Scxhaffer, C.A., Rational design of a potent, long lasting form of interferon: a 40 Kda branched polyethylene glycol conjugated interferon α-2a for the treatment of Hepatitis C, *Bioconjugate Chem.*, 12, 195, 2001.
9. Pradghanaga, S., Wilkinson, I., and Ross, R.J.M., Pegvisomant: structure and function, *Journal of Molecular Endocrinology*, 29, 11, 2002.
10. Chapman, A., Antoniw, P., Spitali, M., West, S., Stephens, S., King. Therapeutic antibody fragments with prolonged half-lives, *Nature Biotechnology*, 1999;17:780–783.
11. Molineux G., Kinstler O., Briddell, B., Hartley, C., McElroy, P., Kerzic, P., Sutherland, w., Stoney, G., Kern, B., Fletcher, F., Cohen, A., Korach, E., Ulich, T., McNiece, I., Lockbaum, P., Miller-Messana, M. A., Gardner, S., Hunt, T., Schwab, G. A new form of Filgrastim with sustained duration in vivo and enhanced ability to mobilize pBPC in both mice and humans, *Experimental Hematology*, 27. 1999; 27:1724–1734.
12. Farese, A. M., D.B. Casey, R.M. Vigneulle, N.R. Siegel, R. F. Finn, J. A. Klover, D.Villani-Price, W.G. Smith, J. P. McKearn and T. J. MacVittie. 2001. A Single Dose of PEGylated Leridistim Significantly Improves Neutrophil Recovery in Sublethally Irradiated Rhesus Macaques, *Stem Cells*, 19: 514–521.

13. Yang, Z., Wang, J., Lu, J., Kobayashi, y.,Takakura, T., Takimoto, A., Yoshioka, T., Lian, C., Chen, C., Zhang, d., Zhang, Y., Li , s., sun, x., Tan, Y., Yagi, S., Frenkel, E., Hoffman, R. M. PEGylation confers greatly extended half-life and attenuated immunogenicity to recombinant methionase in primates, *Cancer Research*, 2004; 64:6673–6678.

14. Aubuchowski, A., van Es, T.,Palczuk, N.C., and Davis, F. F. Alteration of immunological properties of bovine serum albumin by covalent attachment of polyethylene glycol, *J. Biol. Chem.*, 1977; 252:3578–3581.

15. Katre N.V., Knauf, M.J., and Laird, W.J., Chemical modification of recombinant interleukin 2 by polyethylene glycol increases its potency in the murine meth A sarcoma model, *Proc. Natl. Acad. Sci. USA*, 84, 1487, 1987.

16. Caliceti, P., Veronese, F.M., and Jonak, Z., Immunogenic and tolerogenic properties of monomethox-ypoly (ethylene glycol) conjugated proteins, *Il Farmaco*, 54, 430, 1999.

17. Knauf, M.J., Bell, D.P., Hirtzer, P., Luo, ZP., Young, J.D., Katres N.V. Relationship of effective molecular size to systemic clearance in rats of recombinant interleukin-2 chemically modified with water-soluble polymers, *J. Biol Chem.*, 1988;263; 29:15064–15070.

18. Clark, R., Olson, K., Fuh, G., Marian, M., Mortensen, D., Teshima, G., Chang, S., Chu, H., Mukku, V., Canova-Davis, E., Somers T., Cronin, M., Winkler, M., Wells J. A. Long-acting growth hormones produced by conjugation with polyethylene glycol, *J. Biol. Chem.*, 1996; 271:21969–21977.

19. Katre N.V., Immunogenicity of recombinant IL-2 modified by covalent attachment of polyethylene glycol, *Journal of Immunology*, 144, 209, 1990.

20. Hinds K.D. and Kim, S.W., Effects of PEG conjugation on insulin properties, *Advanced Drug Delivery review*, 54, 505, 2002.

21. Guerra, P.I., Acklin, C., Kosky, A.A., Davis, J. M., Treuheit, M.J., Brems, D.N. PEGylation prevents N-terminal degradation of megakaryocyte growth and development factor, *Pharmaceutical Research*, 1998;15;12: 1822–1827.

22. Heller, M.C., Carpenter, J.F., and Randolph, T.W., Conformational stability of lyophilized PEGylated proteins in a phase-separating system, *Journal of Pharmaceutical Sciences*, 88, 58, 1999.

23. Brown, Larry R., Commercial challenges of protein drug delivery, *Expert Opinion on Drug Delivery*, 2, 29, 2005.

24. Zalipsky, S., Seltzer, R. and Menon-Rudolph, S., Evaluation of a new reagent for covalent attachment of polyethylene glycol to proteins, *Biotechnology and Applied Biochemistry*, 15(1), 100, 1992.

25. Morpurgo, M. and Veronese, F.M., Conjugates of peptides and proteins to polyethylene glycols, in *Methods in Molecular Biology*, Totowa, NJ, United States, *Bioconjugation Protocols*, 283, 45, 2004, Humana Press Inc.

26. Veronese, F.M., Peptide and protein PEGylation. A review of problems and solutions, *Biomaterials*, 22, 405, 2001.

27. Veronese, F.M. and Harris, J.M., Introduction and overview of peptide and protein PEGylation, *Advanced Drug Delivery Reviews*, 54, 453, 2002.

28. Roberts, M.J., Bentley, MD., and Harris, J.M., Chemistry for peptide and protein PEGylation, *Advanced Drug Delivery Reviews*, 54, 459, 2002.

29. Francis G E; Fisher D; Delgado C; Malik F; Gardiner A; Neale D. PEGylation of cytokines and other therapeutic proteins and peptides: the importance of biological optimization of coupling techniques. International journal of hematology (1998), 68(1), 1–18.

30. Sperinde, J.J., Martens, B.D. and Griffith, L.G., Tresyl-mediated synthesis: Kinetics of competing coupling and hydrolysis reactions as a function of pH, temperature, and steric factors, *Bioconjugate Chemistry*, 10, 213, 1999.

31. Wang, Yu-Sen; Youngster, Stephen; Bausch, James; Zhang, Rumin; McNemar, Charles; Wyss, Daniel F. Identification of the major positional isomer of pegylated interferon alpha-2b. Biochemistry (2000), 39(35), 10634–10640.

32. Saifer, Mark G. P.; Williams, L. David; Sherman, Merry R.; French, John A.; Kwak, Larry W.; Oppenheim, Joost J. Improved conjugation of cytokines using high molecular weight poly(ethylene glycol): PEG-GM-CSF as a prototype. Polymer Preprints (American Chemical Society, Division of Polymer Chemistry) (1997), 38(1), 576–577.

33. Delgado, C., Francis, G.E. and Fisher, D., The uses and properties of PEG-linked proteins, *Critical Reviews in Therapeutic Drug Carrier Systems*, 9, 249, 1992.

34. Dolence, Eric K.; Hu, Chen-Ze; Tsang, Ray; Sanders, Clifton G.; Osaki, Shigemasa. Electrophilic polyethylene oxides for the modification of polysaccharides, polypeptides (proteins) and polymer surfaces. (Surface Engineering Technologies, Division of Innerdyne, Inc., USA). US patent 55650234 1997.

35. Shuai, Xintao; Merdan, Thomas; Unger, Florian; Wittmar, Matthias; Kissel, Thomas. Novel Biodegradable Ternary Copolymers hy-PEI-g-PCL-b-PEG: Synthesis, Characterization, and Potential as Efficient Nonviral Gene Delivery Vectors, *Macromolecules*, (2003), 36(15), 5751–5759.

36. Abuchowsk,i A.; McCoy, J.R.; Palczuk, N.C.; van Es, T.; Davis, F.F., Effect of covalent attachment of polyethylene glycol on immunogenicity and circulating life of bovine liver catalase, *Journal of Biological Chemistry*, (1977), 252(11), 3582–6.

37. Veronese, F.M.; Largajolli, R.; Boccu, E.; Benassi, C.A.; Schiavon, O., Surface modification of proteins. Activation of monomethoxy-polyethylene glycols by phenylchloroformates and modification of ribonuclease and superoxide dismutase, *Applied Biochemistry and Biotechnology*, (1985), 11(2), 141–52.

38. Beauchamp, C.O.; Gonias, S.L.; Menapace, D.P.; Pizzo, S.V., A new procedure for the synthesis of polyethylene glycol-protein adducts; effects on function, receptor recognition, and clearance of superoxide dismutase, lactoferrin, and alpha 2-macroglobulin. Analytical biochemistry (1983), 131(1), 25–33. Journal code: 0370535. ISSN:0003-2697.

39. Zalipsky, S. and Barany, G., Facile synthesis of α-hydroxy ω-carboxymethyl poly(ethylene oxide), *Journal of Bioactive and Compatible Polymers*, 5(2), 227, 1990.

40. Vincent, G. and Philippe, T., Pegvisomant Pfizer/Sensus, *Current Opinion in Investigational Drugs*, (London, England:2000), 5(4), 463, 2004.

41. Dhalluin, Christophe; Ross, Alfred; Leuthold, Luc-Alexis; Foser, Stefan; Gsell, Bernard; Mueller, Francis; Senn, Hans. Structural and Biophysical Characterization of the 40 kDa PEG-Interferon-α2a and Its Individual Positional Isomers, *Bioconjugate Chemistry* (2005), 16(3), 504–517.

42. Lee, Haeshin; Jang, Il Ho; Ryu, Sung Ho; Park, Tae Gwan. N-terminal site-specific mono-PEGylation of epidermal growth factor, *Pharmaceutical Research* (2003), 20(5), 818–825.

43. Kinstler Olaf; Molineux Graham; Treuheit Michael; Ladd David; Gegg Colin, Mono-N-terminal poly(ethylene glycol)-protein conjugates, *Advanced Drug Delivery Reviews* (2002), 54(4), 477–85.

44. Arutselvan, N.; Xiong Cheng-Yi; Albrecht Huguette; DeNardo Gerald L; DeNardo Sally J Characterization of site-specific ScFv PEGylation for tumor-targeting pharmaceuticals, *Bioconjugate Chemistry* (2005), 16(1), 113–21.

45. Morpurgo, Margherita; Veronese, Francesco M.; Kachensky, David; Harris, J. Milton. Preparation and Characterization of Poly(ethylene glycol) Vinyl Sulfone, *Bioconjugate Chemistry* (1996), 7(3), 363–368.

46. Kogan, T.P., Synthesis of substituted methoxypolyethylene glycol derivatives suitable for selective protein modification, *Synthetic Communications*, 22(16), 2417, 1992.

47. Rosendahl, Mary S.; Doherty, Daniel H.; Smith, Darin J.; Bendele, Alison M.; Cox, George N. Site-specific protein PEGylation: Application to cysteine analogs of recombinant human granulocyte colony-stimulating factor, BioProcess International (2005), 3(4), 52–56, 58–60, 62.

48. Hinds, K.D., Protein conjugation, cross-linking, and PEGylation, in *Biomaterials for Delivery and Targeting of Proteins and Nucleic Acids*, Mahato, R.I., Ed. CRC Press LLC, Boca Raton, Fla, 2005, 119–185.

49. Zalipsky, S. and Barany, G., Preparation of polyethylene glycol derivatives with two different functional groups at the termini, *Polymer Preprints (American Chemical Society, Division of Polymer Chemistry)*, 1986, 27(1), 1–2.

50. Felix, A.M. and Bandaranayake, R.M., Synthesis of symmetrically and asymmetrically branched PEGylating reagents, *Theoretical and Applied Science, Journal of Peptide Research*, 63(2), 85, 2004

51. Kaul, G. and Amiji, M., Long-circulating poly(ethylene glycol)-modified gelatin nanoparticles for intracellular delivery, *Pharmaceutical Research*, 19(7), 1061, 2002.

52. Pardridge, W.M., Wu, D. and Sakane, T., Combined use of carboxyl-directed protein PEGylation and vector-mediated blood-brain barrier drug delivery system optimizes brain uptake of brain-derived neurotrophic factor following intravenous administration, *Pharmaceutical Research*, 15(4), 576, 1998.

53. Liu, Y., Franzen, S. and Feldheim, D.L., Preparation and characterization of mixed monolayers of PEG and peptides on gold nanoparticles, in Abstracts of Papers, 229th ACS National Meeting, San Diego, CA, United States, March 13-17, 2005, 2005.

54. Bhadra, D., Bhadra, S. and Jain, N.K., PEGylated peptide-based dendritic nanoparticulate systems for delivery of artemether, *Journal of Drug Delivery Science and Technology*, 15(1), 65–2005.

55. Greenwald, R.B., Pendri, A. and Bolikal, D., Highly water soluble taxol derivatives: 7-polyethylene glycol carbamates and carbonates, *Journal of Organic Chemistry*, 60(2), 331, 1995.

56. Youngster, S., Wang, Yu-Sen, Grace, M., Bausch, J., Bordens, R., Wyss, D. F., Structure, biology and therapeutic implications of PEGylated interferon alpha-2b, *Current Pharmaceutical Design*, 8, 2139, 2002.

57. Fee, C.J. and Van Alstine, J.M., PEG-proteins: Reaction Engineering and separation issues, *Chemical Engineering Science*, 61, 924, 2006.

58. Ken Hinds, Ken; Koh, Jae Joon; Joss,Lisa; Liu, Feng; Baudys, Miroslav and Kim, Sung Wan. Synthesis and Characterization of Poly(ethylene glycol)-Insulin Conjugates, *Bioconjugate Chem.*, 11, 195–201 (2000)

59. Gioacchini, Anna Maria; Carrea, Giacomo; Secundo, Francesco; Baraldini, Mario and Roda1, Aldo. Electrospray Mass Spectrometric Analysis of Poly(ethylene Glycol)-Protein Conjugates. Rapid Communication in Mass Spectrometry, 11, 1219-1222 (1997)

60. Lee H. and Parg, T. G., Preparation and characterization of mono-pegylated epidermal growth factor: evaluation of in vitro biologic activity, *Pharmaceutical Research*, 19, 845, 2002.

61. Wang, Yu-Sen; Youngster, Stephen; Grace, Michael, et. al. Structural and biological characterization of pegylated recombinant interferon alpha-2b and its therapeutic implications, *Advanced Drug Delivery Reviews*, 54 (2002) 547–570.

62. Bass, L., Strategy for Evaluation of Analytical Comparability of PEGylated Biomolecules, presented at International Association for Biologicals: State of the Art Analytical Methods for the Characterization of Biological Products and Assessment of Comparability, NIH, Bethesda, Maryland, June 2006.

63. Manjula, B,. Tsai, A., Upadhya, K., Perumalsamy, P. K., Malavalli, A., Vandegriff, K., Winslow, R. M., Intaglietta, M.J., Prabhakaran, M., Friedman, J.M., Acharya, A.S. Site-specific PEGylation of Hemoglobin at Cys −93 (β): Correlation between the colligative properties of the PEGylated protein and the length of the conjugated PEG chain, *Bioconjugate Chem.*, 14 464–472, 2003.

64. Kinstler O., Molineux G., Treuheit M., Ladd D., Gegg C. Mono-N-terminal poly(ethylene glycol)-protein conjugates, Advanced drug delivery reviews (2002), 54(4), 477–85.

65. Han, S., Thermal/oxidative degradation and stabilization of polyethylene glycol, *Polymer*, 38(2), 317, 1997.

66. Goodson, R.J. and Katre, N.V., Site directed PEGylation of recombinant interleukin-2 at its glycosylation site, *Biotechnology*, 8(4), 343, 1990.

67. Tsutsumi ,Y., Onda, M., Nagata, S., Lee, B., Kreitman R.J., Pastan, I. Site -specific chemical modification with polyethylene glycol of recombinant immunotoxins anti-Tac(Fv) –PE38 (LMB-2)_improves antitumor and reduces animal toxicity and immunogenicity. *Proc. Natl. Acad. Sci.*, 2000; 18,97,15: 8548–8553.

68. Natarajan, A., Xiong, C-Y, Albrecht, H. DeNardo, G. L., DeNardo, S. J., Characterization of Site-Specific ScFc PEGylation for Tumor targeting Pharmaceuticals, *Bioconjugate Chem.*, 2005; 16: 113–121.

69. Cazalis, C. S., Haller, C. A., Sease-Cargo, L., Chaik of, E.L. C terminal Site Specific PEGylation of a truncated thrombomodulin mutant with retention of full bioactivity, *Bioconjugate Chem.*, 2004; 15:1005–1009.

70. Sato, H., Enzymatic procedure for site specific PEGylation of proteins, *Adv. Drug Deliv. Rev.*, 54, 487, 2002.

71. Wang, L., Zhang, Z., Ansger, B., Schultz, P.G. Addition of the keto functional group to the genetic code of Escherichia coli, *PNAS*, 2003, 100; 1:56–61.

72. Kodera,Y., Sekine, T., Yasokochi, T., Kiriu, Yoshihiro, Hiroto, Misao, Matsushima, A., Inada , Y., Stabilization of L-Aspariginase modified with comb shaped poly(ethylene glycol) derivatives, in Vivo and in Vitro, *Bioconjugate Chem.*, 1994;5:283–286.

73. Lee, E.K. and Lee, J.D., Solid-phase, N-terminus-specific, mono-PEGylation of recombinant inter-feron-alpha-2a: Purification, characterization, and bioactivity, presented at Int. Symp. on the Separation of Proteins, Peptides and Polynucleotides, Aachen, Federal Republic of Germany, Oct 19-22, 2004.

74. Fee, C.J., Size-exclusion reaction chromatography (SERC): A new technique for protein PEGylation, *Biotechnology & Bioengineering*, 82(2), 200, 2003.

75. Baran, E.T., Ozer, N. and Hasirci, V., Solid-phase enzyme modification via affinity chromatography, *Journal of Chromatography B*, 794(2), 311, 2003.

76. Felix, A.M., Site-specific poly(ethylene glycol)ylation of peptides, in *Poly(ethylene glycol) ACS Symposium series 680*, Harris, J.M. and Zalipsky, S., Eds., American Chemical Society, Washington, D.C., 1997, 218-238.

77. Monkarsh, S.P., Spence, C., Porter, J.E., Palleroni, A., Nalin, C., Rosen, P. and Bailon, P. (1997). Isolation of positional isomers of mono-poly(ethylene glycol)ylated interferon/α-2a and the determi-nation of their biochemical and biological characteristics, in: Harris, J.M. and Zalipsky, S., (Eds.), Poly(ethylene glycol) ACS Symposium series 680, American Chemical Society, Washington, D.C. 207-216.

78. Delgado, C., Malmsten, M. and Van Alstine, J.M., Analytical partitioning of poly(ethylene glycol)-modified proteins, *Journal of Chromatography B*, 692(2), 263, 1997.

79. Gepasi 3.30, (Bio)chemical Kinetics Simulation Software written by Pedro Mendes, University of Wales Aberystwyth., downloadable at http:/www.gepasi.org.

25 Microwave Technology in Process Optimization

Farah Mavandadi

CONTENTS

25.1 INTRODUCTION

25.1.1 MICROWAVE-ASSISTED ORGANIC SYNTHESIS (MAOS)—AN OVERVIEW

Microwave energy originally applied for the heating of foodstuff by Percy Spencer in the 1940s has found a variety of technical applications in the chemical and related industries since the 1950s. The applications include food processing[1] drying, analytical chemistry (microwave digestion, ashing, extraction), polymer science, biochemistry (protein hydrolysis, sterilization), pathology (histoprocessing, tissue fixation), and medical treatments (diathermy).

Surprisingly, microwave heating was not implemented in chemistry until much later. In inorganic chemistry, microwave technology has been used since the late 1970s, while it has only been implemented in organic chemistry since the mid-1980s. This slow uptake of the technology can be attributed to a lack of understanding of the basics of microwave dielectric heating by chemists

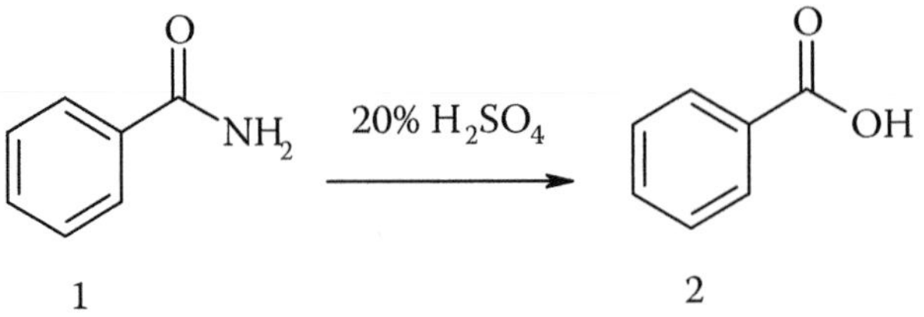

Conventional: 1 h; 90% yield (reflux)
Microwave: 10 min; 99% yield (sealed vessel)

SCHEME 25.1 Albuterol. Earliest report of microwave heating used for hydrolysis of an amide.

coupled with a lack of controllability and reproducibility among equipment used. The first reports on the use of microwave heating to accelerate organic reactions were published by the groups of Richard Gedye (Scheme 25.1)[2] and Raymond J. Giguere and George Majetich in 1986.[3]

These early studies were carried out in sealed Teflon® or glass containers in a domestic household microwave oven without temperature or pressure measurements. The results were often violent explosions due to the rapid uncontrolled heating of organic solvents under closed vessel conditions. The risks associated with the flammability of organic solvents in a microwave field and the lack of available dedicated reactors with adequate temperature and pressure control were major concerns. In the 1990s, several groups started experimenting with "dry-media" (solvent-free) reactions that eliminated the dangers of explosions due to solvent. Here the reagents were preabsorbed onto supports such as silica, alumina, clay, or graphite and heated in a domestic oven under standard open-vessel conditions. These solvent-free conditions were at the time quite popular, and a large number of interesting transformations have been published.[4–7] The technical difficulties, relating to nonuniform heating, mixing, and the precise measurement of temperature, remained unresolved. Some microwave-assisted organic synthesis (MAOS) was also carried out using organic solvents under atmospheric pressure in open vessel conditions. Because the boiling point of the solvent typically limits the reaction temperature that can be achieved under open vessel conditions, high-boiling solvents were frequently used[8,9] that posed some serious challenges in terms of product isolation. The recent availability of commercial microwave equipment[10] intended for organic chemistry, with real-time monitoring of both temperature and pressure, has rapidly moved MAOS from being a laboratory curiosity in the late 1980s to an established technique in organic synthesis, heavily used in both academia and industry. The significant increase in the application of this technology in organic synthesis has been heavily reviewed.[11–15]

The observed rate accelerations and sometimes altered product distributions compared to classical "oil-bath" experiments have led to speculation on the existence of specific or nonthermal microwave effects.[16–18] Historically, such effects were claimed when the outcome of a synthesis performed under microwave conditions was different from that of the conventionally heated counterpart. When reviewing the present literature, it appears that most scientists now agree that in the majority of cases the reason for the observed rate enhancements is a purely thermal/kinetic effect. Even though for the industrial chemist this discussion seems largely irrelevant, the debate on "microwave effects" is undoubtedly going to continue for many years in the academic world. Today, microwave chemistry is as reliable as the vast arsenal of synthetic methods that preceded it. Microwave heating not only reduces reaction times significantly, but is also known to reduce side reactions, increase yields, and improve reproducibility.

25.1.2 MICROWAVE THEORY

25.1.2.1 Microwave Radiation and Dielectric Heating

Microwaves are electromagnetic waves with wavelengths longer than those of infrared light but shorter than those of radio waves. Microwave is a collective name for electromagnetic irradiation with frequency ranges that include ultra-high frequency (UHF; 0.3 to 3 GHz), super-high frequency (SHF; 3 to 30 GHz), and extremely high frequency (EHF; 30 to 300 GHz) signals. Most of the frequencies in this band are dedicated to the use of radar equipment and telecommunication (e.g., dual-band mobile phones operate at 0.9 and 1.8 GHz). To avoid interferences among the different applications, all domestic microwave ovens and microwave reactors for chemical synthesis operate at 2.45 GHz corresponding to a wavelength of 12.2 cm.

Microwave-assisted chemistry is based on the efficient heating of materials by microwave dielectric heating effects, which are dependent on the ability of a specific material (e.g., reagent or solvent) to absorb microwave energy and convert it into heat. The electromagnetic waves of a microwave are composed of an electric and a magnetic component, of which the former is important for interaction with materials. Heating by the electric component of the microwave electromagnetic field occurs through two major mechanisms: dipolar polarization and ionic conduction. When exposed to microwave frequencies, the dipoles of a sample align in the applied electric field. As the alternating electrical field oscillates, the dipole or ion attempts to realign itself with the fluctuating field. In this process, energy is lost in the form of heat due to molecular friction and dielectric loss. The amount of heat generated is directly correlated with the ability of the matrix to align itself with the phase of the microwave field (Figure 25.1).

The energy quantum (0.0016 eV) of the microwave irradiation is totally inadequate for exciting atom–atom bonds or specific parts of a molecule and hence cannot induce chemical reactions, as opposed to ultraviolet or infrared radiation (Table 25.1).[19] When molecules rotate in a matrix, they generate heat by friction. The amount of heat generated by a given reaction mixture is a complex function of its dielectric property, volume, geometry, concentration, viscosity, and temperature.[20] Thus, two samples irradiated at the same power level for the same period of time will most likely end up with rather different final temperatures.

25.1.2.2 Dielectric Properties—Solvents and Microwaves

A dielectric material contains either permanent or induced dipoles. Dielectric polarization arises from the rotation of dipoles in an electric field, as opposed to conduction that results from translational motion of charges. The dielectric constant or relative permittivity is the permittivity

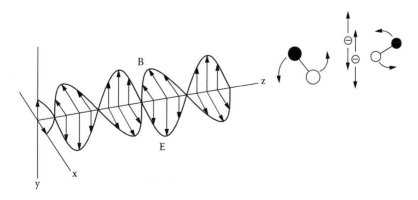

FIGURE 25.1 (See color insert following page 40). Dipolar molecules and ions try to move with an oscillating electric field.

TABLE 25.1
Bond Energies of Different Types of Radiation

Radiation Type	Frequency (MHz)	Quantum Energy (eV)	Bond Type	Bond Energy (eV)
Gamma rays	3.0×10^{14}	1.24×10^6	C–C	3.61
X-rays	3.0×10^{13}	1.24×10^5	C=C	6.35
Ultraviolet	1.0×10^9	4.1	C–O	3.74
Visible light	6.0×10^8	2.5	C=O	7.71
Infrared	3.0×10^6	0.012	C–H	4.28
Microwaves	2450	0.0016	O–H	4.80
Radiofrequency	1	4.0×10^{-9}	H–H	0.04–0.44

Note: See Kappe, C.O. and Stadler, A., in *Microwaves in Organic and Medicinal Chemistry*, Vol. 25, Wiley-VCH, Weinheim, Germany, 2005.

TABLE 25.2
Dielectric Constants of Some Common Solvents at 20°C

Solvent	Dielectric Constant
Carbon tetrachloride	2.2
Benzene	2.3
Chloroform	4.8
Acetone[a]	21.4
Ethanol[a]	25.7
Methanol[a]	33.7
Water[a]	80.4

Note: See also Gabriel, C., Gabriel, S., Grant, E.H., Halstead, B.S.J., Mingos, D.M.P. *Chem. Soc. Rev.*, 27, 213, 1998.

[a] For these polar solvents, ε' is frequency dependent, and the values given in the table refer to the static value ε_s.

of the material relative to that of free space. Compounds that have large permanent dipole moments also have large dielectric constants (Table 25.2), because the dielectric polarization depends primarily on the ability of their dipoles to re-orient in an applied electric field.

The permittivity of a material (ε) is a property that describes the charge storing ability of that substance. Essentially, the ability of a substance to heat in a microwave field is dependent upon two factors: (1) the efficiency with which the substance adsorbs the microwave energy, normally described by its dielectric properties, ε' and (2) the efficiency with which the adsorbed energy can be converted to heat, described by the loss factor, ε''.

A convenient way to evaluate the ability of two closely related substances to convert microwave energy into heat is to compare their respective "loss tangent" values, where the loss tangent is defined as the tangent of the ratio of the loss factor and the dielectric properties (Equation 25.1; Table 25.3). For deeper insight into the mechanism of microwave dielectric heating, the review by Mingos et al.[21] is recommended.

TABLE 25.3
Microwave Absorbing Properties in Relation to Loss Tangents
(tan δ) of Some Common Solvents at 2.45 GHz, 20°C

Solvent	BP (°C)	tan δ	Microwave Absorption
Ethylene glycol	197	1.35	Excellent
Dimethyl sulfoxide (DMSO)	189	0.825	Good
Ethanol	78	0.941	Good
Methanol	63	0.659	Good
1-Methyl-2-pyrrolidone (NMP)	204	0.275	Good
1, 2-Dichlorobenzene	180	0.280	Good
N,N-dimethylformamide (DMF)	154	0.161	Good
Water	100	0.123	Medium
Acetonitrile	81	0.062	Medium
Dichloromethane (DCM)	40	0.042	Poor
Tetrahydrofuran (THF)	66	0.047	Poor
Toluene	110	0.040	Poor

Note: See also Kappe, C.O. and Stadler, A., in *Microwaves in Organic and Medicinal Chemistry*, Vol. 25, Wiley-VCH, Weinheim, Germany, 2005; and Gabriel, C., Gabriel, S., Grant, E.H., Halstead, B.S.J., Mingos, D.M.P. *Chem. Soc. Rev.*, 27, 213, 1998.

$$\tan \delta = \varepsilon''/\varepsilon' \tag{25.1}$$

Today's dedicated pieces of equipment are capable of reliably heating a wide variety of substances, with the use of variable power output and temperature control, so that the problem of selecting the appropriate matrix is more or less invisible to the end user. All types of solvents can be used in MAOS. Polar solvents, such as N,N-dimethylformamide (DMF), N-methylpyrrolidinone (NMP), and ethanol, are good microwave absorbers and will heat efficiently. On the other hand, less-polar or nonpolar solvents such as toluene, dioxane, and tetrahydrofuran (THF) are more or less transparent to the microwave irradiation (possess low tan δ) and will not heat in the pure form. However, most chemical reactions contain enough polar or ionic substances that will absorb the microwaves and generate heat. Nevertheless, when the dielectric properties of the sample are too poor to allow efficient heating by microwave irradiation, the addition of small amounts of polar or ionic additives, with large loss tangent values, can significantly overcome this problem and enable adequate heating of the entire mixture. Fluid salts or ionic liquids, consisting entirely of ions, absorb microwave radiation in a highly efficient manner and are particularly attractive additives because they are relatively inert and stable at temperatures up to 200°C, have a negligible vapor pressure,[22,23] and dissolve to an appreciable extent in a wide range of organic solvents. Energy transfer between the polar molecules that couple with the microwave radiation and the nonpolar solvent bulk is rapid and often provides an efficient means of using nonpolar solvents for synthesis using microwave irradiation.

25.1.2.3 Microwave versus Classical Heating

Chemical reactions, performed using MAOS techniques, are rapid mainly because the reactions are performed at higher temperatures than their conventional counterparts. The modern microwave-based synthesizers can achieve temperatures of up to 250°C and pressures of up to 20 bars, allowing reactions to be carried out at higher temperatures than their reflux counterparts. Microwave dielectric heating, however, has the following advantages compared to classical heating for chemical conversions:

1. The microwave energy is introduced into the chemical reactor remotely; therefore, there is no direct contact between the energy source and the reacting chemicals. Containment materials interact differently from chemicals at the commonly used microwave frequency; hence, by choosing the containment material carefully, the microwaves can be made to pass through the walls of the vessel and heat only the reactants.
2. The introduction of microwave energy into a chemical reaction, which has at least one component capable of coupling with microwaves, can lead to a much higher heating rate than reactions achieved through classical heating. The high temperatures that result when metal powders are exposed to microwaves create hot spots that accelerate the reaction of metals with organic and gaseous substrates.
3. Using a simple, sealed, microwave transparent apparatus (e.g., Teflon or glass), it is possible to rapidly increase the temperature of a reaction in common organic solvents up to 100°C above the conventional boiling point of the solvent, which leads to a thousand-fold acceleration of the reaction rate. For example, methanol, which has a boiling point of 65°C, can be rapidly heated to 160°C and a pressure of 17 atmospheres.

25.1.2.4 Microwave and Thermal Effects

In early literature, there were many claims of a "specific microwave effect" responsible for the observed rate accelerations.[24–26] Later experiments showed some of these early reports to be artifacts,[27] while others are debatable or hard to explain.[28] An attempt to rationalize a possible specific microwave effect has been published by Perreux et al.[29] Most of the reports on specific effects, however, can be rapidly dismissed due to poor temperature control. These inaccuracies in temperature measurements often occur when performing the reactions in domestic ovens, with microtiter plates or on solid supports, where there are inherent difficulties in measuring the temperature accurately. Even with today's specialized equipment, it is very difficult to capture the true temperature of a reaction performed on a dry solid support or in a continuous flow system.

Under some circumstances, the rapid rate of microwave heating can produce heat profiles that are not easily accessible using traditional heating techniques. In such cases, experiments performed using MAOS may well result in a different outcome to conventionally heated reactions, even if the final reaction temperature is the same.[30]

Another phenomenon that might account for some of the claims of specific effects, for reactions run under atmospheric pressure, is the superheating effect.[31] Under microwave irradiation at atmospheric pressure, the boiling points of solvents can be raised up to 26°C above their conventional values. The enhanced boiling point can be maintained in pure solvents for as long as the microwave radiation is applied. Substrates or ions present in the solvent aid in the formation of "boiling nuclei," and the rate at which the temperature of the mixture returns back to the normal boiling point is solvent dependent. It is now well accepted that the major part of rate enhancements observed with MAOS is strictly due to thermal effects, even though the unique temperature profiles accessible by microwave radiation may result in novel outcomes. Although the existence of a "specific microwave effect" cannot be completely ruled out, the effect appears to be of marginal synthetic importance and a rarity.

25.1.3 GETTING STARTED WITH MICROWAVE-ASSISTED ORGANIC SYNTHESIS (MAOS)

MAOS has often been applied to known conventional thermal methodology in the past to accelerate the reaction or increase yield. Nowadays, the technology is applied more and more toward new reactions with unknown outcomes. It is easier to select starting conditions for known thermal reactions than for unknown reactions. Therefore, it is best to perform an unknown reaction in small volume and with careful selection of conditions.

Increasing Temperature $\longrightarrow$

20 °C (room temperature)	30 °C	40 °C	50 °C	60 °C	70 °C	80 °C	90 °C	100 °C
16 h (overnight)	8 h	4 h	2 h	1 h	30 min	15 min	7.5 min	~ 4 min

Decreasing time $\longrightarrow$

FIGURE 25.2 The Arrhenius equation relationship between reaction temperature and time.

25.1.3.1 Selecting Starting Temperature and Time

According to Arrhenius's equation, $k = A*\exp^{(Ea/R*T)}$, the rate of a reaction is doubled for every 10°C rise in temperature. Hence, reactions performed at a 100°C higher temperature would have a reaction rate of 1/1000th of the conventional condition. Arrhenius's rule can be applied to derive the starting temperature and time for a reaction whose conventional conditions are known. For instance, a reaction that takes overnight (16 h) at room temperature (20°C) would be complete in 4 min at ~100°C (Figure 25.2). Theoretically this is an accurate assessment, but it would be prudent to perform reactions at temperatures ±10°C of the Arrhenius derived value. Reaction times are sometimes shorter than the predictions made using Arrhenius's equation. This is probably due to the development of pressure in sealed tubes or due to localized superheating of catalysts and additives within a reaction.

In case of reactions for which there is no precedence of conventional conditions, it is generally safe to run these at double the boiling point of the solvent without surpassing the pressure limit. Five minutes is generally sufficient to give a good indication of a reaction's progress, after which further modifications can be made if necessary.

Substrate and reagent stability should be considered wherever possible. In general, most substrates and reagents have been found to survive high temperatures for the short periods of time characteristic of MAOS. The maximum temperature that can be achieved without exceeding the pressure limit should also be considered.

25.1.3.2 Selecting Solvent

Although different solvents interact very differently with microwaves, because of their diverse polar and ionic properties, it may not be necessary to change the solvent that is specified for a reaction under traditional chemistry conditions. Polar solvents (e.g., DMF, NMP, dimethyl sulfoxide [DMSO], methanol, ethanol, and acetic acid) couple well with microwaves due to their polarity, and hence, the temperature will increase rapidly with these solvents. Nonpolar solvents (e.g., toluene, dioxane, THF) will be heated only if other components in the reaction mixture respond to microwave energy (i.e., if the reaction mixture contains either polar reactants or ions). When using nonpolar solvents, more concentrated reaction mixtures might be preferable. Under such circumstances, the achievable temperature can be quite high. Ionic liquids are reported as new, environmentally friendly, recyclable alternatives to dipolar aprotic solvents for organic synthesis. The dielectric properties of ionic liquids make them highly suitable for use as solvents or additives in MAOS. Ionic liquids consist entirely of ions and therefore absorb microwave irradiation extremely efficiently. Furthermore, they have a low vapor pressure, enhancing their suitability even further. Despite ionic liquids being salts, they dissolve to an appreciable extent in a wide range of organic solvents and can therefore be used to increase the microwave absorption of low-absorbing reaction mixtures.

Solvents can behave differently at elevated temperatures, and most solvents become less polar with increased temperature. Water is maybe the most interesting case. At elevated temperatures, the bond angle in water widens, and its dielectric properties approach those of organic solvents. Water at 250°C has similar dielectric properties as acetonitrile at room temperature. Thus, water can be used as a pseudo-organic solvent at elevated temperatures where organic molecules will dissolve, not only because of the temperature, but also because of the change in dielectric properties. This makes some reactions that normally would not run in water perfectly feasible. Solvents with low boiling points (e.g., methanol, dichloromethane, and acetone), give lower achievable temperatures due to the pressure build-up in the vessel. If a higher absolute temperature is desirable to achieve a fast reaction, it is advisable to change to a closely related solvent with a higher boiling point (e.g., dichloroethane instead of dichloromethane).

25.1.3.3 Working with Gaseous Substrates

Reactions involving gas such as hydrogenation, carbonylation, and amination are conventionally done under catalytic conditions using a gaseous atmosphere of hydrogen gas, carbon monoxide, and ammonia, respectively. Gases, however, are not good microwave absorbers due to the lower frequency of molecule-to-molecule collisions. An alternative to using gases is to use the appropriate "donors" of the gas, which also has the advantage of eliminating the need to use highly flammable and toxic gases (Table 25.4).

Room temperature ionic liquid, N-butyl-N-methylimidazolium hexafluoro-phosphate (Figure 25.3; **Ib**; [BMIM]$^+$[PF$_6$]) has been used as a solvent for catalytic transfer hydrogenations of different

TABLE 25.4
In situ Sources of Gas

Gas	*In Situ* Source
NH$_3$	Conc NH$_4$OH or NH$_4$OAc + AcOH
O$_2$	Oxone
CO	Cr(CO)$_6$ or Mo(CO)$_6$ or DMF or HCOOH + H$_2$SO$_4$
H$_2$	HCOO$^-$NH$_4$$^+$ or cyclohexene
HCl	NaHSO$_3$ + NaCl
CH$_4$	CH$_3$COONa + NaOH
SO$_2$	NaHSO$_3$ + NaOH

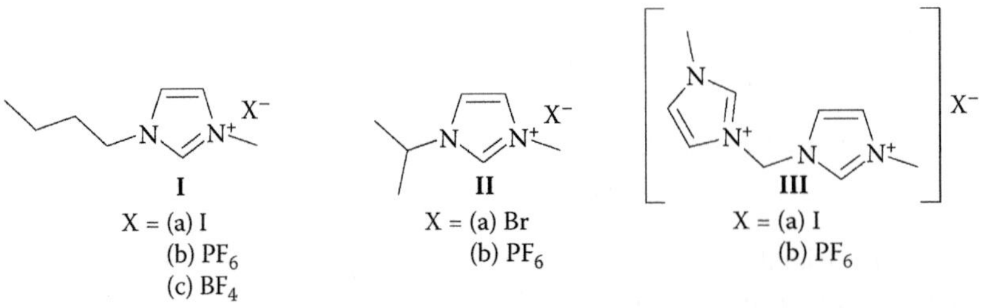

I	II	III
X = (a) I	X = (a) Br	X = (a) I
(b) PF$_6$	(b) PF$_6$	(b) PF$_6$
(c) BF$_4$		

FIGURE 25.3 The structure of some liquids.

SCHEME 25.2(A) Selective microwave assisted reduction of a nitrogroup using tin chloride (Reference 33).

SCHEME 25.2(B) Microwave assisted nitro reduction of a halogenated substrate using Indium (Reference 33).

SCHEME 25.2(C) Microwave assisted carbonylation using Mo(CO)$_6$ as an *in situ* carbon monoxide source (Reference 37).

homo- and heteronuclear organic compounds using microwave heating.[32] Formate salts such as ammonium formate and triethylammonium formate were used as a hydrogen source in the reaction catalyzed by 10% palladium on carbon. Pure products were isolated in moderate to excellent yields by simple liquid–liquid extraction with methyl *tert*-butyl ether (MTBE). Efficient reduction was also achieved within 10 min at 140°C using tin chloride in ethanol (Scheme 25.2A) or indium metal and saturated aqueous ammonium chloride in ethanol (Scheme 25.2B).[33]

Carbonylations using molybdenum hexacarbonyl Mo(CO)$_6$, as a source of carbon monoxide, were reported extensively by Larhed et al.[34–37] in the synthesis of amides and esters from halides (Scheme 25.2C). In their initial studies, pure DME or other nonpolar solvents (e.g., THF, 1,4-dioxane, toluene) afforded precipitation of the solid molybdenum metal on the glass wall of the reaction vessel which resulted in an extreme microwave absorption at that spot and thermal cracking of the Pyrex. After further investigation, a combination of diglyme and 4 M K$_2$CO$_3$(aq) was found to provide a convenient protocol for carbonylations without cracking or overpressurization.

The group also reported[38] DMF in the presence of potassium *tert*-butoxide to be an efficient source of carbon monoxide and dimethyl-amine in palladium-catalyzed aminocarbonylation (Heck carbonylation; Scheme 25.2D). The addition of excess amines to the reaction mixture provided good yields of the corresponding aryl amides. The reaction proceeded smoothly with bromobenzene and more electron-rich aryl bromides, but not with electron-deficient aryl bromides.

Oxone was used in the oxidation of thiol **13** to its sulfoxide **14** or sulfone **15** in water (Scheme 25.2E).[33] Because oxone decomposes in water, the reaction was most successful when the water was added just before microwave heating. The reaction was performed at 150°C which is 100°C above the maximum recommended temperature for oxone (50°C). The sulfoxide was formed as the major product with 0.5 equiv of oxone, whereas the sulfone was formed with 1.5 equiv of the

SCHEME 25.2(D) Microwave assisted carbonylation using DMF as the carbon monoxide source (Reference 38).

SCHEME 25.2(E) Microwave assisted oxidation of a thiol using Oxone (Reference 33).

oxone under the same conditions. A by-product, which was identified as 2-hydroxybenzimidazole, was also formed under the conditions for the sulfone, but not with the sulfoxide.

25.1.3.4 Reaction Optimization

As in conventional synthesis, reaction optimization is crucial for successful MAOS. Similar to traditional chemistry, the success of reactions is as dependent on factors such as solvent and reagent selection as it is upon temperature and time. These parameters can be optimized in a relatively short period of time and the optimal conditions found via MAOS optimization can in most cases be applicable to conventional heating methods for scale-up.

A typical optimization process would involve starting with conventional reaction conditions. The first set of reactions would be performed around the Arrhenius-derived temperature for a set time, typically 5 min, and analyzed. Depending on the outcome of this first set of reactions, the reaction temperature and time could be increased or decreased.

1. *Complete conversion*—The reaction time can be reduced further to optimize the reaction rate. This is particularly important when the conditions are being developed for library synthesis.
2. *Incomplete or no conversion*—The reaction temperature may be increased in 10°C increments, while keeping the time constant at 5 min, until complete conversion or appearance of decomposition. If the threshold temperature for decomposition is reached, then the reaction time may be extended in 5-min increments until complete conversion takes place.
3. *Decomposition and by-products*—The reaction temperature and time should be decreased until minimal or no decomposition or by-product is seen. If the formation of by-product or decomposition is still prominent at the lowest temperature and time achievable, then reaction conditions need to be reviewed and optimized in terms of combination of solvents, additives, and catalysts.

In general, it is preferable to increase the reaction temperature rather than time until decomposition or by-products are observed, because the goal of MAOS is to shorten the reaction time. In case of poor or no yield of the desired product, it may be necessary to explore solvent and reagent combinations in addition to temperature and time. Examples of reaction optimization can be found in Section 25.4.

25.1.4 MAOS IN PROCESS RESEARCH

MAOS is beginning to play a greater role in process research, especially in cases where conventional methods require prolonged reaction times and forced conditions. The unique properties of MAOS allow the otherwise very tedious optimization procedure to be drastically accelerated. Once the proper conditions have been determined, there are several ways to proceed depending on what scale the production is to be performed.

If the product is needed on a small scale, up to 100 g, the reaction can be scaled out rather than scaled up. The excellent reproducibility together with automation can easily produce up to 100 g overnight. Bose et al.[39] have described an alternative for minor scale-up where the use of the microwave-assisted organic reaction enhancement (MORE) technique reduces the need for organic solvents and increases "atom economy" by improving product selectivity and chemical yield, thus, minimizing the need for larger scale-up.

If kilogram quantity of the product is needed, there are two different possibilities for scaling up: batch processing or continuous flow processing.

In a batch process, all the reaction components are combined and held under controlled conditions until the desired process endpoint has been reached. In a batch approach, it is very easy to truly reproduce the conditions used on a smaller scale and simultaneously keep good control of temperature and pressure. To a certain extent, MAOS suffers less from going up scale because the energy is still deposited directly in the reaction mixture, avoiding problems with heat transfers from reactors. Batch solutions also provide direct scalability for heterogeneous mixtures.

Batch reactors have recently been used in the scale-up of the decongestant and anti-asthmatic drug L-ephidrine,[40] the preservative n-butylparaben,[41] and esterification of benzoic acid.[42] Utilizing neat conditions, batch processing has produced as much as 269 g of a dithioketal by ketal exchange[43] and 620 g of an ester by PTC alkylation.[44] Although single-mode cavity microwaves are ideal for small volumes (<1 L), when working with larger volumes these are no longer the best choice and multimode microwaves have to be used. Microwave processes can produce localized high temperatures and pressures, and any scale-up operation must consider these potential dangers and limitations.

In continuous flow-through systems,[45] reagents are pumped through the microwave cavity, allowing only a portion of the sample to be irradiated at a time. A review by Anderson[46] describes scale-up of reactions using continuously stirred tank reactors (CSTRs) and plug flow reactors (PFRs). The review discusses the advantages of continuous flow processing for thermochemical rearrangements, immobilized catalysts, and microwave photochemical, electrochemical, and sonochemical processes. The main drawback is that, for some reactions, not all the components are soluble prior to, or after, microwave irradiation, and this can stop the flow due to blockade of the tubes.

Shieh et al.[47] reported the use of a continuous flow reactor for the esterification of a 50 to 100 g scale of carboxylic acids to methyl esters with dimethyl carbonate and 1,8-Diazabicyclo[5.4.0]undec-7-en (DBU) as catalyst in 20 min (microwave residence time).

A continuous microwave reactor developed at Boehringer Ingelheim Pharmaceuticals used focused single-mode microwave irradiation in scaling up synthetic transformations up to multigram quantities.[48] Representative reactions that were investigated included aromatic nucleophilic substitution (SNAr), esterification, and the Suzuki cross-coupling reaction. In general, the product yields were equivalent to or greater than reactions run under conventional thermal heating conditions. For the nucleophilic aromatic substitution of 4-fluoro-3-nitroaniline **16** with phenethylamine **17**, the flow-through system provided 54% conversion after 5 h, the total irradiation time per milliliter (mL) of reaction mixture was 24 min (Scheme 25.3). As the reaction progressed, however, the SNAr product crystallized from the solution as a fine orange powder, and the resulting particles eventually clogged the lines and frits making it necessary to terminate the reaction before complete consumption of the starting material.

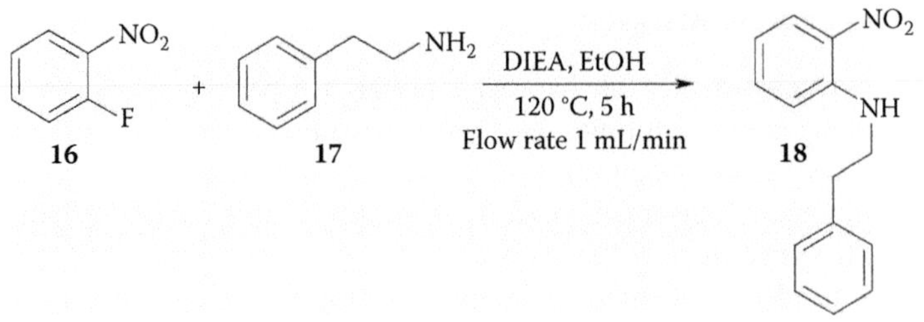

SCHEME 25.3 A nucleophilc aromatic substitution reaction performed in a flow-through microwave system using a flow rate of 1mL/min (Reference 48).

Organ et al.[49] recently developed a capillary-based flow system for conducting microscale microwave synthesis. Excellent conversions were reported in a variety of metal-free cross-coupling and ring-closing reactions, although reactions that had solids in them did not seem to pose a concern, and capillaries coated internally with a thin film of Pd metal were capable of catalyzing reactions. Reagents in separate syringes could also be coinjected into the capillary, mixed, and reacted with none of the laminar flow problems that plague microreactor technology.

Continuous-flow microwave reactors are the solution for production-scale microwave synthesis, but no viable commercial alternatives are available today. The one glaring common problem for production-scale flow-through microwave reactors is the inability to consistently and accurately pump heterogeneous mixtures through the reactor, which *is critical*. It is also impossible to maintain a uniform temperature throughout the reaction mixture and the temperature of the reaction will inevitably increase as the material flows through the cavity. Thus, continuous-flow-through systems, in their current design, require a total re-optimization of the reaction conditions or maybe even a total change of the entire reaction route.

25.1.5 MAOS AND GREEN CHEMISTRY

Introduced in the early 1990s, green chemistry is defined as the utilization of a set of principles that reduces or eliminates the use or generation of hazardous substances in the design, manufacture, and application of chemical products. The concept of green chemistry is enshrined in a set of 12 principles, such as the prevention of waste, the design of energy-efficient processes, the use of safe, environmentally benign solvents where possible, and so forth. Recently, there has been great effort in developing green chemistry-promoting technologies that can reduce or eliminate the use or generation of hazardous substances during the design, manufacture, and use of chemical products. The focus has been on (1) reactions performed in aqueous media, (2) atom-economical synthesis, and (3) recycling catalysts. Almost all reference to "greenness" refers to the efficiency of a reaction escalating energy cost and consumption of natural resources, making energy consumption an important consideration for environmental acceptability and economic viability. Even though MAOS has in the past been used in reference to green chemistry due to its ability to increase product yield and selectivity in short periods of time and often in the absence of solvent, the energy efficiency of such a process was only recently studied by Clark et al.[50] from the Clean Technology Center. The energy consumed in preparing 1 mole of a chemical compound through a Suzuki coupling, Knoevenagel condensation, and Friedel–Crafts acylation was compared for traditional oil bath and microwave reactors. It was observed that there was a reduction in energy demand on switching from oil bath to microwave reactor for all three reactions. For the Suzuki reaction, this reduction was a notable 85-fold.

In the past, MAOS has been carried out under dry or solvent-free conditions, mainly to avoid the hazards of using volatile and flammable organic solvents in domestic microwave ovens. Although these solvent-free techniques claimed to be environmentally friendly, as they avoided the

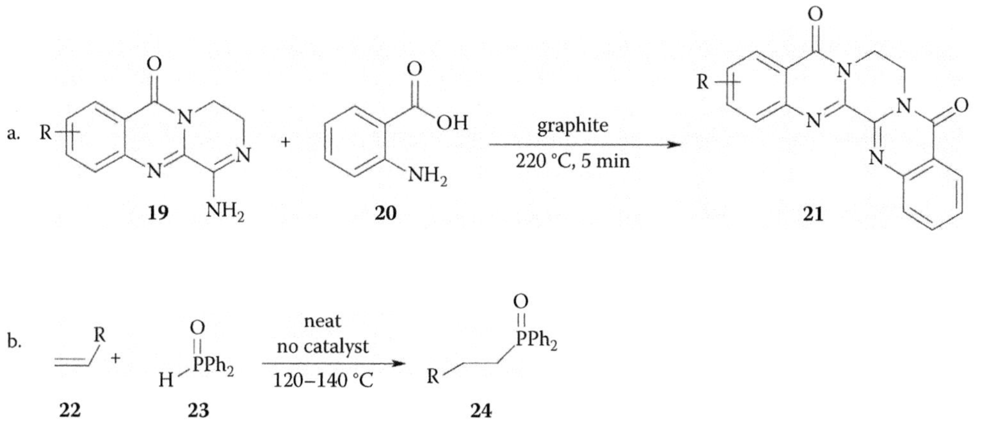

SCHEME 25.4 (a) Microwave-assisted Niementowski reaction (Reference 51). (b) Neat microwave reaction (Reference 52).

use of solvents, this is debatable because solvents were often used to pre-absorb the substrates onto, and wash the products off, the solid supports. For neat solids, it is very difficult to obtain a good temperature control at the surface of the solids, and local "hot spots" may be encountered. This can sometimes give rise to unexpected results and inevitably lead to problems regarding reaction predictability, reproducibility, and control. For certain reactions requiring high temperatures, however, the presence of microwave-absorbing solids can be advantageous. For instance, the best procedure[51] for the preparation of bis-quinazolin-4-ones **21** was found to be via a microwave-assisted Niementowski reaction (Scheme 25.4a), whereby a mixture of the starting amidine **19** and an excess of anthranilic acid **20**, were heated at 220°C, in the presence of graphite. The sealed vials allowed high temperatures to be reached and prevented sublimation of the anthranilic acid. This reaction, when performed in the presence of solvents like NMP or DMF, offered only 37% product and a large amount of by-products.

Neat reactions of liquid substrates can be quite successful. For example, the addition of P(O)-H bonds to alkenes has been accomplished using microwave irradiation in the absence of added solvent or catalyst (Scheme 25.4b).[52] Tandem hydrophosphinylation reactions with alkynes afforded unsymmetrical species such as phosphine oxide—phosphinates.

25.2 MICROWAVE INSTRUMENTATION BASICS: MULTI- AND SINGLE-MODE CAVITIES

Despite several subcategories, there are two fundamentally different constructs of microwave-heating devices—namely, multimode or single mode. The main difference between the two is in the buildup of the energy field within the systems. In both cases, microwaves are generated by a magnetron and led into the reaction chamber, the cavity, through a wave guide. When the microwaves in a multimode apparatus enter the cavity, they are reflected by the walls, generating a three-dimensional stationary pattern of standing waves within the cavity, called modes. The cavity of a domestic microwave oven is designed to have typically three to six different modes intended to provide a uniform heating pattern for general food items. The three-dimensional field pattern consists of areas of high and low field strength, commonly referred to as "hot and cold spots." This results in a drastic variation in the heating efficiency at different positions of the load. Domestic microwave appliances are optimized to give high efficiency for 200- to 1000-g loads, and consequently, they operate less reliably for smaller loads. In addition, the irradiation power is generally controlled by on–off cycles of the magnetron, which makes temperature monitoring in such systems quite difficult. Today's multimode systems, while conceptually similar to the domestic oven, are dedicated to synthesis and feature regulation of the microwave power output, direct temperature control, and real-time monitoring of a reaction mixture via infrared (IR) sensors or fiber-optic probes. The presence of the three-dimensional field pattern, however, renders these cavities better suited for larger sample loads of up to a liter. Moreover, because all the reaction vessels in the multivessel rotors are irradiated simultaneously, these instruments are unsuitable for optimization of individual reaction conditions.

Microwave apparatus utilizing a single-mode or a mono-mode cavity allow only a single mode to be present creating a well-defined heating pattern, which is preferred for small loads or volumes. A properly designed cavity prevents the formation of hot and cold spots within the sample, resulting in a uniform heating pattern. This is very important when microwave technology is used in organic chemistry, because the heating pattern for small samples can be well controlled. This allows for higher reproducibility and predictability of results as well as optimization of yields, which are usually more difficult when using a domestic microwave oven. Much higher field strengths can be obtained, giving rise to more rapid heating. Because reaction vessels are heated individually, such systems are ideal for reaction condition optimization. High-throughput can be achieved by means of integrated robotics that transport each vessel in and out of the microwave cavity, in automation. Diverse designs of single-mode instruments are available from instrument companies with different degrees of sophistication with respect to automation, safety, vessel design, and temperature/pressure monitoring. One of the most important features of single-mode instruments available today is their ability to cool reaction mixtures with compressed air after the irradiation is complete. Such active cooling allows for precise determination of reaction times and avoids degradative product formation, which may occur in case of prolonged passive cooling. The dedicated single-mode instruments today can process volumes ranging from 0.2 to 50 mL.

Both multimode and single-mode instruments can be used to carry out chemical reactions efficiently. The choice of instrumentation depends on the desired application and scale and not on the kind of chemistry to be performed.

25.3 EXAMPLES OF PROCESS OPTIMIZATION WITH MAOS

The short reaction times provided by MAOS make it ideal for rapid reaction scouting and optimization. Most reagents, catalysts, and substrates have been shown to survive temperature extremes for short periods of time. Similar to traditional chemistry, the success of reactions is as dependent on factors such as solvent and reagent selection as it is upon temperature and time.

For instance, the Suzuki–Miyaura coupling of bromofuranone **25** with phenylboronic acid **26** in acetonitrile with sodium carbonate resulted in complete decomposition at 90°C, whereas in toluene with potassium carbonate, a 40% yield of the coupled product **27** was obtained at 140°C (Scheme 25.5A).[53] These conditions appear to be general and enabled halogenated examples to be synthesized in 46 to 63% yields.

A series of substituted liquid aliphatic nitriles **28** were trimerized to their corresponding pyrimidine structures **29** under solvent-free conditions in the presence of catalytic quantities of potassium *tert*-butoxide using a focused microwave reactor (Scheme 25.5B).[54] A small sample of ten commercially available nitriles was used to devise an optimized set of conditions for the formation of 4-aminopyrimidine derivatives. This process was achieved in a parallel fashion using two automated microwave synthesizers to determine the most effective reaction parameters in terms of temperature, time, the effect of single or repeated heating cycles, the concentration and identity of base used, and the result of added cosolvents. Multigram quantities of the corresponding 4-aminopyrimidines have been prepared in high yields and purity following a simple and scalable protocol.

Player et al.[55] from Johnson & Johnson Pharmaceuticals have devised an efficient and versatile method for stereoselective synthesis of (E)-3,3-(diarylmethylene)indolinones **31** by a palladium-

SCHEME 25.5(A) Microwave-assisted Suzuki-Miyaura coupling of bromofuranone with aryl boronic acid (Reference 53).

SCHEME 25.5(B) Microwave-assisted solvent-free trimerization of aliphatic nitriles to pyrimidines (Reference 54).

SCHEME 25.5(C) Microwave-assisted stereoselective tandem Heck-carbocyclization/Suzuki coupling. Conditions described in Table 25.5 (Reference 55).

catalyzed tandem Heck-carbocyclization/Suzuki-coupling sequence (Scheme 25.5C). Factors influencing yield and selectivity included catalyst, coordinating ligand, and solvent. The reaction conditions screened by microwave heating and yields are shown in Table 25.5.

Using the standard Suzuki coupling condition utilizing tetrakis (triphenylphosphine) palladium (0) (entry 1, Table 25.5) as a catalyst source and potassium phosphate as base in THF, no reaction occurred after 14 h at room temperature, but brief microwave irradiation provided the desired diarylidenyl indolinone product in 78% isolated yield. CombiPhos-Pd6 (entry 2, Table 25.5), an air stable mixture of palladium catalyst, was ineffective at promoting this cascade reaction at 25°C, but with microwave heating the reaction proceeded to completion, though not as cleanly as was the case with Pd(PPh$_3$)$_4$. Similarly, palladium acetate (entry 3, Table 25.5) and the highly active carbine ligand also failed to trigger room-temperature transformation, and microwave irradiation was required to force the reaction to completion affording 72% of the indolinone. Interestingly

TABLE 25.5
Suzuki Coupling Conditions Used in Scheme 25.5(C)

Entry	Reaction Condition	Yield
1	Pd(PPh$_3$)$_4$ (0.1 equiv), K$_3$PO$_4$	78%
2	CombiPhos-Pd$_6$ (0.1 equiv), K$_3$PO$_4$ (2.5 equiv), THF, H$_2$O, 100°C, 30 min	65%
3	Pd(OAc)$_2$ (0.1 equiv), K$_3$PO$_4$ (2.5 equiv), THF, 100°C, 30 min (0.1 equiv)	72%

4	Pd(OAc)$_2$ (0.1 equiv), K$_3$PO$_4$ (2.5 equiv), THF, 100°C, 30 min (0.2 equiv)	6%

Buchwald's ligand, 2-(dicyclohexylphosphino)biphenyl, which is usually regarded as a more activating ligand (entry 4, Table 25.5), did not promote the reaction even under microwave conditions.

Kappe et al.[56] reported a microwave-assisted automated screening of conditions for the Biginelli reaction which was then used to synthesize a library of dihydropyrimidines (DHPMs; Scheme 25.5D).

Preliminary runs with solvents such as dioxane and THF proved to be far less effective than a mixture of 3:1 acetic acid (AcOH) and ethanol (EtOH). Both AcOH and EtOH effectively couple with microwave irradiation and had the advantage that the starting materials were soluble under the reaction conditions at elevated temperatures, while the DHPM products **35** were comparatively insoluble at room temperature facilitating product isolation. Having selected the solvent, the next

SCHEME 25.5(D) Microwave-assisted Biginelli reaction (Reference 56).

optimization step was the consideration of a suitable catalyst. Hydrochloric acid (HCl) which is conventionally used in Biginelli reactions was not the most suitable reaction promoter because of inadvertent decomposition of the urea components **33** to ammonia, leading to unwanted by-products. More tolerable Lewis acids such as $Yb(OTf)_3$, $InCl_3$, $FeCl_3$, and $LaCl_3$ were screened. Initial screening results of all these catalysts for the model systems ethyl acetoacetate, benzaldehyde, and urea revealed that 10 mol% Yb(OTf) was the most effective catalyst with the AcOH/EtOH solvent system. Having identified an efficient solvent/catalyst combination, the next issue that was dealt with was reaction temperature and time. After a few optimization cycles, it was found that 120°C proved to be a very efficient reaction temperature. Higher temperatures would lead to decreased yields because of undesirable by-products, while lower reaction temperatures required longer reaction time for complete conversion. For the model system, a total irradiation time of 10 min at 120°C resulted in a 92% isolated yield of pure product. The final DHPM product precipitated directly after the active cooling period.

The efficient use of reaction conditions can be used to obtain the desired chemo-, regio-, or stereoselectivities.[57] For instance, the bromination of quinoline **36** with N-bromosuccinimide (NBS) (Scheme 25.5E)[58] was affected by both the temperature and the solvent selection. The ease of bromination was critically dependent on the polarity of the solvent, whereas the reaction regioselectivity was temperature dependent. At 100°C in acetonitrile, there was selective formation of the isomer **38** after 20 min, with only trace amounts of **37**.

The use of appropriate solvents also allowed the highly regioselective preparation of a series of conformationally constrained bicyclic bisaryl α-amino acids via microwave-assisted Diels–Alder reactions of 9-substituted anthracenes **39** and 2-acetamidoacrylates **40**, in significantly shorter periods of time (1 h versus 48 to 72 h; Scheme 25.5F).[59] With DMF, a polar and highly microwave-absorbing solvent, microwave irradiation at elevated temperatures (200°C/1 h) was found to enhance the meta-regioselectivity **41a–h** and improve reaction yields. Nitrobenzene, which gave good yields of the meta-product under conventional heating, was not the optimal solvent under microwave irradiation.

SCHEME 25.5(E) Microwave-assisted regioselective bromination of quinolines (Reference 58).

SCHEME 25.5(F) Regioselective Diels-Alder reactions under varying reaction conditions (Reference 59).

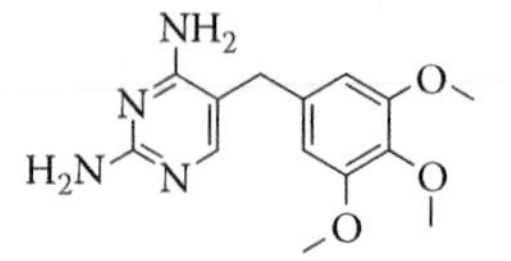

43

FIGURE 25.4 Trimethoprim.

25.4 CASE STUDIES

25.4.1 SYNTHESIS OF TRIMETHOPRIM

Trimethoprim [2,4-diamino-5-(3,4,5-trimethoxybenzyl)pyrimidine, **43**, (Figure 25.4) is a potent and selective inhibitor of bacterial dihydrofolate. It is used alone or in combination with sulfamethoxazole to treat a wide range of clinical bacterial infections.

Several synthetic strategies for the preparation of trimethoprim and its analogs have been reported in the literature. These involve the reaction between 2,4-diaminopyrimidine derivatives and the phenolic Mannich base,[61] Friedel–Crafts reaction using methoxymethylpyrimidine,[62] and cyclization of guanidine with 3-alkoxymethyl-cinnamonitriles[61,63,64] 3-methylthioacrylonitrile,[65] or 3 anilinoacrylonitrile[66–69] as key steps. The method described in a European patent (Scheme 25.6) was selected for total synthesis using microwave irradiation. The first step of this process involves the Michael addition of phenylenediamine **44** to acrylonitrile at elevated temperatures followed by condensation of the nitrile **45** with aldehyde **46**. This room-temperature reaction requires a prolonged reaction time that made it an interesting candidate for MAOS. The final formation of the heterocyclic moiety **43** is performed under reflux in EtOH.

Initial attempts at microwave synthesis using conventional conditions, for the first step, were unsuccessful. However, employing the cyanoethylation procedure described by Cymerman-Craig and Moyle,[70] the microwave irradiated yields were comparable with those from the conventional patented method. Thus, *o*-3,3-(phenylenediimino) dipropanenitrile **45** was obtained with a 70% isolated yield by reacting *o*-phenylenediamine with an eight-fold excess of acrylonitrile and four-fold excess of diethylamine hydrochloride in water under microwave irradiation for 20 min at 160°C. Under milder conditions and using a lower excess of reagents, significant amounts of the starting material remained together with the monocyanomethylated product. Higher temperatures and longer reaction times did not improve yields, probably because of the formation of a tris-N,N,N-cyanoethylated by-product.

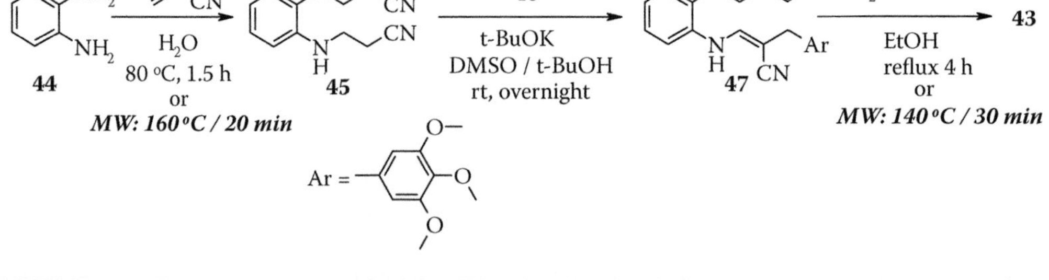

SCHEME 25.6 Early attempts at synthesizing Trimethoprim via MAOS using a patented conventional procedure.

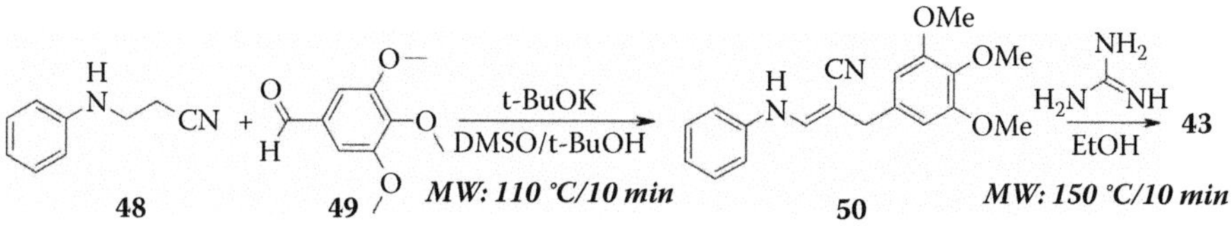

SCHEME 25.7 A shorter higher yielding microwave route to Trimethoprim synthesis.

It was established that the reaction between **45** and 3,4,5-trimethoxybenzaldehyde **46** required cooling in an ice bath in order to obtain reproducible results and a cleaner product, and hence, microwave irradiation was not utilized in this step. Intermediate **47**, thus obtained using conventional methods, was used for the synthesis of **43** under microwave conditions without further purification.[71] Reaction of **47** with a large excess of guanidine under microwave irradiation for 30 min at 140°C afforded trimethoprim **43** in 40% yield, assessed by high-performance liquid chromatography (HPLC). Further optimization[72] of these reaction conditions did not increase the yield of the product.

Because the key reaction in this synthetic sequence did not benefit from microwave rate acceleration, the total amount of time savings using microwave irradiation was not significant as compared to the conventional conditions. Hence, an alternate synthesis pathway starting from commercially available 3-anilinopropionitrile **48** was explored (Scheme 25.7).

In contrast to **47**, intermediate **50** could be readily synthesized under microwave irradiation. Like **47**, intermediate **50** was not purified any further, although it was analyzed using HPLC and hydrogen nuclear magnetic resonance (^{1}H NMR) spectra. The reaction condition was optimized around temperature and time. A 77% yield of **50** was obtained after 5 min at 90°C along with a large amount of unreacted 3-anilinopropionitrile **48**. Complete conversion was, however, observed after 10 min at 110°C, a condition that afforded 83% yield (HPLC) of the desired intermediate **50** along with by-products. Treatment of crude precipitates of **50** from the first step with excess guanidine in EtOH solution for 10 min at 150°C provided trimethoprim **43** (89% yield). Differences in the composition of the crude precipitates of **50** using these two conditions did not affect the yield of the target **43**. Prolonging the reaction time to 20 min slightly improved yields of **43** (91%).[73]

The synthesis of trimethoprim **43** reported here demonstrates the possibility of rapidly optimizing reaction conditions and routes for synthesizing active pharmaceutical ingredients (APIs) using microwave technology. The original synthesis (Scheme 25.5) did not provide significant advantages in terms of time over the conventional patented procedure, so a shorter and higher yielding route (Scheme 25.6) was rapidly sought and optimized as a result of that failure, and the total synthesis time for trimethoprim was reduced from 24 h to 20 min.

25.4.2 Synthesis of Albuterol

Albuterol **51**, (Figure 25.5) is a selective agonist for β_2-adrenergic receptors possessing bronchodilator activity and is widely used in the treatment of bronchial asthma. It was among the highest-selling drugs in the United States for the year 2000.[74]

Numerous routes for the synthesis of albuterol **51**, starting from p-hydroxy acetophenone, have been reported in the literature.[75,76] This study employs the conventional four-step synthesis, reported by Babad et al.[77] starting from salicylaldehyde **52** (Scheme 25.8). Acylation of **52** with 2-bromoacetylchloride or 2-chloroacetylchloride is followed by reaction with tert-butylamine and hydrolysis to obtain aminoketone **56**. Subsequent hydrogenation of **56** under catalytic conditions affords the target compound **51**. Several steps of this synthesis methodology require prolonged heating.

2-Chloroacetylchloride was selected for the Friedel–Crafts acylation of salicylaldehyde **52** because of its ready availability and because it is less expensive than 2-bromoacetylchloride. As anticipated, microwave dielectric heating at 120°C significantly increased the rate of the acylation reaction (30 min versus 18 h) as well as improved yields (70% versus 40%).

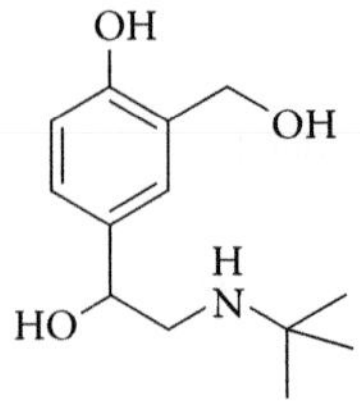

FIGURE 25.5 Albuterol.

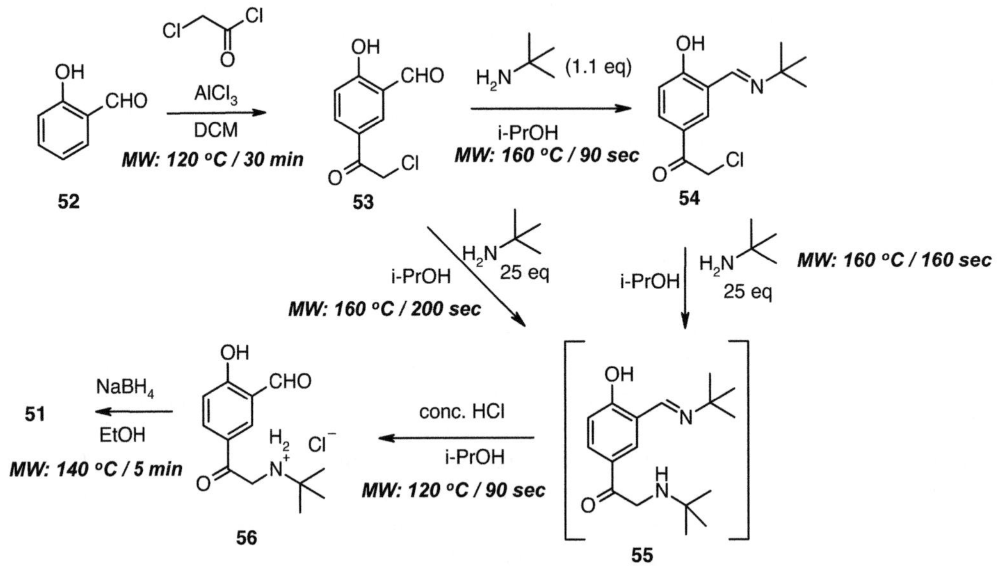

SCHEME 25.8 Microwave-assisted total synthesis of Albuterol.

The second stage of the synthesis involves a three-step sequence: (1) formation of the imine **54**, (2) alkylation to the aminoimine **55**, and (3) *in situ* hydrolysis of the imine moiety to the aldehyde **56**. A series of experiments indicated the alkylation process to be the rate-limiting step in this sequence. Formation of the Schiff base **54**, using *tert*-butylamine (1.1 eq) in *i*-PrOH, which is conventionally rapid,[77] proceeds quantitatively within 90 sec at 160°C.

Although it was reported[77] that 2.2 to 3.0 equivalents of *tert*-butylamine were optimal for converting the imine **54** or 2-chloroacetophenone **53** to the aminoketone **55** within 1 to 2 h of reflux, when repeated these conditions gave only trace amounts of the intermediate **55**[78] after 4 h of reflux. Even after 30 min at 160°C of microwave dielectric heating, this reaction afforded only 40 to 70% conversion[79] with 2.2 to 3.0 eq of *tert*-butylamine, respectively. A large excess (~25 eq) of *tert*-butylamine was found to be crucial for the quantitative conversion of the imine **54** to the intermediate **55** at 160°C in 160 sec. Although compound **55** could be hydrolyzed *in situ*, the excess *tert*-butylamine was removed to facilitate an easier isolation of the aminoketone **56**. Subsequent treatment with concentrated hydrochloric acid (2 eq) in *i*-PrOH resulted in complete hydrolysis of the imine **55** within 90 sec at 120°C using microwave heating. Compound **56** was selectively precipitated (74% isolated yield) from the reaction mixture with cooling, without coprecipitation of *tert*-butylammonium hydrochloride.

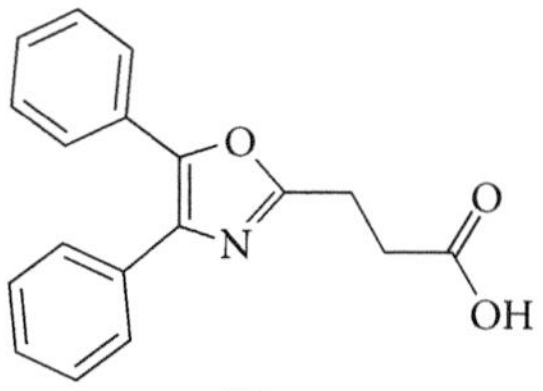

57

FIGURE 25.6 Oxaprozin.

The aminoketone **55** could also be prepared directly from 2-chloroacetophenone **53** using an excess of *tert*-butylamine (25 equiv) and microwave dielectric heating at 160°C for 200 sec. However, it was necessary to use $MgSO_4$ as a water scavenger to prevent oxidative decomposition of **55** and to be able to run the reaction to completion. Subsequent treatment of the aminoketone derivative **56** with $NaBH_4$ in EtOH at 140°C for 5 min resulted in the reduction of both the keto and aldehyde functionalities, providing the target compound **51** in 89% yield.

Microwave dielectric heating was thus used for the total synthesis of albuterol in 62% overall isolated yield compared to 43% reported conventionally.[77] The total synthesis time was also significantly reduced from the conventional 47 h[77] to 40 min.

25.4.3 SYNTHESIS OF OXAPROZIN

Oxaprozin **57** (Figure 25.6) is a nonsteroidal anti-inflammatory drug that has been used to treat symptoms of osteoarthritis and rheumatoid arthritis.[80]

It is traditionally prepared by esterification of benzoin **58** with subsequent cyclization of the ketoester **60** with the nitrogen-donating cyclization agent ammonium acetate (Scheme 25.9).[81,82] The fact that production of oxaprozin requires elevated temperatures for several hours makes its synthesis attractive for use in our investigations using microwaves to decrease the reaction time and increase the overall yield.

The first step in the synthesis of **57**, acylation of benzoin **58** with succinic anhydride **59**, was performed under microwave irradiation in the presence of catalysts such as pyridine[81] or 4-dime-thylaminopyridine (DMAP).[82] Poor yields of **60** were obtained when this reaction was carried out in DMF or toluene at 80 to 110°C for 30 min (Table 25.6, entries 1 and 2). Trifluorotoluene was selected as a substitute solvent for the poorly absorbing toluene in another set of experiments. Under these conditions, yields from esterification of **58** were slightly higher (20 to 30%) than before but were still very low (Table 25.6, entries 3 to 6). A significant increase in yield was seen

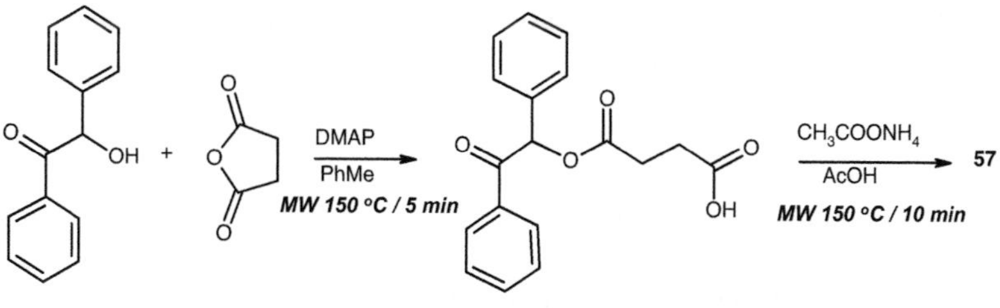

| 58 | 59 | 60 |

SCHEME 25.9 Microwave-assisted synthesis of Oxaprozin. Optimization of conditions for the first step described in Table 25.6.

TABLE 25.6
Optimization of the Benzoyl Acylation Step in the Synthesis of Oxaprozin (Scheme 25.9)

Entry	Benzoin (mmol)	Succinic Anhydride[a] (mmol)	Solvent[b]	Time (min)	Temp. (°C)	HPLC Yield (%)
1	0.1	0.12	DMF	30	110	12
2	0.1	0.12	PhMe	30	80	4
3	0.1	0.12	PhCF$_3$	30	110	27
4	0.1	0.12	PhCF$_3$	30	170	18
5	0.2	0.24	PhCF$_3$	30	110	21
6	0.5	0.6	PhCF$_3$	30	110	28
7	1.0	1.2	PhMe	30	110	66
8	1.0	1.2	PhMe	30	150	82
9	2.0	2.4	PhMe	30	150	88
10	2.0	4.0	PhMe	15	150	89
11	6.0	12.0	PhMe	5	150	96

[a] 20 mol/% dimethylaminopyridine (DMAP) was used in all reactions.
[b] All reactions were run in 2.5 mL of solvent.

when the concentration of the starting material was increased tenfold (Table 25.6, entries 7 to 9). A greater excess of succinic anhydride (Table 25.6, entries 10 and 11) resulted in almost quantitative conversion of benzoin **58** to benzoin hemisuccinate **60** in a shorter period of time (5 to 15 min) at 150°C. Temperatures greater than 150°C led to the formation of by-products.

The two general procedures for cyclization of the keto ester **60** to **57** require heating with ammonium acetate (8 eq) in acetic acid[81] or ammonium acetate (2 eq) in a mixture of acetic acid and formic acid[82] for several (2 to 6) hours. Optimization of the reaction parameters (time, temperature, and ratio of the reaction components) afforded an 86% yield (HPLC) of oxaprozin when **63** was heated, with 8 eq of CH$_3$COONH$_4$ in acetic acid, at 150°C for 10 min. The overall isolated yield of oxaprozin **57** from benzoin **58** under microwave irradiation was 76% (versus 72%[81]) at significantly accelerated reaction speeds (15 min versus 10 h[81]).

25.5 CONCLUSION AND FUTURE TRENDS

Rapid lead generation and optimization were recently facilitated by the emergence of MAOS, and the technique is today one of the major tools for the pharmaceutical chemist where speed of discovery equals competitive advantage in terms of intellectual property and positioning in the market place. MAOS is accepted in discovery chemistry groups, but its acceptance into process development groups has been a slow process, although it has increased substantially during the last 2 years. The development of batch, semi-batch and continuous flow systems will continue, and with time the availability of more dedicated systems for large-scale synthesis will help overcome the process chemist's apprehension toward MAOS. A similar reluctance was expressed by medicinal chemists when dedicated systems were introduced into the market, a situation that has rapidly changed as evident through the burgeoning list of publications as well as the number of conferences and conference contributions on the subject.

MAOS is undoubtedly going to play a major role in chemistry development; this is substantiated by the fact that in most pharmaceutical and biotechnology companies microwave synthesis is the cutting-edge methodology today.

REFERENCES AND NOTES

1. Buffler, C.R. *Microwave Cooking and Processing*; Van Nostrand Reinhold: New York, 1993, pp. 1–68.
2. Gedye, R.; Smith, K.; Westaway, H.A.; Baldisera, L.; Laberge, L.; Rousell, J. *Tetrahedron Lett.*, 1986, *27*, 279–282.
3. Giguere, R.J.; Bray, T.L.; Duncan, S.M.; Majetich, G. *Tetrahedron Lett.*, 1986, *27*, 4945–4958.
4. Loupy, A.; Petit, A.; Hamelin, J.; Texier-Boullet, F.; Jacquault, P.; Mathe, D. *Synthesis*, 1998, 1213–1234.
5. Varma, R.S. *Green Chemistry*, 1999, 1, 43–55.
6. Kidwai, M. *Pure Appl. Chem.*, 2001, *73*, 147–151.
7. Varma, R.S. *Tetrahedron*, 2002, *58*, 1235–1255.
8. Bose, A.K.; Banik, B.K.; Lavlinskaia, N.; Jayaraman, M.; Manhas, M.S. *Chemtech.*, 1997, *27*, 18–24.
9. Bose, A.K.; Manhas, M.S.; Ganguly, S.N.; Sharma, A.H.; Banik, B.K. *Synthesis*, 2002, 1578–1591.
10. www.biotage.com; www.cem.com; www.milestone.com; www.antonpaar.com.
11. Mavandadi, F.; Lidstrom P. *Curr. Topics in Med. Chem.*, 2004, *4*, 773–792.
12. Kappe, C.O. *Angew. Chem. Int. Ed.*, 2004, *43*, 6250–6284.
13. Mavandadi, F.; Pilotti, A. Drug Discovery Today, 2006, 11, 165–174.
14. Lew, A.; Krutzik, P.O.; Hart, M.E.; Chamberlain, A.R. *J. Combi. Chem.*, 2002, *4*, 95–105.
15. Lidström, P.; Tierney, J.; Wathey, B.; Westman, J. *Tetrahedron*, 2001, *57*, 9225–9283.
16. Perreux, L.; Loupy, A. *Tetrahedron*, 2001, *7*, 9199–9223.
17. Kuhnert, N. *Angew. Chem. Int. Ed.*, 2002, *41*, 1863–1866.
18. Strauss, C.R. *Angew. Chem. Int. Ed.*, 2002, *41*, 3589–3590.
19. Kappe, C.O.; Stadler, A. *Microwaves in Organic and Medicinal Chemistry*. 2005, *Vol 25*. Mannhold, R.; Kubinyi, H.; Folkers, G. Eds. Wiley-VCH, Weinham, Germany.
20. Galema, S.A. *Chem. Soc. Rev.*, 1997, *26*, 233–238.
21. Gabriel, C.; Gabriel, S.; Grant, E.H.; Halstead, B.S.J.; Mingos, D.M.P. *Chem. Soc. Rev.*, 1998, *27*, 213.
22. Seddon, K.R. *Kinet. Catal.*, 1997, *37*, 693–697.
23. Welton, T. *Chem. Rev.*, 1999, *99*, 2071–2083.
24. Gedye, R. et al. *Tetrahedron Lett.*, 1986, *27*, 279–282.
25. Giguere, R.J. et al. *Tetrahedron Lett.*, 1986, *27*, 4945–4948.
26. Langa, F. et al. *Contemp. Org. Synth.*, 1997, *4*, 373–386.
27. Gedye, R.N.; Wei, J.B. *Can. J. Chem.*, 1998, *76*, 525–532.
28. Pagnotta, M.; Pooley, C.L.F.; Gurland, B.; Choi, M. *J. Phys. Org. Chem.*, 1993, *6*, 407–411.
29. Perreux, L.; Loupy, A. *Tetrahedron*, 2001, *57*, 9199–9223.
30. Stuerga, D.; Gonon, K.; Lallemant, M. *Tetrahedron*, 1993, *49*, 6229–6234.
31. Baghurst, D.R.; Mingos, D.M.P. *J. Chem. Soc. Chem. Commun.*, 1992, 674–677.
32. Berthold, H.; Schotten, T.; Hönig, H. *Synthesis*, 2002, *11*, 1607–1610.
33. Personal Chemistry, Inc. (Biotage), unpublished data. Available through www.biotagepathfinder.com.
34. Kaiser, N.-F.; Hallberg, A.; Larhed, M. *J. Comb. Chem.*, 2002, *4*, 109–111.
35. Georgsson, J.; Hallberg, A.; Larhed, M. *J. Comb. Chem.*, 2003, *5*(4), 350–352.
36. Wu, X.; Nilsson, P.; Larhed, M. *J. Org. Chem.*, 2005, *70*(1), 346–349.
37. Wu, X.; Larhed, M. *Org. Lett.*, 2005, *7*(15), 3327–3329.
38. Wan, Y.; Alterman, M.; Larhed, M.; Hallberg, A. *J. Org. Chem.*, 2002, *67*(17), 6232–6235.
39. Bose, A.K.; Manhas, M.S.; Ganguly, S.N.; Sharma, A.H.; Banik, B.K. *Synthesis*, 2002, 1578–1591.
40. Shukla, V.B.; Madyar, V.R.; Khadlikar, B.M.; Kulkarni, P.R. *J. Chem. Technol. Biotechnol.*, 2002, *77*, 137–140.
41. Liao, X.; Raghavan, G.S.V.; Yaylayan, V.A. *Tetrahedron Lett.*, 2002, *43*, 45–48.
42. Pipus, G.; Plazl, I.; Koloini, T. *Ind. Eng. Chem. Res.*, 2002, *41*, 1129–1134.
43. Perio, B.; Dozias, M.-J.; Hamelin, J. *Org. Process Res. Dev.*, 1998, *2*, 428.
44. Cleophax, J.; Liagre, M.; Loupy, A.; Petit, A. *Org. Process Res. Dev.*, 2000, *4*, 498.
45. Cablewski, T.; Faux, A.F.; Strauss, C.R. *J. Org. Chem.*, 1994, *59*, 3408–3412.
46. Anderson, N.G. *Org. Proc. Res. & Dev.*, 2001, *5*, 613–621.
47. Shieh, W.-C.; Dell, S.; Repi, O. *Tetrahedron Lett.*, 2002, *43*, 5607–5609.
48. Wilson, N.S. et al. *Org. Process Res. Dev.*, 2004, *8*, 535–538.
49. Corner, E.; Organ, M. *J. Am. Chem. Soc.*, 2005, *127*, 8160–8167.
50. Gronnow, M.J. et al. *Org. Process Res. Dev.*, 2005, *9*, 516–518.

51. Pereira, M. deF. et al. *Tetrahedron Lett.,* 2005, *46,* 3445–3447.
52. Stockland, R.A. Jr. et al. *Org. Lett.,* 2005, *7,* 851–853.
53. Mathews, C.J. et al. *Synlett,* 2005, *3,* 538–540.
54. Baxendale, I.R.; Ley, S. *J. Comb. Chem.,* 2005, *7,* 483–489.
55. Cheung, W.S.; Patch, R.J.; Player, M.R. *J. Org. Chem.,* 2005, *70,* 3741–3744.
56. Stadler, A.; Kappe, C.O. *J. Comb. Chem.,* 2001, *3,* 624–630.
57. DeLa Hoz, A. et al. *Current Organic Chemistry,* 2004, *8,* 903–918.
58. Glasnov, T.N. *et al. J. Org. Chem.,* 2005, *70,* 3865–3870.
59. Yang, B.V.; Doweyko, L.M. *Tetrahedron Lett.,* 2005, *46,* 2857–2860.
60. The three case studies described herein consist of unpublished data from Personal Chemistry, Inc. (Biotage). The reaction conditions for individual steps are available from www.biotagepathfinder.com
61. Roth, B.; Aig, E.; Lane, K.; Rauckman, B.S. *J. Med. Chem.* 1980, 23, 535–541.
62. Manchand, P.S.; Rosen, P.; Belica, P.S.; Oliva, G.V.; Perrotta, A.V.; Wong, H.S. *J. Org. Chem.,* 1992, 57, 3531–3535.
63. Stenbuck, P.; Baltzly, R. *J. Org. Chem.,* 1963, *28,* 1983–1988.
64. Hoffer, M.; Grunberg, E.; Mitrovic, M.; Brossi, A. *J. Med. Chem.,* 1871, *14,* 462–463.
65. Harada, K.; Choshi, T.; Sugino, E. *Heterocycles,* 1994, *38,* 1119–1125.
66. Roth, B.; Aig, E.; Rauckman, B.S.; Strelitz, J.Z.; Phillips, A.P. *J. Med. Chem.,* 1981, *24,* 933–941.
67. Kompis, I.; Then, R.; Boehni, E.; Rey-Bellet, G.; Zanetti, G.; Montavon, M. *Eur. J. Med. Chem.,* 1980, *15,* 17–22. Kompis, I.; Wick, A. *Helv. Chim. Acta,* 1997, *60,* 3025–3034.
68. Perun, T.J.; Rasmussen, R.R.; Hurrom, B.W. U.S. patent: US 4,087,528, May 2, 1978.
69. Koskenniska, L.A.; Manninen, K.J. Eur. Patent Application 0 067 593, May 28, 1981.
70. Cymerman-Craig, J.; Moyle, M. *Organic Synthesis,* 1963, *4,* 205.
71. All attempts to purify compound **47** by crystallization or chromatography failed, probably because of the instability of **47**. Additionally, the product can exist both in "acrylonitrile" form **47** or "cinnamylnitrile" form **47a**. Moreover, *Z* and *E* isomers are possible. These substances have not been fully characterized in the literature.

47   **47a**

72. HPLC analysis of intermediate **47** was not possible because of the instability and high molecular weight of this product.
73. Variations in other reaction parameters did not further improve the yield. The excess of solvents guanine and ethanol and the use of the highest possible temperature are essential. Other solvents, additional bases, and lower temperatures diminish yields of trimethoprim.
74. www.rxlist.com/top200.htm; Source: Scott-Levin, Newton, PA.
75. Tann, C.-H.; Thiruvengadam, T.K.; Chiu, J.; Green, M.; McAllister, T.L.; Colon, C.; Lee, J. U.S. Patent 5,283, 359 February 1, 1994.
76. Collin, D.T.; Hartley, D.; Jack, D.; Lunts, L.H.C.; Press, J.C.; Ritchie, A.C.; Toon, P. *J. Med. Chem.,* 1970, *13,* 674–680.
77. Babad, E.; Carruthers, N.I.; Jaret, R.S.; Steinman, M. *Synthesis,* 1988, 966–968.
78. Because compound **55** is unstable, it was not isolated, and the reaction was controlled by HPLC using weakly acidic buffer (pH 5.5).
79. Measured by the disappearance of the imine derivative **54**.
80. Cummings, D.; Amadio, P. *Am. Fam. Physician,* 1994, *49,* 1197–1202.
81. Weston, G.O. US Patent 4,190,584, February 26, 1980.
82. Breviglieri, G.; Bruno, G.; Contrini, S.; Assanelli, C. U.S. Patent 6,096,896, August 1, 2000.

26 Process Development Considerations for Therapeutic Monoclonal Antibodies in Mammalian Cell Culture

Susan Casnocha, Ronald Fedechko, Paul Mensah, John Mott, and Sandeep Nema

CONTENTS

26.1 INTRODUCTION

During the last decade, a significant trend in the pharmaceutical industry has been the development of monoclonal antibodies (mAbs) as therapeutic agents. Antibodies are polypeptides that bind to specific antigen targets. A major advantage of mAbs as therapeutic agents is their high target specificity, which results in a low side effect profile. Monoclonal antibodies can also be engineered to have additional biological activity beyond binding of the antibody to its antigen. Antibodies can be conjugated to other therapeutic agents (i.e., for targeted delivery of a cytotoxic or radioactive agent to tumor cells). Antibodies, due to their size (~150 kD for IgG) and glycosylation profile, have a long half-life *in vivo*, resulting in a dosing regimen ranging in scale from a week to months.

Monoclonal antibodies have great therapeutic potential and are being pursued both by biotechnology companies and by traditional pharmaceutical companies. A recent report[1] indicates 376 mAb development programs (from preclinical to market), 132 mAbs in clinical development, and 23 approved mAbs (see Table 26.1). The sales figures for 2002 were $5.4 billion (up 37.5% since 2001), and this market share is expected to grow to $16.7 billion by 2008. This growth is fueled by the projected launch of 19 new antibody-based products in the time period 2004 to 2008.

Monoclonal antibodies are predominantly manufactured as recombinant proteins in genetically engineered mammalian cell culture. The challenges in biotechnology differ from those faced by the development of small molecules by the pharmaceutical industry, particularly in the areas of timelines, development costs, and regulatory requirements regarding testing for adventitious agents. In this chapter, mAb process technology will be presented, as well as mAb structure, discovery processes for therapeutic mAbs, formulation, manufacturing, delivery considerations, and future directions.

Throughout this chapter, the major steps in mAb process development are discussed. This process begins with target identification and generation of the mAb. Therapeutic mAbs are generated most commonly through "humanization" of mouse antibodies, immunizing a mouse line genetically engineered with human immunoglobin genes, or screening phage display libraries. The DNA encoding the mAb of interest is then inserted into the genome of a host mammalian cell line. The resulting cell lines (100s to 1000s) are screened and a lead selected based on desirable characteristics, such as cell growth, cell productivity, and quality of the mAb produced. Next, the manufacturing process is developed. Typically, a stirred tank bioreactor is used for production of the secreted mAb (the upstream process), followed by a downstream process consisting of harvest (also called clarification, meaning separation of supernatant from cells) and chromatographic capture and polishing steps. The resulting drug substance is typically a liquid and is formulated into drug product to enhance stability at the desired storage condition.

26.2 STRUCTURE OF MONOCLONAL ANTIBODIES

There are five classes of antibodies: IgA, IgD, IgG, IgE, and IgM. Each class of antibody is characterized by its own location and distribution in the body. IgMs are found primarily in serum and are very large molecules (~850 KD) that have reduced tissue penetration. IgDs are found in serum at very low concentrations and are present on the surface of B cells. Other antibodies such as IgG, IgA, and IgE can diffuse from blood into tissues. IgA is the primary antibody in secretions, such as the mucosal surfaces of the intestine and respiratory tract. IgE antibody is primarily associated with mast cells found in the skin, the mucosa, and connective tissue blood vessels, but it also can be found at low levels in blood or in extracellular fluid. IgG is the most abundant

TABLE 26.1
Approved mAbs

Type	Brand Name	Company	U.S. Approval	Dosage Form	Route	Indication	Expression System	Formulation
Murine	Orthoclone OKT3	Johnson & Johnson	1986	Liquid	IV Injection	Acute kidney transplant rejection	Hybridoma	Monosodium phosphate, dibasic sodium phosphate, sodium chloride, polysorbate 80
	Verluma (Fab)	Boehringer Ingelheim and DuPont Merck	1996	Liquid	IV Injection	Imaging agent for small-cell lung cancer		Phosphate buffer saline
Chimeric	ReoPro	Centocor	1994	Liquid	IV injection and infusion	Reduction of acute blood clot-related complications	SP2/0	Sodium phosphate, sodium chloride, polysorbate 80
	Rituxan	Biogen-IDEC	1997	Liquid	IV infusion	Non-Hodgkin's lymphoma	CHO	Sodium citrate, sodium chloride, polysorbate 80
	Simulect	Novartis	1998	Lyophilized	IV injection and Infusion	Acute kidney transplant rejection	SP2/0	Monobasic potassium phosphate, dibasic sodium phosphate, sodium chloride, sucrose, mannitol, glycine
	Remicade	Centocor	1998	Lyophilized	IV infusion	Rheumatoid arthritis and Crohn's disease	SP2/0	Monobasic sodium phosphate, dibasic sodium phosphate, sucrose, polysorbate 80
	Erbitux	ImClone Systems	2004	Liquid	IV infusion	EGFR-expressing colorectal carcinoma	SP2/0	Monobasic sodium phosphate, dibasic sodium phosphate, sodium chloride
Humanized	Zenapax	Protein Design Labs	1997	Liquid	IV infusion	Prophylaxis of acute organ rejection in patients receiving renal transplants	NSO	Monobasic sodium phosphate, dibasic sodium phosphate, sodium chloride, polysorbate 80
	Synagis	MedImmune/ AstraZeneca	1998	Lyophilized	IM injection	Respiratory syncytial virus (RSV) infection	NSO	Histidine, glycine, mannitol
	Herceptin	Genentech	1998	Lyophilized	IV infusion	Metastatic breast cancer whose tumor over-expresses HER2 protein	CHO	Histidine buffer, trehalose, polysorbate 20

TABLE 26.1 (CONTINUED)
Approved mAbs

Type	Brand Name	Company	U.S. Approval	Dosage Form	Route	Indication	Expression System	Formulation
	Campath	Millennium /ILEX	2001	Liquid	IV infusion	B-cell chronic lymphocytic leukemia	CHO	Dibasic sodium phosphate, monobasic potassium phosphate, sodium chloride, potassium chloride, EDTA, polysorbate 80
	Raptiva	Genentech	2003	Lyophilized	SC	Plaque psoriasis	CHO	Histidine buffer, sucrose, polysorbate 20
	Xolair	Genentech	2003	Lyophilized	SC	Asthma	CHO	Histidine buffer, sucrose, polysorbate 20
	Avastin	Genentech	2004	Liquid	IV infusion	Metastatic carcinoma of colon or rectum	CHO	Monobasic sodium phosphate, dibasic sodium phosphate, trehalose, polysorbate 20
	Lucentis (Fab)	Genentech	2006	Liquid	Intra-vitreal injection	Age-related macular degeneration	*E. coli*	Histidine HCl, Trehalose, Polysorbate 20
	Tysabri	Biogen-IDEC	2004	Liquid	IV infusion	Multiple sclerosis		Dibasic sodium phosphate, monobasic sodium phosphate, sodium chloride, polysorbate 80
Conjugated	Mylotarg (humanized)	Wyeth	2000	Lyophilized	IV infusion	CD33-positive acute myeloid leukemia	NSO	Monobasic sodium phosphate, dibasic sodium phosphate, sodium chloride, dextran 40
	Zevalin (murine)	Biogen-IDEC	2002	Liquid	IV infusion	Follicular or transformed B-cell non-Hodgkin's lymphoma	CHO	Sodium chloride
	CEA-Scan (murine)	Immunomedics	1996	Lyophilized	IV injection or infusion	Imaging agent for colorectal cancer	Mouse ascites	Sucrose, stannous chloride, potassium sodium tartrate, sodium acetate, sodium chloride, glacial acetic acid, hydrochloric acid

Product	Company	Year	Form	Route	Indication	Cell line	Buffer/formulation
ProstaScint (murine)	Cytogen	1996	Liquid	IV injection	Imaging agent for prostate cancer	Hybridoma	Phosphate buffer saline
OncoScint	Cytogen	1992	Liquid	IV injection	Imaging agent for colorectal and ovarian cancer	Hybridoma	Phosphate buffer saline
Bexxar (murine)	Corixa	2003	Liquid	IV infusion	CD20-positive follicular non-Hodgkin's lymphoma	Hybridoma	Phosphate buffer, sodium chloride, maltose, Povidone, ascorbic acid
Humira (Human)	Abbott	2002	Liquid	SC	Rheumatoid arthritis, psoriasis	CHO	Monobasic sodium phosphate, dibasic sodium phosphate, sodium citrate, citric acid, mannitol, sodium chloride, polysorbate 80

Notes: CHO = Chinese hamster ovary (see Urlaub, G. and Chasin, L.A., *Proc. Natl. Acad. Sci. USA*, 77, 4216, 1980); SP2/0 = murine myeloma (see Shulman, M., Wilde, C.D., and Kohler, G., *Nature*, 276, 269, 1978); NSO = murine myeloma (see Barnes, L.M., Bentley, C.M., and Dickson, A.J., *Cytotechnology*, 32, 109, 2004); IV = intravenous; SC = subcutaneous; IM = Intramuscular.

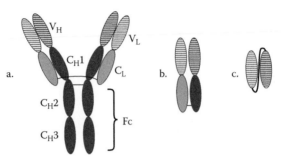

FIGURE 26.1 (See color insert following page 40). (a) The general structure of an IgG antibody. (b) The structure of a Fab fragment containing the variable domains and the C_H1 and C_L constant regions. (c) A single-chain Fv (scFv) fragment. The V_H and V_L domains in a scFv fragment are joined together by a flexible peptide linker encoded by the scFv gene.

antibody in blood plasma and interstitial fluids and has been the antibody isotype of choice for monoclonal therapeutic application[5,6] for several reasons. The IgG subtype has the longest serum half-life of the antibodies classes (half-life can exceed 20 days), and some IgG subclasses or isotypes have Fc effector functions that can help to enhance their therapeutic efficacy.

The IgG antibody is composed of two heavy- and two light-chain proteins. The generic structure is shown in Figure 26.1. The heavy chains and light chains are held together by disulfide bonds. The IgG heavy chain has one variable (V_H) domain, and three constant domains (C_H1, C_H2, and C_H3). The light chain has one variable (V_L) domain and one constant (C_L) domain. The antigen-specific binding affinity is localized in the variable domains within three hypervariable regions called complementarity-determining regions (CDRs). The Fc region consists of the C_H2 and C_H3 domains from two heavy chains (see Figure 26.1). The Fc region has effects on secondary antibody function such as complement binding.

Glycans are attached to antibodies during a multistep maturation process in the Golgi apparatus (a cellular organelle) before the antibody is secreted (see Section 26.6 for additional information). N-linked glycosylation is the most common form. The glycans are linked to the protein backbone via an amide bond of Asparagine at sites in the protein that contain the sequence Asn-X-Ser/Thr. The IgG antibodies have a conserved glycosylation site in the Fc antibody domain.

26.3 DISCOVERY PROCESS FOR THERAPEUTIC MONOCLONAL ANTIBODIES

26.3.1 TARGET AND ANTIBODY IDENTIFICATION

The selection of targets for monoclonal therapeutics represents a significant challenge and requires extensive risk versus benefit analysis. The use of therapeutic mAbs has been limited to cell surface proteins, receptors, and soluble protein targets in blood plasma or interstitial fluids. However, there are engineered antibody fragments in development, called intrabodies, that may penetrate into cells.[7] There are several factors to cell surface target selection. These factors include the presence of the surface target on the appropriate targeted cell lines, and their absence from other healthy cells in the body, and an understanding of the effect of antibody binding on the cell. Antibodies by their nature are divalent and can often activate receptor targets on cells by bringing two receptors in close contact. Depending on the indication, activation of the receptors may or may not be desirable. Many of the targets such as CD20 can have therapeutic applications for several indications such as rheumatoid arthritis and other immunology disorders.

The identification of targets with complete or nearly complete specificity is a key challenge. Herceptin®, used in breast cancer treatment, binds the Her-2 receptor for epidermal growth factor (EGF). This receptor is present only in a subset of breast cancer patients, providing the specificity needed for intervention. However, Her-2 is also expressed in heart tissue, and Herceptin can have cardiac risks. The binding of the mAb to the cancer cell target should prevent tumor growth and, more preferably, kill the tumor cell. Herceptin does both. The binding of the antibody to the Her-2 receptor blocks the growth-promoting effects of EGF on the cell by causing internalization of the receptor, and the binding of Herceptin to Her-2 induces apoptosis or programmed cell death.[8,9] Ideally, a cancer cell target should be present not only on the primary tumor but also on metastasized cells from the primary tumor.

Cell receptors that have been targeted for treatment of various leukemias include CD20 and CD52. The target should be present on the tumor cells but not on hemopoietic stem cells required for the generation of other myeloid and lymphoid blood cells. Ideally, the target antigen should be absent on mature blood cells such as erythrocytes, platelets, and plasma cells. CD20 and CD52 are absent from stem cells. CD20 is present on both pre-B cells and mature B cells, and CD52 is present at high densities on normal lymphocytes. The distribution of these receptors makes them good targets for treating leukemia, but targeting lymphocytes can also have immunosuppressive effects that increase the risk of infections.[10]

Critical factors in antibody choice are choosing the appropriate subclass such as an IgG1, if Fc effector function is desired, or an IgG2 antibody for blocking or neutralizing target function with limited Fc-mediated cell killing, and identifying a monoclonal antibody with the binding specificity and appropriate affinity to the target protein. The anti-CD20 antibody Rituxan® is an IgG1 isotype whose Fc sequence can mediate complement-dependent cytotoxic (CDC) and antibody-dependent cell cytotoxic (ADCC) responses. The ADCC response occurs when antibody binds to target cells followed by neutrophils, macrophages, or natural killer cells binding to the Fc portion of the IgG1 antibody. This directs the macrophage to eliminate the target cell by phagocytosis or the lysis of the target cell by secreted enzymes from the neutrophils or natural killer cells.[11] The CDC response occurs when antibody binds to a cell and triggers the complement cascade. The cascade is comprised of some 26 serum proteins that assemble into a membrane attack complex (MAC) that inserts into the cell's membrane causing osmotic lysis. The CDC and ADCC responses increase the efficacy of Rituxan, Bexxar®, and Zevalin®. Zevalin is the Rituxan antibody labeled with yttrium-90 to generate a chemotoxic affect against targeted lymphocytes. In a similar strategy, Bexxar is labeled with iodine-131.[10] Conjugated antibody can improve efficacy, particularly if the target surface protein is internalized by the cell. Other conjugates are currently under consideration, such as Pseudomonas toxin and cytokines.

26.3.2 MONOCLONAL ANTIBODY GENERATION

26.3.2.1 Historic—Hybridoma

In 1975 Kohler and Milstein[12] showed that B lymphocytes from a mouse spleen could be fused to an immortalized myeloid cell line and grown in culture. This allowed the cloning and isolation of an individual fused cell or hybridoma cell that secretes a unique antibody. An antibody so produced was referred to as a monoclonal antibody. A mouse could be injected with a specific immunogen such as a protein therapeutic target, and B cells producing antibodies against the target could be selected and cultured. Mouse-derived monoclonal antibodies have been approved for human clinical use (i.e., Orthoclone OKT3). However, these antibodies elicit a "human anti-mouse antibody" (HAMA) response that can result in anaphylactic reactions in patients.[13] The antibody response can increase the rate of clearance of the mAb, and murine antibodies have a weak ADCC response in humans.

Improvements have been made in antibody design. Chimeric antibodies have the mouse constant domains exchanged with human domains. Humanized antibodies, in addition to the human constant domain, have modifications in the variable domain making the domain more human-like.[14] It can be challenging to humanize the variable regions while retaining binding affinity. Chimeric and humanized antibodies each provide an additive advantage in reducing the patient's immunogenic

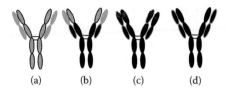

(a)   (b)   (c)   (d)

FIGURE 26.2 (See color insert following page 40). The differences from a murine antibody (a) to a fully human antibody (d). In (b), the constant regions of the murine antibody have been replaced with human constant domains. In (c), the amino acid sequences adjacent to the three hypervariable regions in the variable domains had been humanized by modeling amino acid substitutions that are more commonly found in human antibodies.

response. More recently mice have been constructed to express human antibodies that reduce the challenges involved with constructing chimeric or humanized antibodies (Figure 26.2).

26.3.2.2 Engineered Mice

Engineered mouse strains that produce human antibodies represent a major technological breakthrough that has contributed significantly to the renaissance in mAb therapeutics. The development of mice capable of producing human antibodies requires two major steps. First, the mouse Ig loci must be inactivated resulting in mice incapable of producing mouse Ig antibodies, and second, the human Ig loci must be stably introduced.

The IgH locus encodes the antibody heavy chain, and the Igκ and Igλ loci encode the light chains. Two Ig-deficient mice lines were constructed by inactivation of the murine IgH or the Igκ loci using homologous recombination.[15,16] Cross-breeding of the two mouse lines resulted in a double inactive mouse strain.

Because many human proteins differ from their mouse equivalents, these mouse strains allow for immune responses to be mounted against human proteins. In the situation where the human protein target does not differ from the mouse equivalent or if injection of the target into the mouse is toxic, the "display" approach can offer an alternative method for producing antibodies.

26.3.2.3 Display Technology

The general strategy behind the display libraries is to physically link genotype and phenotype in a way that allows the selection of proteins with a desired property. The protein is "linked" to a DNA sequence that is used to rapidly determine the peptide sequence. The various display methods can provide alternatives for selecting and for modifying human antibodies outside of an animal host. Display libraries of human antibody fragments scFv or Fabs (Figure 26.1) can be created synthetically or derived from human blood cells.[17,18] Display systems can be used to further improve the properties of antibodies through directed molecular evolution consisting of multiple rounds of mutagenesis and selection.

The phage display method was developed in 1985, and it can use single-stranded phages such as fd, f1, and M13, which grow on filamentous strains of *Escherichia coli*.[19] An antibody fragment, such as a scFv, is cloned into a viral vector and translationally fused to the gene III viral surface protein.[20] The sequence of the fused gene can be randomized and the resulting mutant proteins displayed as a phage library. The protein sequence of an individual virus, selected from the library based on its binding properties to an immobilized target ligand, can be determined by DNA sequencing of the viral DNA. Large libraries containing 10^{11} different antibodies have been constructed.[21]

There are several other display technologies. Ribosome display is based upon the capturing of nascent polypeptide and its cognate RNA in ternary complexes ribosome *in vitro* translation reaction.[22] The nascent protein consists of a displayed protein fused to a short C-terminal region that acts to tether the protein while allowing the displayed protein the ability to fold unencumbered by the ribosome. The diversity of the library is generated by degenerative oligomers similar to the phage display libraries. A related method is mRNA display. It relies on the covalent coupling of a RNA–DNA hybrid molecule to its nascent polypeptide.[23]

Yeast display uses the a-agglutinin receptor to display recombinant proteins on the surface of *Saccharomyces cerevisiae*.[24] Yeast display of scFvs or Fabs allows the detection and selection by fluorescence-assisted flow cytometry or by magnetic sorting. In addition, flow cytometry can be used for kinetic characterization of antibody affinity (K_D) as well as K_{off} and K_{on} rates.[25]

Display technologies add a very valuable dimension to the development of therapeutic antibodies derived from human Ig mice or to antibodies developed from synthetic libraries. They provide an alternative for developing therapeutic human antibodies for targets that are either identical or similar to the mouse homologues in humanized transgenic mice.

26.3.3 Screening and Selection

After a target protein has been identified, there are several paths to the development and selection of a lead monoclonal antibody. The paths are generally divided into a whole animal approach such as a humanized mouse or the screening of a combinatorial antibody library using display technologies.

As an example of a whole animal approach, mice carrying human antibody genes are immunized with the target protein. After an appropriate time the spleen is removed and B cells from the spleen are hybridized with cells from established myeloma cell lines. The resultant hybridoma cells are able to grow in culture. Individual hybridoma cell lines are screened for the production of antibody against the target protein. Hybridoma cell lines producing antibodies containing the desired properties are identified and can be used to produce antibody proteins for further study. For mAb candidates for human therapeutic development, molecular cloning of the antibody genes from the hybridoma cell line occurs. Once the antibody genes are cloned, larger quantities of mAb can be produced through transient expression in mammalian cell lines or by the more time-consuming route of developing stable expression cell lines.

An alternative approach is the use of scFv phage display libraries to select a lead antibody. Typically, the target protein is linked to a resin to immobilize the target protein. Phages from the library are then exposed to the resin, allowing those phages from the library to bind the target. The result is an enriched population of phages with affinity to the target. The enriched phage population is expanded by growing the phage, and the selection is repeated several times. During these repetitions, the stringency of the binding conditions can be increased to select for phages with tighter binding affinities. Individual phage candidates are isolated from the library and are subjected to detailed characterization.[26] Lead candidates can be produced in greater quantities in a bacterial expression system as scFv or reengineered Fab fragments. Whole antibodies can be reengineered using the variable regions from the scFv, and expression of these antibodies can be achieved in mammalian cell lines using transient expression techniques or by developing stable expression cell lines as used for the mouse-derived antibodies.

26.4 PROCESS TECHNOLOGY

26.4.1 Cell Line Creation and Cell Bank Testing

26.4.1.1 Expression Systems

To date, commercial antibody expression exclusively uses mammalian cell line expression. These mammalian cell lines have expression cassettes for the antibody heavy and light chain stably integrated into the host cell chromosome. The most commonly used cell lines are derived from Chinese hamster ovary (CHO) cells. About half the approved antibody therapeutics are made in CHO cell lines. The dihydrofolate reductase (dhfr) gene is used as a selectable marker owing to the development of CHO cell lines that are deficient for dhfr genes such as CHO DG44 and CHO

DUKX-B11.[2,27] Expression vectors have been engineered to encode the selectable marker gene, and only those CHO cells that recombine the vector into its genome will survive under the imposed growth conditions. These vectors are also engineered to contain expression units for the heavy and light mAb genes consisting of a promoter, the specific gene, and a polyA transcriptional termination sequence. The engineered vector DNA is capable of autonomous replication in bacteria. After propagation, the vector DNA is isolated from the bacterial culture and is introduced into the mammalian cell line to recombine with the mammalian host cell DNA.

The next most commonly used cell lines are the murine myeloma cell lines such as NS0[4] and Sp2/0.[3] The NS0 cell line is deficient for the glutamine synthetase gene, and the presence of glutamine synthetase (GS) gene on vectors has been used for the selection of stably integrated expression cassettes.[28]

The expression vectors are linearized by cutting with restriction enzymes and introducing the vector into cells by a variety of techniques including electroporation, calcium phosphate, and lipofectin.[29] The linearized vectors are randomly inserted in the chromosome of the host cell. The integration site can have a strong influence on the stability of the insert, and on the expression titers. Vectors are designed to express the selectable markers at a lower level than the antibody chains. This can help in enriching for chromosomal insertions that can support high levels of transcription. (Transcription is the intracellular process by which the DNA is used as a template to encode a primary RNA transcript. The RNA is then translated by the cell into protein.) The antibodies are typically expressed separately with their own promoters and transcriptional termi-nators. The antibody chains can be on the same vector or each can be placed on separate vectors. The selection and amplification of dhfr and GS genes can be augmented by the addition of the dhfr-specific inhibitor methotrexate (MTX) or the GS-specific inhibitor methionine sulfoxamine (MSX), respectively, to help select for cell lines with higher mAb expression.[30]

Cell-specific expression levels range from 10 picogram/cell day (pg/cd) to 60 pg/cd. Volumetric antibody titers have been in the 1 to 2 g/L range, but with improvements in cell line screening, medium development, and cell cultivation development, titers in the 2 to 4 g/L range are becoming more common. Antibody produced in CHO and NS0 have similar glycosylation patterns on the whole (see Section 26.6), but nonhuman glycoforms are also present, and this can be an important factor in cell line selection.[31]

26.4.1.2 Screening of Cell Lines

The transfection of the DNA vector into the mammalian cell line results in a heterogeneous population of cells. This is because the vector DNA integrates into the host chromosomes at random locations in different individual cells. As a result, the expression level of the desired protein can vary dramatically among individual cells. Other important properties, such as stability of expression over numerous generations and the growth properties, can vary among the cells in these populations.

A variety of different strategies can be used to screen pools or individual clones from the transfected cell population. These strategies are designed to initially screen hundreds to thousands of cell lines. Cells from the transfection population are dispersed into 96-well plates. Initial screening of cell lines uses high-throughput assays such as enzyme-linked immunosorbent assay (ELISA). ELISA is an immumological-based assay used in determining the antibody expression level and the preliminary ranking of the cell lines. Cells lines identified from the ELISA screening are scaled up by growth in progressively larger lab-scale plates and flasks. At larger volumes, high-performance liquid chromatography (HPLC) assays can be used to more accurately identify high-productivity cell lines, and growth in flasks can provide the first detailed information on growth profiles and antibody quality. The assays used for initial protein quality include reduced and nonreduced sodium dodecyl sulfate–polyacrylamide gel electrophoresis (SDS-PAGE) to indicate molecular weight of the protein, isoelectrofocusing (IEF) for examining the charge of the protein,

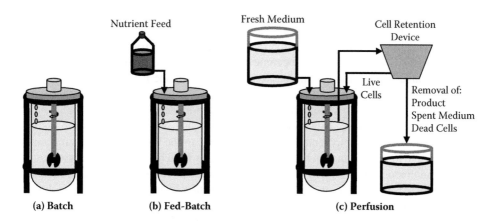

FIGURE 26.3 (See color insert following page 40). Modes of bioreactor operation. (A) Batch bioreactor, (B) fed-batch reactor, and (C) perfusion bioreactor.

and glycosylation analysis. The number of cell lines is reduced to a small number of lead cell lines that are evaluated in bioreactors.

Fluorescence-activated cell sorting (FACS) can be employed to enrich the transfected cell population for high-level expression cells and to clone those individual cells.[32,33] This approach can reduce the effort associated with screening large numbers of cell lines.

26.4.1.3 Cell Banking

Once a final lead clonal cell line has been selected, the cell line is used to create a cell bank. Cell bank generation is performed under good manufacturing practice (GMP) conditions and is the first step to clinical and commercial antibody production. The cell bank is subjected to rigorous testing for sterility (i.e., absence of bacteria, mycoplasma, and fungal contamination), absence of adventitious viruses, and species identity. Other tests can include a battery of molecular tests such as DNA sequencing of the antibody genes and the determination of the number of vector copies in the host's genome.

26.4.2 Upstream Process Development

26.4.2.1 Modes of Upstream Processes

Once a cell line is identified for production of the protein of interest, an upstream (also known as cell cultivation) process must be developed that is robust and scalable. The first industrial process using mammalian cells was the production of the Salk polio virus vaccine in the 1950s. Because cells used by industry early in the development of mammalian production processes required adhesion for proliferation, an important first step in the development of industrial-scale mammalian processes was the development of suitable culture surfaces for attachment. Cells were grown on surfaces in flasks, "cube" bioreactors, roller bottles (large bottles placed horizontally and rotated to maintain liquid contact with cells), or in suspension attached to beads known as microcarriers. However, it was not until the establishment of continuous cell lines, the ability to grow single cells in suspension, and the optimization of media formulations that large-scale mammalian cell culture processes for the production of biotherapeutics, including mAbs, were firmly established. Today, optimization of production processes continues, with suspension or suspension-adapted cell lines in large stirred tanks being the preferred method. The following is a brief synopsis of current bioreactor modes of operation (see also Figure 26.3):

Batch bioreactors—This is the simplest bioreactor process. In this mode, all nutrients needed for cell growth are formulated into the basal medium and added to the production reactor with the cell inoculum at the initiation of the run. The cells grow for a finite period and are harvested after the nutrients become limited and the cell viability decreases to a predetermined level. Although simple to operate, batch mode is least likely to be used as other modes of operation yield higher product titers.

Fed-batch bioreactors—This method begins as a batch mode of operation as cells undergo their initial growth phase. However, run time, cell mass, and product titer are increased with the addition of supplemental nutrients, or feeds, to the bioreactor. These reactors are cost effective to run because higher product titers can be achieved from a single extended reactor run, increasing volumetric productivity. However, volumetric output is fundamentally restricted by the size of the vessel, and formation of toxic cellular metabolites can limit the duration of each reactor run.

Perfusion bioreactors—This mode of bioreactor operation addresses the volumetric output limitations for batch and fed-batch bioreactors through continuous media exchange and toxic metabolite and product removal. Perfusion bioreactors utilize cell separation devices, such as settling tubes, spinner baskets, hollow fiber filters, or acoustical filters, to retain cells within the reactor. Reactor volume stays constant because the volume of spent medium removed is replaced by fresh medium. This symmetry is referred to as the perfusion rate. The design of a cell retention device exploits a physical difference (i.e., particle size) between desired healthy cells and dying cells or cell debris. This continuous feeding, removal of by-products, and removal of dead or dying cells leads to the ability to achieve high steady-state cell concentration and a prolonged production phase (typically >20 days). As such, high volumetric productivity from a relatively small reactor can be achieved when compared to batch and fed-batch processes. Processes utilizing perfusion bioreactors have been successfully commercialized; however, there are a number of drawbacks. The most significant drawbacks include the usage of large volumes of medium, a continuous high-volume harvest stream, and difficulty in lot designations due to the continuous nature of the process. However, using perfusion technology, production rates can be quite high and cost effective due to the extended production phase, reduced capital lay-out for smaller vessels and facilities, and the resulting decreased frequency for breakdown, cleaning, and assembly required to obtain the same volumetric output from a batch or fed-batch process.

26.4.2.2 Medium and Feed Development

Medium and feed development are integral parts of any successful process development program, as cells must be provided sufficient amounts of nutrients so that they can proliferate and yield product at high titer and of consistent quality. During different phases of a bioreactor run (i.e., growth, production), different media formulations may be used to best support either growth or productivity. Early on in the evolution of medium development, supplements such as serum were added to medium to promote growth and viability. As cellular requirements for nutrients became better understood, serum could be replaced by protein-containing substitutes, such as ITS (insulin, transferrin, and selenium) or hydrolysates of animal tissue or proteins. The next step in this evolution came as protein components were replaced by recombinant or nonprotein sources (i.e., soy protein hydrolysates rather than animal hydrolysates, and iron-chelating small molecules rather than the protein iron carrier transferrin).[34] However, there has been increasing pressure from regulatory agencies to completely eliminate all materials of animal origin and create animal component-free (ACF) processes. Today, the gold standard for medium is one that is protein-free, animal component free, and chemically defined. Currently, reagents that make up most chemically defined ACF media include salts (to regulate osmolarity critical for optimal cell growth), amino acids, fatty acids, vitamins, sugars, and nucleic acid precursors.

Basal production media are developed for specific host cell types (i.e., CHO or NS0), and companies that manufacture mAbs can take the approach of utilizing commercially available formulations through media vendors or developing a proprietary formulation. Once a basal medium has been selected, cell line-specific optimization can occur to enhance bioreactor performance. Approaches include ensuring that critical media components are not limiting, and reducing the accumulation of potentially toxic by-products, such as lactate or ammonium, through optimization of carbon and nitrogen sources. After the basal medium composition is determined, then process development can focus on optimizing nutrient feeds.

The nutrient feed is designed to enhance cellular longevity, cell-specific productivity, or product quality. The bioreactor mode of operation (fed batch versus perfusion) will affect the type of feed required, as these reactors are run for very different lengths of time, but the same basic principles are involved. A first approach often employed is the use of a concentrated basal medium as the nutrient feed followed by determination of by-product accumulation and limitation of essential components. The feed composition can then be altered to reduce production of by-products and to maintain essential nutrients.[35,36] Reduction of by-products often involves optimizing the composition and concentration of carbon and nitrogen sources to reduce accumulation of lactate from glucose metabolism and ammonia from degradation of glutamine.[37] Product quality assessments, such as product integrity or glycosylation, are also used to guide feed development. As with basal medium, the gold standard for a feed is animal component-free and chemically defined.

26.4.2.3 Bioreactor Process Development

The design of a bioreactor process is complex due to the many factors that impact bioreactor performance (i.e., temperature, pH, dissolved oxygen tension, dissolved carbon dioxide tension, osmolarity, metabolite accumulation, etc.) and the potential for interaction among these factors (i.e., pH, dissolved CO_2 tension, lactate, and base[38]). In addition, the aspect ratio of the bioreactor, and the type of impeller (flat, pitched, or marine blades) and its position in the tank, and the design of the sparger (which influences the size of sparged air and oxygen bubbles) impact the mixing and aeration within a bioreactor. As the most common reactors used for commercial production are fed-batch stirred tanks, the remainder of this discussion on process development will focus on these reactors.

A critical factor for process development is mixing. Efficient mixing must occur to ensure that nutrients and oxygen are transported to the cells and that the pH and dissolved oxygen are uniform throughout the vessel. The height-to-diameter ratio of the stirred tank (aspect ratio), baffling, and the type of impellers used affect mixing. Ideally, a bioreactor needs both good axial and radial mixing to ensure homogeneity.

Another important aspect of process development is pH control. The basal medium is formulated to contain a buffer compatible with cell growth; the current industry standard is bicarbonate. Bicarbonate is in equilibrium with CO_2 such that bioreactor pH can be lowered by addition of CO_2 and raised by addition of a base (such as NaOH). Particularly at large scale, CO_2 accumulation has been shown to be detrimental to bioreactor performance,[39] and CO_2 levels are lowered by stripping this dissolved gas with sparged nitrogen or air. Because cell growth is dependent on pH, optimization of this parameter allows for maximal cell mass accumulation and increased production of the product of interest.

Dissolved oxygen (DO) is another parameter that can have an effect on cellular growth and productivity through directly influencing the metabolism of the cells. DO levels in the range of 20 to 70% are typical, achieved by sparging with pure oxygen within a controlled oxygen flow-rate range. The flow rate is controlled because high sparge rates can indirectly damage cells through bubble bursts at the air–liquid surface interface[40] or lead to reduced overall mass transfer through buildup of foam at this interface. The impeller agitation rate can also be used to increase the level of dissolved oxygen through increased mass transfer at the air–liquid interface, but the relative contribution to oxygen transfer is reduced in the case of higher-volume bioreactors because the volume-to-surface-area ratio decreases as bioreactor scale increases.

Initial process development efforts usually begin with evaluation of a standard process that is well characterized. During this evaluation, various indicators of bioreactor performance, such as cell mass and productivity, are monitored and then analyzed in order to design further process development studies. Bioreactor parameters optimized often include inoculum cell density, impeller speed, medium pH, nutrient levels, temperature, and so forth.

Bioprocess development in reactors is essential, but it is very labor intensive and costly at traditional bench-top bioreactor scale (typically 1 to 10 L). As a result, scale-down models are used to evaluate some bioreactor parameters. Some models are simple (i.e., shake flasks in the range of 20 to 200 mL), while others are more sophisticated microbioreactor models (0.5 to 30 mL) which include automated fluid transfer (i.e., sampling and nutrient feeding) and noninvasive online monitoring of key parameters (i.e., pH and dissolved oxygen). These sophisticated microbioreactors are being developed and marketed for process development by such companies as BioProcessors. With the higher throughput of these scale-down systems, multifactorial experiments can be run in parallel to shorten timelines for process development.

26.4.2.4 Scalability

Scaling up from a lab-scale bioreactor used for process development to larger pilot- and commercial-scale bioreactors is more complex than multiplying parameters by the scale factor as there are many limitations and interactions of parameters that must be considered. Many guiding principles used for scaling are based on work accomplished in development of microbial processes. A simplified approach is maintaining similar geometry, particularly the aspect ratio (i.e., ratio of vessel height to vessel diameter) of process development and pilot- and commercial-scale bioreactors for mammalian bioreactors. Maintaining similar aspect ratios will ameliorate the effect of scale-up on mass and heat transfer.

Two common approaches biochemical engineers employ in designing scaled-up bioreactor processes are maintaining a constant oxygen-transfer coefficient ($k_L a$), or a constant theoretical power input per unit volume of liquid. One method to determine $k_L a$ uses sulfite oxidation and another uses a technique called gassing out (i.e., monitoring the rate at which oxygen is depleted in an operating bioreactor). The $k_L a$ is determined semi-empirically and is dependent on the geometry of the vessel and the operating conditions of agitation rate and superficial gas velocity. While determining mixing and aeration parameters of a scaled-up process, the biochemical engineer can fine-tune the agitation rate and superficial gas velocity of sparged air or oxygen in order to maintain similar oxygen transfer rates for each scale of the process.

The second method commonly used for bioreactor process scale-up is constant theoretical power input per unit volume. This can be determined using the following formula:

$$\frac{P}{V} = \frac{\rho N_P D^5 n^3}{V}$$

where P = power (W), V = volume (m^3), ρ = density (kg/m^3), N_p = power number, D = diameter of impeller (m), and n = agitation rate (revolutions per second).

The aeration rate will drastically alter the effective density of the liquid in the vessel. Due to this density effect, a fully gassed bioreactor has half the theoretical power input compared to an ungassed vessel. Using the power input/volume determined from process development scale and the bioreactor geometry at large scale, the equivalent large-scale impeller speed can be calculated.

26.4.3 DOWNSTREAM PROCESS DEVELOPMENT

Downstream processing of the conditioned medium produced in bioreactors includes cell separation (or harvest) and protein purification. The goal of downstream process development is to develop a high-yielding, robust, scalable, and reliable process that results in high-purity product. The current industrial practice for the production of pharmaceutical antibodies employs mammalian cell lines. In these systems, the antibody is secreted into the cell culture media with other contaminants. Typical contaminants in the process include DNA, host cell protein (HCP), viruses, aggregated and fragmented product, endotoxins, and residual media components. Separation of the antibody product from these contaminants requires an orthogonal purification process that utilizes various modes of purification. In general, a mAb purification process involves various combinations of filtration and chromatographic steps. Among the key properties of a mAb critical to the design of the optimal process are its size and charge characteristics. Other properties of interest are hydrophobicity and stability. When available, it is important to also note the above properties for major impurities. The first part of this section will describe the various unit operations used in the isolation and purification of mAbs (specifically, IgGs) from mammalian cell processes. Then a sample purification train for mAbs is presented.

26.4.3.1 Unit Operations in Monoclonal Antibody Purification

26.4.3.1.1 Clarification

Clarification, or harvest, is typically the first step in downstream processing. During clarification, the cells, cell debris, and other particulates are removed from the cell culture broth, which contains the product. Clarification can be accomplished by depth filtration or by centrifugation followed by some form of filtration. Tangential flow filtration (TFF), sometimes also referred to as cross-flow filtration (CFF) is also used for clarification. Tangential flow filters are designed either in a flat sheet mode or hollow fiber mode.

The choice of a clarification method is dependent on equipment availability, the cell culture process, including the cell line and cell viability at time of harvest, process economics, and the scale of the process. For relatively small-scale processes (up to 1000 L bioreactor scale), depth filtration may be a more efficient method for clarification. Depth filters are designed with a nominal pore size that allows the liquid to flow through while retaining cells and cell debris. They contain two sections. The first section acts as a pre-filter with the pore size telescoping along the fluid path from larger than nominal to nominal. The second section has constant pore size at the nominal rating of the filter. In some cases, positively charged resins are embedded in the depth filters to provide additional mechanical strength and also some adsorption of negatively charged contaminants such as endotoxin and DNA. Depth filters do not require extensive hardware and are easily set up for operation. Moreover, because they are usually used once and discarded, they do not require cleaning after use. The aforementioned characteristic of the depth filter also makes it less favorable for use in large-scale processes. It is usually not economical to use depth filters at large scales (>5000 L bioreactor), as that typically requires a significant number of filters, which increases the cost per batch of manufacture.

Centrifugation is the preferred clarification method at large scales. Usually, a depth filter is used after the centrifuge to remove any smaller particulates that are not cleared by the centrifuge. In this case, however, a relatively small number of depth filters are needed as the centrifuge clears most of the cell debris.

TFF systems have been used at all process scales. With the appropriate cleaning validation to demonstrate removal of protein and debris from previous use, tangential flow filters can be cleaned and used multiple times. TFF is usually not appropriate for clarification of a very shear-sensitive cell line or a low viability process as the turbulent flow required to recirculate the cells in the system leads to more cell breakage, resulting in release of more impurities in the clarified product.

26.4.3.1.2   Affinity Chromatography

Protein A chromatography is the most widely used affinity chromatography method for purification of monoclonal antibodies (particularly IgG). Protein A is used as the capture step and primarily binds to the Fc region of the mAb. In most cases, the clarified cell culture broth (which is generally near neutral pH) is loaded directly onto the protein A column and the mAb is eluted at low pH (i.e., pH 3.3 to 4.2). Host cell proteins, DNA, endotoxin, and media components flow through the column during the load. Protein A affinity chromatography is very effective in clearing contaminants and typically results in yield and purity greater than 90% and 95%, respectively. However, a small fraction of the protein A ligand leaches into the product and must be removed in subsequent steps. Dynamic binding capacities of mAbs on protein A columns typically range from 20 to 50 g/L of resin.

A drawback in the use of protein A chromatographic media in downstream processing is its high cost (more than ten times the cost of ion exchange media). Considering processing time, the overall cost of downstream processing is generally minimized by packing a smaller protein A column and cycling it multiple times for the purification of a single batch of cell culture. Other options to reduce cost, including the use of protein A mimetics (synthetic forms), have been introduced into the market, but none have attained the popularity of protein A.

26.4.3.1.3   Cation Exchange Chromatography

Cation exchange chromatography resins consist of a negatively charged functional group (typically sulfopropyl [SP], carboxymethyl [CM], or methyl sulfonate [S]) attached to chromatographic media. Positively charged products bind to the cation exchange resin and neutral and negatively charged molecules flow through.

In mAb purification, cation exchange chromatography is used to remove leached protein A, DNA, HCP, and product aggregates. The cation exchange column is typically operated in a pH range of 4.5 to 6.0 and a conductivity of less than 10 mS/cm. Under such conditions, the mAb has a net positive charge and binds to the column. (Most IgG mAbs have an isoelectric point in the range of 7.0 to 9.0.) Leached protein A, product aggregates, and a majority of the HCP have an even higher net positive charge under the aforementioned conditions and bind tighter to the column, whereas the neutral and negatively charged DNA and HCP flow through. The mAb, being the relatively lower positively charged product is eluted first from the column with a high salt step gradient. The tightly bound contaminants are subsequently eluted with an even higher salt buffer in the regeneration phase. Cation exchange chromatography columns can be loaded to a capacity greater than 30 g/L of resin. There is an optimal pH and conductivity at which the maximum dynamic capacity is attained. Recoveries on cation exchange columns are typically 80% or higher.

26.4.3.1.4   Anion Exchange Chromatography

Like cation exchange chromatography, anion exchange chromatography is also a charge-based separation except that negatively charged products bind to the resin while neutral and positively charged products flow through. Anion exchange chromatography resins consist of a positively charged functional group (quaternary ammonium [Q], quaternary amionethyl [QAE] or diethylami-noethyl [DEAE]) attached to chromatographic media.

Anion exchange chromatography is typically used in the flow-through mode in mAb purification. It is typically operated at a pH range of 6.5 to 8.0 and a conductivity of less than 12 mS/cm. The antibody flows through the column while DNA, HCP, and endotoxin bind. The bound contaminants are eluted in the regeneration phase with high salt buffer. As the column operates in the flow-through mode, a significantly higher amount of mAb may be loaded onto the column than in other chromatographic steps. It is, however, important to ensure that the contaminants do not break through (exceed their loading capacity) during the loading step. A typical step yield on an anion exchange is greater than 95%.

26.4.3.1.5　Hydrophobic Interaction Chromatography (HIC)

Hydrophobic interaction chromatography (HIC) consists of immobilized hydrophobic functional groups (such as phenyl, butyl, octyl, etc.) on a chromatographic medium. HIC exploits differences in the hydrophobicity of biomolecules for separation. Proteins bind to hydrophobic columns at relatively high salt concentrations (e.g., 1 M (NH$_4$)$_2$SO$_4$) and are eluted with a low salt buffer. HIC can be used in either the bind and elute mode or the flow-through mode in the purification of antibodies. Using HIC in the bind and elute mode can be challenging because some antibodies precipitate in the high salt buffer required for binding. HIC is effective in clearing aggregates and HCP. The performance of an HIC column is dependent on the type of functional group, the pH, and the type and concentration of the salt used in the process. The interactions and effects of all of these parameters with the mAb are not well understood, and the conditions for the separation must be established experimentally for each process.

26.4.3.1.6　Viral Inactivation and Filtration

Mammalian cell culture broth contains retroviral-like particles, which must be cleared during purification. The chromatographic steps are able to clear some of these viruses, but they are not considered to be robust steps for viral clearance. Hence, there are usually one or two additional steps in the downstream process, designed to specifically clear viruses. Low pH inactivation is most widely used in the purification of mAbs. Low pH inactivation naturally follows the protein A chromatographic step because the product is eluted from the column with a low pH buffer. Typical inactivation pH ranges from pH 3.3 to 3.9 with the preference being at the lower pH provided the mAb is stable under that condition. Inactivation time is usually on the order of 1 h.

Another robust step for removing viruses is nanofiltration using commercially available filters which include the DV series from Pall, the NFP from Millipore, and the Planova™ series from Asahi Kasei.

The capacity of a purification process to clear viruses is demonstrated at a representative small scale using model viruses. The most common model viruses used in this validation study are xenotropic murine leukemia virus (x-MuLV), mouse minute virus (MMV), and reovirus (Reo). The viral clearance capacity of the chromatographic steps, inactivation steps, and the viral filtration step is demonstrated by spiking a known amount of a model virus into the load of each of these unit operations and calculating the efficiency of removal by measuring the remaining viral titer in the product containing fractions.

26.4.3.1.7　Concentration/Diafiltration (Buffer Exchange)

It is sometimes necessary to concentrate a load material or exchange buffers (i.e., diafiltrate) in preparation for the next step during protein purification. Concentrations and diafiltrations are typically done in a cross-flow mode using flat sheet or hollow fiber membranes. The membranes are specified by their molecular weight cutoff (MWCO) and are selected to allow smaller molecular weight component and buffers to go through while retaining the protein. Because antibodies (and in particular IgG) have molecular weights on the order of 150 kDa (or higher for other isotypes), membranes with MWCO of at least 30 to 50 kDa may be used for concentration/diafiltration.

26.4.3.2　Sample Purification Train for Monoclonal Antibodies

Figure 26.4 shows a sample process train for the purification of a monoclonal antibody.[41,42] In this particular scheme, the clarification is accomplished by centrifugation followed by depth filtration (see Section 26.4.3.1.1). The filtered product is then directly loaded onto a protein A column where a majority of the HCP, DNA, and other media components are cleared. The eluted product from the protein A column is then adjusted to a low pH (typically around pH 3.5) with an acid to inactivate viruses in the pool. Upon completion of inactivation (usually 1 h), the pH is adjusted upward (to about pH 4.5 to 6) to be loaded onto a cation exchange chromatography column. Cation exchange chromatography further clears HCP, DNA, aggregates, and other contaminants including

Harvest/Clarification
Centrifugation Followed by Depth Filtration
⇩
Protein A Chromatography
⇩
Low pH Inactivation
⇩
Cation Exchange Chromatography
⇩
Concentration and Diafiltration
⇩
Anion Exchange Chromatography
⇩
Concentration and Diafiltration
(into Formulation Buffer)
⇩
Viral Filtration
⇩
Excipient(s) Addition
(if needed)
⇩
Final Filtration and Dispensing
⇩
Drug Substance

FIGURE 26.4 Sample train for mAb purification.

protein A ligands that leached into the eluted product pool. Because the product on the cation exchanger is eluted at high salt and at relatively low pH, buffer exchange is required to remove the salts and place the product in the optimal buffer (typically at pH 6.5 to 8) for subsequent anion exchange chromatography. The product flows through on the anion exchange column, whereas contaminants such as DNA, HCP, and leached protein A bind to the column. The product from the anion exchange column could then be diafiltered into a formulation buffer and passed through a viral filter to clear viruses. The viral filtration could also be placed after cation exchange chromatography or before anion exchange chromatography. The ideal placement for the viral filter depends on various process requirements including conditions that allow for optimal flux profile. In the scheme shown in Figure 26.4, excipients may then be added to the viral filtered product and the sample dispensed to form drug substance.

26.4.3.3 Downstream Process Scale-Up

Downstream process development is generally done at the laboratory scale and is scaled up for clinical and commercial manufacture. The simplest and most widely used method to scale-up chromatographic columns is to increase the column diameter while maintaining the column height and linear flow rate. This methodology allows for a constant residence time in the column. Filters (i.e., depth, nano- and tangential flow) are scaled up proportionally on a surface-area-to-volume ratio, and the flux (defined as volume of material processed per filter surface area per time $[Lm^{-2}h^{-1}]$) is held constant. In scaling up both chromatographic columns and filters, it is assumed that the make-up of the process material (i.e., product and contaminants concentration) is generally the same at all scales.

TABLE 26.2
Common Degradation Pathways for mAbs

Physical Degradation	Chemical Degradation
Aggregation (soluble and insoluble)[a]	Deamidation
Adsorption	Basic species formation[b]
Denaturation/misfold	Isomerization
	Oxidation[c]
	Deglycosylation
	Disulfide interchange
	C-terminal clipping[d]
	Fragmentation[e]

[a] See Schreiber, G., *Curr. Opin. Struct. Biol.*, 12, 41, 2002.

[b] See Usami, A., et al., *J. Pharm. Biomed. Anal.*, 14, 1122, 1996.

[c] See Kroon, D.J., et al., *Pharm. Res.*, 9, 1386, 1992.

[d] See Harris, R.J., et al., *J. Chromatogr. B Biomed. Sci. Appl.*, 752, 233, 2001.

[e] See Powell, M.F., in *Formulation, Characterization, and Stability of Protein Drugs*, Plenum Press, New York, 1996, pp. 1–140.

26.4.4 FORMULATION OF MONOCLONAL ANTIBODIES

Development of a stable antibody formulation is often challenging due to conflicting requirements of stability, cost of goods, limitations associated with the route of administration, clinical or administration convenience, and lack of robust models to predict long-term stability. One key difference in formulation development of mAbs and other proteins (referred to as NBEs, new biological entities), compared to traditional small molecules (referred to as NCEs, new chemical entities), is focus on physical stability in addition to chemical integrity of the molecule. Table 26.2 summarizes various degradation pathways for mAbs.

The first step in development of the dosage form is preformulation to determine the major degradation pathway and to understand the mechanism of degradation. Oftentimes, the degradation pathway is dependent on the type of stress that is applied (e.g., thermal, light, moisture), and hence, the stress degradation studies may not be predictive of the real-time stability studies. To date, all the commercial mAb formulations are either lyophilized or liquid formulations that are intended to be stored at 2 to 8°C (refrigerated) condition (refer to Table 26.1).

The goal of drug product formulation development is to minimize the various degradation pathways to achieve a minimum shelf life of 2 years at the intended storage condition. Generally, the last step of the mAb production (diafiltration or buffer exchange) is used to formulate the drug product. Drug product manufacturing is then an aseptic fill–finish operation.

To reduce the degradation of drug substance prior to drug product manufacturing, the solution is either stored refrigerated or frozen. Low-temperature treatment may induce aggregation of antibodies. Physical and chemical instability of mAbs during freezing or at low temperature is referred to as 'cold denaturation' and could be due to:

- Freeze concentration (during freezing, as water crystallizes, concentration of mAbs and any formulation stabilizer increases [e.g., 0.15 M sodium chloride in solution reaches 3 M at −10°C]. Buffers may preferentially crystallize leading to pH shifts.)
- Decrease in hydrophobic bond strength at low temperature
- Enhanced concentration of oxygen in unfrozen solution
- Interaction between ice surface and protein molecule

Hence, noncrystallizing sugars (sucrose, trehalose) or polyols (dextran, polyethylene glycol) are added to the formulation to minimize cold denaturation during freeze–thaw or during lyophilization. Also, nonionic surfactants like polysorbates or Pluronics are included to prevent mAb adsorption at the interface of ice, air, or other product contact surfaces. Use of ionic surfactants like SDS is not advisable due to their potential to cause tissue irritation and hemolysis. Preformulation studies are important to identify the pH of maximum stability. For most mAbs, it is in the range of pH 5.0 to 6.5. Common buffers used to control the pH of the solution include phosphate, citrate, histidine, and glycine.

Most of the mAb formulations are either administered via subcutaneous or intravenous route, and hence, the solutions are formulated to be isotonic with blood. Sodium chloride, sugars, and polyols are added to bring the tonicity of the solution to approximately 285 mOsm/kg. Reducing sugars like lactose should not be used in liquid formulation as they can react with side-chain amino groups (especially on lysine) resulting in the brown color adduct (Maillard reaction). Such a reaction will result in loss of positive charges and formation of apparently more acidic species. Sometimes, lactose (reducing sugar), maltose (reducing sugar), and mannitol (crystallizing sugar alcohol) have been added to the lyophilized formulation as bulking agents, but their use should be approached with caution.

The preferred mode to prevent oxidation is to overlay the product with nitrogen or argon. However, dissolved oxygen in the solution and trace metals from the sugars or buffer ingredients could accelerate oxidation. Hence, it may be essential to add antioxidants or chelating agents like ethylene diamine tetraacetic acid (EDTA).

Prior to the selection of formulation, generally a hemocompatibility or local irritation study is performed for drug product intended to be injected either intravenously or subcutaneously, respectively. Route of administration can, in some cases, influence the immunogenicity of the molecule with the subcutaneous (SC) route having higher potential to illicit immune responses than the intravenous (IV) route.

The commercial mAb formulations are summarized in Table 26.1.

26.5 MANUFACTURING CONSIDERATIONS

After a process is developed, it is transferred to a manufacturing site for a production campaign. Due to the high cost of building and maintaining a mammalian product manufacturing facility, contract manufacturers are often employed. Particularly for a start-up biotechnology company, the use of contract manufacturers mitigates the risk of building a facility prior to approval of a product. However, the use of contract manufacturers also translates into less flexibility in scheduling and less flexibility in implementing novel equipment or process elements. Manufacturing scale for mammalian products typically ranges from 500 L (early clinical manufacturing) to 15,000 L (commercial manufacturing) for stirred tank, fed-batch processes.

The timescale for production of mAbs is quite different from small molecules and from recombinant protein products made in prokaryotic cells (i.e., *E. coli*), largely due to the slower growth rate of mammalian cells. An important and time-consuming part of the manufacturing process is the inoculum train, or scaling up the cell mass needed to inoculate a production reactor. The starting point is typically a 1 mL vial containing on the order of 1 to 10 million cells. Considering typical cell doubling times are in the range of 22 to 30 h, scale-up to a 15,000-L production vessel can take 3 to 4 weeks. Improvements in fed-batch processes have resulted in run times up to 3 weeks (although 2 weeks or even shorter is more typical), and perfusion processes run from several weeks to months. A multistep downstream process can add another week to the timeline. All together, a single 500-L fed-batch run, from vial thaw to drug substance, can exceed 6 weeks in duration. Drug product manufacture (including final formulation and vialing) from drug substance adds another week to the timeline.

Overall project timelines starting with discovery of the mAb molecule (i.e., DNA cloned for transfection into a host cell line) to start of clinical trials can vary, but the industry standard is in the range of 18 to 24 months. Some biotechnology companies, relying heavily on platform technology (i.e., generic technology applied to multiple projects), claim timelines of 9 to 12 months. Regardless of the duration, there are four main steps in this process:

- Create a recombinant cell line producing mAb product with desired bioreactor performance characteristics and product quality, and then manufacture and test a master cell bank of this cell line.
- Develop upstream and downstream processes and determine formulation concurrent with safety testing of the cell bank.
- Manufacture the drug substance and drug product.
- Conduct drug product characterization, safety testing, and initial stability studies.

It is common for a process to undergo changes as a mAb moves through development, from preclinical manufacture through commercialization. Some of these changes are made to meet more stringent regulatory requirements (as a product approaches commercialization) and to allow for a higher producing and a more efficient and robust process to minimize the cost of goods. Modifications that may be implemented in the process during development include a change in cell line (e.g., from nonclonal to clonal cell line), culture medium (e.g., from chemically undefined to chemically defined medium), bioreactor operating parameters (e.g., pH, temperature, feeding rate), chromatographic media and column operating parameters, manufacturing scale and site, and formulation. A number of these changes have the potential to impact post-translational modifications, which could in turn impact activity and safety. Hence, product comparability must be established.

The first part of a comparability strategy is focused on analytical comparability, which entails the characterization and purity information, examples of which are shown in Table 26.3. Depending on the degree of analytical comparability, preclinical studies and pharmacokinetics and safety studies in humans may be needed. The complete strategy is typically compiled in a comparability plan document, which is provided to the regulatory agency for feedback.

Before a manufacturing campaign begins, technology transfer occurs. Because the most time-consuming step is the scale-up of cell mass (inoculum train), technology transfer includes characteristics of cell growth in smaller vessels (i.e., t-flasks, shake flasks, spinner flasks, or small volume bioreactors), in order to efficiently execute an inoculum train. Other key elements of the technology transfer are bioreactor parameters, specifications for raw materials (including culture medium, nutrient feeds, resins, buffers, and final formulation components), chromatography parameters, unit operations designed to provide viral inactivation and viral clearance, and storage requirements (vessel type and environmental conditions) for drug substance and drug product.

Once a manufacturing campaign is started, the early phases (inoculum train) can be executed with a once per day sampling to determine cell growth (i.e., cell concentration and viability) and appropriate timing for transfer to a larger vessel to maintain the cultures in an optimal cell density range. An ideal bioreactor process also involves once a day sampling and operating systems check. Post-harvest, with harvest defined as separation of the product containing cell culture supernatant from the cell mass, downstream processing can be run on a shift basis (with predetermined overnight hold steps), or on a continuous basis. This decision is often driven by equipment, facility, and staff scheduling constraints rather than scientific needs.

26.5.1 MANUFACTURING CAPACITY

During the late 1990s, there was a prediction of a significant shortage of manufacturing capacity. This was based on the wealth of the industry's pipeline at that time and the lack of capacity. However, as of late 2006, this shortfall has not materialized. Whether this has been a result of

TABLE 26.3
Typical Characterization Analysis for mAbs

Identity

Isoelectric focusing (IEF)
Peptide map

Characteristics

Appearance (visual)
Particulate count
Protein concentration
pH

Purity

SDS-PAGE (reduced)
SDS-PAGE (nonreduced)
Size exclusion chromatography (for monomer content)
Carbohydrate profile
Deamidated species (%)
Oxidated species (%)

Impurities

DNA
Host cell protein
Leached ligands from chromatographic column (i.e., protein A)
Other process-related impurities

Activity

Binding ELISA

Safety

Endotoxin
Sterility

attrition in the pipeline or new manufacturing capacity coming online is unclear. What does the future hold? The number of products in the pipeline is still large, and while the industry is increasing capacity, is unknown whether this will be enough. Factors that will influence material requirements include dosing and market size, which are predictable. However, intangibles such as attrition of products from the pipeline are not. Detailed tracking of clinical trials will help predict whether a shortfall will indeed occur within the next 5 years, but at this point, concern over manufacturing capacity seems to have waned. Also, if major process improvements occur, this will further alleviate concerns regarding shortages in capacity.

26.5.2 PRODUCT CHARACTERIZATION AND SPECIFICATIONS

Unlike small molecules, proteins (including mAbs) are heterogeneous in nature. Post-translational modifications such as glycsoylation, along with deamidation, and oxidation contribute to heterogeneity. Any of these changes could impact biological activity and safety; therefore, extensive characterization of the product is required to ensure product consistency. Table 26.3 shows typical characterization analysis for a mAb product specification. Acceptance criteria for these tests progress from broader in the early phases of development, when process and product knowledge is minimal, to tighter specifications as the product approaches phase III.

26.6 GLYCOSYLATION

Mammalian cell culture-derived proteins undergo post-translational modifications within the cell before they are secreted into the culture medium. A major posttranslational modification that can impact half-life, bioavailability, and even immunogenicity of therapeutic proteins is glyscoyslation.[48–51] Due to these potential effects, glycosylation is carefully monitored during cell line selection and process and formulation development.

All of the biotherapeutic proteins produced in mammalian systems are glycosylated, having sugars covalently attached to their amino acid backbone. This attachment can occur via two types of linkages, O-linked (sugars attached to a Serine, Threonine, or hydroxyllysine residue in collagen) and N-linked (sugars attached to an asparagine [Asn] residue; consensus sequence for attachment is Asparagines-X-Serine/Threonine where *X* is any amino acid other than Proline).

For reasons such as safety and scalability, nonhuman mammalian cell lines, such as NS0 and CHO cells, are generally used for mAb production. Although these are mammalian cells, they can produce glycoforms that are slightly different from those seen in humans. Glycoforms are monitored and controlled as a process characteristic during development.

26.7 FUTURE DIRECTIONS

Monoclonal antibodies have demonstrated therapeutic value and have established themselves as essential components of the pharmaceutical armamentarium. The large number of mAbs in clinical trials and in development is a testament to the important roles that this class of molecule will play in the future. However, opportunities exist for improving some of the drawbacks associated with therapeutic mAbs as compared with small molecule therapeutics, including size of the molecule, cost to manufacture, and delivery systems.

A variety of efforts are underway to exploit the antibody scaffold through modifications in the intact IgG antibody structure or through the development of antibody fragments. Alterations in the amino acid sequence and the glycosylation patterns of the antibody Fc region have led to more effective ADCC and CDC responses as well as improved serum half-lives. The Fc function on intact mAbs is not always desirable and the large size of a mAb is not always preferable because it limits tissue penetration. A therapeutic antibody fragment, the anti-VEGF Fab molecule Lucentis for macular degeneration has recently been approved. Nanobodies based on the single variable domain antibodies found in camelids and cartilaginous fish are being exploited due to their small size and unique binding properties. Intrabodies may permit targeting of a protein within the cell. Bivalent antibody fragments have applications for both therapy and imaging. A general limitation of antibody fragments is their high rate of clearance from the body. This can be overcome in part by strategies such as protein PEGylation (addition of large polymers to the molecule to increase the molecular weight). The evolution of antibody and antibody fragments in the laboratory is ongoing, and we can be certain that there will be a range of highly diversified protein classes available for a variety of applications derived from the antibody scaffold. The evolution of protein therapeutics will, in turn, impact the way in which proteins are produced by both cell-based and cell-free technology.

The production of mAbs in mammalian cells is an expensive process due to innate costs associated with mammalian cell culture, downstream isolation, and regulatory requirements. Given the current number of monoclonal antibodies in development using current cell culture technology platforms, these platforms will dominate the manufacturing landscape in the foreseeable future. However, several companies have reported significant improvements in expression titers that can potentially help reduce costs provided the product quality is maintained and manufacturing cycle times are not significantly increased. Cell line engineering can achieve improved product quality by altering glycosylation patterns, decreasing development time, and increasing product yields (i.e., through reduction of apoptosis [cell death]). An active area of research by most companies is timeline reduction. This has two advantages: getting a product to market quicker, which will increase the revenue stream for the producing company, and being able to get to key decision points quicker, which will help preserve capital if compounds are found to not be efficacious in early clinical trails. Thus, streamlining and increasing speed do have the potential for reducing overall costs to bring mAbs to market.

Nevertheless, the high cost of mammalian production has generated interest in less costly expression system alternatives. These systems include yeast, algae, moss, plants, and transgenic animals. Although yields from these sources can be very high, concern centers on the potential for adventitious agents or immunogenicity of nonmammalian carbohydrate structures. In general, these systems are not sufficiently developed to compete with mammalian cell culture, but they may provide niche opportunities, such as for antibodies that are difficult to express in mammalian cells. No marketed products are currently produced in these alternative hosts, but the potential for cost benefits coupled with advanced technology to manipulate carbohydrate structures of mAbs may result in increased potential for these alternative systems. Antibody fragments have been successfully expressed in bacterial systems such as *E. coli*, and a significant amount of work continues to be applied to improving antibody fragment expression in *E. coli*.[52] Improvements in alternative expression systems continue at a rapid pace, and in a short time some of these systems may emerge as serious competitors for mammalian-based mAb production.

In addition to development of modified mAb molecules and new expression systems, improved delivery systems for mAbs are under development. Currently, all the approved mAbs are delivered via parenteral (IV, IM, or SC) route due to large dose (hundreds of milligrams) and poor absorption from the gastrointestinal tract (caused by low pH degradation in the stomach and enzymatic cleavage). In order to improve convenience and assure better patient compliance, chronically administered mAbs could be administered at home via a prefilled syringe or cartridge system. A prefilled administration system reduces the manipulations on part of the patient (which could be difficult, for example, for arthritic individuals) and allows better control of sterility. Figure 26.5 shows several devices that improve self-administration and reduce accidental needle stick. These include the following:

1. Prefilled syringe system that could be coupled with an autoinjector
2. Dual-chamber cartridge system for the lyophilized product and pen devices
3. Needle-free systems for injection

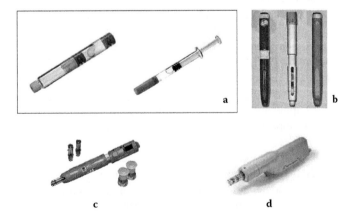

FIGURE 26.5 (See color insert following page 40). Current and future trends in delivery systems for mAbs include (a) dual-chamber cartridge and prefilled syringe with attached needle (courtesy Vetter), (b) pen systems for self-administration, and (c, d) needle-free injection systems (BioJect).

The ultimate goal for the formulation scientist will be to provide a delivery mechanism that is noninvasive. Work is currently ongoing to evaluate transdermal (through skin), oral, and inhalation (pulmonary) delivery, although success at this point in time has been limited.

Several options are being investigated to increase the transport of mAb across the skin—iontophoresis (application of direct current to drive charged drugs), poration technologies (high-frequency pulses of energy to temporarily disrupt the stratum corneum), radio-frequency micro-ablation, mechanical disruption of the skin using microfabricated needles/projections, thermal microporation (application of thermal energy to the skin surface), and ultrasound (also called sonophoresis or phonophoresis). One of the limitations of the technology is that only tens of milligrams of drug could be practically delivered. All these techniques can be combined with chemical permeation enhancers to further improve drug transport.

Pulmonary delivery via the lung is an attractive option due to the large surface area of the alveoli. Dry powder inhalers (DPIs) and metered dose inhalers (MDIs) use formulated and stabilized protein powders and solutions, respectively, and deliver particles in the size range of 1 to 10 μm. Again, poor bioavailability (due to the mucosal barrier) and the ability to deliver only low (milligram) doses of proteins are limitations of this route of administration. These limitations can be overcome by an oral route where large doses can be easily taken by the patient. Current approaches to enhance oral bioavailability of mAbs include conjugating the mAb with molecules that are actively transported across the gut (e.g., transferrin, polyarginine), coadministering the drug with protease inhibitors (e.g., aproteinin, leupeptin), enteric coating the drug so that it escapes exposure to low pH in the stomach, encapsulating mAb in liposomes and cyclodextrin to prevent enzymatic cleavage, adding absorption enhancers or epithelial cell tight junction modulators to allow permeation through the gut wall, or a combination of the above-mentioned approaches.[53]

The technology for therapeutic antibodies has matured since the first murine antibody was approved in 1986. Industry has addressed the initial challenge of the HAMA response by developing methods for identification and development of human antibodies. Now industry is turning toward the business challenges of reducing development costs and timelines, and ultimately, the cost of goods of approved products. Complementing this approach is the scientific development of improved cell lines, expression systems, and processes that will continue to raise the bar for what defines an acceptable commercial process.

ACKNOWLEDGMENTS

The authors would like to thank Dr. Gerald Casperson for review of the manuscript, and Mark Berge, Paul Hermes, and Dr. Sybille Galosy for technical contributions.

REFERENCES

1. Pavlou, A.K. and Belsey, M.J., The therapeutic antibodies market to 2008, *European Journal of Pharmaceutics and Biopharmaceutics*, 59, 389, 2005.
2. Urlaub, G. and Chasin, L.A., Isolation of Chinese hamster cell mutants deficient in dihydrofolate reductase activity, *Proc. Natl. Acad. Sci. USA*, 77, 4216, 1980.
3. Shulman, M., Wilde, C.D., and Kohler, G., A better cell line for making hybridomas secreting specific antibodies, *Nature*, 276, 269, 1978.
4. Barnes, L.M., Bentley, C.M., and Dickson, A.J., Advances in animal cell recombinant protein production: GS-NS0 expression system, *Cytotechnology*, 32, 109, 2000.
5. Goldsby, R.A. et al., *Immunology*, 5th ed., W.H. Freeman, San Francisco, 2002.
6. Manis, J.P., Tian, M., and Alt, F.W., Mechanism and control of class-switch recombination. *Trends Immunol.*, 23, 31, 2002.
7. Stocks, M.R., Intrabodies: production and promise, *Drug Discov. Today*, 9, 960, 2004.
8. Leyland-Jones, B., Trastuzumab: hopes and realities, *Lancet Oncol.*, 3, 127, 2002.

9. Menard, S., et al., Biologic and therapeutic role of HER2 in cancer, *Oncogene*, 22, 6570, 2003.

10. Countouriotis, A., Moore, T.B., and Sakamoto, K.M., Cell surface antigen and molecular targeting in the treatment of hematologic malignancies, *Stem Cells*, 20, 215, 2002.

11. Presta, L.G., Engineering antibodies for therapy, *Curr. Pharm. Biotechnol.*, 3, 237, 2002.

12. Kohler, G. and Milstein, C., Continuous cultures of fused cells secreting antibody of predefined specificity. *Nature*, 256, 495, 1975.

13. Mirick, G.R., et al., A review of human anti-globulin antibody (HAGA, HAMA, HACA, HAHA) responses to monoclonal antibodies. Not four letter words, *Q. J. Nucl. Med. Mol. Imaging*, 48, 251, 2004.

14. Clark, M., Antibody humanization: a case of the "Emperor's new clothes"? *Immunol. Today*, 21, 397, 2000.

15. Jakobovits, A., et al., Analysis of homozygous mutant chimeric mice: deletion of the immunoglobulin heavy-chain joining region blocks B-cell development and antibody production. *Proc. Natl. Acad. Sci. USA*, 90, 2551, 1993.

16. Green, L.L. and Jakobovits, A., Regulation of B cell development by variable gene complexity in mice reconstituted with human immunoglobulin yeast artificial chromosomes, *J. Exp. Med.*, 188, 483, 1998.

17. Knappik, et al., Fully synthetic human combinatorial antibody libraries (HuCAL) based on modular consensus frameworks and CDRs randomized with trinucleotides, *J. Mol. Biol.*, 296, 57, 2000.

18. Amersdorfer, P. and Marks, J.D., Phage libraries for generation of anti-botulinum scFv antibodies, *Methods Mol. Biol.*, 145, 219, 2000.

19. Smith, G.P., Filamentous fusion phage: novel expression vectors that display cloned antigens on the virion surface, *Science*, 228, 1315, 1985.

20. McCafferty, J., et al., Phage antibodies: filamentous phage displaying antibody variable domains. *Nature*, 348, 552, 1990.

21. Vaughan, T.J., et al., Human antibodies with sub-nanomolar affinities isolated from a large non-immunized phage display library, *Nat. Biotechnol.*, 14, 309, 1996.

22. Hanes, J. and Pluckthun, A., *In vitro* selection and evolution of functional proteins by using ribosome display, *Proc. Natl. Acad. Sci. USA*, 94, 4937, 1997.

23. Roberts, R.W. and Szostak, J.W., RNA-peptide fusions for the *in vitro* selection of peptides and proteins, *Proc. Natl. Acad. Sci. USA*, 94, 12297, 1997.

24. Boder, E.T. and Wittrup, K.D., Yeast surface display for screening combinatorial polypeptide libraries. *Nat. Biotechnol.*, 15, 553, 1997.

25. Feldhaus, M.J. and Siegel, R.W., Yeast display of antibody fragments: a discovery and characterization platform. *J. Immunol. Methods*, 290, 69, 2004.

26. Penichet, M.L. and Morrison, S.L., Design and engineering human forms of monoclonal antibodies, *Drug Dev. Res.*, 61, 121, 2004.

27. Urlaub, G., et al., Deletion of the diploid dihydrofolate reductase locus from cultured mammalian cells, *Cell*, 33, 405, 1983.

28. Cockett, M.I., Bebbington, C.R., and Yarranton, G.T., High level expression of tissue inhibitor of metalloproteinases in Chinese hamster ovary cells using glutamine synthetase gene amplification. *Biotechnology (NY)*, 8, 662, 1990.

29. Norton, P.A. and Pachuk, C.J., Methods for DNA introduction into mammalian cells, in *Gene Transfer and Expression in Mammalian Cells*, Makrides, S.C., Ed., Elsevier Science BV, Amsterdam, 2003, pp. 265–277.

30. Wurm, F.M., Production of recombinant protein therapeutics in cultivated mammalian cells, *Nat. Biotechnol.*, 22, 1393, 2004.

31. Jefferis, R., Glycosylation of recombinant antibody therapeutics. *Biotechnol. Prog.*, 21, 11, 2005.

32. Marder, et al., Selective cloning of hybridoma cells for enhanced immunoglobulin production using flow cytometric cell sorting and automated laser nephelometry, *Cytometry*, 11, 498, 1990.

33. Bohm, E., et al., Screening for improved cell performance—selection of subclones with altered production kinetics or improved stability by cell sorting, *Biotech. and Bioeng.*, 88, 699, 2004.

34. Keen, M.J. and Hale, C., The use of serum-free medium for the production of functionally active humanized monoclonal antibody from NS0 mouse myeloma cells engineered using glutamine synthetase as a selectable marker, *Cytotechnology*, 18, 207, 1995.

35. Bibila, T.A. and Robinson, D.K., In pursuit of the optimal fed-batch process for monoclonal antibody production, *Biotechnol. Prog.*, 11, 1, 1995.

36. Sauer, P.W., et al., A high-yielding generic fed-batch cell culture process for production of recombinant antibodies, *Biotechnol. Bioeng.*, 67, 585, 2000.

37. Altamirano, C., et al., Considerations on the lactate consumption by CHO cells in the presence of galactose, *J. Biotechnol.*, 125, 547, 2006.

38. Gramer, M.J. and Ogorzalek, T., A semi-empirical mathematical model useful for describing the relationship between carbon dioxide, pH, lactate, and base in a cell culture process, presented at Cell Culture Engineering X, Whistler, BC, Canada, April 23–28, 2006.

39. Mostafa, S.S. and Gu, X., Strategies for improved dCO_2 removal in large-scale fed-batch cultures, *Biotechnol. Prog.*, 19, 45, 2003.

40. Ma, N., et al., Quantitative studies of cell-bubble interactions and cell damage at different pluronic F-68 and cell concentrations, *Biotechnol. Prog.*, 20, 1183, 2004.

41. Better, M.D. and Horowitz, A.H., Human engineered antibodies to Ep-CAM, US20050009097.

42. Shadle, P.J., et al., Antibody purification, US5429746.

43. Schreiber, G., Kinetic studies of protein–protein interactions, *Curr. Opin. Struct. Biol.*, 12, 41, 2002.

44. Usami, A., et al., The effect of pH, hydrogen peroxide and temperature on the stability of human monoclonal antibody, *J. Pharm. Biomed. Anal.*, 14, 1122, 1996.

45. Kroon, D.J., et al., Identification of sites of degradation in a therapeutic monoclonal antibody by peptide mapping, *Pharm. Res.*, 9, 1386, 1992.

46. Harris, R.J., et al., Identification of multiple sources of charge heterogeneity in a recombinant antibody, *J. Chromatogr. B Biomed. Sci. Appl.*, 752, 233, 2001.

47. Powell, M.F., A compendium and hydropathy/flexibility analysis of common reactive sites in proteins: reactivity at Asn, Asp, Gln, and Met motifs in neutral pH solution, in *Formulation, Characterization, and Stability of Protein Drugs*, Pearlman, R. and Wang, Y.J., Eds., Plenum Press, New York, 1996, pp. 1–140.

48. Chui, D., et al., Genetic remodeling of protein glycosylation *in vivo* induces autoimmune desease, *Proc. Natl. Acad. Sci. USA*, 8, 662, 2001.

49. Jefferis, R. Glycosylation of human IgG antibodies, *BioPharm Int.* 14, 19, 2001.

50. Pendley, C., Schantz, A., and Wagner, C., Immunogenicity of therapeutic monoclonal antibodies, *Curr. Opinion in Mol. Ther.*, 5, 1142, 2003.

51. Raju, T.S., Glycosylation variations with expression systems, *Bioprocess Int.* 1, 44, 2003.

52. Holliger, P. and Hudson, P.J., Engineered antibody fragments and the rise of single domains, *Nat. Biotechnol.*, 23, 1126, 2005.

53. Morishita, M. and Peppas, N.A., Is the oral route possible for peptide and protein drug delivery? *Drug Discovery Today*, 11, 905, 2006.

27 Reaction Progress Kinetic Analysis: A Powerful Methodology for Streamlining Pharmaceutical Reaction Steps

Natalia Zotova, Suju P. Mathew, Hiroshi Iwamura, and Donna G. Blackmond

CONTENTS

27.1 INTRODUCTION

Kinetic studies of complex organic reactions figure prominently in both academic and industrial research. In academia, a proposed multistep catalytic reaction mechanism may be tested by comparing experimentally observed kinetic behavior to that predicted by a mechanistic model. Insights from these kinetic investigations may be used to design further experiments aimed at refining the mechanistic analysis. In pharmaceutical research, the aim of kinetic studies is more practical; obtaining information about the driving forces of a reaction, as well as about its robustness, are important parameters for scaling reactions from the laboratory through to the pilot plant and on to commercial production.

One of the most important developments in pharmaceutical process research of the past several decades has been the introduction of bench-scale experimental tools for efficient and accurate *in situ* monitoring of chemical reactions. Such tools include reaction calorimetry, which measures a property directly proportional to reaction rate (a "differential" method), and spectroscopic methods such as Fourier transform infrared (FTIR) spectroscopy, which measures a property proportional to concentration, or the integral of reaction rate. The temporal profile of a reaction's progress can provide invaluable information, and the development of such experimental protocols has received significant recent attention in the context of "Quality by Design" initiatives for streamlining regulatory procedures in pharmaceutical manufacture.

Interestingly, even with significant advances in experimental and computational tools, kinetic analysis in the 21st century is carried out largely by the same methods used nearly a century ago. These methods focus on linearizing rate equation in order to extract kinetic parameters from the slope and intercepts of the resulting straight line. Reactions involving two or more different substrate concentrations are problematic for this approach, however, and typically must be analyzed in sequence, holding one concentration constant while probing the concentration dependence of the other. This classical kinetic methodology is time consuming, involving a large number of repetitive experiments, and in most cases the reactions are necessarily carried out under concentration conditions far removed from those of practical operation.

We focus here on the new methodology of reaction progress kinetic analysis,[1] an approach that addresses these issues. Reaction progress kinetic analysis streamlines kinetic studies by exploiting the extensive data sets available from accurate *in situ* monitoring of global reaction progress under synthetically relevant conditions, where two or more concentration variables are changing simultaneously in the same manner that they are expected to change during practical operating conditions. This methodology involves the graphical manipulation of a critical minimum set of carefully designed experiments that permits rapid extraction of key information about the reaction's driving forces and its robustness. The advantage of this approach is that this kinetic information may be rapidly obtained even in earliest process research studies and requires little mathematical prowess and no specialized computer programs. These studies may also help inform the direction of further reaction optimization to finalize the synthetic route and prepare for efficient scale-up as well as in further studies to seek fundamental mechanistic understanding.

27.2 KINETIC STUDIES USING *IN SITU* TOOLS

Our experimental technique of choice in many cases is reaction calorimetry.[2] This technique relies on the accurate measurement of the heat evolved or consumed when chemical transformations occur. Consider a catalytic reaction proceeding in the absence of side reactions or other thermal effects. The energy characteristic of the transformation—the heat of reaction, ΔH_{rxn}—is manifested each time a substrate molecule is converted to a product molecule. This thermodynamic quantity serves as the proportionality constant between the heat evolved and the reaction rate (Equation 27.1).

$$q = \Delta H_{rxn} \cdot \left(\frac{reaction}{volume}\right) \cdot rate \qquad (27.1)$$

The quantity of heat evolved up to at any given time during the reaction may be divided by the total heat evolved when all the molecules have been converted to give the fractional heat evolution. When the reaction under study is the predominant source of heat flow, the fractional heat evolution at any point in time is identical to the fraction conversion of the limiting substrate (Equation 27.2). Fraction conversion is then related to the concentration of the limiting substrate via Equation 27.3.

$$fraction\ conversion = f = f_{final} \cdot \frac{\displaystyle\int_0^t q(t)dt}{\displaystyle\int_0^{t(final)} q(t)dt} \qquad (27.2)$$

$$[substrate] = [substrate]_0 \cdot (1 - f) \qquad (27.3)$$

First and foremost in any kinetic study using reaction calorimetry, we must confirm the validity of the method for the system under study by showing that Equation 27.2 holds. Comparing the temporal fraction conversion obtained from the heat flow measurement with that measured by an independently verified measurement technique, such as chromatographic sample analysis or FTIR or nuclear magnetic resonance (NMR) spectroscopy, confirms the use of the calorimetric method.

It is important to point out that the methodology described in this work is not limited to studies using reaction calorimetry but may be employed using any accurate *in situ* experimental technique. These include spectroscopic methods such as FTIR, Raman, and ultraviolet/visible (UV-Vis) spectroscopy.

In our illustration of the graphical manipulations of data using reaction progress kinetic analysis, we will make use of the example of a model reaction, the intermolecular aldol reaction between acetone **1** and aldehyde **2** to form the aldol addition product **3**, mediated by proline **4**, as shown in Scheme 27.1. The demonstration by List, Lerner, and Barbas[3] in 2000 that proline mediates intermolecular aldol reactions with a high degree of asymmetric induction heralded a revolution in the field of organocatalysis, encompassing the discovery of new catalysts and new catalytic transformations.[4]

Figure 27.1 shows how we have employed two techniques to monitor reaction progress in this case. Figure 27.1a and Figure 27.1b show the raw data from the differential (reaction calorimetry: heat flow versus time) and integral (FTIR spectroscopy: absorbance versus time) techniques, and Figure 27.1c illustrates how we convert these data to what we term a "graphical rate equation,"[1] plotting reaction rate versus substrate concentration.

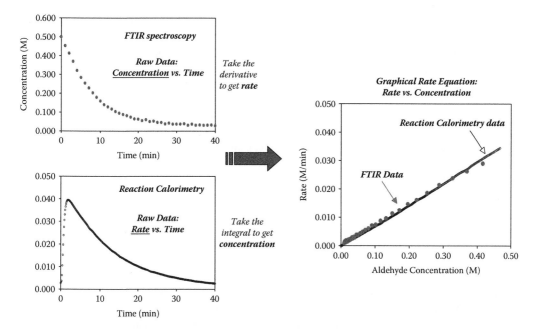

FIGURE 27.1 (See color insert following page 40). Monitoring reaction progress in the proline-mediated aldol reaction of Scheme 27.1 by Fourier transform infrared (FTIR) spectroscopy and reaction calorimetry. The data are manipulated in each case as illustrated in a "graphical rate equation" plotting rate versus substrate concentration for the limiting substrate.

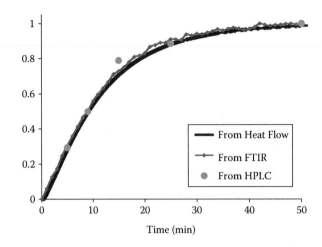

FIGURE 27.2 (See color insert following page 40). Comparison of conversion versus time for the reaction of Scheme 27.1 using high-performance liquid chromatography (HPLC) sampling of product concentration to *in situ* monitoring by Fourier transform infrared (FTIR) spectroscopy and reaction calorimetry.

The agreement between the two *in situ* methods confirms that each gives a valid representation of the reaction progress. We may also wish to compare our *in situ* method to gas chromatography (GC) or high-performance liquid chromatography (HPLC) analysis of conversion as a function of time to provide further validation of the techniques, as shown in Figure 27.2. Sample analysis is typically the most common method of validating *in situ* methodology; even though the number of sample data points available in a sampling study alone is rarely sufficient to carry out the graphical manipulations of reaction progress kinetic analysis, the comparison of Figure 27.2 provides assurance that the data provided by the *in situ* method are valid for this treatment.

The terminology "graphical rate equation" derives from our attempt to relate rate behavior to the reaction's concentration dependences in plots constructed from *in situ* data. Reaction rate laws may be developed for complex organic reactions via detailed mechanistic studies, and indeed much of the research in our group has this aim in mind. In pharmaceutical process research and development, however, it is rare that detailed mechanistic understanding accompanies a new transformation early in the research timeline. Knowledge of the concentration dependences, or reaction driving forces, is required for efficient scale-up even in the absence of mechanistic information. We typically describe the reaction rate in terms of a simplified power law form, as shown in Equation 27.4 for the reaction of Scheme 27.1, even in cases where we do not have sufficient information to relate the kinetic orders to a mechanistic scheme.

$$rate = k \cdot [\mathbf{1}]^x \cdot [\mathbf{2}]^y \cdot [\mathbf{4}]^z \qquad (27.4)$$

Our task is to find values for x, y, and z that hold over the range of concentrations under which the reaction will be carried out. An important further consideration for catalytic reactions is the question of whether the catalyst concentration remains constant over the course of the reaction. The methodology of reaction progress kinetic analysis offers experimental protocols as described below that allow us to probe catalyst robustness as well as substrate dependences.

The key to reaction progress kinetic analysis lies in making use of the reaction stoichiometry. For the aldol reaction of Scheme 27.1, a mass balance tells us that for each molecule of ketone **1** consumed, one molecule of aldehyde **2** is also consumed. The parameter of interest for our analysis is the difference between the initial concentrations of the two substrates. This parameter is a constant in any given experiment, and we call it the excess, or [*e*], with units of molarity (Equation 27.5).

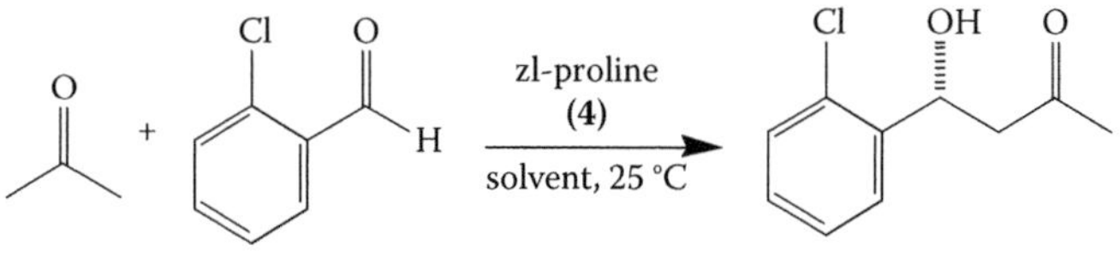

SCHEME 27.1 Proline-mediated aldol reaction.

The quantity [e] allows us to relate the two substrates' concentrations at any point during the reaction (Equation 27.6).

$$[e] = [\mathbf{1}]_0 - [\mathbf{2}]_0 \tag{27.5}$$

$$[\mathbf{1}] = [e] + [\mathbf{2}] \tag{27.6}$$

The excess [e] can be large, small, positive, or negative. When pseudo-zero-order conditions in [**2**] are employed, [e] is » [**1**]. Under conditions of practical synthetic experiments, [e] is usually small, and [**1**] and [**2**] are only slightly different from each other. These synthetically relevant conditions are employed in reaction progress kinetic analysis.

27.3 EXPERIMENTAL PROTOCOLS

The key to the graphical methodology developed in this perspective lies in defining two different experimental protocols: (1) *same excess [e]* experiments and (b) *different excess [e]* experiments.

27.3.1 SAME EXCESS [E] EXPERIMENTS

Catalyst stability is one of the most critical aspects determining the robustness of a process. The treatment described above assumes that the catalyst concentration term, [**4**]^z, is a constant throughout the reaction. We can test this assumption using a second reaction protocol developed in our methodology. These are termed *same excess experiments*. Table 27.1 shows two sets of conditions for carrying out the proline-mediated aldol reaction in Scheme 27.1, employing the same catalyst concentration but different initial concentrations of the two reactants chosen such that the value of the excess, [e], is the same in the two experiments. Note that more conventionally reported parameters (e.g., the number of equivalents of acetone and the catalyst mol%) are not the same for the two experiments.

TABLE 27.1

Comparison of Initial Conditions of Two Reactions Carried Out at Same Excess (e) and Their Concentrations during Reaction

Component	Acetone 1	Aldehyde 2	[e]	Product 3
Exp. 1: initial conditions	2.5 M	0.5 M	2 M	0 M
Exp. 2: initial conditions	2.25 M	0.25 M	2 M	0 M
Exp. 1: 50% conversion	2.25 M	0.25 M	2 M	0.25 M

What is the significance of this choice of conditions? Table 27.1 shows that the initial conditions of experiment 2 are identical to the concentrations of reactants found in experiment 1 at the point when it reaches 50% conversion (Table 27.1 entries highlighted in yellow in color insert). In fact, from this point onward, the two experiments exhibit identical [acetone] at any given [aldehyde] throughout the course of both reactions. The only differences between the two experiments are that the catalyst performed more turnovers in experiment 1, and the reaction vial had a greater concentration of product in experiment 1.

This same [e] experimental protocol leads to a graphical overlay plot that yields valuable kinetic information: if the two experiments described in Table 27.1 are plotted together as reaction rate versus [aldehyde], the two curves will fall on top of one another (overlay) over the range of [aldehyde] common to both only if the rate is not significantly influenced by changes in the overall catalyst concentration within the cycle, including catalyst activation, deactivation, or product inhibition. Overlay in same excess plots, therefore, may be used to confirm catalyst robustness or identify problems such as catalyst deactivation or product inhibition.

27.3.2 DIFFERENT EXCESS [e] EXPERIMENTS

The power law rate equation (Equation 27.4) may be rewritten as Equation 27.7, which "normalizes" the rate by the concentration function of substrate **2**.[5]

Our experimental protocol in this case involves plotting data from reactions carried out at different values of excess [e] as the function shown in Equation 27.7, with rate/[**1**]^x as the y-axis variable and [**2**] as the x-axis variable. Looking back at Equation 27.6, we see that any two reactions with different values of [e] have different concentrations of **1** at any given concentration of **2**. Our task is to find a value of x for which the data from experiments carried out at different excess fall on top of one another—or *overlay*—when plotted as Equation (27.7). From such overlay plots, we learn the order in [substrate] for both **1** and **2**:

- **Overlay** between the two curves reveals that the reaction exhibits order x kinetics in concentration of the "normalized" substrate.
- **The shape of the curve** reveals the reaction order in the concentration of the substrate plotted as the x-axis variable; for example, when the overlaid curves give a straight line, this reveals that $y = 1$, meaning that the reaction exhibits first-order kinetics in [**2**]. The slope of this straight line equals $k \cdot [\mathbf{4}]^z$. Reaction orders other than one may be assessed by the shape of the curve (concave for $y < 1$; convex for $y > 1$); y may also be determined quantitatively by simple fitting of the overlaid curves.

$$normalized\ rate = \frac{rate}{[\mathbf{1}]^x} = k \cdot [\mathbf{2}]^y \cdot [\mathbf{4}]^z$$

$$y-axis : \left(\frac{rate}{[\mathbf{1}]^x}\right) \qquad x-axis : [\mathbf{2}]$$

(27.7)

Let's summarize the reaction protocol set out by our methodology of reaction progress kinetic analysis.

- Two (or more) *different excess* experiments allow determination of the reaction orders of two substrate concentrations.
- Two *same excess* experiments allow determination of catalyst stability over the course of the reaction.

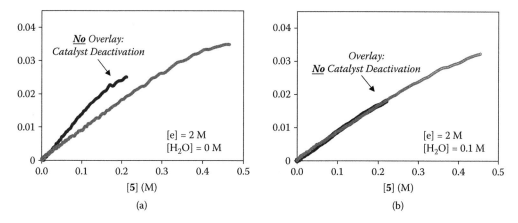

FIGURE 27.3 (See color insert following page 40). Comparison of rate versus [aldehyde] **2** for two reactions of Scheme 27.1 carried out at the same excess [*e*] with initial conditions as given in Table 27.1 (magenta dots: Exp. 1; blue dots: Exp. 2). (a) No water added; (b) 0.1 M water added to the reaction.

Because one of the same excess experiments can be designed based on the conditions of one of the different excess experiments, this protocol describes a mathematical minimum of only three experiments required to obtain a comprehensive kinetic picture of a reaction that will suffice to carry out scale-up of the reaction.

27.3.3 CASE STUDY: ALDOL REACTION

27.3.3.1 Same Excess

Now we are prepared to illustrate these experimental protocols of reaction progress kinetic analysis using data from reaction calorimetric monitoring of the aldol reaction shown in Scheme 27.1. We turn first to the issue of catalyst stability using our same excess protocol. In these aldol reactions, it was noted that the active catalyst concentration can be effectively decreased by the formation of oxazolidinones between proline and aldehydes or ketones,[6] and that addition of water can suppress this catalyst deactivation.[7] Same excess reactions carried out in the absence of water and in the presence of water are shown in Figure 27.3a and Figure 27.3b, respectively.[1b] The plots do not overlay in the absence of water, but they do when water is present. The overlay in these same [*e*] experiments in Figure 27.3b means that the total concentration of active catalyst within the cycle is constant and is the same in the two experiments where water is present.

27.3.3.2 Different Excess

Once conditions are found for obtaining constant catalyst concentration during reactions, we may proceed to our different excess experiments to probe substrate concentration dependences, in full confidence that the complicating influence of a changing active catalyst concentration is absent. Figure 27.4 shows data from three different excess experiments of the reaction in Scheme 27.1 plotted as the rate function (rate/$[1]^x$ versus $[2]$). In Figure 27.4a, we have set $x = 1$, corresponding to first-order kinetics in acetone concentration $[1]$. The fact that the three curves do not overlay tells us that the reaction does not exhibit simple first-order kinetics in $[1]$. The true kinetic order in $[1]$ may be a complex function as is characteristic of catalytic reactions and could require detailed kinetic modeling and further mechanistic analysis to obtain a full mechanistic understanding.

However, the approach we outlined in this work allows us to assess the magnitude of the driving force attributable to $[1]$ and $[2]$ in a way that provides a safe and efficient reaction protocol for scale-up, even in the absence of detailed mechanistic information. We do this by changing the value

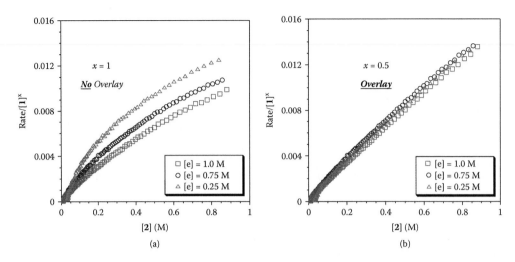

FIGURE 27.4 (See color insert following page 40). Comparison of rate/[1]^x versus [aldehyde] **2** for three reactions of Scheme 27.1 carried out at different excess [e] as shown in the figure legends. [**2**]$_0$ = 1 M, [**4**] = 0.1 M in each case. (a) x = 1; (b) x = 0.5.

of x until the three experimental data sets overlay in plots of Equation 27.7. Figure 27.4b shows that when x = 0.5, the three plots overlay, and the curves become straight lines, meaning that y = 1. This provides a reasonable description of the reaction's concentration driving forces as given in Equation 27.8, where the catalyst driving force has been lumped into k.

Thus, the reaction may be reasonably scaled up assuming that the rate is approximately half order in acetone concentration and first order in aldehyde concentration. This tells us that increasing the initial concentration of acetone has less of an accelerating effect on rate than would a similar increase in the aldehyde. In practical terms, this suggests that operating with an excess of acetone to drive the reaction to completion would have fewer process safety implications than would the alternative of using excess aldehyde.

We may compare our graphical result with the result obtained from solving for x and y by nonlinear regression fitting of the experimental rate curves to Equation 27.8. Carrying this out using the Excel Solver tool and the program Solvstat, we obtain x = 0.46 ± 0.002 and y = 0.90 ± 0.001. This reasonable agreement between our graphical methodology and the more mathematically rigorous approach lends support to the concept that kinetic information from experimental data sets may be extracted equally well from graphical as well as mathematical manipulations.

$$rate = k'\cdot [\mathbf{1}]^{0.5} \cdot [\mathbf{2}] \tag{27.8}$$

The value of x = 0.5 for the kinetic order in acetone [**1**] is not a true kinetic order for this reaction. Rather, this value represents the power-law "compromise" for a catalytic reaction with a more complex catalytic rate law such as Michaelis–Menten kinetics, as shown in Equation 27.9.[8] It may be demonstrated that our power law result giving x 0.5 yields a Michaelis constant of K_M 0.1 M. A reaction exhibiting saturation kinetics in [**1**], where the true kinetic order is close to first order at high concentrations and closer to zero order at low concentrations, can show "apparent" power law kinetics with 0 < x < 1, as shown in Equation 27.9. This suggests that our power law form of Equation 27.8 should be applied only under conditions close to those used to determine the kinetics. This condition is easy to fulfil because reaction progress measurements are typically carried out under synthetically relevant conditions.

$$[\mathbf{1}]^x \approx \frac{[\mathbf{1}]}{K_M + [\mathbf{1}]} \qquad (27.9)$$

We may summarize by outlining a step-by-step procedure to analyze the driving forces and robustness of any isothermal catalytic reaction process:

1. Choose a set of standard conditions of temperature, pressure, solvent, substrate concentrations, and catalyst concentrations for the reaction under study.
2. Choose a second set of conditions of same excess.
3. Choose at least one further set of conditions of different excess.
4. Choose at least one further set of conditions identical to one of the experiments described above, varying only the catalyst concentration.
5. Choose an appropriate *in situ* measurement tool to monitor these reactions under the conditions outlined above.
6. Manipulate the data to produce data triplets (of rate, concentration reactant 1, concentration reactant 2). If the raw data are proportional to concentration, this means differentiating the raw data, and integrating the raw data if they are proportional to rate.
7. Plot same excess data as rate versus [1]; use overlay to probe for catalyst deactivation.
8. Plot different excess data as shown in Equation 27.7; use overlay to determine reaction orders x and y.
9. Plot rate/[catalyst]^z; use overlay to determine z, the order in catalyst concentration.

This simple step-by-step procedure reveals how the use of extensive and accurate *in situ* data combined with simple graphical manipulations can provide a wealth of information of use in both scale-up and in furthering both practical and fundamental mechanistic understanding of complex catalytic reactions in a fraction of the time and number of experiments required by classical kinetic analysis.

27.4 CAVEATS

It is important to emphasize that kinetic analysis of a complex reaction is separate from the critical process safety analysis to which all reaction steps must be subjected for the determination of safe operating conditions. Our use of reaction calorimetry as a kinetic tool does not replace the thermal analysis that is critical to any reaction safety assessment, for estimating exothermicity in terms of adiabatic temperature rise and other parameters. The methodology described here assumes that this safety analysis has determined a range of safe operating temperatures. Our approach applies to isothermal reactions and could be repeated at different temperatures to help in optimizing operating temperature. For reactions that may be carried out using a variable temperature protocol, our methodology can fix the upper limits of the concentration driving forces to be expected when scaling up using such an approach.

A second point concerning this methodology is that it is designed as a hands-on tool to facilitate extraction of information from laboratory data primarily for the use of bench chemists not necessarily at ease with numerical approaches to kinetics, of which there are many available today. It goes without saying that concentration dependences may be determined using a numerical linear regression approach, and many chemists have adopted this approach. What our methodology offers is the opportunity to extract this information without the need to become familiar with commercial software and numerical analysis, and without the need to propose *a priori* a series of elementary steps through which the reaction proceeds. The graphical approach we illustrate here may have a greater likelihood of being embraced by synthetic chemists whose primary task is synthetic methodology applied to route selection. The step-by-step, algorithm-like approach is analogous to that used by the design of experiments studies that are now widely adopted even in very early pharmaceutical research for a broad scoping of reaction conditions. Our kinetic analysis aims to aid in reaction optimization in a similar way within each reaction step in a process.

27.5 SUMMARY

We show in this work that by applying the methodology of reaction progress kinetic analysis to data acquired by accurate and continuous monitoring of a complex multistep reaction with an *in situ* probe, we are able both to provide a quantitative assessment of the reaction orders for two separate substrate concentrations as well as to delineate conditions under which the stability of the catalyst and the robustness of the process is insured. Even without knowledge of the reaction mechanism or of the nature of the catalytic intermediate species, this information is sufficient to allow safe and efficient scale-up of the reaction as well as to provide a basis for further optimization aimed both at efficient production and detailed mechanistic understanding.

ACKNOWLEDGMENTS

Funding from the Engineering and Physical Science Research Council (EPSRC), AstraZeneca, Pfizer, and Mitsubishi Pharma is gratefully acknowledged.

REFERENCES AND NOTES

1. (a) Blackmond, D.G., *Angew. Chemie Int. Ed.* 2006, *44*, 4302; (b) Mathew, J.S.; Klussmann, M.; Iwamura, H.; Valera, F.; Futran, A.; Emanuelsson, E.A.C.; Blackmond, D.G. *J. Org. Chem.* 2006, *71*, 4711.
2. (a) Beezer, A.E. *Thermochimica Acta* 2001, *380*, 205–208; (b) Mathot, V.B.F. *Thermochimica Acta* 2000, *355*, 1–33; (c) Kemp, R.B.; Lamprecht, I. *Thermochimica Acta* 2000, *348*, 1–17; (d) Ferguson, H.F.; Frurip, D.J.; Pastor, A.J.; Peerey, K.M.; Whiting, L.F. *Thermochimica Acta* 2000, *363*, 1; (e) Regenass, W. *J. Thermal Anal.* 1997, *49*, 1661.
3. List, B.; Lerner, R.A.; Barbas, C.F. III, *J. Am. Chem. Soc.* 2000, *122*, 2395.
4. (a) Seayad, J.; List, B. *Org. Biomol. Chem.* 2005, *3*, 719; (b) List, B. *Acc. Chem. Res.* 2004, *37*, 548; (c) Notz, W.; Tanaka, F.; Barbas, C.F. III *Acc. Chem. Res.* 2004, *37*, 580; (d) Dalko, P.I.; Moisan, L. *Angew. Chem. Int. Ed.* 2004, *43*, 5138.
5. An analogous equation may be written to normalize the rate by [2]; it is most practical to carry out the normalization using the limiting substrate.
6. List, B.; Hoang, L.; Martin, H.J. *PNAS* 2004, *101*, 5839–5842.
7. Nyberg, A.I.; Usano, A.; Pihko, P.M. *Synlett* 2004,1891.
8. Billo, E.J. *Excel for Chemists*, Wiley-VCH, New York, 2001.
9. A full kinetic study of the proline-mediated aldol reaction based on a detailed catalytic reaction mechanism will be published separately.

28 Trends in Outsourcing

John Lu and Ichiro Shinkai

CONTENTS

28.1 INTRODUCTION

According to a recent *Chemical & Engineering News* report, 2005 showed no major change in either sales growth or outlook for the global pharmaceutical industry. Global sales for 2004 grew 7% to reach a record $550 billion, according to IMS Health. Growth in 2004 was slightly lower compared with the 9% growth seen in 2003, when the worldwide drug market reached US$492 billion. In 2004, the U.S. market grew 8%—the first drop below double-digit growth since 1995. With the major U.S. pharmaceutical companies focusing on discovery research and marketing efforts, there are many opportunities for outsourcing the production of drug substances and key intermediates. Recent trends indicate that process development work has also been outsourced to contract laboratories.

28.2 ROLE OF CHINA

Along with India, China has been a favorable emerging country for the pharmaceutical industry since its access into the World Trade Organization in 2001. The clear message for the pharmaceutical industry is that companies need to have a strategy that incorporates China as both a market and a supply source. As an emerging market, China offers the following.

China's population of 1.3 billion (compared to 300 million in the United States) makes up more than 20% of the world's population. Current gross domestic product (GDP) is US$1.4 trillion and is projected to grow to US$43 trillion by 2050 at an annual rate of 8 to 10%. With 6% of that GDP expected to be spent on health care, China's pharmaceutical market is expected to be the only one that continues double-digit growth into 2010:

- 2000: US$5 billion (ranked eleventh in worldwide pharmaceutical sales)
- 2005: US$12 billion (ranked ninth)
- 2010: US$24 billion projected (ranked seventh)
- 2020: Projected to be top ranked alongside the U.S. market

In addition, China is already well established as a low-cost supply source, often with shorter lead times. China has a vast pool of qualified technical labor. Currently, there are 13 million students enrolled in undergraduate programs and another 500,000 in postgraduate programs. In 2004, there were 120,000 chemistry degree candidates, and 29,000 graduated with degrees in chemistry. The previous year, there were more than 20,000 returnees who brought with them postgraduate degrees and technical expertise, as well as years of experience and training with Western chemical manufacturing companies and multinational pharmaceutical companies.

China's manufacturing capacities are also considerable. Before the government mandated that all chemical manufacturers supplying to pharmaceutical companies be certified by the State Food and Drug Administration (SFDA) in good manufacturing practice (GMP) regulations compliance, there were more than 7,000 chemical manufacturer suppliers to the pharmaceutical industry. Currently, there are 5,000 chemical manufacturers, with 4,000 already certified by the Chinese SFDA in GMP compliance.

Finally, the appeal of sourcing in China is based on China's low cost structure for all phases of drug development. The costs of conducting clinical trials in China is approximately 20 to 25% of those in the United States. A full-time equivalent (FTE) in China costs about US$40,000 to US$100,000 per year in the Shanghai area and even less in inland regions such as Chengdu. In the United States, an FTE costs upward of US$200,000 to US$250,000 per year. In addition to low overhead costs, the hourly wages in China are about 25% those of Western countries.

In sum, compared to the US$800 to US$900 million necessary, on average, to develop a drug over 8 to 10 years in the United States, it takes only US$120 to US$150 million over 5 to 8 years by some estimates to develop a drug in China.

Roche has a research and development (R&D) center in China with 40 employees. Novartis built an R&D center in China in 2006. Eli Lilly and Company has an R&D contract for chemistry services with a Chinese custom research laboratory that is under exclusive service to Lilly. Other pharmaceutical companies also have extensive FTE arrangements with custom research laboratories in China (one example is Merck's collaboration with WuXi PharmaTech).

So why is it that few pharmaceutical companies have aggressively outsourced to China given its attractively low-cost and fast-turnaround positioning?

The primary concern is over intellectual property (IP) protection. Although the Chinese government, as a condition for entering into the World Trade Organization, has promulgated laws on IP protection, public understanding of the value of IP is still a relatively new concept, as is public awareness of the need to protect IP and of the laws governing it. Therefore, although the laws have been established, their prominence and enforcement are still in their infancy.

Chinese confidential disclosure agreements, too, may not be as strong as those in the Western countries, but they nevertheless offer another protection against IP loss. The burden rests on the customer to perform the necessary due diligence to find the right partner, however. Research such as plant visits, meeting with the management, and familiarization with the manufacturer's strengths and weaknesses is strongly recommended before executing confidential disclosure agreements. Gathering information on the company's financial data and technical expertise as well as obtaining a list of its commercial products and customers will help to determine the company's financial health and potential growth. Checking the manufacturer's reputation in the marketplace is another important part of research. Most important, however, is a visit to the site to inspect the facilities and examine the company's recordkeeping. Meeting with managers in person provides a better understanding of the direction in which the company may be heading.

Another major concern about sourcing from China is that the quality standards among Chinese chemical manufacturers are still not up to those of U.S. pharmaceutical companies. Although the overall quality level is improving very quickly, and the SFDA is stepping up oversight on GMP compliance, the concept of current GMP is still foreign to many Chinese manufacturers, from top management down to workers.

Frequent visits to and informal audits of the plants will certainly help to educate the manufacturer about expectations and the GMP standard. Involving them in discussions with customers will also emphasize the need to upgrade quality standards. But any improvement in the quality standard will also need the full support of upper management. As a result of the many returnees from the Western countries, Chinese management talent is becoming more forward thinking (i.e., Western style and pro-customer) at Chinese companies. And with the privatization of many previously state-owned companies, the forces of market competition will also drive managers to be more customer oriented.

Language and culture barriers, as well as awareness and interpretation of Chinese laws continue to be deterrents for some companies to source aggressively from China. Having a local partner or establishing an office in China will certainly help to alleviate this problem.

Although the appeal of sourcing from China is its low cost structure, the question remains about how long China can sustain this favorable cost position. Owing to recent and ongoing economic expansion, the cost of labor, energy, and raw materials has already increased significantly. But because the cost structure in China was so low to begin with, Chinese companies have been able to absorb these recent cost increases and remain very competitive. It remains to be seen how many similar increases can be absorbed by China's chemical manufacturing industry.

In addition, speculation continues about whether the central government of China will yield to foreign diplomatic pressure and "float" more freely its renminbi (RMB) currency, which has been regarded as undervalued (to support China's export business). Although China's central government no longer pegs its RMB to U.S. dollars but to a variety of other currencies, it has limited the movement of RMB to a very narrow range. Therefore, for now at least, all indications are that the Chinese government will continue to keep a tight rein on the fluctuation of its RMB to maintain a balance between China's import/export values and diplomatic pressures from other countries.

28.3 ROLES OF AGENTS

Agents of trading companies doing business in the pharmaceutical industry have been more focused on its manufacturing partners because of the long history they have had with them. A manufacturer would ask the agent to promote certain products. The agent would then approach various potential customers to introduce these products, mainly via the purchasing department. But with the spin-off of the pharmaceutical business from the trading company to a subsidiary agent, the focus has now shifted to the customer, as shown in Figure 28.1.

By frequently contacting pharmaceutical companies to find out the requirements from the purchasing/sourcing department and the trends in the types of chemistries and molecules that may be coming from R&D departments and published in the literature, the agent can approach various

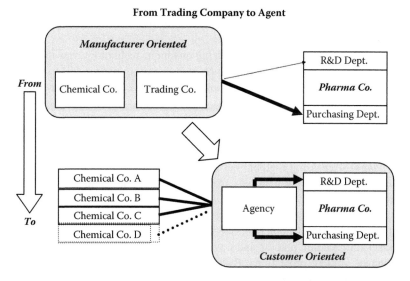

FIGURE 28.1 (See color insert following page 40). From trading company to agent.

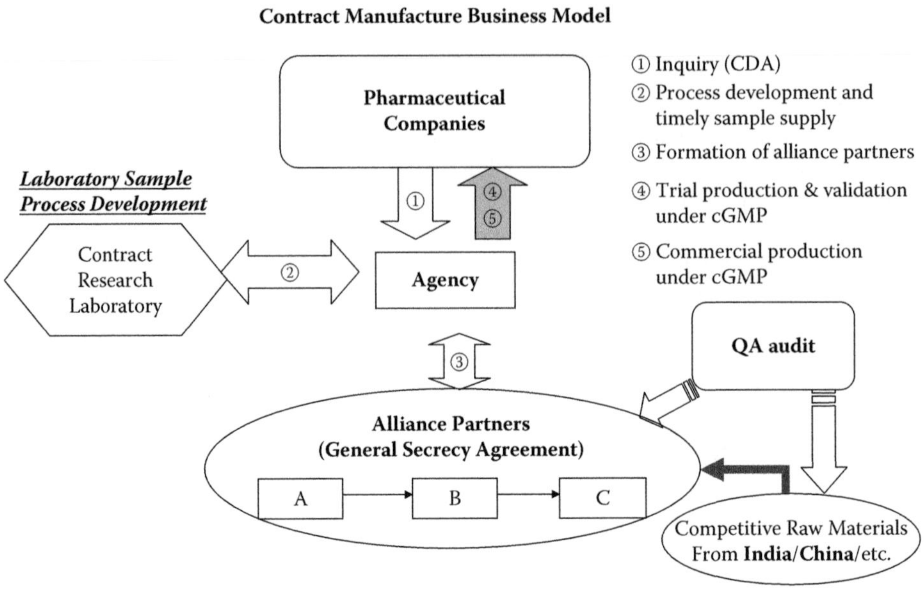

Contract Manufacture Business Model

FIGURE 28.2 (See color insert following page 40). Contract manufacture business model (CDA, confidential disclosure agreement; cGMP, current good manufacturing practice; QA, quality assurance).

potential manufacturers to find out which facilities and processes may be readily available and what may need to be developed.

With the ever-changing climate and requirements of the pharmaceutical industry, such as the current focus on sourcing from China (and India), the agent has also adapted its business model to meet the needs of its customers (Figure 28.2). No longer just providing sourcing and logistics services, the agent now also provides various value-added services:

1. *Process development*—For projects for which the client may not have developed a commercially viable process, the agent uses FTEs at contract research laboratories to develop, transfer, and scale up this process with the selected manufacturing partners (Figure 28.3).
2. *Supply-chain management*—By capitalizing on unique technology, raw material position, and any excess capacity that each alliance manufacturing partner has to offer, the agent uses global coordination to minimize total production costs as well as lead times and risks (Figure 28.4).
3. *Quality assurance audits*—As part of due diligence service to clients, agents frequently visit and audit manufacturing sites to train staff and upgrade the plants' quality systems to the customer's standard before a client audit.
4. *Cost savings*—Accessing a global network, the agent can help alliance partners source for more competitive raw materials from low-cost countries such as China and India (Figure 28.5).

The agent's objective in a customer-service-oriented model is to propose the most cost-effective solution that takes into account process and raw materials as well as quality systems, all predicated on strict compliance with IP protections. By providing speed and science to the client, the agent helps clients free their own resources to concentrate on "core" activities such as drug discovery, clinical development, and marketing. Needless to say, speed is a crucial element in satisfying pharmaceutical companies, particularly in the early stages of projects. Agents should identify the

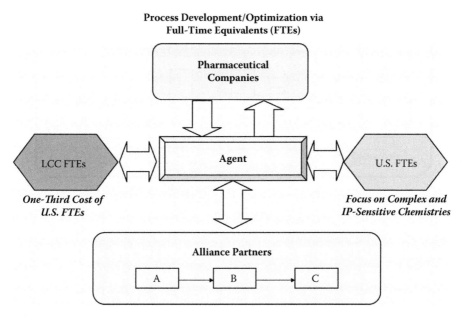

FIGURE 28.3 (See color insert following page 40). Process development/optimization via full-time equivalents (IP, intellectual property).

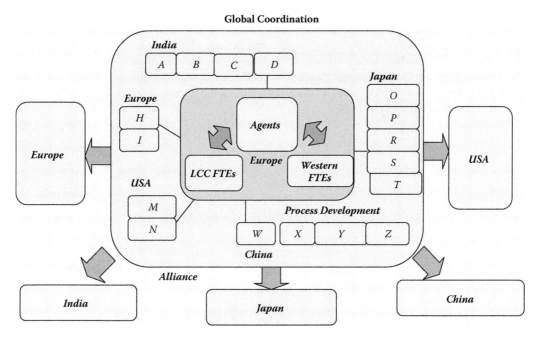

FIGURE 28.4 (See color insert following page 40). Global coordination (FTEs, full-time equivalents).

optimal viable process with the shortest possible lead time and implement actual production. The proposal of new processes should be based on science so as to be convincing. Yet the agent must not disregard the business mindset, because pharmaceutical companies and chemical manufacturers are always under strong economic pressure both for their own survival and for their mutual benefit in an increasingly competitive marketplace. Agents are continuously trying their utmost to contain manufacturing costs and minimize business risks by using existing multipurpose plants.

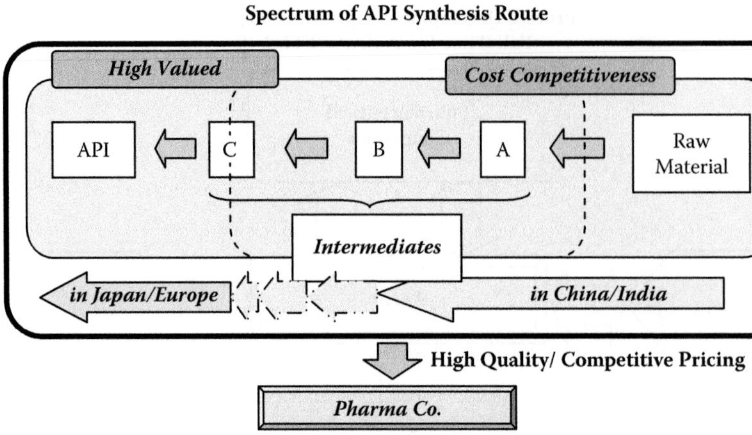

FIGURE 28.5 (See color insert following page 40). Spectrum of active pharmaceutical ingredient (API) synthesis route.

All of these agent services are valuable and help differentiate an agent from other strong competitors, either chemical companies or trading companies. As we all recognize with Merck's recent announcement to recall Vioxx® (Merck & Co., Inc., Whitehouse Station, NJ), the pharmaceutical business always incurs a great amount of risk. Contract manufacturing is confirmed to be an intelligent solution that benefits both pharmaceutical companies and chemical manufacturers. Despite different challenges in the industry (such as fewer new drug applications, increasingly restrictive regulatory affairs, fierce competition, and generic challenges), we are still optimistic about the future. Promising indicators are the still-increasing number of early-stage development projects (up to phase 2) and the importance of specialized scientific expertise, which is becoming central in the selection of processes and manufacturers as the chemical structures of target compounds become increasingly complex.

29 Sourcing Pharmaceutical Products in China and India

David Robins and Steve Hannon

CONTENTS

29.1 INTRODUCTION

China and India have demographic similarities and differences. They are the two largest populations on earth, comprising 40% of the world's total population, and are aggressively being propelled from third-world countries to economic rivals of the United States and the European Union. China and India are both very active in pharmaceuticals and the production of pharmaceutical actives, intermediates, and raw materials. Both have young, highly educated workforces that are driven by entrepreneurial zeal. Both represent consumer markets of enormous size, yet their manufacturing and research and development (R&D) sectors are underdeveloped relative to the projected market demand for pharmaceutical products internally. Wages in China and India, especially for those with technical and college educations, are growing at breakneck speed. Potential in both countries is currently limited by a shortage of experienced, trained, and skilled professionals in technical, accounting, and business management fields. Both India and China are benefiting enormously from U.S. and European expatriates returning to their home countries.

There are also differences. India is the world's largest democracy; China is the world's largest single-party government. India's democracy follows the parliamentary model, so political matters are particularly messy and unpredictable. China's government, while gradually becoming open to market forces, is, to outsiders, still opaque and poorly understood. China's population growth has stabilized; India's population is still growing at an unsustainable rate. India has a large percentage of English speakers in the technical and professional fields, whereas China is still transitioning from Mandarin/Russian to Mandarin/English multilingualism. India has a British-style legal system, whereas China is still transitioning from a judiciary controlled by the ruling party to an independent system. India's legal system is notoriously slow and ponderous, but China's, equally slow, is unpredictable. India is focusing on private investments, so factories and hotels are isolated islands of order and tranquility, whereas China has collectively focused on infrastructure on a scale never before seen in human history. India has invested capital through private capital markets. China has focused capital spending through state-owned banks, only now opening to international capital markets.

Recent media coverage has graphically compared and contrasted India and China yet fails to capture the dynamic situation. For example, English is the primary language behind Mandarin in Chinese schools. India is recognizing that without infrastructure, the productivity of factories and R&D centers is wasted. China is working to bring its legal system up to Western standards, for without predictable legal support, the high-technology industries cannot thrive. The classic analysis popular with the business school and consulting industry is that China has a robust manufacturing economy, whereas India has better-developed R&D and creative discovery infrastructure. This generalization, however, ignores the strength of some of the more mature Indian manufacturers that are now investing their capital in manufacturing resources in the United States and European Union. It also ignores the Chinese government's commitment to and support of high-tech R&D such as manned space flight and cultivation of mammalian cells for biopharmaceuticals. The rate of growth and change in these two economies makes any forecasts misleading and meaningless; by the time the print is dry, the situation has changed.

Three predictions or principles, however, necessarily color all business considerations related to working in India or China:

- Neither India nor China has a sustainable balance between consumer demand for products and services and supply that can be met by current manufacturing and R&D resources, so we can expect that manufacturing and R&D in both countries will grow faster than in the United States or European Union, and this shortcoming will cut into U.S./EU exports and jobs until rough parity is achieved, commensurate with the median lifestyle and gross domestic product (GDP).
- China has fueled growth in manufacturing and R&D (promoted by the central committee and its directives to the capital banks) to support the internal Chinese market for pharmaceutical products, not for export as is popularly believed. This internal market is so profitable that projects for Western customers are often dropped because their profit margins are so poor. Conversely, India has focused on export of pharmaceutical products, primarily because capital funding is privately provided, and rates of return for the export market (historically the United States and European Union) are so great. The Indian middle class is only now developing buying power after almost two decades of export growth in products to the West. Although this historical difference between India and China (export versus internal markets) in economic terms is likely to disappear in 1 to 2 years the cultural legacy of this historical difference will likely influence business decisions for some time to come.

- Both India and China are very youthful societies, but for very different reasons. India has seen astronomical population growth fueled by improvements in agriculture and health care. China has similarly benefited from these technological improvements, but owing to government restrictions on births as well as the devastation to the 40- to 50-year-old age group wrought by China's cultural revolution and the Great Leap Forward, the population in China is relatively stable yet very youthful. Neither situation is ideal for long-term stability. In the short term, however, both countries have scores of motivated, educated young people, but the workforces lack experience and middle managers with the expertise to mentor and shape these energetic youth. Young people are forced to make choices and decisions with little guidance or direction. Hence, we see enormous energies expended at very little cost by Western standards, yet everyone has a horror story about projects gone catastrophically wrong.

A further strategic note for U.S./EU pharmaceutical companies is that China is the fastest-growing market for pharmaceuticals in the world. India's market is growing more quickly than that in the West but not as fast as China's. The market for Western products in India or China dwarfs the current markets in the United States, European Union, or Japan and will continue to do so provided that growth rates do not stall or fall. General Motors sold more cars in China in 2005 than in the United States. An old axiom of doing business is that selling products in a country is far easier if those products are made in that country (employing local citizens/consumers). So it should come as no surprise that Western pharmaceutical companies that invested in plants in India and China decades ago are now at the forefront of establishing asymptotic sales growth of their products.

With this background, we can explore some detailed differences, recommended practices, and case studies of custom projects in India and China within the context of some typical questions.

29.2 INTELLECTUAL PROPERTY AND PATENT RIGHTS

"I am worried about protecting intellectual property when I take projects to India or China. How can I protect my technology?" Both India (1995) and China (2001) have now signed the World Trade Organization protocols for protection of intellectual property and recognition of patents. Developing countries are granted 10 years to bring their laws into compliance with the Trade-Related Aspects of Intellectual Property Rights (TRIPS) agreement. India passed TRIPS-compliant legislation in 2005. China expects to have TRIPS-enabling legislation in force in 2008. In reality, the concern is with the practical protection of technology: how can companies prevent employees from taking information to the manufacturer next door who does not have a confidential disclosure agreement with the inventor? Typical controls are to have each employee sign a confidentiality agreement as a condition of employment and for employers to take legal action against former employees who violate the terms. India has much stronger legal institutions through which to enforce employment guidelines, but all legal procedures are time consuming and laborious. Chinese legal institutions are much less sophisticated. Typically, employers black out critical information in records to keep employees from seeing the total picture.

29.3 LAW, LEGAL SYSTEMS, COURTS, AND LAWYERS

"Can I protect my intellectual property rights and business contracts in a court of law?" In keeping with its British-style government, India has adopted a model of legal jurisprudence that is recognizable to most Western customers. It is agonizingly slow and ponderous, however. China's legal system, which has traditionally been part of the ruling party apparatus, has only recently achieved independence. The judges are often holdovers from the party days, however, and there are not enough Western-trained lawyers. The law is still evolving and case law is skimpy at best. The net result is that Chinese legal decisions are poorly understood and often unpredictable. The civil system is also slow.

29.4 CURRENCY EXCHANGE AND MONETARY CONTROLS

India's rupee is freely floated on world currency markets. China's yuan (renminbi) is tightly controlled by the government and formerly pegged to the U.S. dollar. Recently, there has been some movement to use a variety of currencies to float the yuan, but the movement is miniscule and unpredictable.

29.5 LOGISTICS AND INFRASTRUCTURE

"How can I get my products delivered on schedule? Can I accurately deliver goods for just-in-time delivery?" India has favored private investment rather than public expenditures, which rely on taxation, are never popular, and often place a burden on the poor who operate in a cashless society. Public spending has not kept pace with private spending, leading to immaculate facilities that can only be reached on unpaved paths shared by all. China has a century of socialist thought that has supported the expenditure of enormous sums for public works, from parks and monuments to roads, railways, airports, and other transportation infrastructure. Toll roads are well designed and expanding rapidly to reach all parts of China.

29.6 OWNERSHIP OF FACILITIES

"Who owns the businesses I work with, and do they have the resources to conduct business?" Most manufacturers in India are private, family owned, or recently publicly traded stock companies. Investments are conservative, borrowing from banks is not undertaken lightly, and speculative trading is uncommon. Ownership is clear and straightforward. As might be expected in a well-developed legal system such as that in India, legal documents of incorporation and ownership are available. In China, ownership is only several decades old. Because private individuals historically lacked the needed capital to buy a manufacturing business, government privatization of manufacturers has resulted in the use of very inventive techniques to distribute manufacturing assets to individuals. Thus, ownership is often murky, and real valuations for companies are hard to establish. Most manufacturing sites are under joint ownership, with the share of ownership determined by the amount of business managed by each owner. Because the legal system is weak, documentation is not very detailed, and deals are often based on personal agreements, not legal documents.

29.7 GOVERNMENT AND BUREAUCRACY

"Both India and China have big, bureaucratic governments: one is democratic with a socialist twist, and the other is communist with a capitalist flavor. How do business procedures differ in the two systems?" It is remarkable how bureaucracies adopt similar systems regardless of their underpinnings. Both India and China have duties on importation that are staggering if the raw materials need to be imported for a project. Both require "proper paperwork" to move any goods, and the airlines are very cautious about shipping hazardous goods. Chinese marketers are notorious for creating fictitious paperwork to circumvent regulations for air shipments.

29.8 HOLIDAYS AND WORK SCHEDULES

India and China have lunar calendars for holidays, as well as holiday schedules that differ from each other and from that of the West. India tends to have many different holidays of several days' duration. China tends to have fewer holidays with time off, but although fewer, they are of longer duration. For example, spring festival (popularly known as Chinese New Year) is officially a 2-week holiday. May and fall festivals are officially 1 week. There are few holidays in between. Personal vacation time in India and China is uncommon. Work weeks are officially 40 hours, although abuses are common. China's phenomenal construction boom is fueled by a 24–7 shift schedule.

29.9 BUSINESS PRACTICES

Both India and China rely on personal relationships to develop business relationships despite the widespread belief that the Internet has replaced them. Written agreements are considered less important than they are in the West, and face-to-face negotiations are prioritized. Price is the last item in any negotiation. Without a working relationship, the lowest price will be elusive. Dinners, banquets, and meals are important contributors to the development of any relationship. Indians have a longer history in dealing with the West for business, so they are a little more flexible in adapting to Western social practices. India has adopted many of the business practices of Britain and thus both legal and accounting appear familiar. The stock market in India has some credibility. China's accounting practices are far from transparent. It will take some time with foreign investors to introduce Western accounting to China. A cash transaction or barter economy still exists in much of China. Both Indian and, to a greater extent, Chinese business people expect to negotiate any transaction. These negotiations are often intense and emotional, but once concluded, the parties are partners.

29.10 INTERNAL VERSUS EXTERNAL MARKETS

As indicated earlier, the Indian pharmaceutical market is primarily focused on export, and the Indian government's tax and duty policies favor foreign sales. The internal market for pharmaceuticals has not grown as rapidly as the export market, although this gap is closing owing both to the rise in the middle class and to the trade policies and intellectual property policies of the World Trade Organization. Chinese development policy for pharmaceuticals has favored manufacturing for the internal Chinese market. Western medicines are introduced into China by official government policy. Before market reforms, this introduction took place through government quotas and orders for specific drugs. Now it is done through processes developed by government-supported R&D institutes and influenced by the availability of capital from banks to support the manufacturers. In either case, the government supports introducing the latest in Western medicines for the population. The growing Chinese middle class is fueling the fastest sales growth for pharmaceuticals in the world today. Western innovators are encouraged to introduce their products into the market: the first launch gets exclusivity in the market and protection from competitors through the drug approval process. The ready availability of the key raw materials as well as the drugs themselves is catapulting China into an export paradise. Likewise, the restrictions on exporting patented medicines from India and the growing middle class are accelerating internal market growth in India.

29.11 EXPORT SHIPPING AND PAPERWORK

Both India and China have adopted international standards for export shipments. The biggest concerns about exportation arise from the lack of controls and inspections to assure that the paperwork accurately describes the products. This concern is greater in China than in India, where exports are actually inspected before shipping, and are often delayed until the inspector releases the shipment.

29.12 IMPORT DUTIES AND PAPERWORK

Both India and China have duty schedules favoring internal goods over imports, with substantial duties. These schedules have a dramatic effect on early research, as there is no access to the widely available catalog houses for small quantities. India and China have considerable incentives to source all raw materials locally for their processes. If they must buy imported raw materials, any cost savings from reduced labor and overhead costs are often completely wiped out.

29.13 RESEARCH AND DEVELOPMENT CAPABILITIES

India's private companies have invested heavily in the equipment and infrastructure to do world-class R&D. Historically, this investment has been made by each manufacturer, so companies have fully integrated R&D through commercial manufacturing capability. The limitations are usually related to size, with larger companies fully equipped and staffed to handle almost any project and smaller companies restricted to core competencies. Government-owned or government-funded research centers are not as well dispersed. There are a few very good R&D centers, but they are not focused on industrial development. China has historically not distinguished between government and industry. Until privatization, they were part of the same team. So R&D could take place at one institution and be transferred to another, often totally unrelated, manufacturer by "executive decision." With privatization, separate organizations remain, now brought together by common business objectives and collaboration. Manufacturers have very primitive R&D, both process and analytical. Kilo labs and pilot plants at the manufacturing site are rare. R&D personnel from the institute travel to the manufacturing site and stay in dormitories during the process development and transfer phase. The significant analytical instrumentation needed to develop modern synthetic processes exists primarily in the R&D institutes. Western customers need to appreciate that the ownerships of the R&D institutes and the manufacturing sites are usually different, but it is actually possible to make this situation work successfully.

29.14 STAFF TURNOVER

India and China have very mobile, young workforces of trained professionals. The best only want to live in Mumbai or Shanghai, not in the less cosmopolitan areas preferred for the factories. Consequently, staff turnover is extremely high, and retention of key personnel is critical. Salaries for critical positions are rising rapidly.

29.15 TRAINING AND EDUCATION

Indian and Chinese workforces are large and educated. Both countries have very young populations of trained scientists, reflecting the age demographics of the countries. Although youthful enthusiasm and vigor are important, experience and training are also crucial and in short supply. India has more mid-career, experienced professionals, but their numbers are still inadequate for the explosive growth the country is currently experiencing. China is even more in need of experienced professionals. The education from the major universities is competitive on a worldwide basis. Instrumentation and equipment usually lag behind what is more common in the commercial marketplace. India has a long history of English-language training; English use in China is growing rapidly.

29.16 WAGES AND PRODUCTIVITY

A common misperception is that if wages in India or China are one-tenth U.S., EU, or Japanese salaries, then the prices of products and services should also be one-tenth of those elsewhere. In reality, the productivities are inadequate to realize this cost advantage. Several factors contribute. One is the lack of access to the catalog materials common in the West owing to the extreme duties imposed. Also, the equipment and instrumentation are neither as good nor as common as they are in the West. Again, this disparity is partly due to duties. Infrastructure problems also contribute to lost productivity. One experienced pharmaceutical outsourcing professional noted almost 10 years ago that realizing a 20% savings from India or China is doing well. This estimate is still fairly accurate today. The cost of introducing current Western standards for occupational health and safety practices, environmental controls, and current good manufacturing practices (cGMPs) will challenge the manufacturers, just as the introduction of more and better equipment, facilities, and infrastructure will improve productivity.

29.17 MARKET INTELLIGENCE

Indian businesspeople are quite savvy in developing accurate market intelligence. This expertise is likely due in large part to the facility with which Indian citizens can and do travel to the United States, European Union, and Japan. Chinese citizens are not yet so well traveled, owing partly to the costs of travel and partly to travel restrictions posed by terrorism and anti-immigration controls. Until recently, Chinese citizens could not easily get passports for foreign travel; however, travel is now growing rapidly, and we can expect that market intelligence and awareness will increase.

Inquiries for goods in China multiply exponentially because of the multitude of agents—the Chinese marketing companies separate from the manufacturers—who do not identify the customer to protect their information source. Western customers often prefer to deal with Chinese manufacturers through Western trading companies to avoid dealing with the Chinese market directly. Thus, inquiries are received in multiples from the same source. Indian companies are much more aware of this problem and tend to avoid the multiplier effect. Also, most Western companies are comfortable dealing with known Indian suppliers, and when the Indian manufacturer selects an agent to represent it, the representation is exclusive.

29.18 LANGUAGE AND COMMUNICATIONS

Traditionally, one of the strengths in working with Indian manufacturers is the widespread use of English among technical and managerial staff. This ease of conversing is further supported by the extensive number of Indian students educated in English-speaking countries and the repatriation of these scientists and managers as the job market in India expands. The information technology and call center industries are also supporting the use of English (colloquial English as well) among newer graduates from Indian universities. One should always keep in mind that English is still a second language for most Indians, and there are always potential problems in interpretation, especially when using more colloquial terms and expressions. Even within the use of English, there are cultural differences and nuances that need to be abided by to maximize working collaborations. Local languages and dialects number in the hundreds in India, with English becoming the national language of business and science.

English is the most popular language in China after Mandarin. Because most Chinese citizens also speak the local language or dialect (there are hundreds), English must be a third language for students. The growth of English speaking is explosive and among young students: they see it as central for their future careers. One reporter from Xinhua, the Chinese news agency covering the Asia-Pacific Economic Cooperation meeting in Busan, South Korea, was asked if he thought Busan was a good candidate for the 2020 Summer Olympics. The Chinese reporter commented that no one in Busan spoke English, so it would not be a good candidate. As China moves rapidly toward joining the international world of business and science, the use of English will accelerate. The practical use of English for communication among business managers or scientists is still a long way off and incomparable to communication in India. Exactly how far off will depend on whether the rapid industrialization of China continues at its present pace.

29.19 CHEMICAL NAMES AND NOMENCLATURE

India has adopted the chemical names and nomenclature rules used in the West. Although it is tempting to say that this choice was made because the names and nomenclature are English based, that is not true—our names and nomenclature are originally German. It is certainly true that it was the common use of English (proper British English) in the schools that introduced Western chemical names and nomenclature systems.

China has a long history of public education that predates the growth of English as the international language of business and science. So it should come as little surprise that the Chinese language system has Chinese terms for chemicals. At the simplest level, much as we would call acetic acid by its common name, vinegar, the Chinese also have names and nomenclature built around common terms. In sourcing products and services in China, it is important to have chemists trained in both Western and Chinese nomenclature to translate. The use of Chemical Abstracts Service numbers is a great help, because these numbers are recognized in both systems.

29.20 ENVIRONMENTAL REGULATIONS AND COMPLIANCE

China is today's poster child for environmental problems. The latest spill into the Songhua River after an explosion and fire at a chemical plant in 2005 is the most visible example of the problems. No Western customer wants to find out that a manufacturing partner has unacceptable environmental problems. The better informed and managed manufacturers in China accept responsibility for the safeguarding of the people and environment in their areas. It is critical for customers to know the Chinese manufacturers that they are collaborating with and to help them recognize and correct unacceptable practices. It is worthwhile to understand that the most recent chemical spills in China are the result of industrial accidents. There is a dearth of experienced industrial engineers and risk-assessment personnel experienced in detecting and correcting pollution sources, directing remediation, and determining the causes of industrial accidents that are often the source of the pollution. China has a government office with the authority to shut down plants with unacceptable environmental practices. As usual, they are too few and often too late. This shortcoming will need to change, and change rapidly.

India currently has a better record of environmental compliance, although one need only look at the covers of back issues of *Chemical & Engineering News* to find pictures of polluted effluent from Indian manufacturers. The industrial accident followed by the airborne dispersal of methyl isocyanate at Bhopal was not so long ago and was perhaps the event that triggered the Indian industry to better manage its processes and effluent.

29.21 COMMON GOOD MANUFACTURING PRACTICE REGULATIONS AND COMPLIANCE

India has benefited from years of selling pharmaceutical products to the West because the sales usually have a prerequisite GMP audit by the customer or by the U.S. Food and Drug Administration. The Europeans have historically not included GMP audits in the approval process, but this policy is changing. Thus, Indian plants have a long history of GMP compliance through their foreign trade. There is no compliance audit (i.e., independent Indian government office) for internal sales in India or for sales to countries without compliance functions.

China implemented an internal audit process in 2002 for all plants selling human pharmaceuticals. No plant may manufacture or sell a pharmaceutical for use in humans in China without the approval of the State Drug Administration (SDA). The SDA certifies each plant every 2 years, and every pharmaceutical manufacturer is required to display an approval letter. Compared to the GMP requirements of the U.S. Food and Drug Administration, the SDA GMP practices are adequate in some areas but lacking in others. For example, all final active pharmaceutical ingredient (API) finishing areas have segregated manufacturing suites, with gowning and degowning requirements for staff and cleanable surfaces. Batch record keeping and inventory and analytical records are recognizable as GMP. Labeling is similarly familiar. Staff are required to have GMP training. Process validation, cleaning validation (because most processes are in dedicated facilities, this inspection is less critical), and up-to-date analytical methodology are areas needing improvement. The control of cross-contamination in intermediate processing areas is poor. Nevertheless, the GMP compliance is not the area in China in need of the most improvement (see the next topic).

29.22 OCCUPATIONAL HEALTH AND SAFETY REGULATIONS AND COMPLIANCE

Indian companies can claim that the long history of selling products to the West has given them a hefty lead in health and safety compliance. Indian plants with such a history are quite sophisticated in meeting current standards for health and safety compliance. This compliance is also aided by the private ownership of most companies and their need for insurance to cover losses and liabilities, enforced through the legal system. At least a generation of chemical plants exist in India that are owned by Western companies, were built to Western standards by Western companies, and employ managers responsible for enforcing the health and safety standards of the parent company.

Health and safety are the areas in most need of improvement in China. Many of the current environmental problems mentioned above can be traced to accidents. The annual number of workers killed or injured on the job is staggering. There are several reasons for the problems in these areas. China lacks India's long association with Western customers and partners. The internal Chinese market dominates the current market, and thus, there is little incentive to implement costly Western standards. There is significant tension between the different levels of government in China. The central government is often at odds with local governments that want the jobs, taxes, and growth. Recent accidents and disasters on a scale seldom seen or reported previously will likely increase the central government's enforcement powers. International reaction and political and economic pressures will force changes as well. From the opposite direction, local resident populations are fighting back against local governments often corrupted by graft and payoffs and dependent on the taxes from local manufacturers for their revenues. The triple pressure from central authority, international customers, and local residents will bring changes. Without knowledgeable and experienced hands to shape them, however, these changes may not be as effective as they might be. Compared to India with its long history of Western collaboration, China has less than a decade of experience with Western manufacturing standards. Until very recently, sole foreign ownership was not allowed and was relegated to minority ownership in joint ventures, meaning that Western partners have been overruled when deciding to implement designs and standards. Rarely Chinese joint venture plants have accepted the design and safety standards of their minority partners, and these are model facilities.

29.23 PROFITABILITY

"Can I do business in India and China?" or rather "Can I afford not to do business in India or China?" Although population figures for China and India are unreliable, these countries each have roughly one-fifth of the world's population. According to 2005 estimates from the Central Intelligence Agency's *World Factbook*, China's population growth rate is 0.58%, with 1.3 billon people living in an area roughly the size of the United States. The median age is about 32 years. India's population is estimated at 1.1 billon living in an area roughly one-third that of the United States, with a growth rate of 1.4%. The median age is about 25 years. The GDP in purchasing power parity (PPP, not actual GDP) was estimated at 7.3 trillion for China and 3.3 trillion for India for 2004. Before the recent recalculation of Chinese GDP growth, the annual real GDP growth in China for 2004 was estimated at 9.1%, while India's was estimated at 6.2%. Chinese GDP has been quadrupling since 1978, and the Chinese GDP is ranked with most European countries. Recently, the Chinese revised the 2004 GDP numbers to include the service industry. The revised GDP growth for 2004 was approximately 17%. The missing 300 billion is larger than the entire GDP of many countries. China will officially try to slow the GDP growth to a mere 9% for 2006, with consumer price index growth of 2 to 3%.

With populations and markets as large as these, India and China cannot be ignored by pharmaceutical companies. In China, home of well-regarded traditional medicines, Western medicines now outsell traditional remedies three- to four-fold. The Chinese growth in Western pharmaceutical sales is the fastest in the world (estimated at 15% annually or US$24 billion) and is predicted to rank fifth in 2010. India's pharmaceutical market is estimated at US$5 billion, with more than US$1.2 billion exported as APIs, representing more than 60% of the total API output.

Most of the Western medicines sold in India and China are what we term generics; pharmaceutical innovation, although starting, is in its infancy in both countries. India will likely be first to bring a new chemical entity to market. China's official government policy is to bring the latest in Western medicines to market to meet consumers' growing demand. In India, market forces will drive the growth. At this time, both countries are focused on preventing and treating communicable diseases. India is next in line for the eradication of polio, and avian flu and severe acute respiratory syndrome are global manifestations of a failure to develop controls and treatments for such diseases. Nevertheless, recent estimates project that China will rapidly become the center for senility treatment. As a result of a rapidly improving standard of living, life expectancy is increasing, and an unprecedented number of Chinese citizens are reaching an age at which diseases of mental deterioration are diagnosed.

India will continue to strengthen its lead as a supplier of generic APIs, with broadening to include dosage-form manufacture as well. China will continue to focus on its internal markets. Both India and China thus will be centers of manufacture for generic drugs and raw materials and intermediates to support generic APIs. Custom manufacturing of raw materials and intermediates for new chemical entities will follow that of generics, much as they did in Italy when it was the center for generic API manufacture. India has already used its capital and market strength both to buy Western pharmaceutical manufacturing sites and expand into China.

China has traditionally used state-owned companies for the distribution of Western-type pharmaceuticals, serving over 64,000 hospitals and numerous retail pharmacies. This state-owned system is gradually being privatized. India is reducing the layers of bureaucracy in its health-care system and price controls are being relaxed. India is introducing private health insurance to help fund the population's growing demand for pharmaceuticals. India is also expanding its outreach system to people unable to take advantage of the market conditions in the larger cities and affluent areas. International insurance companies are now permitted to offer their services to employers in India through joint ventures with local firms. Foreign companies can now own pharmaceutical development and manufacturing sites completely.

India and China represent the most promising areas of growth for pharmaceutical companies. The corresponding growth of the manufacturing sectors in both India and China to serve their markets offers an enormous manufacturing base for Western companies looking for capable and competitive sources of key raw materials and intermediates. It is easy to predict that the companies doing business in India and China (buying products, employing people) will have the least difficulty introducing their own products into these markets. It has worked this way in every other geographical market to date, and there is no reason to believe India and China will be different.

Index

A

Printed and bound by CPI Group (UK) Ltd, Croydon, CR0 4YY

24/10/2024

01778285-0007